SYMMETRIES IN SCIENCE III

SYMMETRIES IN SCIENCE III

Edited by
Bruno Gruber
Southern Illinois University
Carbondale, Illinois
and
Francesco Iachello
Yale University
New Haven, Connecticut

PLENUM PRESS · NEW YORK AND LONDON

Library of Congress Cataloging in Publication Data
Symposium Symmetries in Science III (1988: Vorarlberg, Austria)
Symmetries in science III / edited by Bruno Gruber and Francesco Iachello.
p. cm.
"Proceedings of the Symposium Symmetries in Science III, held July 25-28, 1988, in Schloss Hofen, Lochau, Vorarlberg, Austria"—T.p. verso.
Symposium co-sponsored by Southern Illinois University at Carbondale and others.
Bibliography: p.
Includes index.
ISBN 978-1-4612-8082-8
1. Symmetry—Congresses. 2. Symmetry (Physics)—Congresses. I. Gruber, Bruno, 1936- . II. Iachello, F. III. Southern Illinois University at Carbondale. IV. Title. V. Title: Symmetries in science three. VI. Title: Symmetries in science 3.
Q172.5.S95S96 1988 89-30747
500—dc19 CIP

Proceedings of the symposium Symmetries in Science III, held July 25–28, 1988, in Schloss Hofen, Lochau, Vorarlberg, Austria

The work contained in Section II, pages 377 to 503 relates to Department of the Navy grant N00014-88-J-9010 issued by the Office of Naval Research. The United States Government has a royalty-free licence throughout the world in all copyrightable material contained therein.

ISBN-13: 978-1-4612-8082-8 e-ISBN-13: 978-1-4613-0787-7
DOI: 10.1007/978-1-4613-0787-7

Softcover reprint of the hardcover 1st edition 1989

A Division of Plenum Publishing Corporation
233 Spring Street, New York, N.Y. 10013

FOREWORD

The Symposium "Symmetries in Science III" was held at Schloss Hofen, Vorarlberg, Austria, during the period July 25-28, 1988. Some 70 invited scientists from 24 countries attended the Symposium and presented reviews, research papers and reports, as well as participated in workshops. The format of the Symposium was deliberately kept international and interdisciplinary, and also provided an opportunity for presentations of symmetry applications in fields other than in the conventional disciplines.

The Symposium was co-sponsored by

Southern Illinois University at Carbondale
Land Vorarlberg,
U.S. Office of Naval Research (Special Session), and
IBM Austria.

Support by these institutions is gratefully acknowledged.

Several individuals have contributed significantly to the success of Symposium III. They are

Dr. John C. Guyon, President, Southern Illinois University, at Carbondale,
Dr. Martin Purtscher, Landeshauptman des Landes Vorarlberg,
Dr. Guntram Lins, Landesrat des Landes Vorarlberg, and
Dr. Hubert Regner, Direktor, Landesbildungszentrum Schloss Hofen.

I wish to thank these individuals for their support of Symposium III, also in the name of the participating scientists. The Symposium provided these scientists with an opportunity to get together and exchange scientific information in an efficient manner, and in a very pleasant environment.

Bruno Gruber

CONTENTS

PART I: REVIEWS

PART II: SPECIAL SESSION
(Sponsor: U.S. Office of Naval Research, London)

PART III: RESEARCH PAPERS AND REPORTS

PART I: REVIEWS

FUNDAMENTAL SYMMETRIES AND QUANTUM ELECTRODYNAMICS

A. O. Barut

Department of Physics
University of Colorado
Boulder, CO 80309

ABSTRACT

An essay on the symmetries of the Maxwell-Dirac system and on all the symmetries that follow from it.

I. The Importance of Electromagnetism

It can be said that perhaps 99.9% of all everyday physical phenomena concerning the constitution of matter and radiation in physics, chemistry and biology, can in principle be explained by a single simple theoretical system, namely the coupled Maxwell-Dirac equations (or their limiting cases, the Maxwell-Schrödinger, or the Maxwell-Lorentz equations). This is a tremendous achievement of the last hundred years of theoretical physics. This system describes phenomena at distances from about 10^{-16} cm (electron positron scattering at high energies) to radar probing of astronomical distances and energies from 10^{-5} eν (Lamb shift) to 40^{12} eν (e^+e^- accelerators).

The remaining .1% of the phenomena (nature leaves always a small door open to deeper levels) concern the rare events in radioactivity and cosmic rays, which led to the introduction of new particles, at first seemingly outside the Maxwell-Dirac system, beginning in 1932, namely the neutron and the neutrino. With this date the particle physics begins and with it we have the new phenomenon of forceful production in the laboratory of all sorts of new particles, some of which already occur in cosmic rays, some perhaps produced for the first time by man. Gravitation presumably plays little role in the formation of nuclei, atoms and molecules, but is, of course, dominant in the formation of celestial bodies.

Because the electromagnetic theory, and its extension after Maxwell-Lorentz, to include positron, electron spin and the wave properties of the electron, hence quantumelectrodynamics, works so well and so universally, we must first see if it can also account for these remaining phenomena of radioactivity, the properties of the nucleus and its disintegration products, before introducing completely new theories, new forces and new particles. This expectation is based on the realization that, as we shall see it in more detail, quantumelectrodynamics that we practice it today is in fact incomplete and does not give us the full information about the electromagnetic behavior at short distances or high energies, in fact an unknown territory--terra incognita--that must be explored.

We have divided physical phenomena according to their energetic appearances into weak, strong, electromagnetic and gravitational. They also differ in their range (short-ranged or long-ranged interactions) or in being microscopic or macroscopic. However, electromagnetism can and does manifest itself in widely different strengths. For example, the "chemical force" between two neutral atoms is very weak compared to Coulomb force, is charge independent, hence seemingly non-electromagnetic, but has revealed itself, after quantum mechanics, to be a residual weak manifestation of electromagnetism when atomic structure is taken into account. This small residual force is of course of vital importance for the whole of chemistry and biophysics. Another example is the α-radioactivity, which is also very weak in general, and has also revealed itself not as a new force, but a largely electromagnetic and quantum phenomenon due to the tunneling of the α-particle through the electromagnetic barrier of the α-nucleus system. In both these instances we can describe the process as a new interaction vertex with appropriate new coupling constants: For α-decay, the interaction $g_\alpha \bar{\psi}_A \psi_B \phi_\alpha$ representing the vertex A ⟶ B + α. For chemical force the interaction $g_{chem}\ \bar{\psi}_1 \bar{\psi}_2 \psi_3 \psi_4$ representing the reaction H + H ⟶ H + H, for example.

These are examples of a true unification in which the new coupling constants g_α, g_{chem} are completely eliminated in terms of the underlying electromagnetic coupling constant e. We could have put these interactions together with the electromagnetism into a larger system, and construct a gauge field theory with a broken symmetry. There is in fact a gauge theory of chemical forces plus electromagnetic forces. Instead g_{chem} and g_α are calculated in terms of e. Coming back to the four fundamental interactions of physics, weak, electromagnetic, strong and gravitational, the unification attempts of the recent decades have concentrated on putting all these separate forces side by side into a larger system and attribute the differences to symmetry breaking. The alternative that we wish to explore is to derive and calculate these interactions from electromagnetism which would be an already unified theory with possibly a single coupling constant e. We should remember that Newton's unification of terrestrial and celestial gravity, and Maxwell's unification of electricity and magnetism have a single coupling constant; the magnetic coupling constant is calculated in terms of e and c, $g_{magn} = e/c$, and c is known by independent measurements. These are true unifications.

With these remarks on the fundamental significance and possible universal role of electromagnetism I shall discuss the following topics in symmetry:

1) Symmetries of and from the Maxwell-Dirac System
2) Symmetries of electromagnetic matter: Two or many body systems that bind electromagnetically, and lead eventually to macroscopic symmetries
3) Symmetries of particle physics. The extrapolation of QED to short distances and a possible phase transition of QED at short distances to strong interactions.
4) Symmetries of the electron itself, the most basic of all particles.

II. Symmetries We Have Learned from Electrodynamics

Space-Time Symmetries. Maxwell's equations gave us the notion of the invariance under Lorentz transformations (Waldemer Voigt (1887) and Hendrik A. Lorentz), this in turn the relativistic particle equations (H. Poincaré (1904), and finally the physical interpretation of simultaneity and a new definition of inertial frames (A. Einstein (1905)). Thus the special relativistic symmetries of space-time originate from electrodynamics.

The electromagnetic field concept is in fact intimately related with the structure of space and time. The electromagnetic fields were originally thought to be, like in any other wave phenomena, waves in a medium, in an aether which fills and defines the whole space. This cannot be a rigid aether (Michelson-Morley experiment), but a deformable aether is perfectly relativistic. After the special relativity, we have gotten used to talk about waves without a medium (like ocean waves without the ocean) at the expense of introducing new physical quantities, e.g. potentials $A_\mu(x)$, whereas in a medium $A_\mu(x)$ would be simply the displacement of the aether from its equilibrium position. Either way, it is the wave operator, $\Box$, that makes the space-time of particle physics as we know it now.

Discrete Symmetries. The wave equation, or more generally the Maxwell's equations define also invariance under space-reflection (Parity P), and invariance under time reflection (time reversal T). There is one other important discrete symmetry, particle-antiparticle conjugation C which comes from the properties of the Dirac current $j^\mu(x)$ on the right hand side of the Maxwell's equations:

$$F^{\mu\nu}{}_{,\nu} = - j^\mu = - e\bar{\psi}\gamma^\mu\psi$$

$$*F^{\mu\nu}{}_{,\nu} = 0 \qquad (1)$$

Without introducing any new coupling constants we can work with an electric charge - magnetic charge symmetry by replacing the second equation in (1) by

$$*F^{\mu\nu}{}_{,\nu} = - k^\mu \qquad (2)$$

where k^μ is the current of magnetic monopoles. The coupling constant g in k^μ is determined by e by the charge quantization relation $eg = n/2$; $n = 0, 1, 2 \ldots$. Since no free magnetic monopoles have been discovered, we work usually in the sector $n = 0$.

Currents. The left hand side of Maxwell's equations have not changed since Maxwell even in quantum electrogynamics. The right hand side, the form of the current j^μ, describing the matter, has undergone considerable change, however, from the macroscopic currents of Maxwell, to the current of classical point charges of Lorentz, to the distributed currents of Schrödinger, and finally to the currents of spinning electrons of Dirac. The Dirac current describes both electrons and positrons, and to every motion of an electron there is another symmetric motion of the positron (particle-antiparticle symmetry C).

The electrodynamics gives us thus the full relativistic invariance (Poincaré symmetry with space and time reflections, P and T) as well as particle-antiparticle conjugation C due to the properties of the electronic current.

There is no indication anywhere else that the proper relativistic invariance is broken. As to the discrete symmetries P, C and T their apparent violation in some processes is best understood from the peculiar structure of the underlying particles, like neutrino and K_0- mesons.

Conformal Symmetry. Electrodynamics gave us also the conformal symmetry, a symmetry which contains in addition to the Poincaré Symmetry of Lorentz transformations and translations, also the dilations and inversions of coordinates. This 15-parameter symmetry group is first a property of the free electromagnetic field, but can be expanded to full electromagnetism if the current j_μ has special transformation properties.

The additional symmetry operations can be interpreted as the change of scales from point to point.

Self Consistent Coupling Between Field and Matter. Eq's. (1) are only one-half of electrodynamics. The other half describes the dynamics of the source-current, the Dirac equation

$$\{\gamma^\mu(i\partial_\mu - eA_\mu) - m\}\,\psi = 0\;, \tag{3}$$

or, a classical equation for particles, if the source is a point charge.

As H.A. Lorentz taught us, the electromagnetic field $F_{\mu\nu}$ is produced by the source, which in turn is driven by the field, hence both must be treated self-consistently. The way to guarantee this self-consistency is to derive both equations (1) and (3) from a single variational principle: The action of electrodynamics is

$$W = \int \{\bar{\psi}(\gamma^\mu i\partial_\mu - m)\bar{\psi} - ej^\mu A_\mu - \tfrac{1}{4}F_{\mu\nu}F^{\mu\nu}\}\, d^4x \tag{4}$$

where the first term is the kinetic energy of the source, the last the kinetic energy of the field, and the middle term the interaction between the field and matter. According to Leibnitz we live in the best of all possible worlds, and the extremum of the action W gives us not only both equations (1) and (3), but also the conservation laws and symmetries of electrodynamics that we talked about, and the consistency of these conservation laws with the time-evolution of the system.

III. Symmetries of the Electromagnetic Matter

Electromagnetic matter consists of the bound states and other clusters of basic particles formed solely by their electromagnetic interactions. Thus starting with electrons, muons and protons we have the formation of positronium, muonium, H-atoms, as well as, of course, other atoms, molecules, biological molecules, up to crystals and condensed matter. The structure of proton and nuclei plays a very little role (e.g., hyperfind structure).

I shall first show how the symmetry and dynamical groups of the two-body problem follows from the basic action of electrodynamics, and how we obtain in the limit the dynamical groups of the one-body problem in a potential--a problem that has been widely studied since the 1960s. In particular, the postulated infinite-component wave equations can now be given a field theoretic derivation based on electrodynamics. In order to derive the equations and symmetries of the two-body system from field theory we consider two matter fields ψ_1 and ψ_2 as the source of the current and the action (4) becomes

$$\begin{aligned} W = \int dx[&\bar{\psi}_1(\gamma^\mu i\partial_\mu - m)\psi_1 + \bar{\psi}_2(\gamma^\mu i\partial_\mu - m_2)\psi_2 \\ &-(e_1\bar{\psi}_1\gamma^\mu\psi_1 + e_2\bar{\psi}_2\gamma^\mu\psi_2)A_\mu - \tfrac{1}{4}F_{\mu\nu}F^{\mu\nu}] \end{aligned} \tag{5}$$

If we choose a gauge such that $A^\mu, \mu = 0$, then the first Maxwell eq. (1) becomes $\Box A^\mu = j_\mu$ and can be solved for A_μ

$$\Box A_\mu = \int dy\, D(x-y)\,[e_1\bar{\psi}_1\gamma_\mu\psi_1 + e_2\bar{\psi}_2\gamma_\mu\psi_2] \tag{6}$$

where $D(x-y)$ is the Green's function of the wave operator $\Box$. The potential A_μ in (6) -- it is the Lienard-Wiechert potential--can be called the self-field of the electron: The Dirac current, $e\bar{\psi}\gamma^{\mu}\psi$, is assumed to produce a field, like any other current, which acts not only on the other particles, but also on itself.

Inserting A_{μ} of (6) into W (also in the term $F_{\mu\nu}F^{\mu\nu}$) we obtain a total interaction action (kinetic parts being unchanged)

$$W^{int} = \int dx dy\ [j_{1\mu}(x) + j_{2\mu}(x)] D(x-y) [j_1^{\mu}(y) + j_2^{\mu}(y)] \qquad (7)$$

representing two mutual interactions and two self-interactions of currents. There are two variational principles for our action (5)-(7).

(*i*) If we vary W with respect to ψ_1 and ψ_2 separately, we obtain coupled nonlinear Hartree-type equations. I call such a system a "separated two-body quantum system"

(*ii*) We can vary W with respect to the composite field

$$\Phi(x_1x_2) = \psi_1(x_1)\psi_2(x_2) \qquad (8)$$

only. We now get a linear equation for Φ only (plus some extra-terms coming from self-interactions). This is the quantum mechanical two-body equation in configuration space, well-known in the standard nonrelativistic many-body problem. Once such a wave equation for $\Phi(x_1x_2)$ is written, the solution is no longer factorizable. I call such a system a "nonseparated quantum 2-body system." The peculiar long-range correlations of quantum 2-body systems (e.g. Einstein-Podolsky-Rosen problem) are due to this configuration space wave functions.

I think both types of systems, separated and nonseparated, occur in quantum systems. For example, in the H_2-molecules, the two electrons are not separated, their wave functions are in the tensor product space and must be antisymmetrized with output to the exchange of the two electrons; they mix. But the two protons are separated. We do not antisymmetrize the wave function with respect to the protons. The Born-Oppenheimer method treats protons and electrons on different footing. I think this is also physically so: an H_2-molecule is an individual single system defined by the positions of the nuclei; we may use the probabilities for the distribution of the two electrons.

The two-body relativistic equation for the composite field $\Phi(x_1x_2)$ is obtained as follows. We first express the mutual interaction action in (7) in terms of Φ:

$$W^{int} = \int dx_1 dx_2 \bar{\Phi}(x_1x_2)\gamma^{\mu} \otimes \gamma_{\mu}\Phi(x_1x_2)D(x_1x_2) \qquad (9)$$

Φ is a 16-component spinor with two spinor indices $\Phi_{\alpha\alpha'}$, so this equation has to be understood in the tensor product space of two Dirac Spin algebra, i.e.

$$\bar{\Phi}_{\alpha\alpha'}\ \gamma^{\mu}_{\alpha\beta} \otimes \gamma_{\mu\alpha'\beta'}\ \Phi_{\beta\beta'}$$

The self-energy terms are a bit complicated; we shall indicate them separately at the end.

In order to write the two kinetic energy terms in eq. (5) in terms of Φ, we multiply the first term in W with the normalization integral $\int d\vec{x}_2\ \bar{\psi}_2\ \gamma \cdot n \psi_2$ and the second term with the integral $\int d\vec{x}_1\ \bar{\Psi}_1\ \gamma \cdot n\ \Psi_1$. In this way also the kinetic energy terms can be written in terms of

Φ and $\bar{\Phi}$. We now vary the action with respect to $\bar{\Phi}$ and obtain the two-body equation:

$$\{(\gamma^\mu i\partial_\mu - m_1) \otimes \gamma\cdot n + \gamma\cdot n \otimes (\gamma^\mu i\partial_\mu - m_2) - e_1 e_2 \frac{\gamma^\mu \times \gamma_\mu}{r_\perp} + V_{self}\}\ \Phi = 0 \quad (10)$$

Here we always write in the tensor product the spin matrices and indices of the particle 1 first, those of the second particle second, e.g. $\gamma_\mu \times \gamma^\mu$. Further, n^μ is a time-like four-vector and $r_\perp = [((x_1-x_2)\cdot n)^2 - (x_1-x_2)^2]$ is the relativistic distance. We can choose $n^\mu = (1000)$, then $r_\perp = r$. Self energy terms V_{self} we shall explain later. Equation (10) is fully covariant, and more importantly contains a single time. In order to see this we introduce center of mass and relative coordinates by the usual transformations.

$$P = p_1 + p_2,\quad r = x_1 - x_2,\quad p = (1-a)p_1 - ap_2,\qquad R = ax_1 + (1-a)x_2$$

and obtain the equation

$$\{\Gamma^\mu P_\mu - g^\mu p_\mu - m_1 I \odot \gamma\cdot n - m_2\, \gamma\cdot n \odot 1 - e_1 e_2 \frac{\gamma^\mu \times \gamma^\mu}{r} + V_{self}\}\ \Phi = 0 \quad (11)$$

where $\Gamma^\mu = a\gamma^\mu \otimes \gamma\cdot n + (1-a)\gamma\cdot n \otimes \gamma^\mu$

$$g^\mu = \gamma^\mu \otimes \gamma\cdot n - \gamma\cdot n \otimes \gamma^\mu$$

Now we see indeed that component of g^μ parallel to n^μ vanishes identically. In fact separating all four-vectors into a component parallel to n^μ and another perpendicular to n^μ, e.g.

$$\Gamma^\mu_\parallel = (\gamma\cdot n)n^\mu,\quad \gamma^\mu_\perp = \gamma^\mu - (\gamma\cdot n)n^\mu\ ,$$

we obtain

$$\{P_\parallel - \alpha_\perp\cdot P_\perp - g_\perp\cdot p_\perp - m_1 \gamma\cdot n \otimes I - m_2 I \otimes \gamma\cdot n - e_1 e_2 \frac{1}{r_\perp}(1 - \alpha_{1\perp}^\mu \cdot \alpha_{2\perp}^\mu)$$

$$+ \tilde{V}_{self}\}\ \Phi = 0 \quad (12)$$

$P_\parallel$ is the Hamiltonian

$$\alpha^\mu_\perp = a\gamma\cdot n\gamma^\mu \otimes I + (1-a)\ I \otimes \gamma^\mu_\perp \gamma\cdot n$$

$$g^\mu_\perp = \gamma\cdot n\ \gamma^\mu_\perp \otimes I - I \otimes \gamma^\mu_\perp\ \gamma\cdot n$$

Or, for $n^\mu = (1000)$,

$$\{P_0 - \vec{\alpha}\cdot\vec{P} - \vec{g}\cdot\vec{p} - m_1\beta_1 - m_2\beta_2 - \frac{e_1 e_2}{r}(1 - \vec{\alpha}_1\cdot\vec{\alpha}_2) + \tilde{V}_{self}\}\ \Phi = 0$$

$$\vec{\alpha} = a\vec{\alpha}_1 + (1-a)\vec{\alpha}_2,\quad \vec{g} = \vec{\alpha}_1 - \vec{\alpha}_2 \quad (13)$$

For Coulomb problem the operator

$$\Theta = r P_0$$

has a simpler spectrum than $P_0 = H$ itself:

$$\Theta = r\vec{\alpha}\cdot\vec{P} + \vec{g}\cdot r\vec{p} + (m_1\beta_1 + m_1\beta_2)r + e_1 e_2(1 - \vec{\alpha}_1\cdot\vec{\alpha}_2) + r\tilde{V}_5 \quad (15)$$

Dynamical Symmetries. For a fixed center of mass momentum $\mathcal{P}$(e.g. $\mathcal{P} = 0$), the operators $\hbar$, $\hbar p$ are in the Lie algebra of SO(4,2), the well-known dynamical group of the H-atom without spins. The coefficients $\alpha\cdot\mathcal{P}$, g, $(m_1\beta_1 + m_2\beta_2)$, $\alpha_1\cdot\alpha_2$, . . . are in the Lie algebra of the tensor product of two Dirac representations of SO(4,2). Thus, the full dynamical group of the two-body problem with spins is, as expected, $SO(4;2)_{orbital} \otimes SO(4,2)_{Dirac} \otimes SO(4,2)_{Dirac}$.

Neglecting self-energy terms V_{self} which for atoms contains small terms like Lamb-shift and spontaneous emission, we can pass easily to the limits of first to one-body Dirac equation, then to spinless case, and finally to nonrelativistic Kepler problems.

If one of the particles is heavy, $m_2 \to \infty$, $\overset{(2)}{\gamma_0} \sim 1$, $\overset{(2)}{\vec{\gamma}} \to 0$, $a = {m_1}/M \to 0$, the Hamiltonian of the first particle in the center of mass frame with $P_0 = p_0^{(1)} + m_2$

$$\Theta^{(1)} = \hbar P_0^{(1)} = \vec{\alpha}\cdot\hbar\vec{p}_1 + \hbar\beta m_1 + e_1 e_2 \tag{15}$$

i.e. the one-body Dirac Hamiltonian.

Furthermore, eq. (11) written concisely as

$$\{\Gamma^\mu \mathcal{P}_\mu + K\}\ \Phi = 0 \tag{16}$$

where K is a matrix, is an infinite-component wave equation which, as is also well-known, describe composite systems realistically.

To summarize, the dynamical group approach to quantum systems, in particular the infinite component wave equation, can be derived from first principles from an underlying electron dynamics field theory. In this way the parameters of the infinite component wave equations are determined in terms of the masses and coupling constants of the basic constituents.

Finally we remark that the full dynamical algebra of our system (13) or (16), when the generators of the Poincaré group $P_0, \vec{P}_1$ . . . are included, will be an infinite-dimensional algebra of the Kac-Moody type. This is because, for a composite system we have a highly reducible representation of the Poincaré group representing the infinitely many mass and spin states, and the generators of the dynamical group connect these different mass states.

From Microscopic Symmetry to Macroscopic Symmetry

The previous method of deriving equations for 2-body system from field theory can be extended to 3 or more particles. But which many body systems do actually lead to stable (or nearly stable) bound states is a question of the sizes of the parameters; the stability of atoms depends on the nuclear charge Z, for example.

But a new situation occurs when we go from atoms to molecules. Consider the simplest molecule H_2 consisting of two hydrogen atoms. It does not seem to be possible to understand H_2 starting from a 4-body Schrödinger equation representing two protons and two electrons. Rather, as we have already mentioned, we have spontaneously broken the permutation symmetry for the two-protons, that is "separate" the two protons and apply quantum mechanics only to the electrons. The Born-Oppenheimer method fixes the positions of the protons, treats the two electrons quantum mechanically, and then considers the small oscillations of the two protons separately. In doing this we take into account that a molecule is a definite single objective quantum system, and not a probability distribution. The individuality of quantum systems is established, I think, at this level. Another way of expressing the Born-Oppenheimer procedure is to say that the distance R between the protons obeys at first a superselection rule, i.e. it is not quantized. After solving the problem with fixed R we allow for oscillations of the protons around their center of mass but without mixing. In fact it is possible to treat the two-body problem in quantum mechanics without quantizing the center of mass: only relative coordinates are quantized; one obtains the same result as the usual theory. Super-selection rules are the proper way to deal with nonquantized dynamic variables.

Continuing further from molecules to more complex systems, I have discussed in "Symmetry in Science II" the question whether we can derive crystal symmetry from first principles, i.e. from an N-body Schrödinger equation for $\psi(x_1 \ldots x_N, R_1 \ldots R_m)$ for N electrons and M nuclei. The answer, I believe, is no. The positions of the nuclei, i.e. the crystal symmetry, h are determined by essentially classical equilibrium or group theoretical arguments. The large permutation and rotation symmetries of nuclei is broken down to a smaller crystal symmetry. But once crystal symmetry is given, we can then quantize the electrons in this given environment in which the electron clouds spread and mix.

These are, I think, limitations to the unquestioned use of quantum rules to the structure of matter.

IV. The Extrapolation of Electrodynamics to Short Distances

We have discussed the formation of two- or more-body bound states in electrodynamics corresponding to atomic and molecular structure. In these instances the dominant force is the mutual interaction between the particles; the self-fields give for these states only small corrections (Lamb shift, spontaneous emission). The size of the atomic structures are determined by the constants: m(mass of the electron), α(fine structure constant), c and $\hbar$, and nuclear charge Z.

There is, however, a second regime where the self potentials dominate and the mutual forces between the particles are small corrections. To see this we go back to the general action (7). The self-interaction terms are:

$$W_{int}^{self} = e_1^2 \int dxdy\, \bar{\psi}_1(x)\gamma^\mu\psi_1(x)\, D(x-y)\bar{\psi}_1(y)\gamma_\mu\psi_1(y)$$

$$+ e_2^2 \int dxdy\, \bar{\psi}_2(x)\gamma^\mu\psi_2(x) D(x-y)\bar{\psi}_2(y)\gamma_\mu\psi_2(y) \qquad (17)$$

When we pass to the composite field $\bar{\Phi}(x_1 x_1)$ defined by eq. (8), we can evaluate the self potential V_{self} in eq. (10) and rewrite eq. (10) now completely as

$$\{[\gamma^\mu(p_{1\mu}-e\, A_\mu^{(1)})-m_1] \otimes \gamma\cdot n+\gamma\cdot n \otimes [\gamma^\mu(p_{2\mu}-e_2\, A_\mu^{(2)})-m_2]\}\, \Phi\,(x_1,x_2) = 0$$

$$A_\mu^{(1)} = \frac{1}{2}e_2\frac{\gamma_\mu\gamma\cdot n}{r} + A_\mu^{(1)\,self}$$

$$A_\mu^{(2)} = \frac{1}{2}e_1\frac{\gamma\cdot n\gamma_\mu}{r} + A_\mu^{(2)\,self} \qquad (18)$$

where the self-potentials are given by

$$A_\mu^{(1)\,self}(x_1) = \int dzdu\; \mathcal{D}(x_1-z)\bar{\Phi}(z,u)\gamma_\mu \otimes \gamma\cdot n\; \Phi(z,u)$$

$$A^{(2)\,self}(x_2) = \int dzdu\; \mathcal{D}(x_2-u)\bar{\Phi}(z,u)\gamma\cdot n \otimes \gamma_\mu \Phi(z,u)$$

In the center of mass frame, the Hamiltonian form of (18) is

$$\{\vec{\alpha}_1\cdot(\vec{p}_1-e_1\vec{A}_1)+\beta_1 m_1+\vec{\alpha}_2\cdot(\vec{p}_2-e_2\vec{A}_2)+\beta_2 m_2+e_1 V_1+e_2 V_2\}\Phi(x_1,x_2)$$
$$= E\Phi(x_1,x_2) \tag{19}$$

where

$$\vec{A}_1 = \frac{e_2}{2}\frac{\vec{\alpha}_2}{\hbar} - \frac{e_1}{2}\int dzdu\; \mathcal{D}(x_1-z)\Phi^+(z,u)\vec{\alpha}_1\Phi(z,u)$$

$$\vec{A}_2 = \frac{e_1}{2}\frac{\vec{\alpha}_1}{\hbar} - \frac{e_2}{2}\int dzdu\; \mathcal{D}(x_2-z)\Phi^+(z,u)\vec{\alpha}_2\Phi(z,u)$$

$$V_1 = \frac{e_2}{2}\frac{1}{\hbar} - \frac{e_1}{2}\int dzdu\; \mathcal{D}(x_1-z)\Phi^+(z,u)\,\Phi(z,u)$$

$$V_2 = \frac{e_1}{2}\frac{1}{\hbar} - \frac{e_2}{2}\int dzdu\; \mathcal{D}(x_2-z)\Phi^+(z,u)\,\Phi(z,u)$$

Equation (18) is exact from the point of view of our classical relativistic self consistent field theory, and from the point of view of interpreting ψ as an objective representation of electronic matter. But equation (18)-(19) are rather complicated. We write it for the case when one of the particles is very heavy, as in H-atom, and its field is represented by a fixed external Coulomb potential.

$$[\gamma^\mu(i\partial_\mu - eA_\mu^{ext})-m]\,\psi = e^2\gamma^\mu\psi(x)\int dy\bar{\psi}(x)\gamma^\mu\psi(x)\;\mathcal{D}(x-y)\bar{\psi}(y)\gamma_\mu\psi(y) \tag{20}$$

This is now a single nonlinear integrodifferential equation for a single particle in an external field. If the self-energy term on the right hand side of this equation is treated iteratively, around the stationary solutions of the external field problem, then it is possible to obtain all the radiative corrections of quantum electrodynamics, i.e.

(*i*) spontaneous emission
(*ii*) Lamb shift
(*iii*) vacuum polarization
(*iv*) anomalous magnetic moment,

without second quantization of fields. Thus quantized electromagnetic field on the one hand, and self-field on the other hand are two dual ways of dealing with radiative processes. But now we can extrapolate eq. (20) to short distances, whereas we cannot do this in the perturbative QED--it would mean to be able to sum infinitely many Feynman diagrams.

Nonlinear equations of the type (20) have another regime in which the nonlinearity dominates over the external field; we may get new type of solutions corresponding to a self-focusing, or self-organization of the system. Such localized solutions are known for equations having soliton solutions. It has then been conjectured that electrodynamics should exhibit a phase transition at short distances to new self-organized states. To be more specific the electron-positron system (e^+,e^-), for example, which we know to form the atomic positronium at large distances due to mutual interaction, should also form new states at short distances due to their own self fields. Now one of the effects of self energy is to give to the particles an anomalous magnetic moment whose value depends on the external field itself, self-consistently. In many models with an anomalous magnetic moment it is possible, in fact to show that new states of (e^+e^-) occur at distances of the order of the classical electron radius (α/m), which is also the typical hadronic distance. The masses of such states are multiples of $2m/\alpha \sim 140$ MeV.

This basic idea of a phase transition in QED at short distances gives us a possibility to re-examine the new particles, heavy leptons, mesons and baryons, as new composite states of electromagnetism. Although the dynamics is difficult and not completely solved, it is possible to completely understand the kinematics, that is the classification of particles and their quantum numbers on the basis of two fundamental absolutely stable particles only, the electron and its neutrino. All other particles, according to this view, are composite and unstable--but two of them apparently with an extremely long life-time, proton and muon-neutrino.

V. The Symmetries of the Electron

Finally, I discuss some new results concerning the structure of the electron itself, the most fundamental of all particles (see Sect. IV). It is in the structure of the electron that we must look for the origin of the many rather mysterious qualities of the electron; the spin, the charge, Pauli exclusion principle, the existence of antiparticles, and the existence of its partners, electron neutrino on the one hand and heavy leptons (μ,τ) on the other hand. The electron, for all these properties, is more than just a point particle, or an irreducible representation of the Poincaré group. Most importantly, the symmetry between heavy leptons, that is the identical behavior of e, μ, τ leptons excepting their mass, seems to defy any explanation so far. This problem is known as the existence of three families of leptons each with their own neutrinos. In standard models all these leptons are assumed to be elementary. The structure of the electron and its self field may give us a possibility that these heavy leptons are in fact in some sense "excited states" of the electron itself.

An intuitive picture of the Dirac electron can be obtained by a classical model which gives us very naturally the origin of spin and antiparticles; it may also lead to "excited" states. This classical, but of course covariant, model is most simply described in terms of an invariant time parameter τ by two pairs of conjugate variables: (x_μ, p_μ), the usual coordinates and momenta, and the internal spin variables $(\bar{z}, z)$, where $z_\alpha(\tau)$ are classical 4-component spinors with $\bar{z} = z^+\gamma^0$ its injugate. The theory is defined by the action

$$W = \int d\tau (i\lambda \dot{\bar{z}} z + p_\mu \dot{x}^\mu - p_\mu \bar{z}\gamma^\mu z + eA_\mu \bar{z}\gamma^\mu z) \qquad (21)$$

up to a total time derivative. It is thus formulated on a larger phase-space but it is a symplectic Hamiltonian system. The two fundamental constants are λ, with the dimension of action, and e · $(c=1)$. The mass enters later as the value of the integral of motion $H = \bar{z}\gamma^\mu z(p_\mu - eA_\mu)$, the "Hamiltonian with respect to τ" (or the mass). When this theory is quantized--either canonically by replacing Poisson-brackets with commutators, or by a Schrödinger quantization, or by a path integral quantization--one obtains the Dirac equation. But the concepts of spin and antiparticles are already present in the theory (21) as follows. If we solve the equations of motion resulting from (21) for a free particle, we find that the natural motion of the particle in space time is a helix, around a fictitious center of mass which moves linearly like a relativistic particle. The frequency and radius of the helical motion are $2m$ and $\frac{1}{2}m$, respectively. Now the spin of the particle turns out to be simply the orbital angular momentum of the helical motion with respect to the center of mass. Furthermore, particles and antiparticles correspond to the positive and negative frequencies of helical motion, or right and left helical motions.

We thus see that such a microscopic dynamical system as described by eq. (21) has many remarkable symmetry properties which are then transformed to complex systems that they form. We should like to emphasize that no "force" is necessary to keep the particle in a helical path. The existence of internal variables does it automatically. This is a beautiful example of Heinrich Hertz's "forceless mechanics," where a geometry in a higher dimensional space implies forces in ordinary space-time. Another way to put this is to say that electron has other coordinates than just position and momenta.

It remains to be seen whether the internal structure of the electron can give us a deeper understanding of the existence of the heavy leptons, and why every lepton comes with its own neutrino.

CONCLUSIONS

The electrodynamics has been, since its conception over a hundred years ago, a most enduring theory. It has unified an enormous range of phenomena under one simple set of laws. These are the hallmarks of true scientific knowledge: general validity, extreme simplicity, freedom from arbitrary parameters. Quantumelectrodynamics is rightly called the best theory that physicists have ever built. In contrast the current formulations of the physics of nuclei and particles have shown such as complexity, hundreds of fundamental objects, many new forces and models with dozens of parameters, that we are undoubtedly far from a basic understanding of these phenomena. It is generally believed that these phenomena have nothing to do with electromagnetism.

In this essay I have tried to show not only the central role of electromagnetism in our understanding of the structure of all ordinary matter, but also the exciting possibility that the behavior of electromagnetic interactions at short distances is very likely to be quite different than at large distances, and that they undergo a phase transition and might very well explain the occurrence of multiples of new particle states at high energies. Simplicity may again be restored as the most important feature of the scientific endeavour.

REFERENCES

More technical details of the topics discussed in this essay may be found in the following reviews and in references listed therein.

A.O. Barut, Physica Scripta, 35, 229(1987).
A.O. Barut, Annalen der Physik, 43, 83(1986).
A.O. Barut, Physica Scripta, T21, 18(1988).
A.O. Barut, Foundations of Physics, 17, 549(1987), 18, 95(1988).

See also

A.O. Barut, in Symmetry in Science II (edit. by B. Gruber), Plenum Press, 1987; p. 37.

RECENT PROGRESS IN IMPLEMENTING THE TENSOR OPERATOR CALCULUS *,**

L. C. Biedenharn

Department of Physics
Duke University
Durham, NC 27706

and

Department of Physics
University of Texas
Austin, TX 78712

R. Le Blanc

Department of Physics
McGill University
Montreal, Quebec, Canada H3A 2T8

J. D. Louck

Theoretical Division
Los Alamos National Laboratory
Los Alamos, NM 87545

1. INTRODUCTION

We are indebted to the organizers of the conference and especially Professor Bruno Gruber for this opportunity to review recent progress in implementing the tensor operator calculus. This subject is fundamental for the symmetry approach to quantum physics, and–as is well-known–has extensive applications to nuclear structure physics, nuclear collective motion and quark models in particle physics,to name only a few of the more important examples.

The explicit construction, algebraically, of all tensor operators for the compact Lie groups would, in itself, provide a very large class of implementable physical models realizing group symmetries and adapted to large scale calculations on complex, many-body, physical systems. One realizes at once, however, that such an undertaking is a formidable task, feasible only if the problem can be broken into smaller pieces treated recursively. For the family of groups $U(n), n = 2, 3, \cdots$, the canonical reduction: $U(n) \supset U(n-1) \supset \cdots \supset U(2)$ allows the tensor operators to be symmetry adapted to this chain in a modular fashion: this is the concept of the projective unit tensor operators (also called reduced Wigner coefficients), which we discuss in Section 2. We will confine our review to $U(n)$ although the techniques can in principle be extended

* Invited paper presented at the Symmetries in Science III Conference, July 24-28, 1988, at Bregenz, Austria.

** Supported, in part, by the National Science Foundation and the Department of Energy.

to all compact Lie groups. We will focus our attention on three general aspects of the subject to which there have been significant recent contributions:

(i) The relation between *characteristic null space* and the enumeration of a basis for *unit tensor operators*;

(ii) The use of *vector coherent state* methods and the extension of this technique to tensor operators;

(iii) The nature of the *operator pattern labelling* for tensor operators.

The relation between characteristic null space and the enumeration of unit tensor operators in $U(n)$, topic (i) above, is inherently a detailed study[0] of the properties of the Littlewood-Richardson numbers (the multiplicity of the irreps occurring in a Kronecker product of irreps). We discuss in Section 2 a significant result obtained by Baclawski,[1] and its implications for the canonical labelling of unit tensor operators in $U(n)$. The basic problem at issue, still not fully resolved for $n \geq 4$, is to determine to what extent tensor operator structures are essentially unique and not merely well defined. (Certainly, one knows that the algebraic constructs will embody phase conventions—which is trouble enough!—but if the structure itself were to be "author-dependent", this would be most unfortunate.)

The induction process, generating representations of a group from representations of a sub-group is a well-known technique in group representation theory. Something of a break-through in this field has been achieved in the last few years by the introduction of "vector coherent states", our second topic. This development is associated primarily with the work of Deenen and Quesne,[2] Hecht,[3,4] Rowe,[5] and Le Blanc,[6] and is reviewed in Section 3. The more recent[7,8] extension of the VCS technique to tensor operators is discussed in Sections 4 and 5. The VCS construction for tensor operators, as will be clear from these sections, has the very great advantage of *building* $U(n)$ *objects by* $U(n-1)$ *constructs*, exactly the modular procedure one would have hoped for.

The explicit results (for general $U(n)$) obtained by VCS methods in Section 5 is then discussed in Section 6 in terms of the operator pattern labelling, topic (iii) above.

Finally in Section 7 we discuss several open problems including some brief remarks about "quantum groups".

Group representations and tensor operators have been a rich source for the development of special functions. Space limitations dictated that we eliminate our intended survey of the considerable progress made recently in this field, especially by Milne and Gustafson among others. We have also not been able to review very recent work by Klimyk[9] and by Alisauskas.[10] These authors take a very different approach to tensor operator construction by abandoning the canonical (orthogonal) operators in favor of non-orthogonal operators.

2. PROGRESS IN DETERMINING CANONICAL OPERATOR STRUCTURES FOR $U(n)$

Let us begin by defining an explicit space on which our operators are to act. For the group $U(n)$, define the Hilbert space:

$$\mathcal{H} \equiv \sum_{[m]} \oplus \mathcal{H}([m]), \tag{2.1}$$

constructed as the direct sum of carrier spaces $\mathcal{H}([m])$ for each unirrep of $U(n)$, taken once and only once, subject, however, to the equivalence relation:

$$\mathcal{H}([m_{1n}m_{2n}\ldots m_{nn}]) \cong \mathcal{H}([m_{1n}+k, m_{2n}+k, \ldots, m_{nn}+k]), \quad k \text{ integral.} \tag{2.2}$$

The imposition of the equivalence relation shows that we actually have constructed a model space[11] for the simple group $SU(n)$. The formal extension from $SU(n)$ to $U(n)$ affords great simplifications in manipulating the irrep changes induced by $SU(n)$ tensor operators. Since $U(n)$ tensor operators are at the same time also $SU(n)$ tensor operators, it is reasonable to continue using the convenient language of $U(n)$ tensor operators.

Consider now all linear transformations of $\mathcal{H}$ into $\mathcal{H}$, that is, to say matrices in the space $\mathcal{H}\times\mathcal{H}^{\text{dual}}$. What can one say as to a $U(n)$ symmetry adapted classification of all such operators?

It is well-known that one can classify these operators by their transformation properties under $U(n)$ into *tensor operator sets* labelled by a Young frame:

$$[M] \quad = \quad [M_{1n}M_{2n}\ldots M_{nn}], \tag{2.3}$$

and by a Gel'fand-Weyl pattern[12,13]

$$(M) \quad \equiv \quad \begin{pmatrix} M_{1n} & & M_{2n} & & \cdots & & M_{nn} \\ & M_{1,n-1} & & \cdots & & M_{n-1,n-1}, & \\ & & & \vdots & & & \\ & & & M_{11} & & & \end{pmatrix}, \tag{2.4}$$

with M_{ij} integral and obeying the constraints:

$$M_{ij} \geq M_{i,j-1} \geq M_{i+1,j}.$$

Any such tensor operator can be multiplied by an arbitrary function of the invariant operators in $U(n)$, so it is useful to limit the classification to *unit* tensor operators. We have still not achieved at this stage a unique specification, but it is clear that the labelling, so far, is canonical, that is, free of arbitrary choice to within equivalence. (The choice of order in enumerating the variables by $1, 2, \cdots, n$ is indeed arbitrary but any such choice is equivalent to any other under the permutation sub-group S_n.)

To specify the unit tensor operator sets further, one introduces the *shifts* $\{\Delta_{in}\}$ induced by the tensor operator $\mathcal{O}$:

$$\mathcal{O}: \quad [m_{in}] \to [m'_{in}] \quad \equiv \quad [m_{in} + \Delta_{in}]. \tag{2.5}$$

The shift $\Delta = (\Delta_{1n}, \Delta_{2n}, \cdots, \Delta_{nn})$ can be any weight of the operator irrep labelled by $[M]$.

Tensor operators belonging to *uniform* shifts, $\Delta_{in} = k$ $(\forall i, k)$, are the special class of operators considered in the Poincaré- Birkhoff-Witt theorem on the universal enveloping algebra of a Lie group.

Just as the weights of a vector in $\mathcal{H}[m]$ do not fully specify the vector, so also the shifts are an incomplete labelling of a unit tensor operator. It was conjectured early on that the way to complete the labelling was to extend the shifts to a pattern—called *the operator pattern* (Γ) —built in analogy to the Gel'fand-Weyl pattern (M).

THEOREM:[14] (Biedenharn, Giovannini and Louck 1967)

Unit tensor operators in $U(3)$ are canonically labelled by three patterns:
(i) A Young frame: $[M_{13}, M_{23}, M_{33}]$ specifying an irrep;
(ii) A unique vector of the irrep specified by a Gel'fand-Weyl pattern (M); and
(iii) A unique operator component specified by an operator pattern (Γ) whose weights are the shifts induced by the action of the operator.

A somewhat different view of tensor operators, as generalizations of the universal enveloping algebra, was given by Flath. A very interesting characterization (and canonical labelling) of the set of all tensor operators in $U(3)$ can be given as *an algebra realized in the universal enveloping algebra of* $so(8)$. Moreover:

THEOREM:[15] (Flath and Biedenharn 1985)

The algebra of tensor operators in $U(3)$ *contains no proper two-sided ideals.*

It was entirely unexpected that the $SU(3)$ algebra of tensor operators would be a 'single entity' in this way. By contrast the algebra of tensor operators for $U(4)$ appears to be enormously more complicated.

There is a quite different approach to a canonical labelling for operators based on the notion of null space. For tensor operators the relevant concept is that of *characteristic null space* (the additional requirement that the operator annihilates complete irrep spaces and not just individual vectors in the space). To what extent do these two distinct approaches to a canonical labelling coincide?

THEOREM:[16] (Louck and Biedenharn 1970)

The canonical labelling by operator patterns and by characteristic null space agree for $U(3)$, *and for all unit tensor operators in* $U(n)$ *belonging to irreps* $[k\ \dot{0}\ -k]$.*

Since the notion of characteristic null space is clearly independent of arbitrary choices, and moreover can be proved to admit a total ordering, it would appear that this approach could be definitive for all $U(n)$. Unfortunately, this hope cannot be realized:

THEOREM:[1] (Baclawski, 1984)

There exist distinct tensor operators in $U(n), n \geq 4$, *which belong to the same irrep and induce the same shifts, yet have identical characteristic null spaces.*

In the course of proving this theorem Baclawski developed a very effective technique for determining the characteristic null space.

THEOREM:[1] (Baclawski, 1984)

Consider a $U(n)$ *tensor operator* $\mathcal{O}$ *with Young frame* $[M]$ *and shift* Δ. *Let* d *be the multiplicity of this operator. Then there exists a sequence of dominant weights (irreps)* $b_1, b_2, \ldots, b_d$ *such that for every irrep* $[\lambda]$ *the multiplicity of the tensor operator* $\mathcal{O}$ *acting on the irrep space* $\mathcal{H}([\lambda])$ *is precisely the number of indices* i *such that* $[\lambda] \geq b_i$. *The sequence is unique to within order.*

The determination of what we shall call the "Baclawski sequence" $\{b_i\}$ or the "Baclawski points" (in the sense of the intertwining number null space diagram) makes possible a very straightforward determination of the characteristic null spaces. The procedure is most easily understood from examples.

Before proceeding to these examples let us remark that the b_i in the sequence $\{b_i\}$ are *not necessarily distinct*; there may be repetitions. Moreover, the multiplicity at a $[\lambda]$ irrep corresponding to a b_i need not necessarily be unity. The order relation $[\lambda] \geq b_i$ is to be understood as a *partial ordering* (see below). Finally we remark that the characteristic null space is actually best defined in terms of the Baclawski sequence itself. The reason is that there might exist "accidentally" irreps annihilated by the operator, and hence in the null space, but not connected in any known way

* A dot over a numeral means that the numeral is repeated as many times as necessary to complete the irrep labels.

with the null space multiplicity determined by the Baclawski sequence. (Examples are known in $U(2)$.)

As an example for determining the sequence $\{b_i\}$ let us consider the $\langle[210]\rangle$ tensor operator in $U(3)$, with $\Delta = (111)$. There are two operator patterns $\begin{pmatrix} 1 \\ 20 \\ 210 \end{pmatrix}$ and $\begin{pmatrix} 1 \\ 11 \\ 210 \end{pmatrix}$, so that $d = 2$. If we regard the operator patterns as state labels, then the associated Young-Weyl tableaux (also called "semi-tableaux") are:

$$\begin{pmatrix} 1 \\ 20 \\ 210 \end{pmatrix} \longleftrightarrow \begin{array}{|c|c|}\hline 1 & 2 \\ \hline 3 \\ \cline{1-1}\end{array} \implies b_1 = (1\ 0),$$

$$\begin{pmatrix} 1 \\ 11 \\ 210 \end{pmatrix} \longleftrightarrow \begin{array}{|c|c|}\hline 1 & 3 \\ \hline 2 \\ \cline{1-1}\end{array} \implies b_2 = (0\ 1).$$

The rule to determine the b_i from the associated tableau is this:

(a) the first entry is the number of unpaired 1's, where a 1 with a 2 on a row strictly below can be associated together and considered a "pair".

(b) the second entry is the number of unpaired 2's (pairing with respect to 3's),

(c) and similarly for the remaining $n - 1$ entries (for $U(n)$).

The b_i denote irreps which are points in the intertwining number null space diagram. In assigning the irrep labels to b_i, the first entry corresponds to $(m_{1n} - m_{2n})$, the second entry to $(m_{2n} - m_{3n})$, and similarly. Thus $b_1 = (10)$ corresponds to the $SU3$ irrep $[100]$; $b_2 = (01)$, to the irrep $[110]$. The partial order relation considers irreps $[\lambda]$ to be written in the form: $[\lambda] = (\lambda_{1n} - \lambda_{2n}, \lambda_{2n} - \lambda_{3n}, \cdots, \lambda_{n-1,n} - \lambda_{n,n})$. Then $[\lambda] \geq b = (b_{in})$ means that every difference is non-negative, that is: $(\lambda_{in} - \lambda_{i+1,n}) - b_{in} \geq 0,\ i = 1, 2, \cdots, n-1$.

Since b_1 and b_2 in this example are not comparable (neither is equal to or greater than the other), it follows that for $[\lambda] = b_i$, the multiplicity is unity. Clearly the points $(0\ x) \geq (0\ 1)$ if $x \geq 1$, but $(0\ x) \not\geq (1\ 0)$. Hence the *line* $(0\ x), x \geq 1$, has multiplicity 1. Similarly for the line $(x\ 0), x \geq 1$. The point $(0\ 0)$ has multiplicity 0, whereas the point $(1\ 1)$ has multiplicity 2. We have thus easily determined the *null space intertwining number diagram* for $\langle[210]\rangle, \Delta = (111)$ in $U(3)$.

This method generalizes such diagrams to $U(n)$ straightforwardly.

Given this information on the characteristic null space for $\langle[210]\rangle, \Delta = (111)$ how would one assign the correct operator pattern? The tensor operator $\langle[210]\rangle$ for $\Delta = (111)$ includes the generators, which are known to annihilate only the identity irrep, equivalent to the point (00). Thus the generators belong to the smaller of the two null spaces above. But this still doesn't answer the question: which operator pattern?

A hint as to the answer comes from a property of the Baclawski points: deleting the n's from the semi-tableaux associated to the points yields the Baclawski points for $U(n-1)$ tensor operators. In this sense, the set $\{b_i\}$ is *hereditary*.

Applying this to the $\langle[210]\rangle$, $\Delta = (111)$ operators (that is, removing the 3's in the example above), we see that: $b_1 \to (1), b_2 \to (0)$. Clearly b_2 inherits the smaller null space and indicates that the corresponding operator pattern, $\begin{pmatrix} 1 \\ 11 \\ 210 \end{pmatrix}$, should be assigned to the generators as the tensor operators having the smaller null space.

A more direct way to assign operator pattern labels would be most desirable. By contrast, every element of a Gel'fand-Weyl pattern can be obtained[16] as the eigenvalue of a constructible operator.

Recently an early conjecture[16] as to how to assign operator patterns has been proved.[18] To explain this result it is necessary to recall the definition of a *projective (tensor) operator.*

The concept of a projective operator has its origin in the elementary observation that each tensor operator in $U(n)$ determines a set of tensor operators in the subgroup $U(n-1)$. Since the unit tensor operators in $U(n-1)$ are a basis for all tensor operators in $U(n-1)$, we may expand the $U(n)$ unit tensor operator:

$$\left\langle \begin{matrix} (\Gamma)_{n-1} \\ [M]_n \\ (M)_{n-1} \end{matrix} \right\rangle = \sum_{(\gamma)_{n-2}} \begin{bmatrix} (\Gamma)_{n-1} \\ [M]_n \\ (\gamma)_{n-1} \end{bmatrix} \left\langle \begin{matrix} (\gamma)_{n-2} \\ [M]_{n-1} \\ (M)_{n-2} \end{matrix} \right\rangle, \tag{2.6}$$

where

(a) (subgroup condition): $\gamma_{i,n-1} = M_{i,n-1}$,

(b) the object in the square brackets on the right hand side denotes a *unit projective operator in the* $U(n)/U(n-1)$ *space* (thus a $U(n-1)$ invariant operator),

(c) $\langle [M]_{n-1} \rangle$ is a $U(n-1)$ unit tensor operator with upper operator pattern $(\gamma)_{n-2}$ and the lower (state label) pattern $(M)_{n-2}$ inherited from the $U(n)$ labels on the LHS.

(d) The operator pattern $(\gamma)_{n-2}$ in the $U(n-1)$ tensor operator also comprises the $n-2$ rows of the lower pattern $(\gamma)_{n-1}$ of the projective operator.

The projective operator in (2.6) acts on the factor space $U(n)/U(n-1)$ and, to be fully explicit, the matrix elements of the projective operators take the form

$$\left\langle \begin{matrix} [m+\Delta(\Gamma)]_n \\ [m+\Delta(\gamma)]_{n-1} \end{matrix} \right| \left| \begin{bmatrix} (\Gamma)_{n-1} \\ [M]_n \\ (\gamma)_{n-1} \end{bmatrix} \right| \left| \begin{matrix} [m]_n \\ [m]_{n-1} \end{matrix} \right\rangle, \tag{2.7}$$

where the $U(n-1)$ shifts are $\Delta(\gamma)$ and the $U(n)$ shifts, $\Delta(\Gamma)$.

Explicit matrix elements for projective operators, (2.7), are known for all *elementary* operators,[17] that is, tensor operators having irrep labels $[\dot{1}_k\ \dot{0}_{n-k}]$. It can easily be shown that for these matrix elements the limit $m_{nn} \to -\infty$ exists. Writing this limit in ket vector form, we have the operator relation:

$$\begin{aligned} &\lim_{m_{nn}\to-\infty} \begin{bmatrix} (\Gamma)_{n-1} \\ [\dot{1}_k \dot{0}_{n-k}] \\ (\gamma)_{n-1} \end{bmatrix} \left| \begin{matrix} m_{1n} & & \cdots & & m_{nn} \\ & m_{1,n-1} & \cdots & m_{n-1,n-1} & \end{matrix} \right\rangle \\ &= \prod_{i=1}^{n-1} (\delta^{\Gamma_{i,n-1}}_{\gamma_{i,n-1}}) \cdot \begin{bmatrix} (\Gamma)_{n-1} \\ (\gamma)_{n-1} \end{bmatrix}_{ext} \left| \begin{matrix} m_{1n} & \cdots & m_{n-1,n} & \\ & m_{1,n-1} & \cdots & m_{n-1,n-1} \end{matrix} \right\rangle. \end{aligned} \tag{2.8}$$

Note that—because of the δ- function constraints—the two operator patterns for the extended projective operator (defined by this limit) have irrep labels in $U(n-1)$ in common, and thus the operator could also be written in standard projective operator form with the shared $U(n-1)$ irrep labels

$[\Gamma_{1,n-1}\cdots\Gamma_{n-1,n-1}] = [\gamma_{1,n-1}\cdots\gamma_{n-1,n-1}] \equiv [M_{1,n-1}\cdots M_{n-1,n-1}]$ displayed as a single row. The operator is denoted as *ext* (for "extended") since it acts, by definition, on *a new space of* $U(n)/U(n-1)$ *ket vectors in which the label* m_{nn} *is deleted.*

This result may appear *ad hoc*, but it is well defined: *the pattern calculus rules*[17] *give all matrix elements for all elementary extended projective operators.*

What is most remarkable is that the matrix elements of these extended elementary projective operators *are none other than* $(6j)$ *coefficients.*

This result has now been proved in general.

THEOREM:[18] (Louck and Biedenharn 1988)

(a)*Matrix elements of* $U(n)$ *projective tensor operators in the limit* $m_{nn}\to-\infty$ *become matrix elements of extended projective operators which may be written symbolically in operator form as*

$$\lim\left(\begin{bmatrix}(\Gamma)_{n-1}\\ [M]_n\\ (\gamma)_{n-1}\end{bmatrix}\right) = \prod_{i=1}^{n-1}\left(\delta^{\Gamma_{i,n-1}}_{\gamma_{i,n-1}}\right)\cdot\begin{bmatrix}(\Gamma)_{n-2}\\ [\Gamma]_{n-1}\\ (\gamma)_{n-2}\end{bmatrix}_{ext}. \tag{2.9}$$

(b) *Matrix elements of these extended projective operators are* $(6j)$ *coefficients in* $U(n-1)$ *given by:*

$$\left\langle\begin{matrix} m'_{1n} & \cdots & m'_{n-1,n} & \\ & m'_{1,n-1} & \cdots & m'_{n-1,n-1}\end{matrix}\right|\begin{bmatrix}(\Gamma)_{n-2}\\ [M]_{n-1}\\ (\gamma)_{n-2}\end{bmatrix}_{ext}\left|\begin{matrix} m_{1n} & \cdots & m_{n-1,n} & \\ & m_{1,n-1} & \cdots & m_{n-1,n-1}\end{matrix}\right\rangle$$

$$= \begin{bmatrix} [m_{1n}\ \ \cdots\ \ m_{n-1,n}] & [\dot{0},-w] & [m_{1,n-1}\ \ \cdots\ \ m_{n-1,n-1}] \\ \begin{pmatrix}[M]_{n-1}\\ (\Gamma)_{n-2}\end{pmatrix} & [\dot{0}] & \begin{pmatrix}[M]_{n-1}\\ (\gamma)_{n-2}\end{pmatrix} \\ [m'_{1,n}\ \ \cdots\ \ m'_{n-1,n}] & [\dot{0},-w] & [m'_{1,n-1}\ \ \cdots\ \ m'_{n-1,n-1}] \end{bmatrix} \tag{2.10}$$

where $m'_{i,n} = m_{i,n} + \Delta_{i,n}(\Gamma), m'_{i,n-1} = m_{i,n-1} + \Delta_{i,n-1}(\gamma)$ *and*
$w = \sum_{i=1}^{n-1}(m_{i,n} - m_{i,n-1}) = \sum_{i=1}^{n-1}(m'_{i,n} - m'_{i,n-1})$.

We have written the $(6j)$ coefficient here in the form of a $(9j)$ coefficient since the coupling relationships are put more clearly in evidence. (For the identity irrep $[\dot{0}]$ the $(9j)$ reduces to a $(6j)$.) Note that the irrep $[\dot{0},-w]$ is *multiplicity free so that no operator pattern is needed for each of the three horizontal couplings. In sharp contrast to this the two operator patterns* (Γ) *and* (γ) *are essential to define uniquely the two non-trivial vertical couplings in the* $(9j)$ *coefficient.*

Remarks:

(a) It was demonstrated [16,7] quite early that all *elementary projective operators* (operators belonging to irreps of the form $[\dot{1}_k\ \dot{0}_{n-k}]$) *for all* $U(n)$ *obey the result stated in the theorem.*

(b) We will show below that the VCS results (to be developed in Section 5) obey (2.9) as stated in the theorem.

(c) The theorem given above shows that the $(3j)$ coefficients (vector addition coefficients), the projective tensor operators, and the $(6j)$ coefficients are all structurally inter-related via limits. Equally importantly, the meaning of an operator pattern becomes more accessible now that one can explicitly give, via the theorem, a way to determine the operator pattern labels.

Because, however, the labelling is achieved through a limiting operation one cannot claim to have established all operator patterns categorically. (Indeed, one can arbitrarily mix tensor operators (having the same state labels (M)) using invariant operators as coefficients, without affecting the limiting (operator pattern) results, provided only that the mixing becomes diagonal in the limit.)

Although the Baclawski result shows that null space is itself not categoric, still null space considerations very greatly restrict the freedom to mix operators and considerably extend the number of operators having a unique labelling via limits.

3. RÉSUMÉ OF THE VCS CONSTRUCTION[8]

Consider a Lie group G with subgroup H, with Lie algebra $\mathbf{g}$ and $\mathbf{h}$, respectively. The Mackey-Wigner induction technique[19] constructs representations of G from representations of H. The "vector coherent state" (VCS) construction is a special case of this general technique in which two conditions are imposed: (a) the rank of $\mathbf{g}$ and $\mathbf{h}$ are the same; (b) the (complexified) Lie algebra of G splits into the three disjoint subalgebras: $\mathbf{g} = \mathbf{n}_+ + \mathbf{h} + \mathbf{n}_-$, where $\mathbf{n}_+(\mathbf{n}_-)$ are nilpotent algebras of raising (lowering) operators carrying a (possibly reducible) representation of $\mathbf{h}$: $[\mathbf{h}, \mathbf{n}_\pm] = \mathbf{n}_\pm$. The aim of the VCS procedure is to construct all irreps of G from knowledge of all irreps of H.

Let us specialize to the unitary Lie group $G = U(n)$ with the Lie algebra $\mathbf{g} = \{E_{ij}; 1 \leq i, j \leq n\}$, obeying the standard commutation relations: $[E_{ij}, E_{k\ell}] = \delta_{jk}E_{i\ell} - \delta_{i\ell}E_{kj}$.

For the subgroup H we take the Lie group $H = U(n-1) \times U(1)$, so that the Lie algebra consists of four subsets,

(a) *the $u(n-1)$ subalgebra* *: $\{C_{\alpha,\beta}\} = \{E_{\alpha,\beta};\ 1 \leq \alpha,\ \beta \leq n-1\}$.

(b) *the u(1) subalgebra*: $\{W\} = \{E_{nn}\}$,

(c) *the raising operators*: $\{A_\alpha\} = \{E_{\alpha n};\ 1 \leq \alpha \leq n-1\}$ and

(d) *the lowering operators*, $\{B_\alpha\} = \{E_{n\alpha};\ 1 \leq \alpha \leq n-1\}$.

The unitary irreps (unirreps) of $U(n)$ are well-known to be uniquely specified (to within equivalence) by the Young frame $[m] = [m_{1n} \cdots m_{nn}]$. The carrier space of $[m]$ will be denoted by the set of orthonormal vectors $\{|(m)\rangle\}$, where (m) is a Gel'fand-Weyl pattern.

The VCS approach focuses on a special *subset* of the vectors in $[m]$: the vectors annihilated by the raising operators, $\{A_\alpha\}$. Since: $(A_\alpha|(m)\rangle = 0) \Rightarrow (W|(m)\rangle = m_{nn}|(m)\rangle)$, it follows that the vectors in this subset have maximal labels in row $n-1$. That is: $m_{i,n-1} = m_{i,n}$. This special subset of vectors— called the "intrinsic space"—corresponds exactly to vectors carrying the irrep $[m_{1n} \cdots m_{n-1,n}] \times [m_{nn}]$ of the subgroup $H = U(n-1) \times U(1)$.

The idea underlying VCS theory is to construct the *same irrep* $[m]$ but over a new basis consisting of vectors from the intrinsic space, multiplied by holomorphic vector functions (boson polynomials acting on the vacuum ket).

To construct this new carrier space, let us define a vector of the new basis by

* We adopt the convention that Latin indices run from 1 to n while Greek indices run from 1 to $n-1$.

$$|(m)\rangle_{VCS} \equiv \sum_{(\mu)\in\mathcal{H}} \langle(\mu)|e^{z\cdot A}|(m)\rangle\ |0\rangle \otimes |(\mu)\rangle, \tag{3.1a}$$

where

(a) The summation is over the intrinsic space $(\mu) \in [m]$ with $W|\mu\rangle = m_{nn}|\mu\rangle$. (We denote this by: $(\mu) \in \mathcal{H}$.)

(b)

$$z\cdot A \equiv \sum_{\alpha=1}^{n-1} z_\alpha A_\alpha. \tag{3.1b}$$

(In (1b) the $\{z_\alpha\}$ are boson creation operators commuting with all $\{E_{ij}\}$, and acting on the vacuum ket $|0\rangle$.)

This notation is designed to make clear that vectors in the new basis are labelled by the same labels (m) as in the old basis, but that the new basis is built on the direct product of boson ket vectors with the intrinsic space subset of vectors of the original basis. (The boson ket vectors span the factor space $SU(n)/U(n-1)$.)

Let us now demonstrate that the basis $\{|(m)\rangle_{\mathrm{VCS}}\}$ carries exactly the same irrep as before. Let g belong to $G = U(n)$. Then we define the action:

$$g|(m)\rangle = \sum_{(m')} D^{[m]}_{(m'),(m)}(g)|(m')\rangle, \tag{3.2a}$$

with $g \rightarrow D(g)$ defining the irrep $[m]$ on the original basis $\{|(m)\rangle\}$.

To show that the new basis (3.1a) carries this same irrep, $D(g)$, we define the action:

$$g \circ |(m)\rangle_{VCS} \equiv \sum_{(\mu)\in\mathcal{H}} \langle(\mu)|e^{z\cdot A} g|(m)\rangle|0\rangle \otimes |(\mu)\rangle. \tag{3.2b}$$

It follows from (3.2a) that

$$g \circ |(m)\rangle_{VCS} = \sum_{(m')} D^{[m]}_{(m'),(m)}(g)|(m')\rangle_{VCS}\ , \tag{3.2c}$$

which establishes the desired result.

Consider now the action of the Lie algebra $\mathbf{g}$ on the basis $\{|(m)\rangle_{VCS}\}$. From (3.2b) we see that

$$\begin{aligned}\mathbf{g} \circ |(m)\rangle_{VCS} &= \sum_{(\mu)\in\mathcal{H}} \langle(\mu)|e^{z\cdot A}\mathbf{g}|(m)\rangle\ |0\rangle \otimes |(\mu)\rangle, \\ &= \sum_{(\mu)\in\mathcal{H}} \langle(\mu)|(e^{z\cdot A}\mathbf{g}e^{-z\cdot A})e^{z\cdot A}|(m)\rangle\ |0\rangle \otimes |(\mu)\rangle. \end{aligned}\tag{3.2d}$$

Using the Baker-Campbell-Hausdorff identity to evaluate $e^{z\cdot A}\mathbf{g}e^{-z\cdot A}$, as $\mathbf{g}$ runs over the generators E_{ij}, we find the VCS realization:

$$\mathbf{g} \circ |(m)\rangle_{VCS} \equiv \Gamma(\mathbf{g})|(m)\rangle_{VCS},$$

with

$$\Gamma(A_\alpha) = \partial_\alpha \equiv \frac{\partial}{\partial z_\alpha}, \tag{3.3a}$$

$$\Gamma(C_{\alpha\beta}) = \mathcal{E}_{\alpha\beta} - z_\beta \partial_\alpha, \tag{3.3b}$$

$$\Gamma(B_\alpha) = z_\gamma \mathcal{E}_{\gamma\alpha} - z_\alpha \mathcal{E}_{nn} - z_\alpha z_\gamma \partial_\gamma, \tag{3.3c}$$

$$\Gamma(W) = \mathcal{E}_{nn} + z_\alpha \partial_\alpha. \tag{3.3d}$$

where, for clarity, we have denoted by $\mathcal{E}_{\alpha\beta}$ the operators $E_{\alpha\beta}$ acting on the intrinsic subspace $\{|(\mu)\rangle\}$.

Note especially that in the VCS realization, the generators of the $U(n-1)\otimes U(1)$ subgroup consist of the direct sum of the (intrinsic) $U(n-1)\otimes U(1)$ generators, $\mathcal{E}_{\alpha\beta}$ and $\mathcal{E}_{nn}$, plus a variant of the Jordan-Schwinger boson realization: $-z_\beta\partial_\alpha$. *The two realizations commute, so that vectors in* $|(m)\rangle_{VCS}$ with sharp $U(n-1)\otimes U(1)$ *labels are defined by* $U(n-1)$ *vector coupling.*

In the defining relation (3.1a), the matrix elements that occur, $\langle(\mu)|e^{z\cdot A}|(m)\rangle$, are polynomials in the variables $\{z_\alpha\}$, since $(z\cdot A)^k|(m)\rangle$ vanishes for sufficiently large k. The form of this matrix element is familiar from coherent state work, but the construction in (3.1a) is more general, since the expansion is over the vectors of the intrinsic space, as well as over vectors in the "coherent state" basis.

Note that $\{\partial_\alpha\}$, like $\{A_\alpha\}$, carries the $U(n-1)$ irrep $[1\dot{0}]$. Similarly $\{z_\alpha\}$ belongs to $[\dot{0},-1]$ in $U(n-1)$. The boson operators $\{z_\alpha\},\{\partial_\alpha\}$, and 1 realize a Heisenberg-Weyl algebra under commutation.

Let us now give a more explicit form for the vectors $|(m)\rangle_{VCS}$. Using the fact that $\{z_\alpha\}\in[\dot{0},-1]$ in $U(n-1)$, we see that the direct product $[\dot{0},-1]\times\cdots\times[\dot{0},-1]$ of w such irreps defines uniquely the irrep $[\dot{0}-w]$ as polynomials in $\{z_\alpha\}$ of degree w. Let us denote the vectors of this irrep, normalized in the standard (boson) way, by:

$$\left|\begin{matrix}[\dot{0},-w]\\(\lambda)\end{matrix}\right\rangle \equiv B^{[\dot{0},-w]}_{(\lambda)}(z)|0\rangle, \tag{3.4a}$$

where $[\dot{0},-w]$ is an irrep in $U(n-1)$, (λ) denotes the associated Gel'fand-Weyl $U(n-2)$ pattern, and $B(z)$ is a normalized boson polynomial in $\{z_\alpha\}$.

The vectors in the VCS basis having prescribed $U(n-1)\otimes U(1)$ labels are then:

$$\left[\left|\begin{matrix}[\dot{0},-w]\\(\cdot)\end{matrix}\right\rangle\times\left|\begin{matrix}m_{1n}\ldots m_{n-1,n}\\(\cdot)\end{matrix}\right\rangle\right]_{(\nu)}\otimes|m_{nn}\rangle, \tag{3.4b}$$

where we have denoted the $U(n-1)$ vector coupling—yielding the $U(n-1)$ state $|(\nu)\rangle$—by the bracket, with $(\cdot)$ denoting quantum numbers summed over in effecting the coupling.

Acting on this set of vectors with the raising operator $\{\partial_\alpha\}$ successively lowers the degree of the polynomial $B(z)$ to zero, so that the highest weight vectors of the $U(n)$ irrep are given by the vectors of the intrinsic space. *It follows that the* $U(n)$ *irrep is* $[m]$, *with these labels being supplied by the intrinsic state vectors alone.*

Although the states (3.4b) are normalized in $U(n-1)\otimes U(1)$, they are not necessarily normalized correctly in $U(n)$. To obtain the proper relative normalization, we recall that both bases $\{|(m)\rangle\}$ and $\{|(m)\rangle_{VCS}\}$ carry the same representation. Using (3.2c) shows that:

$$\begin{aligned}\mathbf{g}\circ|(m)\rangle_{VCS} &= \sum_{(m')}\langle(m')|\mathbf{g}|(m)\rangle|(m')\rangle_{VCS}\\ &\equiv \Gamma(\mathbf{g})|(m)\rangle_{VCS},\end{aligned} \tag{3.5}$$

where $\langle (m')|\mathbf{g}|(m)\rangle$ are the *standard matrices of the generators* of $U(n)$.

By direct computation, using (3.5), one obtains the following explicit form for the VCS vectors of a general $U(n)$ unirrep:

$$|(m)\rangle_{VCS} = (-1)^{\varphi}K(m)\left[\left|\begin{matrix}[\dot{0},-w]\\(\cdot)\end{matrix}\right\rangle \times \left|\begin{matrix}[\mu]\\(\cdot)\end{matrix}\right\rangle\right]_{(m)_{n-1}} \otimes |m_{nn}\rangle, \quad (3.6)$$

where

(a) $|\binom{\dot{0},-w}{\cdot}\rangle$ is defined by the normalized boson polynomial, (3.4a), belonging to the $U(n-1)$ irrep $[\dot{0},-w]$,

(b) the label $w = \sum_{i=1}^{n-1}(m_{in} - m_{i,n-1})$,

(c) $|\,{}^{[\mu]}_{\cdot}\rangle \otimes |m_{nn}\rangle$ denotes an intrinsic state vector carrying the $U(n-1)\times U(1)$ irrep $[m_{1n}m_{2n}\cdots m_{n-1,n}] \times [m_{nn}]$.

(d) The bracket $[\cdots\times\cdots]$, as in (3.4b), denotes $U(n-1)$ vector coupling. The notation $(m)_{n-1}$ denotes the lowest $n-1$ rows of the $U(n)$ Gel'fand-Weyl pattern (m).

(e) The normalization factor $K(m)$ is given by:

$$K(m) = \left|\left(\prod_{\alpha=1}^{n-1}\frac{(p_{\alpha n}-p_{nn}-1)!}{(p_{\alpha,n-1}-p_{nn})!}\right)^{1/2}\right|, \quad (3.7)$$

where the *partial hooks* are defined by $p_{ij} \equiv m_{ij} + j - i$.

Note that $K(m)$ involves only the invariant labels of $U(n)$ and $U(n-1)$. We can promote $K(m)$ to a Hermitian operator, $\mathbf{K}$, with matrix elements defined above. The generators of $U(n-1)\times U(1)$, that is, $\Gamma(E_{\alpha\beta})$ and $\Gamma(E_{nn})$, commute with $\mathbf{K}$.

(f) The phase factor $(-1)^{\varphi}$ is defined from (3.5) by the phase conventions of the standard realization of $U(n)$ (cf. ref. 8).

(g) The orthonormal dual vectors, ${}_{VCS}\langle (m)|$, are Hermitian conjugates to (3.6) *but with* $\mathbf{K}$ *replaced by* $\mathbf{K}^{-1}$. (The inverse is well-defined as a consequence of the betweenness constraints on the m_{ij}.)

Remark:

It will be recognized from this résumé that the VCS construction achieves nothing new in terms of $U(n)$ representations. The novelty of the VCS construction, and its importance, lies in a different direction, where one sees that the VCS construction yields *a new realization,* $\Gamma(\mathbf{g})$, *of the* $U(n)$ *Lie algebra* and more importantly, *a new construction of* $U(n)$ *carrier spaces, defined entirely within the context of the* $U(n-1)$ *vector addition coefficients.*

In other words, the VCS construction allows us to obtain $U(n)$ results with $U(n-1)$ methods, and is accordingly a most useful technique in a recursive approach to the set of all $U(n)$ groups and their associated operators.

4. EXTENSION OF THE VCS CONSTRUCTION TO TENSOR OPERATORS

The construction of a canonical basis for tensor operators relies heavily on two properties: the *equivariance property* and the *derivation property.* (For definiteness we specialize the discussion below to $U(n)$.)

(a) *Equivariance:*

Let t denote a tensor operator in $U(n)$. Then t is a *set* of operators $t = \{t(m)\}$ obeying the defining relation for equivariance,

$$g : t(m) \to t'(m) \equiv g(t(m))g^{-1} = \sum_{(m')} D^{[m]}_{(m'),(m)}(g)t(m'), \tag{4.1}$$

for $g \in U(n)$ and $(m), (m')$ Gel'fand-Weyl patterns in $U(n)$ labelling a basis for the irrep $[m]$.

(b) *The derivation property:*

Let $\mathcal{O}$ belong to the Lie algebra $u(n)$. A realization of $u(n)$ by linear operators has the derivation property iff

$$\mathcal{O}(AB) \quad = \quad (\mathcal{O}A)B + A(\mathcal{O}B), \tag{4.2}$$

where AB denotes the tensor product of two representations of $U(n)$, either or both of which may be tensor operators (using the equivariance property).

It is easily verified that the VCS realization of $U(n)$—with generators given by eqs. (3.3) and state vectors by eq. (3.6)—has the derivation property on a tensor product of VCS states. This elementary, but important, observation proves that the matrix realization of unit tensor operators for $U(n)$ can indeed be implemented in the VCS framework.

Consider a generic VCS state, eq. (3.6). Since it is the transformation properties which are essential, we can replace, using equivariance, intrinsic space vectors by ***normalized unit tensor operators*** in $U(n-1) \otimes U(1)$. Such tensor operators are denoted by

$$\left\langle \begin{matrix} (\Gamma)_{n-2} \\ [M_{1n} \ldots M_{n-1,n}] \\ (\mu)_{n-2} \end{matrix} \right\rangle \otimes \langle M_{nn} \rangle, \tag{4.3}$$

where, for clarity, we denote Young frames in $U(n-1)$ by $[\ldots]_{n-1}$ and Gel'fand-Weyl (or operator) patterns in $U(n-1)$ by $(\ldots)_{n-1}$. Since the VCS construction promotes the $U(1) \times U(n-1)$ labels to $U(n)$ labels, we must require $M_{i,n} \geq M_{nn}$ $(i = 1.\ldots, n-1)$ for consistency.

The associated $U(n)$ tensor operator in the VCS construction then has the form

$$\begin{aligned} \left\langle \begin{matrix} (\Gamma)_{n-1} \\ [M_{1n} \ldots M_{nn}] \\ (M)_{n-1} \end{matrix} \right\rangle_{VCS} &= K \begin{pmatrix} [M]_n \\ [M]_{n-1} \end{pmatrix} \\ &\times \left(\left[B^{\begin{smallmatrix}[\dot{0}, -W] \\ (\cdot)\end{smallmatrix}}(z) \times \left\langle \begin{matrix} (\Gamma)_{n-2} \\ [M_{1n} \ldots M_{n-1,n}] \\ (\cdot)_{n-2} \end{matrix} \right\rangle \right]_{(M)n-1} \otimes \langle M_{nn} \rangle \right), \end{aligned} \tag{4.4}$$

where the square bracket once again denotes a vector coupling in $U(n-1)$. By construction the operators defined in (4.4) transform equivariantly as the unirrep $[M]$ of $U(n)$ with Gel'fand-Weyl pattern (M).

We shall discuss in Section 6 the operator pattern labels (Γ) which appear in (4.4).

The VCS operators given in (4.4)—which are built in analogy to the generic VCS vectors (3.6)—are not the most general since we have used only polynomials in $\{z_\alpha\}$ and not more general polynomials over both $\{z_\alpha\}$ and $\{\partial_\alpha\}$. The VCS realization of the $U(n)$ generators given by (3.3) shows that, as an operator $\{\partial_\alpha\}$ indeed transforms as $[1\ \dot{0}]$ in $U(n-1)$, but in $U(n)$ is actually a member of the tensor operator set $[1\ \dot{0} - 1]$.

Technical modifications are required in order to use polynomials over both $\{z_\alpha\}$ and $\{\partial_\alpha\}$ in more general tensor operators.

There are special simplifications which occur for polynomials over either $\{z_\alpha\}$ or $\{\partial_\alpha\}$ separately. The VCS realization (3.3) shows that in this case the $U(n)$ operator irrep labels $[M]$ are determined by the intrinsic space labels belonging to $U(n-1) \times U(1)$: *maximally tied* for $\{z_\alpha\}$, or *minimally tied* for $\{\partial_\alpha\}$. For brevity we omit further details (see ref. 8).

It is an important result that the two classes of VCS tensor operator—the maximally tied and the minimally tied classes—*have fully explicit matrix elements completely defined in terms of* $U(n-1)$ *constructs* (to within a *multiplicative* ratio of K factors which are $U(n-1)$ invariant quantities.) This is a quite remarkable result since it was entirely unexpected that any such general property could possibly hold true.

For brevity we give only the results for the maximally tied operators ($\Rightarrow$ minimal shift $\Delta_{nn} = M_{nn}$) in the next section. What is important point is the *structure* of these matrix elements, and the omitted operators have the same general structure.

5. EXPLICIT RESULTS FOR ALL $U(n)$ TENSOR OPERATORS WITH MINIMAL SHIFT Δ_{nn}

To give explicit matrix elements for all unit tensor operators in all $U(n)$ is a formidable undertaking, and to reduce this task to manageable proportions it is essential to use recursive techniques tailored to the decomposition $U(n) \supset U(n-1) \supset \ldots \supset U(2)$. Thus one assumes full knowledge of all $U(n-1)$ constructs in proceeding to the next level $U(n)$. We have used this approach to define, in eq. (2.6) projective tensor operators which act in the projective space $U(n)/U(n-1)$. By constructing projective tensor operators at each successive $U(k) : U(k-1)$ level we can assemble $U(n)$ tensor operators out of modular units, a process which may be termed "imbrication".

The evaluation of the maximal and minimal shift VCS operators (Section 4) when taken between VCS realizations of the state vectors (Section 3) leads to the following explicit results.[8]

Minimal Shift Matrix Elements:

$$\left\langle \begin{matrix} m'_{1,n} & & \cdots & & m'_{n,n} \\ & m'_{1,n-1} & \cdots & m'_{n-1,n-1} & \end{matrix} \middle| \begin{bmatrix} (\Gamma)_{n-1} \\ [M]_n \\ (\gamma)_{n-1} \end{bmatrix} \middle| \begin{matrix} m_{1,n} & & \cdots & & m_{n,n} \\ & m_{1,n-1} & \cdots & m_{n-1,n-1} & \end{matrix} \right\rangle$$

$$= \quad (-1)^{\phi} \times \left(\frac{w^{(f)}!}{w^{(i)}!W!} \right)^{1/2} \quad \times \quad \frac{K\left({[m_{1,n} \cdots m_{n,n}] \atop [m_{1,n-1} \cdots m_{n-1,n-1}]} \right) K\left({[M]_n \atop [\gamma]_{n-1}} \right)}{K\left({[m'_{1,n} \cdots m'_{n,n}] \atop [m'_{1,n-1} \cdots m'_{n-1,n-1}]} \right)} \quad \times$$

$$\times \quad \begin{bmatrix} [m_{1,n} \cdots m_{n-1,n}] & [\dot{0}, -w^{(i)}] & [m_{1,n-1} \cdots m_{n-1,n-1}] \\ (\Gamma)_{n-1} & [\dot{0}, -W] & (\gamma)_{n-1} \\ [m'_{1,n} \cdots m'_{n-1,n}] & [\dot{0}, -w^{(f)}] & [m'_{1,n-1} \cdots m'_{n-1,n-1}] \end{bmatrix}, \tag{5.1}$$

where:

(i) $\Gamma_{i,n-1} = M_{i,n} \quad (i = 1, \cdots, n-1)$. This constraint is the condition for minimal shift ($\Delta_{nn} = M_{nn}$).

(ii) $m'_{i,n} = m_{i,n} + \Delta_{i,n}(\Gamma), m'_{i,n-1} = m_{i,n-1} + \Delta_{i,n-1}(\gamma)$;

(iii) $w^{(i)} = \sum_{i=1}^{n-1}(m_{i,n} - m_{i,n-1})$, $w^{(f)} = \sum_{i=1}^{n-1}(m'_{i,n} - m'_{i,n-1})$ and $W = \sum_{i=1}^{n-1}(M_{in} - \gamma_{i,n-1})$. An important constraint is that: $w^{(f)} = w^{(i)} + W$;

(iv) $\phi = \phi([m]_n) + \phi([M]_n) - \phi([m']_n) - \phi([m]_{n-1}) - \phi([\gamma]_{n-1}) + \phi([m']_{n-1})$, with $\phi([m]_j) \equiv 1/2 \sum_{1 \leq k \leq \ell \leq j}(m_{kj} - m_{\ell j})$.

(v) $\begin{bmatrix} \cdots \\ \cdots \\ \cdots \end{bmatrix}$ is a $(9j)$ coefficient in $U(n-1)$.

Remarks:

(a) The essential point about these maximal/minimal shift classes of $U(n)$ matrix elements is that they have a simple structure in terms of three constitutive blocks:

(i) A $(9j)$ coefficient in $U(n-1)$,
(ii) a ratio of K factors, and
(iii) a ratio of w factorials arising from multiplicity-free totally symmetric $U(n-1)$ irreps (cf. eq. 3.6).

(b) Only for $U(3)$ are these results fully explicit since only here are the $(9j)$ coefficients for $U(2)$ known in detail.

(c) It is useful to note, however, that conceptually all $(3j), (6j) \ldots$ *coefficients in $U(n)$ are well-defined only by using operator patterns.*

In particular, the $(6j)$ coefficients involve *six* $U(n)$ irrep labels (geometrically the lines of a tetrahedron) and *four* operator patterns (geometrically associated with the four faces (coupling triangles) of the tetrahedron.

It is the explicit adjunction of the operator patterns in the $(3nj)$ coefficients that makes these objects well-defined. For example, the coefficient appearing in (5.1) would in general require six operator patterns to make each of the nine lines ("triangles") unique. Because irreps of the form $[\dot{0}, -w^{(i)}]$ are multiplicity-free, no operator patterns are necessary for the three horizontal lines (or for the stretched triangle corresponding to the middle vertical line). However, operator patterns $(\Gamma)_{n-1}$ and $(\gamma)_{n-1}$ are required for the remaining two vertical lines.

(d) The results for the maximal/minimal shift matrix elements permit a simpler derivation of the K-factor relation (3.7) than the method sketched in Section 3. Since (i) matrix elements for all elementary operators are explicitly known for all $U(n)$, and (ii) all elementary operators have either $\Delta_{nn} = 0$ (minimal) or $\Delta_{nn} = 1$ (maximal), clearly the K-factors themselves are determined by the pattern calculus rules of ref. (17). In practice, this provides a quick and easy verification of (3.7).

6. DETERMINATION OF OPERATOR PATTERNS

The results obtained in Section 5 can be used to provide a striking, and simple, exemplification in all $U(n)$ of the theorem on limits developed in section 2.

Consider the result in (5.1). The only place the limiting parameter m_{nn} occurs is in the K-factors. From the explicit form of the K- factors, eq. (3.7) we see that in the limit $m_{nn} \to -\infty$,

$$\lim_{m_{nn} \to -\infty} \left(\frac{K\begin{pmatrix} [m^{(i)}]_n \\ [m^{(i)}]_{n-1} \end{pmatrix}}{K\begin{pmatrix} [m^{(f)}]_n \\ [m^{(f)}]_{n-1} \end{pmatrix}} \right) \longrightarrow \frac{1}{(m_{nn})^{\alpha}} \tag{6.1}$$

with

$$\alpha = \sum_{k=1}^{n-1}(m_{kn}^{(f)} - m_{kn}^{(i)}) - \sum_{k=1}^{n-1}(m_{kn-1}^{(f)} - m_{kn-1}^{(i)}) \geq 0, \tag{6.2}$$

implying that the limit is equal to

$$\begin{cases} 1, \text{if } \sum_{k=1}^{n-1}(m_{kn}^{(i)} - m_{kn}^{(i)}) - \sum_{k=1}^{n-1}(m_{k,n-1}^{(f)} - m_{k,n-1}^{(i)}) = w^{(f)} - w^{(i)} = 0, \\ 0, \quad \text{otherwise.} \end{cases} \tag{6.3}$$

This condition for a non-vanishing limit can be rewritten in the terms of the shift $\Delta_{in}(\Gamma) = m_{in}^{(f)} - m_{in}^{(i)}$ and $\Delta_{in-1}(\gamma) = m_{i,n-1}^{(f)} - m_{i,n-1}^{(i)}$, so that we have, for the non-vanishing condition in eq. (6.3), the relation

$$\sum_{i=1}^{n-1} \Delta_{in}(\Gamma) - \sum_{i=1}^{2} \Delta_{i,n-1}(\gamma) = 0. \tag{6.4}$$

Using the definition of the shifts, we see that

$$\sum_{i=1}^{n-1} \Delta_{in}(\Gamma) \quad = \quad \sum_{i=1}^{n-1} \Gamma_{i,n-1} = \sum_{i=1}^{n-1} M_{i,n} \tag{6.5}$$

where the last equality results from the constraint that $\Gamma_{i,n-1} = M_{i,n}$ (the condition for minimal Δ_{nn} shift).

Similarly:

$$\sum_{i=1}^{n-1} \Delta_{i,n-1}(\gamma) \quad = \quad \sum_{i=1}^{n-1} \gamma_{i,n-1}. \tag{6.6}$$

Thus the condition for a non-vanishing limit takes the form

$$\sum_{ik=1}^{n-1} (\Gamma_{in} - \gamma_{i,n-1}) \quad = \quad 0, \tag{6.7}$$

which—because of the betweenness conditions on Gel'fand-Weyl pattern labels—implies that *each term in* (6.7)*is positive* and therefore

$$\Gamma_{in} - \gamma_{i,n-1} = 0, \qquad i = 1, 2, \cdots, n-1. \tag{6.8}$$

The conditions in (6.8) are precisely the δ- function conditions that occur in the limit theorem, eq. (2.9).

The condition $w^{(f)} = w^{(i)}$ from (6.3) has the important consequence that $W = 0$ in (5.3) so that the $(9j)$ coefficient reduces to a $(6j)$ coefficient in agreement with the theorem of Section 2.

Conversely, the desired conclusion that the operator pattern in eq. (5.3) has been *correctly assigned* follows from the above results. Since the lower pattern in these equations is generic, we may choose the lower pattern to be maximal, that is, $\gamma_{i,n-1} = M_{i,n}(i = 1 \cdots n-1)$, and then take the limit $m_{nn} \to -\infty$. We find from eqs. (6.3)-(6.7) that $\Gamma_{i,n-1}$ must be $M_{i,n}(k = 1 \cdots n)$. This allows us to conclude that the VCS tensor operators in eq. (5.3) has at least a limiting component that carries the assigned operator labels, but it does not establish that the tensors before the limit correspond uniquely to the assigned operator patterns.

To complete the proof, we need only remark that (5.1) shows that the K-factors contain the only dependence on m_{nn}. Thus the entire dependence on m_{nn} of (5.3) is contained in a *known multiplicative factor*. The inverse limit to (5.3) thus exists, and this establishes the desired result.

(More generally the limit $m_{nn} \to -\infty$ yields a $U(n-1)(6j)$ coefficient, for which the subsequent limit $m_{n-1,n-1} \to -\infty$ then yields a $U(n-1)/U(n-2)$ space projective operator, etc. *This sequence of limits validates, step-by-step, the entire operator pattern.*

The K-factors, which are crucial to this result, are typical of the VCS approach and show once again the significance of the VCS construction.

7. CONCLUDING REMARKS:

(a) We have shown in the preceding sections that the vector coherent state approach to tensor operators very nicely yields explicit results for two large classes of tensor operator matrix elements for $U(n)$, using only $U(n-1)$ constructs.

The question that immediately occurs is this: can VCS methods yield all $U(n)$ tensor operators explicitly? Since VCS methods do, in fact, yield all *elementary* tensor operators explicitly (this, as we have noted earlier, is a special case of the two classes already evaluated), it is clear that using coupling techniques one can, in principle, obtain all $U(n)$ operators. This observation shows that VCS methods can work in principle, but it does *not* answer the real question as to whether $U(n-1)$ constructs alone will suffice or not. It would be of considerable interest to establish this answer, especially if affirmative.

(b) Let us turn now to the subject of the existence of a canonical operator labelling. The progress reviewed in Section 2 seems strongly indicative that such a labelling may exist for general $U(n)$ tensor operators. The nature of "operator pattern space", however, is still far from clear, and it is here that one would like to see further progress. The fact that operator patterns can only be assigned via limits strongly suggests that we are faced with an analog of a quantified structure and its classical limit.

Operator patterns in $U(n)$ can be coupled by means of the $(6j)$ coefficients in $U(n)$; this is the content of the product laws developed in ref. 16). It is important to recognize that this coupling explicitly shows that *operator patterns do not arise from a group structure.* An example will make this clear.

Consider the operator pattern $\begin{pmatrix} 1 \\ 20 \\ 210 \end{pmatrix}$ denoting the non- generator adjoint operator in $SU(3)$, that is, the octet D-operator. We may construct this operator by coupling the elementary operators $\langle[100]\rangle$ and $\langle[110]\rangle$, using the appropriate $(6j)$ coefficient, and summing over operator patterns.

We find, in particular, that the $(6j)$ coefficient coupling the operator pattern $\begin{pmatrix} 0 \\ 00 \\ 100 \end{pmatrix}$ of $\langle[100]\rangle$ to the operator pattern of $\begin{pmatrix} 1 \\ 11 \\ 110 \end{pmatrix}$ of $\langle[110]\rangle$ yielding the operator pattern $\begin{pmatrix} 1 \\ 20 \\ 210 \end{pmatrix}$ of $\langle[210]\rangle$ *does not vanish.* This is quite impossible if the operator patterns result from a group structure, since it would imply that the $U(2)$ subgroup product, $[00] \times [11] \to [20]$ would be non-vanishing.

We conclude that *whatever structure underlies the space of operator patterns it is not that of a group.*

There is, however, a way to restore the $U(3) : U(2)$ group-subgroup structure: take the limit for $m_{33} \to -\infty$. When this limit it taken—as we have proved more generally in Section 2—the $(6j)$ coefficients limit to square bracket coefficients. For the present case this limit is the projective $U(3)/U(2)$ space matrix element:

$$\left\langle \begin{matrix} 210 \\ 20 \\ 1 \end{matrix} \middle| \begin{bmatrix} 1 \\ 10 \\ 100 \\ 00 \\ 0 \end{bmatrix} \middle| \begin{matrix} 100 \\ 11 \\ 1 \end{matrix} \right\rangle$$

multiplied by a $(6j)$ coefficient in $U(2)$. (The three operator patterns given above are inverted in the notation for this coefficient.)

The projective matrix element given here ***vanishes***, as it must, since $U(3) : U(2)$ group-subgroup branching laws must be obeyed by such operators. This behavior is typical: the limit $m_{nn} \to -\infty$, ***restores*** the $U(n) : U(n-1)$ group-subgroup relations for the operator patterns involved.

This result becomes even clearer if we consider the VCS realization of the commutation relations given in eqs. (3.3). The limit $m_{nn} \to -\infty$ is equivalent to letting the operator $\mathcal{E}_{nn}$ become 'large'; renormalizing in the standard way (by $(m_{nn})^{-1}$) we see that this limit is none other than a ***group contraction***: the Lie algebra of $U(n)$ contracts into a semi-direct product of the $(n-1)$ dimensional Heisenberg-Weyl algebra and the Lie algebra of $U(n-1)$.

We can put this situation more suggestively by recalling that the quantum $\to$ classical limit relation has been defined in a general algebraic context as a ***deformation*** of the algebra structure. The algebra of tensor operators is known to be associative and unital, and, (as we have mentioned in Section 2) is a generalization of the universal enveloping algebra $U(\mathbf{g})$. This algebra can itself be generalized to incorporate products changing operator patterns (this is the generalization from the algebra of $(3j)$ operators to $(6j)$ operators).

An interesting application of these ideas uses the limit relation of eq. (2.10) applied to the $U(3)$ group. This relation shows that the limit $(m_{33} \to -\infty)$ yields all $(6j)$ coefficients in $U(2)$. More importantly, this relation shows that we have ***an algebra of invariant operators*** (the operators whose matrix elements are $(6j)$ coefficients in $U(2)$) ***acting on the space of ket vectors*** $\{\left| \begin{smallmatrix} m_{13} & & m_{23} & \\ & m_{12} & & m_{22} \end{smallmatrix} \right\rangle\}$.

Such an algebra has been discussed in detail in reference (21), as "W- algebra". It is of interest to note[21] that the space on which these operators act, the set of ket vectors $\{\left| \begin{smallmatrix} m_{13} & & m_{23} & \\ & m_{12} & & m_{22} \end{smallmatrix} \right\rangle\}$, can be considered as a space of ***"quantized" triangles***[22]. In the limit that one of the sides of the triangle becomes large (the limit $m_{22} \to -\infty$) the ***algebra becomes precisely the algebra of tensor operators in*** $U(2)$. Thus we have the suggestive result that ***W-algebra is a deformation of the algebra of tensor operators in*** $U(3)$.

These remarks are relevant because they indicate a possible connection with a very important new development in mathematical physics: the concept of a ***quantum group***. This subject—the culmination of a long development, starting originally with the quantum inverse scattering method—now joins together topics as disparate as lattice models in statistical mechanics and current algebras. Clearly these new developments cannot be summarized—or even properly defined—here, but the review of Drinfel'd[23] can be strongly recommended. (The informal survey of Bergman[24] is also recommended for the introductory concepts that Drinfel'd omits).

It is important to state, however, that ***a quantum group is not a group,*** but a sort of quantized generalization. Since recent examples of quantum groups are, in fact, deformations of the universal enveloping algebra of $U(n)$ it seems reasonable to pose the question: can the algebras built on transformations of operator pattern space—such as W-algebras—be subsumed under the concept of quantum groups?

References

0. J. D. Louck and L. C. Biedenharn, "Some Properties of the Intertwining Number of the General Linear Group", in *Science and Computers, Adv. Math. Supp. Studies,* **10** (1986), 265-311.
1. K. Baclawski, *Adv. in Appl. Math.* **5** (1984) 416-432.
2. J. Deenen and C. Quesne, *J. Math. Phys.* **25** (1984) 2638, 2354.
3. K. T. Hecht, *The Vector Coherent State Method and Its Application to Problems of Higher Symmetries, Lecture Notes in Physics* **290**, Springer-Verlag, Berlin, (1987).
4. K. T. Hecht, R. Le Blanc and D. J. Rowe, *J. Phys. A: Math. Gen.* **20** (1987) 2241.
5. D. J. Rowe, R. Le Blanc and K. T. Hecht, *J. Math. Phys.* **29** (1988) 287.
6. R. Le Blanc and K. T. Hecht, *J. Phys. A: Math. Gen.* **20** (1987) 4613;
7. R. Le Blanc, *J. Phys. A: Math. Gen.* **20** (1987) 5015.
8. R. Le Blanc and L. C. Biedenharn, "Implementation of a $U(n)$ Wigner-Racah calculus in a vector coherent state Hilbert space," to be published in J. Phys. A: Math. Gen.
9. A. U. Klimyk and I. I. Kachurik, "Infinitesimal operators of group representations in noncanonical bases," to be published in J. Math. Phys.
10. S. Alisauskas,"On biorthogonal and orthonormal Clebsch-Gordan coefficients of $SU(3)$: Analytical and algebraic approaches," to be published in J. Math. Phys.
11. I. M. Gel'fand and A. V. Zelevinsky, *Societé Math. de France, Astérique, hors série* (1985) 117-128.
12. G. E. Baird and L. C. Biedenharn *J. Math. Phys.* **4** (1963) 1449.
13. J. D. Louck, *Amer. Journ. of Phys.***38** (1970) 3.
14. L. C. Biedenharn, J. D. Louck, and A. Giovannini *J. Math. Phys.* **8** (1967) 691.
15. D. Flath and L.C. Biedenharn, *Can. J. Math* **37** (1985) 710.
    L. C. Biedenharn and D. Flath, *Commun. Math. Phys.* **93** (1984) 143.
16. J. D. Louck and L. C. Biedenharn *J. Math. Phys.***11** (1970) 2368.
17. L. C. Biedenharn and J. D. Louck, *Commun. Math. Phys.* **8** (1968) 89.
18. J. D. Louck and L. C. Biedenharn, to be submitted for publication.
19. J. D. Louck and L. C. Biedenharn *J. Math. Phys.* **14** (1973) 1336.
20. G. W. Mackey, *Induced representations of groups and Quantum Mechanics*, W. A. Benjamin, Inc., NY, (1968).
21. L. C. Biedenharn and J. D. Louck, "The Racah-Wigner Algebra in Quantum Theory", Vol. 9 *Ency. of Math, and Appl.* (G.-C. Rota, Ed.), Addison-Wesley (Reading MA) 1981. Cf. Chap. 4, "W-algebra: An Algebra of Invariant Operators".
22. B. Hasslacher and M. J. Perry, *Phys. Lett.* **103B** (1981), 21-24.
23. V. G. Drinfel'd, " Hopf algebras and the quantum Yang-Baxter equations," *Sov. Math. Dokl.* **32** (1985) 254; "Quantum Groups" ICM publications (1986), 798.
24. G. M. Bergman, *Contemporary Mathematics* **43** (1985) 25-48.

ON A GLOBAL DIFFERENTIAL GEOMETRIC APPROACH TO THE RATIONAL MECHANICS OF DEFORMABLE MEDIA

E. Binz* and D. Socolescu**

***Fakultät für Mathematik und Informatik, Universität Mannheim**
Seminargebäude A. 5
****Fachbereich Mathematik, Universität Kaiserslautern**
Erwin-Schrödinger-Straße, 6750 Kaiserslautern

0. Introduction

In the past the rational mechanics of deformable media was largely concerned with materials governed by linear constitutive equations. In recent years, the theory has expanded considerably towards covering materials for which the constitutive equations are inherently nonlinear, and/or whose mechanical properties resemble in some respects those of a fluid and in others those of a solid (cf[Tr,No],[Le,Fi]).

In the present article we formulate a satisfactory global mathematical theory of moving deformable media, which includes all these aspects.

As we shall see, in our theory the stress tensor is neither necessarily local nor symmetric. In fact it does not even determine the equations of motion. It is a more general object, namely, the stress form, which governs the motion. Typical for our considerations is the study of the motion of a soap bubble, i.e. of a closed, deformable, two-dimensional material surface in $\mathbb{R}^3$. It is intuitively clear that this complex motion can be described as the superposition of two different ones. These are on one hand the "elastic" deformation of the soap bubble in "radial" direction, and the "instantaneous", "viscous" fluid flow of the same soap bubble along its surface, that is "transversally" to its "elastic" deformation on the other.
For our general case let us assume that at any instant the deformable medium in $\mathbb{R}^n$ forms a manifold and that the diffeomorphism type of this manifold does not change. Hence these manifolds are all diffeomorphic to a fixed one, which we denote by M.
As we shall show, this fascinating representative problem of mechanics of continua as well as the general problem of motion of a deformable medium

leads to a dynamical system on a suitably chosen infinite-dimensional manifold. In order to explain the main ideas of our global approach we introduce at first the differential geometric framework.
The manifold M is supposed to be smooth, compact, oriented and of dimension less or equal to n-1. The ambient euclidean space $\mathbb{R}^n$ is assumed to be equipped with a fixed scalar product $\langle\ ,\ \rangle$.
Hence an instantaneous configuration of the medium is given by a smooth embedding of M into $\mathbb{R}^n$. Therefore the configuration space is $E(M,\mathbb{R}^n)$, the space of all smooth embeddings of M into $\mathbb{R}^n$. As shown in [Bi,Fi], $E(M,\mathbb{R}^n)$ can be given a smooth principal bundle structure. More precisely let Diff M be the group of smooth diffeomorphisms of M, and define the action Φ of Diff M on $E(M,\mathbb{R}^n)$ as follows

(0.1) $$\Phi(j,g) = j \circ g, \qquad \forall\ j \in E(M,\mathbb{R}^n),\ g \in \text{Diff } M.$$

Let us denote the quotient of $E(M,\mathbb{R}^n)$ by this action by $U(M,\mathbb{R}^n)$, and identify it with the set of all smooth submanifolds of M in $\mathbb{R}^n$ diffeomorphic with M. Further denote by Π the projection of $E(M,\mathbb{R}^n)$ onto $U(M,\mathbb{R}^n)$. Endowed with the C^∞-topology, $E(M,\mathbb{R}^n)$, $U(M,\mathbb{R}^n)$ and Diff M become Frèchet manifolds. The quadruple $(E(M,\mathbb{R}^n),\Pi,U(M,\mathbb{R}^n),\text{Diff } M)$ is then a principal bundle with Diff M as its structure group. Hence the fibres of this principal bundle have the form

(0.2) $$j \circ \text{Diff } M \quad , j \in E(M,\mathbb{R}^n).$$

In the particular case of the soap bubble we now visualize the two motions described above as follows :
The "instantaneous" fluid flow along its surface is described by a curve in one of the fibres of the above principal bundle, while the "radial" deformation is given by a curve which is transverse to the fibres of $E(M,\mathbb{R}^n)$.

Each configuration $j \in E(M,\mathbb{R}^n)$ yields a Riemannian metric $m(j)$, assuming on any pair of tangent vectors $v,w \in TM$ the value

(0.3) $$m(j)(v,w) := \langle dj\ v, dj\ w \rangle ,$$

where the scalar product is to be taken pointwise.
The "instantaneous" metrical properties of the moving body are described in this metric. Suppose now that the deformable medium is moving. We furnish the description of its motion by assuming that we know the work done by the forces acting upon M. It is in this work that all the constitutive information on the medium is coded. We therefore call it the constitutive law. The fluid component of the medium is expressed through the dependence of the work on an extra parameter. Accordingly the constitutive law is then given by

(0.4) $$F : C^\infty(M,\mathbb{R}^n) \times E(M,\mathbb{R}^n) \times C^\infty(M,\mathbb{R}^n) \longrightarrow \mathbb{R} ,$$

where F is linear in the third argument, the first factor in the cartesian product is the space of extra parameters and furthermore the trivial tangent bundle $TE(M,\mathbb{R}^n) = E(M,\mathbb{R}^n) \times C^\infty(M,\mathbb{R}^n)$ is the phase space of motions in $E(M,\mathbb{R}^n)$.
We concentrate on those constitutive laws which admit an integral representation. More precisely, we assume that F is given by

$$(0.5) \qquad F(k)(j,h) = {}_M\!\int \langle \varphi_F(j,k),h\rangle \mu(j), \qquad \forall\ j\in E(M,\mathbb{R}^n),\ h,k\in C^\infty(M,\mathbb{R}^n),$$

with $\varphi_F : TE(M,\mathbb{R}^n) \longrightarrow C^\infty(M,\mathbb{R}^n)$ being a smooth map called the force density. The equation of motion on $E(M,\mathbb{R}^n)$ described by a smooth curve

$$(0.6) \qquad \sigma : (-\lambda,\lambda) \longrightarrow E(M,\mathbb{R}^n), \qquad \lambda > 0,$$

is given then by

$$(0.7) \qquad \begin{aligned} F(\dot\sigma(t))(\sigma(t),h) &= {}_M\!\int \langle \varphi_f(\sigma(t),\dot\sigma(t)),h\rangle \mu(\sigma(t)) \\ &= {}_M\!\int \rho(\sigma(t))\langle \ddot\sigma(t),h\rangle \mu(\sigma(t)) \qquad \forall\ h \in C^\infty(M,\mathbb{R}^n), \end{aligned}$$

where ρ is the mass density (cf. section 2).

We note that in (0.7) we have assumed for simplicity that the constitutive law F depends on the "velocity" $\dot\sigma(t)$, i.e. $k=\dot\sigma(t)$. Interpreting h as a virtual displacement, (0.7) is just d'Alembert's principle of virtual work, which was formulated for the mechanics of continua by [He]. But (0.7) implies easily

$$(0.8) \qquad \rho(\sigma(t))\ddot\sigma(t) = \varphi_F(\sigma(t),\dot\sigma(t)), \qquad \forall\ t\in (-\lambda,\lambda).$$

To obtain a more refined form, let us denote by "T" and "$\perp$" respectively the tangential and the normal component with respect to $\sigma(t)(M)$. The equation of motion (0.8) splits into the coupled system

$$(0.9) \qquad \begin{cases} \nabla(\sigma(t))_{Z(t)}Z(t) + \dot Z(t) + W(\sigma(t),\dot\sigma(t)^\perp)Z(t) + [(\dot\sigma(t)^\perp)^\cdot]^T \\ \quad = \rho^{-1}(\sigma(t))\ Y(\sigma(t),\dot\sigma(t)), \\ \\ [(\dot\sigma(t))^\perp)^\cdot]^\perp = \\ \quad \rho^{-1}(\sigma(t))\varphi_F(\sigma(t),\dot\sigma(t))^\perp - [d\dot\sigma(t)^\perp Z(t)]^\perp - S(\sigma(t))(Z(t),Z(t)). \end{cases}$$

Here $\nabla(\sigma(t))$ denotes the Levi-Civita connection of $m(\sigma(t))$, the metric given by $\sigma(t)$, $S(\sigma(t))$ is the second fundamental tensor, $Z(t)$ and $Y(\sigma(t),\dot\sigma(t))$ belong to ΓTM. Furthermore $W(j,N)$ is the unique bundle map of TM associated with a smooth map $N : M \longrightarrow \mathbb{R}^n$ satisfying

$$(0.10) \qquad \langle dj\ X(p),N(p)\rangle = 0, \qquad \forall\ X\in \Gamma TM,\ p\in M$$

and which is determined by

$$(0.11) \qquad dj\ W(j,N)Y = (dN,Y)^T.$$

Among the force densities acting on M we distinguish between internal forces and external ones. Of a special interest is the study of the motion of the deformable medium M subjected to an internal force density. Clearly, internal physical properties of the moving medium are described by constitutive laws invariant under the translation group $\mathbb{R}^n$. Evidently, the $\mathbb{R}^n$-invariant configurations are differentials of embeddings. We hence identify

(0.12) $$E(M,\mathbb{R}^n)/_{\mathbb{R}^n} \text{ with } \{dj \mid j \in E(M,\mathbb{R}^n)\}$$

and more generally

(0.13) $$C^\infty(M,\mathbb{R}^n)/_{\mathbb{R}^n} \text{ and } \{dh \mid h \in C^\infty(M,\mathbb{R}^n)\}.$$

The phase space for the $\mathbb{R}^n$-invariant motion is hence

(0.14) $$T(E(M,\mathbb{R}^n)/_{\mathbb{R}^n}) = \{dj \mid j \in E(M,\mathbb{R}^n)\} \times \{dh \mid h \in C^\infty(M,\mathbb{R}^n)\}.$$

We require that the internal constitutive law F admits the representation

(0.15) $$F = F_{\mathbb{R}^n} \circ Td ,$$

where

(0.16) $$F_{\mathbb{R}^n} : C^\infty(M,\mathbb{R}^n)|_{\mathbb{R}^n} \times E(M,\mathbb{R}^n)|_{\mathbb{R}^n} \times C^\infty(M,\mathbb{R}^n)|_{\mathbb{R}^n} \longrightarrow \mathbb{R}$$

is a parameter depending one form (the parameter varies in the front factor in (0.16)) and Td is the tangent map of the differential

(0.17) $$d : E(M,\mathbb{R}^n) \longrightarrow E(M,\mathbb{R}^n)/_{\mathbb{R}^n} .$$

To get a detailed description of the motion of the deformable medium, we assume now that $F_{\mathbb{R}^n}$ itself has an integral representation

(0.18) $$F_{\mathbb{R}^n}(dk)(dj,dl) = {}_M\!\int \alpha(dj,dk)dl\ \mu(j), \qquad \forall\ j \in E(M,\mathbb{R}^n),\ k,l \in C^\infty(M,\mathbb{R}^n),$$

where α is an $\mathbb{R}^n$-valued one-form, the so-called stress form, depending itself on an extra parameter, i.e.

(0.19) $$\alpha : E(M,\mathbb{R}^n)|_{\mathbb{R}^n} \longrightarrow A^1(M,\mathbb{R}^n).$$

The stress form α decomposes naturally at each

$$(dk,dj) \in TE(M,\mathbb{R}^n)|_{\mathbb{R}^n} = C^\infty(M,\mathbb{R}^n)|_{\mathbb{R}^n} \times E(M,\mathbb{R}^n)|_{\mathbb{R}^n}$$

into

(0.20) $$\alpha(dj,dk) = c_\alpha(dj,dk)\ dj + dj\ C_\alpha(dj,dk) + dj\ B_\alpha(dj,dk) ,$$

with $C_\alpha : TM \longrightarrow TM$ and $B_\alpha : TM \longrightarrow TM$ being smooth, strong bundle endomorphisms, which are respectively skew- and selfadjoint with respect to m(i) and $c_\alpha \in C^\infty(M,so(n))$. Here so(n) denotes the Lie algebra of the group of all proper rotations SO(n). In case that M is of codimension 1, the equations of motion (0.9) read as

$$(0.21)\quad \begin{cases} \nabla(\sigma(t))_{Z(t)}Z(t)+\dot{Z}(t)+2\cdot\epsilon(\sigma(t),\dot{\sigma}(t))W(\sigma(t)Z(t) \\ \qquad -\epsilon(\sigma(t),\dot{\sigma}(t))\ \mathrm{grad}_{\sigma(t)}\epsilon(\sigma(t),\dot{\sigma}(t)) \\ \qquad = -\rho^{-1}(\sigma(t))\mathrm{div}_{\sigma(t)}T_\alpha(d\sigma(t),d\dot{\sigma}(t))-2\ W(\sigma(t))U_\alpha(d\sigma(t),d\dot{\sigma}(t)), \\ \dot{\epsilon}(\sigma(t),\dot{\sigma}(t)) = \rho^{-1}(\sigma(t))\ \mathrm{tr}\ (B_\alpha(d\sigma(t),d\dot{\sigma}(t))\ W(\sigma(t)) \\ \qquad - d\epsilon(\sigma(t),\dot{\sigma}(t))\ Z(t) + \mathfrak{h}(\sigma(t))(Z(t),Z(t)). \end{cases}$$

Here $\epsilon(\sigma(t),\dot{\sigma}(t)) \in C^\infty(M,\mathbb{R})$, $U_\alpha(d\sigma(t)) \in \Gamma TM$, $W(\sigma(t))$ is the Weingarten map, $\mathrm{div}_{\sigma(t)}$ is the divergence taken with respect to $m(\sigma(t))$, $\mathfrak{h}(\sigma(t))$ is the second fundamental form, tr denotes the trace and $T_\alpha(dj,dk)$ is the so-called stress tensor, defined as

$$(0.22)\qquad T_\alpha(dj,dk)(X,Y) = m(j)((B_\alpha+C_\alpha)(dj,dk)X,Y)\ , \quad \forall\ X,Y \in \Gamma TM.$$

Each $\alpha \in A^1(M,\mathbb{R}^n)$, and hence the parameter depending stress form splits relative to an embedding $i \in E(M,\mathbb{R}^n)$ into

$$(0.23)\qquad \alpha = dh + \beta\ ,$$

where $h \in C^\infty(M,\mathbb{R}^n)$, the so called integrable part of α, is uniquely determined up to a constant. Moreover h splits into parts tangential and normal to j(M), i.e.

$$(0.24)\qquad h = di\ X_h + h^\perp$$

(with $h^\perp = \Theta_h\cdot N(i)$, $\Theta_h \in C^\infty(M,\mathbb{R}^n)$, in case of dim M = n−1) for a well determined vector field $X_h \in \Gamma TM$. Using the Hodge decomposition

$$(0.25)\qquad \begin{cases} X_h = X_h^0 + \mathrm{grad}_i\psi_h\ , \\ \mathrm{div}_i X_h^0 = 0\ , \end{cases}$$

we thus obtain immediately

$$(0.26)\qquad \alpha(X) = di\nabla(i)_X X_h + diW_h(i)X + S(i)(X_h,X) + (dh^\perp(X))^\perp + \beta(X), \quad \forall\ X \in \Gamma TM.$$

This allows us to read off the coefficients in (0.22) as

$$(0.27)\qquad \begin{cases} c_\alpha di = (dh^\perp)^\perp + S(i)(X_h,\cdot) + c_\beta di\ , \\ C_\alpha = \frac{1}{2}[\nabla(i)X_h^0 - \tilde{\nabla}(i)X_h^0] + C_\beta\ , \\ B_\alpha = \frac{1}{2}L_{X_h^0} + \mathrm{grad}_i\psi_h + W_h(i) + B_\beta\ . \end{cases}$$

$W_h(i)$ denotes here the strong smooth bundle map of TM given by

$$(0.28) \qquad diW_h(i)\, X := (dh^{\perp}(X))^T , \qquad \forall\, X \in \Gamma TM .$$

$\tilde{\nabla}(i)X_h$ is the adjoint of $\nabla(i)X_h$ with respect to m(i) formed fibrewise, so that each $v_p \in T_pM$ is sent into $\tilde{\nabla}(i)X_h(v_p)$, $\forall\, p \in M$. Moreover

$$(0.29) \qquad L_{X_h} : TM \longrightarrow TM$$

is the strong smooth bundle endomorphism of TM defined by the Lie derivative $L_{X_h}(m(i))$ via the equation

$$(0.30) \qquad m(i)(L_{X_h}X,Y) := L_{X_h}(m(i))(X,Y) , \qquad \forall\, X_h,X,Y \in \Gamma TM.$$

Using now the definition of the Laplace–Beltrami operator $\Delta(i)$

$$(0.31) \qquad div_i(\nabla(i)X_h) = \Delta(i)X_h = -tr\, \nabla^2(i)X_h ,$$

and introducing $R(i)X_h$ via

$$(0.32) \qquad m(i)(R(i)X_h,Y) = Ric(m(i))(X_h,Y) , \qquad \forall\, Y \in \Gamma TM ,$$

where Ric(m(i)) denotes the Ricci tensor of m(i), we obtain in the case of codimension 1 the formulas

$$(0.33) \qquad \begin{cases} div_i\, B_{dh} = \frac{1}{2}\,\Delta(i)X_h + \frac{1}{2}\, R(i)X_h \\ \qquad\qquad + W(i)\, grad_i\Theta_h + \Theta_h\, grad_i H(i) , \\ div_i\, C_{dh} = \frac{1}{2}\,\Delta(i)X_h - \frac{1}{2}\, R(i)X_h - \frac{1}{2}\, grad_i div_i X_h , \\ tr\, B_{dh} = -\,\Delta(i)\, \psi_h + tr(\Theta_h W(i)) . \end{cases}$$

a fixed embedding i.

Here the unnormalized mean curvature H(i) is defined to be tr W(i) . Next we introduce the notion of structural viscosity. To this end we consider on the one hand the decompositions (0.23), (0.24) and (0.25) for the stress form $\alpha(dk,dj)$, which now depends on an additional parameter dk with $k \in C^{\infty}(M,\mathbb{R}^n)$. On the other hand, we use the decomposition for k, i.e.

$$(0.34) \qquad k = djX_k + k^{\perp},$$

$$(0.35) \qquad \begin{cases} X_k = X_k^0 + grad_j \psi_k , \\ div_j X_k^0 = 0 . \end{cases}$$

Even though dk is determined only up to a constant, X_k^0 depends uniquely on dk. This allows us to relate $X_k^0$and X_h^0 uniquely to each other by

$$(0.36) \qquad X_h^0(dj,dk) = \nu(dj,dk)X_k^0 + \hat{X}_h(dj,dk) \, ,$$

where $\nu(dj,dk) \in C^\infty(M,\mathbb{R})$ and $\hat{X}(dj,dk) \in \Gamma TM$ is pointwise orthogonal to X_k^0. We call the function $\nu(dj,dk)$, the coefficient of structural viscosity. Accordingly we call these deformable media, whose constitutive laws depend only on $k^\perp$, frictionless deformable media, while the deformable media, whose constitutive laws depend on the whole of k, will be called frictional ones.

Furnished with the structure developed so far, we deduce next the equations of motion of a deformable medium M subjected to a general constitutive law

$$F : E(M,\mathbb{R}^n) \times C^\infty(M,\mathbb{R}^n) \times C^\infty(M,\mathbb{R}^n) \longrightarrow \mathbb{R}.$$

To do this, we assume that F splits into

$$(0.37) \qquad F = F_{ext} + F_{int}$$

and that F_{int} is of the form

$$(0.38) \qquad F_{int} = d^* F_{\mathbb{R}^n} \, .$$

Furthermore, we require that F_{ext} and $F_{\mathbb{R}^n}$ both admit integral representation and denote the resulting force densities by φ_{ext} and φ_{int} respectively. Using Hodge's decomposition, we obtain for all $j \in E(M,\mathbb{R}^n)$ and all $k \in C^\infty(M,\mathbb{R}^n)$

$$(0.39) \qquad \begin{cases} \varphi_{int}(j,k) = dj \, grad_j \tau_{int}(j,k) + dj \, Y^0_{int}(j,k) + \varphi^\perp_{int}(j,k) \, , \\ \varphi_{ext}(j,k) = dj \, grad_j \tau_{ext}(j,k) + dj \, Y^0_{ext}(j,k) + \varphi^\perp_{ext}(j,k) \, , \end{cases}$$

and in turn

$$\begin{aligned} (0.40) \qquad \varphi(j,k) &= dj \, (grad_j \tau_{int}(j,k) + grad_j \tau_{ext}(j,k)) \\ &\quad + dj \, (Y^0_{int}(j,k) + Y^0_{ext}(j,k)) + \varphi^\perp_{int}(j,k) + \varphi^\perp_{ext}(j,k) \\ &:= dj \, grad_j \, \tau(j,k) + dj \, Y^0(j,k) + \varphi^\perp(j,k) \, , \\ &\qquad \forall \; j \in E(M,\mathbb{R}^n), \;\; k \in C^\infty(M,\mathbb{R}^n). \end{aligned}$$

Hence the equations of motion in case of dim M = n−1 are

$$(0.41)\quad \begin{cases} \nabla(\sigma(t))_{Z(t)}Z(t) + \dot{Z}(t) + 2\cdot\epsilon(\sigma(t),\dot\sigma(t))\, W(\sigma(t))Z(t) \\ -\ \epsilon(\sigma(t),\dot\sigma(t))\ \mathrm{grad}_{\sigma(t)}\epsilon(\sigma(t),\dot\sigma(t)) \\ = \rho^{-1}(\sigma(t))(\mathrm{grad}_{\sigma(t)}\tau(\sigma(t),\dot\sigma(t)) - \Delta(\sigma(t))[\nu(d\sigma(t),d\dot\sigma(t))Z^0(t) \\ +\ \hat{Z}_h(d\sigma(t),d\dot\sigma(t)) + \mathrm{grad}_{\sigma(t)}\psi(\sigma(t),\dot\sigma(t))] \\ -\ W(\sigma(t))[\mathrm{grad}_{\sigma(t)}\Theta_h(d\sigma(t),d\dot\sigma(t)) + 2(W(\sigma(t))X_h - \mathrm{grad}\ \Theta_h)] \\ -\ \Theta_h(d\sigma(t),d\dot\sigma(t))\mathrm{grad}_{\sigma(t)}H(\sigma(t)]) \ , \\ \\ \dot\epsilon(\sigma(t),\dot\sigma(t)) = \rho^{-1}(\sigma(t))[-\tau_{int}(d\sigma(t),d\dot\sigma(t))\ H(\sigma(t)) \\ -\ dH(\sigma(t))[\nu(d\sigma(t),d\dot\sigma(t))\ Z^0(t) + \hat{Z}_h(d\sigma(t),\dot\sigma(t))] \\ +\ \mathrm{div}_{\sigma(t)}\nu(d\sigma(t),d\dot\sigma(t))W(\sigma(t))Z^0(t) \\ +\ \mathrm{div}_{\sigma(t)}W(\sigma(t))\hat{X}_h(d\sigma(t),d\dot\sigma(t)) \\ -\ \Theta_h(d\sigma(t),d\dot\sigma(t))\ \mathrm{tr}\ W(\sigma(t))^2] + \mathfrak{h}(\sigma(t))\ (Z(t),Z(t)) \\ -\ d\epsilon(\sigma(t),\dot\sigma(t))\ Z(t) + \kappa_{ext}(\sigma(t),\dot\sigma(t))] \end{cases}$$

where $\varphi_{ext}{}^{\perp}(\sigma(t),\dot\sigma(t)) = \kappa_{ext}(\sigma(t),\dot\sigma(t))\ N(\sigma(t))$.
In case the motion follows a fixed surface $i(M)\subset\mathbb{R}^n$ given by a fixed embedding $i \in E(M,\mathbb{R}^n)$, the equation (0.41) reduces to

$$(0.42)\quad \begin{cases} \nabla(i)_{X(t)}X(t) + \dot{X}(t) = \rho^{-1}(X(t))[-\mathrm{grad}_i\tau_{int}(X(t),\dot X(t)) \\ \qquad -\ \Delta(i)[\nu(X(t),\dot X(t))\ X^0(t) + \hat{X}_h(X(t),\dot X(t)) \\ \qquad +\ \mathrm{grad}_i\psi(X(t),\dot X(t))] - W(i)[\mathrm{grad}_i\ \Theta_h(X(t),\dot X(t)) \\ \qquad +\ 2\ (W(\sigma(t))X_h - \mathrm{grad}\ \Theta_h)] - \Theta_h(X(t),\dot X(t))\ \mathrm{grad}_iH(i)\ , \\ \\ 0 = \rho^{-1}(X(t))\ (-\tau_{int}(X(t),\dot X(t))\cdot H(i) - dH(i)[\nu(X(t),\dot X(t))X^0(t) \\ \qquad +\ \mathrm{div}_i\nu(X(t),\dot X(t))W(i)X^0(t) + \mathrm{div}_iW(i)\hat{X}_h(X(t),\dot X(t)) \\ \qquad +\ \hat{X}_h(X(t),\dot X(t))] - \Theta_h(X(t),\dot X(t))\ \mathrm{tr}\ W(i)^2) \\ \qquad +\ \mathfrak{h}(i)(X(t),X(t)) + \kappa_{ext}(X(t),\dot X(t))\ , \end{cases}$$

where $X(t)$ is the push-forward of $Z(t)$ by $g(t) \in \mathrm{Diff}\ M$, i.e.

$$(0.43)\qquad X(t) = Tg(t)\ Z(t)\ g(t)^{-1}\ , \qquad \forall\ t \in (-\lambda,\lambda)\ .$$

At the end of the paper we remark how to introduce a volume active pressure $\pi(dj,dk)$, which allows us to decompose $F(dj,dk)$ into

$$(0.44)\qquad F(dj,dk) = \bar{F}(dj,dk) - \pi(dj,dk)\cdot DV(j)\ ,$$

where $V(j)$ denotes the volume of $j(M)$.
$\pi(dj,dk)\cdot DV(j)$ is the work used against the infinitesimal volume change by $DV(j)$. Let us point out that $\pi(dj,dk)$ is not identical with $\tau_{int}(dj,dk)$, the former is a real, the latter a smooth function.

We have omitted to discuss the influence of thermodynamics to the deformations of the medium. We will do these studies in a forthcoming paper.

1. The space of configurations as a principal bundle

As already mentioned a configuration of the moving deformable medium M is described by a smooth embedding of M into an euclidean ambient space $\mathbb{R}^n$. In the present paper we assume that the dimension dim M of the manifold M satisfies the inequality

(1.1) $$\dim M \leq n-1.$$

Let us recall that a smooth embedding $j : M \longrightarrow \mathbb{R}^n$ is a smooth map satisfying the following conditions (i) the tangent map

(1.2) $$Tj(p) : T_pM \longrightarrow \{j(p)\} \times \mathbb{R}^n$$

of j at $p \in M$ is injective for any $p \in M$,

(ii)

(1.3) $$j : M \longrightarrow j(M) \subset \mathbb{R}^n$$

is a homeomorphism.

We point out at this occasion that the tangent map

(1.4) $$Tj : TM \longrightarrow T\mathbb{R}^n = \mathbb{R}^n \times \mathbb{R}^n$$

splits naturally into

(1.5) $$Tj = (j,dj),$$

where

(1.6) $$dj = pr_2 \circ Tj$$

and

(1.7) $$pr_2 : \mathbb{R}^n \times \mathbb{R}^n \longrightarrow \mathbb{R}^n$$

is the projection onto the second factor along the first one, i.e.

(1.8) $$pr_2(a,b) = b \qquad \text{for all pairs } (a,b) \in \mathbb{R}^n \times \mathbb{R}^n .$$

Thus dj represents locally nothing else but the Frèchet differential of the local representative of j.

We denote by $E(M,\mathbb{R}^n)$ the set of all smooth embeddings of M into $\mathbb{R}^n$. Hence $E(M,\mathbb{R}^n)$ is the set of all configurations of our moving deformable medium. If we equip $E(M,\mathbb{R}^n)$ with the Whitney C^∞- topology (cf. [Gui,Go]), then it becomes an infinite dimensional Frèchet manifold.

The reason is the following one :

Consider the set $C^\infty(M,\mathbb{R}^n)$ of all smooth maps of M into $\mathbb{R}^n$, which is endowed with Whitney's C^∞- topology and note that together with the pointwise defined operations of addition and multiplication with scalars $C^\infty(M,\mathbb{R}^n)$ becomes a complete metrizable, locally convex space, a so-called Frèchet space.

Since as shown in [Hi] $E(M,\mathbb{R}^n)$ is open in $C^\infty(M,\mathbb{R}^n)$, it hence carries the structure of a Frèchet manifold (cf. [Bi,Fi]).
Using now the differential calculus in locally convex spaces constructed either in [Ba], [Gu], [Mi] or [Fr,Kr], it is evident that the tangent space $T_j E(M,\mathbb{R}^n)$ at $j \in E(M,\mathbb{R}^n)$ is nothing else but $C^\infty(M,\mathbb{R}^n)$.
Therefore the tangent bundle $TE(M,\mathbb{R}^n)$ is trivial, i.e.

$$TE(M,\mathbb{R}^n) = E(M,\mathbb{R}^n) \times C^\infty(M,\mathbb{R}^n). \tag{1.9}$$

We note that $TE(M,\mathbb{R}^n)$ is the phase space of motions in $E(M,\mathbb{R}^n)$.
Next we introduce the principal bundle structure of $E(M,\mathbb{R}^n)$, which is crucial for our formalism to describe the motion of the deformable medium M.
Following [Bi,Fi] we first describe the group action. Let Diff M be the group of all smooth diffeomorphisms of M equipped with the C^∞- topology. Diff M is a Frèchet manifold, in which the operations are smooth in either one of the above mentioned notions of differentiability.
Consequently we call Diff M a differentiable group. The tangent space at the identity in Diff M is naturally identified with ΓTM, the set of smooth vector fields on M.
The operation Φ of Diff M on $E(M,\mathbb{R}^n)$ given by

$$\Phi(j,g) = j \circ g \qquad \forall\ j \in E(M,\mathbb{R}^n),\ g \in \text{Diff } M, \tag{1.10}$$

is smooth.
Consequently $E(M,\mathbb{R}^n)$ can be represented as

$$E(M,\mathbb{R}^n) = \bigcup_{i \in E(M,\mathbb{R}^n)} i \circ \text{Diff } M \quad , \tag{1.11}$$

i.e. as the collection of smooth fibers. The quotient $U(M,\mathbb{R}^n)$ of $E(M,\mathbb{R}^n)$ by Diff M

$$U(M,\mathbb{R}^n) = E(M,\mathbb{R}^n)/_{\text{Diff } M} \tag{1.12}$$

inherits a smooth Frèchet manifold and it is naturally identified with the collection of all smooth submanifols of M in $\mathbb{R}^n$ which are diffeomorphic to M. Let us denote the quotient map, i.e. the projection of $E(M,\mathbb{R}^n)$ onto $U(M,\mathbb{R}^n)$, by Π. The quadruple ($E(M,\mathbb{R}^n)$, Π, $U(M,\mathbb{R}^n)$, Diff M) is then a principal bundle in the sense of [Gr,H,V].
Each configuration $j \in E(M,\mathbb{R}^n)$ induces a Riemannian metric $m(j)$ on M defined by

$$m(j)(X,Y) := \langle djX, djY \rangle \qquad \forall\ X,Y \in \Gamma TM. \tag{1.13}$$

Denoting now by $\mathfrak{M}(M)$ the set of all smooth Riemannian metrics on M equipped with the C^∞- topology, we obtain then a natural map

(1.14) $$m : E(M,\mathbb{R}^n) \longrightarrow \mathfrak{M}(M).$$

Noting that $\mathfrak{M}(M)$ is an open cone in $S^2(M)$, the Frèchet space of smooth symmetric two tensors on M, we deduce that $\mathfrak{M}(M)$ is also a Frèchet manifold and that the map m is smooth (cf. [Sch]).
Moreover the tangent bundle $T\mathfrak{M}(M)$ of $\mathfrak{M}(M)$ is trivial, i.e.

(1.15) $$T\mathfrak{M}(M) = \mathfrak{M}(M) \times S^2(M).$$

For later use we calculate at this point the derivative $Dm(j)(k)$ of m at $j \in E(M,\mathbb{R}^n)$ in the direction of $h \in C^\infty(M,\mathbb{R}^n)$. Due to the $\mathbb{R}$ - bilinearity of m in the variable $j \in E(M,\mathbb{R}^n)$ we have

(1.16) $$Dm(j)(h)(X,Y) = \langle djX, dhY \rangle + \langle dhX, djY \rangle, \quad \forall\, X,Y \in \Gamma TM.$$

Splitting h into
$$h = dj\, X_h + h^\perp$$
with $X_h \in \Gamma TM$ and $h^\perp$ being pointwise normal to $j(M)$ in $\mathbb{R}^n$, equation (1.16) turns into

(1.17) $$Dm(j)(h)(X,Y) = \langle djX,d(djX_h)(Y)\rangle + \langle d(djX_h)X,djY\rangle + \langle djX,dh^\perp Y\rangle + \langle dh^\perp X,djY\rangle .$$

Since $d(djX_h)Y = dj\nabla(j)_Y X_h + S(i)(X_h,Y)$, where $\nabla(i)$ is the Levi-Civita connection of $m(j)$, (1.17) turns into

(1.18) $$\begin{aligned} Dm(j)(h)(X,Y) &= m(j)(X,\nabla(j)_Y X_h) + m(j)(\nabla(j)_X X_h,Y) \\ &\quad + \langle djX,dh^\perp Y\rangle + \langle dh^\perp X,djY\rangle \\ &= L_{X_h}(m(j))(X,Y) + \langle djX,dh^\perp Y\rangle + \langle dh^\perp X,djY\rangle , \end{aligned}$$

with $L_{X_h}(m(j))$ being the Lie derivative of $m(j)$.
In case $h^\perp = \Theta_h \cdot N(j)$ then

(1.19) $$Dm(j)(h)(X,Y) = L_{X_h}(m(j))(X,Y) + 2\cdot\Theta_h\cdot\mathfrak{h}(j)(X,Y) ,$$

with $\mathfrak{h}(j)$ the second fundamental form of j defined by

$$\mathfrak{h}(j)(X,Y) := m(j)(W(j)X,Y) = \langle dN(j)X,djY\rangle .$$

Here $W(j)$ denotes the Weingarten map of j given by

(1.20) $$djW(j)X = dN(j)X\ , \qquad \forall\ X \in \Gamma TM\ .$$

If $H(j)$ denotes $\operatorname{tr} W(j)$, then

(1.21) $$\operatorname{tr} Dm(j)(h) = 2\cdot\left(\operatorname{div}_j X_h + \Theta_h \cdot H(j)\right)\ .$$

By $\operatorname{div}_j X_h$ we denote the divergence of X_h formed with respect to $m(j)$. This means,

$$\operatorname{div}_j X_h = \operatorname{tr} \nabla(j) X_h\ .$$

The function $\frac{H(j)}{\dim M}$ is called the mean curvature of j, while $H(j)$ denotes the unnormalized mean curvature.

2. The metric $\mathfrak{G}_E$ on the configuration space $E(M,\mathbb{R}^n)$

In order to define a metric on $E(M,\mathbb{R}^n)$, which is adapted to the mass distribution of our moving deformable medium M, we first introduce a density map

(2.1) $$\rho : E(M,\mathbb{R}^n) \longrightarrow C^\infty(M,\mathbb{R}),$$

which is supposed to be smooth in either sense of the above mentioned notions of differentiability. In addition we require that ρ fullfills a continuity equation, namely

(2.2) $$D\rho(j)(h) = -\tfrac{1}{2}\,\rho(j)\operatorname{tr}_j Dm(j)(h), \quad \forall\ j \in E(M,\mathbb{R}^n),\ h \in C^\infty(M,\mathbb{R}^n),$$

where tr_j denotes the trace taken with respect to $m(j)$. Using the fact that the derivative at j in the direction of any $h \in C^\infty(M,\mathbb{R}^n)$ of the Riemannian volume form $\mu(j)$ has the form

(2.3) $$D\mu(j)(h) = \tfrac{1}{2}\,\mu(j)\operatorname{tr}_j Dm(j)(h)$$

as shown in [Bi,1], it follows that (2.2) is indeed the continuity equation. Consequently the total mass $m(j)$ attached to any $j \in E(M,\mathbb{R}^n)$ via the formula

(2.4) $$m(j) = \int_M \rho(j)\,\mu(j)$$

is constant in j.
The existence of such a function ρ can be established as follows :
Let $i \in E(M,\mathbb{R}^n)$ be any embedding and denote by O_i its connected component

in $E(M,\mathbb{R}^n)$. Then for any $j \in O_i$ the differential dj is related to di in the following way

$$(2.5) \qquad dj = g \circ di \circ f,$$

where $g \in C^\infty(M,SO(n))$ and f is a strong bundle isomorphism of TM which is fibrewise positive with respect to m(i). One easily verifies that the Riemannian volume forms $\mu(j)$ and $\mu(i)$ of j and i respectively are related by

$$(2.6) \qquad \mu(j) = \det f \cdot \mu(i) .$$

We then set

$$(2.7) \qquad \rho(j) := \rho(i) \cdot \det f^{-1}$$

with $\rho(i)$ chosen such that

$$(2.8) \qquad \rho(i)(p) > 0 , \qquad \forall\, p \in M,$$

and note that this map satisfies the continuity equation (2.2).
Next we introduce the metric $\mathfrak{G}_E$ on $E(M,\mathbb{R}^n)$ by the formula

$$(2.9) \qquad \mathfrak{G}_E(j)(h,k) := {}_M\!\int \rho(j) \langle h,k\rangle\, \mu(j) , \quad \forall\, h,k \in C^\infty(M,\mathbb{R}^n).$$

Due to (2.2) $\mathfrak{G}_E$ is constant in j. Therefore the geodesics of $\mathfrak{G}_E$ are straight line segments as shown in [Bi,1].

3. The constitutive law and the general equations of motion

By a constitutive law we understand a smooth parameter depending one form, the so-called work

$$(3.1) \qquad F : C^\infty(M,\mathbb{R}^n) \times E(M,\mathbb{R}^n) \times C^\infty(M,\mathbb{R}^n) \longrightarrow \mathbb{R}.$$

The first factor $C^\infty(M,\mathbb{R}^n)$ in the above cartesian product is the parameter space. We often will regard F as a map

$$F : C^\infty(M,\mathbb{R}^n) \longrightarrow A^1(E(M,\mathbb{R}^n),\mathbb{R}).$$

The domain of F is the parameter space, the range the collection of all smooth one forms on $E(M,\mathbb{R}^n)$ with values in $\mathbb{R}$. To handle this abstract notion (3.1) we require an integral representation for F given by

$$(3.2) \qquad F(k)(j,h) = {}_M\!\int \langle \varphi_F(j,k),h\rangle\, \mu(j) , \quad \forall\, j \in E(M,\mathbb{R}^n),\ h,k \in C^\infty(M,\mathbb{R}^n),$$

where $\varphi_F : TE(M,\mathbb{R}^n) \longrightarrow C^\infty(M,\mathbb{R}^n)$, the so-called force density, is assumed to be a smooth map. We point out that F(k)(j,h) varies linearly only in h and that it furthermore depends on the maps j and k globally.

The equation of motion on $E(M,\mathbb{R}^n)$ described by a smooth curve

(3.3) $$\sigma : (-\lambda,\lambda) \longrightarrow E(M,\mathbb{R}^n) ,$$

for some positive real λ, is given by

(3.4) $$F(\dot\sigma(t))(\sigma(t),h) = \mathcal{B}(\sigma(t))(\ddot\sigma(t),h) = {}_M\!\int \langle \varphi_F(\sigma(t),\dot\sigma(t)),h\rangle \, \mu(\sigma(t)) = {}_M\!\int \rho(\sigma(t)) \langle \ddot\sigma(t),h\rangle \, \mu(\sigma(t)) \quad \forall\, h\in C^\infty(M,\mathbb{R}^n),$$

where for the sake of simplicity we have taken $k=\dot\sigma(t)$.
It is obvious that this equation implies

(3.5) $$\rho(\sigma(t))\, \ddot\sigma(t) = \varphi_F\, (\sigma(t),\dot\sigma(t)), \quad \forall\, t\in (-\lambda,\lambda).$$

We note that (3.5) is a second order differential equation on $E(M,\mathbb{R}^n)$ and not on parts of $\mathbb{R}^n$.
We rewrite it now according to the principal bundle structure of $E(M,\mathbb{R}^n)$, by proceeding as follows :
At first we note that $\dot\sigma(t)$ admits in $\mathbb{R}^n$ the pointwise splitting

(3.6) $$\dot\sigma(t) = d\sigma(t)\, Z(t) + \dot\sigma(t)^\perp ,$$

where $Z(t)\in \Gamma TM$ is uniquely determined and $\dot\sigma(t)^\perp$ is, according to the definition, pointwise perpendicular to $\sigma(t)(M)$ for each $t\in (-\lambda,\lambda)$.
Consequently $\ddot\sigma(t)$ is given by

(3.7) $$\ddot\sigma(t) = d\dot\sigma(t)\, Z(t) + d\sigma(t)\, \dot Z(t) + (\dot\sigma(t)^\perp)' , \quad t\in (-\lambda,\lambda),$$

where

(3.8) $$\begin{aligned} d\dot\sigma(t)\, Z(t) &= d(d\sigma(t)\, Z(t))\, Z(t) + d\dot\sigma(t)^\perp\, Z(t) \\ &= d\sigma(t)\, \nabla(\sigma(t))_{Z(t)}\, Z(t) + S(\sigma(t))\, (Z(t),Z(t)) \\ &\quad + d\sigma(t)\, W(\sigma(t),\dot\sigma(t)^\perp)\, Z(t) + (d\dot\sigma(t)^\perp\, Z(t))^\perp . \end{aligned}$$

Here $\nabla(\sigma(t))$ means the Levi – Cività connection of $m(\sigma(t))$, and $W(\sigma(t),\dot\sigma(t))$, the Weingarten map of $\sigma(t)$ is defined as follows :
Let $N : M \longrightarrow \mathbb{R}^n$ be any vector field along $j\in E(M,\mathbb{R}^n)$, such that

(3.9) $$\langle dj\, X(p), N(p)\rangle = 0 , \quad \forall\, X\in \Gamma TM,\ p\in M.$$

Then the Weingarten map $W(j,N)$ of j given by N is the uniquely determined bundle map of TM for which

(3.10) $$dj\, W(j,N)\, Y = (dN\, Y)^T .$$

In the particular case where M is oriented and of dimension equal to n-1 and N coincides with the unit normal vector field N(j) of j(M) in $\mathbb{R}^n$, then W(j,N(j)) is nothing else but the Weingarten map, denoted in this particular case just by W(j).
Let us turn back to $\ddot{\sigma}(t)$. Obviously (compare cf. [Bi,6])

$$\begin{aligned}(3.11)\qquad \ddot{\sigma}(t) &= d\sigma(t)\ \nabla(\sigma(t))_{Z(t)}Z(t) + d\sigma(t)\ W(\sigma(t),\dot{\sigma}(t)^{\perp})\ Z(t)\\ &+ ((\dot{\sigma}(t)^{\perp})^{\cdot})^{T} + d\sigma(t)\ \dot{Z}(t) + S(\sigma(t))\ (Z(t),Z(t))\\ &+ (d\dot{\sigma}(t)^{\perp}\ Z(t))^{\perp} + ((\dot{\sigma}(t)^{\perp})^{\cdot})^{\perp}\ .\end{aligned}$$

For unifying terminology we set

$$(3.12)\qquad ((\dot{\sigma}(t)^{\perp})^{\cdot})^{T} := d\sigma(t)\ U(\sigma(t),\dot{\sigma}(t))$$

for each $t \in (-\lambda,\lambda)$ and a well defined vector field $U(\sigma(t),\dot{\sigma}(t)) \in \Gamma TM$.
If we split now $\varphi_F(\sigma(t),\dot{\sigma}(t))$ into a tangential and normal part respectively then for all $t \in (-\lambda,\lambda)$

$$\begin{aligned}(3.13)\qquad \varphi_F(\sigma(t),\dot{\sigma}(t)) &= \varphi_F(\sigma(t),\dot{\sigma}(t))^{T} + \varphi_F(\sigma(t),\dot{\sigma}(t))^{\perp}\\ &= d\sigma(t)\ Y_F(\sigma(t),\dot{\sigma}(t)) + \varphi_F(\sigma(t),\dot{\sigma}(t))^{\perp}\end{aligned}$$

holds for a uniquely determined $Y_F(\sigma(t),\dot{\sigma}(t)) \in \Gamma TM$.
The equation of motion splits thus into the coupled system

$$(3.14)\qquad \left\{\begin{aligned} &\nabla(\sigma(t))_{Z(t)}Z(t) + \dot{Z}(t) + W(\sigma(t),\dot{\sigma}(t)^{\perp})\ Z(t) + U(\sigma(t),\dot{\sigma}(t))\\ &\qquad = \rho^{-1}(\sigma(t))\ Y_F(\sigma(t),\dot{\sigma}(t)),\\[1ex] &((\dot{\sigma}(t)^{\perp})^{\cdot})^{\perp} = \rho^{-1}(\sigma(t))\ \varphi_F(\sigma(t),\dot{\sigma}(t))^{\perp}\\ &\qquad - (d\dot{\sigma}(t)^{\perp}\ Z(t))^{\perp} - S(\sigma(t))\ (Z(t),Z(t))\ .\end{aligned}\right.$$

The first equation multiplied on both sides by $d\sigma(t)$ yields an equation of vectors which are tangential to $\sigma(t) \circ \mathrm{Diff}\ M$, while the second one is an equation of vectors in $C^{\infty}(M,\mathbb{R}^n)$, which are normal to $\sigma(t) \circ \mathrm{Diff}\ M$.

Hence the above coupled system (3.14) is a splitting of the equation of motion according to the principal bundle structure of $E(M,\mathbb{R}^n)$ as mentioned above.
In the particular case dim M = n-1 we obtain

$$(3.15)\qquad \dot{\sigma}(t)^{\perp} = \epsilon(\sigma(t),\dot{\sigma}(t))\ N(\sigma(t))$$

$$(3.16)\qquad \varphi_F(\sigma(t),\dot{\sigma}(t))^{\perp} = \kappa_F(\sigma(t),\dot{\sigma}(t))\ N(\sigma(t))$$

for well determined $\epsilon(\sigma(t),\dot{\sigma}(t)),\ \kappa_F(\sigma(t),\dot{\sigma}(t)) \in C^{\infty}(M,\mathbb{R})$.
Hence

$$(3.17)\qquad (\dot{\sigma}(t)^{\perp})^{\cdot} = \dot{\epsilon}(\sigma(t),\dot{\sigma}(t))\ N(\sigma(t)) + \epsilon(\sigma(t),\dot{\sigma}(t))\ N(\sigma(t))^{\cdot}.$$

It remains now to calculate $N(\sigma(t))'$. To this end we prove at first that

$$\text{(3.18)} \qquad DN(j)(\tau \cdot N(j)) = - dj \, grad_j \tau , \qquad \forall \tau \in C^\infty(M,\mathbb{R}) .$$

Indeed let

$$j(t) = j + t \cdot \tau \cdot N(j) , \qquad \forall t \in \mathbb{R}.$$

Then we get

$$\text{(3.19)} \qquad \begin{aligned} \frac{d}{dt} \langle dj(t),N(j(t))\rangle \Big|_{t=0} &= 0 \\ &= \langle \tau \cdot djW(j),N(j)\rangle + d\tau + \langle dj,DN(j)(\tau \cdot N(j))\rangle. \end{aligned}$$

Using that $\langle N(j),N(j)\rangle = 1$, (3.18) then follows.
Next consider a smooth curve $\gamma(t) \in \mathrm{Diff}\, M$, $t \in (-\lambda,\lambda)$, $\lambda>0$, and note that

$$\text{(3.20)} \qquad N(j \circ \gamma(t)) = N(j) \circ \gamma(t) .$$

Differentiating (3.20) we get

$$\text{(3.21)} \qquad DN(j)(dj\, X) = dN(j)\, X = dj\, W(j)\, X ,$$

where $X = \dot\gamma(0)$. From (3.6), (3.15), (3.18) and (3.21) it follows that

$$\text{(3.22)} \qquad \begin{aligned} N(\sigma(t))' &= DN(\sigma(t))\, (d\sigma(t)\, Z(t) + \epsilon(\sigma(t),\dot\sigma(t))\, N(\sigma(t)) \\ &= d\sigma(t)\, \big(W(\sigma(t))\, Z(t) - grad_{\sigma(t)}\epsilon(\sigma(t),\dot\sigma(t))\big) , \end{aligned}$$

and hence

$$\text{(3.23}_1\text{)} \qquad U(\sigma(t),\dot\sigma(t)) = \epsilon(\sigma(t),\dot\sigma(t))\big(W(\sigma(t))Z(t) - grad_{\sigma(t)}\epsilon(\sigma(t),\dot\sigma(t))\big)$$

$$\text{(3.23}_2\text{)} \qquad ((\dot\sigma(t)^\perp)')^\perp = \dot\epsilon(\sigma(t),\dot\sigma(t))\, N(\sigma(t)).$$

Moreover

$$\text{(3.24)} \qquad \begin{aligned} d(\dot\sigma(t)^\perp)\, Z(t) = {} & d\epsilon(\sigma(t),\dot\sigma(t))\, Z(t)\, N(\sigma(t)) \\ & + \epsilon(\sigma(t),\dot\sigma(t))\, d\sigma(t)\, W(\sigma(t))\, Z(t) \\ & - d\sigma(t)\, grad_{\sigma(t)}\epsilon(\sigma(t),\dot\sigma(t)). \end{aligned}$$

Thus (3.14) rewrites as

$$\text{(3.25)} \qquad \left\{ \begin{aligned} & \nabla(\sigma(t))_{Z(t)}\, Z(t) + \dot Z(t) + 2\, \epsilon(\sigma(t),\dot\sigma(t))\, W(\sigma(t))\, Z(t) \\ & \quad - \epsilon(\sigma(t),\dot\sigma(t))\, grad_{\sigma(t)}\epsilon(\sigma(t),\dot\sigma(t)) = \rho^{-1}(\sigma(t))\, Y(\sigma(t),\dot\sigma(t)), \\ & \dot\epsilon(\sigma(t),\dot\sigma(t)) = \rho^{-1}(\sigma(t))\, \kappa_F(\sigma(t),\dot\sigma(t)) + \mathfrak{h}(\sigma(t))\, (Z(t),Z(t)) \\ & \qquad - d\epsilon(\sigma(t),\dot\sigma(t))\, Z(t) . \end{aligned} \right.$$

We refer to (3.25) as the general equations of motion of a deformable medium.

Let us now split $Y_F(\sigma(t),\dot\sigma(t))$ with respect to $m(\sigma(t))$ according to the Hodge decomposition into

$$(3.26) \qquad Y_F(\sigma(t),\dot\sigma(t)) = Y^o{}_F(\sigma(t),\dot\sigma(t)) + \operatorname{grad}_{\sigma(t)} \tau_F(\sigma(t),\dot\sigma(t)),$$

where $\tau(\sigma(t),\dot\sigma(t)) \in C^\infty(M,\mathbb{R}^n)$ and

$$(3.27) \qquad \operatorname{div}_{\sigma(t)} Y^o{}_F(\sigma(t),\dot\sigma(t)) = 0 ,$$

i.e. the divergence of $Y^o{}_F(\sigma(t),\dot\sigma(t))$ taken with respect to $m(\sigma(t))$ vanishes, $\tau_F(\sigma(t),\dot\sigma(t)) \in C^\infty(M,\mathbb{R})$, and $\operatorname{grad}_{\sigma(t)}$ means the gradient taken with respect to $m(\sigma(t))$.

Using (3.26) we rewrite the first equation (3.25) as

$$(3.25') \qquad \begin{aligned} &\nabla(\sigma(t))_{Z(t)} Z(t) + \dot Z(t) + 2\, \epsilon(\sigma(t),\dot\sigma(t))\, W(\sigma(t)\, Z(t) \\ &- \epsilon(\sigma(t),\dot\sigma(t)) \operatorname{grad}_{\sigma(t)}\epsilon(\sigma(t\ ,\dot\sigma(t)) \\ &= \rho^{-1}(\sigma(t))\, [\, \operatorname{grad}_{\sigma(t)}\tau_F(\sigma(t),\dot\sigma(t)) + Y^o{}_F(\sigma(t),\dot\sigma(t))]. \end{aligned}$$

The above Hodge decomposition of $Y_F(\sigma(t),\dot\sigma(t))$ yields a decomposition of φ_F into

$$(3.28) \qquad \varphi_F(\sigma(t),\dot\sigma(t)) = \varphi^0_F(\sigma(t),\dot\sigma(t)) + \varphi'_F(\sigma(t),\dot\sigma(t)) ,$$

where

$$(3.29) \qquad \begin{cases} \varphi^0_F(\sigma(t),\dot\sigma(t)) = d\sigma(t) \operatorname{grad}_{\sigma(t)}\tau_F(\sigma(t),\dot\sigma(t)), \\ \varphi'_F(\sigma(t),\dot\sigma(t)) = d\sigma(t)\, Y^o{}_F(\sigma(t),\dot\sigma(t)) + \varphi_F(\sigma(t),\dot\sigma(t))^\perp. \end{cases}$$

We note that the tangential part of the force density φ'_F is divergence free. Its corresponding work, i.e. the one form F', will be called the reduced constitutive law.

4. The motion along a fixed surface $i(M) \subset \mathbb{R}^n$

Let us consider again the coupled system (3.14) describing the motion of the deformable medium M. Noting that the embedding $\sigma(t)$ varies with t and that the submanifolds $\sigma(t_1)(M)$ and $\sigma(t_2)(M)$ of $\mathbb{R}^n$ differ generically from each other for different $t_1,t_2 \in (-\lambda,\lambda)$, we obtain that the first equation (3.14) describes the instantaneous motion of the deformable medium along the submanifold M.

In this section we assume that the submanifolds $\sigma(t)(M)$ of $\mathbb{R}^n$ are identical for all $t \in (-\lambda,\lambda)$.

As a visualising example we image a fluid moving on a sphere of fixed radius .

More generally let $i \in E(M,\mathbb{R}^n)$ be fixed.

Thus $i(M)$ is a submanifold of $\mathbb{R}^n$ on which a deformable medium moves according to a constitutive law to be specified below. A configuration of this motion is an embedding j of M onto i(M) and is hence of the form

$$(4.1) \qquad j = i \circ g,$$

for some $g \in$ Diff M.
Consequently the configuration space is $i \circ$ Diff M . It remains now to specify the constitutive law on T(i $\circ$ Diff M) , the phase space of the motions on i(M). To this end let us first study the nature of a tangent vector h to i $\circ$ Diff M at $i \circ g$, i.e. $h \in T_{iog}$ $i \circ$ Diff M. If we denote by R_g the right translation by g, i.e.

$$(4.2) \qquad \begin{cases} R_g : \text{Diff } M \longrightarrow \text{Diff } M, \\ g' \mapsto g \circ g', \end{cases}$$

then the tangent map $TR_g(id)$ sends any tangent vector $X \in \Gamma TM$ at Id Diff M into a tangent vector in T_gDiff M. Moreover

$$(4.3) \qquad TR_g : \Gamma TM \longrightarrow T_g\text{Diff } M$$

is obviously surjective. Regarding $i \circ g \circ$ Diff M as a submanifold of $E(M,\mathbb{R}^n)$, any $h \in T_{iog} E(M,\mathbb{R}^n)$ tangential to $i \circ g \circ$ Diff M is thus of the form

$$(4.4) \qquad h = d(i \circ g)\, X_h\, ,$$

for a uniquely defined vector field $X_h \in \Gamma TM$.
Thus we have a natural bijection

$$(4.5) \qquad d(i \circ g) : \Gamma TM \longrightarrow T_{iog}\, i \circ \text{Diff } M \subset T_{iog}\, E(M,\mathbb{R}^n)\, ,$$

sending each $X \in \Gamma TM$ into $d(i \circ g)\, X \in T_{iog}$ $i \circ$ Diff M.
By (4.5) we see that T ($i \circ$ Diff M) is trivialized via right translations as

$$(4.6) \qquad T(\, i \circ \text{Diff } M\,) = i \circ \text{Diff } M \times \Gamma TM\, .$$

Clearly

$$(4.7) \qquad d(i \circ g)\, X = di \circ TgX \quad .$$

We now introduce the constitutive law via the smooth map

$$(4.8) \qquad F : di\, \Gamma TM \times i \circ \text{Diff } M \times C^\infty(M,\mathbb{R}^n) \longrightarrow \mathbb{R}\, ,$$

which is linear in $h \in C^\infty(M,\mathbb{R}^n)$. We note that in analogy to (3.1) we put the parameter space $di\, \Gamma TM$ as the first factor of the domain of definition.

The justification for choosing the third factor as being $C^\infty(M,\mathbb{R}^n)$ instead or ΓTM will be given later.
As in the preceding section we require an integral representation for the constitutive law reading as

$$(4.9) \qquad F(d(i \circ g)X)(i\circ g,h) = \int \langle \varphi_F(i \circ g,d(i \circ g)X),h\rangle \ \mu(i \circ g)$$
$$\forall\ h \in C^\infty(M,\mathbb{R}^n),\ X \in \Gamma TM,\ g \in \text{Diff } M,$$

where the force density

$$(4.10) \qquad \varphi_F : i \circ \text{Diff } M \times \Gamma TM \longrightarrow C^\infty(M,\mathbb{R}^n)$$

is a smooth map.
The equation of motion on $i \circ \text{Diff } M$ described by a smooth curve

$$(4.11) \qquad \sigma : (-\lambda,\lambda) \longrightarrow i \circ \text{Diff } M\ , \quad \lambda > 0,$$

subjected to the above constitutive law is hence

$$(4.12) \qquad \rho(\sigma(t))\ddot{\sigma}(t) = \varphi_F(\sigma(t),\dot{\sigma}(t)),$$

or, equivalently,

$$(4.13) \qquad \begin{cases} \rho(\sigma(t))(\ddot{\sigma}(t))^T = (\varphi_F(\sigma(t),\dot{\sigma}(t)))^T\ , \\ \rho(\sigma(t))(\ddot{\sigma}(t))^\perp = (\varphi_F(\sigma(t),\dot{\sigma}(t)))^\perp\ , \end{cases} \qquad t \in (-\lambda,\lambda).$$

We obviously have

$$(4.14) \qquad (\varphi_F(\sigma(t),\dot{\sigma}(t)))^T = d\sigma(t)\ Y_F(\sigma(t),\dot{\sigma}(t))\ ,$$

for a well defined vector field $Y_F(\sigma(t),\dot{\sigma}(t)) \in \Gamma TM$. Since a solution $\sigma(t)$ of (4.13) has the form

$$(4.15) \qquad \sigma(t) = i \circ g(t)\ ,$$

for a smooth map

$$g : (-\lambda,\lambda) \longrightarrow \text{Diff } M\ ,$$

it follows that

$$(4.16) \qquad \begin{aligned} \dot{\sigma}(t) &= d(\ i \circ g(t))\ Z(t) \\ &= di \circ T_g(t)\ Z(t)\ , \end{aligned}$$

where $Z(t) \in \Gamma TM$ is well determined, and consequently

$$(4.17) \qquad \begin{aligned} \ddot{\sigma}(t) &= (d(\ i \circ g(t))\ Z(t))^{\cdot} \\ &= d(\ i \circ g(t))^{\cdot}\ Z(t) + d(\ i \circ g(t))\ \dot{Z}(t) \\ &= d(\ i \circ g(t))\ \nabla(\ i \circ g(t))_{Z(t)} Z(t)\ + \\ &\quad\ d(\ i \circ g(t))\ \dot{Z}(t) + S(\ i \circ g(t))\ (Z(t),Z(t))\ . \end{aligned}$$

Using (4.17), from (4.13) we obtain, $\forall\, t\in(-\lambda,\lambda)$,

$$(4.18)\qquad \begin{cases} \rho(\, i\circ g(t))(\nabla(\, i\circ g(t))_{Z(t)}Z(t) + \acute{Z}(t)) = Y_F(g(t),Z(t)), \\ \rho(\, i\circ g(t))\; S(\, i\circ g(t))\; (Z(t),Z(t)) = \varphi_F(g(t),Z(t))^{\perp} , \end{cases}$$

where we have used the notation

$$(4.19)\qquad f\,(g(t),Z(t)) := f\,(\, i\circ g(t),\; d(\, i\circ g(t))\; Z(t))\;.$$

By comparing (4.18) with (3.14) we observe that the last equations are obtained from the first ones by setting $\dot{\sigma}(t)^{\perp} = 0$, in accordance with the fact that $i(M)$ is a fixed surface.
We note that we can remove the instantaneous connection $\nabla(i\circ g(t))$ and the instantaneous second fundamental tensor $S(i\circ g(t))$ in (4.18) by using the push-forward of $Z(t)$ by $g(t)\in \mathrm{Diff}\, M$, that is we introduce $X(t)$ by

$$(4.20)\qquad X(t) := Tg(t)Z(t)g(t)^{-1}\;, \qquad t\in(-\lambda,\lambda)\;.$$

Using (4.20) we obtain on one hand

$$(4.21)\qquad \begin{cases} \dot{\sigma}(t) = di\; X(t)g(t) = d(\, i\circ g(t))\; Z(t)\;, \\ \ddot{\sigma}(t) = di\; \acute{X}(t)g(t) + d(\, di\; X(t))\; \dot{g}(t)\;. \end{cases}$$

On the other hand the equation

$$(4.22)\qquad \dot{\sigma}(t) = di\; \dot{g}(t) = di\; X(t)g(t)$$

yields

$$(4.23)\qquad \dot{g}(t) = X(t)g(t)\;.$$

(4.21) and (4.23) imply

$$(4.24)\qquad \ddot{\sigma}(t)g(t)^{-1} = di\; \acute{X}(t) + di\; \nabla(i)_{X(t)}X(t) + S(i)(X(t),X(t))\;.$$

Setting now

$$(4.25)\qquad \begin{cases} Y_F(g(t),X(t)) := Y_F(g(t),Z(t))\; g(t)\;, \\ \varphi_F(g(t),X(t))^{\perp} := (\varphi_F(g(t),Z(t)))^{\perp}\; g(t)\;, \qquad t\in(-\lambda,\lambda) \end{cases}$$

and observing that the map

$$\rho : E(M,\mathbb{R}^n) \longrightarrow C^{\infty}(M,\mathbb{R})$$

does depend by construction on dj rather than on j itself the system (4.18) turns into

$$
(4.26)\quad \begin{cases} \rho(d\sigma(t))(\nabla(i)_{X(t)}X(t) + \dot{X}(t)) = Y_F(g(t),X(t)) , \\ \rho(d\sigma(t))S(i)(X(t),X(t)) = \varphi_F(g(t),X(t))^{\perp} . \end{cases}
$$

Let us note that in the case when we would require that $F(d(i \circ g))$ would act on ΓTM rather than on $C^\infty(M,\mathbb{R}^n)$, we would obtain only the equations (4.13), by missing the observation that the normal forces are up to the density ρ of a geometric nature.

The next step is to decompose $Y_F(g(t),X(t))$ with respect to $m(i \circ g(t))$ according to Hodge uniquely into

$$
(4.27)\quad \begin{cases} Y_F(g(t),X(t)) = \operatorname{grad}_i \tau_F(g(t),X(t)) + Y^0{}_F(g(t),X(t)) , \\ \operatorname{div}_i Y^0{}_F(g(t),X(t)) = 0, \end{cases}
$$

where $\tau_F(g(t),X(t)) \in C^\infty(M,\mathbb{R})$.
Thus the first equation (4.26) becomes

$$
(4.26') \qquad \begin{aligned} \rho(X(t))(\nabla(i)_{X(t)}X(t) &+ \dot{X}(t)) \\ &= \operatorname{grad}_i \tau_F(g(t),X(t)) + Y^0(g(t),X(t)) . \end{aligned}
$$

If we require

$$
(4.28) \qquad \sigma((-\lambda,\lambda)) \subset i \circ \operatorname{Diff}_{\mu(i)}M ,
$$

where $\operatorname{Diff}_{\mu(i)}M$ is the subgroup of all elements in $\operatorname{Diff} M$ which leave $\mu(i)$ invariant, then $X(t)$ has to be divergence free for all $t \in (-\lambda,\lambda)$. This is due to the fact that

$$
(4.29) \qquad T_{id} \operatorname{Diff} M = \{ X \in \Gamma TM \mid \operatorname{div}_i X = 0 \} .
$$

In this case $\sigma(t)$ has to satisfy the system of equations

$$
(4.30)\quad \begin{cases} \rho\, (\nabla(i)_{X(t)}X(t) + \dot{X}(t)) \\ \qquad = \operatorname{grad}_i \tau_F(g(t), X(t)) + Y^0{}_F(g(t),X(t)), \\ \rho\, S(i)\, (X(t),X(t)) = \varphi_F(\sigma(t),X(t))^{\perp} , \\ \operatorname{div}_i X(t) = 0 , \end{cases}
$$

with $\rho : i \circ \operatorname{Diff} M \longrightarrow \mathbb{R}$ being a constant function.

5. The $\mathbb{R}^n$ – invariance of internal constitutive laws

Let us assume now that the motion of the deformable medium M is subjected to an internal constitution law F, which admits an integral representation. The fact that the corresponding force density is an internal one requires it to be independent of the region in $\mathbb{R}^n$ in which the deformable medium moves. Hence an internal force density has to be invariant under the translation or, more precisely, under the action of the translation group $\mathbb{R}^n$ of $\mathbb{R}^n$ (cf.[Bi,4]).

Let us describe next this action of $\mathbb{R}^n$ on $TE(M,\mathbb{R}^n)$.

At first we recall that the translation group $\mathbb{R}^n$ of the vector space $\mathbb{R}^n$ is the underlying abelian group of the $\mathbb{R}$-vector space $\mathbb{R}^n$.

The action

(5.1) $$r : C^\infty(M,\mathbb{R}^n) \times \mathbb{R}^n \longrightarrow C^\infty(M,\mathbb{R}^n)$$

on $C^\infty(M,\mathbb{R}^n)$ is given by

(5.2) $$r(h,z) = h + z \;, \qquad \forall\, h\in C^\infty(M,\mathbb{R}^n),\; z\in \mathbb{R}^n \;,$$

where by $h + z$ we mean the map defined via

(5.3) $$(h + z)(p) = h(p) + z \;, \qquad \forall\, p\in M \;.$$

Hence $z\in \mathbb{R}^n$ is naturally identified with the constant map in $C^\infty(M,\mathbb{R}^n)$ assuming z as its value.

Clearly in the particular case where $h = j \in E(M,\mathbb{R}^n)$

(5.4) $$r(j,z) = j + z \;, \qquad \forall\, z\in E(M,\mathbb{R}^n),$$

belongs to $E(M,\mathbb{R}^n)$. Hence r reduces to

(5.5) $$r : E(M,\mathbb{R}^n) \times \mathbb{R}^n \longrightarrow E(M,\mathbb{R}^n).$$

The tangent map Tr of r is given by

(5.6) $$Tr(h,z)(k,u) = (h+z,\, k+u) \;, \qquad \forall\, k\in T_hC^\infty(M,\mathbb{R}^n),\; h\in C^\infty(M,\mathbb{R}^n),\; z,u\in \mathbb{R}^n.$$

Hence r induces an action on $TC^\infty(M,\mathbb{R}^n)$ defined by

(5.7) $$Tr : TC^\infty(M,\mathbb{R}^n) \times T\mathbb{R}^n \longrightarrow TC^\infty(M,\mathbb{R}^n), \quad ((h,k),(z,u)) \mapsto (h+z,\, k+u) \;,$$

where $k\in T_hC^\infty(M,\mathbb{R}^n) = C^\infty(M,\mathbb{R}^n)$ and $u\in T_z\mathbb{R}^n = \mathbb{R}^n$, and respectively on $TE(M,\mathbb{R}^n)$ given by

(5.8) $$Tr : TE(M,\mathbb{R}^n) \times T\mathbb{R}^n \longrightarrow TE(M,\mathbb{R}^n) \;, \quad ((j,k),(z,u)) \mapsto (j+z,\, k+u) \;,$$

where $k\in T_jE(M,\mathbb{R}^n) = C^\infty(M,\mathbb{R}^n)$ and $u\in T_z\mathbb{R}^n = \mathbb{R}^n$.

Given now a parameter depending smooth constitutive law

(5.9) $$F : C^\infty(M,\mathbb{R}^n) \longrightarrow A^1(E(M,\mathbb{R}^n),\mathbb{R})$$

and continuing to write

(5.10) $$F(j,k) \text{ instead of } F(k,j)\ ,\ \ j \in E(M,\mathbb{R}^n)\ ,\ k \in C^\infty(M,\mathbb{R}^n)$$

we form next

(5.11) $$F \circ Tr((j,k),(z,u)) : C^\infty(M,\mathbb{R}^n) = T_{j+z}E(M,\mathbb{R}^n) \longrightarrow \mathbb{R}\ .$$

The requirement

(5.12) $$F\ (j+z,\ k+u) = F\ (j,k) \qquad \forall\ j\in E(M,\mathbb{R}^n),\ k\in C^\infty(M,\mathbb{R}^n),\ z,u\in \mathbb{R}^n,$$

does then yield the type of constitutive law we want to work with.
In order to construct the desired type of $\mathbb{R}^n$-invariant constitutive laws, we consider the quotients of the actions r and Tr. To this end we note that the map

(5.13) $$d : C^\infty(M,\mathbb{R}^n) \longrightarrow \{\ dh\ \mid\ h\in C^\infty(M,\mathbb{R}^n)\}$$

has the property that

(5.14) $$d^{-1}(dh) = \{\ h+z\ \mid\ z\in \mathbb{R}\ \}.$$

Hence if we quotient out the action of $\mathbb{R}^n$ on $C^\infty(M,\mathbb{R}^n)$ we obtain a bijection again called d

(5.15) $$d : C^\infty(M,\mathbb{R}^n)\big|_{\mathbb{R}^n} \longrightarrow \{\ dh\ \mid\ h\in C^\infty(M,\mathbb{R}^n)\}.$$

We equip $\{\ dh \mid h\in C^\infty(M,\mathbb{R}^n)\ \}$ with the uniquely determined topology making d to a homeomorphism to $C^\infty(M,\mathbb{R}^n)\big|_{\mathbb{R}^n}$ carrying the quotient topology. Note that both topological spaces are Frechet manifolds.
Next we identify them via d. Hence we have identified also the two Fréchet manifolds $E(M,\mathbb{R}^n)\big|_{\mathbb{R}^n}$ and $\{\ dj \mid j\in E(M,\mathbb{R}^n)\ \}$ yielding

(5.16) $$TE(M,\mathbb{R}^n)\big|_{\mathbb{R}^n} = \{\ dj\ \mid\ j\in E(M,\mathbb{R}^n)\ \} \times \{\ dh\ \mid\ h\in C^\infty(M,\mathbb{R}^n)\ \}.$$

Therefore we obtain the following

Lemma 5.1 :
Given a smooth map

(5.17) $$F_{\mathbb{R}^n} : C^\infty(M,\mathbb{R}^n)\big|_{\mathbb{R}^n} \times E(M,\mathbb{R}^n)\big|_{\mathbb{R}^n} \times C^\infty(M,\mathbb{R}^n)\big|_{\mathbb{R}^n} \longrightarrow \mathbb{R}\ ,$$

linear in the third argument, then the resulting (parameter depending) one form F given by

(5.18) $$F = F_{\mathbb{R}^n} \circ (d, Td)$$

is a (parameter depending) $\mathbb{R}^n$–invariant one form on $E(M,\mathbb{R}^n)$.
Here Td is the tangent map of

(5.19) $$d : E(M,\mathbb{R}^n) \longrightarrow E(M,\mathbb{R}^n)\big|_{\mathbb{R}^n} .$$

Remark 5.1 :
In the following we write

(5.20) $$d^* F_{\mathbb{R}^n} \quad \text{instead of} \quad F_{\mathbb{R}^n} \circ (d, Td) .$$

Remark 5.2 :
The above lemma allows us to study the constitutive laws of the type $d^* F_{\mathbb{R}^n}$ rather than constitutive laws invariant under Tr.

6 . On the characterization of $\mathbb{R}^n$–valued one–forms relative to embeddings

Let throughout this section $i \in E(M,\mathbb{R}^n)$ be a fixed smooth embedding and $\alpha \in A^1(M,\mathbb{R}^n)$ be a fixed smooth $\mathbb{R}^n$–valued one–form. We follow [Bi,2]. As the first observation we formulate the following

Proposition 6.1 :
Let $\alpha \in A^1(M,\mathbb{R}^n)$ and $i \in E(M,\mathbb{R}^n)$ be given.
Then the following decomposition holds

(6.1) $$\alpha = dh + \beta ,$$

where $h \in C^\infty(M,\mathbb{R}^n)$, the so–called integrable part of α, is uniquely determined up to a constant. Moreover this decomposition is maximal in the sense that the integrable part of β is a constant.

Proof :
Let $e_1,\dots,e_n$ be an orthonormal basis of $\mathbb{R}^n$. Then we get

(6.2) $$\alpha(X) = \sum_{s=1}^{n} \alpha^s(X)\, e_s , \qquad \forall\, X \in \Gamma TM ,$$

for an uniquely determined family $\alpha^1,\dots,\alpha^n$ of smooth $\mathbb{R}$– valued one–forms on M, i.e. $\alpha^s \in A^1(M,\mathbb{R})$, s= 1,...,n. Clearly

(6.3) $$\alpha^s(X) = \langle \alpha(X), e_s \rangle , \qquad \forall\, X \in \Gamma TM,\ s= 1,\dots,n.$$

In addition α^s , s= 1,...,n, can be represented as

(6.4) $$\alpha^s(X) = m(i)\,(Y_s, X)\,, \qquad \forall\, X \in \Gamma TM,$$

for a well defined $Y_s \in \Gamma TM$. This vector field splits according to Hodge's decomposition uniquely into

(6.5) $$\begin{cases} Y_s = \mathrm{grad}_i\, \tau_s + Y^0_s\,, \\ \mathrm{div}_i\, Y^0_s = 0\,, \end{cases}$$

where $\tau_s \in C^\infty(M,\mathbb{R}^n)$ and $Y^0_s \in \Gamma TM$.
Hence

(6.6) $$\alpha^s(X) = d\tau_s(X) + m(i)(Y^0_s, X)\,, \qquad \forall\, X \in \Gamma TM\,.$$

Next we define the integrable part h of α by

(6.7) $$h := \sum_{s=1}^{n} \tau_s\, e_s\,,$$

and the non-integrable part β by

(6.8) $$\beta\,(X) := \sum_{s=1} m(i)\,(Y^0_s, X)\, e_s\,, \qquad \forall\, X \in \Gamma TM\,.$$

Inserting (6.6) into (6.2) and using (6.7) and (6.8) yields the decomposition (6.1). It remains only to show that (6.1) does not depend on the choice of the basis of $\mathbb{R}^n$. To this end let $\bar{e}_1,\dots,\bar{e}_n \in \mathbb{R}^n$ be another orthonormal basis of $\mathbb{R}^n$ and define $\bar{\alpha}$, $\bar{\tau}$, $\bar{Y}^0$, $\bar{h}$ and $\bar{\beta}$ accordingly.
Then

(6.9) $$\begin{aligned} \bar{\alpha}\,(X) &= \langle \alpha(X), \bar{e}\, \rangle \\ &= \langle dh(X), \bar{e}\, \rangle + \langle \beta(X), \bar{e}\, \rangle \\ &= \langle \sum_{s=1}^{n} d\tau_s(X)\, e_s, \bar{e}\, \rangle + \langle \sum_{s=1}^{n} m(i)\,(Y^0_s, X)\, e_s, \bar{e}\, \rangle \\ &= m(i)\,(\, \sum_{s=1}^{n} \mathrm{grad}_i\, \tau_s\, \langle e_s, \bar{e}\, \rangle\,,\, X) \\ &\quad + m(i)\,(\, \sum_{s=1}^{n} Y^0_s\, \langle e_s, \bar{e}\, \rangle\,,\, X) \\ &= \langle d\bar{h}(X)\,,\, \bar{e}\, \rangle + \langle \bar{\beta}(X)\,,\, \bar{e}\, \rangle \\ &= m(i)\,(\mathrm{grad}_i\, \bar{\tau}\,, X) + m(i)\,(\bar{Y}^0, X). \end{aligned}$$

Since on one hand

$$(6.10) \qquad \sum_{s=1}^{n} \operatorname{grad}_i \tau_s \langle e_s, \bar{e} \rangle = \operatorname{grad}_i \Big(\sum_{s=1}^{n} \tau_s \langle e_s, \bar{e} \rangle \Big),$$

on the other hand

$$(6.11) \qquad \operatorname{div}_i \Big(\sum_{s=1}^{n} Y_s^0 \langle e_s, \bar{e} \rangle \Big) = \sum_{s=1}^{n} (\operatorname{div}_i Y_s^0) \langle e_s, \bar{e} \rangle = 0,$$

we conclude due to the uniqueness of Hodge's decomposition the following relations

$$(6.12) \qquad \begin{cases} \sum_{s=1}^{n} \operatorname{grad}_i (\tau_s \langle e_s, \bar{e} \rangle) = \operatorname{grad}_i \bar{\tau}, \\ \sum_{s=1}^{n} Y_s^0 \langle e_s, \bar{e} \rangle = \bar{Y}^0 . \end{cases}$$

Consequently the uniqueness of the decomposition (6.1) follows, namely

$$(6.13) \qquad \begin{cases} dh(X) = \sum_{=1}^{n} \langle dh(X), \bar{e} \rangle \bar{e} = \sum_{=1}^{n} \langle d\bar{h}(X), \bar{e} \rangle \bar{e} = d\bar{h}(X), \\ \beta(X) = \sum_{=1}^{n} \langle \beta(X), \bar{e} \rangle \bar{e} = \sum_{=1}^{n} \langle \beta(X), \bar{e} \rangle \bar{e} = \bar{\beta}(X), \\ \qquad \forall X \in \Gamma TM . \end{cases}$$

Let us detail now the decomposition (6.1).
For this purpose we note first that h can be given the form

$$(6.14) \qquad h = di\, X_h + h^{\perp},$$

where $X_h \in \Gamma TM$ is well defined and $h^{\perp}$ denotes the pointwise formed component of h normal to i(M).
Using the fact that X_h splits into

$$(6.15) \qquad \begin{cases} X_h = X_h^0 + \operatorname{grad}_i \psi_h, \\ \operatorname{div}_i X_h^0 = 0, \end{cases}$$

where $\psi_h \in C^\infty(M,\mathbb{R})$, $X_h^0 \in \Gamma TM$, we deduce from (6.14) that

$$(6.16) \qquad dh(X) = di\, \nabla(i)_X X_h^0 + di\, (\nabla(i)_X \operatorname{grad}_i \psi_h + W_h(i)\, X) + S(i)\, (X_h^0, X) + (dh^{\perp}(X))^{\perp} .$$

For the sake of readability we remind that $W_h(i)$ defined via

$$(6.17) \qquad di\, W_h(i)\, X := (dh^{\perp}(X))^T ,$$

is a smooth strong bundle endomorphism of TM, which is selfadjoint with respect to m(i).

Let us show next that the divergence-free part X_h^0 of X_h is uniquely determined by h. To this end we use the fact that according to the above proposition the integrable part h of α is uniquely determined up to a constant, i.e.

$$(6.18) \qquad h' = h + z ,$$

for some $z \in \mathbb{R}^n$. Regarding z as a constant map in $C^\infty(M,\mathbb{R}^n)$ we write it in the form

$$(6.19) \qquad z = di\, X_z + z^{\perp} .$$

But the vector field Z on $\mathbb{R}^n$ assigning to any $z \in \mathbb{R}^n$ the vector $Z(z) = Z \in \mathbb{R}^n$ is the gradient of some map $\varphi \in C^\infty(M,\mathbb{R})$ and hence

$$(6.20) \qquad X_z = \operatorname{grad}_i (\varphi \circ i) .$$

Therefore

$$(6.21) \qquad \begin{aligned} h' &= di\, X_h^0 + di\, \operatorname{grad}_i (\psi_h + \varphi \circ i) + h^{\perp} + z^{\perp} \\ &= di\, X_{h'}^0 + di\, \operatorname{grad}_i \psi_{h'} + h'^{\perp} \end{aligned}$$

or, equivalently

$$(6.22) \qquad \begin{aligned} &di\, (X_{h'}^0 - X_h^0) + di\, \operatorname{grad}_i(\psi_{h'} - \psi_h - \varphi \circ i) \\ &= h^{\perp} + z^{\perp} - h'^{\perp} = 0 . \end{aligned}$$

But (6.22) implies

$$(6.23) \qquad X_{h'}^0 - X_h^0 + \operatorname{grad}_i (\psi_{h'} - \psi_h - \varphi \circ i) = 0 .$$

Using once more the uniqueness of Hodge's decomposition we conclude then (cf.[Bi,2])

Proposition 6.2 :

Let $\alpha \in A^1(M,\mathbb{R}^n)$ and $i \in E(M,\mathbb{R}^n)$ be given, and denote by $h \in C^\infty(M,\mathbb{R}^n)$ the integral part of α, which is uniquely detemined up to a constant

$$\alpha = dh + \beta .$$

Splitting h into

$$h = di\, X_h + h^{\perp} ,$$

where $X_h \in \Gamma TM$ is well defined, then the divergence-free part X_h^0 of X_h is uniquely determined.

Next we characterize $\alpha \in A^1(M,\mathbb{R}^n)$ relative to $i \in E(M,\mathbb{R}^n)$ from a quite different point of view. To this end let us introduce the following two tensor T_α on M

$$(6.24) \qquad T_\alpha(X,Y) := \langle \alpha(X), di\, Y \rangle , \qquad \forall\, X,Y \in \Gamma TM .$$

Clearly T_α is smooth. Next we denote by

$$(6.25) \qquad P : TM \longrightarrow TM$$

the unique smooth strong bundle endomorphism for which

$$(6.26) \qquad T_\alpha(X,Y) = m(i)\,(PX,Y) ,$$

and by $\tilde{P}$ the fibre-wise formed adjoint of P with respect to m(i). The symmetric and the antisymmetric part of T_α, the tensors T_α^s and T_α^a respectively have the form

$$(6.27) \qquad T_\alpha^s\,(X,Y) = m(i)\,(\tfrac{1}{2}(P + \tilde{P})\, X,Y) ,$$

$$(6.28) \qquad T_\alpha^a\,(X,Y) = m(i)\,(\tfrac{1}{2}(P - \tilde{P})\, X,Y) .$$

Setting now

$$(6.29) \qquad \begin{cases} B_\alpha := \frac{1}{2}\,(P + \tilde{P}) , \\ C_\alpha := \frac{1}{2}\,(P - \tilde{P}) , \end{cases}$$

we obtain that

$$(6.30) \qquad \alpha(X) = \alpha'(X) + di\, C_\alpha X + di\, B_\alpha X , \qquad \forall\, X \in \Gamma TM.$$

Clearly $\alpha'(X)(p)$ is a vector in the normal space of TiT_pM , $\forall\, p \in M$. Hence there is a unique smooth map

$$(6.31) \qquad c_\alpha \in C^\infty(M,so(n)) ,$$

where so(n) denotes the Lie algebra of the group of all proper rotations of SO(n), such that

(6.32) $\alpha'(X) = c_\alpha diX$,

with

(6.33) $c_\alpha(N(j)) \perp \mathrm{Ker}\, c_\alpha$, $\forall\, X \in \Gamma TM$.

We may now state the following

Proposition 6.3 :

Let $\alpha \in A^1(M,\mathbb{R}^n)$ and $i \in E(M,\mathbb{R}^n)$ be given. Then there exist two uniquely determined smooth, strong bundle endomorphisms

$$C_\alpha : TM \longrightarrow TM$$

and

$$B_\alpha : TM \longrightarrow TM ,$$

which are skew- and respectively selfadjoint with respect to m(i), and a uniquely determined map $c_\alpha \in C^\infty(M,so(n))$, such that the following relation holds

(6.34) $\alpha(X) = c_\alpha\, di\, X + di\, C_\alpha X + di\, B_\alpha X$, $\forall\, X \in \Gamma TM$.

Remark 6.1 :

Given $\alpha \in A^1(M,\mathbb{R}^n)$ and $i \in E(M,\mathbb{R}^n)$, then

(6.35) $\delta T^a_\alpha = 0$ if $\delta\alpha = 0$.

Indeed, let us consider the one form $\langle i,\alpha\rangle \in A^1(M,\mathbb{R}^n)$, which assigns to any $X \in \Gamma TM$ the real function $\langle i,\alpha(X)\rangle$.
Since

(6.36) $\delta\langle i,\alpha\rangle = T^a_\alpha$ if $\delta\alpha = 0$,

(6.35) then follows immediately.

Next we link the two characterizations of $\mathbb{R}^n$-valued one-forms relative to embeddings, as expressed by the two propositions above. To this end let $\alpha \in A^1(M,\mathbb{R}^n)$ and $i \in E(M,\mathbb{R}^n)$ be given.
Using (6.1) and (6.14) $\alpha(X)$ turns into

(6.37)
$$\alpha(X) = di\, \nabla(i)_X X_h + di\, W_h(i)\, X + S(i)(X_h,X) + (dh^\perp(X))^\perp + \beta(X) .$$

Inserting (6.37) in (6.24) we get

(6.38)
$$\begin{aligned} T_\alpha(X,Y) &= \langle\, \alpha(X), di\, Y \,\rangle \\ &= \langle\, di\, \nabla(i)_X X_h + W_h(i)\, X, di\, Y \,\rangle + T_\beta(X,Y) \\ &= m(i)\, (\nabla(i)_X X_h,Y) + m(i)\, (W_h(i)\, X,Y) + T_\beta(X,Y) , \\ &\qquad \forall\, X,Y \in \Gamma TM. \end{aligned}$$

Therefore

$$(6.39)\qquad T^s_\alpha(X,Y) = \frac{1}{2}[\, m(i)\,(\nabla(i)_X X_h, Y) + m(i)\,(\nabla(i)_Y X_h, X)\,] + m(i)\,(W_h(i)\,X,Y) + T^s_\beta(X,Y)$$
$$= \frac{1}{2} L_{X_h}\,(m(i))\,(X,Y) + m(i)\,(W_h(i)\,X,Y) + T^s_\beta(X,Y)\,,$$

$$(6.40)\qquad T^a_\alpha(X,Y) = \frac{1}{2}[\, m(i)\,(\nabla(i)_X X_h, Y) - m(i)\,(\nabla(i)_Y X_h, X)] + T^a_\beta(X,Y),$$

Rewriting the Lie derivative $L_{Z_h}(m(i))$ of $m(i)$ in the direction of Z_h with the help of the Theorem of Fischer and Riesz as

$$(6.41)\qquad L_{Z_h}\,(m(i)\,(X,Y) = m(i)\,(L_{Z_h}\,X,Y) \qquad \forall\; Z_h \in \Gamma TM\,,$$

by a uniquely determined strong smooth bundle endomorphism

$$(6.42)\qquad L_{Z_h} : TM \longrightarrow TM\,,$$

from (6.34) and (6.37) we infer the following formulas for c_α, C_α and B_α :

$$(6.43)\qquad c_\alpha\, di\, X = (dh^\perp(X))^\perp + S(i)\,(X_h,X) + c_\beta\, di\, X\,,$$

$$(6.44)\qquad C_\alpha\, X = \frac{1}{2}[\, \nabla(i)\, X_h - \tilde{\nabla}(i)\, X_h\,]\, X + C_\beta\, X\,,$$

$$(6.45)\qquad B_\alpha X = \frac{1}{2}[\, \nabla(i)\, X_h + \tilde{\nabla}(i)\, X_h\,]\, X + W_h(i)\, X + B_\beta\, X$$
$$= (\, \frac{1}{2} L_{X_h} + W_h(i) + B_\beta\,)\, X\,.$$

Instead of $\nabla(i)_v X_h$ we often write $\nabla(i)X_h(v)$ for any $v \in T_pM$. Similarily we use $\tilde{\nabla}(i)X_h(v)$ instead of $\tilde{\nabla}(i)_v X_h$.
Using next the Hodge decomposition of X_h, i.e. (6.15) and taking into account that

$$(6.46)\qquad m(i)\,((\nabla(i)\,\mathrm{grad}_i\,\psi_h - \tilde{\nabla}(i)\,\mathrm{grad}_i\,\psi_h)\,X,Y) = 0\,,$$

we obtain finally the following

Proposition 6.4 :

Let $\alpha \in A^1(M,\mathbb{R}^n)$ and $i \in E(M,\mathbb{R}^n)$ be given. Then the following relations hold

$$\alpha = dh + \beta\,,$$
$$\alpha(X) = c_\alpha\, di\, X + di\, C_\alpha X + di\, B_\alpha X\,, \qquad \forall\; X \in \Gamma TM,$$

where the integrable part $h \in C^\infty(M,\mathbb{R}^n)$ is uniquely determined up to a constant , $c_\alpha \in C^\infty(M,so(n))$ is a uniquely determined, $C_\alpha : TM \longrightarrow TM$ is a uniquely determined smooth, strong and skew-adjoint bundle endomorphism, $B_\alpha : TM \to TM$ is a uniquely determined smooth, strong and selfadjoint bundle endomorphism.

Writing

$$h = di\, X_h + h^{\perp} ,$$

where $X_h \in \Gamma TM$, and using Hodge's decomposition

$$\text{(6.47)} \qquad \begin{cases} X_h = X_h^0 + \operatorname{grad}_i \psi_h , \\ \operatorname{div}_i X_h^0 = 0 , \end{cases}$$

we obtain finally

$$\text{(6.48)} \qquad c_\alpha\, di = (dh^{\perp})^{\perp} + S(i)\,(X_h, \,.\,) + c_\beta\, di ,$$

$$\text{(6.49)} \qquad C_\alpha = \tfrac{1}{2} [\, \nabla(i)\, X_h^0 - \tilde{\nabla}(i)\, X_h^0 \,] + C_\beta ,$$

$$\text{(6.50)} \qquad B_\alpha = \tfrac{1}{2}\, L_{X_h^0 + \operatorname{grad}_i \psi_h} + W_h(i) + B_\beta .$$

Hence

$$\text{(6.51)} \qquad \begin{aligned} \operatorname{tr} B_\alpha &= \operatorname{div}_i X_h + \operatorname{tr} W_h(i) + \operatorname{tr} B_\beta \\ &= -\,\Delta(i)\, \psi_h + \operatorname{tr} W_h(i) + \operatorname{tr} B_\beta , \end{aligned}$$

where $\Delta(i)$ is the Laplace-Beltrami operator of $m(i)$.

Let us calculate now the covariant divergence of B_α and C_α. To this end we recall at first the covariant divergence $\operatorname{div}_i A$ of a smooth strong bundle endomorphism

$$\text{(6.52)} \qquad A : TM \longrightarrow TM .$$

Let $e_1,\dots,e_m$ be a moving orthonormal frame of TM, and set

$$\text{(6.53)} \qquad \operatorname{div}_i A = \sum_{r=1}^{m} \nabla(i)_{e_r} (A)\, e_r .$$

At first we compute $\operatorname{div}_i \nabla(i)\, X_h$. Using the equation

$$\text{(6.54)} \qquad \begin{aligned} & m(i)\, (\nabla(i)_{e_r}(\nabla(i)\, X_h)\, e_r, Y) \\ &= m(i)\, (\nabla(i)_{e_r}(\nabla(i)_{e_r} X_h) - \nabla(i)_{\nabla(i)_{e_r} e_r} X_h, Y), \\ & \qquad\qquad\qquad \forall\, Y \in \Gamma TM, \end{aligned}$$

we get

$$\text{(6.55)} \qquad \operatorname{div}_i(\nabla(i)\, X_h) = \Delta(i)\, X_h,$$

where $\Delta(i) X_h$ is the Laplace-Beltrami operator of $m(i)$ applied to X_h which by definition is $-\operatorname{tr} \underset{\sim}{\nabla(i)^2}\, X_h$.

In order to compute $\operatorname{div}_i \tilde{\nabla}(i)\, X_h$ we consider the equations

(6.56) $$m(i)\,(\nabla(i)_{e_r}(\tilde{\nabla}(i)\,X_h)(e_r),Y)$$
$$= e_r\,(m(i)\,(\tilde{\nabla}(i)\,X_h(e_r),Y) - m(i)\,(\tilde{\nabla}(i)\,X_h\,(\nabla(i)_{e_r}e_r),Y)$$
$$- m(i)\,(\tilde{\nabla}(i)\,X_h(e_r),\nabla(i)_{e_r}Y)$$
$$= m(i)\,(e_r,\nabla(i)_{e_r}\nabla(i)_Y X_h) - m(i)\,(e_r,\nabla(i)_{\nabla(i)_{e_r}Y}\,X_h)$$
$$= m(i)\,(e_r,\nabla(i)_{e_r}\,(\nabla(i)\,X_h)\,Y)\,,$$

(6.57) $$m(i)\,(\nabla(i)_Y(\tilde{\nabla}(i)\,X_h)(e_r),e_r)$$
$$= m(i)\,(e_r,\nabla(i)_Y\nabla(i)_{e_r}X_h) - m(i)\,(\,e_r,\nabla(i)_{\nabla(i)_Y e_r}X_h)$$
$$= m(i)\,(e_r,\,\nabla(i)_Y(\nabla(i)\,X_h)(e_r))$$

and find

(6.58) $$\sum_{r=1}^{n}[\,m(i)\,(\nabla(i)_{e_r}(\tilde{\nabla}(i)X_h)(e_r),Y) - m(i)\,(\nabla(i)_Y(\tilde{\nabla}(i)X_h)(e_r),e_r)]$$
$$= \mathrm{Ric}\,(m(i))(Y,X_h)\,,$$

where $\mathrm{Ric}(m(i))$ denotes the Ricci tensor of $m(i)$. Hence

(6.59) $$m(i)\,(\mathrm{div}_i(\tilde{\nabla}(i)X_h),Y) = \mathrm{tr}\,\nabla(i)_Y(\nabla(i)X_h) + \mathrm{Ric}(m(i))(X,X_h).$$

But (6.59) yields

(6.60) $$\mathrm{div}_i(\tilde{\nabla}(i)X_h) = \mathrm{grad}_i\mathrm{div}_i\,X_h + R(i)\,X_h\,,$$

where $R(i)X_h$ is defined via

(6.61) $$m(i)\,(R(i)X_h,Y) = \mathrm{Ric}(m(i))\,(X_h,Y)\,,\qquad \forall\ Y \in \Gamma TM.$$

From (6.49),(6.50),(6.55) and (6.60) we deduce

(6.62) $$\mathrm{div}_i L_{X_h} = \Delta(i)\,X_h + R(i)\,X_h + \mathrm{grad}_i\mathrm{div}_i X_h\,,$$

(6.63) $$2\,\mathrm{div}_i\,C_h = \Delta(i)\,X_h - R(i)\,X_h - \mathrm{grad}_i\mathrm{div}_i\,X_h$$

and consequently

(6.64) $$\mathrm{div}_i\,(\tfrac{1}{2}\,L_{X_h} + C_h) = \Delta(i)\,X_h\,,$$

(6.65) $$\mathrm{div}_i\,(\tfrac{1}{2}\,L_{X_h} - C_h) = R(i)\,X_h + \mathrm{grad}_i\mathrm{div}_i\,X_h\,.$$

Let us restrict our attention to the case where M has codimension 1. Since M is oriented we have a positively oriented unit normal vector field $N(i)$ along i. Hence $h \in C^\infty(M,\mathbb{R}^n)$ splits uniquely into

(6.66) $$h = di\,X_h + \Theta_h\,N(i)\,,$$

where $X_h \in \Gamma TM$, $\Theta_h \in C^\infty(M,\mathbb{R})$.

Thus

$$(6.67) \qquad W_h(i) = W(i) \qquad \text{if } \Theta_h = 1 \quad .$$

Defining the unnormalized mean curvature H(i) of i by

$$(6.68) \qquad \operatorname{tr} W(i) = H(i)$$

we immediately find

$$(6.69) \qquad \begin{aligned} & m(i)\,(\,\operatorname{div}_i(\Theta_h\, W(i))\,,\, Y\,) \\ &= \sum_{r=1}^{n-1} m(i)\,(\nabla(i)_{e_r}(\Theta_h\, W(i))\, e_r, Y) \\ &= m(i)\,(\operatorname{grad}_i \Theta_h, W(i)\, Y) + m(i)\,(\Theta_h \operatorname{div}_i W(i), Y) \end{aligned}$$

and hence

$$(6.70) \qquad \operatorname{div}_i(\Theta_h\, W(i)) = W(i)\operatorname{grad}_i \Theta_h + \Theta_h \operatorname{div}_i W(i)\,.$$

On the other hand by Codazzi's equation (cf. [Kl])

$$(6.71) \qquad \begin{aligned} \sum_{r=1}^{m} m(i)\,(\nabla(i)_{e_r}(W(i))\, e_r, Y) &= \sum_{r=1}^{n-1} m(i)\,(\nabla(i)_Y(W(i))\, e_r, e_r) \\ &= m(i)\,(\operatorname{grad}_i H(i), Y) \end{aligned}$$

and consequently

$$(6.72) \qquad \operatorname{div}_i(\Theta_h\, W(i)) = W(i)\operatorname{grad}_i \Theta_h + \Theta_h \operatorname{grad}_i H(i)\,.$$

From (6.50), (6.62), (6.63),(6.67),(6.68) and (6.72) we infer then

$$(6.73) \qquad \operatorname{div}_i (B_{dh} + C_{dh}) = \Delta(i)\, X_h + W(i)\operatorname{grad}_i \Theta_h + \Theta_h \operatorname{grad}_i H(i)$$

$$(6.74) \qquad \begin{aligned} \operatorname{div}_i (B_{dh} - C_{dh}) &= R(i)\, X_h + \operatorname{grad}_i \operatorname{div}_i X_h \\ &\quad + W(i)\operatorname{grad}_i \Theta_h + \Theta_h \operatorname{grad}_i H(i)\,, \end{aligned}$$

$$(6.75) \qquad \begin{aligned} \operatorname{div}_i B_{dh} &= \tfrac{1}{2}\Delta(i)\, X_h + \tfrac{1}{2} R(i)\, X_h \\ &\quad + W(i)\operatorname{grad}_i \Theta_h + \Theta_h \operatorname{grad}_i H(i)\,. \end{aligned}$$

We conclude this section by proving the following

Lemma 6.5 :

Let $\alpha \in A^1(M,\mathbb{R}^n)$ and $i \in E(M,\mathbb{R}^n)$ be given.
If α has no integrable part, i.e. $\alpha = \beta$, then

$$(6.76) \qquad \operatorname{div}_i (C_\beta + B_\beta) = 0\,.$$

Proof :

Denoting by $\bar{e}_1,\ldots,\bar{e}_n$ an orthonormal basis of $\mathbb{R}^n$ we have, in accordance with (6.8) ,

$$
\begin{aligned}
T_\beta(X,Y) &= \langle \beta(X),di\ Y\rangle \\
&= m(i)\ ((B_\beta + C_\beta)\ X,Y) \\
&= \sum_{s=1}^{n} m(i)\ (Y_s^0,X)\ \langle \bar{e}_s,di\ Y\rangle \\
&= m(i)\ (\sum_{s=1}^{n} \langle \bar{e}_s,di\ Y\rangle\ Y_s^0,X)\ , \qquad \forall\ X,Y \in \Gamma TM.
\end{aligned} \tag{6.77}
$$

If now $e_1,\ldots,e_m$ is a moving orthonormal frame in TM, then we get

$$
\begin{aligned}
m(i)\ (\mathrm{div}_i(B_\beta + C_\beta),Y) &= \sum_{r=1}^{m} m(i)\ (\nabla(i)_{e_r}(B_\beta + C_\beta)\ e_r,Y) \\
&= \sum_{r=1}^{m}\sum_{s=1}^{n} m(i)\ (\nabla(i)_{e_r} \langle \bar{e}_s,di\ Y\rangle\ Y_s^0,\ e_r)
\end{aligned} \tag{6.78}
$$

By interchanging the summation the assertion then follows.

7. The equations of motion of a deformable medium subjected to an internal constitutive law

Let M be a moving deformable medium of codimension 1, i.e. $\dim M=n-1$, and assume that its motion is due only to an internal constitutive law F. As we have already seen F is $\mathbb{R}^n$-invariant and admits the representation

$$F = d^* F_{\mathbb{R}^n}$$

where

$$F_{\mathbb{R}^n} :\ C^\infty(M,\mathbb{R}^n)\big|_{\mathbb{R}^n} \times E(M,\mathbb{R}^n)\big|_{\mathbb{R}^n} \times C^\infty(M,\mathbb{R}^n)\big|_{\mathbb{R}^n} \longrightarrow \mathbb{R}$$

is a smooth map. Next we assume that $F_{\mathbb{R}^n}$ has an integral representation

$$F_{\mathbb{R}^n}(dk)(dj,dl) := {}_M\!\int \alpha(dj,dk)\cdot dl\ \mu(j)\ , \quad \forall\ j \in E(M,\mathbb{R}^n),\ k,l \in C^\infty(M,\mathbb{R}^n), \tag{7.1}$$

where α is an $\mathbb{R}^n$-valued one form, the so-called stress-form, i.e.

$$\alpha :\ TE(M,\mathbb{R}^n)\big|_{\mathbb{R}^n} \longrightarrow A^1(M,\mathbb{R}^n). \tag{7.2}$$

We note that the integrand

$$\alpha(dj,dk)\cdot dl \tag{7.3}$$

in (7.1) is defined in the following way :

Let first represent $\alpha(dj,dk)$ and dl according to Proposition 6.3 as

$$\alpha(dj,dk) = c_\alpha(dj,dk)\cdot dj + dj\cdot C_\alpha(dj,dk) + dj\cdot B_\alpha(dj,dk)$$

and

$$dl = c_{dl}\cdot dj + dj\cdot C_{dl} + dj\cdot B_{dl}$$

respectively. Then we set

$$\alpha(dj,dk)\cdot dl := \operatorname{tr} B_\alpha(dj,dk)\cdot B_{dl} + \operatorname{tr} C_\alpha(dj,dk)\cdot C_{dl} + \operatorname{tr} c_\alpha(dj,dk)\cdot c_{dl}\,. \tag{7.4}$$

Using now (5.20) and (7.1), it is easy to see that the internal constitutive law F admits an integral representation by a force density φ_F, which depends on (the coefficients of) the stress form α. Indeed, to this end we have to solve the equation

$$\int_M \langle\varphi_F(dj,dk),l\rangle\mu(j) = \int_M \alpha(dj,dk)\cdot dl\ \mu(j)\,, \quad \forall\ j \in E(M,\mathbb{R}^n),\ k,l \in C^\infty(M,\mathbb{R}^n). \tag{7.5}$$

Before doing so, however, we point out that α in (7.5) is not uniquely determined. In fact we have (cf.[Bi,4]).

Theorem 7.1 :
Given $\alpha \in A^1(M,\mathbb{R}^n)$ and $j \in E(M,\mathbb{R}^n)$ with dh as the integrable part of α then

$$\int \alpha\cdot dk\ \mu(j) = \int dh\cdot dk\ \mu(j)\,, \tag{7.6}$$

expressing the fact that the non-integrable part β of $\alpha = dh + \beta$ is orthogonal to $C^\infty(M,\mathbb{R}^n)\big|_{\mathbb{R}^n}$, regarded as a subspace of $A^1(M,\mathbb{R}^n)$.

Proof :
By Proposition 6.1 we have, $\forall\ X \in \Gamma TM$,

$$dhX = \sum_{s=1}^{n} m(j)\ (V_s,X)e_s,$$

$$\beta(X) = \sum_{s=1}^{n} m(j)(Y_s^0,X)e_s\,,$$

where $V_s = \operatorname{grad}_j\tau_s$, $\operatorname{div}_j Y_s^0 = 0$, $s = 1,\ldots,n$. Then from (6.24), (6.26) and (6.29) it follows

$$(B_{dh} + C_{dh})X = \sum_{s=1}^{n} m(j)\ (V_s,X)Y_s \tag{7.7}$$

and

$$(B_\beta + C_\beta)X = \sum_{s=1}^{n} m(j)\ (Y_s^0,X)Y_s\,. \tag{7.8}$$

where $Y_s = V_s + Y_s^0$, $s = 1,\dots,n$, satisfies the equation

$$m(j)\,(Y_s,X) = \langle \alpha(X), e_s \rangle\ , \qquad \forall\ X \in \Gamma TM.$$

Hence

$$(7.9) \qquad (B_{dh} + C_{dh}) \circ (B_\beta + C_\beta)X = \sum_{s=1}^{n} \sum_{s'=1}^{n} m(j)\,(Y_s^0,X)\ m(j)\,(V_{s'},Y_s)\ Y_{s'}\ .$$

Therefore if $\bar{e}_1,\dots,\bar{e}_{n-1}$ is an orthonormal moving frame on M

$$(7.10) \qquad \begin{aligned} tr(B_{dh} + C_{dh}) \circ (B_\beta + C_\beta) &= tr(B_{dh} \circ B_\beta + C_{dh} \circ C_\beta) \\ &= \sum_{r=1}^{n-1} \sum_{s'=1}^{n} \sum_{s=1}^{n} m(j)\,(Y_s^0,\bar{e}_r) m(j)\,(V_{s'},Y_s) m(j)\,(Y_{s'},\bar{e}_r) \\ &= \sum_{s=1}^{n} m(j)\,(Y_s^0,V_s)\ . \end{aligned}$$

Next we form

$$(7.11) \qquad \begin{cases} c_{dh}\ djX = \sum_{s=1}^{n} m(j)\,(V_s,X)\cdot\langle e_s,N(j)\rangle\cdot N(j) \\ c_{dh}\ N(j) = \sum_{s=1}^{n} \langle e_s,N(j)\rangle\cdot dj\ V_s\ . \end{cases}$$

Similar equations hold for c_β with V_s replaced by Y_s^0 for each s.
Hence

$$(7.12) \qquad \begin{cases} c_{dh} \circ c_\beta\ dj\ X = \sum_{s'=1}^{n} \sum_{s=1}^{n} m(j)\,(Y_s^0,X)\cdot\langle e_s,N(j)\rangle\cdot\langle e_{s'},N(j)\rangle\cdot V_s\ , \\ c_{dh} \circ c_\beta\ N(j) = \sum_{s'=1}^{n} \sum_{s=1}^{n} m(j)\,(V_s,Y_{s'})\cdot\langle e_s,N(j)\rangle\cdot\langle e_{s'},N(j)\rangle\cdot N(j)\ . \end{cases}$$

Therefore

$$(7.13) \qquad tr\ c_{dh} \circ c_\beta = 2\cdot \sum_{s=1}^{n} m(j)\,(Y_s^0,V_s)$$

and hence

$$(7.14) \qquad tr(B_{dh} \circ B_\beta + C_{dh} \circ C_\beta + c_{dh} \circ c_\beta) = 3\cdot \sum_{s=1}^{n} m(j)\,(Y_s^0,V_s)\ .$$

But (7.14) implies

$$(7.15) \qquad {}_M\!\int dh\cdot\beta\ \mu(j) = 3\cdot \sum_{s=1}^{n} {}_M\!\int m(j)\,(Y_s^0,V_s)\ \mu(j) = 0$$

since Y_s^0 and V_s are L_2-orthogonal, $s = 1,\dots,n$.

Let us now turn back to (7.5).
At first we study the equation

$$(7.16) \qquad \langle \varphi_F, l \rangle = tr\ (\ B_\alpha \circ B_{dl} + C_\alpha \circ C_{dl} + c_\alpha \circ c_{dl}\ )\ ,$$

where for the sake of simplicity we omited the arguments of α and φ_F. On one hand we note that $l \in C^\infty(M,\mathbb{R}^n)$ can be represented as

$$(7.17) \qquad l = dj\, X_l + \Theta_l\, N(j)$$

and that it will be sufficient to consider only those l for which $\Theta_l = 1$, i.e.

$$(7.18) \qquad l = dj\, X_l + N(j)\ .$$

On the other hand according to Proposition 6.4 we have for a fixed orthonormal moving frame $e_1,\ldots,e_{n-1} \in \Gamma TM$

$$(7.19) \qquad \begin{aligned} \operatorname{tr} B_\alpha \circ B_{dl} &= \sum_{s=1}^{n-1} \tfrac{1}{2}\, m(j)\, ((B_\alpha \circ (\nabla(j))\, X_l + \tilde{\nabla}(j)X_l)\, e_s, e_s) \\ &\quad + \sum_{s=1}^{n-1} m(j)\, (B_\alpha \circ W(j)e_s, e_s) \\ &= \sum_{s=1}^{n-1} m(j)\, (B_\alpha \circ \nabla(j)_{e_s} X_l, e_s) - \sum_{s=1}^{n-1} \langle N(j), S(j)(e_s, B_\alpha e_s)\rangle \\ &= \operatorname{div}_j B_\alpha X_l - m(j)(\operatorname{div}_j B_\alpha, X_l) \\ &\quad + \langle \operatorname{tr}(B_\alpha \circ W(j)) \cdot N(j), N(j)\rangle\ . \end{aligned}$$

We calculate next in the same way $\operatorname{tr} C_\alpha \circ C_{dl}$ and obtain, using the fact that

$$(7.20) \qquad \operatorname{tr} W(j) \circ C_\alpha = 0,$$

$$(7.21) \qquad \operatorname{tr} C_\alpha \circ C_{dl} = \operatorname{div}_j C_\alpha X_l - m(j)(\operatorname{div}_j C_\alpha, X_l)\ .$$

It remains now to calculate $\operatorname{tr} c_\alpha \circ c_{dl}$. Recalling to this purpose that

$$(7.22) \qquad c_{dl}\, dj\, Y = S(j)\,(X_l, Y)\,, \qquad \forall\, Y \in \Gamma TM,$$

and writing $c_\alpha\, N(j)$ with the help of a field, say U_α, in ΓTM as

$$(7.23) \qquad c_\alpha\, N(j) = dj\, U_\alpha$$

we then get

$$\begin{aligned} \operatorname{tr} c_\alpha \circ c_{dl} &= \sum_{s=1}^{n-1} \langle c_\alpha \circ c_{dl} dje_s, dje_s\rangle + \langle c_{dl} \circ c_\alpha N(j), N(j)\rangle \\ &= \sum_{s=1}^{n-1} \langle c_\alpha S(j)(X_l, e_s), dje_s\rangle + \langle c_{dl} dj\, U_\alpha, N(j)\rangle \\ &= -\sum_{s=1}^{n-1} m(j)(W(j)X_l, e_s)\, \langle c_\alpha N(j), dje_s\rangle \\ &\quad - m(j)(W(j)U_\alpha, X_l) \\ &= -\sum_{s=1}^{n-1} m(j)(W(j)X_l, e_s) \cdot m(j)(U_\alpha, e_s) - m(j)(W(j)U_\alpha, X_l) \\ &= -\,2\, m(j)\, (W(j)U_\alpha, X_l)\ . \end{aligned}$$

The equation for φ_F becomes then

$$(7.24) \qquad {}_M\int \langle \varphi_f, l\rangle \mu(j) = {}_M\int [\operatorname{div}_j((B_\alpha + C_\alpha)X_l) - \langle dj(\operatorname{div}_j(B_\alpha + C_\alpha) + 2W(j)U_\alpha) + \operatorname{tr}(B_\alpha \circ W(j))N(j), l\rangle]\ \mu(j).$$

Using Gauss' theorem this equation yields

Proposition 7.2 :

Let $F_{\mathbb{R}^n}$ admits an integral representation given by

$$F_{\mathbb{R}^n}(dk)(dj,dl) = {}_M\int \alpha(dj,dk)\cdot dl\ \mu(j)\ , \qquad \forall\ j \in E(M,\mathbb{R}^n),\ k,l \in C^\infty(M,\mathbb{R}^n),$$

where the stress form

$$\alpha : TE(M,\mathbb{R}^n)\big|_{\mathbb{R}^n} \longrightarrow A^1(M,\mathbb{R}^n)$$

splits into

$$\alpha = c_\alpha \cdot dj + dj\cdot C_\alpha + dj\cdot B_\alpha\ .$$

Then F admits an integral representation with a force density φ_F given at (dj,dk) by

$$(7.25) \qquad \varphi(dj,dk) = -\ dj\ \operatorname{div}\left(B_\alpha(dj,dk) + C_\alpha(dj,dk)\right) + 2\ W(j)U_\alpha(dj,dk)) + \operatorname{tr}\ (B_\alpha(dj,dk)\circ W(j))\cdot N(j)\ .$$

Since α in (7.1) can be replaced by its integrable part as expressed in (7.6) a redundancy occurs in (7.23). Let us therefore rewrite this equation by replacing $\alpha(dj,dk)$ by its integrable part $dh(dj,dk)$. To this end we first rewrite $U_{dh}(dj,dk)$ in terms of $h(dj,dk)$ as done by [Bi,5] and restated in the following :

Lemma 7.3 :
Let $h = dj\ X_h + \Theta_h\cdot N(j)$ for any $h \in C^\infty(M,\mathbb{R}^n)$ then U_{dh} , as defined in (7.23) takes the form

$$(7.26) \qquad U_{dh} = W(j)X_h - \operatorname{grad}_j\Theta_h\ .$$

Proof :
Equation (7.23) reads in the case under consideration as

$$c_{dh}\ N(j) = dj\ U_{dh}\ .$$

By (6.44) we have moreover

$$c_{dh}\ dj\ X = S(j)(X_h,X) + d\Theta_h(X)\cdot N(j) = -\ m(j)(W(j)X_h - \operatorname{grad}_j\Theta_h, X)\cdot N(j)\ .$$

Therefore

$$(7.27)\qquad c_{dh}^{2}\, N(j) = c_{dh}\cdot dj\, U_{dh} = -\, m(j)(W(j)X_h - \operatorname{grad}_j \Theta_h, U_{dh})\, N(j)$$

implying (7.26).
In view of (7.6) and (7.26) we obtain

Corollary 7.8 :
Since for any $(dj,dk) \in TE(M,\mathbb{R}^n)|_{\mathbb{R}^n}$ the stress-form α splits uniquely into

$$\alpha(dj,dk) = dh(dj,dk) + \beta(dj,dk)$$

with $h(dj,dk)$ the integrable and $\beta(dj,dk)$ the non-integrable part respectively, equation (7.25) turns into

$$(7.28)\qquad \varphi(dj,dk) = -\operatorname{div}\left(B_{dh}(dj,dk) + C_{dh}(dj,dk)\right) + 2\, W(j)^2 X_h - 2\, W(j)\operatorname{grad}_j \Theta_h$$

where

$$h = dj\, X_h + \Theta_h\, N(j)$$

with $X_h \in \Gamma TM$ and $\Theta_h \in C^\infty(M,\mathbb{R})$.

Using now the equations of motion (3.25) of the deformable medium M we immediately obtain the following

Main theorem :

Let M be a moving deformable medium of codimension 1 subjected to an internal constitutive law

$$F : C^\infty(M,\mathbb{R}^n) \times E(M,\mathbb{R}^n) \times C^\infty(M,\mathbb{R}^n) \longrightarrow \mathbb{R}\,,$$

with

$$F = d^* F_{\mathbb{R}^n}\,,$$

and

$$F_{\mathbb{R}^n} : C^\infty(M,\mathbb{R}^n)|_{\mathbb{R}^n} \times E(M,\mathbb{R}^n)|_{\mathbb{R}^n} \times C^\infty(M,\mathbb{R}^n)|_{\mathbb{R}^n} \longrightarrow \mathbb{R}\,,$$

a smooth map admitting an integral representation given by the so-called stress form

$$\alpha : TE(M,\mathbb{R}^n)|_{\mathbb{R}^n} \longrightarrow A^1(M,\mathbb{R}^n)\,.$$

This stress form α decomposes according to (6.34) at each

$$(dj,dk) \in TE(M,\mathbb{R}^n)\big|_{\mathbb{R}^n} = E(M,\mathbb{R}^n)\big|_{\mathbb{R}^n} \times C^\infty(M,\mathbb{R}^n)\big|_{\mathbb{R}^n}$$

into

$$\alpha(dj,dk) = c_\alpha(dj,dk)\cdot dj + dj\cdot C_\alpha(dj,dk) + dj\cdot B_\alpha(dj,dk) .$$

The equation of motion on $E(M,\mathbb{R}^n)$ described by a smooth curve

$$\sigma : (-\lambda,\lambda) \longrightarrow E(M,\mathbb{R}^n), \qquad \lambda > 0 ,$$

is then given by

$$(7.29) \qquad {}_M\int \rho(\sigma(t)) \langle \ddot{\sigma}(t),l\rangle \, \mu(\sigma(t)) = {}_M\int \alpha(d\sigma(t),d\dot{\sigma}(t))\cdot dl \, \mu(\sigma(t)) = {}_M\int \langle \varphi_F(d\sigma(t),d\dot{\sigma}(t)),l\rangle \, \mu(\sigma(t)),$$

where the force density φ_F satisfies (7.25). Using the fact that (7.29) implies

$$(7.30) \qquad \rho(\sigma(t))\ddot{\sigma}(t) = \varphi_F(d\sigma(t),d\dot{\sigma}(t)) , \qquad \forall\, t \in (-\lambda,\lambda)$$

and writing $\dot{\sigma}(t)$ in $\mathbb{R}^n$ as

$$\dot{\sigma}(t) = d\sigma(t)Z(t) + \epsilon(\sigma(t),\dot{\sigma}(t))N(\sigma(t)) , \qquad \forall\, t \in (-\lambda,\lambda) ,$$

where $Z(t) \in \Gamma TM$ and $\epsilon(\sigma(t),\dot{\sigma}(t)) \in C^\infty(M,\mathbb{R})$, the equation of motion (7.30) splits into the coupled system

$$(7.31) \qquad \left\{ \begin{aligned} & \nabla(\sigma(t))_{Z(t)}Z(t) + \dot{Z}(t) + 2\,\epsilon(\sigma(t),\dot{\sigma}(t))W(\sigma(t))Z(t) \\ & - \epsilon(\sigma(t),\dot{\sigma}(t))\,\mathrm{grad}_{\sigma(t)}\epsilon(\sigma(t),\dot{\sigma}(t)) \\ & = \rho^{-1}(\sigma(t))\,[-\,\mathrm{div}_{\sigma(t)}T_\alpha(d\sigma(t),d\dot{\sigma}(t)) - 2\,W(\sigma(t))U_\alpha)], \\ & \\ & \dot{\epsilon}(\sigma(t),\dot{\sigma}(t)) = \rho^{-1}(\sigma(t))\,\mathrm{tr}\,B_\alpha(d\sigma(t),d\dot{\sigma}(t))\,W(\sigma(t)) \\ & \qquad - d\epsilon(\sigma(t),\dot{\sigma}(t))Z(t) + \mathfrak{h}(\sigma(t))(Z(t),Z(t)) , \end{aligned} \right.$$

where the smooth two tensor $T_\alpha(dj,dk)$, the so-called stress tensor, is defined by

$$(7.32) \qquad T_\alpha(dj,dk)(X,Y) = m(j)(B_\alpha+C_\alpha)(dj,dk)X,Y), \qquad \forall\, X,Y \in \Gamma TM$$

and $\mathfrak{h}(\sigma(t))$ denotes the second fundamental form of $\sigma(t)$.

Remark :
(7.31) corresponds to Cauchy's law in the mechanics of continua.

Using Corollary 7.8 equations (7.31) turn into

Corollary :
Using the splitting of the stress-form into an integrable dh and a non-integrable part β respectively, (7.31) reads

$$
(7.33)\quad \begin{cases} \nabla(\sigma(t))_{Z(t)}Z(t) + Z(t) + 2\,\epsilon(\sigma(t),\dot{\sigma}(t))W(\sigma(t))Z(t) \\ \quad - \epsilon(\sigma(t),\dot{\sigma}(t))\ \mathrm{grad}_{\sigma(t)}\epsilon(\sigma(t),\dot{\sigma}(t)) \\ \quad = \rho^{-1}(\sigma(t))\ [-\ \mathrm{div}_{\sigma(t)}T_{dh}(d\sigma(t),d\dot{\sigma}(t)) \\ \qquad - 2\,W(\sigma(t))(W(\sigma(t))X_h - \mathrm{grad}_{\sigma(t)}\Theta_h)], \\ \dot{\epsilon}(\sigma(t),\dot{\sigma}(t)) = \rho^{-1}(\sigma(t))\ \mathrm{tr}\ B_{dh}(d\sigma(t),d\dot{\sigma}(t))\ W(\sigma(t)) \\ \qquad - d\epsilon(\sigma(t),\dot{\sigma}(t))Z(t) + \mathfrak{h}(\sigma(t))(Z(t),Z(t)) . \end{cases}
$$

Example 7.9 :

Let α be given by

$$
(7.34)\qquad \alpha(dj,dk) = -\ \tau(dj,dk)\cdot dj\ , \qquad \forall\ j\in E(M,\mathbb{R}^n),\ k\in C^{\infty}(M,\mathbb{R}^n),
$$

where

$$
(7.35)\qquad \tau : TE(M,\mathbb{R}^n)\big|_{\mathbb{R}^n} \longrightarrow C^{\infty}(M,\mathbb{R}^n)
$$

is a smooth map. Then we have

$$
(7.36)\quad \begin{cases} B_{\alpha}(dj,dk) = \tau(dj,dk)\cdot \mathrm{Id}_{TM}\ , \\ C_{\alpha}(dj,dk) = 0\ , \\ c_{\alpha}(dj,dk) = 0\ . \end{cases} \qquad \forall\ (dj,dk)\ \in\ TE(M,\mathbb{R}^n)\big|_{\mathbb{R}^n}\ ,
$$

Therefore the stress tensor T_{α} and the force density φ_F become

$$
(7.37)\qquad T_{\alpha}(dj,dk)(X,Y) = \tau(dj,dk)\cdot m(j)(X,Y)\ , \qquad \forall\ X,Y\ \in\ \Gamma TM\ ,
$$

and

$$
(7.38)\qquad \varphi_F(dj,dk) = dj\ \mathrm{grad}\ \tau(dj,dk) + \tau(dj,dk)\cdot H(j)\ N(j)\ .
$$

Then the Main theorem reduces to the following

Proposition 7.10 :

Let the hypotheses of the Main theorem hold and assume that α is given by (7.34). Then the equations of motion of the deformable medium M are given by

$$
(7.39)\quad \begin{cases} \nabla(\sigma(t))_{Z(t)}Z(t) + \dot{Z}(t) + 2\cdot\epsilon(\sigma(t),\dot{\sigma}(t))\ W(\sigma(t))Z(t) \\ - \epsilon(\sigma(t),\dot{\sigma}(t))\ \mathrm{grad}_{\sigma(t)}\epsilon(\sigma(t),\dot{\sigma}(t)) \\ \qquad = \rho^{-1}(\sigma(t))\ \mathrm{grad}_{\sigma(t)}\tau(d\sigma(t),d\dot{\sigma}(t))\ , \\ \dot{\epsilon}(\sigma(t),\dot{\sigma}(t)) = \rho^{-1}(\sigma(t))\ \tau(d\sigma(t),d\dot{\sigma}(t))\cdot H(\sigma(t)) \\ \qquad - d\epsilon(\sigma(t),\dot{\sigma}(t))Z(t) + \mathfrak{h}(\sigma(t))\ (Z(t),Z(t))\ . \end{cases}
$$

We call (7.39) the equations of motion of a perfect deformable medium. The corresponding constitutive law will be refered too as the perfect constitutive law.
The above corollary relates with the motion induced by a reduced constitutive law introduced earlier as shown by the following

Corollary 7.11 :
Let F be a smooth $\mathbb{R}^n$-invariant constitutive law splitting into the sum

$$F = F^0 + F' , \tag{7.40}$$

where F^0 and F' are smooth $\mathbb{R}^n$-invariant constitutive laws admitting both stress forms α^0 and α' respectively. Assume moreover that α^0 and α' have the decompositions

$$\begin{cases} \alpha^0(dj,dk) = \tau(dj,dk)\cdot dj , \\ \alpha'(dj,dk) = c_\alpha'(dj,dk)\cdot dj + dj\cdot C_\alpha'(dj,dk) + dj\cdot B_\alpha'(dj,dk) , \\ \qquad \forall\ (dj,dk) \in TE(M,\mathbb{R}^n)|_{\mathbb{R}^n} \end{cases} \tag{7.41}$$

with

$$\operatorname{tr} B_\alpha'(dj,dk) = 0 . \tag{7.42}$$

Then the motion

$$\sigma : (-\lambda,\lambda) \longrightarrow E(M,\mathbb{R}^n)$$

induced by F satisfies the equations

$$\begin{cases} \nabla(\sigma(t))_{Z(t)}Z(t) + \dot{Z}(t) + 2\,\epsilon(\sigma(t),\dot\sigma(t))W(\sigma(t))Z(t) \\ \quad - \epsilon(\sigma(t),\dot\sigma(t))\ \operatorname{grad}_{\sigma(t)}\epsilon(\sigma(t),\dot\sigma(t)) \\ \quad = \rho^{-1}(\sigma(t))\ [\operatorname{grad}_{\sigma(t)}\tau(d\sigma(t),d\dot\sigma(t)) - \operatorname{div}_{\sigma(t)}T_\alpha'(d\sigma(t),d\dot\sigma(t)) \\ \qquad - 2\ W(\sigma(t))\ U_\alpha)] \\ \dot\epsilon(\sigma(t),\dot\sigma(t)) = \rho^{-1}(\sigma(t))\ [\operatorname{tr}(B_\alpha'(d\sigma(t),d\dot\sigma(t))\cdot W(\sigma(t))) \\ \qquad - \tau(d\sigma(t),d\dot\sigma(t))\ H(\sigma(t))] \\ \qquad - d\epsilon(\sigma(t),\dot\sigma(t))\ Z(t) + \mathfrak{h}(\sigma(t))\ (Z(t),Z(t)) . \end{cases} \tag{7.43}$$

8. The structural viscosity

As known the notion of viscosity was first introduced by Newton as "the resistance which arises from the lack of slipperiness of the parts of the liquid". He made the assumption that the viscosity "is proportional to the velocity with which the parts of the liquid are separated from one another". As a measure of the viscous resistance one has introduced the coefficient of viscosity ν.

In this section we introduce the notion of structural viscosity, as done in [Bi,5], i.e. the notion of viscosity within our apparatus.

Let again M be a moving deformable medium of codimension 1 and assume that its motion is due only to an internal constitutive law F. As shown F is $\mathbb{R}^n$-invariant and admits the representation

$$F = d^* F_{\mathbb{R}^n}$$

with

$$F_{\mathbb{R}^n} : C^\infty(M,\mathbb{R}^n)|_{\mathbb{R}^n} \times E(M,\mathbb{R}^n)|_{\mathbb{R}^n} \times C^\infty(M,\mathbb{R}^n)|_{\mathbb{R}^n} \longrightarrow \mathbb{R}$$

a smooth map. In addition we assume that $F_{\mathbb{R}^n}$ admits a stress form

$$\alpha : TE(M,\mathbb{R}^n)|_{\mathbb{R}^n} \longrightarrow A^1(M,\mathbb{R}^n) .$$

According to Proposition 6.1 the $\mathbb{R}^n$-valued one-form α admits the decomposition

$$\alpha(dj,dk) = dh(dj,dk) + \beta(dj,dk),$$

where the integrable part $h \in C^\infty(M,\mathbb{R}^n)$ is uniquely determined up to a constant. Next we give h and k the equivalent forms

$$h(dj,dk) = dj\, X_h(dj,dk) + h^\perp(dj,dk)$$

and respectively

$$k = dj\, X_k + k^\perp .$$

Moreover we set, using the fact that M has codimension 1,

$$h^\perp(dj,dk) = \Theta_h(dj,dk)\, N(j) . \tag{8.1}$$

We split now according to Hodge's theorem X_h and X_k into

$$X_h = X_h^0 + \operatorname{grad}_j \psi_h ,$$
$$\operatorname{div}_j X_h^0 = 0 ,$$

and

$$X_k = X_k^0 + \operatorname{grad}_j \psi_k ,$$
$$\operatorname{div}_j X_k^0 = 0$$

respectively. Let us remind of proposition 6.2, expressing in particular that X_h^0 and X_k^0 are uniquely determined by α and dk respectively.
We relate next X_h^0 and X_k^0 uniquely to each other by setting

$$X_h^0(dj,dk) = \nu(dj,dk) X_k^0 + \hat{X}_h(dj,dk) , \tag{8.2}$$

where $\nu(dj,dk) \in C^\infty(M,\mathbb{R})$ and $\hat{X}_h(dj,dk) \in \Gamma TM$ is pointwise orthogonal to X_k^0, and call the function $\nu(dj,dk)$ coefficient of structural viscosity.
Accordingly we call the deformable media, whose constitutive laws depend only on $k^\perp$, frictionless deformable media, while the deformable media

whose constitutive laws depend on the whole of k, will be called frictional ones. Taking now into account (8.1) and (8.2), Proposition 6.4 becomes

Proposition 8.1 :

Let $\alpha \in A^1(M,\mathbb{R}^n)$ and $j \in E(M,\mathbb{R}^n)$ be given and assume that M has codimension 1. Then the following relations hold :

$$\begin{cases} \alpha = dh + \beta , \\ \alpha(X) = c_\alpha \, dj \, X + dj \, C_\alpha \, X + dj \, B_\alpha \, X , \quad \forall X \in \Gamma TM , \\ h(dj,dk) = dj \, X^0_h(dj,dk) + dj \, \mathrm{grad}_j \Psi_h + \Theta_h(dj,dk) \, N(j) , \\ \mathrm{div}_j \, X^0_h = 0 , \\ X_k = X^0_k + \mathrm{grad}_j \Psi_k , \\ \mathrm{div}_j \, X^0_h = 0 , \\ X^0_h = \nu(dj,dk) X^0_k + \hat{X}_k(dj,dk) , \end{cases}$$

(8.3)
$$c_\alpha(dj,dk)\cdot dj = (dh^\perp(dj,dk))^\perp + \nu(dj,dk) S(j)(X^0_k, \,.\,) + S(j)(\hat{X}, \,.\,) + S(j)(\mathrm{grad}_j \Psi_h, \,.\,) + c_\beta(dj,dk)\cdot dj ,$$

(8.4)
$$C_\alpha(dj,dk) = \tfrac{1}{2} [\nabla(j) \, \nu(dj,dk) X^0_k - \tilde{\nabla}(j) \, \nu(dj,dk) X^0_k] + \tfrac{1}{2} [\nabla(j) \hat{X}_h(dj,dk) - \tilde{\nabla}(j) \hat{X}_h(dj,dk)] + C_\beta(dj,dk) ,$$

(8.5)
$$B_\alpha(dj,dk) = \tfrac{1}{2} L_{\nu(dj,dk) X^0_k} + \tfrac{1}{2} L_{\hat{X}_h(dj,dk)} + \tfrac{1}{2} L_{\mathrm{grad}_j \Psi_h(dj,dk)} + \Theta_h(dj,dk) \, W(j) + B_\beta(dj,dk) , \quad \forall \, (dj,dk) \in TE(M,\mathbb{R}^n)|_{\mathbb{R}^n} .$$

If now $\nu(dj,dk)$ is a constant map in $C^\infty(M,\mathbb{R})$, then we get

(8.6)
$$C_\alpha(dj,dk) = \tfrac{1}{2} \nu(dj,dk) \, (\nabla(j) \, X^0_k - \tilde{\nabla}(j) \, X^0_k) + \tfrac{1}{2} [\nabla(j) \hat{X}_h(dj,dk) - \tilde{\nabla}(j) \hat{X}_h(dj,dk)] + C_\beta(dj,dk) ,$$

(8.7)
$$B_\alpha(dj,dk) = \tfrac{1}{2} \nu(dj,dk) \, L_{X^0_k} + \tfrac{1}{2} L_{\hat{X}_h(dj,dk)} + \tfrac{1}{2} L_{\mathrm{grad}_j \Psi_h(dj,dk)} + \Theta_h(dj,dk) \, W(j) + B_\beta(dj,dk) , \quad \forall \, (dj,dk) \in TE(M,\mathbb{R}^n)|_{\mathbb{R}^n} .$$

Using now the definition (7.32) of the stress tensor we get

(8.8)
$$T_\alpha(dj,dk)(X,Y) = m(j)((\tfrac{1}{2} \nu(dj,dk) \, L_{X^0_k} + \tfrac{1}{2} \nu(dj,dk) (\nabla(j) X^0_k - \tilde{\nabla}(j) X^0_k) + \tfrac{1}{2} L_{\hat{X}_h(dj,dk)} + \tfrac{1}{2} [\nabla(j) \hat{X}_h(dj,dk) - \tilde{\nabla}(j) \, \hat{X}_h(dj,dk)] - \tfrac{1}{2} L_{\mathrm{grad}_j \Psi_h(dj,dk)} + \Theta_h(dj,dk)\cdot W(j) + B_\beta(dj,dk) + C_\beta(dj,dk)) \, X \, , Y) .$$

It is this equation which motivates us to call the function $\nu(dj,dk)$ coefficient of structural viscosity.

9. The equations of motion of a deformable medium subjected to a general constitutive law

Let us suppose that the motion of the deformable medium M is governed by the smooth constitutive law

$$F : C^{\infty}(M,\mathbb{R}^n) \times E(M,\mathbb{R}^n) \times C^{\infty}(M,\mathbb{R}^n) \longrightarrow \mathbb{R} ,$$

and assume that F splits into

$$F = F_{ext} + F_{int} , \tag{9.1}$$

where the internal constitutive law F_{int} and the external one F_{ext} are also smooth. As shown F_{int} is $\mathbb{R}^n$-invariant and admits the representation

$$F_{int} = d^{*} F_{\mathbb{R}^n}$$

with

$$F_{\mathbb{R}^n} : C^{\infty}(M,\mathbb{R}^n)\big|_{\mathbb{R}^n} \times E(M,\mathbb{R}^n)\big|_{\mathbb{R}^n} \times C^{\infty}(M,\mathbb{R}^n)\big|_{\mathbb{R}^n} \longrightarrow \mathbb{R} .$$

Finally we assume that both F_{ext} and $F_{\mathbb{R}^n}$ admit integral representations and denote by φ_{ext} and φ_{int} the corresponding force densities. According to (3.13) and (3.26) we have the splittings

$$\varphi_{F_{int}}(j,k) = dj\ \mathrm{grad}_j \tau_{int}(j,k) + dj\ Y^0{}_{int}(j,k) + \varphi_{F_{int}}(j,k)^{\perp} , \tag{9.2}$$

$$\varphi_{F_{ext}}(j,k) = dj\ \mathrm{grad}_j \tau_{ext}(j,k) + dj\ Y^0{}_{ext}(j,k) + \varphi_{F_{ext}}(j,k)^{\perp} , \tag{9.3}$$

$$\forall\ j \in E(M,\mathbb{R}^n),\ k \in C^{\infty}(M,\mathbb{R}^n) ,$$

and hence

$$\varphi_F(j,k) = dj\ \mathrm{grad}_j \tau(j,k) + dj\ Y^0(j,k) + \varphi_F(j,k)^{\perp} , \tag{9.4}$$

where

$$\begin{aligned} &\tau_{int}(j,k) + \tau_{ext}(j,k) = \tau(j,k) , \\ &Y^0{}_{int}(j,k) + Y^0{}_{ext}(j,k) = Y^0(j,k) , \\ &\varphi_{F_{int}}(j,k)^{\perp} + \varphi_{F_{ext}}(j,k)^{\perp} = \varphi_F(j,k)^{\perp} , \\ &\qquad \forall\ j \in E(M,\mathbb{R}^n),\ k \in C^{\infty}(M,\mathbb{R}^n). \end{aligned} \tag{9.5}$$

Since $\varphi_{F_{int}}$ is $\mathbb{R}^n$– invariant, its tangential and normal parts grad $\tau_{int}+Y^0_{int}$ and $\varphi_{F_{int}}^{\perp}$ respectively are also $\mathbb{R}^n$–invariant. Using now the $\mathbb{R}^n$–invariance of the Laplace–Beltrami operator Δ, we infer that τ and consequently grad τ are $\mathbb{R}^n$–invariant. Accordingly we split F_{int} into

(9.6) $$F_{int} = F^0_{int} + F'_{int} ,$$

where F^0_{int} has the force density $\varphi^0_{F_{int}}$ defined by

(9.7) $$\varphi^0_{F_{int}}(j,k) = dj\ \mathrm{grad}_j \tau_{int}(dj,dk)$$

and F'_{int} admits the stress form

(9.8) $$\begin{cases} \alpha : TE(M,\mathbb{R}^n) \longrightarrow A^1(M,\mathbb{R}^n) \\ \alpha'(dj,dk) = c_\alpha'(dj,dk)\cdot dj + dj\ C_\alpha'(dj,dk) + dj\ B_\alpha'(dj,dk)\ . \end{cases}$$

Hence

(9.9) $$\alpha(dj,dk) = c_\alpha'(dj,dk)\cdot dj + dj\ C_\alpha'(dj,dk) + dj\ (-\tau_{int}(dj,dk)\ Id_{TM} + B_\alpha'(dj,dk))\ .$$

Using (7.40) we then find

(9.10) $$\begin{aligned} \varphi_{F_{int}}(j,k) &= \varphi^0_{F_{int}}(j,k) + \varphi'_{F_{int}}(j,k) \\ &= dj\ \mathrm{grad}_j\tau_{int}(dj,dk) + dj\ Y^0_{int}(dj,dk) + \varphi_{F_{int}}(dj,dk)^{\perp} \\ &= dj\ \mathrm{grad}_j\tau_{int}(dj,dk) - dj(\mathrm{div}_j B_\alpha'(dj,dk) + \mathrm{div}_j C_\alpha'(dj,dk) \\ &\quad + 2\ W(j)\ (W(j)X_h - \mathrm{grad}\ \Theta_h)) + [\mathrm{tr}\ (B_\alpha'(dj,dk)\cdot W(j)) - \\ &\quad \tau_{int}(dj,dk)\cdot H(j)]\ N(j)\ , \end{aligned}$$

(9.11) $$T_\alpha(j,k)(X,Y) = -\tau_{int}(dj,dk)\ m(j)(X,Y) + m(j)((B_\alpha' + C_\alpha')(dj,dk)X,Y)\ ,\ \forall\ X,Y \in \Gamma TM\ .$$

On the other hand α' splits according to Proposition 6.4 at (dj,dk) into

$$\alpha'(dj,dk) = dh\ (dj,dk) + \beta(dj,dk)\ ,$$

where $h(dj,dk) \in C^\infty(M,\mathbb{R}^n)$. Writing

$$h(dj,dk) = dj\ X_h(dj,dk) + \Theta_h(dj,dk)\ N(j)\ ,$$
$$k(dj,dk) = dj\ X_k(dj,dk) + \Theta_k(dj,dk)\ N(j)\ ,$$

with $\Theta_h, \Theta_k \in C^\infty(M,\mathbb{R})$, we obtain by (9.10) and Proposition 6.4

$$(9.12) \qquad \varphi_{F_{int}}(j,k) = dj \ grad_j \tau_{int}(dj,dk) - dj(\triangle(j) \ X_h$$
$$+ W(j) grad_j \Theta_h(dj,dk) + \Theta_h(dj,dk) grad_j H(j)$$
$$+ 2 \ W(j)(W(j)X_h - grad \ \Theta_h))$$
$$+ (tr \ ([\ \tfrac{1}{2} L_{X_h(dj,dk)} + \Theta_h(dj,dk) \ W(j)] \circ W(j))$$
$$- \tau_{int}(dj,dk) \ H(j)) \ N(j) \ .$$

Decomposing now $X_h(dj,dk)$ and X_k by the Hodge theorem into

$$(9.13) \qquad \begin{cases} X_h(dj,dk) = X_h^0(dj,dk) + grad_j \ \psi_h(dj,dk) \ , \\[2mm] div_j \ X_h^0 = 0 \ , \end{cases}$$

$$(9.14) \qquad \begin{cases} X_k(dj,dk) = X_k^0(dj,dk) + grad_j \ \psi_k(dj,dk) \ , \\[2mm] div_j \ X_k^0 = 0 \ , \end{cases}$$

introducing the structural viscosity $\nu(dj,dk) \in C^\infty(M,\mathbb{R})$ via

$$(9.15) \qquad X_h^0(dj,dk) = \nu(dj,dk) \ X_k^0 + \hat{X}_h(dj,dk)$$

and noting that

$$(9.16) \qquad \tfrac{1}{2} \cdot tr \ (L_Z \cdot W(j)) = div_j W(j)Z - dH(j)Z$$

from (9.2), (9.3) and (9.11) we obtain

$$(9.17) \qquad \varphi_F(j,k) = \varphi_{F_{ext}}(j,k) + dj \ (grad_j \tau(j,k) - \triangle(j) \ [\ \nu(dj,dk)X_k^0$$
$$+ \hat{X}_k(dj,dk) + grad_j \psi_h(dj,dk)] - W(j) \ [grad_j \Theta_h(dj,dk)$$
$$+ 2 \ (W(j)X_h - grad \ \Theta_h)] - \Theta_h(dj,dk) \ grad_j H(j)]$$
$$- (\tau_{int}(dj,dk) \ H(j) + dH(j) \ [\nu(dj,dk) \ X_k^0 + \hat{X}_h(dj,dk)]$$
$$- div_j \ \nu(dj,dk)W(j)X_k^0 - div_j W(j) \ \hat{X}_h(dj,dk)$$
$$- \Theta_h(dj,dk) \ tr \ W(j)^2) \ N(j),$$

$$(9.18) \qquad T_\alpha(dj,dk)(X,Y) = -\tau_{int}(dj,dk) \ m(j) \ (X,Y)$$
$$+ \tfrac{1}{2} \ (L_{\nu(dj,dk)} \ X_k^0 + L_{\hat{X}_k(dj,dk)} + L_{grad_j \psi_h(dj,dk)})$$
$$(m(j))(X,Y) + m(j)(W_h(j)X,Y)$$
$$+ m(j)(C_\alpha'(dj,dk) \ X,Y) \ .$$

Consequently we may state the following theorem, based on (9.17), (9.18) and (3.25)

Theorem 9.1 :

Let $F : C^\infty(M,\mathbb{R}^n) \times E(M,\mathbb{R}^n) \times C^\infty(M,\mathbb{R}^n) \longrightarrow \mathbb{R}$ be a smooth constitutive law admitting the splitting (9.1), i.e.

$$F = F_{int} + F_{ext} ,$$

where both the internal and the external constitutive laws, F_{int} and F_{ext} respectively, admit integral representations with the respective force densities $\varphi_{F_{int}}$ and $\varphi_{F_{ext}}$.

Then the general equations of motion of a deformable medium are given by

$$(9.19)\quad \begin{cases} \nabla(\sigma(t))_{Z(t)}Z(t)+\dot{Z}(t)+2\epsilon(\sigma(t),\dot{\sigma}(t))W(\sigma(t))Z(t) \\ \quad - \epsilon(\sigma(t),\dot{\sigma}(t))\ \mathrm{grad}_{\sigma(t)}\epsilon(\sigma(t),\dot{\sigma}(t)) \\ \quad = \rho^{-1}(\sigma(t))(\mathrm{grad}_{\sigma(t)}\tau_{int}(\sigma(t),\dot{\sigma}(t)) - \Delta(\sigma(t))\ [\nu(d\sigma(t),d\dot{\sigma}(t))Z^0(t) \\ \quad + \hat{X}_h(d\sigma(t),d\dot{\sigma}(t)) + \mathrm{grad}_{\sigma(t)}\Psi_h(\sigma(t),\dot{\sigma}(t))] \\ \quad - W(\sigma(t))[\mathrm{grad}_{\sigma(t)}\Theta_h(d\sigma(t),d\dot{\sigma}(t)) + 2\ (W(\sigma(t))X_h(\sigma(t),\dot{\sigma}(t)) \\ \quad - \mathrm{grad}\ \Theta_h)(\sigma(t),\dot{\sigma}(t))] - \Theta_h(d\sigma(t),d\dot{\sigma}(t))\mathrm{grad}_{\sigma(t)}H(\sigma(t)) \\ \quad + \rho^{-1}(\sigma(t))Y_{ext}(\sigma(t),\dot{\sigma}(t)) , \\ \\ \dot{\epsilon}(\sigma(t),\dot{\sigma}(t)) = \rho^{-1}(\sigma(t))[-\tau_{int}(d\sigma(t),d\dot{\sigma}(t))\ H(\sigma(t)) \\ \quad - dH(\sigma(t))[\nu(d\sigma(t),d\dot{\sigma}(t))\ Z^0(t) + \hat{X}_h(d\sigma(t),\dot{\sigma}(t))] \\ \quad + \mathrm{div}_{\sigma(t)}\nu(d\sigma(t),d\dot{\sigma}(t))W(\sigma(t))Z^0(t) \\ \quad + \mathrm{div}_{\sigma(t)}W(\sigma(t))\hat{X}_h(d\sigma(t),d\dot{\sigma}(t)) \\ \quad - \Theta_h(d\sigma(t),d\dot{\sigma}(t))\ \mathrm{tr}\ W(\sigma(t))^2] + \mathfrak{h}(\sigma(t))\ (Z(t),Z(t)) \\ \quad - d\epsilon(\sigma(t),\dot{\sigma}(t))\ Z(t) + \kappa_{ext}(\sigma(t),\dot{\sigma}(t))] , \end{cases}$$

where $\varphi^{\perp}_{ext}(\sigma(t),\dot{\sigma}(t)) = \kappa_{ext}(\sigma(t),\dot{\sigma}(t))\ N(\sigma(t))$.

The motion of a deformable medium along a fixed surface $i(M) \subset \mathbb{R}^n$ is given by

$$(9.20)\quad \begin{cases} \nabla(i)_{X(t)}X(t) + \dot{X}(t) = \rho^{-1}(X(t))[-\mathrm{grad}_i\tau_{int}(X(t),\dot{X}(t)) \\ \qquad - \Delta(i)[\nu(X(t),\dot{X}(t))\ X^0(t) + \hat{X}_h(X(t),\dot{X}(t)) \\ \qquad + 2\ (W(i)X_h - \mathrm{grad}\ \Theta_h)] - \Theta_h(X(t),\dot{X}(t))\ \mathrm{grad}_iH(i) \\ \qquad + \rho^{-1}(X(t))\ Y_{ext}(X(t),\dot{X}(t)) \\ \\ 0 = \rho^{-1}(X(t))\ (-\tau_{int}(X(t),\dot{X}(t))\cdot H(i) \\ \quad - dH(i)[\nu(X(t),\dot{X}(t))X^0(t) + \hat{X}_h(X(t),\dot{X}(t))] \\ \quad + \mathrm{div}_i\nu(X(t),\dot{X}(t))W(i)X^0(t) + \mathrm{div}_iW(i)\hat{X}_h(X(t),\dot{X}(t)) \\ \quad - \Theta_h(X(t),\dot{X}(t))\mathrm{tr}W(i)^2) + \mathfrak{h}(i)(X(t),X(t)) \\ \quad + \kappa_{ext}(X(t),\dot{X}(t)), \end{cases}$$

where $X(t)$ is the push-forward (4.20) of $Z(t)$ by $g(t) \in \text{Diff}\, M$.
As an example let us consider the stress form

$$\alpha(dj,dk) = -\tau_{int}(dj,dk)\cdot dj + \nu\, dj\cdot L_{X_k} \tag{9.21}$$

with a constant $\nu \in \mathbb{R}$. Then the motion along $i(M)\subset\mathbb{R}^n$ is governed by

$$\begin{aligned}\rho(X(t))(\nabla(i)_{X(t)}X(t) + \dot{X}(t)) \\ = -\,\text{grad}_i\tau_{int}(g(t),X(t)) - \nu\cdot\Delta(i)\, X(t) - \nu\cdot\text{Ric}(i)\, X(t)\end{aligned} \tag{9.22}$$

in case $\text{div}_i X(t) = 0$. Thus the equation (9.22), a Navier-Stokes type of equation, is an approximation to (9.20).

Remark:
As done in [Bi,5] a type of pressure $\pi(dj,dh)$ can be introduced by forming the L^2-component of $\text{tr}\, B_\alpha(dj,dk)\circ W(j)$ (in 7.25) along $H(j)$ yielding the decomposition

$$\text{tr}\, B_\alpha(dj,dk) = -\,\pi(dj,dk)\cdot H(j) + \text{tr}\, \bar{B}_\alpha(dj,dk)\ . \tag{9.23}$$

This allows us to split $F(dj,dk)$ into

$$F(dj,dk) = \bar{F}(dj,dk) - \pi(dj,dk)\cdot DV(j) \tag{9.24}$$

where $V(j)$ is the volume of $j(M)$. Both of these equations hold for all dj, dk and dh. Motivated by the last equation we call $\pi(dj,dk)$ the volume active pressure. $\pi(dj,dk)\cdot DV(j)$ is the work needed to change the volume by $DV(j)$. Clearly this type of pressure also exists in the realm of section 3 and is clearly not identical (in general) with $\tau_{int}(dj,dk)$.

Acknowledgement

We are indebted to M. Epstein, who has told us how to use d'Alembert's principle as a constitutive law years ago. We are especially thankfull to J. Sniatycki, who showed us the present formulation of d'Alembert's principle used in our treatment and who discussed with us some fundamental aspects of the Navier-Stokes equation.
The first author thanks the University of Calgary for its kind hospitality during his stay in fall 1987 when parts of these notes were written.

References

[Ba] A. Bastiani : Applications différentiables et variétés différentiables de dimension infinie, J. Anal. Math., 13, 1-114, 1964

[Bi,1] E. Binz : Two natural metrics and their covariant derivatives on a manifold of embeddings, Mh. Math., 89, 275-288, 1980

[Bi,2] E. Binz : The description of $\mathbb{R}^n$-valued one forms relative to an embedding, Mannheimer Manuskripte No. 74, Universität Mannheim, 1987

[Bi,3] E. Binz : One forms on $E(M,\mathbb{R}^n)$ with integral representation, Mannheimer Manuskripte No. 79, Universität Mannheim, 1988

[Bi,4] E. Binz : On the notion of the stress tensor associated with $\mathbb{R}^n$-invariant constitutive laws admitting integral representations, (to appear in Reports on Mathematical Physics), 1988

[Bi,5] E. Binz : Viscosity and volume active pressure, Mannheimer Manuskripte No. 82, Universität Mannheim, 1988

[Bi,6] E. Binz : Natural Hamiltonian systems on spaces of embeddings, Proceedings of the XV international conference on differential geometric methods in theoretical physics, ed. H. D. Doebner, J. D. Hennig, World Scientific, Singapore, New Jersey, Hong Kong, 1987

[Bi,Fi] E. Binz, H. R. Fischer : The manifold of embeddings of a closed manifold, Lecture Notes Phys., 139, Springer-Verlag, Berlin, 1981

[Fr,Kr] Frölicher, Kriegl : Linear spaces and differentiation theory, John Wiley, Chichester England, 1988

[Gr,H,V] W. Greub, S. Halperin, J. Vanstone : Connections, Curvature and Cohomology, I, II, Acad. Press, New York, 1972-73

[Gu] J. Gutknecht : Die C^∞_Γ - Struktur auf der Diffeomorphismengruppe einer kompakten Mannigfaltigkeit, Diss. ETH 5879, Zürich, 1977

[Go,Gui] M. Golubitski, V. Guillemin : Stable mappings and their singularities, Graduate Texts of Mathematics, Springer GTM 14, Berlin, 1973

[He] E. Hellinger : Die allgemeinen Ansätze der Mechanik der Kontinua, Enzykl. Math. Wiss. 4/4, 1914

[Hi] M. W. Hirsch : Differential Topology, Springer GTM, Berlin, 1976

[Kl] W. Klingenberg : Riemannian Geometry, De Gruyter Studies in Math., Berlin, 1982

[Le,Fi] M. J. Leitman, G. M. C. Fisher : The linear theory of visco-elasticity, Handbuch der Physik, VI a/3, Editors C. Truesdell, S. Flügge, Springer-Verlag, Berlin, 1965

[Mi] P. Michor : Manifolds of differentiable mappings, Shiva Math. Series, Shiva Publ., Kent, UK, 1980

[Sch] Th. Schnyder-Peter : Zur Struktur der Immersionen einer Mannigfaltigkeit in einem euklidischen Raum, Universität Zürich, 1983

[Tr,No] C. Truesdell, W. Noll : The non-linear field theories of mechanics, Handbuch der Physik, III/3, Editors C. Truesdell, S. Flügge, Springer-Verlag, Berlin, 1973

BERRY'S CONNECTION, THE CHARGE-MONOPOLE SYSTEM AND THE GROUP THEORY OF THE DIATOMIC MOLECULE

A. Bohm*

School of Physics and Astronomy
Tel Aviv University
Tel Aviv 69978, Israel

I. INTRODUCTION

When we study complicated quantum physical systems, we divide them into their parts and study these parts separately. The parts can be constituents, but the parts can also be other subsystems, like certain collective motions (e.g. rotations about the center of mass). In either case the space of physical states of the subsystem is a factor space of the tensor product space for the whole system. One can then study the structure of the subsystem for each definite state of the remainder of the whole system.

The most successful, and essentially generic example of this procedure is the Born-Oppenheimer approximation[1)] of molecules. The whole (diatomic) molecule consists of the fast motion of the electrons and the slow collective motion (rotations and oscillations) of the molecule as a whole, described by the internuclear variables. One first deals with the motion of the fast variables keeping the slow variables fixed but arbitrary. This leads to the electronic energy spectra. After the dynamics of the fast variables is considered as resolved, one allows the "slow variables" to vary and describes the subsystem of the collective motions by an effective Hamiltonian or an effective action. In this way one arrives at the well known spectra of molecules. The electronic levels ε_n are split into vibrational excitations $E_{j,\nu;n}$ ($\nu = 1,2,3,\cdots$) which are in turn split into rotational bands $E_{j,\nu;n}$ ($j = 0,1,2,\cdots$).

In the original Born-Oppenheimer theory the internuclear distance $\vec{x}$ (the slow distance variable) was not considered as a dynamical variable.

*On leave from Physics Department, The University of Texas at Austin, Austin, Texas 78712. Supported by a Fulbright grant.

That something can go wrong with this "adiabatic" limit of the Born-Oppenheimer procedure was first noticed in the spectra of molecules and interpreted as the occurrence of an effective vector potential.[2)] The fast motion induces a vector potential $\vec{A}$ in the dynamics of the slow motion, which was called the "molecular Bohm-Aharanov effect" by Mead.[2)] Berry[3a)] then demonstrated the generality of this phenomenon and showed that (Abelian) magnetic monopole fields occur near a degeneracy of the quantum levels in the space of the slow variables. These induced vector potentials A which can also be non-Abelian,[3b)] have therefore been called[4)] Berry's connection, a line integral of this connection is Berry's phase. How this Berry connection appears in the Born-Oppenheimer procedure when the "slow" variables are considered as quantum mechanical observables and not as fixed parameters is shown in section II.

In section III we review the properties of a symmetric top (molecule) and a dumbbell with flywheel and in section IV we describe the group theory (Spectrum Generating Group) for these systems. This will prepare us for a detailed study of the Berry connections for the specific case of rotations in diatomic molecules which is presented in section VI. Moody, Shapere and Wilczek recently investigated the relation between these diatomic molecules and the charge-monopole system.[5)] We will therefore present in section V the charge-monopole and the dyon system[6)] in a way which displays the analogy with the dumbbell system of section III, before we derive in section VI the Berry connection for the dumbbell.

The material presented here is mainly a review of various results that can be found in the literature and only its presentation is new emphasizing the algebraic aspects and the relation to group theory.

II. THE BORN-OPPENHEIMER APPROXIMATION AND BERRY'S CONNECTION

The collective model for molecules is based on the Born-Oppenheimer procedure. In the Born-Oppenheimer procedure the physical system is divided into two parts. The first part consists of the motion of the electron cloud around the nuclei, the second part is the motion of the nuclei. The electron variables are called the fast variables, the variables of the nuclei are called the slow variables.

The complete Hamiltonian is written as

$$H = \frac{\vec{P}^2}{2\mu} + \frac{\vec{\mathcal{p}}^2}{2m} + V(\vec{X},\vec{\mathcal{r}}) \qquad (2.1)$$

here

$(\vec{p},\vec{x})$ are the slow variables, $(\vec{P}\ \vec{X})$ the corresponding operators (2.2)

$(\vec{\mathcal{p}},\vec{\mathcal{r}})$ are the fast variables (operators) (2.3) (which need not be linear momentum and coordinate but could as well be spin etc.). One first deals with the motion of the fast variables (electronic degrees of freedom) keeping the slow variables fixed. This motion is described by the sub-Hamiltonian

$$h(\vec{x}) = \frac{\vec{p}^2}{2m} + V(\vec{x},\vec{r}) \tag{2.4}$$

which depends parametrically on the slow coordinates $\vec{x}$. The eigenvalues $\varepsilon_n(\vec{x})$ and the eigenvectors $|n;(\vec{x})\rangle$ with property

$$h(\vec{x})\,|n;(\vec{x})\rangle = \varepsilon_n(\vec{x})\;|n;(\vec{x})\rangle \tag{2.5}$$

depend upon the parameter $\vec{x}$.

To consider $\vec{x}$ as a fixed parameter means to assume that it commutes with $\vec{P}$ and thus to assume that $h(\vec{x})$ commutes with $\vec{P}$. Therefore it also means to assume that H and $h(\vec{x})$ commute. Under this assumption $h(\vec{x})$ and H can be diagonalized together. We denote the common eigenvector by $|N,n;\vec{x}\rangle$ and have:

$$h(\vec{x})\,|N,n;\vec{x}\rangle = \varepsilon_n(\vec{x})\;|N,n;\vec{x}\rangle \tag{2.6}$$

$$H\;|N,n;\vec{x}\rangle = E_{N,n}\;|N,n;\vec{x}\rangle \tag{2.7}$$

$\varepsilon_n(\vec{x})$ are the eigenvalues of the electronic Hamiltonian $h(\vec{x})$ for any fixed value of the slow coordinates $\vec{x}$ and n stands for the set of electronic quantum numbers. N stands for the set of quantum numbers $N = j,\nu,\cdots$ which characterize the slow motions of the nuclei, or the collective motion of the molecule.

After (2.6) has been solved, $\varepsilon_n(\vec{x})$ is inserted on the l.h.s. of (2.7) and one obtains the eigenvalue equation

$$\left(\frac{\vec{P}^2}{2\mu} + \varepsilon_n(\vec{X})\right)\;|N,n,\vec{x}\rangle = E_{N,n}\;|N,n,\vec{x}\rangle \tag{2.7'}$$

in which $\varepsilon_n(\vec{x})$ is treated as the potential energy. This equation determines the energy values $E_{j,\nu,\cdots n}$ of the collective excitations for every fixed value of n. The Born-Oppenheimer procedure leads thus to the well known spectra of molecules as shown in Fig. II.1.

Each electronic level with given value n is split into vibrational excitations $E_{j,\nu,n}$ $\nu = 1,2,3,\cdots$ and rotational bands $j = 0,1,2,\cdots$.

Some time ago it was noticed[2)] that the above described "adiabatic limit" of the Born-Oppenheimer approximation may be insufficient. As will be shown below by a straightforward calculation, instead of (2.7') for the effective Hamiltonian of the collective motion one should take for the effective Hamiltonian the expression

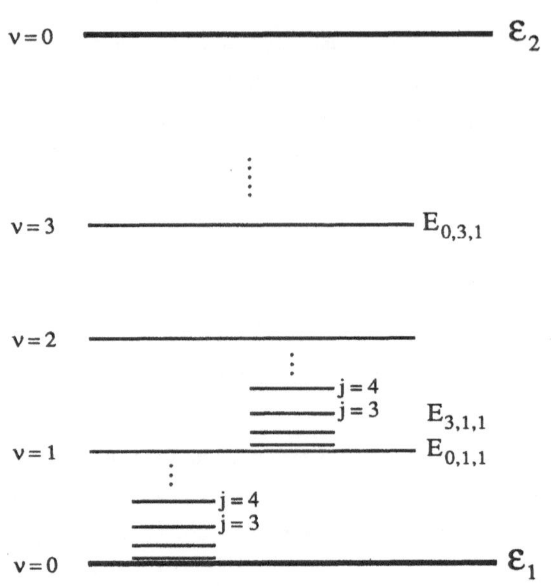

Figure 1. Schematics of typical molecular spectra.

$$H_{eff} = \frac{1}{2\mu} (\vec{P} - \vec{A}(X))^2 + \tilde{\varepsilon}_n(X) = \frac{1}{2\mu} \vec{\pi}^2 + \tilde{\varepsilon}_n(x) \tag{2.8}$$

where

$$\begin{aligned} \pi_i &= P_i - A_i , \\ A_i(\vec{x}) &= \langle n,\vec{x}| i \frac{\partial}{\partial x^i} |n,\vec{x}\rangle , \end{aligned} \tag{2.9}$$

and $|n,\vec{x}\rangle$ is the eigenvector of the Hamiltonian for the fast motion (2.5). $A_i(x)$ is thus a vector potential induced by the fast motion into the slow motion, $\tilde{\varepsilon}_n(\vec{x})$ is like $\varepsilon_n(\vec{x})$ in (2.7') a scalar potential induced by the fast motion.

The gauge potential is sometimes also written in a different way

$$A = A_i(x)dx^i = i \langle n,\vec{x}|d|n,\vec{x}\rangle \tag{2.9'}$$

where d denotes differentiation in $\vec{x}$. A change of phase

$$|n,\vec{x}\rangle \to |n,\vec{x}\rangle e^{-i\beta(\vec{x})} \tag{2.10}$$

induces a gauge transformation

$$\begin{aligned} A \to & \; ie^{i\beta(\vec{x})}\langle\vec{x},n|d(|n,\vec{x}\rangle e^{-i\beta(x)}) = \\ & = i\langle\vec{x},n|d|n,\vec{x}\rangle + e^{i\beta}\langle\vec{x},n|n,\vec{x}\rangle e^{-i\beta} d\beta(x) = A + d\beta(\vec{x}) . \end{aligned} \tag{2.11}$$

The gauge potential (2.9) or (2.9') is called Berry's connection. The corresponding field

$$\vec{B} = \vec{\nabla} \wedge \vec{A} \qquad \text{or} \qquad B = dA = i(d\langle n\vec{x}|)d|n\vec{x}\rangle \tag{2.12}$$

is called the Berry curvature

With (2.8) and (2.9) --which will be derived below-- it has been shown that as a consequence of the fast motion (electrons in the present case) the canonical momentum operator of the slow motion (collective vibrations and rotations) is replaced by the covariant momentum

$$\pi_i = P_i - A_i \tag{2.8}$$

where A_i is the Berry's connection (2.9) induced by the fast motion. As a consequence of the influence of the fast motion upon the slow dynamics the π_i will obey anomalous quantum mechanical commutation relations.

APPENDIX II. DERIVATION OF EQUATIONS (2.8) AND (2.9)

The derivation of (2.8) and (2.9) makes use only of fundamental notions of quantum theory: In the Hilbert space $\mathcal{H} = \mathcal{H}^{\text{slow}} \times \mathcal{H}^{\text{fast}}$ of physical states of the whole physical system (fast and slow motion) one can introduce basis systems of (generalized) eigenvectors of different complete systems of commuting observables. We choose the following two c.s.c.o.'s

$$X_i \; r_i \; \cdots \tag{2.13}$$

and

$$X_i \; \cdots \; h(X) \tag{2.14}$$

X_i are the "slow" position operators (positions of the nuclei in the molecule), r_i are the "fast" position operators (positions of the electrons) $h(X_i)$ is the hamiltonian of the fast motion given by (2.4) and $\cdots$ stands for possible other observables needed to form a c.s.c.o.. The eigenvectors of (2.13) are denoted by

$$|x_i \; r_i\rangle \tag{2.15}$$

and the eigenvectors of (2.14) are denoted by

$$|x_i, \; n\rangle \; . \tag{2.16}$$

That $h(X)$ and X_i commute follows from (2.4) because p_i and r_i commute with X_i. The eigenvectors (2.16) have the property

$$h(X_i)|x_i,n\rangle = \varepsilon_n(x_i)|x_i,n\rangle \tag{2.17}$$

$$X_i|x_i,n\rangle = x_i|x_i,n\rangle \quad . \tag{2.18}$$

The basis systems (2.15) and (2.16) can be expanded with respect to each other

$$|x_i,n\rangle = \int d^3x' \; d^3r' \; |x_i'r_i'\rangle \; \langle x_i'r_i'|x_i,n\rangle \tag{2.19}$$

and the tansition coefficients $\langle x_i'r_i'|x_i,n\rangle$ have the property

$$\langle x_i' r_i' | x_i, n\rangle = \delta^3(x'_i - x_i)\langle r_i' | n\rangle_{(x_i)} \tag{2.20}$$

where the reduced transformation coefficient $\langle r_i' | n\rangle_{(x_i)}$ depends upon x_i.

In the adiabatic limit the evolution of the "slow" positions $X_i(t)$ is viewed as a slow prescribed classical motion so that X_i in $V(X_i, r_i)$ is a number, $V(x_i, r_i)$. As a consequence of this, P_i (the "slow" momenta) commute with V, $[P_i, V(x_i, r_i)] = 0$. Consequently we have also: $[H, h(x_i)] = 0$.

Thus in the adiabatic limit (the total hamiltonian) H and the (electronic) hamiltonian $h(X_i)$ can be diagonalized simultaneously. Therefore in this limit there exists a c.s.c.o .

$$H \cdots h \tag{2.21}$$

whose simultaneous eigenvectors we call

$$|N,n\rangle \tag{2.22}$$

They have the property

$$H|N,n\rangle = E_{N,n}|N,n\rangle \tag{2.23}$$

$$h|N,n\rangle = \varepsilon_n|N,n\rangle \quad . \tag{2.24}$$

We will consider first the adiabatic limit in which the basis system of eigenvectors (2.22) of $\mathcal{H}$ exists. We can expand these (approximate) basis vectors with respect to the (exact) basis vectors $|x_i, n\rangle$

$$|N,n\rangle = \int d^3x' \; |x_i', n\rangle \langle x_i' n | N\, n\rangle \tag{2.25}$$

The wave function

$$\Psi_{Nn}(x_i, r_i) = \langle x_i r_i | N,n\rangle \tag{2.26}$$

can therefore be written as

$$\begin{aligned} \Psi_{Nn}(x_i, r_i) &= \int d^3x' \; \langle x_i r_i | x_i' n\rangle \; \langle x_i' n | Nn\rangle \\ &= \int d^3x' \; \delta^3(x_i - x_i')\langle r_i | n\rangle_{(x_i)} \; \langle x_i' n \; Nn\rangle \\ &= \langle r_i | n\rangle_{(x_i)} \; \langle x_i \; n | Nn\rangle \\ &= \phi_n(\vec{x}, \vec{r}) \; \Phi_n^N(\vec{x}) \quad . \end{aligned} \tag{2.27}$$

In here we have used (2.20). $\phi_n(\vec{x}, \vec{r}) = \langle r_i | n\rangle_{(x_i)}$ are the (electronic) wave functions of the fast motion for the fixed value $\vec{x}$ of the slow coordinate. The $\Phi_n^N(\vec{x}) = \langle x_i n | Nn\rangle$ are the wave functions of the slow motion for the fixed value ε_n of the fast hamiltonian.

We now start with (2.23) and calculate

$$\langle x_i r_i | H | N,n\rangle = E_{N,n} \; \langle x_i r_i | Nn\rangle \tag{2.28}$$

Using (2.1), (2.4) and (2.26) we obtain

$$\langle x_i r_i|(\frac{1}{2\mu}\vec{P}^2 + h(X_i))|N,n\rangle = E_{Nn}\,\psi_{Nn}\,(x_i,r_i) \tag{2.29}$$

Using for the ket $|N,n\rangle$ the expression (2.25) this becomes

$$\int \langle x_i r_i|\frac{1}{2\mu}\vec{P}^2 + h(X_i)\,|x_i',n\rangle\,\langle x_i'n|N,n\rangle\, d^3x' .$$

Now we use (2.20) and

$$\langle x_i r_i|\frac{1}{2\mu}\vec{P}^2|\psi\rangle = -\frac{1}{2\mu}\vec{\nabla}^2\,\langle x_i r_i|\psi\rangle$$

and (2.17). Then (2.29) becomes after using (2.27) on the r.h.s.:

$$(-\frac{1}{2\mu}\vec{\nabla}^2 + \varepsilon_n(x_i))\,\langle r_i|n\rangle_{x_i}\,\langle x_i n|Nn\rangle = E_{Nn}\,\langle r_i|n\rangle_{x_i}\langle x_i n|Nn\rangle \tag{2.30}$$

$\vec{\nabla}$ is the differentiation with respect to x_i and acts on both factors on the l.h.s. If $\langle r_i|n\rangle_{x_i}$ depends slowly upon x_i so that $\vec{\nabla}\langle r_i|n\rangle_{x_i}$ can be neglected then we obtain from (2.30):

$$(-\frac{1}{2\mu}\vec{\nabla}^2 + \varepsilon_n(x_i))\phi_n^N(\vec{x}) = E_{N,n}\,\Phi_n^N(\vec{x}) , \tag{2.31}$$

as the Schrödinger equation for the slow moving degrees of freedom. $\varepsilon_n(x_i)$ given by (2.17), or in terms of wave functions (using (2.26) and (2.27)) by

$$(-\frac{1}{2m}\vec{\nabla}_r^2 + V(\vec{x},\vec{r}))\,\phi_n(\vec{x},\vec{r}) = \varepsilon_n(\vec{x})\,\phi_n(\vec{x},\vec{r}) , \tag{2.32}$$

acts as effective potential for the slow motion described by (2.31). The electrons (or fast motion) remain in the n-th energy level $\varepsilon_n(\vec{x})$ while $\vec{x}$ slowly changes and $E_{N,n}$ takes the energy values $E_{N_1,n}$ $E_{N_2,n}$ ... with n-fixed.

We will now derive the equation that replaces (2.31) if one does not make the assumptions of the adiabatic limit. One can then no more introduce the basis system (2.22) and we denote the eigenvectors of H by ψ_n^E

$$H\,\psi_n^E = E\,\psi_n^E . \tag{2.33}$$

Each ψ^E corresponds to a given $|N,n\rangle$ of the adiabatic limit

$$\psi^E \leftrightarrow |N,n\rangle \tag{2.34}$$

It can be obtained from $|N,n\rangle$ by approximation methods and is again labeled by n,ψ_n^E. We will now assume that every eigenstate of H (energy state of the whole system (molecule)) belongs to a definite state of the fast (electronic motion). This means that to ψ_n^E only the n-th eigenvectors of $n(X_i)$ contribute:

$$\psi^E = \psi^E = \int d^3x \ |x_i,n\rangle \langle x_i,n|\psi^E\rangle \tag{2.35}$$

This is an approximation which is good for the ground state but not for the excited states of molecules. In general, for a given value E one has to take in (2.35) the sum over all n which will result in a non-Abelian gauge potential $\vec{A}_{n'n}$ in place of $\vec{A}_n$ of (2.48).

With (2.35) we calculate

$$\langle x_i,r_i|H|\psi_n^E\rangle = \int d^3x' \ \langle x_i,r_i| \ \frac{1}{2\mu}\vec{P}^2 + h(X_i)|\vec{x}'n\rangle \ \langle\vec{x}'n|\psi_n^E\rangle \quad .$$

For the l.h.s. we use (2.33) and on the r.h.s. we follow the procedure that led from (2.29) to (2.30) using (2.20). Then we obtain

$$E\langle\vec{r}|n\rangle_{\vec{x}} \ \langle\vec{x},n|\psi_n^E\rangle = (-\frac{1}{2\mu}\vec{\nabla}^2 + \varepsilon_n(\vec{x})) \ \langle\vec{r}|n\rangle_{\vec{x}} \ \langle\vec{x},n|\psi_n^E\rangle \quad . \tag{2.36}$$

This equation has the same form as (2.30). However we will now not make the assumption that $\langle\vec{r}|n\rangle_{\vec{x}}$ depends slowly upon $\vec{x}$. Then the differentiation $\vec{\nabla}$ on the r.h.s. is applied to both functions $\langle\vec{r}|n\rangle_{\vec{x}}$ and $\langle\vec{x},n|\psi_n^E\rangle$ of $\vec{x}$.

Before we proceed we follow the convention and introduce a vector, denoted $|n;\vec{x}\rangle$, in the space $\mathcal{H}^{fast}$. This vector can be defined using the reduced transition matrix element $\langle\vec{r}|n\rangle_{\vec{x}}$ and its properties can then be established from this definition. Vice versa, one can define it by its properties (and some phase conventions) and then establish its connection with the reduced matrix element. As conventionally one does the latter, we here follow the former procedure.

We define the vector $|n;\vec{x}\rangle$ of $\mathcal{H}^{fast}$ by

$$\langle\vec{r}|n\rangle_{\vec{x}} \equiv \langle\vec{r}|n;\vec{x}\rangle \quad . \tag{2.37}$$

The vector $|n;\vec{x}\rangle$ is, like the vector $|n\rangle$ of (2.22), a normalized vector and is not to be mistaken for $|\vec{x},n\rangle$ of (2.17), (2.18), which is a generalized basis vector of $\mathcal{H} = \mathcal{H}^{slow} \times \mathcal{H}^{fast}$. It is a vector labeled by the quantum number n (electronic energy quantum number or in general quantum number of the fast motion) and depends parametrically upon $\vec{x}$, the variable of the slow (collective) motion.

This vector has the following property:

$$h(\vec{X})|n;\vec{x}\rangle = h(\vec{x})|n;\vec{x}\rangle \tag{2.38}$$

where $h(\vec{X})$ is the operator $h(\vec{X})$ with the operator $\vec{X}$ taken at its values $\vec{x}$ ($h(\vec{x}')$ with $\vec{x}' \neq \vec{x}$ is not defined on $|n;\vec{x}\rangle$); it has also the property

$$e^{-i\vec{\phi}\vec{J}^{fast}} \ |n;\vec{x}_0\rangle = |n;\vec{x}\rangle \tag{2.39}$$

where $e^{-i\vec{\phi}\vec{J}^{fast}}$ is the operator that represents the transformation (rotation if $\vec{J}$ is angular momentum):

$$\vec{x}_0 \to \vec{x} = R(\vec{\phi})\vec{x}_0 \tag{2.40}$$

in the space $\mathcal{H}^{fast}$.

As in $\mathcal{H}^{slow}$ we have

$$e^{-i\vec{\phi}\vec{J}^{slow}} |\vec{x}_0\rangle = |\vec{x}\rangle \quad , \tag{2.41}$$

(2.39) means that the parametric dependence in $|n,\vec{x}\rangle$ of $\mathcal{H}^{fast}$ transforms in the same way as the generalized eigenvalues in $|\vec{x}\rangle$ of $\mathcal{H}^{slow}$.

[To establish these properties we calculate

$$\begin{aligned}
\varepsilon(\vec{x})\ \delta^3(\vec{x}' - \vec{x})\langle\vec{r}'|n;\vec{x}\rangle &= \langle\vec{x}'\vec{r}'|h(\vec{X})\ |\vec{x};n\rangle \\
&= \langle\vec{x}'\vec{r}'|h(\vec{X})\ |\vec{x}\rangle \times\ |n;\vec{x}\rangle \\
&= \langle\vec{x}'| \times \langle\vec{r}'|h(\vec{x})|\vec{x}\rangle \times |n;\vec{x}\rangle \\
&= \langle\vec{x}'|\vec{x}\rangle\ \langle\vec{r}'|h(\vec{x})|n;\vec{x}\rangle \quad .
\end{aligned}$$

Comparing the r.h.s. with the l.h.s. and using the completeness of $|\vec{r}'\rangle$ establishes (2.38).

The transformation $R(\vec{\phi})$ is represented in $\mathcal{H} = \mathcal{H}^{fast} \times \mathcal{H}^{slow}$ by

$$e^{-i\vec{\phi}\vec{J}} = e^{-i\vec{\phi}\vec{J}^{fast}} \times e^{-i\vec{\phi}\vec{J}^{slow}} \quad .$$

where J, J^{fast}, J^{slow} are the representatives of the generator of $R(\vec{\phi})$ in $\mathcal{H}$, $\mathcal{H}^{fast}$ and $\mathcal{H}^{slow}$, respectively. We calculate

$$\langle\vec{x}'|\vec{x}\rangle\ \langle\vec{r}'|n;\vec{x}\rangle = \langle\vec{x}'|\vec{x}\rangle\ \langle\vec{r}|n\rangle_{\vec{x}}$$

$$\begin{aligned}
\langle\vec{x}'\vec{r}'|\vec{x},n\rangle &= \langle\vec{x}'\vec{r}'|e^{-i\phi J^{fast}}\ |\vec{x}_0,n\rangle \\
&= \langle\vec{x}'|\times\langle\vec{r}'|e^{-i\phi J^{fast}}\ |\vec{x}_0\rangle \times |n;\vec{x}_0\rangle \\
&= \langle\vec{x}'|e^{-i\phi J^{slow}}|\vec{x}_0\rangle\ \langle\vec{r}'|e^{-i\phi J^{fast}}\ |n;\vec{x}_0\rangle \\
&= \langle\vec{x}'|\vec{x}\rangle\ \langle\vec{r}'|e^{-i\phi J^{fast}}\ |n;\vec{x}_0\rangle \quad .
\end{aligned}$$

Comparing the r.h.s. with the l.h.s. establishes (2.39).]

We now proceed with the calculation of (2.36). Multiplying (2.36) with $\overline{\langle\vec{r}|n\rangle_{\vec{x}}} = \overline{\langle\vec{r}|n;\vec{x}\rangle}$ and integrating over $\vec{r}$ using the fact that $\int d^3r\ \overline{\langle\vec{r}|n;\vec{x}\rangle}\ \langle\vec{r}|n;\vec{x}\rangle = 1$ we obtain

$$\int d^3r\ \overline{\langle\vec{r}|n;\vec{x}\rangle}\ (-\frac{1}{2\mu}\ \vec{\nabla}^2 + \varepsilon_n(\vec{x}))\langle\vec{r}|n;\vec{x}\rangle\ \langle\vec{x},n|\psi_n^E\rangle = E\langle\vec{x};n|\psi_n^E\rangle \tag{2.42}$$

where the differentiation $\vec{\nabla}$ is applied to the function $\langle\vec{r}|n,\vec{x}\rangle$ and to the function $\langle\vec{x},n|\psi_n^E\rangle$ of $\vec{x}$. We define the wave function

$$\psi_n^E(\vec{x}) = \langle\vec{x},n|\psi_n^E\rangle \quad . \tag{2.43}$$

Then (2.42) is rewritten as

$$\langle n;\vec{x}|(-\frac{1}{2\mu}\vec{\nabla}^2 + \varepsilon_n(\vec{x}))|n;\vec{x}\rangle\ \psi_n^E(\vec{x}) = E\psi_n^E(\vec{x}) \tag{2.44}$$

where $\vec{\nabla}$ acts on the vector $|n;\vec{x}\rangle$ which depends parametrically upon $\vec{x}$, and it also acts on the function $\psi_n^E(\vec{x})$ which stands to the right of it. Working this out in detail using the product rule one obtains

$$[-\frac{1}{2\mu}\vec{\nabla}^2 - \frac{2}{2\mu}(\langle n;\vec{x}|\vec{\nabla}|n;\vec{x}\rangle)\vec{\nabla} - \frac{1}{2\mu}\langle n;\vec{x}|\vec{\nabla}^2|n,\vec{x}\rangle$$

$$+ \varepsilon_n(\vec{x})]\ \psi_n^E(\vec{x}) = E\psi_n^E(\vec{x}) \tag{2.45}$$

We subtract and add in [ ] of (2.45) the two terms

$$-\frac{1}{2\mu}(\vec{\nabla}\langle n;\vec{x}|)(\vec{\nabla}|n,\vec{x}\rangle) - \frac{1}{2\mu}\langle n;\vec{x}|\vec{\nabla}|n;\vec{x}\rangle\ \langle n;\vec{x}|\vec{\nabla}|n;\vec{x}\rangle$$

then (2.45) can be written as

$$[-\frac{1}{2\mu}(\vec{\nabla} + \langle n;\vec{x}|\vec{\nabla}|n,\vec{x}\rangle)^2 + \tilde{V}_n(\vec{x})]\ \psi_n^E(\vec{x}) = E\psi_n^E(\vec{x}) \tag{2.46}$$

where $\tilde{V}_n(x)$ is defined by

$$\tilde{V}_n(\vec{x}) \equiv \varepsilon_n(\vec{x}) + \frac{1}{2\mu}((\vec{\nabla}\langle n;\vec{x}|)\vec{\nabla}|n;\vec{x}\rangle + (\langle n,\vec{x}|\vec{\nabla}^2|n,\vec{x}\rangle)^2)\ . \tag{2.47}$$

We define

$$\vec{A}_n(\vec{x}) \equiv i\langle n;\vec{x}|\vec{\nabla}|n;\vec{x}\rangle \tag{2.48}$$

then (2.46) take the desired form

$$[\frac{1}{2\mu}(\frac{1}{i}\vec{\nabla} - \vec{A}_n(\vec{x}))^2 + \tilde{V}_n(\vec{x})]\ \psi_n^E(\vec{x}) = E\psi_n^E(\vec{x})\ . \tag{2.49}$$

This is the Schrödinger equation for the wave function $\psi_n^E(\vec{x})$ of the slow motion (for every fixed value n of the fast motion). It does not only have a scalar potential $\tilde{V}_n(\vec{x})$ --which is given by (2.47)-- induced by the fast motion, but it also has a vector potential $\vec{A}_n(\vec{x})$ given by (2.48) which is also induced by the fast motion. The effective hamiltonian for slow motion is therefore given by (2.8) with the induced scalar potential $\tilde{\varepsilon}_n(\vec{x}) = \tilde{V}_n(\vec{x})$ given by (2.47). (Note the two terms added to $\varepsilon_n(\vec{x})$ as compared to (2.7').

III. THE SYMMETRIC TOP AND DUMBBELLS WITH AND WITHOUT SPIN

The particular form of the effective Hamiltonian for the collective motion depends upon the particular molecule that one considers. We give here examples of molecules that will be of relevance for later comparison with the charge-monopole system.

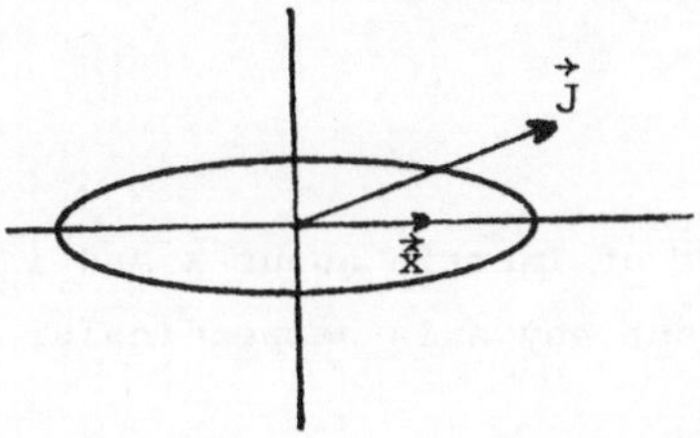

Figure III.1. Symmetric top (angular momentum $\vec{J}$ and symmetry axis $\hat{\vec{x}}$).

(a) The Rigid Symmetric Top.

Let the symmetry axis of the symmetric top be given by the unit vector $\hat{\vec{x}} = \frac{\vec{x}}{x}$ and let

I_A be the moment of inertia about $\hat{\vec{x}}$

I_B be the moment of inertia about any axis perpendicular to $\hat{\vec{x}}$.

When the symmetric top is rigid it can perform --in addition to the center of mass translations, which we ignore here-- only rotations about the center of mass. Let $\vec{\omega}$ be the angular frequency (with which every point of the top rotates), then the angular momentum of the symmetric top can be given as

$$\vec{J} = I_B\vec{\omega} + \frac{I_A - I_B}{I_A} (\hat{\vec{x}} \cdot \vec{J})\hat{\vec{x}} \tag{3.1}$$

and the Hamiltonian is given by

$$H = \frac{1}{2I_B} (\vec{J}^2 - \frac{I_A - I_B}{I_A} (\frac{\vec{x}}{x} \cdot \vec{J})^2) \tag{3.2}$$

$\hat{\vec{x}} \cdot \vec{J} = \frac{\vec{x}}{x} \cdot \vec{J}$ is the component of angular momentum along the body fixed symmetry axis $\hat{\vec{x}}$.

[Derivation of (3.1) and (3.2):

The unit vector $\hat{\vec{x}}$ rotates with the angular frequency $\vec{\omega}$, i.e.:

$$\dot{\hat{\vec{x}}} = \vec{\omega} \wedge \hat{\vec{x}} \tag{3.a$_1$}$$

$\vec{\omega}$ has a component along $\hat{\vec{x}}$ and a component perpendicular to $\hat{\vec{x}}$ which one can see by taking the cross product of (3.a$_1$) with $\hat{\vec{x}}$:

$$\hat{\vec{x}} \wedge \dot{\hat{\vec{x}}} = \vec{\omega} - \hat{\vec{x}}\,(\hat{\vec{x}} \cdot \vec{\omega}) \quad . \tag{3.a$_2$}$$

The angular momentum of any rigid body is given by

$$\vec{J} = I \cdot \vec{\omega} \tag{3.a$_3$}$$

where I is the tensor of inertia. Using for $\vec{\omega}$ (3.a$_2$) we can write this as

$$\vec{J} = I \cdot (\vec{x}(\vec{\omega} \cdot \hat{\vec{x}}) + \hat{\vec{x}} \wedge \dot{\hat{\vec{x}}})$$

$$= I_A(\vec{\omega} \cdot \hat{\vec{x}})\hat{\vec{x}} + I_B(\hat{\vec{x}} \wedge \dot{\hat{\vec{x}}}) \tag{3.a$_4$}$$

because I_A was the component of the moment of inertia about $\hat{\vec{x}}$ and I_B was the component of the moment of inertia about any axis perpendicular to $\hat{\vec{x}}$, that means in particular about $\hat{\vec{x}} \wedge \dot{\hat{\vec{x}}}$.

Inserting $\hat{\vec{x}} \wedge \dot{\hat{\vec{x}}}$ from (3.a$_2$) again into (3.a$_4$) we obtain

$$\vec{J} = I_B\vec{\omega} + (I_A - I_B)(\vec{\omega} \cdot \hat{\vec{x}})\hat{\vec{x}} \quad . \tag{3.a$_5$}$$

From (3.a$_3$) we obtain (making use of the fact that I_A is the component of I about $\hat{\vec{x}}$)

$$\hat{\vec{x}} \cdot \vec{J} = \hat{\vec{x}} \cdot I \cdot \vec{\omega} = I_A\, \hat{\vec{x}} \cdot \vec{\omega} \tag{3.a$_6$}$$

and inserting this into (3.a$_5$) we obtain the following general formula for the angular momentum of the symmetric top

$$\vec{J} = I_B\vec{\omega} + \frac{I_A - I_B}{I_A}\,(\hat{\vec{x}} \cdot \vec{J})\hat{\vec{x}} \quad . \tag{3.1}$$

The general formula for $\vec{\omega}$ expressed in terms of the angular momentum is obtained from this as:

$$\vec{\omega} = \frac{1}{I_B}\,\vec{J} - \frac{I_A - I_B}{I_A I_B}\,(\hat{\vec{x}} \cdot \vec{J})\hat{\vec{x}} \quad . \tag{3.a$_7$}$$

The energy of a rotating rigid body is in general given by

$$H = \frac{1}{2}\,\vec{\omega} \cdot \vec{J} = \frac{1}{2}\,\vec{\omega} \cdot I \cdot \vec{\omega}. \tag{3.a$_8$}$$

Multiplying (3.a$_7$) by $\vec{J}$ we obtain (after $\vec{x}$, $\hat{\vec{x}}$ and $\vec{J}$ are taken to be operators) for the energy operator of the symmetric top molecule

$$H = \frac{1}{2I_B}\,(\vec{J}^2 - \frac{I_A - I_B}{I_A}\,(\hat{\vec{x}} \cdot \vec{J})^2)$$

or

$$H = \frac{1}{2I_B}\,(\vec{J}^2 - \frac{I_A - I_B}{I_A}\,(\frac{\vec{x}}{x} \cdot \vec{J})^2) \tag{3.2}$$

(b) The Rigid Dumbbell With a Flywheel On Its Axis.

For the symmetric top the component of $\vec{J}$ along the axis $\hat{\vec{x}}$ can take many values (we will see which ones later). For certain cases it can be that this component has a fixed value. For instance, if we have a dumbbell that consists of two masspoints (the nuclei of a diatomic molecule) and an electron revolving rapidly about the dumbbell axis with fixed component of angular momentum Λ, then $\hat{\vec{x}} \cdot \vec{J} = \Lambda$. (This situation is realized in mole-

Figure III.2. Dumbbell with flywheel (consisting of masspoint m_2 and rapidly rotating beat m_1, I_A)

$$I_B = \mu x^2, \quad \mu = \frac{m_1 \cdot m_2}{m_1 + m_2}.$$

cules of the Hund type (a).) Another example, shown in Fig. III.2, which leads to the same model, is a dumbbell that does not consist of two masspoints but of beats (one beat and a masspoint or two beats), which rapidly spin about the axis $\hat{x}$. (Its relativistic generalization is probably the hadronic di-quarks.) Let us specialize the symmetric top to this particular case:

$$I_B = \mu x^2 \quad (\mu = \text{reduced mass}) \quad I_A << I_B \quad . \tag{3.3}$$

From

$$\dot{\vec{x}} = \vec{\omega} \wedge \vec{x}$$

it follows (by taking the cross product with $\vec{x}$) that

$$\vec{\omega} = \frac{1}{x^2} \vec{x} \wedge \dot{\vec{x}} + \frac{1}{x^2} (\vec{\omega} \cdot \vec{x})\vec{x} \quad . \tag{3.4}$$

Insert this into (3.1) and we obtain:

$$\vec{J} = I_B\left(\frac{1}{x^2} \vec{x} \wedge \dot{\vec{x}} + \frac{1}{x^2} \vec{x}(\vec{\omega} \cdot \vec{x})\right) + \frac{I_A - I_B}{I_A} \frac{\vec{x} \cdot \vec{J}}{x^2} \vec{x} \quad . \tag{3.5}$$

If we further use

$$\vec{\omega} \cdot \vec{x} = \frac{1}{I_A} \vec{x} \cdot \vec{J}$$

which follows from ($3.a_6$), then we obtain

$$\vec{J} = I_B \frac{1}{x^2} \vec{x} \wedge \dot{\vec{x}} + \frac{\vec{x} \cdot \vec{J}}{x^2} \vec{x} \tag{3.6}$$

Or if we use (3.3) we have

$$\vec{J} = \mu\vec{x} \wedge \dot{\vec{x}} + (\hat{x} \cdot \vec{J})\hat{x} \ , \qquad \hat{x} = \frac{\vec{x}}{x} \quad . \tag{3.7}$$

This equation holds as operator equation if the observables are quantum mechanical. With $\hat{x} \cdot \vec{J} = \Lambda$ we write this:

$$\vec{J} = \mu\vec{x} \wedge \dot{\vec{x}} + \Lambda\hat{x} \tag{3.7a}$$

For the Hamiltonian of this dumbbell with spin we obtain from (3.2):

$$H^{rot} = \frac{1}{2\mu x^2} (\vec{J}^2 - \Lambda^2) + \left(\frac{1}{2I_A} \Lambda^2\right) \quad .$$

The last term, $\frac{1}{2I_A}\Lambda^2$, is a constant for a given state of the fast motion (a given electronic state or a given state of rapid spin motion), and can be included in the energy ε_n of the fast motion and is therefore omitted. The Hamiltonian is then given as

$$H^{rot} = \frac{1}{2\mu x^2}(\vec{J}^2 - \Lambda^2) \quad . \tag{3.8}$$

For the special case that $\Lambda = 0$ (there is no flywheel on the axis or the beat is not spinning but is just a mass point) we obtain the usual rotator Hamiltonian

$$H^{rot} = \frac{1}{2\mu x^2}\vec{J}^2 \tag{3.9}$$

with the angular momentum perpendicular to the internuclear axis.

(c) The Vibrating and Rotating Dumbbell

In the examples that we have considered so far, the slow, collective, motion was a rigid rotation. We will now consider an example in which the slow collective motion consists of two "parts": rotation and vibration.

Figure III.3. Vibrating rotator consisting of masspoint m_2 and rapidly rotating beat m_1, I_A.

It is a diatomic molecule which can perform vibrations --or other radial motions-- and rotations (see Fig. III.3).

The energy of the radial vibrator (along the $\vec{x}$-direction) is given by

$$H^{vib} = \frac{1}{2\mu}P_r^2 + \frac{\mu\omega^2}{2}x^2 \qquad x^2 = \vec{x}^2 \tag{3.10}$$

where P_r is the radial momentum operator.[7)] It is given in terms of the canonical momenta P_i and position x_i by

$$P_r = \frac{1}{2}\{\frac{x_i}{x}, P_i\} = \frac{x_i}{x}P_i - \frac{i}{x} \tag{3.11}$$

(P_r is the component of $\vec{P}$ along the direction $\hat{x} = \vec{x}/x$; the anticommutator in (3.2) takes into account the non-commutativity of x_i and P_i. One can see that

$$\vec{P}^2 = P_r^2 + \frac{1}{x^2}\vec{L}^2 \; ; \qquad \vec{L} = \vec{x} \wedge \vec{p} \;) \; . \tag{3.12}$$

The radial momentum operator P_r is conjugate to the radius operator $x = (x_i x_i)^{\frac{1}{2}}$:

$$[x, P_r] = i1 \tag{3.13}$$

and $P_r^2/2\mu$ is the operator of kinetic energy of the radial motion.

The harmonic oscillator potential $(\mu\omega^2/2)x^2$ in (3.10) comes from the potential energy $\varepsilon_n(x)$ which was induced into the slow (nuclear) dynamics by the fast system (electrons) cf. (2.7'). (3.10) therefore describes an idealized situation, because in addition to the leading x^2 term in $\varepsilon_n(x)$, there will always be additional terms proportional to higher powers of x. So the radial motion will in general not be given by (3.10) but by

$$H^{\text{radial}} = \frac{1}{2\mu} P_r^2 + V(x) \tag{3.14}$$

where V(x) is different for different values of the "fast" quantum numbers n.

If the dumbbell also performs rotations about its center of mass with the energy operator for the angular motion given by (3.8) then the energy operator of the vibrating rotator is given by the sum of these two Hamiltonians (if one neglects the interaction between the vibrational and rotational motions):

$$H^{\text{dumbbell}} = \frac{1}{2\mu x^2}\left(\vec{J}^2 - \left(\frac{\vec{J}\cdot\vec{x}}{x}\right)^2\right) + \frac{P_r^2}{2\mu} + V(x) \; . \tag{3.15}$$

IV. THE SPECTRUM GENERATING GROUP OF THE DUMBBELL AND THE CHARGE-MONOPOLE SYSTEM--REPRESENTATIONS OF E(3), SO(3,1) and SO(4)

The observables X_i, J_i of the symmetric top and the dumbbell in section III fulfill the commutation relations

$$[J_i, J_j] = i\varepsilon_{ijk} J_k \tag{4.1}$$

$$[J_i, X_j] = i\varepsilon_{ijk} X_k \tag{4.2}$$

$$[X_i, X_j] = 0 \; . \tag{4.3}$$

These are the commutation relation (c.r.) of the group E(3). We will give a review of the representations of E(3) because their mathematical properties are relevant for the physical properties of the dumbbell and symmetric top molecule, as well as for the charge-monopole system. E(3) will turn out to be the Spectrum Generating Group of these physical systems. In fact there are three groups which could serve as a spectrum generating group;

these are E(3), SO(3,1) and SO(4). We will review their property together.

We will denote by SG any of these three groups, the defining c.r. of SO(3,1) and SO(4) differ from (4.1) ··· (4.3) by the c.r.

$$[X_i, X_j] = \pm i\lambda^2 \varepsilon_{ijk} J_k \tag{4.4$_\pm$}$$

in place of (4.3). In here λ is a real parameter with the same dimension as the X_i, the + on the r.h.s. is for SO(4) and the − for SO(3,1).

As SG contains $SO(3)_{J_i}$ as a subgroup, an irreducible representation (irrep.) space of SG can thus be reduced with respect to SO(3) and the angular momentum vectors $|jj_3\rangle$ with

$$\vec{J}^2 |jj_3\rangle = j(j+1)|jj_3\rangle \qquad J_3|jj_3\rangle = j_3|jj_3\rangle$$

are basis vectors of the irrep space.

The irreducible representations of SG (not only the unitary ones but a much larger class[10)]) are characterized by a pair of numbers (k,c) where

$$k = \text{integer or half-integer} \tag{4.5}$$

and represents the lowest value of j in the irrep (k,c). c can take any complex value. For unitary representations of E(3) we denote this pair of numbers by (k,r) where r is a real number. These numbers are connected with the eigenvalues of the Casimir operators of the groups in the following way:

$$\left(\frac{1}{\lambda^2} X_i X_i - J_i J_i\right)|(k,c)jj_3\rangle = (-k^2 - c^2 + 1)|(k,c),jj_3\rangle \tag{4.6}$$

$$\frac{1}{\lambda} X_i J_i |(k,c)jj_3\rangle = k\, ic|(k,c)jj_3\rangle \,. \tag{4.7}$$

For E(3):

$$X_i X_i |(k,r)jj_3\rangle = r^2|(k,r)jj_3\rangle \tag{4.8}$$

$$X_i J_i |(k,r)jj_3\rangle = k\, r|(k,r)jj_3\rangle \quad . \tag{4.9}$$

(The irrep (k,r) of E(3) is obtained from the irrep (k,c) of SO(3,1) by the contraction $\lambda \to 0$, $ic \to \infty$ such that $ic\lambda \to r$.) k completely specifies the reduction of the irreps of E(3) and SO(3,1) with respect to SO(3). This reduction is given by

$$\mathcal{H}(k,\cdot) \xrightarrow[SO(3)_{J_i}]{} \sum_{j=k,k+1\cdots}^{\infty} \oplus R^{(j)} \tag{4.10}$$

where • stands either for r in case of E(3) or for c in case of SO(3,1). The symbol $\xrightarrow[SO(3)]{}$ means that, when one restricts oneself to actions of the J_i then $\mathcal{H}(k,\cdot)$ becomes the direct sum on the r.h.s. of (4.10). This reduction is depicted by the K-type (K stands for maximal compact subgroup). The K-type is a diagram of dots in which each dot represents a value of j that occurs on the r.h.s. of (4.10). The value of j characterizes the irrep of K = SO(3) (represents angular momentum).

K-type is a mathematical property; to it corresponds in the physical interpretation the energy (or other) spectrum. To each dot in the K-type corresponds an energy level of the physical system that is described by the representation (k,•) of SG (or in other words whose space of physical states is $\mathcal{H}(k,\cdot)$).

(For SO(4) the sum on the r.h.s. of (4.10) is finite and the highest j is an integer or half integer k_1 and (k,k_1) characterize the irreps of SO(4). The K-type of Fig. IV.1 is also finite and there are a finite number of energy levels in Fig. IV.1(b).)

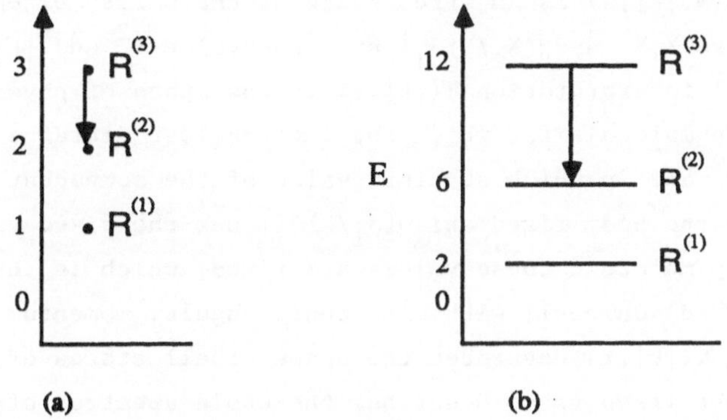

Figure IV.1: (a) K-type (weight diagram) of the irrep (k = 1,c) of SO(3,1) or of the irrep (k = 1,r) of E(3). (b) Energy diagram of the rigid rotator.

The energy values of e.g. the symmetric top with Hamiltonian (III,2) are obtained using (4.8) and (4.9) (assuming E(3) as the SG) as:

$$E_j = \frac{1}{2I_B}\left(j(j+1) - \frac{I_A - I_B}{I_A} k^2\right), \quad j = k,\ k+1,\ k+2 \cdots \tag{4.11}$$

The basis vectors $|(k,\cdot)jj_3\rangle$ are not eigenvectors of the parity operator U_p. Parity is defined in the usual way by:

$$\begin{aligned} U_p X_i U_p &= -X_i \\ U_p J_i U_p &= J_i \end{aligned} \tag{4.12}$$

from which follows that

$$U_p X_i J_i U_p = -X_i J_i \quad . \tag{4.13}$$

As a consequence of (4.13) and of (4.9) it follows that U_p transforms a vector with k into a vector with the value -k. U_p thus transforms from the space $\mathcal{H}(k,\cdot)$ to the space $\mathcal{H}(-k,\cdot)$ and in order to obtain a representation space of SG and parity one has to take the direct sum (parity doubling):

$$\mathcal{H}(|k|\cdot) \equiv \mathcal{H}(k,\cdot) \oplus \mathcal{H}(-k,\cdot) \quad . \tag{4.14}$$

A parity eigenvector is thus given by

$$||k|\cdot jj_3\pm\rangle = \frac{1}{\sqrt{2}}(|(k,\cdot)jj_3\rangle \pm |(-k,\cdot)jj_3\rangle) \ . \tag{4.15}$$

As (4.11) contains only k^2 each energy level is doubly degenerate (in addition to the usual $(2j + 1)$ degeneracy), except for the case $k = 0$.

The parity operator has the property

$$U_p||k|\ jj_3\pm\rangle = (-1)^j(\pm 1)||k|\ jj_3\pm\rangle \tag{4.16}$$

The space $\mathcal{H}(|k|,r)$ is an irrep space of the parity extended E(3). In it the operators X_iX_I and $|(X_i/X)J_i|$ have the value r^2 and $|k|$ respectively. In the physical interpretation $\mathcal{H}(|k|,r)$ is the space of physical states of a symmetric top molecule for which the internuclear distance has a fixed value $X_iX_i = r^2$ and in which absolute value of the component of angular momentum along the body fixed axis $|(X_i/X)J_i|$ has the fixed value k. If for a symmetric top molecule these values are fixed, which is the case for a rigid (r^2 = fixed) dumbbell with electronic angular momentum component $\Lambda = |k|$, then $\mathcal{H}(|k|,r)$ describes the space of all states of this molecule. The group, whose irrep space describes the whole spectrum of a physical system, has been called spectrum generating group.[8] SG = extended E(3) is thus the spectrum generating group of this molecule and the irrep space (4.14) is its space of physical states.

As the molecules come in parity eigenstates (4.16) the angular momentum component along $\vec{X}$, which is given by k, does not have a definite value. The irrep space $\mathcal{H}(k,r)$ of the unextended E(3) is not a space of physical states (except when $k = 0$), one must always take the direct sum (4.14) and the linear combination (4.15).

In place of the extended E(3) we could have considered the extended SO(3,1) or extended SO(4). It describes a physical system whose intrinsic positions (dipole operators) do not commute, $(4.4_\pm)$.[9]

The basis system $|(k\ r)jj_3\rangle$ of the irrep space $\mathcal{H}(k\ ,r)$ is one in which the following complete system of commuting observables (c.s.c.o.) is diagonal

$$J_3, \vec{J}^2, \quad X_iX_i, \quad X_iJ_i \quad . \tag{4.17}$$

Another basis system is given by the eigenvectors of the c.s.c.o.:

$$X_1, X_2, X_3, X_iJ_i \quad . \tag{4.18}$$

This basis system is denoted as

$$|x_i, k\rangle = |\hat{x}_i, (k, r)\rangle = |\theta\phi, (k, r)\rangle \tag{4.19}$$

in here $\hat{x}_i = x_i/r$ and $r = \sqrt{x_i x_i}$

$$x_1 = r\sin\theta\cos\phi; \quad x_2 = r\sin\theta\sin\phi; \quad x_3 = r\cos\theta$$

These basis vectors are usually normalized as

$$\langle k' x_i' | x_i k\rangle = \delta_{k'k}\ \delta^3(\vec{x}' - \vec{x}) \quad . \tag{4.20}$$

Whereas the phases for the basis vectors $|(k, r)jj_3\rangle$ are fixed (by convention for phase choices of matrix elements) the phases of the (generalized) eigenvectors (4.19) are unspecified. It is clear that every vector obtained from (4.19) by multiplying it with a phase factor that depends upon x_i or the polar angles θ, ϕ is also a system of eigenvectors with the same normalization (4.20).

The transition matrix elements between the basis vectors $|(k,r)jj_3\rangle$ and the basis vectors $|\theta\phi(k, r)\rangle$ depend upon these phase factors and choosing the phases of these matrix elements in a particular way fixes the phase factors of the basis system.

For one particular choice of phase factors the transition matrix elements are given by[11)]

$$\langle (k\ r)jm|\theta\phi(k,r)\rangle = \sqrt{\frac{2j+1}{4\pi}}\ \frac{1}{r}\ D^j_{mk}(\phi,\theta,-\phi) \tag{4.21}$$

where $D^j_{mk}(\phi,\theta,\psi)$ are the SU(2) representation functions.[12)]

After the phases of the basis system (4.19) are fixed by (4.21) (because the D function is completely fixed including phases) one can introduce a new set of basis vectors by e.g.:

$$|\theta\phi(k\ r)\rangle' = e^{-2ik\phi}|\theta\phi(k\ r)\rangle \quad . \tag{4.22}$$

For these basis vectors the transition coefficients are given by

$$\langle (k\ r)jm|\theta\phi(k\ r)\rangle' = \sqrt{\frac{2j+1}{4\pi}}\ \frac{1}{r}\ D^j_{mk}(\phi,\theta,\phi) \quad . \tag{4.23}$$

The basis system (4.22) is as good a system of (generalized) eigenvectors of the c.s.c.o. (4.18) as the original basis sytem (4.19). To see why we would use the one or the other, we will study the specific case $j = \frac{1}{2}$, $k = \pm\frac{1}{2}$, $m = \pm\frac{1}{2}$. For this specific case the D-function[12)]

$$D^{\frac{1}{2}}_{mk}(\phi,\theta,\psi) = e^{-im\phi}\ d^{j=\frac{1}{2}}_{mk}(\theta)e^{-ik\psi} \tag{4.24}$$

leads to the following values for the transition matrix elements (complex conjugate of the wave function):

$$\langle (k,r)\ j = \tfrac{1}{2} m | (k,r)\theta\phi\rangle = \begin{matrix} & k=\tfrac{1}{2} & k=-\tfrac{1}{2} \\ m=\tfrac{1}{2} & \cos\frac{\theta}{2} & -e^{-i\phi}\sin\frac{\theta}{2} \\ m=-\tfrac{1}{2} & e^{i\phi}\sin\frac{\theta}{2} & \cos\frac{\theta}{2} \end{matrix} \sqrt{\frac{2}{4\pi}}\,\frac{1}{r} \tag{4.25}$$

In the matrix on the r.h.s. of (4.25) m labels the rows and k the columns. For the wave function (4.23) one obtains instead

$$\langle (k,r)\ j = \tfrac{1}{2} m | (k,r)\theta\phi\rangle = \begin{pmatrix} e^{-i\phi}\cos\frac{\theta}{2} & -\sin\frac{\theta}{2} \\ \sin\frac{\theta}{2} & e^{i\phi}\cos\frac{\theta}{2} \end{pmatrix} \sqrt{\frac{1}{2\pi}}\,\frac{1}{r} \tag{4.26}$$

We first consider the representation space $\mathcal{H}(k = \frac{1}{2}, r)$. The wave functions of (4.25),

$$\begin{pmatrix} \cos\frac{\theta}{2} \\ e^{i\phi}\sin\frac{\theta}{2} \end{pmatrix}_{\theta=\pi} = \begin{pmatrix} 0 \\ e^{i\phi} \end{pmatrix} \tag{4.27}$$

are ill defined at $\theta = \pi$, because at the point $\theta = \pi$, ϕ can have any value. However the wave functions (4.26) are well defined at $\theta = \pi$, because

$$\begin{pmatrix} e^{-i\phi}\cos\frac{\theta}{2} \\ \sin\frac{\theta}{2} \end{pmatrix}_{\theta=\pi} = \begin{pmatrix} 0 \\ 1 \end{pmatrix} \tag{4.28}$$

and thus have one value at the point $\theta = \pi$.

On the other hand at $\theta = 0$ the wave functions of (4.26) are ill defined because

$$\begin{pmatrix} e^{-i\phi}\cos\frac{\theta}{2} \\ \sin\frac{\theta}{2} \end{pmatrix}_{\theta=0} = \begin{pmatrix} e^{-i\phi} \\ 0 \end{pmatrix} \tag{4.29}$$

However at $\theta = 0$ (4.25) is well defined because

$$\begin{pmatrix} \cos\frac{\theta}{2} \\ e^{i\phi}\sin\frac{\theta}{2} \end{pmatrix}_{\theta=0} = \begin{pmatrix} 1 \\ 0 \end{pmatrix} . \tag{4.30}$$

Thus near $\theta = 0$ one should use the basis vectors $|\theta\phi(k \geq 0, r)\rangle$ and near $\theta = \pi$ one should use the basis vectors $|\theta,\phi(k \geq 0,r)\rangle$. No basis system exists which is well defined over the whole sphere S_2.

For the representation space $(k = -\frac{1}{2}, r)$ the situation is the same so that one can define parity eigenvectors

$$||k|,r;jj_3\pm\rangle = \frac{1}{\sqrt{2}}\left(|(k,r)jj_3\rangle \pm |(-k,r)jj_3\rangle\right) \tag{4.31a}$$

and

$$||k|,r;jj_3\pm\rangle' = \frac{1}{\sqrt{2}}\left(|(k,r)jj_3\rangle' \pm |(-k,r)jj_3\rangle'\right) \tag{4.31b}$$

where the unprimed vectors (41.31a) are well defined on the upper half of the sphere and the primed vectors (4.31b) are well defined on the lower half of the sphere.

V. THE ELECTROMAGNETIC CHARGE-MONOPOLE SYSTEM

Here we review the properties of the charge-monopole system.[13] It consists of a non-relativistic charged particle of mass m_2 and electric charge e which is moving in the field of a magnetic monopole of mass m_1 and strength g.

The magnetic field of the monopole is given by

$$\vec{B} = \frac{g}{4\pi}\frac{\vec{x}}{x^3} \ . \tag{5.1}$$

The vector potential in the standard gauges can be defined by

$$A_x^{em} = 0 \quad A_\theta^{em} = 0 \quad eA_\phi^{em} = \frac{eg}{4\pi x}\frac{1-\cos\theta}{\sin\theta} \quad \text{for} \quad 0 \le \theta < \frac{\pi}{2} + \delta \tag{5.2a}$$

$$A_x'^{em} = 0 \quad A_\theta'^{em} = 0 \quad eA_\phi'^{em} = \frac{-eg}{4\pi x}\frac{1+\cos\theta}{\sin\theta} \quad \text{for} \quad \frac{\pi}{2} - \delta < \theta \le \pi \tag{5.2b}$$

$$\text{with} \quad 0 \le \delta \le \frac{\pi}{2} \ .$$

In the overlap region the two vector potentials differ by the gauge transformation

$$A_\ell - A'_\ell = \frac{2g}{4\pi x \sin\theta} \ . \tag{5.2c}$$

There are other forms of the vector potential that yield the field (5.1). We will always assume that the vector potential has the property

$$A_x \equiv \vec{A} \cdot \frac{\vec{x}}{x} = 0 \tag{5.3}$$

because a vector potential of the form $\vec{A} = A(x)\vec{x}$ is not of much interest as it would lead to a field $\vec{B} = \vec{\nabla} \wedge \vec{A} = 0$.

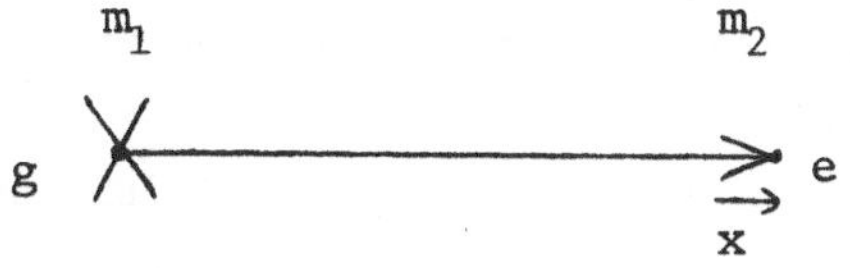

Figure V.1

With (5.1) the conventional Lorentz force equation reads

$$\mu\ddot{x}_i = \frac{eg}{4\pi} \varepsilon_{ijk} x_j \dot{x}_k \frac{1}{x^3} \quad . \tag{5.4}$$

We write this equation in terms of radial and angular variables. Using

$$x_i = x\hat{x}_i \qquad \hat{\vec{x}}^2 = 1$$

(5.4) is equivalent to

$$\ddot{x} = x \, \dot{\hat{x}}_i \dot{\hat{x}}_i \tag{5.5}$$

and

$$\frac{d}{dt} (\varepsilon_{ijk} x_j \mu \dot{x}_k - \frac{eg}{4\pi} \hat{x}_i) = 0 \ . \tag{5.6}$$

The radial equation (5.5) is the same as that of a free non-relativistic particle. If $g = 0$ then (5.6) gives the conservation of the usual orbital angular momentum. With $g \neq 0$ equation (5.6) says that the conserved angular momentum has the additional piece $\frac{eg}{4\pi} \hat{x}_i$ and is given by:

$$J_i = \varepsilon_{ijk} x_j \mu \dot{x}_k - \frac{eg}{4\pi} \hat{x}_i \quad . \tag{5.7}$$

Taking the scalar product of (5.7) with $\hat{x}_i$ gives the angular momentum component along the axis $\vec{x}$ which connects charge and monopole.

$$\frac{eg}{4\pi} = -\hat{x}_i J_i \quad . \tag{5.8}$$

The Dirac quantization condition specifies that

$$\frac{eg}{4\pi} = \text{integer or half integer} \quad .$$

This condition, first given by Dirac in 1931, can be derived in many different ways. The first derivation that we shall give here makes use of the correspondence between the charge monopole system and the dumbbell with spin of section III and uses the representation theory of E(3) of section IV which describes the dumbbell (and therewith also the charge-monopole) system. Later in section VI we shall present the usual topological considerations that lead to the Dirac quantization condition.

The preceding equations are valid for the classical charge-monopole system but also if x_i and J_i are operators. Then the position operator x_i and the operator J_i of (5.7) fulfill the commutation relations of E(3) given by (4.1) ··· (4.3). Thus the algebra of observables of the charge monopole system contains --like the algebra for the molecules in section III-- the (enveloping) algebra of E(3). We compare (5.7) with the angular momentum of the symmetric top, (3.1), or with the angular momentum of the dumbbell with rapidly rotating beats (3.7). In the representation space $\mathcal{H}(k,r)$ of E(3) $\hat{\vec{x}} \cdot \vec{J}$ becomes k and (3.7) takes the form

$$J_i = \mu(\vec{x} \wedge \dot{\vec{x}})_i + k\hat{x}_i \quad . \tag{5.9}$$

Comparing (5.9) with (5.7) we see that they are identical if

$$- \frac{eg}{4\pi} = k \tag{5.10}$$

k, the angular momentum of the rapidly rotating beat, characterizes the irreps of E(3). From the representation theory of the group E(3) follows that k can take only integer or half-integer values. Therewith we have seen that group-representation theoretical considerations lead to the Dirac quantization condition.

So far we have seen that the charge-monopole system and the dumbbell molecule with flywheel (or rotating beat) of section III have the same E(3) in common. We shall now show that there is a much closer relation between these two systems, as was already hinted at by the similarities between Figure V.1 and Figure III.2. For this purpose we compare their Hamiltonians.

The Hamiltonian of the charge-monopole system is given by

$$H = \frac{1}{2\mu} \vec{\pi}^2 \tag{5.11}$$

where

$$\vec{\pi} = \vec{p} - \frac{e}{c} \vec{A}^{e.m.} = \mu\dot{\vec{x}} \tag{5.12}$$

and $\vec{A}^{e.m.}$ is the vector potential of the monopole. p_i are the canonical momenta which together with x_i fulfill the c.r.

$$[x_i,x_j] = 0 \qquad [p_i,p_j] = 0 \qquad [p_i,x_j] = -i\delta_{ij} \tag{5.13}$$

The covariant momenta π_i defined by (5.12) fulfill in contrast the c.r.

$$[\pi_i,x_j] = -i\delta_{ij} \tag{5.14}$$

and

$$[\pi_i,\pi_j] = -i\epsilon_{ijk}eB_k = -i\epsilon_{ijk} \frac{eg}{4\pi} \frac{1}{x^2} \hat{x}_k = -i\epsilon_{ijk} \frac{1}{x^2} S_k^{int} \quad . \tag{5.15}$$

These commutation relations make sense only if $x^2 = r^2 \neq 0$; i.e. the representation spaces $(k_0, r = 0)$ are excluded. $S_i^{int} = \frac{eg}{4\pi} \hat{x}_i$ is the angular momentum of the electromagnetic field.

From (5.7) one calculates $\vec{J}^2$ in a straightforward calculation using the c.r. (5.14)

$$\vec{J}^2 = (\vec{x}^2\vec{\pi}^2 + 2i\vec{x}\vec{\pi} - x_j x_k \pi_k \pi_j + k_0^2) \quad . \tag{5.16}$$

Dividing this by $2\mu\vec{x}^2$ (= $2I_B$ of the dumbbell with rotating beat) then one obtains from (5.16)

$$\frac{1}{2\mu} \vec{\pi}^2 = \frac{1}{2\mu x^2} (\vec{J}^2 - k_0^2) - \frac{1}{2\mu x^2} (i\vec{x}\vec{\pi} - (\vec{x}\vec{\pi})^2) \quad . \tag{5.17}$$

As for the monopole potential we always assume that (5.3) holds, it follows from the definition of $\vec{\pi}$ in (5.12) that

$$\vec{x} \cdot \vec{\pi} = \vec{x} \cdot \vec{p} \tag{5.18}$$

so that the second bracket on the r.h.s. of (5.17) becomes

$$\frac{1}{2\mu x^2} (i\vec{x}\vec{\pi} - (\vec{x}\vec{\pi})^2) = \frac{1}{2\mu x^2} (i\vec{x} \cdot \vec{p} - (\vec{x} \cdot \vec{p})^2) \quad . \tag{5.19}$$

From the definition of the radial momentum operator (3.11) follows that

$$P_r^2 = \frac{1}{x^2} (\vec{x} \cdot \vec{p})^2 - i \frac{1}{x^2} (\vec{x} \cdot \vec{p}) \quad . \tag{5.20}$$

Inserting this for the second bracket on the r.h.s. of (5.17), we obtain for the Hamiltonian of the charge monopole system

$$H \equiv \frac{1}{2\mu} \vec{\pi}^2 = \frac{1}{2\mu x^2} (\vec{J}^2 - k_0^2) + \frac{1}{2\mu} P_r^2 \quad . \tag{5.21}$$

The Hamiltonian of the charge-monopole system is thus the same -- after the identification (5.10)-- as the dumbbell Hamiltonian in the representation space $\mathcal{H}(|k|,r)$, except that the charge-monopole Hamiltonian has in addition the kinetic energy term of the free radial motion.

$$H^{\text{free}} = \frac{1}{2\mu} P_r^2 \quad . \tag{5.22}$$

This term was expected to be present because of the free radial motion in the charge-monopole system given by equation (5.5). The charge-monopole system is thus a combination of two systems, the system of rotational motion and the system of free motion. It has, therefore, greater similarity with the rotating vibrating dumbbell of section III(c), which is also a combination of rotational and radial motion, only that the Hamiltonian of the radial motion is given by (3.10) or (3.14), and not by (5.22).

An example of a physical system that combines rotational motion of a charged monopole system with non-free radial motion is the dyon system.[6) A dyon system is a system of two particles of masses m_1 and m_2 (as shown in Figure V.1) each of which has an electrical charge e_1 and e_2 respectively and a magnetic monopole strength g_1 and g_2 respectively. The equation of motion for this dyon system is

$$\mu\ddot{x}_i = \frac{1}{4\pi} (e_1e_2 + g_1g_2) \frac{x_i}{x^3} - \frac{1}{4\pi} (e_1g_2 - e_2g_1) \epsilon_{ijk}x_j\dot{x}_k \frac{1}{x^3} \tag{5.23}$$

The Hamiltonian is then given by

$$H = \frac{1}{2\mu x^2} [\vec{J}^2 - k^2] + \frac{1}{2\mu} P_r^2 + \frac{1}{4\pi} (e_1e_2 + g_1g_2) \frac{1}{x} \quad . \tag{5.24}$$

This dyon Hamiltonian differs from the monopole Hamiltonian (5.21) by the Coulomb potential term for the radial motion. If the radial interaction would not be given by the Coulomb potential but by any other function, say $\varepsilon_n(x)$, then the Hamiltonian would be

$$H = \frac{1}{2\mu x^2}(\vec{J}^2 - k_0^2) + \frac{1}{2\mu} P_r^2 + \varepsilon_n(x) = \frac{1}{2\mu}\vec{\pi}^2 + \varepsilon_n(x) \quad . \tag{5.25}$$

In here $\vec{\pi}$ is the covariant derivative given by (5.12) with the vector potential $A^{e.m.}$ given by (5.2) in which eg is replaced by $(e_1g_2 - e_2g_1)$.

We have therewith seen that the charge-monopole system is identical with the molecular dumbbell that has a rapidly rotating beat, except that the molecular Hamiltonian for this case does not have the free Hamiltonian of the radial motion. Instead of the simple charge-monopole system one can consider a dyon like system (in which the radial interaction is not necessarily given by the Coulomb interaction but by an arbitrary $\varepsilon_n(x)$. Then its Hamiltonian is identical with the Hamiltonian of a dumbbell with one or two rapidly rotating beats which can also perform radial motion of a rather general kind, including radial oscillations.

What is the origin of these monopole or dyon like systems in molecular physics? In order to answer this question we will have to return to section II where we saw that, in general, the fast motion (of the electrons or of one electron or of the spin of one electron or of the rapidly rotating beat) induces a Berry connection $\vec{A}$ and a scalar potential $\varepsilon(x)$ into the slow, collective, motion. We therefore suspect that the fast motion for the dumbbell with flywheel will be of such kind that it induces a Berry connection of the form (5.2).

VI. THE BERRY CONNECTION FOR THE DUMBBELL WITH SPIN

In this section we want to compare the vector potential of the charge monopole system with the induced Berry connection for the dumbbell with flywheel. The fast motion of the dumbbell with flywheel is the electrons rapid revolving motion about the dumbbell axis or (for the half integer angular momentum) the spinning of the outer electron (or the spinning of one of the beats of the dumbbell). The slow degrees of freedom are the directions of the dumbbell axis $\vec{x} = (\theta,\phi)$. The slow changes in $\vec{x}$ are not supposed to cause transitions between the states with different electron angular momentum along the dumbbell axis.

The relevant part of the Hamiltonian for the fast motion $h(\vec{x})$ is given by

$$h(\vec{x}) = \mu\vec{x} \cdot \vec{J} \tag{6.1}$$

where $\vec{J}$ is the angular momentum operator and μ is a strength constant which is large (compared to the energy splitting caused by the slow motion).

The eigenvector of the Hamiltonian (6.1) depends in general upon $\vec{x} = (\phi,\theta)$.

$$h(\vec{x})\,|(k,r)\,;\,\theta,\phi\rangle = \mu k|(k,r);\theta,\phi\rangle \quad . \tag{6.2}$$

In here k, the eigenvalue of $\vec{x}\cdot\vec{J}$, takes the values $k = -\Lambda$ or $k = +\Lambda$. In particular if the flywheel effect comes from one spinning electron then $k = +\frac{1}{2}$ or $k = -\frac{1}{2}$.

In quantum mechanics a physical state is described by a ray of (unit) vectors differing by a phase factor. Thus every

$$|(k,r);\theta,\phi\rangle\; e^{-i\beta(\hat{x})} \qquad (\hat{x} = \theta,\phi)$$

with β a real function of the slowly varying parameters is an equally valid representative for the physical state.

The Berry connection is according to (2.9) and (2.9') given by

$$\begin{aligned} A^{kk'} &= i\langle(kr)\theta\phi|d|(k'r)\theta\phi\rangle \\ &= i\langle(kr)\theta\phi|\frac{\partial}{\partial\theta}\,|(k'r)\theta\phi\rangle d\theta + i\langle(kr),\theta,\phi|\frac{\partial}{\partial\phi}\,|(k'r)\theta\phi\rangle d\phi \\ &= A_\theta^{kk'}\,d\theta + A_\phi^{kk'}\,d\phi \quad . \end{aligned} \tag{6.3}$$

A change of the phase β,

$$|(kr)\theta\phi\rangle \rightarrow |(k\;r)\theta\phi\rangle\; e^{-i\beta(\theta,\phi)}$$

induces a gauge transformation in the Berry connection

$$\begin{aligned} A \rightarrow A' &= e^{i\beta(\hat{x})}\;\langle(k,r)\theta,\phi|(d|(k'r)\theta\phi\rangle\; e^{-i\beta(\theta,\phi)}) \\ &= i\langle(r,k)\theta\phi|d|(r,k')\theta\phi\rangle + ie^{i\beta}\langle(rk)\theta\phi|(rk')\theta\phi\rangle\; e^{-i\beta}\; d(-i\beta) \\ &= A + d\beta(\theta,\phi) \quad . \end{aligned} \tag{6.4}$$

We choose one particular vector of the ray for the case that $\vec{x}$ is pointing into the 3-direction $\theta = 0$, $\phi = 0$. This vector is $|(kr)00\rangle$. Let R denote the rotation that transforms $\begin{pmatrix}0\\0\\1\end{pmatrix}$ into $\vec{x}$:

$$R\begin{pmatrix}0\\0\\1\end{pmatrix} = \begin{pmatrix}\hat{x}_1\\\hat{x}_2\\\hat{x}_3\end{pmatrix} = \begin{pmatrix}\sin\theta\,\cos\phi\\\sin\theta\,\sin\phi\\\cos\theta\end{pmatrix} \tag{6.5}$$

Its inverse, which takes $\vec{x}$ into the 3-axis, is obtained as a rotation about the 3-axis by the angle $-\phi$, $R_3(-\phi)$, followed by a rotation about the 2-axis by an angle $-\theta$, $R_2(-\theta)$, and followed again by a rotation about the 3-axis by an angle ϕ, $R_3(\phi) = R_3^{-1}(-\phi)$:

$$R^{-1} = R_3^{-1}(-\phi)R_2(-\theta)R_3(-\phi) \quad . \tag{6.6}$$

In the space of the electronic states this rotation is represented by the operator[10)]

$$U(R) = U(\phi,\theta,-\phi) = e^{-i\phi J_3}\, e^{-i\theta J_2}\, e^{i\phi J_3} \tag{6.6'}$$

Thus

$$|(kr);\theta\phi\rangle = e^{-i\phi J_3}\, e^{-i\theta J_2}\, e^{i\phi J_3}\, |(kr);00\rangle \tag{6.7}$$

is one of the vectors which represents the state when $\hat{x} = \hat{x}(t)$ is pointing in the direction given by $(\theta(t),\phi(t))$. The parameters $\vec{x} = (\theta,\phi)$ in the fast vectors (6.7) are according to (2.39) the same as the generalized eigenvalues of the basis system (4.19) in the space of the slow motion.

This vector (6.7) is well defined at $\theta = 0$, it is however not well defined at $\theta = \pi$ (because for $\theta = \pi$ different values of ϕ represent the same direction $\vec{x}$). We will, therefore, use it only in the region which excludes $\theta = \pi$, i.e. for $0 \le \theta < \frac{\pi}{2}$. From the vector (6.7) we calculate the Berry connection using (6.3).

The calculation proceeds as follows (in place of (kr) we write just k)

$$\begin{aligned}
A_\theta^{kk'} &= \langle 00k| e^{-i\phi J_3} e^{i\theta J_2} e^{i\phi J_3}\, i\frac{\partial}{\partial\theta}\, e^{-i\phi J_3} e^{-i\theta J_2} e^{i\phi J_3} |k'00\rangle \\
&= e^{-ik\phi} e^{+ik'\phi}\, \langle 00k| e^{i\theta J_2}\, i\frac{\partial}{\partial\theta}\, e^{-i\theta J_2} |k'00\rangle \\
&= e^{-ik\phi} e^{ik'\phi} \langle 00k| e^{i\theta J_2} J_2 e^{-i\theta J_2} |k'00\rangle
\end{aligned}$$

$$A_\theta^{kk'} = e^{-ik\phi} e^{ik'\phi} \langle 00k| J_2 |k'00\rangle \ . \tag{6.8}$$

In the first line we have used the fact that

$$J_3 |k'00\rangle = \vec{J}_i \vec{x}_i |k'00\rangle = k' |k'00\rangle \ .$$

For the ϕ-component of the induced vector potential we calculate

$$\begin{aligned}
A_\phi^{kk'} &= \langle \theta\phi k| i\frac{\partial}{\partial\phi} |k'\theta\phi\rangle \\
&= \langle 00k\, e^{-i\phi J_3} e^{i\theta J_2} e^{i\phi J_3}\, i\frac{\partial}{\partial\phi}\, (e^{-i\phi J_3} e^{-i\theta J_2} e^{i\phi J_3}) |k'00\rangle \\
&= \langle 00k| e^{-i\phi J_3} e^{i\theta J_2} e^{i\phi J_3} \times \\
&\quad \times (J_3 e^{-i\phi J_3} e^{-i\theta J_2} e^{i\phi J_3} - e^{-i\phi J_3} e^{-i\theta J_2} e^{i\phi J_3} J_3 |k'00\rangle \\
&= \langle 00k| (e^{-i\phi J_3} e^{i\theta J_2} J_3 e^{-i\theta J_2} e^{i\phi J_3} - J_3) |k'00\rangle
\end{aligned} \tag{6.9}$$

We use

$$\begin{aligned}
e^{-i\phi J_3} e^{i\theta J_2} J_3 e^{-i\theta J_2} e^{i\phi J_3} &= U(\phi,-\theta,0) J_3\, U^{-1}(\phi,-\theta,0) \\
&= J_i R_{i3}\, (\phi,-\theta,0) \\
&= -\cos\phi\, \sin\theta\, J_1 - \sin\phi\, \cos\theta\, J_2 + \cos\theta\, J_3 \ ,
\end{aligned}$$

where we have used for the rotation matrix

$$R(\alpha,\beta,\gamma) = \begin{pmatrix} \cos\alpha\cos\beta\cos\gamma - \sin\alpha\sin\gamma & -\cos\alpha\cos\beta\sin\gamma - \sin\alpha\cos\gamma & \cos\alpha\sin\beta \\ \sin\alpha\cos\beta\cos\gamma + \cos\alpha\sin\gamma & -\sin\alpha\cos\beta\sin\gamma + \cos\alpha\cos\gamma & \sin\alpha\sin\beta \\ -\sin\beta\cos\gamma & \sin\beta\sin\gamma & \cos\beta \end{pmatrix} .$$

With this we obtain

$$A_\phi^{kk'} = \langle 0,0,k|-\sin\theta(\cos\phi\ J_1 + \sin\phi\ J_2) + (\cos\theta - 1)J_3|k'00\rangle \quad . \tag{6.10}$$

To calculate this further we need to know the matrix elements of the angular momentum operator in the fast (electronic) states $\langle 00k|J_i|k'00\rangle$.

We shall first discuss the case that the space of the slow motion is given by the irreducible representation space (k,r) of $E(3)_{X_iJ_i}$ where k has a fixed value k = (0, which is trivial) $+\frac{1}{2}$ or $-\frac{1}{2}$ or +1 or -1 ··· etc. Then the matrices $A^{kk'}$ are one-by-one and according to (6.8) and (6.10) given by

$$A_\theta = 0 \qquad \text{and} \qquad A_\phi = -k(1 - \cos\theta) \tag{6.11}$$

or

$$A = A_\theta d\theta + A_\phi d\phi = -k(1 - \cos\theta)d\phi \quad .$$

We will now consider the dumbbell with angular momentum component Λ along the dumbbell axis (diatomic molecule with a rapidly rotating electron (flywheel) or a spinning electron). The component of angular momentum along the dumbbell axis Λ in the space of slow collective motion is identical with the electron angular momentum $|k|$ along this axis. The space of the slow motion is given by (4.14)

$$\mathcal{H}(|k|,r) = \mathcal{H}(|k|,r) + \mathcal{H}(-|k|,r) \quad . \tag{6.12}$$

Thus k and k' in the preceding equations, in particular in (6.8) and (6.10) is $+|k|$ or $-|k|$.

We distinguish two cases

1) $|k| > \frac{1}{2}$; then $k,k' = \pm1, \pm3/2, \pm\cdots$

and the matric elements

$$\langle 00k|J_2\ \ |k00\rangle = 0 \quad . \tag{6.13}$$

Therefore we obtain

$$A_\theta^{kk'} = \begin{pmatrix} 0 & 0 \\ 0 & 0 \end{pmatrix} \tag{6.14}$$

$$A_\phi^{kk'} = |k| \begin{pmatrix} -(1-\cos\theta) & 0 \\ 0 & (1-\cos\theta) \end{pmatrix} \begin{matrix} k = |k| \\ k = -|k| \end{matrix} \qquad \text{columns: } k' = |k|,\; k' = -|k| \tag{6.15}$$

(k' labels the columns as indicated and k labels the rows).

2) For $|k| = \frac{1}{2}$ a special situation arises, because the matrix elements of J_2 and J_1 can be different from zero. k and k' are $+\frac{1}{2}$ or $-\frac{1}{2}$ and we parametrize

$$\langle 00;k|J_2|k'00\rangle = \kappa\tfrac{1}{2}\,\sigma_2 = \kappa\tfrac{1}{2}\begin{pmatrix} 0 & -i \\ i & 0 \end{pmatrix} \qquad \text{columns: } k' = \tfrac{1}{2},\; k' = -\tfrac{1}{2} \tag{6.16}$$

From (6.8) we then obtain for A_θ:

$$A_\theta^{kk'} = \tfrac{1}{2}\begin{pmatrix} 0 & -ie^{-i\phi}\kappa \\ ie^{i\phi}\kappa & 0 \end{pmatrix} \begin{matrix} k = \frac{1}{2} \\ k = -\frac{1}{2} \end{matrix} \qquad \text{columns: } k' = \tfrac{1}{2},\; k' = -\tfrac{1}{2} \tag{6.17}$$

And from (6.10) we obtain

$$A_\phi^{kk'} = \tfrac{1}{2}\begin{pmatrix} -(1-\cos\theta) & -\sin\theta\, e^{-i\phi}\kappa \\ -\sin\theta\, e^{i\phi}\kappa & (1-\cos\theta) \end{pmatrix} \tag{6.18}$$

For $\kappa = 1$ the matrices $\langle 0,0,k|J_i|k'00\rangle$ (of (6.16)) form the $j = \frac{1}{2}$ representation of the rotation group.

The Berry connections given above are in a particular gauge, which was fixed by the choice of the particular vector (6.7). The vector (6.7) was well defined in the region that excludes $\theta = \pi$, the Berry connections above are therefore well defined in the region

$$\mathcal{R} = \{(r\theta\phi)\,|\,0 < r;\ 0 \le \theta < \tfrac{\pi}{2} + \delta,\ 0 \le \phi < 2\pi\} \tag{6.19}$$

$$\text{where} \quad 0 < \delta \le \tfrac{\pi}{2} \ .$$

This is the same region in which the vector potential of the charge-monopole system is given by (5.2a). (5.2) gives the Cartesian coordinates of the vector potential. In order to compare the above results with it we have to use the Cartesian coordinates, $A_\theta^\perp$, $A_\phi^\perp$ which are defined by

$$A = A_\theta^\perp\, r d\theta + A_\phi^\perp\, r \sin\theta\, d\phi \ . \tag{6.20}$$

From (6.11) we therefore obtain

$$A_\theta^\perp = 0 \qquad A_\phi^\perp = -\frac{k(1-\cos\theta)}{r\sin\theta} \ . \tag{6.21}$$

This is identical with (5.2a) if one makes the identification (5.10):

$$-\frac{eg}{4\pi} = k \tag{5.10}$$

(6.11) and therewith (6.21) gives the Berry connection for the particular case in which the representation space is given by $\mathcal{H}(k,r)$ of $E(3)_{X_i J_i}$. The space of physical states of the dumbbell is given by (6.12) and the physical states are superpositions of eigenvectors with eigenvalue $|k|$ and $-|k|$. The vector potential of the charge monopole system is thus not exactly identical with the Berry connection of the dumbbell with flywheel, but only with the Berry connection of one half of the linear combinations of physical states. (Physical states of the dumbbell are parity eigenstates given by (4.31).)

The Berry connection (6.15) is then a generalization of the vector potential of the charge monopole system. The Cartesian components of (6.15) are given by

$$A_{\phi}^{\perp kk'} = |k| \begin{array}{c} \begin{array}{cc} k' = +|k| & k' = -|k| \end{array} \\ \begin{pmatrix} -\frac{(1-\cos\theta)}{r\sin\theta} & 0 \\ 0 & \frac{1-\cos\theta}{r\sin\theta} \end{pmatrix} \end{array} \begin{array}{l} k = +|k| \\ k = -|k| \end{array} \quad . \tag{6.22}$$

These are the components in the basis defined by (4.6) ··· (4.9). To obtain the matrix elements in the physical basis of parity eigenstates (4.15):

$$|(|k|r);\theta,\phi;\pi = \pm 1\rangle = \frac{1}{\sqrt{2}} \left(|(kr)\theta\phi\rangle \pm |(-k,r)\theta\phi\rangle\right) \tag{6.23}$$

we have to calculate

$$\langle\pi'|A|\pi\rangle = \sum_{\substack{k,k'= \\ \pm|k|}} \langle\pi'|k\rangle A^{kk'} \langle k'|\pi\rangle \tag{6.24}$$

where the transition matrix $\langle\pi|k\rangle$ is according to (6.23) given by

$$\langle k|\pi\rangle = \frac{1}{\sqrt{2}} \begin{array}{c} \begin{array}{cc} \pi = + & \pi = - \end{array} \\ \begin{pmatrix} 1 & 1 \\ 1 & -1 \end{pmatrix} \end{array} \begin{array}{l} k = +|k| \\ k = -|k| \end{array} \tag{6.25}$$

(π labels the columns and k the rows, as indicated).

In this way we obtain for the Cartesian components of the Berry connection

$$\langle\pi'|A_{\phi}^{\perp}|\pi\rangle = \begin{array}{l} \\ \pi' = + \\ \pi' = - \end{array} \begin{array}{c} \begin{array}{cc} \pi = + & \pi = - \end{array} \\ \begin{pmatrix} 0 & -\frac{(1-\cos\theta)}{r\sin\theta} \\ -\frac{(1-\cos\theta)}{r\sin\theta} & 0 \end{pmatrix} \end{array} \tag{6.26}$$

A similar calculation has to be done for (6.17) and (6.18) to obtain the matrices of the Berry connection in the parity basis for the case $|k| = \frac{1}{2}$.

If we choose instead of (6.7) a vector which differs from it by a phase factor, we obtain the Berry connection in a different gauge. As (6.7) is ill defined for $\theta = \pi$ we have to choose a different gauge for the region that includes $\theta = \pi$:

$$\mathcal{R}' = \{r\theta\phi | 0 < r;\ \frac{\pi}{2} - \delta < \theta \le \pi \qquad 0 \le \phi < 2\pi\} \qquad 0 < \delta \le \frac{\pi}{2} .$$

For this region it is conventional to choose instead of (6.7) the vector:

$$|(kr)\theta\phi\rangle' = e^{-i\phi J_3} e^{-i\theta J_2} e^{-i\phi J_3} |(kr),00\rangle = e^{-i2k\phi} |(kr),\theta,\phi\rangle \quad . \tag{6.7'}$$

This vector is well defined for $\theta = \pi$, because

$$|(kr)\ \theta = \pi,\phi\rangle' = e^{-i\phi k} e^{-i\phi J_3} e^{-i\pi J_2} |(kr)00\rangle$$
$$= e^{-i\phi k} e^{-i\phi J_3} |(-k,r)00\rangle$$
$$= e^{-i\phi k} e^{+i\phi k} |(-k,r)00\rangle = |(-k,r)00\rangle \ .$$

It is, however, not well defined at $\theta = 0$. The Berry connection for the region $\mathcal{R}'$ is given in terms of the vectors (6.7')

$$A'^{kk'} = i\ '\langle(kr)\theta\phi|d|(kr)\theta\phi\rangle' \quad .$$

A straightforward calculation like the one given above leads to the following expressions for the Berry connection. For the irreducible representation space $\mathcal{H}(k,r)$ we obtain

$$A'_\theta = 0 \qquad A'_\phi = k(1 + \cos\theta) \quad . \tag{6.11'}$$

For the dumbbell with spin in the space $\mathcal{H}(|k|,r)$ we have now in place of (6.14) and (6.15):

$$A_\theta'^{kk'} = 0 \tag{6.14'}$$

$$A_\phi'^{kk'} = |k| \begin{pmatrix} (1 + \cos\theta) & 0 \\ 0 & -(1 + \cos\theta) \end{pmatrix} \tag{6.15'}$$

And for the special case $|k| = \frac{1}{2}$ we obtain in place of (6.17) and (6.18)

$$A_\theta'^{kk'} = \frac{1}{2} \begin{pmatrix} 0 & -ie^{i\phi}\kappa \\ -ie^{-i\phi}\kappa & 0 \end{pmatrix} \tag{6.17'}$$

$$A_{\phi}^{'kk'} = \frac{1}{2} \begin{pmatrix} (1+\cos\theta) & -\sin\theta\, e^{i\phi}\kappa \\ -\sin\theta\, e^{-i\phi}\kappa & -(1+\cos\theta) \end{pmatrix} \tag{6.18'}$$

The Cartesian components of (6.11') are given by (cf. (6.20)):

$$A_{\theta}^{'\perp} = 0 \qquad A_{\phi}^{'\perp} = k\,\frac{1+\cos\theta}{r\sin\theta}\ . \tag{6.21'}$$

They are identical with the vector potential for the charge monopole system given by (5.2b).

We can again introduce a basis system of parity eigenvectors like (6.23) and calculate the matrices $\langle\pi'|A|\pi\rangle$. We do it this time for the case $|k| = \frac{1}{2}$ given by (6.17') and (6.18'). Using (6.24) we obtain:

$$'\langle\pi'|A_{\theta}|\pi\rangle' = \frac{1}{2}\kappa \begin{pmatrix} \sin\phi & i\cos\phi \\ -i\cos\phi & -\sin\phi \end{pmatrix} \tag{6.27}$$

and

$$'\langle\pi'|A_{\phi}|\pi\rangle' = \begin{pmatrix} -\sin\theta\cos\phi\,\kappa & (1+\cos\theta) \\ (1+\cos\theta) & \sin\theta\cos\phi\,\kappa \end{pmatrix} \tag{6.28}$$

As according to (6.7') the vectors $|(kr);\theta\phi\rangle$ and $|(kr);\theta\phi\rangle'$ differ by the phase factor $e^{-i\beta} = e^{-i2k\phi}$, the Berry connections differ according to (6.4) by the gauge transformation

$$A' = A + 2k\,d\phi\ . \tag{6.29}$$

The curvature two-form of A and A' is the same. It is obtained according to (2.12) from (6.11) or (6.11'):

$$B = dA' = dA = -k\sin\theta\, d\theta \wedge d\phi\ . \tag{6.30}$$

The integral divided by 2π of B over the sphere S^2 surrounding the monopole must be an integer (first Chern class) N:

$$N = \frac{1}{2\pi}\int B = -\frac{k}{2\pi}\int \sin\theta\, d\theta \wedge d\phi = -\frac{k}{2\pi}4\pi = -2k\ . \tag{6.31}$$

From this one obtains again (using (5.10)) the Dirac quantization condition

$$k = \text{integer or half-integer}\ . \tag{6.32}$$

We have thus obtained the same condition (6.32) and (4.5) from topological considerations and from the representation theory of the spectrum generating group.

We have seen in this section that the fast motion, which causes an angular momentum component along a given axis, induces into the slow motion of this axis a vector potential that is of the same form as the vector potential of a magnetic monopole or a dyon system. For a physical system whose Hamiltonian respects parity, the Berry connection is at least a 2×2 matrix like e.g. (6.15) or (6.26), the ordinary vector potential of the magnetic monopole (6.11) holds only in half of the space. The magnetic monopole and the dyon system are characterized by two specific forms of the induced scalar potential ((5.11) and (5.24) respectively), a molecule with Berry connection of the monopole type is unlikely to have such induced scalar potentials (Coulomb potentials) but has more general potential as in (5.25) which is predominantly of the oscillator type (3.10). As a consequence of the induction of the vector potential, the canonical momenta turn into covariant momenta π_i which fulfill anomalous commutation relations (5.15).

APPENDIX VI.

The Berry connection is often calculated in a different way using the wave functions (4.25) and (4.26).[14] We restrict ourselves here to the case $k = \frac{1}{2}$ and the region $\mathcal{R}$ of (6.19). In order to calculate (6.3) we insert a complete set of angular momentum basis vectors $|j = \frac{1}{2}, m(k,r)\rangle$ and obtain

$$\begin{aligned} A^{kk'} &= i \langle (k)\theta\phi | d | (k)\theta\pi\rangle \\ &= i \sum_m \langle (k)\theta\phi | \; j = \tfrac{1}{2} m\rangle \; d \; \langle j = \tfrac{1}{2} | m \; (k')\theta\phi\rangle \quad . \end{aligned} \tag{A.1}$$

The wave functions $\langle k\theta\phi | j = \frac{1}{2}\; m\rangle$ are according to (4.25):

$$\begin{aligned} \langle (k = \tfrac{1}{2}r)\theta\phi | j = \tfrac{1}{2}, m\rangle &= (\cos\tfrac{\theta}{2},\; e^{-i\phi}\sin\tfrac{\theta}{2}) \\ \langle (k = -\tfrac{1}{2}r)\theta\phi | j = \tfrac{1}{2}, m\rangle &= (-e^{+i\phi}\sin\tfrac{\theta}{2},\; \cos\tfrac{\theta}{2}) \end{aligned} \tag{A.2}$$

the first entry is for $m = +\frac{1}{2}$, the second for $m = -\frac{1}{2}$.

Writing (A.1) in matrix notation using the matrix elements on the r.h.s. of (4.25) we obtain

$$\begin{aligned} A^{k=\frac{1}{2},k'=\frac{1}{2}} &= i(\cos\tfrac{\theta}{2},\; e^{-i\phi}\sin\tfrac{\theta}{2})\; d \begin{pmatrix} \cos\tfrac{\theta}{2} \\ e^{+i\phi}\sin\tfrac{\theta}{2} \end{pmatrix} \\ &= i(\cos\tfrac{\theta}{2},\; e^{-i\phi}\sin\tfrac{\theta}{2}) \begin{pmatrix} -\sin\tfrac{\theta}{2}\, d\,\tfrac{\theta}{2} \\ e^{+i\phi}\cos\tfrac{\theta}{2}\, d\,\tfrac{\theta}{2} + ie^{+i\phi}\, d\phi\, \sin\tfrac{\theta}{2} \end{pmatrix} \end{aligned} \tag{A.3}$$

and consequently

$$A^{k=\frac{1}{2},k'=\frac{1}{2}} = i(-\cos\frac{\theta}{2}\sin\frac{\theta}{2}\, d\,\frac{\theta}{2} + \sin\frac{\theta}{2}\cos\frac{\theta}{2}\, d\,\frac{\theta}{2} + i\sin^2\frac{\theta}{2}\, d\phi)$$

$$= -\sin^2\frac{\theta}{2}\, d\phi = -\tfrac{1}{2}(1-\cos\theta)d\phi\ . \qquad (A.4)$$

As a second example we calculate

$$A^{k=-\frac{1}{2},k'=\frac{1}{2}} = i(-e^{i\phi}\sin\frac{\theta}{2},\ \cos\frac{\theta}{2})\ d\begin{pmatrix}\cos\frac{\theta}{2}\\ e^{i\phi}\sin\frac{\theta}{2}\end{pmatrix}$$

$$= i(-e^{i\phi}\sin\frac{\theta}{2};\ \cos\frac{\theta}{2})\begin{pmatrix}-\sin\frac{\theta}{2}\, d\,\frac{\theta}{2}\\ +ie^{i\phi}\, d\phi\,\sin\frac{\theta}{2} + e^{i\phi}\cos\frac{\theta}{2}\, d\,\frac{\theta}{2}\end{pmatrix}$$

$$= ie^{i\phi}\,\tfrac{1}{2}\, d\theta - e^{i\phi}\, d\phi\,\tfrac{1}{2}\sin\theta$$

The result of all calculations (for the region $\theta \neq \pi$) is summarized in the following matrices.

$$\begin{matrix} & k'=\frac{1}{2} & k'=-\frac{1}{2} & \\ A_\theta^{kk'} = \frac{1}{2} & \begin{pmatrix} 0 & -ie^{-i\phi} \\ ie^{i\phi} & 0 \end{pmatrix} & & \begin{matrix} k=\frac{1}{2} \\ k=-\frac{1}{2}\end{matrix}\end{matrix} \qquad (A.5)$$

$$A_\phi^{kk'} = \frac{1}{2}\begin{pmatrix} -(1-\cos\theta) & -\sin\theta\, e^{-i\phi} \\ -\sin\theta\, e^{i\phi} & (1-\cos\theta)\end{pmatrix} \qquad (A.6)$$

where A_ϕ and A_θ are defined by

$$A = A_\phi\, d\phi + A_\theta\, d\theta \qquad (A_r = 0)\ .$$

The Cartesian coordinates $A^\perp$, defined by

$$A = A_\theta^\perp\, r d\theta + A_\phi^\perp\, r\sin\theta\, d\phi$$

are then given by, e.g.:

$$A_\theta^{\perp kk'} = \frac{1}{2r}\begin{pmatrix} 0 & -ie^{-i\phi} \\ +ie^{i\phi} & 0\end{pmatrix} \qquad (A.5a)$$

$$A_\phi^{\perp kk'} = \frac{1}{2r}\begin{pmatrix} \frac{(1-\cos\theta)}{\sin\theta} & e^{-i\phi} \\ -e^{i\phi} & \frac{1-\cos\theta}{\sin\theta}\end{pmatrix} \qquad (A.6a)$$

The results (A.5) and (A.6) are in agreement with the expressions (6.17) and (6.18) derived above for $\kappa = 1$.

VII. CONCLUSION

On the occasion of the 50th anniversary of the introduction by Dirac of the concept of the magnetic monopole he wrote[15]: "I am inclined now to believe that monopoles do not exist." The electromagnetic type of monopoles and dyons have indeed never been seen. But from our above considerations we know now that physical systems that have the same dynamics as monopoles do occur in nature. Often a complicated physical system is naturally and conveniently divided into two parts. This dissection is usually not in terms of constituents but in terms of motions. An example is the rapid motion of the electrons in a molecule and the slow collective motion of the molecule as a whole (the motion of the nuclei). If the electronic motion gives rise to an angular momentum component along the body axis then the dynamics of the body as a whole is determined --among others-- by a monopole vector potential. As a consequence of this the intrinsic momenta π_i do not commute, (5.15).

The rapid rotation around the axis does not have to come from electronic motion. A possible physical system which gives rise to the same intrinsic dynamics is a thin tube in which two (or one, or more) beats ("dyons") perform rapid rotation about the tube axis with value of angular momentum about the tube axis given by k. The collective motion of the tube as a whole is then described by intrinsic variables which obey anomalous commutation relations of the kind given by (5.15).

We have considered in these lectures only the effect of the rapid rotation in position space. Rapid rotation can also occur in momentum space (Zitterbewegung), in which case position is replaced by momentum in (6.1).[16] One would, therefore, also expect non-commuting intrinsic position operators.

A relativistic model with these properties has recently been suggested for the description of hadron spectra and hadron structure.[17] In this case the relativistic extended object is a flux tube in which the beats are given by quark and antiquark (for mesons) and quark and diquark (for baryons). The relativistic vibrations and rotations of the flux tube are described in analogy to the molecular collective motion. If one proceeds in complete analogy to the naive Born-Oppenheimer approximation one will be led to the usual canonical commutation relations (of the relativistic string) for the intrinsic position operators ξ_μ ($\mu = 0,1,2,3,\cdots$) and the intrinsic momentum operators π_μ. If one takes into account that the rapid spinning motion of the quarks (dyons) can induce Berry connections one would expect that the momentum and position operators of the flux string do not commute. The commutation relations that have been suggested are the following:

$$[\xi_\mu,\xi_\nu] = -\frac{i}{(Mc)^2}\Sigma_{\mu\nu} , \qquad (7.1a)$$

$$[\pi_\mu,\pi_\nu] = -\frac{i}{(\alpha' Mc)^2}\Sigma_{\mu\nu} , \qquad (7.1b)$$

$$[\xi_\mu,\pi_\nu] = -i\,\breve{g}_{\mu\nu}\frac{2McH}{\alpha'(Mc)^2} , \qquad (7.1c)$$

$$[\Sigma_{\mu\nu},\Sigma_{\rho\sigma}] = -i(\breve{g}_{\mu\rho}\Sigma_{\nu\sigma} + \breve{g}_{\nu\sigma}\Sigma_{\mu\rho} - \breve{g}_{\mu\sigma}\Sigma_{\nu\rho} - \breve{g}_{\nu\rho}\Sigma_{\mu\sigma}) . \qquad (7.2)$$

In here $\Sigma_{\mu\nu}$ is the operator of the spin tensor, P_μ is the center of mass momentum, $M = (P_\mu P^\mu)^{\frac{1}{2}}$, $\hat{P}_\mu = P_\mu M^{-1}$, $\breve{g}_\mu{}^\rho = \eta_\mu{}^\rho - \hat{P}_\mu \hat{P}^\rho$ (projector onto the plane perpendicular to P_μ), H is the relativistic Hamiltonian, $\frac{1}{2\pi\alpha'}$ is the string tension which is a phenomenological parameter and c is the velocity of light.

The analogy between (7.1b) and (5.24) is readily seen and so is the reciprocity of π_μ and ξ_μ. If this model is successful (present results for hadron spectra and transition rates are encouraging), then not only molecules but also hadrons "contain" monopoles, albeit not of the type that had been conceived by Dirac.

ACKNOWLEDGEMENT

I should like to thank Y. Aharanov, Y. Dothan, R. Levine, and participants of the Workshop on Algebraic Methods in Molecular Physics for valuable discussions of subjects related to this paper. I would also like to express my gratitude to L. Boya and T. Imbo for reading the manuscript and for suggesting improvements.

REFERENCES

1. M. Born and J. Oppenheimer, Ann. Phys. 84:457 (1927).
2. G. Herzberg and M.C. Longuet-Higgins, Discuss.Faraday Soc.35:77 (1963); C.A. Mead and D.G. Truhlar, J. Chem. Phys. 70:2284 (1979); C.A. Mead, Chem. Phys. 49:33 (1980)
3. M. Berry, Proc. Roy. Soc.A392:45 (1984); F. Wilczek and A. Zee, Phys.Rev. Lett. 52:2111 (1984).
4. R. Jackiw, Intern. Journ. Mod. Phys. A3:285 (1988).

5. J. Moody, A. Shapere, and F. Wilczek, Phys. Rev. Lett. 56:893 (1986).
6. J. Schwinger, Science 165:757 (1969); D. Zwanziger, Phys. Rev. 176: 1480 (1968); D6:458 (1972).
7. See e.g., A Bohm, Quantum Mechanics, Sect. VII.2, Springer, New York (1986).
8. A. O. Barut and A. Bohm, Phys. Rev. 139B:1107 (1965); Y. Dothan, M. Gell-Mann, and Y. Ne'eman, Phys. Rev. Lett. 17:148 (1965).
9. Though these cases have been discussed in the literature, e.g., F. Iachello, R. D. Levine, et al., J. Chem. Phys. 77:3047 (1982), we shall not consider them here.
10. M. A. Naimark, Linear Representations of the Lorentz Group, New York (1964); A. Bohm, Quantum Mechanics (2nd edition), Appendix V3.
11. For a derivation see A. Bohm and R. B. Teese, J. Math. Phys. 17:94 (1976), Appendix A.
12. E.g., L. C. Biedenharn and J. D. Louck, Angular Momentum in Quantum Physics, Addison-Wesley (1981), Chapter 3.
13. E.g., P. Goddard and D. I. Olive, Rep. Prog. Phys. 41:1357 (1978); S. Coleman, (Erice lectures 1981) The Unity of the Fundamental Interactions, H. Zichichi editor, Plenum Press (1983).
14. M. Stone, Phys. Rev. D33:1191 (1986).
15. P. Dirac, Letter to Abdus Salam of 11 November 1981.
16. Y. Aharanov, Private communications.
17. A. Bohm, P. Kielanowski, M. Kmiecik, and M. Loewe, Tel Aviv University preprint TAUP 1637-88.

SYMMETRY MODELS IN SOLID STATE CHEMISTRY

William O.J. Boo

Chemistry Department
University of Mississippi
University, MS 38677

One aspect of solid state chemistry which has flourished in recent years is the ability of chemists to synthesize new and exciting materials. Unfortunately, the synthesis of most advanced materials has been directed toward applications; as a result, solid state chemistry has become a collection of unrelated topics. Furthermore, chemistry departments have been slow in accepting the solid state as a part of chemistry. Part of the problem is that chemists are obsessed with molecules. Even the models of crystal structures which have been used for almost 75 years are an outgrowth of molecular models (i.e. ball-and-stick models). Our laboratory, has devoted many years to the systematic preparation and characterization of fundamental inorganic systems and we have concluded that the structural behavior of non-molecular solids may be explained better through symmetry models. Our models employ two aspects of symmetry: (1) fundamental structures are represented by tilings and/or packings and (2) dynamic behaviors involving structural changes are governed by laws of symmetry. Results which support these conclusions are presented in this paper.

It is well known to solid state scientists that when crystals of complex composition are cooled, concomitant ordering often occurs; and as this ordering sets in, the symmetry of the crystal is lowered. Perhaps the best known example is cooperative magnetic ordering in paramagnetic crystals. The antiferromagnetic ordering of V^{2+} ions in KVF_3[1,2] is illustrated on Figure 1. Jahn-Teller cooperative ordering is another common example. When Cr^{2+} and Cu^{2+} ions are octahedrally coordinated, they become "Jahn-Teller ions". Compounds containing these nonspherical ions will usually have a critical temperature below which they become cooperatively oriented. Jahn-Teller ordering of Cr^{2+} ions in $KCrF_3$[3] is illustrated on Figure 2. Both KVF_3 and $KCrF_3$ have the cubic perovskite structure at high temperatures; but below their respective ordering temperatures, their structures are tetragonal.

Many other kinds of concomitant ordering occur in solids. In first-row transition metal fluorides, these become increasingly abundant as one progresses from binary, to ternary, to quaternary component systems. Ordering events are usually accompanied by other phenomena such as nonstoichiometry, solid-solid phase transitions, multiphases, modulations and superstructures, small crystal distortions and domain structures. These complications make it difficult, if not impossible, to study ordered structures by single crystal diffraction techniques. As a consequence, these inte-

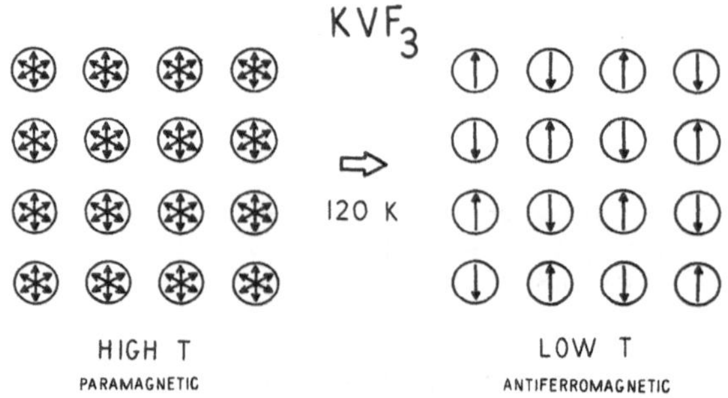

Figure 1. As KVF_3 is cooled, magnetic ordering sets in at 120 K.

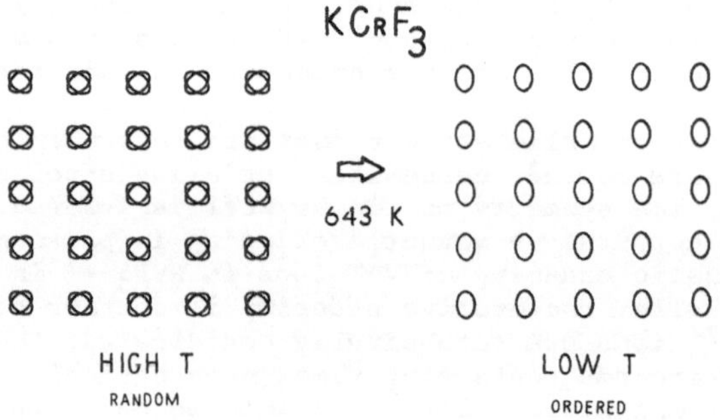

Figure 2. As $KCrF_3$ is cooled, Jahn-Teller cooperative ordering sets in at 643 K.

resting materials, and their dramatic structural properties, have been mostly ignored.

We have applied space-age technology to systematically synthesize and study fundamental solid state systems. This includes ultra pure synthesis employing vacuum encapsulation by electron beam welding techniques, high precision X-ray powder diffraction, magnetic measurements from 4 to 300 K, and thermal studies up to 1500 K.

Our efforts have focused on an interesting group of compounds of general formula $A^{I}_{x}M^{II}_{x}M^{III}_{1-x}F_3$ (where A = Na, K, Rb, or Cs; x = 0 to 1; and M = V, Cr, Mn, Fe, Co, Ni, Cu, or Zn). This general description includes the binary compounds $M^{III}F_3$ (x = 0), the ternary compounds $AM^{II}F_3$ (x = 1), and a variety of quaternary compounds (1 > x > 0). These compounds are often described as perovskite-like because they consist of MF_3 networks of octahedra (usually corner sharing) with A^+ ions located in holes within the network.

In ionic compounds, anions (e.g. F^-) usually pack together like spheres; whereas, cations (e.g. M^{2+} and M^{3+}) tend to distance themselves as far from each other as possible. These experimental facts are the basis of Pauling's rules.[4] Pauling did not have access to present day synthetic materials or he probably would have concluded, as we have, that there are two fundamental structural types. In Type I structures, polygons are formed by connecting the centers of neighboring anions with line segments. The polygons form either tilings (2 dimension) or packings (3 dimension), or both. In Type II structures, the geometrical forms are defined by cations. "Ideal" two dimensional structures include those in which regular, quasi-regular, or semi-regular tilings are formed.[5] "Ideal" three dimensional structures include those in which ideal packings are formed (packings which consist of regular, quasi-regular, and/or semi-regular polyhedra).[5] There are also Type I,II structures in which both sets of conformations are defined.

All of the ideal packings have cubic symmetry. The ideal tilings are either hexagonal or square. Furthermore, if layers of atoms with hexagonal or square symmetry are stacked, they form three dimensional hexagonal and tetragonal systems, respectively (in some instances, stacking also generates cubic systems). We conclude, therefore, that three ideal crystal systems exist in nature: cubic, hexagonal, and tetragonal (the latter two qualify only if they are layer structures). All of the structures of the $A^{I}_{x}M^{II}_{x}M^{III}_{1-x}F_3$ compounds fall into these categories.

At high temperatures, the MF_3 compounds have the ReO_3 structure; $NaMF_3$, KMF_3, and $RbMF_3$ have the cubic perovskite ($CaTiO_3$) structure. Except for the A^+ ions, the MF_3 and AMF_3 structures are identical and are both two and three dimensional Type I,II structures (Figure 3). At high temperatures, the $A^{I}_{x}M^{II}_{x}M^{III}_{1-x}F_3$ compounds (A = K, Rb, or Cs; x = 0.20 - 0.30) have the hexagonal tungsten bronze (A_xWO_3) structure, which is a two dimensional Type I,II structure (Figure 4); the $K^{I}_{x}M^{II}_{x}M^{III}_{1-x}F_3$ compounds (x = 0.40 - 0.60) have the tetragonal tungsten bronze (K_xWO_3) structure, which is a two dimensional Type I,II structure (Figure 5); and the $A^{I}_{x}M^{II}_{x}M^{III}_{1-x}F_3$ compounds (A = Rb, or Cs; x = 0.40 - 0.60) have the cubic modified pyrochlore structure, which is a three dimensional Type II structure (Figure 6). As these compounds are cooled, various kinds of concomitant ordering set in which, in every case, lowers their symmetries.

Joe Rosen states in his book: "under the correspondence (degree of symmetry <---> entropy) the special symmetry evolution principle and the second law of thermodynamics are isomorphic".[6] Hence, symmetry and entropy

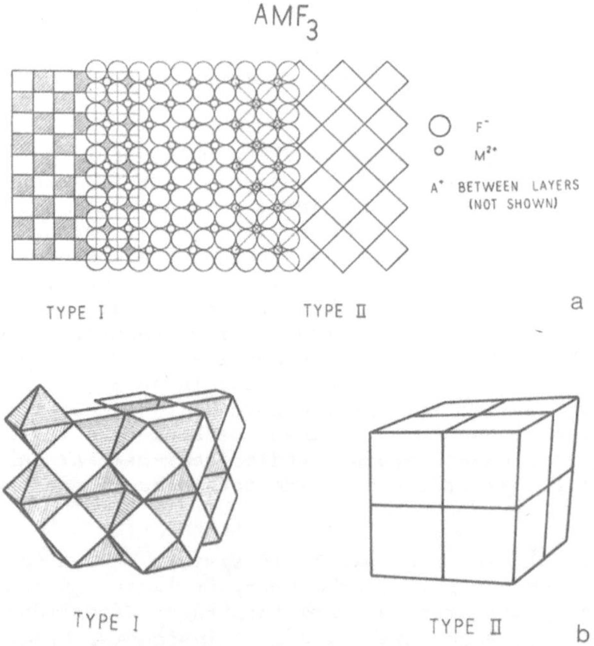

Figure 3. The perovskite structure. (a) As a two-dimensional TYPE I,II structure, the F^- ions are located at the vertices of the quasi-regular tiling (TYPE I) and the M^{2+} ions are located at the vertices of the regular tiling (TYPE II). (b) As a three-dimensional TYPE I,II structure, the F^- ions are located at the vertices of a packing of octahedra and cuboctahedra (TYPE I), and the M^{2+} ions are located at the vertices of a packing of cubes (TYPE II).

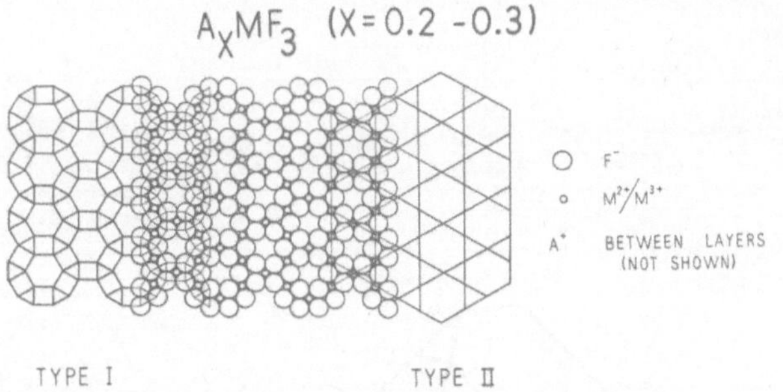

Figure 4. The hexagonal tungsten bronze structure as a two-dimensional TYPE I,II structure.

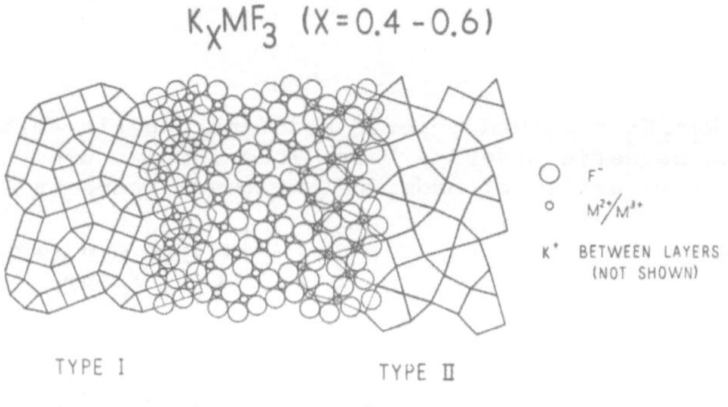

Figure 5. The tetragonal tungsten bronze structure as a two-dimensional TYPE I,II structure.

describe the same phenomena in nature. Having established this relationship, we have devised three useful rules about symmetry and order.

(1) In nature, a system will assumed the most symmetrical configuration possible.
(2) Symmetry and randomness are directly related whereas symmetry and order are opposites.
(3) The symmetry of a real system is usually lowered as its temperature is decreased.

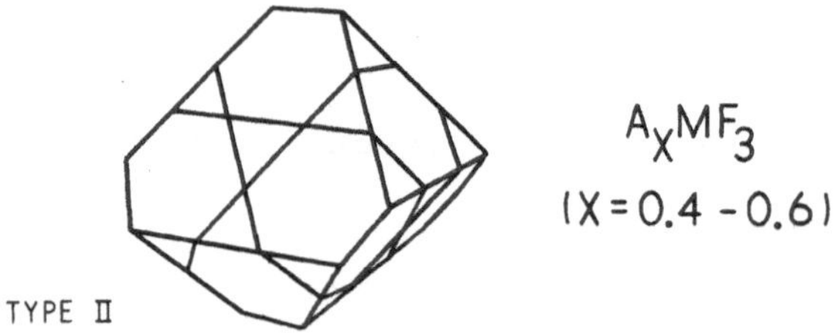

Figure 6. The modified pyrochlore structure. As a three-dimensional TYPE II structure, the M^{2+} and M^{3+} ions are located at the vertices of a packing of tetrahedra and truncated tetrahedra.

In the $A^{I}_{x}M^{II}_{x}M^{III}_{1-x}F_3$ compounds, ionic ordering usually sets in above room temperature, and magnetic ordering below room temperature. The following lists summarize the variety of ordering phenomena found in these compounds:

Ionic ordering

(1) ordering of partially filled A^+ sites

(2) electronic ordering (M^{2+} and M^{3+} are homonuclear)

(3) ordering of cations (M^{2+} M^{2+} are not homonuclear)

(4) Jahn-Teller ordering

Magnetic ordering

Long range

(1) antiferromagnetic
(2) ferrimagnetic
(3) ferromagnetic
(4) metamagnetic

Short range

(1) linear chains
(2) dimers
(3) trimers

Examples of ionic ordering are illustrated on Figures 2, 7, 8, and 9. Numerous examples of magnetic ordering have been described elsewhere.[7-14]

There are also solid-solid phase changes other than those involving ordering phenomena. We have narrowed the list of transitions observed in the $A^{I}_{x}M^{II}_{x}M^{III}_{1-x}F_3$ compounds to four categories:

(1) order-disorder
(2) collapsing or changing of the MF_3 parent lattice without breaking M-F bonds
(3) restructuring of the MF_3 parent lattice (M-F bonds are broken and reformed)
(4) changes involving defect structures (cations occupy anion sites and/or vice versa).

For all of these transitions, the high temperature structure appears to be higher in symmetry than the low temperature structure. In the first and second types, the symmetry group of the low temperature form is a subgroup of the high temperature form. In the third and fourth types, there may exist no symmetry relationship between the high and low temperature phases.

The symmetries of all structures are limited by two kinds of constraints: geometrical and natural. In crystals, the only allowed rotational symmetries are 2, 3, 4, or 6 fold. The cubic system provides the upper limit of symmetry for three dimensional crystals; the hexagonal and square systems for two dimensional crystals. These are examples of geometrical constraints. Some examples of natural constraints are: chemical bonds, shapes of ions or atoms, chemical composition, relative magnitudes of ionic charges, relative sizes of ions, and signs of ionic charges. As a crystal is cooled, natural constraints are the cause of solid-solid phase transitions which result in the lowering of crystal symmetry. We must also acknowledge the existence of external constraints such as temperature and pressure. As is the case with gases, crystals demonstrate ideal behavior at high temperatures and low pressures.

Several examples of phase changes which occur as a consequence of natural constraints have already been given. When paramagnetic crystals like KVF_3 are cooled, the coupling of weak covalent bonds results in the magnetically ordered state (Figure 1). It is because of their nonspherical shapes that Jahn-Teller ions become ordered (Figure 2). In all inorganic crystals, the structures formed are highly dependent upon composition; however, there are subtle composition effects which give rise to concomitant ordering such as partially filled sites (Figure 7). The difference in magnitude of the ionic charges of M^{2+} and M^{3+} gives rise to their order in Figure 8. The relative sizes of ions in $NaZnF_3$ causes the perovskite lattice to collapse around Na^+ ions; and at still lower temperatures, the Na^+ ions become cooperatively ordered (Figure 9). Finally, in most substances, the chemical composition, the sign and magnitude of ionic charges, and the relative sizes of ions place severe constrains on possible crystal structures. In the case of $K_{0.50}VF_3$ all of these constraints are overcome at high temperatures. One of the high temperature forms of $K_{0.50}VF_3$ has the hexagonal tungsten bronze structure. In order to adjust for an apparent incorrect stoichiometry, K^+ ions occupy F^- sites and some vanadium sites are empty. When $K_{0.50}VF_3$ is cooled to room temperature, the conversion from the hexagonal to the tetragonal form is sluggish and occurs over a period of several days (Figure 10).

We have been purposely vague regarding the quantification of symmetry. Quantification implies that there exists a means of comparing any two structures to determine which has the higher degree of symmetry. It is

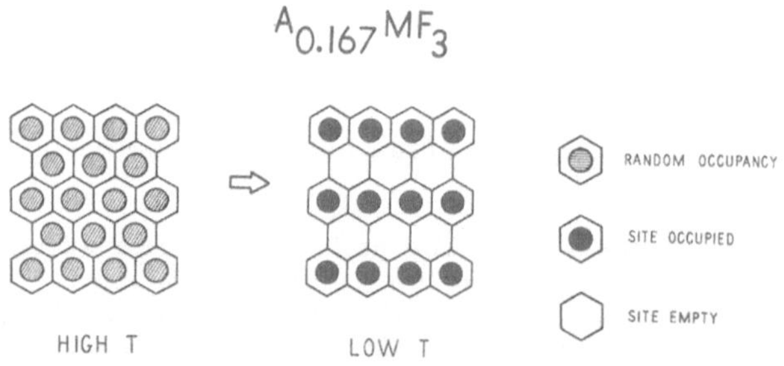

Figure 7. Ordering of partially filled A^+ sites in $A_{0.167}MF_3$ (A = K, Rb, or Cs; M = V, or Cr)

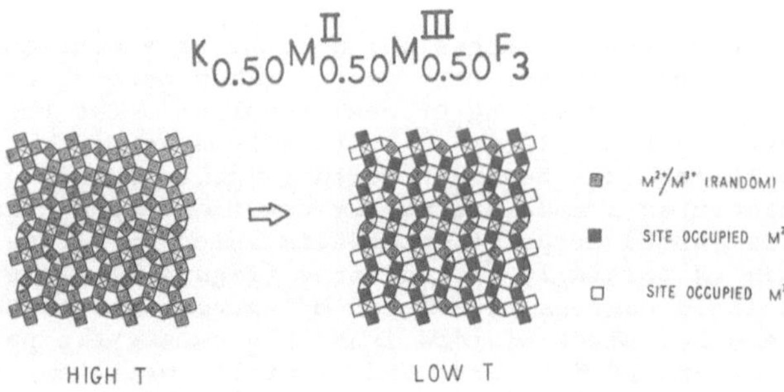

Figure 8. Ordering of M^{2+}/M^{3+} ions in $K_{0.50}M^{II}_{0.50}M^{III}_{0.50}F_3$. If M^{2+} and M^{3+} are homonuclear (e.g. V^{2+} and V^{3+}) the ordering is electronic. If M^{2+} and M^{3+} are not homonuclear the ordering is ionic.

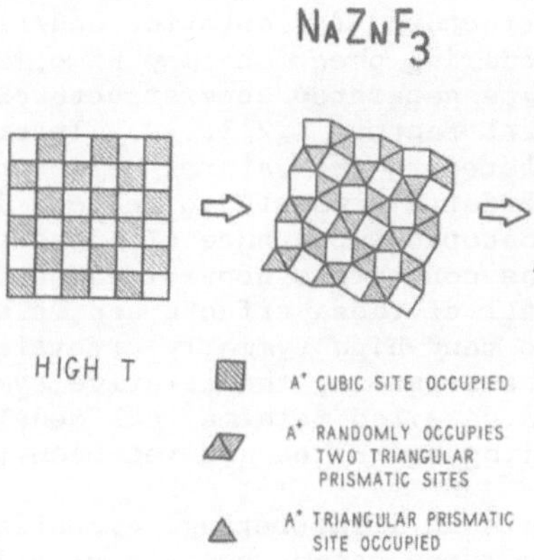

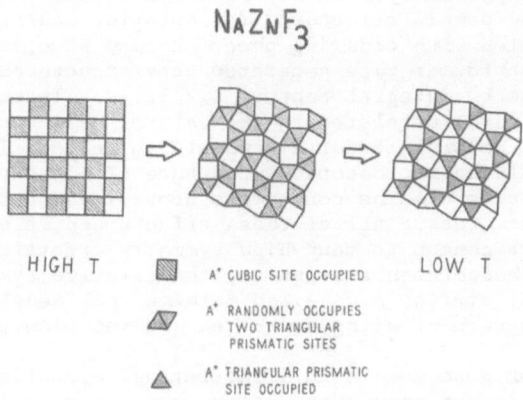

Figure 9. Ordering in $NaZnF_3$.

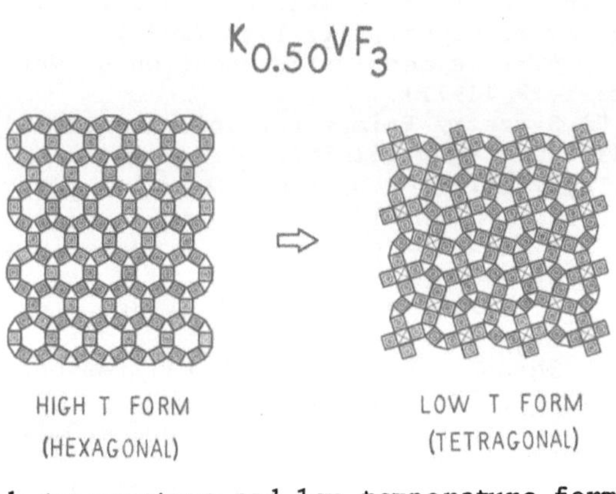

Figure 10. High temperature and low temperature forms of $K_{0.50}VF_3$.

possible to set up a hierarchy of symmetries based on a set of mathematical rules, but nature may or may not abide by these rules. There is some safe ground such as the comparison of groups and their subgroups; the former being of higher symmetry than the latter. The $A_x^I M_x^{II} M_{1-x}^{III} F_3$ compounds, however, demonstrate that there is more to be considered than just point group symmetry. In addition to lowering of point group symmetry, ordering phenomena are often accompanied by more subtle effects such as modulations, superstructures, and domain structures (orientation and/or composition). Modulations associated with ordering phenomena may be commensurate or incommensurate. The former type generates superstructures which increase the unit cell volume by integral factors 1,2,3... . Incommensurate modulations have periods unrelated to the natural period of the crystal. Orientation domains occur in ordered crystals to relieve internal strains and give the crystal the macroscopic appearance of high symmetry. The creation of composition domains converts a nonstoichiometric crystal into small stoichiometric areas. All of these effects are related to the degree of symmetry, and are common to many high symmetry crystals. Although we may make qualitative statements regarding the relative symmetry of a crystal in different states, a detailed mathematical model for quantification that is in agreement with nature has not yet been proposed.

The importance of good models in illustrating, visualizing, and understanding solid-state structures cannot be overemphasized. Geometrical symmetry and the special symmetry evolution principle provide simple guidelines for understanding the structures and behavior of high symmetry phase systems. Synthetic materials like the $A_x^I M_x^{II} M_{1-x}^{III} F_3$ compounds are important because they are real systems which demonstrate ideal structural behavior and, therefore, provide a foundation for understanding many advanced solid state materials of the near future.

REFERENCES

1. Y.S. Hong, R.F. Williamson, and W.O.J. Boo, J. Chem. Ed., 57, 583 (1980).
2. R.F. Williamson, and W.O.J. Boo, Inorg. Chem., 16, 646 (1977).
3. J.C. Cousseins and A. de Kozak, C.R. Acad. Sci., 263, 1533 (1966).
4. L. Pauling, Acta. Cryst., 51, 1010 (1929).
5. R. Williams, "The Geometrical Foundation of Natural Structure", Dover, New York (1972).
6. J. Rosen, "A Symmetry Primer for Scientists", Wiley-Interscience, New York (1983), pages 161-167.
7. Y.S. Hong, R.F. Williamson, and W.O.J. Boo, Inorg. Chem., 19, 2229 (1980).
8. Y.S. Hong, R.F. Williamson, and W.O.J. Boo, Inorg. Chem., 20, 403 (1981).
9. Y.S. Hong, R.F. Williamson, and W.O.J. Boo, Inorg. Chem., 21, 3898 (1982).
10. E. Banks, M. Shone, Y.S. Hong, R.F. Williamson, and W.O.J. Boo, Inorg. Chem., 21, 3894 (1982).
11. E. Banks, M. Shone, R.F. Williamson, and W.O.J. Boo, Inorg. Chem., 22, 3339 (1983).
12. R.M. Metzger, N.E. Heimer, C.S. Kuo, R.F. Williamson, and W.O.J. Boo, Inorg. Chem., 22, 1060 (1983).
13. Y.S. Hong, K.N. Baker, R.F. Williamson, and W.O.J. Boo, Inorg. Chem., 23, 2787 (1984).
14. R.F. Williamson, E.S. Arafat, K.N. Baker, C.H. Rhee, J.R. Sanders, T.B. Scheffler, H.S. Zeidan, and W.O.J. Boo, Inorg. Chem., 24, 482 (1985).

SPONTANEOUS SYMMETRY BREAKING IN UNIFIED THEORIES WITH GAUGE GROUP $SO(10)$

F. Buccella
Dipartimento di Scienze Fisiche
Universitá di Napoli

INTRODUCTION

This work is based on research started several years ago in Geneva and continued in Naples, where it is still in progress. This study took advantage of previous work on critical orbits[1] and on the construction of positive definite invariants which vanish in the direction of the desired vacuum.[2] Part of the matter here presented may be found in the Ph.D. Thesis by Guy Anastaze in Geneva and in the theses of Lorella Cocco, Tiziana Tuzi and Luigi Rosa in Naples.[3] The spontaneous symmetry breaking of $SO(10)$ into the gauge group $G = SU(3)XSU(2)XU(1)$ of the strong, weak and electromagnetic interactions, requires VEV s in at least two irreducible representations, since all the G singlets belonging to the various irreducible representations considered have a larger little group.

The VEV s in the two different irreducible representations are not necessarily of the same order of magnitude and therefore, between the two scales, one has an intermediate symmetry. If this intermediate symmetry is just $SU(5)$, one recovers the same relationship of the $SU(5)$ unified theory between the low energy gauge coupling constants and the lifetime of the nucleon, which disagrees with the experimental lower limit on $\tau(P \rightarrow e^+ + \pi^0)$.

In the case of a different intermediate symmetry the lepto-quarks responsible for nucleon decay may be heavier than in the $SU(5)$ case. Also the experimental neutrino results,[4] providing hints for neutrino mixings and masses, give another reason to investigate $SO(10)$ models where the neutrinos are expected to take masses smaller by various orders of magnitude than the other fundamental fermions. In fact their mass, through the well-known sea-saw mechanism,[5] is inversely proportional to the scale where B-L symmetry is spontaneously broken. We shall determine in the various models considered with intermediate symmetry different from $SU(5)$, the values of the scales in terms of the experimental values of $\alpha/\alpha_S(M_W)$ and $sin^2\theta_W(M_W)$. Some of the models considered will appear rather intriguing from a phenomenological point of view, since they would predict a sufficiently large value for the mass of the lepto-quarks and Majorana masses of the right-handed neutrinos around $10^{10} - 10^{11}$ Gev, which is the proper order of magnitude to get masses for ν_τ and ν_μ suitable to explain the related phenomena.

In the next section we shall describe the G singlets in the low representations of $SO(10)$. In the following we shall construct the invariant potentials with absolute minimum in the direction desired for the various vacua. Finally we shall discuss the phenomenological consequence of the renormalization group equations for the various models considered.

1. SYMMETRY PROPERTIES OF THE $SU(3)XSU(2)XU(1)$ SINGLETS

We write now the singlets under $SU(3)XSU(2)XU(1)$ contained in the lowest representations of $SO(10)$.

In the 16 representation the state

$$\varphi_+ = (+++++) \tag{1}$$

which is singlet under the $SU(5)$ of the generators commuting with

$$M_{12} + M_{34} + M_{56} + M_{78} + M_{910} \tag{2}$$

In the 45 every combination:

$$p(T_{12} + T_{34} + T_{56}) + q(T_{78} + T_{910}) \tag{3}$$

For $p = q$ it is a $SU(5)XU(1)$ singlet, for $p = 1/3$ and $q = -1/2$ it is just the hypercharge commuting with $SU(3)_c XSU(2)_L XU(1)_Y XU(1)_{B-L}$. In the 54 the tensor:

$$2(S_{11} + S_{22} + S_{33} + S_{44} + S_{55} + S_{66}) - 3(S_{77} + S_{88} + S_{99} + S_{1010}) \tag{4}$$

which is invariant under $SO(6)XSO(4) \equiv SU(4)_{PS} XSU(2)_L XSU(2)_R$. In the 126 the state

$$\Psi_+ = \varphi_+ \varphi_+ \tag{5}$$

is evidently again invariant under the same $SU(5)$ of φ_+. In the 210 any linear combination:

$$\begin{aligned} a(T_{1234} + T_{1256} + T_{3456}) + b(T_{1278} + T_{12910} + T_{3478} \\ + T_{34910} + T_{5678} + T_{56910}) + c\, T_{78910} \end{aligned} \tag{6}$$

which is always invariant under $SU(3)_c XSU(2)_L XU(1)_Y XU(1)_{B-L}$ for $a = b = c$, it is invariant under $SU(5)XU(1)$ for $a = b = 0$ under $SO(6)XSO(4)$ and for $b = c = 0$ it is also invariant under the discrete left-right symmetry generator D.

In the 10, 120 and 320 representations there are no $SU(3)XSU(2)XU(1)$ singlets. From the present analysis one sees that at least two of the representations just listed are needed to get the spontaneous breaking of $SO(10)$ into $SU(3)XSU(2)XU(1)$ since all the singlets under G in the irreducible representations considered have a larger symmetry.

2. CONSTRUCTION OF INVARIANT POTENTIALS WITH ABSOLUTE MINIMUM IN THE DIRECTION OF THE DESIRED VACUUM

In order to have a stable minimum we shall require that, be ϕ and Ψ the scalars belonging to the two irreducible representations needed to break $SO(10)$ into

$SU(3) X SU(2) X U(1)$, that both the parts depending only on ϕ or Ψ, V_ϕ or V_Ψ respectively and the one containing both the fields, $V_{\phi\Psi}$ get their absolute minimum in correspondence of the desired vacuum. The method described here will consist in the construction of positive definite non-trivial invariants, which vanish in the direction of the desired vacuum. By assuming that these invariants have positive coefficients in the expression of the potential, we shall get our goal.

Let us first consider the invariant potentials depending only on a single representation (for the complex cases, 16 and 126, we shall of course consider 16 + 16* and 126 + 126* respectively). For the case of the 16 we have only one non-trivial invariant, which may be put in the form:

$$\|(\varphi_+ \varphi_+)_{10}\| \tag{7}$$

It is evidently positive definite and vanishes in the $SU(5)$ invariant direction since the 10 does not contain any $SU(5)$ singlet. So, if the coefficient of this invariant in V_ϕ is positive, the absolute minimum is in the $SU(5)$ invariant direction. In the case of the 45 one has only one non-trivial invariant, which may be put in the form:

$$\|(\phi_A \phi_A)_{54}\| \tag{8}$$

Also this invariant is positive definite and vanishes in the direction with $SU(5) X U(1)$ invariance ($p = q$) since the 54 does not contain any $SU(5)$ singlet. With the scalars of the 54 one can build two non-trivial invariants, the cubic $(\phi_S \phi_S)_{54}$ x ϕ_S and $(\|(\phi_S \phi_S)_{54}\|)$; one can build the positive definite invariant

$$\|(\phi_S \phi_S)_{54} - k\, \phi_S\| \tag{9}$$

This invariant would vanish if $k = C \begin{matrix} 54 \\ (1,1,1) \end{matrix} \begin{matrix} 54 \\ (1,1,1) \end{matrix} \begin{matrix} 54 \\ (1,1,1) \end{matrix} \langle\phi_S\rangle$ in the direction $SO(6) X SO(4)$ invariant. There are four independent non-trivial invariants built with the states of the 126, namely

$$\begin{aligned} &\|(\Psi_+ \Psi_+)_{54}\|, \; \|(\Psi_+ \Psi_+)_{1050}\| \\ &\|(\Psi_+ \Psi_+)_{4125}\| \text{ and} \\ &(\Psi_+\Psi_+\Psi_+\Psi_+)_1 + (\Psi_-\Psi_-\Psi_-\Psi_-)_1 \end{aligned} \tag{10}$$

The first three invariants vanish in the $SU(5)$ invariant direction since the product of two $SU(5)$ singlets of the 126 has its only component on the maximal weight representation contained in the product of two 126, namely the 2772.

Also the last invariant vanishes, together with all its first and second derivatives in that direction; but it may take either sign (in fact changes sign under the transformation $\Psi_\pm \rightarrow exp(\pm i\pi/4)\Psi_\pm$). So if the coefficients of the first three invariants are positive and the coefficient of the fourth one is sufficiently small with respect to the others, we may expect the minimum to be in the $SU(5)$ invariant direction. It would be always the case whenever we may write the potential in the form:

$$\begin{aligned} &\|(\Psi_+ \Psi_+)_{54} + \eta_1 e^{i\delta_1}(\Psi_- \Psi_-)_{54}\| \\ &\|(\Psi_+ \Psi_+)_{1050}\| \\ &\|(\Psi_+ \Psi_+)_{4125} + \eta_2 e^{i\delta_2}(\Psi_- \Psi_-)_{4125}\| \end{aligned} \tag{11}$$

Finally for the case of the 210 one has one cubic and three quartic non-trivial independent invariants:

$$(\phi\,\phi)_{210} \times \phi \text{ and}$$
$$||(\phi\,\phi)_{45}||\ ||(\phi\,\phi)_{54}||\ ||(\phi\,\phi)_{210}||\ (\text{or } ||(\phi\,\phi)_{1050}||) \tag{12}$$

The invariants $||(\phi\,\phi)_{45}||$, $||(\phi\,\phi)_{210}||$ and $||(\phi\,\phi)_{1050}||$, as well as the cubic one vanish in correspondence to the $SO(6)XSO(4)$ invariant direction: ($a = b = 0$ in Eq. 6). In fact there is no $SO(6)XSO(4)$ singlet in the 45 and 1050 representations and the only $SO(6)XS0(4)$ singlet of the 210 is such that

$$C\begin{matrix}210 & 210 & 210\\ (1,1,1) & (1,1,1) & (1,1,1)\end{matrix} = 0 \tag{13}$$

The invariant $||(\phi\,\phi)_{45}||$ also vanishes in the $SU(3)_c XSU(2)_L XSU(2)_R XU(1)_{B-L} XD$ invariant direction ($b = c = 0$). In this direction one can also construct the invariant

$$||(\phi\,\phi)_{210} - k'\,\phi|| \tag{14}$$

with $k' = C\begin{matrix}210 & 210 & 210\\ (15,1,1)_1 & (15,1,1)_1 & (15,1,1)_1\end{matrix}\langle\phi\rangle$

So with positive (and large with respect to the others) coefficients in V_ϕ for $||(\phi\,\phi)_{45}||$ and $||(\phi\,\phi)_{210} - k'\,\phi||$, the absolute minimum is expected in the $SU(3)XSU(2)XSU(2)XU(1)XD$ invariant direction.

With a larger value of k", namely

$$k'' = C\begin{matrix}210 & 210 & 210\\ [1,0] & [1,0] & [1,0]\end{matrix}\langle\phi\rangle \tag{15}$$

(in the square brackets we write the behaviour under $SU(5)\ XU(1)$), the invariant vanishes in the $SU(5)XU(1)$ invariant direction ($a = b = c$). In this direction vanishes also $||(\phi\,\phi)_{54}||$ for the already outlined reason that the 54 does not contain any $SU(5)$ singlet. Since, as long as we know, $||(\phi\,\phi)_{45}||$ and $||(\phi\,\phi)_{210}||$ are maxima in the same direction, a potential with positive coefficient for $||(\phi\,\phi)_{210} - k''\,\phi||$ and $(\phi\,\phi)_{54}$ and negative for $||(\phi\,\phi)_{45}||$ and $||(\phi\,\phi)_{210}||$ is expected to take its absolute minimum in the $SU(5)XU(1)$ invariant direction.

The invariant $||(\phi\,\phi)_{54}||$ vanishes also in two particular directions of the $SU(3)XSU(2)XSU(2)XU(1)$ two-dimensional stratus obtained in Eq. (6) with $b = 0$, namely for $c = \pm\sqrt{2}\,a$. One has been able to show that, under certain conditions on the parameters it is possible to get the minimum, in the G invariant subset of the directions of the 210, in this direction.

The following pairs of irreducible representations have been up to now considered:

$$45 + 16$$
$$54 + 126$$
$$210 + 126$$

Let us study the relative mixed terms in the potentials in the various cases. There is only one third degree invariant lienar in the 16, 16 and 45 and only one non-trivial of degree four, respectively $(\phi_A \phi_+)_{16} \times \phi_-$ and $\|(\phi_A \phi_+)_{16}\|$. We may consider the positive invariant

$$\|(\phi_A \varphi_+)_{16} - k''' \varphi_+\| \tag{16}$$

We are interested in the case when the $SU(5)$ which is the invariance of φ_+ does not coincide with the $SU(5)$, which, together with $U(1)$, is the little group of ϕ but the two $SU(5)$ have in common $SU(3)_c XSU(2)_L$, namely for ϕ_A one has to choose the flipped version $(SU(5)XU(1))'$ (in our notation $p = -q$). So in this circumstance, we take

$$k''' = C \begin{matrix} 45 & 16 & 16 \\ [1,0] & [1,+5] & [1,+5] \end{matrix} \langle SU(5)XU(1) \text{ singlet } |(SU(5)XU(1))' \text{ singlet } \rangle \langle \phi \rangle \tag{17}$$

and with a positive sign also $V_{\phi\Psi}$ would be minimum in correspondence to $\phi_{\text{flipped}} \varphi_+$. For the second case 54+126 one has only one non-trivial quartic invariant

$$\|(\phi_S \Psi_+)_{126}\| \tag{18}$$

which vanishes with ϕ_S and Ψ_+ $SO(6)XSO(4)$ and $SU(5)$ invariant respectively with common $SU(3)XSU(2)XU(1)$ little group, since the only G singlet of the 126 transforms as [1,-10] under $SU(5)XU(1)$ while Ψ_+ and ϕ_S transform as [1,+10] and [24,0] respectively. Again with a positive sign to multiply that invariant also $V_{\phi\Psi}$ gets its absolute minimum in correspondence with ϕ_S and Ψ_+.

Finally for the case of the 126 and 210 representations one has one cubic invariant linear in the 126, 126 and 210, and five non-trivial quartic invariants quadratic in the 210 and linear in the 126 and 126 and three quadratic in the 126 (or in the 126) and in the 210.

We may choose as independent invariants the following:

$$\begin{aligned} &\|(\phi \Psi_+)_{10}\| \; \|(\phi \Psi_+)_{120}\| \; \|(\phi \Psi_+)_{320}\| \\ &\|(\phi \Psi_+)_{126} + k \Psi_+\| \; (\Psi_+ \Psi_-)_{45} \times (\phi \phi)_{45} \\ &(\phi \Psi_\pm)_{10} \times (\phi \Psi_\pm)_{10} \\ &(\phi \Psi_\pm)_{120} \times (\phi \Psi_\pm)_{120} \\ &\text{and} \\ &(\phi \Psi_\pm)_{320} \times (\phi \Psi_\pm)_{320} \end{aligned} \tag{19}$$

If ϕ and Ψ_+ have a common $SU(3)XSU(2)XU(1)$ algebra $(\phi \Psi_\pm)_{10}$, $(\phi \Psi_\pm)_{120}$ $(\phi \Psi_\pm)_{320}$ vanish since the 10, 120 and 320 representations do not contain singlets under G and therefore so do the first three and last six invariants. Of course only the first three, the norms, are positive because the last six change sign under the transformation:

$$\Psi_\pm \to e^{i\pi/2} \Psi_\pm \,.$$

Therefore $V_{\phi\Psi}$ is certainly minimum in the direction where ϕ and Ψ have a common $SU(3)xSU(2)XU(1)$ symmetry if it can be put as combination of the invariants.

$$\begin{aligned} &\|(\phi \Psi_+)_{10} + \epsilon_1 e^{i\rho_1} (\phi \Psi_-)_{10}\| \\ &\|(\phi \Psi_+)_{120} + \epsilon_2 e^{i\rho_2} (\phi \Psi_-)_{120}\| \\ &\|(\phi \Psi_+)_{320} + \epsilon_3 e^{i\rho_3} (\phi \Psi_-)_{320}\| \end{aligned} \tag{20}$$

with positive coefficients. As long as for the non-homogeneous invariant

$$\|(\phi \Psi_+)_{126} - k^{iv}\Psi_+\| \tag{21}$$

with ϕ and Ψ_+ $SU(3)XSU(2)XU(1)$ singlets the only nonvanishing component of $(\phi \Psi_+)_{126}$ is in the direction Ψ_+ (the only G singlet) and depends only on the component of ϕ along the $SU(5)$ invariant direction, since Ψ_+ is a $SU(5)$ singlet; so we have:

$$(\phi \varphi_+)_{126} = C \begin{matrix} 210 \\ [1,0] \end{matrix} \quad \begin{matrix} 126 \\ [1,+10] \end{matrix} \quad \begin{matrix} 126 \\ [1,+10] \end{matrix} \langle \phi \, SU(5) \, \text{singlet} \rangle \tag{22}$$

So for every ϕ, which is a $SU(3)XSU(2)XU(1)$ singlet one can find a k^{iv} such that the non-homogeneous invariant vanishes.

3. DETERMINANTS OF THE HIGH SCALES FROM THE GAUGE COUPLING CONSTANTS AT M_W

With an intermediate symmetry, different from $SU(5)$, the predictions for the low energy coupling constants dictated by the group renormalization equations depend on the scales M_W and M_R of the spontaneous symmetry breaking of $SO(10)$ and the intermediate symmetry respectively:

$$\begin{aligned} sin^2\theta_W(M_W) &= 3/8 - 55\alpha(M_W)/24\pi[(1 - 1/110) ln\, M_R/M_W + c_1\, ln\, M_X/M_R] \\ \alpha/\alpha_S(M_W) &= 3/8 - 33\alpha(M_W)/8\pi[(1 + 1/66) ln\, M_R/M_W + c_2\, ln\, M_X/M_R] \end{aligned} \tag{23}$$

where we have separated, in the coefficients of $ln\, M_R/M_W$, the gauge boson and scalar contributions and the values of the c_i, again split in the same way, are reported in Table I, together with the range of values possible for $sin^2\theta_W(M_W)$ assuming $\alpha_S(M_W) = 0,1$. We give also the value of the masses, measured in terms of the scale of $SU(5)$ breaking in the minimal mode, which in the various models would give for $sin^2\theta_W(M_W)$ the value .228 coming from the recent analysis (6) of the neutral current reactions and from the values of the masses of the intermediate weak bosons $W_\pm$ and Z_0. From Table I we see that it is not possible to predict the experimental values of $sin^2\theta_W$ and α/α_S at the scale M_W with intermediate symmetry $[SU(5)XU(1)]$. In the case of $SO(6)XSO(4)$ we have rather broad range of allowed values for $sin^2\theta_W$: in presence of D symmetry and predicted value for M_X is not sufficiently high to comply with the lower limit on $\tau(P \to e^+\Pi^0)$. Narrower ranges are expected for $SU(3)XSU(2)XSU(2)XU(1)$ where one also gets the intriguing prediction of a rather low value for M_R. A Yukawa coupling of the fermions to the scalars of the $126 + 126*$ is allowed, which implies a Majorana mass for the right-handed neutrinos given by the product of the Yukawa coupling f_i to the i-th family times the $SU(5)$ invariant VEV $\langle \Psi_\pm \rangle$. By applying the sea-saw expression for the (mainly) left-handed neutrinos we get, for the case with $SU(3)XSU(2)XSU(2)XU(1)$ intermediate symmetry:

$$\begin{aligned} m(\nu_\tau) &= \frac{g(M_R) m_\tau^2 m_t^2}{f_3\, M_R\, m_b^2} = 4\, eV \frac{g(M_R)}{f_3} \frac{m_t^2}{(50GeV)^2} \\ m(\nu_\mu) &= \frac{g(M_R) m_\tau^2 m_c^2}{f_2\, M_R\, m_b^2} = 3 \cdot 10^{-3}\, eV\, \frac{g(M_R)}{f_2} \end{aligned} \tag{24}$$

which compares fairly with the proposed experimental values.

TABLE I

HIGGS	45 + 16	54 + 126	210+126	210+126	210+126	210+126
INTERMEDIATE SYMMETRY	[SU(5)XU(1)]'	SO(6)XSO(4)XD	SO(6)XSO(4)	SU(3) XSU(2)XSU(2) XU(1) XD	SU(3)X SU(2)L XSU(2) XU(1)	[SU(5)XU(1)]'
SCALARS OF THE 16 or126 AT THE SCALE M_R	[10,+1]	(10*,1,3) +(10,3,1)	(10*,1,3)	(1,1,3, +2) +(1,3,1,+2)	(1,1,3,+2)	[50,+2]
c_1	12/5-1/110-3/55	-2/5-1/110-2/55	-2/5-1/110+39/55	2/5-1/110+7/55	2/5-1/110+3/22	12/5-1/110-6/11
c_2	4/3+1/66-1/33	2/3-1/66+2/33	2/3+1/66+1/33	2/3+1/66+5/33	3/2+1/66+5/66	4/3+1/66-10/33
ALLOWED VALUE For $\sin^2\theta_W$	.078-.214	.214-.475	.214-.306	.214-.273	.214-.260	.089-.214
$\frac{M_R}{M_{SU(5)}}$		.21	$1.3\cdot10^{-2}$	$1.1\cdot10^{-3}$	$1.8\cdot10^{-4}$	
$\frac{M_X}{M_{SU(5)}}$		1.7	6.4	4.4	19	

REFERENCES

1. L. Michel and L. A. Radicati, Ann. Phys. (N.Y.) 66:758 (1971).
   M. Abud and G. Sartori, Phys. Lett. B104:147 (1981) and Ann. Phys. (N.Y.) 150:307 (1983).
2. F. Buccella, H. Ruegg and C. A. Savoy, Phys. Lett. 94B:491 (1980).
   F. Buccella and H. Ruegg, Nuovo Cimento 67A:61 (1982).
   O. Kaymakcalan, W. D. McGlinn, L. Michel, L. O'Raifeartaigh and K. C. Wali, Nucl. Phys. B267:203 (1986).
   L. O'Raifeartaigh and J. Burzlaff, 3rd Capri Symposium on Elementary Particles, May 1985.
3. G. Anastaze, F. Buccella and J. P. Derendinger, Zeitschrift für Phys. C20:269 (1983).
   F. Buccella, L. Cocco and C. Wetterich, Nucl. Phys. B243:273 (1984).
   F. Buccella, L. Cocco, A. Sciarrino and T. Tuzi, Nucl. Phys. B274:559 (1986).
   F. Buccella and L. Rosa, Zeitschrift für Phys. C36:425 (1987).
   M. Abud, F. Buccella, L. Rosa and A. Sciarrino (to be submitted to Nucl. Phys. B).
4. M. Baldo Ceolin, Capri Symposium on Elementary Particles, May 1988.
5. M. Gell-Mann, P. Ramond and R. Slansky in "Supergravity" eds. D. Z. Freedman and P. van Nieuwenhuizen (North Holland, 1980).
6. U. Amaldi, A. Böhm, L. S. Durkin, P. Langacker, A. K. Mann, W. J. Marciano, A. Sirlin and H. H. Williams, Phys. Review D36:1385 (1987).
   G. Costa, J. Ellis, G. L. Fogli, D. V. Nanopoulos and F. Zwirner, Nucl. Phys. B297:244 (1988).

SYMMETRIES AND STATISTICAL MODELS

Vittorio Cantoni

Dipartimento di Matematica
Università di Milano
Via Saldini 50, Milano, Italy

INTRODUCTION

A *symmetry*, in the strictest original sense, can be defined as a distance-preserving transformation of a metric space, and a subset of the space is *symmetric* with respect to such a transformation if it is mapped onto itself. The *symmetry group* of the subset is the group constituted by all the isometric transformations of the space with respect to which the subset is invariant.

By extension, the term "symmetry" has come to designate, in mathematical physics, any transformation, defined within a definite model, that preserves some (not necessarily metric) structure or relation occurring in the model.

Here we shall be concerned with a natural metric structure that can be defined in the state space of any statistical system, i.e. of any model of physical system for which identical experiments repeated under identical external conditions determine statistical distributions that are not, in general, concentrated at a single value. (If all the distributions were concentrated at a single value, our considerations would still make sense, but the metric would be degenerate and devoid of informational content.)

No restrictive assumption on the specific structure of the model is required to show that the space-time symmetries of the system (which are symmetries in the broader sense) determine symmetries in the restricted sense (i.e. isometries) in the state space $\mathcal{S}$. In particular, Galilei or Lorentz invariance implies isometric action of the corresponding group on $\mathcal{S}$. Further isometries of $\mathcal{S}$ are connected with the possible evolutions of the physical system, under suitable assumptions of reversibility.

Among existing theories, relativistic quantum mechanics provides the most relevant illustration of such general considerations. From the point of view adopted here, the basic structure of the set $\mathcal{S}$ of pure states is the distance determined by the Fubini-Study Riemannian metric (or its generalization to the infinite-dimensional complex projective Hilbert space), which is directly related to the statistical distributions of the observables. The representations of the Poincaré group and of the evolution operators by unitary transformations of the underlying Hilbert space correspond to isometric actions on $\mathcal{S}$, in conformity with our general scheme.

As models of state spaces, complex projective spaces possess symmetry properties which admit a physical interpretation and are sufficient to characterize these spaces at least in the finite-dimensional cases.

A NATURAL METRIC IN THE STATE SPACE

Our descriptive framework is drawn from a system of axioms introduced by Mackey (1963), designed to analyze the foundations of the quantum-mechanical formalism. However, of Mackey's axioms we shall adopt only the ones that possess a physical interpretation of indisputable generality.

The scheme consists of a set $\mathcal{S}$ of *states*, a set $\mathcal{O}$ of real-valued *observables*, and a function $p(A, \alpha, E)$ representing the probability that the measurement of the observable $A \in \mathcal{O}$ on the state $\alpha \in \mathcal{S}$ gives a result in the Borel set E of the real numbers $\mathbb{R}$. Thus p is defined on $\mathcal{O} \times \mathcal{S} \times \mathcal{B}$ (where $\mathcal{B}$ is the family of the Borel sets of $\mathbb{R}$), has values in the closed interval $[0, 1]$, and is assumed to be such that $p(A, \alpha, \mathbb{R}) = 1$, $p(A, \alpha, \phi) = 0$ (where ϕ denotes the empty set), and that $p(A, \alpha, \bigcup E_h) = \sum p(A, \alpha, E_h)$ for every discrete family of disjoint Borel sets. Moreover, it is natural to assume that $\alpha = \beta$ whenever $p(A, \alpha, E) = p(A, \beta, E)$ for every $A \in \mathcal{O}$ and every $E \in \mathcal{B}$, and that $A = B$ whenever $p(A, \alpha, E) = p(B, \alpha, E)$ for every $\alpha \in \mathcal{S}$ and every $E \in \mathcal{B}$.

These data are sufficient to determine a metric in the state space $\mathcal{S}$. First we associate with every observable A the two-point function

$$(1) \quad T_A(\alpha, \beta) = \left| \int_{\mathbb{R}} d\sqrt{\alpha_A \beta_A} \right|^2$$

where α_A is the probability measure on $\mathbb{R}$ defined *by* $\int_E d\alpha_A = p(A, \alpha, E)$, β_A is defined analogously, and $\sqrt{\alpha_A \beta_A}$ denotes the measure on $\mathbb{R}$ defined by

$$\int_E d\sqrt{\alpha_A \beta_A} = \int_E \sqrt{\frac{d\alpha_A}{d\sigma} \frac{d\beta_A}{d\sigma}}\, d\sigma$$

(σ being any finite measure on $\mathbb{R}$ with respect to which α_A and β_A are absolutely continuous). Next we define the two-point function

$$(2) \quad T(\alpha, \beta) = \inf_{A \in \mathcal{O}} T_A(\alpha, \beta).$$

Finally we set

$$(3) \quad d(\alpha, \beta) = 2 \arccos \sqrt{T(\alpha, \beta)}.$$

In Cantoni (1975), the function $T(\alpha, \beta)$ has been referred to as the "generalized transition probability" because, in quantum mechanics, it actually coincides with the transition probability between pure states (see also Gudder, 1978, 1979; Hadjisavvas, 1982). In full generality, however, there seems to be no reason why it should represent a transition probability in any physical sense; rather, it should be regarded as a quantitative evaluation of the "resemblance" of the states α and β. This is particularly evident in classical statistical mechanics, where the states are represented by probabilities densities on the phase space Ω, and one has (Hadjisavvas, 1982)

$$(4) \quad T(\alpha, \beta) = \left| \int_{\Omega} \sqrt{\rho_\alpha \rho_\beta}\, d\mu \right|^2$$

(ρ_α and ρ_β being the probability densities of α and β, and μ the Liouville measure), so that $T(\alpha, \beta)$ gives an evaluation of the overlap of ρ_α and ρ_β.

Since T is symmetric, has values in the interval [0, 1], and is equal to 1 if and only if $\alpha = \beta$, the distance function d defined by (3) is also symmetric, not greater than π, and zero if and only if $\alpha = \beta$. It can be shown also that it satisfies the triangle inequality (Cantoni, 1985), so that it endows $\mathcal{S}$ with a bounded metric. Like T, it provides a criterion of comparison between states: the smaller the distance, the greater the resemblance.

In the special case where $\mathcal{S}$ represents the pure states of a quantum-mechanical system, $\mathcal{S}$ is a complex projective space, and it follows from the analysis of Cantoni (1985) that d coincides with the geodesic distance determined by the Riemannian structure of $\mathcal{S}$.

ISOMETRIES CONNECTED WITH SPACE-TIME SYMMETRIES

If it is assumed that the system possesses a space-time symmetry group $\mathcal{G}$, in the usual sense that there exists a class of physically equivalent reference frames related by $\mathcal{G}$, there is a natural action of the group on the space $\mathcal{S}$ of all the realizable states and on the set $\mathcal{O}$ of all the observables. In fact, if a state α is prepared in the reference frame R by means of a certain experimental arrangement, and $g \in \mathcal{G}$ transforms R into an equivalent frame R', another state $\alpha' \equiv g\alpha$ can be prepared, in principle, by transferring the experimental arrangement from R to R'. Similarly, to an observable A measured in R by a certain apparatus there corresponds another observable $A' \equiv gA$ measured by the same apparatus transferred to R'.

The action of $\mathcal{G}$ on $\mathcal{S}$ determined by the maps

$$g: \mathcal{S} \to \mathcal{S}$$
$$\alpha \to g\alpha$$

is isometric. In fact, since the physical equivalence of the reference frames entails the relation

$$p(A, \alpha, E) = p(gA, g\alpha, E),$$

from the definitions (1) and (2) one obtains

$$T_A(\alpha, \beta) = T_{gA}(g\alpha, g\beta),$$

and

$$T(\alpha, \beta) = \inf_{A \in \mathcal{O}} T_A(\alpha, \beta) = \inf_{gA \in g\mathcal{O}} T_{gA}(g\alpha, g\beta) =$$
$$= \inf_{A' \in \mathcal{O}} T_{A'}(g\alpha, g\beta) = T(g\alpha, g\beta),$$

where the bijective character of the maps g: $\mathcal{O} \to \mathcal{O}$ has been taken into account. Thus $T(\alpha, \beta) = T(g\alpha, g\beta)$ and, from (3),

$$(5) \quad d(\alpha, \beta) = d(g\alpha, g\beta) \qquad (g \in \mathcal{G}).$$

As mentioned in the previous section, when $\mathcal{S}$ represents the pure states of a quantum-mechanical system, T coincides with the transition probability. To satisfy the requirements of a relativistic theory, the distance d (or, equivalently, T) must be invariant under an action of the Poincaré group, and this is indeed the case when the group is represented by unitary transformations of the underlying Hilbert space.

Suppose now that, from the point of view of a single reference frame, one can regard the generic state α as the outcome of a preparation ending at time t_0 and the generic observable A as a measurement process starting at time t_0'. Assume, moreover, that for every $\tau \geqslant 0$ the environmental conditions can be controlled during the time interval $(t_0, t_0' \equiv t_0 + \tau)$, and denote by m the generic choice of a value of τ together with an assignment of environmental conditions. Then to each state α there corresponds another state $m\alpha$ which is the outcome of the preparation of α followed by the evolution under m, and to each observable A there corresponds another observable mA representing the same measurement process as A, but starting with a delay τ after the system has evolved under m. It is natural to assume that the set of all physically realizable "movements" m of the system is represented by a semigroup $\mathcal{M}$ (Mielnik, 1977, 1980), with actions on $\mathcal{S}$ and $\mathcal{O}$ subject to the identity $p(A, m\alpha, E) = p(mA, \alpha, E)$. It is then easy to show (Cantoni, 1982) that $T(\alpha, \beta) \leqslant T(m\alpha, m\beta)$ for all $m \in \mathcal{M}$, or, equivalently, that

(6) $d(\alpha, \beta) \geqslant d(m\alpha, m\beta)$.

In the case of classical statistical mechanics the above relation is always an equality, because the integral (4) is invariant under the underlying canonical transformations associated with the movements of the system. In quantum mechanics the relation (6) is again an equality, on account of the unitary character of the evolution operators.

In greater generality, a sufficient condition in order that relation (6) hold as an equality is the assumption that the movements $\mathcal{M}$ constitute a group, rather than merely a semigroup. This can be interpreted as an assumption of reversibility, in the sense that to each regulation m of the external conditions acting on the physical states one can associate a second regulation m^{-1} restoring the initial situation (with m^{-1} independent of the state).

Thus, for reversible systems, the state space $\mathcal{S}$ possesses a group of isometries representing the movements of the physical system, and it is natural to assume that this group acts *transitively* on $\mathcal{S}$: actually, the condition of transitivity can be regarded as the very definition of what is meant by the assertion that distinct elements of $\mathcal{S}$ represent states of the *same* physical system.

More stringent symmetry assumptions will lead us very close to a characterization of the model for the set of pure states in quantum mechanics.

SYMMETRY PROPERTIES OF THE SET OF PURE STATES IN QUANTUM MECHANICS

As metric spaces (with the geodesic distance determined by the Fubini-Study Riemannian metric or its generalization to the infinite-dimensional case), complex projective spaces are two-point homogeneous (in the sense that they possess an isometry group that acts transitively on all pairs of equidistant points). In addition, they possess a symplectic structure which is invariant under the isometry group, and the distance function is bounded.

Conversely, among the two-point homogeneous metric spaces with a bounded distance function, finite-dimensional complex projective spaces are characterized by the existence of a symplectic form invariant under a strongly transitive group of isometries (i.e. a group of isometries with respect to which the space is two-point homogeneous (Avez; Buzzanca, 1984). It is plausible that such a characterization can be extended to the infinite-dimensional case, though a complete proof does not seem to be available yet.

In models of the state space of a physical system, the boundedness of the distance is, as we have seen, a general property implicit in the definition, while two-point homogeneity is a symmetry property (in the strict sense) expressing, in addition to reversibility, a sufficiently high degree of "mobility" of the system under its possible evolutions. As to the existence of an invariant symplectic form (which is an additional symmetry in the broader sense), it corresponds, in quantum mechanics, to the fact that the expectation values of all the observables behave as the observables of a classical (possibly infinite-dimensional) Hamiltonian system, with the Poisson brackets corresponding to the quantum commutators (Cantoni, 1977b).

REFERENCES

Avez, A., private communication.

Buzzanca, C., 1984, *Rend. Sem. Mat. Univers. Politecn. Torino*, 42:117.

Cantoni, V., 1975, *Commun. Math. Phys.*, 44:125.

Cantoni, V., 1976, *Commun. Math. Phys.*, 50:241.

Cantoni, V., 1977a, *Commun. Math. Phys.*, 56:189.

Cantoni, V., 1977b, *Rend. Accad. Naz. Lincei*, 62:628.

Cantoni, V., 1982, *Commun. Math. Phys.*, 87:153.

Cantoni, V., 1985, *Helv. Phys. Acta*, 58:956.

Cantoni, V., and Logli, A., 1988, *Boll. Unione Mat. It.*, in press.

Gudder, S. P., 1978, *Commun. Math. Phys.*, 63:265.

Gudder, S. P., 1979, "Stochastic Methods in Quantum Mechanics", North Holland, Amsterdam.

Hadjisavvas, N., 1982, *Commun. Math. Phys.*, 83:43.

Mackey, G. W., 1963, "Mathematical Foundations of Quantum Mechanics", Benjamin, New York.

Mielnik, B., 1977, *Rep. Math. Phys.*, 12:331.

Mielnik, B., 1980, *J. Math. Phys.*, 21:44.

INFINITELY GENERATED SUPERMANIFOLDS, SUPER LIE GROUPS AND HOMOGENEOUS SUPERMANIFOLDS

Roberto Cianci
Dipartimento di Matematica, Università di Genova

1. Introduction

In this paper we wish to discuss some aspects of the theory of Super Lie Groups (SL groups for short) and homogeneous supermanifolds. Briefly, a SL Group is a supermanifold which is also a group.

Such groups play a fundamental role in physical theories of fields, since super Lie algebras are widely used by physicists for the construction of supersymmetric gauge fields theories.

First of all, in this paper we discuss the main mathematical concepts which constitute the basis of the theory of supermanifold; secondly, we shall give a rigorous definition of SL Groups and, by means of this, shall discuss some results concerning super Lie algebras, the exponential map and some theorems on sub-SL Groups.

In section 4 we discuss the " supermanifold version " of the fundamental Lie theorems; we shall prove the " third Lie Theorem " analogue which allows us to construct a SL Group in terms of its tangent space in the origin (roughly speaking, the SL algebra).

Finally we discuss some results concerning homogeneous supermanifolds.

2. Mathematical Preliminaries

Let Γ^L the exterior algebra Grassmann over $\mathbb{R}^L$: $\Gamma^L = \bigoplus_{i=0}^{L} \omega^{(i)}$ where $\omega^{(i)}$ denotes the set of the i-forms on $\mathbb{R}^L$. Γ^L is naturally $\mathbf{Z}_2$ graded by considering even and odd forms:

$$\Gamma^L = \Gamma_0^L \oplus \Gamma_1^L$$

Remark. In Γ^L we consider also sums of forms of different degree; for $L = 2$ consider, for instance, the forms $\theta = 2 + 2\mathbf{e}_1 \wedge \mathbf{e}_2$ and $\omega = \frac{1}{2} - \frac{1}{2}\mathbf{e}_1 \wedge \mathbf{e}_2$ which satisfy $\theta\omega = 1$.

Paper partially supported by the National Group for Mathematical Physics of the Italian Research Council and by the Italian Ministry of the Public Education through the research project "Geometria e Fisica".

This grading gives rise to a $\mathbf{Z}_2$ graded commutative structure: we have:

$$\Gamma_0^L \Gamma_0^L \subset \Gamma_0^L \ , \ \Gamma_1^L \Gamma_1^L \subset \Gamma_0^L \ , \ \Gamma_0^L \Gamma_1^L \subset \Gamma_1^L \ , \ \Gamma_1^L \Gamma_0^L \subset \Gamma_1^L$$

$$\text{and} \qquad ab = (-1)^{pq}\, ba \qquad \text{if } a \in \Gamma_p^L,\ b \in \Gamma_q^L.$$

Now we wish to give a $\mathbf{Z}_2$ graded commutative Banach algebra structure to Γ^L [1]; to this and, consider a basis $\mathbf{e}_1, \mathbf{e}_2, ..., \mathbf{e}_L$ of $\mathbb{R}^L$and denote with M_L the set of finite sequences of positive integer: $\mu = (\mu_1, ..., \mu_k)$ with $1 \leq \mu_1 < \mu_2 ... < \mu_k \leq L$; ϕ denotes the empty sequence. Defining, for each $\mu \in M_L$,

$$\beta_\mu = \mathbf{e}_{\mu_1} \wedge \mathbf{e}_{\mu_2} \wedge ... \wedge \mathbf{e}_{\mu_k} \quad \text{and} \quad \beta_\phi = 0,$$

each element x of Γ^L can be uniquely written as

$$x = \sum_{\mu \in M_L} x_\mu \beta_\mu$$

where $x_\mu \in \mathbb{R}$. By considering the products $\beta_\mu \wedge \beta_\nu$ one can also define the symbols $f_{\mu\nu}^\rho$ by means of equations $\beta_\mu \wedge \beta_\nu = f_{\mu\nu}^\rho \beta_\rho$; it holds

$$f_{\mu\nu}^\rho = \begin{cases} \text{degr}\,(\Pi) & \text{if } \mu \cup \nu = \rho \\ 0 & \text{otherwise} \end{cases}$$

where Π denotes the permutation which sends $\mu\nu \to \rho$. By putting $\text{degr}(\mu) \equiv |\mu| = 0$ if k is even and 1 if k is odd, we get the following equalities:

$$f_{\mu\nu}^\rho = f_{\mu\nu}^\rho (-1)^{|\mu\nu|} \qquad f_{\nu\mu}^\rho f_{\alpha\rho}^\gamma = f_{\alpha\nu}^\rho f_{\rho\mu}^\gamma$$

Theorem. 2.1 The norm

$$\| x \| \equiv \sum_{\mu \in M_L} | x_\mu |$$

makes Γ^L into a *Banach algebra*; in other word, it holds:

$$\| 1 \| = 1 \quad \text{and} \quad \| ab \| \leq \| a \| \| b \|$$

for each a e b.

Proof. The first assertion is trivial; the second one follows from:

$$\| ab \| = \| \sum_{\mu \in M_L} a_\mu \beta_\mu \sum_{\nu \in M_L} b_\nu \beta_\nu \| = \sum_{\rho \in M_L} | \sum_{\mu,\nu \in M_L} a_\mu b_\nu f_{\nu\mu}^\rho |$$
$$\leq \sum_{\mu \in M_L} | a_\mu | \sum_{\nu \in M_L} | b_\nu | = \| a \| \| b \|$$

since more terms are present in the last member. □

An example. Consider the algebra Γ^2 generated on $\mathbb{R}^2$ by $1, \mathbf{e}_1, \mathbf{e}_2$ by means of the usual wedge product: an element $a \in \Gamma_2$ can be written as $a = A + B\mathbf{e}_1 + C\mathbf{e}_2 + D\mathbf{e}_1 \wedge \mathbf{e}_2$ where $A, B, C, D \in \mathbb{R}$; $\Gamma_0^2 = \{A + D\mathbf{e}_1 \wedge \mathbf{e}_2\}$, $\Gamma_1^2 = \{B\mathbf{e}_1 + C\mathbf{e}_2\}$.

In the previous discussion the value of L and consequently, the dimension of the algebra Γ^L is not very important; for technical reasons, which will been clarified in the following, we shall consider the inductive limit of Γ^L for $L \to \infty$; we shall denote with B^∞ this limit.

This allow us to consider a more axiomatic approach to supermanifold theory which is based on the following axioms [2]:

Definition 2.1 *A* $\mathbf{Z}_2$ *-graded commutative Banach algebra over* $\mathbb{R}$ *or* $\mathbb{C}$ *is a graded algebra* $Q = Q_0 \oplus Q_1$ *s.t.*

a) there exists an identity element 1 *satisfying* $\| 1 \| = 1$, $\| ab \| \leq \| a \| \| b \|$,

b) $b_r b_s = (-1)^{rs} b_s b_r, \quad b_r \subset Q_r, b_s \subset Q_s$,

c) $\| b_0 + b_1 \| = \| b_0 \| + \| b_1 \|, \quad b_i \subset Q_i$.

Definition 2.2 *A Banach Grassmann algebra (BG algebra) over* $\mathbb{R}$ *or* $\mathbb{C}$ *is a* $\mathbf{Z}_2$ *-graded commutative Banach algebra where*

a) for each continuous Q_0*-linear map* $f: Q_i \to Q_j$ *there is one and only one element* $v \in Q_{i+j}$ *s.t. for each* $b \in Q_i$*, we have* $f(b) = vb$ *with* $\| v \| = \| f \|$,

b) by denoting with Q_0' *the subalgebra generated by even powers of elements of* Q_1 *we have* $Q_0 = \mathbb{R} \oplus Q_0'$*, (for the case of* $\mathbb{R}$*) and* $\| \alpha + b \| = | \alpha | \, \| b \|$.

The norm $\| f \|$ is the usual norm for functions in Banach spaces.

Proposition. 2.1 Let Q a BG algebra; if $xa = 0$ for each $x \in Q_1$ then $a = 0$. This is a trivial consequence of item a) of definition 2.2.

Theorem. 2.2 Γ^L is not a BG algebra while B^∞ is.

Proof. The form $a = dx^1 \wedge ... \wedge dx^L$ does not verify proposition 2.1. Concerning B^∞ we first prove prop. 2.1. Let $xa = 0$ for each $x \in Q_1$.
Write $x = x_0 + \sum_i x_i dx^i + \sum_{ij} x_{ij} dx^i \wedge dx^j + \sum_{ijk} x_{ijk} dx^i \wedge dx^j \wedge dx^k + ...$ By putting $a = dx^1$, one gets $x_0, x_i = 0$, $x_{ij} = 0, ...$ for $i, j, ... \neq 1$; by a straightforward calculation one gets the proof of prop. 2.1. Now consider a map $f: Q_0 \to Q_r$; in this case item a) of def. 2.1 is trivial: $f(b) = f(b) = bf(1) = bv = vb$; in the case $f: Q_1 \to Q_1$ one has, for $a, b, c \in Q_1$, $f(abc) = abf(c) = f(-acb) = -acf(b)$ and $a[bf(c) + cf(b)] = 0$; this implies $bf(c) = -cf(b) = f(b)c$. By putting $b = \sum b^\mu \beta_\mu$ and $c = \sum c^\nu \beta_\nu$ we get $b^\mu c^\nu [\beta_\mu f(\beta_\nu) - \beta_\nu f(\beta_\mu)] = 0$. If we choose $b^\mu = \delta^\mu_\alpha$ and $c^\nu = \delta^\nu_\gamma$ we get $f(\beta_\mu) = \lambda_\mu \beta_\mu$, $f(\beta_\nu) = \lambda_\nu \beta_\nu$ (ν not summed) which, in turn, implies $\lambda_\mu = \lambda_\nu = \lambda$ and therefore $f(c) = f(\sum c^\mu \beta_\mu) = \sum c^\mu \lambda \beta_\mu = \lambda c$. The proof for the case $f: Q_1 \to Q_0$ is similar. Concerning completeness we observe that B^∞ is isomorphic to l_1 which is, of course, complete. We have only to proof $\| f \| = \| v \|$ if $f(b) = vb$. Since

$$\| f \| = \sup_b \frac{\| vb \|}{\| b \|} \leq \frac{\| b \| \| v \|}{\| b \|} \leq \| v \|$$

if $f: Q_0 \to Q_r$ we get $\| f \| = \| v \|$ by choosing $b = 1$; if $f: Q_1 \to Q_r$ we set $v = \tilde{u}_i + v_i$ where $\tilde{u}_i$ does not contain dx^i. Since $f(dx^i) = \tilde{u}_i dx^i$ we get $\| f \| \geq \| \tilde{u}_i \|$ and $\sup_{i \in \mathbf{N}} \| \tilde{u}_i \| \leq \| f \| \leq \| v \|$. Now, for each i we find a value $\mu(i)$ s.t., setting $v = \sum v^\mu \beta_\mu$, i does not appear after $\mu(i)$. Hence $\| v_i \| \leq \sum_{\mu(i)}^\infty | v^\mu |$ and

$$\lim_{\mu(i) \to \infty} \sum_{\mu(i)}^{\infty} | v^\mu | = 0.$$

Now one easily gets $\| v \| = \| f \|$. □

A useful concept is the so called " body " map $\varepsilon: Q \to \mathbb{R}$; it is the projection of $Q = \mathbb{R} \oplus Q'_0 \oplus Q_1$ on its first factor. The algebra B^∞ admits the ordered basis β_μ; in the following we shall consider only algebra with this property which we shall call Special Banach Grassmann Algebra (SBG).

Theorem. 2.3 An element $a \in Q$ is invertible if and only if $\varepsilon(a) \neq 0$ [3].

Proof. The expression

$$a^{-1} = \varepsilon(a)^{-1} \sum_{n=0}^{\infty} (-1)^n \left(\frac{a - \varepsilon(a)}{\varepsilon(a)} \right)^n .$$

is convergent and satisfies $aa^{-1} = 1$. The inverse implication is standard. □

Define now the $m+n$ Cartesian products $Q^{m+n} = Q \times ... \times Q$; the norm $\| Y \| = \sum_{A=1}^{m+n} \| Y^A \|$ makes Q^{m+n} into a Banach Q-module and a Hausdorff space with respect to the topology induced by this norm. This Banach module is also graded: $Q^{m+n} = Q_0^{m+n} \oplus Q_1^{m+n} \equiv Q^{m,n} \oplus Q^{\hat{m},\hat{n}}$. The typical element of the even part $Q^{m,n}$ can be written as $X^A = (x^i, \theta^\alpha)$ where $x^i = (x^1, ..., x^m) \in Q_0^m$ and $\theta^\alpha = (\theta^1, ..., \theta^n) \in Q_1^n$ (here $A = 1, ..., m+n$).

The Banach Q_0-module $Q^{m,n}$ is also called vector superspace (vss for short).

Definition 2.3 [2] *A function $f: U \subset Q^{m,n} \to Q^{p,q}$ is said to be supersmooth if its Fréchet differential is C^∞ and Q_0-linear: that is*

$$(Df)_{(x)}(aY) = a(Df)_{(x)}(Y)$$

for all point $x \in U$, all $a \in Q_0$ and all $Y \in Q^{m,n}$.

A function admitting the following convergent expansion

$$f(X^A) = c_0 + X^A\, c_A + \frac{1}{2} X^A\, X^B\, c_{AB} + \frac{1}{6} X^A\, X^B\, X^C\, c_{ABC} + ...$$

is called superanalytical (SA for short)and satisfies the requirement of definition 2.3.

This definition implies that the " derivatives " of f with respect to X^A is well defined; see the interesting paper [4], in which a beautiful discussion on this crucial point is done. This is very important, since other approaches which adopt the algebra B^L, $L \leq \infty$, like the original one introduced by Rogers in [1] do not give a good definition of odd partial derivatives. Actually, a top form $dx^1 \wedge ... \wedge dx^L$ can always been added to an odd derivatives without changing anything.

We notice that, even if f is C^∞ in the sense of Fréchet, we do not necessarily succeed in defining the directional derivatives with respect to X^A; this because, X^A, despite its synthetic symbol has a complicate intrinsic structure; this matter is very similar to the problem of defining holomorphic functions on $\mathbb{R}^2$: only if Cauchy Riemann conditions are verified we can speak of derivative with respect to z. Definition 2.3 plays, in our framework, the same role played, in complex analysis, by Cauchy Riemann conditions; the analogy with this mathematical branch does not stop here [5].

3. Supermanifolds.

The simpler way to define supermanifolds (S manifolds for short) is to consider particular Banach manifolds [6] which are well suited to our definition of supersmooth function.

Following this idea we have

Definition 3.1 *Let M a Banach manifold; a (m,n)-dimensional super atlas on M is the couple $\mathcal{A} = \{(U_\alpha, \psi_\alpha)\}$; U_α is an open covering on M and $\psi_\alpha : U_\alpha \to V_\alpha \subset Q^{m,n}$ coordinate maps s.t. the transition functions $\psi_\alpha \bullet \psi_\mu^{-1} : \psi_\mu (U_\mu \cap U_\alpha)$ are supersmooth for all U_α.*

Definition 3.2 *Two atlases $\mathcal{A}$ and $\mathcal{A}'$ are said to be compatible (or equivalent) if the functions $\psi'_\alpha \bullet \psi_\mu^{-1} : \psi_\mu (U'_\mu \cap U_\alpha)$ are supersmooth for all U'_α and U_μ.*

(In other words, if $\mathcal{A} \cup \mathcal{A}'$ is still an atlas.)

Definition 3.3 *A (m,n) S differential structure for a Banach manifold M is an equivalence class of atlases of M.*

Definition 3.4 *A (m,n)-dimensional supermanifold M is a Banach manifold with a S differential structure.*

As in the theory of differential manifolds, two different, inequivalent atlases make the *same* Banach manifold M into two different S differential manifolds.

Definition 3.5 *Two S differential structure S and S' for a Banach manifold M are said to be equivalent if there exists a continuous isomorphism $\phi : M \to M'$ which is supersmooth when expressed on the atlases S and S'.*

The previous discussion implies that the only functions which can live in our theory are the supersmooth ones.

Remark. Due to the anticommuting property of the coordinates θ^α, supersmooth functions are very simple as far as the odd functional dependence is concerned: they are polynomials

$$\phi^L(x^i, \theta^\alpha) = \phi^L(x^i) + \sum_{\{\alpha\}} \sum_{k \le m} \phi^L_{(k)}(x^i)\theta^{\alpha_1} ... \theta^{\alpha_k} .$$

This formula is also known as the " superfield expansion ".

Now we wish to construct the tangent space $T_u(M)$ to a point $u \in M$; to this end, we consider the space C_u of the curves with a " first order contact " in u; in other words, by considering two maps $\gamma_1(t)$ and $\gamma_2(t) : I \subset \mathbb{R} \to M$ we say that they are equivalent in u if $\gamma_1(0) = \gamma_2(0) = u$ and

$$\frac{d}{dt} X^A(\gamma_1(t))|_{t=0} = \frac{d}{dt} X^A(\gamma_2(t))|_{t=0} .$$

This equivalence relation, of course independent of the coordinate system locally used, allows to define $T_u(M)$ as $\frac{C_u}{\sim}$. In this case the maps $\gamma \in [\gamma]$ give origin, in u to the vector

$$Y^A = (Y^i, Y^\alpha) = \frac{d}{dt} X^A(\gamma(t))|_{t=0} = \left[\frac{d}{dt} x^i(t), \frac{d}{dt} \theta^\alpha(t)\right] |_{t=0} .$$

In analogy with the standard case, one can define $T(M)$ as the bundle of the tangent vectors on M.

Remark. By denoting, as usual, the vectors as $Y = Y^A \left(\frac{\partial}{\partial x^A}\right)_u$ we notice that a tangent vector has component in $Q^{m,n}$; in other words, while $\frac{\partial}{\partial x^i}$ is a tangent vector since the real number 1 belongs to Q_0, for the same reason $\frac{\partial}{\partial \theta^\alpha}$ is not.

The space of the Q_0-linear maps from $T_u(M)$ to Q is defined as the dual space $T^\dagger_u(M)$; a basis is constituted by the maps $dx^A)_u$, $A = 1, ..., m+n$ defined by equations

$\frac{\partial}{\partial x^A})_u \left(dx^B\right)_u = \delta_A^B$. A cotangent vector of $T_u^\dagger(M)$ is $\omega = \sum_{A=1}^{m+n} dx^A)_u\, \omega_A$ for $\omega_A \in Q^{m+n}$.

The action of the vectors on the forms is the following:

$$\omega(Y) = Y(\omega) = Y^A \frac{\partial}{\partial x^A}(dx^B\, \omega_B) = Y^A\, \omega_A .$$

In a similar way, by taking into account the theory of Banach manifolds, one can easily generalize the usual concepts of *tensorial product* $\otimes$, *wedge product* $\wedge$, *exterior differential* d: $\Lambda^k(M) \to \Lambda^{k+1}(M)$, *and Lie derivatives:* by considering a smooth tangent vector field $v(x)$ on M one introduce the *flow* $\phi_t(x)$ associated to $v(x)$ (here t denotes a real parameter); it is the *local, abelian one dimensional group of automorphisms* of M which satisfies

$$\phi_t(\phi_s(x)) = \phi_{t+s}(x) \qquad v(x) = \phi_t'(x)$$

for real parameters t and s.

Now, for every tangent field $v(x)$ we define

$$\mathcal{L}_v u \equiv \lim_{t \to 0} \frac{-\phi_{t*} u + u}{t}$$

as the *Lie derivative* of the field $u(x)$ with respect to the motion ϕ_t. Of course, for each local abelian one dimensional group of automorphisms of M we can determine a tangent vector field $v(x)$ tangent to ϕ_t; on the contrary, for each tangent vector field $v(x)$, there exists *locally* an abelian one dimensional group of automorphisms of M tangent to $v(x)$. Similarly, by duality, we get the Lie derivative also for each cotangent field $\omega(x)$; the following relation holds:

$$\mathcal{L}_v \omega = v \rfloor d\omega + d\left(v \rfloor \omega\right).$$

4. Super Lie Groups

In this section we discuss some main results concerning SL groups [7, 3, 8].

Definition 4.1 *A Super Lie Group $\mathcal{G}$ is SA manifold s.t.*

a) it is also an abstract group

b) the composition rule $\lambda: \mathcal{G} \times \mathcal{G} \to \mathcal{G}$ is superanalytical.

Consider the tangent space at the identity e of a (m,n)-dimensional SL group $\mathcal{G}$. $T_e(\mathcal{G})$ is formed by the vectors $V = \sum_{i=1}^m v^i \left(\frac{\partial}{\partial x^i}\right)_e + \sum_{\alpha=1}^n v^\alpha \left(\frac{\partial}{\partial x^\alpha}\right)_e$ where $v^i, v^\alpha \in Q^{m,n}$.

Definition 4.2 *The map $L_a: \mathcal{G} \to \mathcal{G}$, for $a \in \mathcal{G}$, given by $L_a b = ab$ is said the left-transport on $\mathcal{G}$.*

By standard techniques, we can define also the " push forward " map $L_{a*}: T_b(\mathcal{G}) \to T_{ab}(\mathcal{G})$; this is also an " internal " SL group automorphism since it is SA. Define now the vector fields $D_A(a) \equiv L_{a*} \frac{\partial}{\partial x^A})_e$ for each $a \in \mathcal{G}$, and for all $A = 1, ..., m+n$. The fields D_A are said the left-invariant vector fields of $\mathcal{G}$.

Theorem. 4.1

a) The sums $\sum_{A=1}^{m+n} V^A D_A(a)$ generate $T_a(\mathcal{G})$ for each $a \in \mathcal{G}$ in a unique way if $V^A \in Q^{m,n}$,

b) $L_{a*} D_A(b) = D_A(c)$ for $c = ab$.

Proof. Obvious.

Consider two vector fields $u = \sum u^A \frac{\partial}{\partial x^A}$ and $v = \sum v^B \frac{\partial}{\partial x^B}$ on $\mathcal{G}$. The Lie derivative $\mathcal{L}_u v = [u, v] = uv - vu$ will now be calculated:

$$\begin{aligned}\mathcal{L}_u v &= \sum u^A \frac{\partial}{\partial x^A} \sum v^B \frac{\partial}{\partial x^B} - \sum v^B \frac{\partial}{\partial x^B} \sum u^A \frac{\partial}{\partial x^A} \\ &= \sum u^A v^B \left(D_B D_A - (-1)^{AB} D_A D_B\right) \equiv \sum u^A v^B \left[D_B, D_B\right]_\pm\end{aligned}$$

Notice that the Lie derivative is given, exactly as Banach manifold theory requires, by the usual commutator; its grading nature arises as a consequence of the fact to write the terms $u^A v^B$ on the left of the other terms.

The following theorem is straightforward [9]:

Theorem. 4.2 The quantities C_{BA}^E defined by equations $[D_B, D_A]_\pm = \sum_E C_{BA}^E D_E$, $E = 1, ..., m + n$, are called " the structural constants of the SL group $\mathcal{G}$ "; they are graded antisymmetric $C_{BA}^E = -(-1)^{AB} C_{BA}^E$ with values in Q_{A+B+C}.

Definition 4.3 *A Super Lie module W is a free finite dimensional $\mathbf{Z}_2$ graded left Q-algebra such that:*

a) $Q_r W_s \subset W_{r+s}$, $\quad [W_r, W_s] \subset W_{r+s}$

b) $[cX, Y] = c[X, Y]$ *for all* $c \in Q$

c) $(-1)^{rt}[X, [Y, Z]] + (-1)^{ts}[Z, [X, Y]] + (-1)^{rs}[Y, [Z, X]] = 0$

Here $X \in W_r$, $Y \in W_s$, $Z \in W_t$.

Remark. W is infinite dimensional when regarded as a $\mathbb{R}$ vector space.

Theorem. 4.3 $T_e(\mathcal{G})$ is the even sector of a SL module W with respect to the [ ] product; W is *decomposable,* that is $W = Q \otimes g$ for some graded Lie algebra g, if and only if the structural constants C_{BA}^E are real.

Proof. The first assertion is trivial; if W is decomposable the theorem follows by elementary properties of g.l. algebras. Conversely, by defining $g = \left\{\sum_{A=1}^{m+n} v^A D_A\right\}$, $a \in \mathbb{R}$, a graded Lie algebra structure arises from the properties of the Lie bracket. □

The $\{D_A\}$ are a Q-module basis for W; we make W into a Banach algebra by setting

$$\| X \| = \sum_A \| X^A \| \quad \text{for } X = \sum_A X^A D_A$$

and by checking that

$$\| [X, Y] \| = \sum_E \| \sum_{A,B} Y^B X^A C_{AB}^E \| \leq M \| X \| \| Y \|$$

where $M = \sum_{ABE} |C_{AB}^E|$.

Our main aim is to prove the following theorem [8]:

Theorem. 4.4 *Let Q a SBG algebra, and W an SL module over it; then there exists a unique connected, simply connected SL group $\mathcal{G}$ whose Lie module is isomorphic to W. The uniqueness is up to equivalence classes of superanalytic structures on G.*

Lemma. We may associate to any SL module W over a SBG algebra Q, a Banach Lie Group in a standard way.

We introduce a Banach Lie algebra $\tilde{g}$ underlying W. Setting $\Xi_{A\mu} = \beta_\mu D_A$ we define real numbers $F^{\alpha}_{\mu\nu\rho}$ and $C^{E\mu}_{AB}$ as $\beta_\mu\beta_\nu\beta_\rho = F^{\alpha}_{\mu\nu\rho}\beta_\alpha = f^{\tau}_{\mu\nu} f^{\alpha}_{\tau\rho}\beta_\alpha$ and $C^{E}_{AB} = C^{E\mu}_{AB}\beta_\mu$ respectively; $\tilde{g}$ is the $\mathbb{R}$-linear span of the $\Xi_{\mu A}$'s; we have

$$X = \sum_{\mu A} b^{\mu A} \Xi_{\mu A} \quad \text{and} \quad \| X \| = \sum_{\mu A} \| b^{\mu A} \| < +\infty$$

$\tilde{g}$ is graded according to $\deg(\Xi_{A\mu}) = \deg A + \deg \mu$; in the following we shall mean deg A when A, for instance, is an exponent. In $\tilde{g}$ there is the product

$$[\Xi_{A\mu}, \Xi_{B\nu}] = E^{C\rho}_{A\mu B\nu}\, \Xi_{C\rho} \quad \text{where} \quad E^{C\rho}_{A\mu B\nu} = (-1)^{A\nu} C^{C\alpha}_{AB} F^{\rho}_{\mu\nu\alpha}.$$

We introduce also a map $\alpha: \tilde{g} \to W$ as $\alpha(a^{A\nu}\Xi_{A\nu}) = a^A D_A$ where $a^A = a^{A\nu}\beta_\nu$. The even part g of $\tilde{g}$ is a Banach Lie Algebra isomorphic to W_0; now we wish to find a BL group whose algebra is g. Since not every BL algebra is the algebra of some BL group [10], we have to represent g into an enlargable BL algebra.

Let A a associative B algebra over $\mathbb{R}$ with an ordered basis $\{e_i\}$. If a is $a = \sum_i a_i e_i$ we set $\| a \| = \sum_i | a_i |$; now define $A \otimes A$ as the $\| \,\|$-completion of the algebraic tensor product and assume that the norm of $t = \sum_{ik} t^{ik} e_i \otimes e_k \in A \otimes A$ is $\| t \| = \sum_{ik} | t^{ik} |$. In a natural way we get the tensor algebra over A of the controvariant tensors $T = \mathbb{R} \oplus A \oplus ... \oplus A_l \oplus ...$ where $A_l = A \otimes ... \otimes A$ l times. Since any A_l is normed in analogy to $A \otimes A$, T becomes a B algebra with respect to the norm $\| t \| = \sum \| t_q \|$ if $t = t_0 + ... + t_q$.

Consider now the following faithful representation $\gamma: A \to \text{End}(A)$ given by $\gamma(a)c = ac$ for all $c \in A$. This is called the *left regular representation* of A. Finally, notice that we can get a BL algebra, called A_L by setting in A the following product $[a, c] = ac - ca$.

The Poincaré Birkoff Witt theorem [11] allows us to identify the universal enveloping algebra U of A with a closed subalgebra of T.

The following theorem is crucial:

Theorem. 4.5 The universal enveloping algebra U of A is a B algebra, and, therefore A_L is a BL algebra; moreover A has a faithful, norm preserving representation into $(\text{End}(U))_L$.

Proof. According to the Poincaré Birkoff Witt theorem there exists a map $u: A \to U_L$ which is a Lie isomorphism of A onto its image [12]; u is also norm preserving; by regarding $\gamma: U \to \text{End}(U)$ as a faithful, norm preserving Lie representation $\gamma: U_L \to \text{End}(U)_L$ we set $\eta = \gamma \dot{u}$ and obtain a norm-preserving map $\eta: A \to \text{End}(U)_L$ that is also a BL isomorphism. □

The crucial fact is that $\text{End}(U)_L$is the Lie algebra of $GL(U)$, the Banach Lie group of bounded linear automorphisms of U. Standard results in BL theory ensures us that there exists a BL group whose algebra is A. For the convenience of the reader, we shall state now, without proof, the following main theorem.

Theorem. 4.6 [6] If H is BL group with Lie algebra h, and s is a sub BL algebra of h, then there exists a unique (up to isomorphisms) sub BL group of H whose Lie algebra is isomorphic to s.

It follows that [13] there exists a unique (up to isomorphisms) connected, simply connected BL group whose Lie algebra g is isomorphic to g.

Summing up we have

Theorem. 4.7 Given any SL module W, there exists a unique (up to BL isomorphisms) connected, simply connected BL group $\mathbf{G}$ whose Lie algebra is BL isomorphic to the even sector of W.

The group G admits a unique analytic structure [14]. G denote the topological group underlying $\mathbf{G}$.

Our main task is to give a SA structure to the BL group $\mathbf{G}$ which turns $\mathbf{G}$ into a SL group $\mathcal{G}$.

Denote with N the family of nuclei of the topological group G; for a definition of nucleus and of *local* topological group see [15, 16].

Theorem. 4.8 Let $\mathbf{G}$ a BL group whose Lie algebra g is isomorphic to the even part of an SL module W; there exists a $U \in N$ which can carry a unique SA structure $\mathcal{S}$ which makes U into a local SL group for which $T_e(U) \sim W_0$.

Proof. For $a \in N$ consider BL canonical coordinates ϕ: if $a = \exp(a^{A\nu}\Xi_{A\nu})$, put $\phi(a) = a^{A\nu}$; by setting $\psi = \alpha \bullet \phi: V \to Q^{m,n}$ we get a local SA structure $\mathcal{S}$ which makes N into a local SL group. By denoting with Λ and $\tilde{\Lambda}$ the representations of the group product λ in coordinates ϕ and ψ respectively, by using the Campbell formula in $\mathbf{G}$ and the fact that α is a Lie isomorphism we get a Campbell formula also in $\mathcal{G}$ with respect to $\mathcal{S}$:

$$z^C = \tilde{\Lambda}(a^A, b^B) = a^C + b^C + \frac{1}{2}\sum_{A,B} b^B a^A C^C_{AB} + \dots$$

where $z = \lambda(a, b)$. This equation implies the S-analicity of $\tilde{\Lambda}$ and the relation: $T_e(U) \sim W_0$ where T_e is taken with respect to $\mathcal{S}$. The proof that the SA structure induced by ψ is unique (up to equivalence) is easily obtained; actually, any further SA structure should give origin to T_e which is, by hypothesis, isomorphic to W_0. □

Now we extend to the whole group G the SA structure $\mathcal{S}$ defined on U.

Theorem. 4.9 Let $\mathbf{G}$ a connected BL group with underlying topological group G; let $\mathcal{S}$ an SA structure defined on nucleus of G. The SA structure $\mathcal{S}$ can be extended to all G obtaining a unique (up to equivalence of SA structure) SL group $\mathcal{G}$.

Outline of the proof. First of all, we extend the structure $\mathcal{S}$ to all of G. By considering the open $U \in N$ of Theorem 4.8, we define the atlas:

$$\mathcal{A} = \left\{(U_b, \psi_b) | U_b = Ub,\ \psi_b(x) = \psi(xb^{-1}), x \in U\right\}.$$

This is an atlas whose local coordinate functions are labeled by the group elements b; it can shown that for each $x \in U_a \cap U_b$ the maps $\psi_a(x) \to \psi_b(x)$ are one-to-one and SA; furthermore the product λ in G is SA when expressed in local coordinates. Concerning the SA structures, if $\tilde{\mathcal{S}}$ is another SA structure which makes G into a different SL group $\tilde{\mathcal{G}}$ whose Lie module is isomorphic to W_0, one can show that $\tilde{\mathcal{S}}$ and $\mathcal{S}$ are equivalent.

We wish to point out that, following definition 3.5, $\tilde{\mathcal{S}} \sim \mathcal{S}$ means that there exists a continuous isomorphism $\phi: \mathcal{G} \to \tilde{\mathcal{G}}$ which is supersmooth when expressed on the atlases of $\mathcal{A}$ and $\tilde{\mathcal{A}}$. This map is an automorphism only in the categories of topological and BL groups but it is *not* a SL automorphism. It follows that $\mathcal{G}$ and $\tilde{\mathcal{G}}$ are different as SL groups: it is not true, that any continuous automorphism is SA; in other words, the same topological group can carry different SA structures.

For instance, by considering the abelian structure of $Q : \lambda(a,b) = a+b$ where $\phi(a) = a^{\nu}$, we get a unique, up to BL automorphism, BL group $\mathbf{Q}$ whose underlying set is Q. On the contrary, Q can carry the following equivalent, incompatible SA structures:

$$\mathcal{A} = \{(Q,\psi)|\psi(a) = (a_0, a_1)\}, \qquad \tilde{\mathcal{A}} = \{(Q,\tilde{\psi})|\tilde{\psi}(a) = (a_0 - \frac{1}{2}\varepsilon(a_0), a_1)\}.$$

Two different SL groups $\mathcal{Q}$ and $\tilde{\mathcal{Q}}$ arise. The " transition function " $\varsigma = \psi \bullet \tilde{\psi}^{-1}$ is not SA, while the map $\xi : \mathcal{Q} \to \tilde{\mathcal{Q}}$ given by $\xi = \tilde{\psi}^{-1} \bullet \psi$ is an SA isomorphism making $\mathcal{Q}$ and $\tilde{\mathcal{Q}}$ equivalent.

Theorem. 4.10 Let G a connected topological group carrying at least one SA structure making it into a SL group $\mathcal{G}$; Then there is a one-to-one correspondence between isomorphism classes of SL modules and equivalence classes of SA structure on G.

Remark. Since given a graded Lie algebra g we can construct the Lie module $W = Q \otimes g$ we have that, for each graded Lie algebra g there exists a unique connected, simply connected SL group $\mathcal{G}$ for which $T_e\mathcal{G} = W_0$.

We have completely proved Theorem 4.4.

5. Results on SL groups.

In this section we shall give some results concerning SL groups and their subgroups. This will be useful for the construction of Homogeneous Supermanifolds.

We denote with $gl(m+n)$ the $(m+n) \times (m+n)$ matrices with values in Q. $gl(m+n)$ is $\mathbf{Z}_2$ graded in the following way:

$$gl(m+n) = gl(m,n) \oplus gl(\tilde{m},\tilde{n})$$

where $Y \in gl(m,n)$ if $Y_A^B \in Q_{A+B}$ and $Y \in gl(\tilde{m},\tilde{n})$ if $Y_A^B \in Q_{A+B+1}$. On the other hand, $gl(m+n)$ is a graded SL module with respect to the graded commutator $[X,Y] = XY - (-1)^{XY}YX$.

In order to study the invertibility of the matrices in $gl(m+n)$, we denote with $sgl(m+n)$ the submodule of $gl(m+n)$ constructed by the matrices with vanishing body: $sgl(m+n) = \{A \in gl(m+n)|\varepsilon(A) = 0\}$.

Theorem. 5.1 For each $A \in sgl(m+n)$ and any real number $\beta \in (0,1)$ we can find a real number $\alpha \leq 0$ such that $\| A \| \leq \alpha\beta^n$ for every positive integer n.

Proof. Since every element $x \in Q$ can be splitted into the sum $x = y + z$ where y is sum of strictly nilpotent terms and $\| z \| \leq 1$ [2] we apply this result to the terms of the matrix $X = Y + Z \in sgl(m+n)$. This means that for each E and D there exists a positive number $N(E,D)$ such that $(Y_E^D)^{N(E,D)} = 0$ Taking $N = (m+n)^2 \times \mathrm{Max}_{E,D} \{N(E,D)\}$ $\quad E,D = 1,...,m+n$ we deduce that, for each integer $s \geq 0$, all the terms containing at least N Y's in the expansion $X^{N+s} = (B+C)^{N+s}$ vanish independently of the order in which the Y's appear.

We have

$$\begin{aligned} X^{N+s} = & Y^{N-1}Z^{s+1} + Y^{N-2}Z^{s+1}Y + \dots \binom{N+s}{N-1} \text{terms} + \\ & Y^{N-2}Z^{s+2} + Y^{N-3}Z^{s+1}YZ + \dots \binom{N+s}{N-2} \text{terms} + \end{aligned}$$

and

$$\| X^{N+s} \| \leq \sum_{r=0}^{N-1} \binom{N+s}{r} \| Y \|^r \| Z \|^{N+s-r} .$$

We can now use lemma 2.7 of [1] to get the proof. □

Define $GL(m,n)$ the SL group of invertible matrices in $gl(m,n)$.

Lemma. The map Exp: $sgl(m+n) \to 1 + sgl(m+n)$ given by $\operatorname{Exp} X = \sum_{n=0}^{\infty} \frac{X^n}{n!}$ is an SA bijection with the SA inverse: $\log : 1 + sgl(m+n) \to sgl(m+n)$ given by $\log(1+X) = \sum_{n=1}^{\infty} \frac{(-X^n)}{n}$.

Proof. These series are convergent by Theorem 6.1; it is easy to show that log is the inverse map of Exp; on the other hand, these maps are SA since they are analytic and admit Q_0 linear Frechét differentials. □

The following theorem deals with the matrices of $GL(m,n)$.

Theorem. 5.2 A matrix $A \in gl(m,n)$ belongs to $GL(m,n)$ if and only if its *body matrix* is non singular.

Proof. Let $A_0 = \varepsilon(A)$; if A is non singular $AA^{-1} = 1$ and, therefore, $A_0 A_0^{-1} = 1$.

Conversely, if A_0 is nonsingular we set:

$$A^{-1} = A_0^{-1} \sum_{k=0}^{\infty} (-A_0^{-1} A)^k .$$

On one hand, this series is convergent by Theorem 6.1 and, on the other hand, the relations $AA^{-1} = A^{-1}A = 1$ hold. □

Now we wish to study the sub SL groups. In order to do that, we need a theorem concerning free graded (left) Q-modules. Suppose one considers a free graded (left) Q-module W;

Definition 5.1 *Let W a free graded Q-module. The couple (V,j) will be called a free submodule of W if:*

a) V is a free graded Q-module

b) $j: V \to W$ is a Q_0-linear injective BL homomorphism.

Theorem. 5.3 Let (V,i) be a free SL submodule of W; then there exists a free graded Q-module U such that $W = i(V) \oplus U$.

Proof. Set $m+n = \dim W$ and $p+q = \dim V$. We prove this theorem by showing that any matrix B_i^A associated to a Q_0-linear injection $i: V \to W$ can be completed to a matrix of $GL(m,n)$; here $i = 1, ..., p+q$; $A = 1, ..., m+n$. Since i is injective, $v^i A_i^A = 0$ implies $v^i = 0$ and, therefore, there exists a matrix Y_A^k such that $B_i^A Y_A^k = \delta_i^k$. Now,

By denoting with $\tilde{B} = \varepsilon(B)$ and $\tilde{Y} = \varepsilon(Y)$ one easily gets:

$$\tilde{B}_{i_0}^{A_0} \tilde{Y}_{A_0}^{k_0} = \delta_{i_0}^{k_0}, \quad \text{and} \quad \tilde{B}_{i_1}^{A_1} \tilde{Y}_{A_1}^{k_1} = \delta_{i_1}^{k_1}$$

The subscripts to the indices denote the sectors where the indices run. Suppose now that there exists a $c^{i_0} \neq 0$ such that $c^{i_0} \tilde{B}_{i_0}^{k_0} = 0$; since $c^{i_0} \tilde{B}_{i_0}^{k_0} \tilde{Y}_{k_0}^{l_0} = c^{l_0} = 0$ we get that the " vectors " $\tilde{B}_{i_0}^{k_0}$ are independent; in the same way we can show that also the " vectors " $\tilde{B}_{i_1}^{k_1}$ are independent: in other words, the rank of $\tilde{B}_i^A$ is $p+q$. Then, there exist real vectors $\mathbf{b}_{p+1} = \left(b_{p+1}^1, ..., b_{p+1}^m\right)$, ..., $\mathbf{b}_m = \left(b_m^1, ..., b_m^m\right)$ which complete $\tilde{B}_{i_0}^{k_0}$ to a basis of $\mathbb{R}^m$; with the same procedure we get the following matrix which satisfies our requirement:

$$C = \begin{pmatrix} B_1^1 & \dots & B_1^m & B_1^{m+1} & \dots & B_1^{m+n} \\ \dots & \dots & \dots & \cdot & \dots & \cdot \\ B_p^1 & \dots & B_p^m & B_p^{m+1} & \dots & B_p^{m+n} \\ b_{p+1}^1 & \dots & b_{p+1}^m & 0 & \dots & 0 \\ \dots & \dots & \dots & 0 & \dots & 0 \\ b_m^1 & \dots & b_m^m & 0 & \dots & 0 \\ B_{m+1}^1 & \dots & B_{m+1}^m & B_{m+1}^{m+1} & \dots & B_{m+1}^{m+n} \\ \dots & \dots & \dots & \cdot & \dots & \cdot \\ B_{m+q}^1 & \dots & B_{m+q}^m & B_{m+q}^{m+1} & \dots & B_{m+q}^{m+n} \\ 0 & \dots & 0 & b_{m+q+1}^{m+1} & \dots & b_{m+q+1}^{m+n} \\ 0 & \dots & 0 & \cdot & \dots & \cdot \\ 0 & \dots & 0 & b_{m+n}^{m+1} & \dots & b_{m+n}^{m+n} \end{pmatrix}.$$

□

Now we give some definitions concerning sub SL groups.

Definition 5.2 *Let $\mathcal{G}$ a SL group. The couple $(\mathcal{H}, i)$ will be called a sub SL group if:*

a) $\mathcal{H}$ is an SL group,

b) $i: \mathcal{H} \to \mathcal{G}$ is a supersmooth one-to-one immersion,

c) $i: \mathcal{H} \to \mathcal{G}$ is a group homomorphism.

Now we shall show that a given sub SL group can carry a unique S differential structure.

Lemma. Let M a S manifold and (P, i) a sub S manifold of M; let $\psi: N \to M$ a supersmooth map from a Supermanifold N into M such that $\psi(N) \subset i(P)$. If $\psi_0: N \to P$ given by $\psi_0 = i^{-1} \bullet \psi$ is continuous, then it is supersmooth.

Now we give some details on the linkage between the SA structure of a SL group $\mathcal{G}$ and the BL structure of its underlying BL group $\mathbf{G}$.

Since $Q^{m,n}$ is also a B space any SL group $\mathcal{G}$ can be also regarded as a BL group; we identify $Q^{m,n}$ with a B space as follows. If $\{\beta_\mu, \mu \in J\}$ is a basis of Q (here J denotes some ordered set) and $v^A \in Q^{m,n}$, we set $v^A = v^{A\mu} \beta_\mu$.

Now we define a map $c: \mathcal{G} \to G$ which, as we shall see, is closely related to the map α of Sect. 4.

Let $\{(U_\beta, \psi_\beta)\}$ and $\{(U_\beta, \phi_\beta)\}$ suitable atlases of $\mathcal{G}$ and G respectively; (we have chosen the same open covering $\{U_\beta\}$ for $\mathcal{G}$ and G).

We define $\phi_\beta \bullet c \bullet \psi_\beta^{-1}(z^A) = z^{A\mu}$. We can easily show that:

a) $c_*(\partial v^A/\partial z^A) = \partial v^{A\mu}/\partial z^{A\mu}$

b) $c_*^{-1}(\partial v^{A\mu}/\partial z^{A\mu}) = \partial v^A/\partial z^A$

c) $c_* = \alpha^{-1}$.

c_* is a $\mathbb{R}$ linear analytic BL isomorphism of W_0 onto g.

Theorem. 5.4 Let $(\mathcal{H}, i)$ a sub SL group of a SL group $\mathcal{G}$; The SA structure of $\mathcal{H}$ is unique.

Proof. First of all introduce the BL groups $\mathbf{G}$ and $(\mathbf{H}, j)$ underlying $\mathcal{G}$ and $\mathcal{H}$ respectively; we notice that $j = c \bullet i \bullet c^{-1}$. Let $\mathcal{A} = \{(U_\beta, \psi_\beta)\}$ an SA atlas for $\mathcal{G}$ and $\mathcal{B} = \{(V_\mu, \phi_\mu)\}$, $\mathcal{C} = \{(Y_\nu, \chi_\nu)\}$ two SA atlases for $\mathcal{H}$. The transition functions $\chi_\nu \bullet \phi_\mu^{-1}$ are continuous since the BL structure of $\mathbf{H}$ is unique.

Consider now an $x \in V_\mu \cap Y_\nu$ such that $i(x) \in U_\beta$; the map i can be represented on $\mathcal{A}$ $\mathcal{B}$ and $\mathcal{C}$ as $i_{\beta\mu} = \psi_\beta \bullet i \bullet \phi_\mu^{-1}$ and $\tilde{i}_{\beta\nu} = \psi_\beta \bullet i \bullet \chi_\nu^{-1}$. The relation $i_{\beta\mu} = \tilde{i}_{\beta\nu} \bullet (\chi_\nu \bullet \phi_\mu^{-1})$ holds. By using the previous Lemma we can show that the functions $(\chi_\nu \bullet \phi_\mu^{-1})$ are SA and, therefore, $\mathcal{B}$ and $\mathcal{C}$ are compatible. □

Our next purpose is to show the following theorem.

Theorem. 5.5 Let $\mathcal{G}$ a SL group with SL module W; let $T \subset W$ a SL submodule of W; then there exists a unique connected sub SL group $(\mathcal{H}, i)$ of $\mathcal{G}$ such that the SL modulus of $\mathcal{H}$ is isomorphic to T.

To this end, we shall need some partial results.

First of all, we set up the definition of the exponential map, which, in a partially constructive framework, has been already introduced in Sect. 5.

Let $\mathcal{G}$ a SL group and $\mathbf{G}$ its underlying BL group. If V is a neighborhood of O in the BL algebra of $\mathbf{G}$ we denote with $\exp: V \to \mathbf{G}$ the exponential map in $\mathbf{G}$.

Definition 5.3 *The exponential map* $\mathrm{Exp}: U \to \mathcal{G}$ *from some neighborhood* U *of* O *in the SL modulus* W_0 *of* $\mathcal{G}$, *is* $\mathrm{Exp} = c^{-1} \bullet \exp \bullet c_*$.

Notice that U has to be chosen as $c_*^{-1}V$. Since it is always possible to choose V such that exp is a diffeomorphism between V and $\exp(V)$, it is possible, putting $U = c_*^{-1}V$, to construct an exponential map Exp which is a homeomorphism of U onto $c^{-1}(\exp V)$ and SA on all W.

Lemma. Let $\psi: \mathcal{H} \to \mathcal{G}$ an SL homeomorphism; then $\psi \bullet \mathrm{Exp} = \mathrm{Exp} \bullet \psi_*$.

Proof. Construct the map $\phi: \mathbf{H} \to \mathbf{G}$ as $\phi = c \bullet \psi \bullet c^{-1}$ which is a BL homomorphism; equation $\phi \bullet \exp = \exp \bullet \phi_*$ completes the proof.

Lemma. Let $\mathcal{G}$ a SL group, $(\mathcal{H}, i)$ a sub SL group; denote by $W_\mathcal{G}$ and $W_\mathcal{H}$ the corresponding SL modules; then $(W_\mathcal{H}, i_*)$ is a SL submodule of $W_\mathcal{G}$.

Proof. It is sufficient to reconstruct the classical proof to get the equation $[i_*X, i_*Y] = i_*[X, Y]$ for each $X, Y \in W_\mathcal{H}$; now, i_* is Q_0-linear since i is SA; on the other hand, the injectivity of i follows from the fact that i is an immersion; finally, Q_0-linearity implies that i_* is norm preserving. □

Now we prove Theorem 6.5. Denote with $\{D_A, A = 1, ..., \dim\mathcal{G}\}$ a basis of $W_\mathcal{G}$ and with C_{AB}^D the structural constants of $\mathcal{G}$; denote with $\mathbf{G}$ the BL group underlying $\mathcal{G}$. The map $c_*: W_\mathcal{G} \to \tilde{g}$ is a BL isomorphism onto the BL superalgebra $\tilde{g}$, whose even sector g is the BL algebra of $\mathbf{G}$. A basis of $\tilde{g}$ is $\{D_{A\mu} = c_*(\beta_\mu D_A)\}$.

Let us denote by $\{E_i = B_i^A D_A,\ A = 1, ..., \dim T\}$ a basis of the SL submodule

$T \subset W_{\mathcal{G}}$ with Lie bracket $[E_i, E_j] = a_{ij}^k E_k$; the set $\tilde{h} = c_\star(T)$ is a sub BL algebra of $\tilde{g}$ with basis $\{R_{i\mu} = c_\star(\beta_\mu E_i)\}$. We denote with $\{R^0_{i\mu}\}$ the even vectors of this basis. Now, BL algebras theory [6] ensures us that there exists a connected sub BL group $(\mathbf{H}, j)$ of $\mathbf{G}$ such that $j_\star(h) = \tilde{h}_0$; h is the BL algebra of $\mathbf{H}$ admitting the basis $\left\{P_{i\mu} \equiv j_\star^{-1} R^0_{i\mu}\right\}$. One can show that h and $\tilde{h}_0$ have the same structural constants:

$$d^{k\rho}_{i\mu,j\nu} = (-1)^{i\nu} F^{\rho}_{\mu\nu\tau}\, a^{k\tau}_{ij} .$$

By using Q^{p+q} as underlying space, we construct a SL module E isomorphic to T ($\dim T = p + q$). If P_i denotes the canonical basis of Q^{p+q}, we construct $\alpha: h \to Q^{p,q}$ by putting $\alpha(v^{i\mu} P_{i\mu}) = v^i P_i$ where $v^i = v^{i\mu}\beta_\mu$.

$Q^{p,q}$ with the commutator $[u, v] = \alpha\left[\alpha(u)^{-1}, \alpha(v)^{-1}\right]$ becomes the even sector of an SL module E with structural constants e^j_{ik} satisfying:

$$d^{k\rho}_{i\mu,j\nu} = (-1)^{i\nu} F^{\rho}_{\mu\nu\tau}\, e^{k\tau}_{ij} .$$

One can show that $e^{k\tau}_{ij} = a^{k\tau}_{ij}$, that is E and T are SL isomorphic. According to the results of Section 5, we put on $\mathbf{H}$ an SA structure which makes it into a SL group $\mathcal{H}$. The only crucial point is to choose, among all possible SA structures that $\mathbf{H}$ can carry, just the one for which $\mathcal{H}$ is a sub SL group of $\mathcal{G}$. We set $W_{\mathcal{G}} = E$ and $i: \mathcal{H} \to \mathcal{G}$ as $i = c^{-1} \bullet j \bullet c$. Now we prove that i is SA. To this end, on the nuclei U and V of $\mathcal{H}$ and $\mathcal{G}$, where normal coordinates ψ, ϕ are used, for an element $a^i \in \phi(V)$, we get

$$\begin{aligned}
\psi \bullet i \bullet \phi(a^i) &= \psi \bullet i \bullet \mathrm{Exp}(a^i P_i) = \psi \bullet c^{-1} \bullet j \bullet \exp \bullet c_\star(a^i P_i) \\
&= \psi \bullet c^{-1} \bullet \exp(a^{i\mu} j_\star P_{i\mu}) = \psi \bullet \mathrm{Exp} \bullet c_\star^{-1}(a^{i\mu} R^0_{i\mu}) . \\
&= \psi \bullet \mathrm{Exp}(a^i B_i^A D_A) = B_i^A
\end{aligned}$$

Since $\mathcal{H}$ is connected i is SA also on all $\mathcal{H}$. $(\mathcal{H}, i)$ is a sub SL group of $\mathcal{G}$ whose SA structure is unique. T and $W_{\mathcal{G}}$ are SL isomorphic since

$$T = c_\star^{-1}(\tilde{h}) = c_\star^{-1} \bullet j_\star \bullet \alpha^{-1}(W_{\mathcal{H}}) = i_\star(W_{\mathcal{H}}).$$

□

6. Homogeneous Supermanifolds

The main goal of this section is to show how to put an SA structure on the topological homogeneous manifold $\mathcal{G}/\mathcal{H}$ where $\mathcal{G}$ and $\mathcal{H}$ are SL groups. The main theorem is the following.

Theorem. 6.1 Let $(\mathcal{H}, i)$ a closed sub SL group of the SL group $\mathcal{G}$; let $\pi: G \to G/H$ be the projection on the cosets $\{\pi(a) = \{a\mathcal{H}\}\}$. Then G/H admits one and only one SA structure for which $\pi: \mathcal{G} \to \mathcal{G}/\mathcal{H}$ is SA.

First of all, we consider the subspaces $\mathcal{D}_a \equiv L_{a\star} i_\star T_e(\mathcal{H})$ where L_a for $a \in \mathcal{G}$ is the left transport map.

Theorem. 6.2 For each $a \in \mathcal{G}$, the subspaces $\mathcal{D}_a$ are SL isomorphic to $T_e(\mathcal{H})$; moreover the set $\{\mathcal{D}_a | a \in \mathcal{G}\}$ is an SA involutive differential system.

Proof. $\mathcal{D}_a$ is naturally $\mathbf{Z}_2$ graded since it is a subset of $T_a(\mathcal{G})$; moreover it is a Q_0 module since $L_{a\star}$ is Q_0 linear. On the other hand, the existence of $L_{a^{-1}\star}$ and the very definition of the $\mathcal{D}_a$'s ensure that $L_{a\star}$ is an isomorphism making $\mathcal{D}_a$ isomorphic, as a Q_0

module to $i_* T_e(\mathcal{H})$. Concerning completeness if $\{X_i\}$ denotes a Cauchy sequence in $\mathcal{D}_a$, $\{Y_i \equiv L_{a^{-1}*} X_i\}$ is a Cauchy sequence in $i_* T_e(\mathcal{H})$ which is a Banach space. Therefore, the latter sequence admits a limit element Y; now, by defining $X \equiv L_{a*} Y \in \mathcal{D}_a$ it is easy to show that $\mathcal{D}_a$ and $i_* T_e(\mathcal{H})$ are isomorphic also as B spaces. Finally, consider two vectors $A, B \in \mathcal{D}_a$; by considering elements $X, Y \in i_* T_e(\mathcal{H})$ such that $A = L_{a*} X$ and $B = L_{a*} Y$ we get:

$$[A, B] = [L_{a*} X, L_{a*} Y] = L_{a*} [X, Y] \in \mathcal{D}_a$$

□

This theorem can also be seen in the following way; by defining $\mathcal{H}_a = L_a(i\mathcal{H})$ one gets $\mathcal{D}_a = T(\mathcal{H}_a)$.

Now we consider a topology on G/H; a classical theorem [17] ensures that, given a group G and a closed subgroup H, there exists a topology on G/H such that $\pi: G \to G/H$ is continuous. We shall use this topology.

Consider equation $W_{\mathcal{G}} = i_* W_{\mathcal{H}} \oplus \mathbf{F}$ and define projections π_1 and π_2 on the first and second term respectively. Consider also a local coordinate system (U, ϕ) of $\mathcal{G}$ such that $e \in U$; using the involutive character of $\mathcal{D}_a$, as will as the supermanifold version of Frobenius integrability theorem [18], we conclude that the integral manifolds of the system $\{\mathcal{D}_a | \forall a \in \mathcal{G}\}$ are given, in U, by the slices $S_x = \pi_2 \bullet \phi(x)$ =const. In particular, the identity e is an element of the slice $U \cap i(\mathcal{H})$.

Theorem. 6.3 Two points $s, t \in U$ are in the same coset if and only if they lie in the same slice.

Proof. If $\pi(s) = \pi(t)$ then $t^{-1} s \in i(\mathcal{H}) \cap U_1 = S_0 \cap U_1$ or, in other words, $s \in t(S_0 \cap U_1)$. U_1 is suitable subset of U. Now, $t(S_0 \cap U_1) = L_t(S_0 \cap U_1) \subset \mathcal{H}_t$ and, therefore, in view of the fact that $T(\mathcal{H}_t) = \mathcal{D}_t$, we have that $t(S_0 \cap U_1)$ is a connected part of an integral manifold of $\{\mathcal{D}_a\}$.

Conversely, if s and t lie in the same slice, there are a $p \in U$ and $\alpha, \beta \in i(\mathcal{H})$ such that $s = p\alpha$ and $t = p\beta$. It follows that $\pi(s) = \pi(t)$. □

By considering the slice S in $\phi(U)$ where $\pi_1 \phi(x) = 0$ we can define the map $\tilde{\phi}^{-1} \equiv \pi \bullet \phi^{-1}|_S : S \to \pi(U)$; this map is one-to-one, C^0 and open. The inverse map $\tilde{\phi}: \pi(U) \to S \subset \mathbf{F}$ is a local system of coordinates for $\mathcal{G}/\mathcal{H}|_{\pi(U)}$.

Now we give an SA structure to G/H by starting from the local one $(\pi(U), \tilde{\phi})$. To this end, we shall use the left transport law. First of all, we notice that L_a induces on G/H a map $\mathcal{L}_a$ given by $\pi \bullet L_a \bullet \pi^{-1}$. By defining $\mathcal{W}_a = \mathcal{L}_a \pi(U)$ it is easily to show that

a) $\{\mathcal{W}_a | a \in G\}$ is onto G/H,

b) $\{aH\} \in \mathcal{W}_b$ and $\{bH\} \in \mathcal{W}_a$ if $\pi(a) = \pi(b)$.

The maps $\tilde{\phi}_a = \tilde{\phi} \bullet \mathcal{L}_{a^{-1}} : \mathcal{W}_a \to \mathbf{F}$ are well defined for all $\mathcal{W}_a$. Finally we show that $\mathcal{A} = \left\{(\mathcal{W}_a, \tilde{\phi}_a)\right\}$ is an atlas for $\mathcal{G}/\mathcal{H}$; after having observed that $\cup_a \mathcal{W}_a = G/H$ we have only to show that $\tilde{\phi}_b \bullet \tilde{\phi}_a^{-1} : \tilde{\phi}_a(\mathcal{W}_a \cap \mathcal{W}_a) \to \tilde{\phi}_b(\mathcal{W}_b)$ is SA. However, this is easy since it is sufficient to check it only for $\tilde{\phi}_b \bullet \tilde{\phi}_a^{-1} : \mathcal{W} \to \tilde{\phi}_b(\mathcal{W})$ in view of the left transport law properties. The main theorem 6.1 is completely proved.

REFERENCES

1. A. Rogers, J. Math. Phys. **21** (1980), p. 1352.
2. A. Jadczyk, and K. Pilch, Commun. Math. Phys. **78** (1981), p. 373.

3. U. Bruzzo, and R. Cianci, Class. Quantum Grav. **1** (1984), p. 213.
4. M. Rothstein, Trans. of A.M.S. **297** (1986), p. 159.
5. A. Rogers, J. Math. Phys. **26** (1985), p. 385.
6. S. Lang, "Differential Manifolds," Addison-Wesley, Reading, Mass., 1972.
7. A. Rogers, J. Math. Phys. **22** (1981), p. 939.
8. U. Bruzzo, and R. Cianci, Lett. in Math. Phys. **8** (1984), p. 279.
9. R. Cianci, J. Math. Phys. **25** (1984), p. 451.
10. V.J. Van Est, and T.J. Korthagen, Indag. Math. **26** (1964), p. 15.
11. N. Jacobson, "Lie Algebras," Interscience, New York, 1962.
12. C.E. Rickart, "General Theory of Banach Algebras," Van Nostrand, Princeton, 1960.
13. P. De la Harpe, in "Lecture Notes in Mathematics," Springer-Verlag, Berlino, 1972.
14. G. Birkoff, Trans. of A.M.S. **43** (1938), p. 61.
15. P.M. Cohn, "Lie Groups," Cambridge University Press, Cambridge, 1957.
16. L. Pontriagin, "Topological Groups," Princeton University Press, Princeton, 1958.
17. F.W. Warner, "Foundations of Differentiable Manifolds and Lie Groups," Scott, Foresman & Co., Gleview, IL, 1971.
18. U. Bruzzo, and R. Cianci., Jour. of Phys. A **18** (1985), p. 417.

LIE ALGEBRAS AND RECURRENCE RELATIONS II

Philip Feinsilver

Department of Mathematics
Southern Illinois University, Carbondale, Illinois 62901 USA

This work is based on Lie Algebras and Recurrence Relations I (L.A.R.R.I). Here we first review the basic material presented in L.A.R.R.I and then we explore some topics in detail. Contents of L.A.R.R.II are:

I. Review of RNL Theory
II. The Hamiltonian Flow
III. Examples. The Lorentz Case
IV. Matrix Elements
V. Topics in the Multivariate Theory

The topics discussed here relate primarily to representation theory or, potentially at least, to mathematical physics.

I. INTRODUCTION AND REVIEW OF BASIC CONSTRUCTIONS (RNL THEORY)

A connection between Lie algebras and orthogonal polynomials may be found by way of the three-term recurrence for the polynomials. Let H_n be a sequence of orthogonal polynomials with recurrence relation

(1.1) $$xH_n = H_{n+1} + a_n H_n + b_n H_{n-1}, \qquad n \geq 1$$

$H_0 = 1$, $H_1 = x$, $b_n > 0$, $n \geq 1$. We can define the operators R, N, L -- raising, neutral, lowering -- by

(1.2) $$RH_n = H_{n+1}, \qquad NH_n = a_n H_n, \qquad LH_n = b_n H_{n-1}$$

and ask about their Lie-algebraic structure. The basic facts are:

(1.3) **Finite-Dimensionality Theorem.** Let $\mathcal{L}$ = {Lie algebra generated by R,N,L}, $\mathcal{L}_0$ = {Lie algebra generated by R,L}. Then

i) $\dim \mathcal{L}_0 < \infty \Leftrightarrow b_n$ is a polynomial in n of degree ≤ 2

ii) $\dim \mathcal{L} < \infty$ if, in addition, a_n is a polynomial of degree ≤ 1.

Sketch of the Proof: Compute $[L,R]H_n = (b_{n+1}-b_n)H_n$. Denote $b_{n+1} - b_n$ by b'_n. Then $[[L,R],R]H_n = (b'_{n+1}-b'_n)H_{n+1} = b''_n RH_n$. Repeated bracketing with $[L,R]$ builds up arbitrary powers of b''_n, so this must be constant. ii) follows similarly.

We are thus interested in recurrences of the form

$$xH_n = H_{n+1} + 2\alpha nH_n + n(t+(n-1)\beta)H_{n-1} \tag{1.4}$$

where we choose $a_0 = 0$, since this amounts to a fixed shift in the variable x, and we take $b_0 = 0$, so that $LH_0 = 0$, i.e. H_0 is a "vacuum state" acting as a cyclic vector for the representation of the algebra $\mathcal{L}$. We denote H_0 by Ω throughout. We thus have $H_n = R^n\Omega$.

A thorough study from the viewpoint of the theory of orthogonal polynomials of recurrences of the type (1.4) is included in the study [AI], see also [AA],[AW]. L.A.R.R.I ([F1]) consists of the principal exposition of our approach. Here we will carry out an in-depth study of some features of the RNL theory. The $\mathcal{L}$-modules we are studying are of the type studied by Harish-Chandra, Verma (see, e.g. [D]), and by many researchers (notably B. Gruber).

Introduce the operator

$$\rho = [L,R] . \tag{1.5}$$

By (1.2) and (1.4), we have $\rho H_n = (t+2n\beta)H_n$ and

$$[L,\rho] = 2\beta L, \qquad [\rho,R] = 2\beta R . \tag{1.6}$$

Thus $\mathcal{L}_0$ (cf. (1.3)) is generically sl(2) and the various real forms su(2), su(1,1), sl(2,$\mathbb{R}$) have corresponding classes of orthogonal polynomials which we call generically "of Bernoulli type" -- associated to various types of binomial distributions. The algebra $\mathcal{L}$, for $a,\beta \neq 0$, includes the identity operator since $\rho/2\beta$ and $N/2\alpha$ differ by the operator tI. $\alpha = 0$ gives a pure sl(2). For $\beta = 0$, $\alpha \neq 0$, one has a solvable algebra, the oscillator algebra (and corresponding Poisson distributions of probability theory); for $\alpha=\beta=0$, we reduce to the (nilpotent) Heisenberg-Weyl algebra (Gaussian distributions). $q = 0$ has corresponding Laguerre polynomials (with gamma distributions as weight functions) and the algebraic structure can be expressed using a solvable Lie algebra (the finite-difference algebra, [F2]).

The next step comes by asking how to relate the algebraic and Hilbert space structures. In analogy to Fock space we require that R and L be adjoints, N self-adjoint. This gives, $\langle\ \ \rangle$ denoting inner product:

$$(1.7) \qquad \langle RH_{n-1}, H_n\rangle = \|H_n\|^2 = \langle H_{n-1}, LH_n\rangle = b_n\|H_{n-1}\|^2.$$

We denote $\|H_n\|^2$ by γ_n. (1.7) yields

$$(1.8) \qquad \gamma_n = n! \prod_{j=1}^{n} (t+(j-1)\beta) = n!\beta^n(\tau)_n \qquad (\beta \neq 0)$$

and we normalize $\gamma_0 = 1$. τ denotes t/β; $(\tau)_n = \tau(\tau+1)\cdots(\tau+n-1)$. This, (1.7), explains the condition $b_n > 0$, $n \geq 1$, stated after (1.1). Denote the Hilbert space generated by the H_n by $\mathcal{H}$.

Now for the next main observation. Let $X = R+N+L$. On the H_n's, X acts as multiplication by x, so gives a self-adjoint operator on $\mathcal{H}$. If we start with an abstract Hilbert space with orthogonal basis ψ_n with operators $\hat{R}$, $\hat{N}$, $\hat{L}$ defined as in (1.2) acting on the ψ_n, with $\hat{L}\psi_0 = 0$, then if $\hat{R}$ and $\hat{L}$ are adjoints it follows that $\hat{X} = \hat{R}+\hat{N}+\hat{L}$ is unitarily equivalent to X by the spectral theorem. We will study particular realizations of this equivalence. Namely, introduce the "derivative operator" V:

$$(1.9) \qquad VH_n = nH_{n-1}.$$

Observe that $[V,R] = I$, and they form a boson pair. We have the boson realization

$$(1.10) \qquad R = R, \qquad N = 2\alpha RV, \qquad L = tV+\beta RV^2, \qquad \rho = t+2\beta RV.$$

We could start with the boson realization where R = multiplication by ξ, $V = d/d\xi$. Thus the problem is to find eigenfunctions of the operator (with $D = d/d\xi$):

$$(1.11) \qquad X = \xi + 2\alpha\xi D + tD + \beta\xi D^2.$$

One finds the desired solutions to $X\omega = x\omega$:

$$(1.12) \qquad \omega(\xi,x) = \exp\left[-\frac{\alpha+q}{\beta}\xi\right] {}_1F_1\left[\frac{x+(\alpha+q)\tau}{2q}\ ;\ \tau;\ \frac{2q\xi}{\beta}\right]$$

where $q^2 = \alpha^2-\beta$. One identifies $\omega(\xi,x)$ with the intertwining kernel giving the unitary equivalence between the spaces generated by the ξ^n and H_n:

$$(1.13) \qquad \omega(\xi,x) = \sum_{n=0}^{\infty} \xi^n H_n(x)/\gamma_n.$$

Using (1.8) and (1.4), one readily checks that (1.13) satisfies $X\omega = x\omega$. The variable x in (1.1), (1.4) is now seen as a spectral variable. (1.12) and (1.13) yield the explicit forms

(1.14) $$H_n(x) = (\tau)_n(-\alpha-q)^n {}_2F_1\left[-n,\ \frac{x+(\alpha+q)\tau}{2q}\ ;\ \tau;\ \frac{2q}{\alpha+q}\right].$$

These (generically ${}_2F_1$) polynomials are the Meixner and Meixner-Pollaczek polynomials ([AI],[C]) including as special cases Krawtchouk, Laguerre, Poisson-Charlier, and Hermite polynomials. These latter correspond to various of the parameters α, β, q going to zero, thus corresponding to particular Lie algebra contractions. It is interesting that this approach gives a Lie-algebraic basis for the classical distributions of probability theory and their associated limit theorems. E.g., if $\alpha = 0$, taking the limit $\beta \to 0$ is a contraction to the Heisenberg-Weyl algebra (cf. (1.10)) corresponding to the classical central limit theorem of DeMoivre-Laplace.

For more general semi-simple cases, consider raising operators $E_{-\alpha}$, corresponding to negative roots, lowering operators E_α, corresponding to positive roots, and H_α generating the Cartan subalgebra. Then with a Hilbert space set up so that E_α and $E_{-\alpha}$ are adjoints, with H_α self-adjoint, spectral theory applied to the operator $\sum_{\alpha \text{ pos}} (E_{-\alpha} + H_\alpha + E_\alpha)$ gives an appropriate extension of the present theory. Below, section V, we discuss certain aspects of the higher rank case.

In section II we present the Hamiltonian viewpoint. In III we give examples, including a discussion of the "Lorentz case" -- $\beta > 0$, $q^2 < 0$, corresponding to the continuous binomial distribution (Pollaczek polynomials [AI],[P]). This appears to be a relativistic analog of the Gaussian distribution, with $\beta = 1/c^2$, but the physics of the situation eludes me. In IV we compute matrix elements for the generic case. In V, two topics in the multivariate case are presented -- (1) matrix elements and (2) generalized Clebsch-Gordan coefficients.

II. THE HAMILTONIAN FLOW

For the canonical version of the theory we have the boson realization $R = R$, $N = 2\alpha RV$, $L = tV+\beta RV^2$. We have

(2.1) $$X = R + N + L = R(1 + 2\alpha V + \beta V^2) + tV.$$

Thus, define the *characteristic polynomial*

(2.2) $$\pi(\theta) = 1 + 2\alpha\theta + \beta\theta^2$$

(2.3) $$X = R\pi(V) + tV\ .$$

Considering the spectral form of the theory, with X acting as multiplication by x, we have the generating function (or coherent state):

$$(2.4)\qquad \sum_{0}^{\infty} \frac{s^n}{n!} H_n(x) = e^{sR}\Omega,$$

and we see that V acts as multiplication by s.

Now put V as a function of $z = \frac{d}{dx}$. So

$$(2.5)\qquad \left[V\left(\frac{d}{dx}\right), x\right] = [V(z), x] = V'(z)$$

by usual boson calculus, where ' denotes the derivative of V as a function of z. And (2.1) yields

$$(2.6)\qquad [V,x] = V'(z) = 1 + 2\alpha V + \beta V^2 = \pi(V).$$

I.e., V, as a function of z, satisfies a Riccati equation. This can be solved, with the normalizing conditions $V(0) = 0$, $V'(0) = 1$, giving

$$(2.7)\qquad V(z) = \tanh qz/(q - \alpha \tanh qz).$$

Introduce U(v), the function inverse to V: $U(V(z)) = z$. Then $U'(v) = \pi(v)^{-1}$, by (2.6). We thus have from (2.3):

$$(2.8)\qquad R = (x - tV)U'(V).$$

Let H(z) satisfy $H'(z) = V(z)$. Then we have

$$(2.9)\qquad x - tV = e^{-tH} x e^{tH},$$

$$(2.10)\qquad R = e^{-tH} xU'(V) e^{tH}, \quad V = e^{-tH} V e^{tH},$$

since $[H,V] = 0$. Thus, H acts as a Hamiltonian and (2.10) is the corresponding Hamiltonian flow. Think of z as generalized momentum. Then since $V = H'(z) = \frac{dH}{dz}$, we call V the velocity operator. Supporting this interpretation, note that in (2.1) we have $\frac{dX}{dt} = V$.

In (2.4), R acts as $\frac{d}{ds}$, so that, with V corresponding to multiplication by s,

$$(2.11)\qquad e^{sR}\Omega = e^{-tH} e^{xU(s)} e^{tH} = e^{xU(s) - tM(s)}$$

where $M(s) = H(U(s))$. This is the generating function for the polynomials $H_n(x)$. We remark that differentiating (2.6) and integrating back the resulting equation for V", yields the relation

$$(2.12)\qquad \pi(V(z)) = e^{2\alpha z + 2\beta H(z)}.$$

With $H(0) = 0$, from $H' = V$ one finds (cf. (2.7))

(2.13) $$H(z) = -\frac{\alpha}{\beta}z + \frac{1}{2\beta}\log \pi(V(z))$$

(2.14) $$H(z) = -\frac{\alpha}{\beta}z + \frac{1}{\beta}\log \frac{q\,V(z)}{\sinh qz}$$

(2.15) $$M(v) = -\frac{\alpha}{\beta}U(v) + \frac{1}{2\beta}\log \pi(v).$$

III. EXAMPLES. THE LORENTZ CASE

1. We first consider the operator RL. We have

(3.1) $$RL\,H_n = b_nH_n, \qquad \text{with } b_n = n(t+(n-1)\beta).$$

This gives the exponential

(3.2) $$e^{sRL}H_n = e^{sb_n}H_n$$

and on a general element $\eta = \sum c_nH_n$ in the domain of e^{sRL}

(3.3) $$e^{sRL}\sum c_nH_n = \sum c_ne^{s\beta(n^2-n)}e^{stn}H_n.$$

Put $r = e^{2s\beta}$. Then

(3.4) $$e^{sRL}\eta = \sum c_nr^{\binom{n}{2}}r^{\tau n/2}H_n$$

(recalling $\tau = t/\beta$) and, with $\|\eta\|^2 = \sum |c_n|^2\gamma_n = 1$, the expectation value

(3.5) $$\langle e^{sRL}\eta,\eta\rangle = \sum |c_n|^2\gamma_nr^{\binom{n}{2}}r^{\tau n/2}$$

theta-series. This is interesting to note,, but further investigations along these lines have yet to be made.

2. The operator V^*V is a good candidate for the square of the physical velocity. First we recall the following fact from Hilbert space theory:

(3.6) *Proposition*: Let ψ_n be a complete orthogonal system of eigenfunctions of an operator T: $T\psi_n = \lambda_n\psi_n$. Then $\|T\| = \sup_n |\lambda_n|$.

Proof: For $\varphi = \sum c_n\psi_n$, $(\|T\varphi\|/\|\varphi\|)^2 = \left[\sum |c_n|^2|\lambda_n|^2\gamma_n\right]/\sum |c_n|^2\gamma_n$, where $\gamma_n = \|\psi_n\|^2$. Thus, $\sup_{\|\varphi\|=1} \|T\varphi\| \le \sup |\lambda_n|$, since $\|T\varphi\|^2$ is a convex combination of the $|\lambda_n|^2$ for $\|\varphi\| = 1$. On the other hand, we can find a sequence $\varphi_{n'} = \psi_{n'}/\gamma_{n'}^{1/2}$ such that $\|T\varphi_{n'}\| \to \sup |\lambda_n|$.

We calculate, replacing T of the Proposition with V^*V,

$$\langle V^*VH_n, H_n\rangle = \lambda_n\gamma_n \tag{3.7}$$

$$\langle V^*VH_n, H_n\rangle = \langle VH_n, VH_n\rangle = n^2\gamma_{n-1} \tag{3.8}$$

$$\lambda_n = n^2\gamma_{n-1}/\gamma_n = n^2/b_n \tag{3.9}$$

where $b_n = n(t+(n-1)\beta)$ and $\gamma_n = b_1b_2\cdots b_n$. Thus $\|V^*V\| = \sup_{n\geq 1} |n^2/b_n|$. We have: $\lambda_0 = 0$, $\lambda_1 = 1/t$, $\lambda_2 = 2/(t+\beta)$, ..., $\lambda_n \to 1/\beta$ as $n \to \infty$. Assuming $\beta > 0$,

$$\|V^*V\| = 1/\min(t,\beta), \qquad \beta > 0. \tag{3.10}$$

That is, with $\tau = t/\beta$, $\tau < 1$ is a tachyon sector, where $|V|$ can be as large as $t^{-1/2}$, but for $\tau \geq 1$, it looks like we are in a relativistic sector, with $|V| \leq \beta^{-1/2}$, hence the suggestion $\beta = 1/c^2$.

3. Recall -- (1.5), (1.10) -- the operator

$$\rho = [L,R] = t + 2\beta RV. \tag{3.11}$$

From $\rho\Omega = t\Omega$, we can say that the value of the time parameter comes from the representation via ρ. Similarly, with $X = R+N+L$, the spectral theorem permits us to interpret X as position in the ordinary sense of having numerical values. It is natural to look at $\begin{bmatrix} X \\ \rho \end{bmatrix}$ as a space-time vector. We consider the case $\alpha = 0$ (cf. (2.2)).

Let $Y = \frac{1}{i}(R-L)$. This is another natural operator to look at on $\mathcal{H}$. Let us calculate:

$$X(s) = e^{isY}Xe^{-isY}, \qquad \rho(s) = e^{isY}\rho e^{-isY}, \tag{3.12}$$

where $X = R+L$. Checking that $[R-L,\rho] = -2\beta(R+L)$, $[R-L,R+L] = -2\rho$, we have

$$\frac{dX}{ds} = -2\rho(s), \qquad \frac{d\rho}{ds} = -2\beta X(s) \tag{3.13}$$

$$\begin{aligned} X(s) &= X\cosh 2s\sqrt{\beta} - \beta^{-1/2}\rho\sinh 2s\sqrt{\beta} \\ \beta^{-1/2}\rho(s) &= \beta^{-1/2}\rho\cosh 2s\sqrt{\beta} - X\sinh 2s\sqrt{\beta}. \end{aligned} \tag{3.14}$$

The conclusion drawn is that $\begin{bmatrix} X \\ \beta^{-1/2}\rho \end{bmatrix}$ is a Lorentz vector for $\beta > 0$. Again, $\beta^{-1/2} = c$ gives the natural form $\begin{bmatrix} X \\ c\rho \end{bmatrix}$. Thus we call $\beta > 0$, $\alpha = 0$ the Lorentz case. This corresponds to the continuous binomial distribution of Pollaczek (see [AI],[AW]).

4. The continuous binomial distribution corresponds to the Hamiltonian $H(z) = \log \frac{\cos \theta}{\cos(\theta+z)}$ (see [Du], p. 170). This is not in normalized form $H(0) = V(0) = 0$, $V'(0) = 1$. Centering and scaling gives the forms

$$(3.15) \qquad H(z) = -\frac{\sin 2\theta}{2} z + (\cos^2\theta) \log \left[\frac{\sec z}{1 - \tan\theta \tan z}\right]$$

$$(3.16) \qquad V(z) = \frac{\tan z}{1 - \tan\theta \tan z}$$

$$(3.17) \qquad \pi(v) = 1 + 2 \tan\theta\, v + \sec^2\theta\, v^2.$$

Thus, $\alpha = \tan\theta$, $\beta = \sec^2\theta$, $q^2 = \alpha^2 - \beta = -1$. (The theory of the associated orthogonal polynomial systems may be found in [AI],[AW],[C]). We consider also the case $\alpha = 0$, $\beta = c^{-2}$, $q = ic^{-1}$. From (2.7), (2.13)-(2.14):

$$(3.18) \qquad V(z) = c \tan(z/c), \qquad V' = 1 + \beta V^2$$

$$(3.19) \qquad H(z) = c^2 \log \sec(z/c)$$

which is a scaled version of (3.15)-(3.16) for $\theta = 0$. Now we see readily that the non-relativistic limit $c \to \infty$, $\beta \to 0$ yields the Gaussian case $H(z) = z^2/2$, $V(z) = z$.

IV. MATRIX ELEMENTS

In this section we look at matrix elements for functions of the operator $X = R + N + L$.

1. Define the functions φ and Ψ as follows. For $\gamma_n = n!\beta^n(\tau)_n$ (cf. (1.8)), set

$$(4.1) \qquad e^{t\varphi(x)} = \sum_0^\infty \frac{x^n}{n!}\frac{\gamma_n}{n!} = (1 - \beta x)^{-\tau}$$

$$(4.2) \qquad \varphi(x) = \begin{cases} -\beta^{-1} \log(1 - \beta x), & \beta \neq 0 \\ x, & \beta = 0. \end{cases}$$

(We generically take $\beta \neq 0$; the results for $\beta = 0$ follow by taking limits.)

Next, define the exponential vectors, or coherent states:

$$(4.3) \qquad \Psi_f = e^{fR}\Omega = \sum_0^\infty \frac{f^n}{n!} R^n\Omega$$

which has, by (2.11), the spectral form:

$$\psi_f = \sum_0^\infty \frac{f^n}{n!} H_n(x) = e^{xU(f)-tM(f)}. \tag{4.4}$$

From (4.1), via the orthogonality relations $\langle R^n\Omega, R^m\Omega\rangle = \delta_{nm}\gamma_n$

$$\langle \psi_f, \psi_g\rangle = \langle e^{fR}\Omega, e^{gR}\Omega\rangle = e^{t\varphi(fg)}. \tag{4.5}$$

We will denote this quantity by ψ_{fg} for convenience.

We calculate the matrix elements of X:

$$\langle X\psi_f, \psi_g\rangle = \langle (R+N+L)\psi_f, \psi_g\rangle \tag{4.6}$$

$$\langle R\psi_f, \psi_g\rangle = \frac{\partial}{\partial f}\langle \psi_f, \psi_g\rangle = tg\varphi'(fg)e^{t\varphi(fg)} = \frac{tg}{1-\beta fg}\psi_{fg} \tag{4.7}$$

$$\langle L\psi_f, \psi_g\rangle = \langle \psi_f, R\psi_g\rangle = \frac{tf}{1-\beta fg}\psi_{fg} \tag{4.8}$$

$$\langle N\psi_f, \psi_g\rangle = 2\alpha\langle RV\psi_f, \psi_g\rangle = 2\alpha f\langle R\psi_f, \psi_g\rangle = \frac{2t\alpha fg}{1-\beta fg}\psi_{fg}, \tag{4.9}$$

using $Ve^{fR}\Omega = f\psi_f$. And we have

$$\langle X\psi_f, \psi_g\rangle = t\,\frac{f+g+2\alpha fg}{1-\beta fg}\psi_{fg}. \tag{4.10}$$

We define Ψ_{fg} so that $\langle X\psi_f, \psi_g\rangle = t\Psi_{fg}\psi_{fg}$, i.e.

$$\Psi_{fg} = \frac{f+g+2\alpha fg}{1-\beta fg}. \tag{4.11}$$

(4.12) *Proposition.* $\langle e^{sX}\psi_f, \psi_g\rangle = \exp[t(H(s)+\varphi(V(s)\Psi_{fg})+\varphi(fg))]$.

Proof: We differentiate both sides with respect to s, noting that they agree at $s = 0$. First, note that as in (4.6)-(4.10), cf. (2.3), $X\psi_g = (R\pi(V)+tV)e^{gR}\Omega = \left[\pi(g)\frac{\partial}{\partial g}+tg\right]\psi_g$. We denote $\langle e^{sX}\psi_f, \psi_g\rangle$ by E for short. Thus:

$$\begin{aligned}\frac{\partial E}{\partial s} &= \langle Xe^{sX}\psi_f, \psi_g\rangle = \langle e^{sX}\psi_f, X\psi_g\rangle \\ &= \left[\pi(g)\frac{\partial}{\partial g}+tg\right]E.\end{aligned} \tag{4.13}$$

We also compute, by basic calculus, writing just Ψ for Ψ_{fg},

$$\frac{\partial\Psi}{\partial g} = \frac{\pi(f)}{(1-\beta fg)^2}. \tag{4.14}$$

Apply the operator $\pi(g)\frac{\partial}{\partial g}+tg$ to the right-hand side of (4.12) to get, with V for V(s):

$$t\left[g+\pi(g)\left[V\frac{\pi(f)}{(1-\beta fg)^2}\frac{1}{1-\beta V\Psi}+\frac{f}{1-\beta fg}\right]\right]E \tag{4.15}$$

noting that, by (4.2), $\varphi'(x) = (1-\beta x)^{-1}$. Combining terms in (4.15) gives

$$(4.16) \qquad t\left[\Psi + \frac{\pi(f)\pi(g)}{(1-\beta fg)^2}\,\frac{V}{1-\beta V\Psi}\right]E.$$

Now apply $\frac{\partial}{\partial s}$ to the right-hand side of (4.12):

$$(4.17) \qquad t(H' + V'\Psi\varphi'(V\Psi))E, \qquad \text{which gives}$$

$$(4.18) \qquad t\left[V + \frac{\pi(V)\Psi}{1-\beta V\Psi}\right]E.$$

We must check that (4.16) and (4.18) agree. Equating them and multiplying through by $1-\beta V\Psi$:

$$(4.19) \qquad \Psi - \beta V\Psi^2 + \frac{\pi(f)\pi(g)}{(1-\beta fg)^2}V \stackrel{?}{=} V - \beta V^2\Psi + \pi(V)\Psi.$$

Writing out $\pi(V) = 1 + 2\alpha V + \beta V^2$, $\beta V^2\Psi$ cancels, and Ψ cancels from both sides of (4.19). Next V cancels out, leaving:

$$(4.20) \qquad -\beta\Psi^2 + \frac{\pi(f)\pi(g)}{(1-\beta fg)^2} \stackrel{?}{=} 1 + 2\alpha\Psi.$$

One checks the relation

$$(4.21) \qquad \pi(\Psi) = \frac{\pi(f)\pi(g)}{(1-\beta fg)^2}$$

and this is the same as (4.20).

2. From (4.12) comes the result:

(4.22) *Proposition.* $\langle \sum_0^\infty \frac{s^n}{n!} H_n(X)\psi_f, \psi_g \rangle = e^{t\varphi(s\Psi_{fg})}\Psi_{fg}$.

Proof: By (4.4), for the left-hand side:

$$(4.23) \qquad \langle e^{XU(s)-tM(s)}\psi_f, \psi_g\rangle = e^{-tM(s)}\langle e^{XU(s)}\psi_f, \psi_g\rangle$$

$$(4.24) \qquad \langle e^{XU(s)}\psi_f, \psi_g\rangle = e^{tM(s)}e^{t\varphi(s\Psi_{fg})}e^{t\varphi(fg)}$$

by (4.12), with $M(s) = H(U(s))$. Cancelling e^{tM} gives the result.

By the definition of φ, the right-hand side of (4.22) equals

$$(4.25) \qquad (1 - s\beta\Psi_{fg})^{-\tau}\Psi_{fg} = \left[\sum \frac{s^n}{n!}\beta^n(\tau)_n\Psi_{fg}^n\right]\Psi_{fg}$$

$$= \left[\sum \frac{s^n}{n!}\left[\frac{\Gamma_n}{n!}\Psi_{fg}^n\right]\right]\Psi_{fg}.$$

It follows that

$$\text{(4.26)}\qquad \frac{\langle H_n(X)\psi_f,\psi_g\rangle}{\langle\psi_f,\psi_g\rangle} = \frac{\gamma_n}{n!}\,\Psi_{fg}^n .$$

From this we calculate the matrix elements of $\omega(\xi,X) = \sum_0^\infty \frac{\xi^n H_n(X)}{\gamma_n}$ to give the

$$\text{(4.27)}\qquad \textit{Exponential Formula.}\quad \frac{\langle\omega(\xi,X)\psi_f,\psi_g\rangle}{\langle\psi_f,\psi_g\rangle} = e^{\xi\Psi_{fg}}.$$

This relation follows immediately from the definition of ω by (4.26). It is remarkable that the reduced matrix elements $\langle\omega\psi_f,\psi_g\rangle/\langle\psi_f,\psi_g\rangle$ are independent of t. We see also that in a sense ω plays the role of an exponential function for this calculus.

3. We now calculate the matrix elements

$$\text{(4.28)}\qquad \langle e^{sX}H_n,H_m\rangle$$

which, e.g., in the angular momentum case give Jacobi polynomials. Writing (4.12) in the form

$$\text{(4.29)}\qquad \langle e^{sX}\sum \frac{f^n}{n!}H_n,\ \sum\frac{g^m}{m!}H_m\rangle = e^{tH(s)}(1-\beta V\Psi)^{-\tau}(1-\beta fg)^{-\tau}$$

shows that the required matrix elements are the coefficients in the expansion of the right-hand side in powers of f and g. One readily checks the following

$$\text{(4.30)}\qquad \textit{Proposition.}\quad (1-s)^{y-a}(1-(1-u)s)^{-y} = \sum_{n=0}^\infty \frac{s^n}{n!}(a)_n\ {}_2F_1(-n,y;a;u).$$

Using (4.11) in (4.29) yields

$$\text{(4.31)}\qquad e^{tH}(1-\beta fg-\beta V(f+g+2\alpha fg))^{-\tau}.$$

Apply $\left[\frac{\partial}{\partial f}\right]^n\Big|_{f=0}$ to (4.31):

$$\text{(4.32)}\qquad e^{tH}\beta^n(\tau)_n(1-\beta Vg)^{-\tau-n}(V+(1+2\alpha V)g)^n$$

$$\text{(4.33)}\qquad e^{tH}\beta^n(\tau)_n V^n(1-\beta Vg)^{-\tau-n}\left[1+\frac{1+2\alpha V}{V}\,g\right]^n .$$

Using (4.30) with $y = -n$, $a = \tau$, $s = \beta Vg$, $u = \pi(V)/\beta V^2$, gives the coefficient of $g^m/m!$:

$$\text{(4.34)}\qquad e^{tH}\beta^{n+m}(\tau)_n V^{n+m}(\tau)_m\ {}_2F_1\left[-m,-n;\tau;\frac{\pi}{\beta V^2}\right].$$

For $m \geq n$, one has the transformation formula

$$\text{(4.35)}\qquad {}_2F_1(-m,-n;\tau;y) = \frac{m!\,y^n}{(m-n)!(\tau)_n}\ {}_2F_1(-n,1-n-\tau;1+m-n;y^{-1}).$$

Setting $m = n+N$, $N \geq 0$, gives the matrix elements

$$(4.36) \qquad \langle e^{sX} H_n, H_{n+N} \rangle = e^{tH} \pi(V)^n V^N \frac{\gamma_{n+N}}{N!} \, {}_2F_1\left[-n, 1-n-\tau; N+1; \frac{\beta V^2}{\pi(V)}\right].$$

From the formulas of section II one can check that $q^2 V(s)^2 = \pi(V(s)) \sinh^2 qs$, giving $\beta V^2/\pi(V)$ in the alternate form $\beta \sinh^2 qs/q^2$. In particular, for the case $q \to 0$, the argument becomes βs^2.

The matrix elements for the case $\beta = 0$ are found by taking the limit $\beta \to 0$ in (4.36) to yield

$$(4.37) \qquad e^{tH} \pi(V)^n V^N \frac{\gamma_{n+N}}{N!} \, {}_1F_1\left[-n; N+1; -t \frac{\sinh^2 \alpha s}{\alpha^2}\right],$$

where $\pi(V) = 1 + 2\alpha V$, $\gamma_k = k! t^k$.

V. TOPICS IN THE MULTIVARIATE THEORY

1. For the multivariable theory, we take a commuting family of self-adjoint operators X_j, each of which is of RNL-type. That is, we have $X_j = R_j + N_j + L_j$, with $[X_j, X_k] = 0$. In general, the requirement that the operators R_j, N_j, L_j form a finite-dimensional Lie algebra leads to the form, $1 \leq k \leq d$:

$$(5.1) \qquad R_k = R_k, \quad N_k = a^{\mu}_{\lambda k} R_\lambda V_\mu, \quad L_k = b_k (t + R_\lambda V_\lambda) V_k$$

where R_k, V_k are standard boson operators, $[R_k, V_j] = \delta_{jk}$ and we use the following

(5.2) *Summation Convention.* Repeated Greek indices are assumed to be summed over, regardless of location; Latin indices are not summed over, unless explicitly indicated by a summation sign.

The generic Lie algebra generated by $\{R_j, L_j\}_{1 \leq j \leq d}$ is sl(d+1) (shown to me by B. Gruber), with N's corresponding to a u(d) algebra. The derivation of the form (5.1) is giving in L.A.R.R.I ([F1]) and from another point of view in [F4]. The form (5.1) corresponds to the recurrence

$$(5.3) \qquad x_j H_n = H_{n+e_j} + a^{\mu}_{\lambda j} n_\mu H_{n+e_\lambda - e_\mu} + b_j n_j (t + |n| - 1) H_{n-e_j},$$

where $n = (n_1, \ldots, n_d)$ is a multi-index, e_k denote the standard basis for $\mathbb{N}^d$, and $|n| = \sum_{j=1}^{d} n_j$. A generating function for the H_n's is given in [F3] using a multivariate generalization of (4.30). The H_n's are given in terms of Lauricella polynomials.

The Riccati equation (2.6), $V' = \pi(V)$, generalizes to a Riccati system of projective type, studied in [F4] and considered in the work of [AH], [AHW].

From (5.1)

$$X_j = tb_jV_j + R_\lambda(\delta_{\lambda j} + a^\mu_{\lambda j}V_\mu + b_jV_\lambda V_j). \tag{5.4}$$

With V_j a function of the variables $z = (z_1,\ldots,z_d)$, $z_j = \partial/\partial x_j$, we have the system

$$\frac{\partial V_j}{\partial z_k} = \delta_{jk} + a^\mu_{jk}V_\mu + b_kV_jV_k \tag{5.5}$$

$$R_k = (X_\lambda - tb_\lambda V_\lambda)W_{\lambda k} \tag{5.6}$$

where W_{jk} is the matrix inverse to $\partial V_j/\partial z_k$. From (5.6) we see that the Hamiltonian $H(z)$ satisfies

$$\frac{\partial H}{\partial z_k} = b_kV_k\,. \tag{5.7}$$

so that the time-dependence of R_k is given by a Hamiltonian flow as in (2.10), with W replacing U'.

Define the exponential vectors, with $f = (f_1,\ldots,f_d)$,

$$\psi_f = e^{f_\lambda R_\lambda}\Omega. \tag{5.8}$$

The orthogonality of the H_n's, with R_k and L_k adjoints gives γ_n in the form

$$\gamma_n = \|H_n\|^2 = b_1^{n_1}\cdots b_d^{n_d}\,n_1!\cdots n_d!(t)_{|n|}. \tag{5.9}$$

$$\langle\psi_f,\psi_g\rangle = e^{t\varphi(fg)} = (1-b_\lambda f_\lambda g_\lambda)^{-t} \tag{5.10}$$

$$e^{t\varphi(x)} = (1-b_\lambda x_\lambda)^{-t}. \tag{5.11}$$

(Note that the scaling of t is different from that used in the one-variable discussion.)

We calculate the Ψ-functions, cf. (4.6)-(4.11). For clarity, we will write bfg to denote $b_\lambda f_\lambda g_\lambda$.

$$\langle R_j\psi_f,\psi_g\rangle = \frac{\partial}{\partial f_j}\langle\psi_f,\psi_g\rangle = \frac{tb_jg_j}{1-bfg}\,\psi_{fg} \tag{5.12}$$

$$\langle L_j\psi_f,\psi_g\rangle = \langle\psi_f,R_j\psi_g\rangle = \frac{tb_jf_j}{1-bfg}\,\psi_{fg} \tag{5.13}$$

$$\langle N_j\psi_f\psi_g\rangle = a^\mu_{\lambda j}f_\mu\langle R_\lambda\psi_f,\psi_g\rangle = a^\mu_{\lambda j}f_\mu\,\frac{tb_\lambda g_\lambda}{1-bfg}\,\psi_{fg}. \tag{5.14}$$

Adding (5.12)-(5.14) gives

$$\langle X_j\psi_f,\psi_g\rangle = tb_j\,\frac{f_j+g_j+a^\mu_{\lambda j}b_\lambda g_\lambda f_\mu}{1-bfg}\,\psi_{fg} \tag{5.15}$$

yielding, with $\langle X_j\psi_f,\psi_g\rangle = tb_j\Psi_{j,fg}$,

(5.16) $$\Psi_{j,fg} = \frac{f_j + g_j + a^{\mu}_{\lambda j}b_\lambda b_j^{-1}g_\lambda f_\mu}{1 - bfg}.$$

An exact analogue of (4.12) can be derived that will give the generating function for the matrix elements $\langle e^{s_\lambda X_\lambda}H_n, H_m\rangle$ which will appear as types of generalized hypergeometric functions (we defer the details to a future paper).

2. We conclude the present work with an application of the R-L structure to the calculation of some generalized Clebsch-Gordan coefficients. In the case $\alpha = 0$, $\beta = -1/2$, these give the usual C-G coefficients of the theory of angular momentum.

We consider the tensor product of m copies of the same algebra, generated by $(R_1,L_1),\ldots,(R_m,L_m)$ respectively. We are thus taking the simplest multivariate case, where $[R_j,R_k] = [R_j,L_k] = [L_j,L_k] = 0$, all $j \neq k$, $1 \leq j$, $k \leq m$. The corresponding polynomials $H_n = H_{n_1} \otimes H_{n_2} \otimes \cdots \otimes H_{n_m}$ and $\gamma_n = \gamma_{n_1} \cdots \gamma_{n_m}$, product of the separate γ_{n_j}. Now define, with $\rho_j = [L_j,R_j]$,

(5.17) $$R = R_1 + \cdots + R_m, \qquad L = L_1 + \cdots + L_m, \qquad \rho = \rho_1 + \cdots + \rho_m$$

with $\rho_j\Omega = t_j\Omega$, Ω a cyclic vacuum vector for the whole representation. We want to find a subspace giving a "one-variable" representation of the R,L,ρ algebra. I.e. we want a vacuum vector Ω_T such that $\rho\Omega_T = T\Omega_T$ and $L\Omega_T = 0$. Then we build up a subspace with basis vectors $R^N\Omega_T$, $N \geq 0$. We proceed to construct Ω_T.

(5.18) *Proposition.* Let $V(x_1,\ldots,x_m)$ denote the Vandermonde determinant $\prod_{i<j}(x_i - x_j)$. Then $\sum_{j=1}^{m} \frac{\partial V}{\partial x_j} = 0$.

Proposition (5.18) follows readily by taking logarithmic derivatives. Furthermore, from (5.18) we conclude that for any integer $I > 0$, V^I satisfies $\sum \frac{\partial}{\partial x_j} V^I = 0$. We write

(5.19) $$V(x_1,\ldots,x_m)^I = \sum c_n x^n/n!,$$

$n = (n_1,\ldots,n_m)$, using the multi-index notations $x^n = x_1^{n_1}\cdots x_m^{n_m}$, $n! = n_1!\cdots n_m!$. With $R = (R_1,\ldots,R_m)$, we have

(5.20) *Proposition.* $\tilde{\Omega}_T = \sum c_n R^n\Omega/\gamma_n$ satisfies $L\tilde{\Omega}_T = 0$.

Proof: From (5.18), (5.19) we know that $\sum_j \sum_n c_{n+e_j} x^n/n! = 0$, i.e. for each n, $\sum_j c_{n+e_j} = 0$. By construction, $L_j\tilde{\Omega}_T = \sum c_{n+e_j} R^n\Omega/\gamma_n$, since $L_j R^n\Omega = n_j(t_j + (n_j-1)\beta)R^{n-e_j}\Omega = b_{n_j} R^{n-e_j}\Omega$, and γ_n is the product $\gamma_{n_1} \cdots \gamma_{n_m}$ of the individual squared norms, each of which in turn is the product of the coefficients b_{n_j}.

Define $C = \|\tilde{\Omega}_T\|$, i.e. $C^2 = \sum c_n^2/\gamma_n$. Let

(5.21) $$\Omega_T = C^{-1} \sum c_n R^n\Omega/\gamma_n.$$

Continuing, observe next that V^I is homogeneous in $(x_1,\ldots,x_m)$ of degree $\binom{m}{2}I$, since there are $\binom{m}{2}$ factors in V. Thus, $\rho R^n\Omega = \sum (t_j + 2\beta n_j)R^n\Omega = \left[\sum t_j + 2\beta|n|\right]R^n\Omega$ gives a factor of $\sum t_j + 2\beta\binom{m}{2}I$ for each $R^n\Omega$ occurring in the expansion of V^I. Thus,

(5.22) $$\rho\Omega_T = T\Omega_T, \qquad \text{where } T = \sum t_j + 2\beta\binom{m}{2}I.$$

If we specify T *a priori*, then we can calculate the necessary value of I

(5.23) $$I = \frac{T - \sum t_j}{\beta m(m-1)}.$$

Define the normalized states

(5.24) $$\psi_N(T) = \gamma_N(T)^{-1/2} R^N\Omega_T, \qquad N \geq 0,$$

where $\gamma_N(T)$ denotes the one-variable norm-squared with parameter T, where β is understood as given. The multinomial theorem gives $R^N = \sum \binom{N}{p} R^p$, with $\binom{N}{p} = \frac{N!}{p_1!\cdots p_m!}$, p denoting $(p_1,\ldots,p_m)$. Thus, with ℓ denoting a multi-index,

(5.25) $$\psi_N(T) = (C\sqrt{\gamma_N})^{-1} \sum_{\ell} \sum_{p} \binom{N}{p} c_\ell \gamma_\ell^{-1} R^{\ell+p}\Omega.$$

With the change of index $\ell+p = n$, we find the coefficient of $\psi_n = R^n\Omega/\|R^n\Omega\|$ in $\psi_N(T)$:

(5.26) $$C_{nt}^{NT} = (\gamma_n/\gamma_N)^{1/2} C^{-1} \sum_{p} \binom{N}{p} c_{n-p}\gamma_{n-p}^{-1}$$

with $t = (t_1,\ldots,t_m)$. These are our C-G coefficients. We remark that the $\psi_N(T)$ are eigenstates of the Casimir operator $\rho(\rho - 2\beta) - 4\beta RL$ with common eigenvalue $T(T-2\beta)$. Finally, note that the above construction works in a similar fashion for the product of m copies of a given multivariate algebra.

In conclusion, we remark that this work illustrates our theory and techniques that are useful in the study of certain aspects of representation theory, the theory of orthogonal polynomials, and probability theory/quantum theory.

(Note. References to recent work very much in the same spirit as our section V (but somewhat more sophisticated), using vector coherent states and boson operator techniques in angular momentum theory may be found in the article by L. Biedenharn in this volume.)

REFERENCES

[AA] G.E. Andrews and R.A. Askey, Classical orthogonal polynomials, in "Lecture Notes in Math." Springer-Verlag, 1171, 36 (1985).

[AH] R.L. Anderson and J. Harnad, Superposition principles for matrix Riccati equations, J. Math. Physics, 24, 5, 1062 (1983).

[AHW] R.L. Anderson, J. Harnad, and P. Winternitz, Systems of ordinary differential equations with nonlinear superposition principles, Physica D, 4, 164, (1982).

[AI] R. Askey and M. Ismail, Recurrence relations, continued fractions, and orthogonal polynomials, Memoirs of AMS, 300, (1984)

[AW] R. Askey and J. Wilson, Some basic hypergeometric orthogonal polynomials that generalize Jacobi polynomials, Memoirs of AMS, 319, (1985).

[C] T.S. Chihara, An introduction to orthogonal polynomials, Mathematics and Applications Series, Gordon and Breach, 13, (1978).

[D] J. Dixmier, "Algèbres enveloppantes", Gauthier-Villars, Paris, (1974).

[Du] C.F. Dunkl, Orthogonal polynomials and a Dirichlet problem related to the Hilbert transform, Indag. Math. 47, 147-171 (1985).

[F1] P. Feinsilver, Discrete Analogues of the Heisenberg-Weyl algebra, Mh. Math., 104, 89, (1987).

[F2] P. Feinsilver, Lie Algebras and recurrence relations I, submitted to Acta Applicandae Math., (1988).

[F3] P. Feinsilver, Heisenberg algebras in the theory of special functions, conference proceedings "Phase Space", "Springer Lect. Notes in Physics", 278, 423-425, (1987).

[F4] P. Feinsilver, Bernoulli systems in several variables, Conference Proceedings, Probability Measures on Groups, "Springer Lecture Notes", Vol. 1064, 86 (1984).

[P] F. Pollaczek, Sur une famile de polynômes orthogonaux qui contient les polynômes d'Hermite et de Laguerre comme cas limites, comptes Rend. Acad. Sci., Paris, 230, 1563 (1950).

DIFFEOMORPHISM GROUPS AND LOCAL SYMMETRIES:
SOME APPLICATIONS IN QUANTUM PHYSICS

Gerald A. Goldin
Depts. of Mathematics and Physics
Rutgers University
New Brunswick, N.J. 08903 USA

David H. Sharp
Theoretical Division
Los Alamos National Laboratory
Los Alamos, N.M. 87545 USA

INTRODUCTION

This paper reviews results about unitary representations of diffeomorphism groups. Such groups occur naturally in a number of different physical contexts. Their areas of application include quantum mechanics and quantization methods, quantum field theory and local current algebra, general relativity and quantum gravity, and hydrodynamics and the quantization of fluids. In the case when the underlying manifold is a circle, they are related to Kac-Moody algebras, the Virasoro algebra, and the quantum mechanics of strings.

A diffeomorphism is a smooth, invertible transformation from one manifold onto another, or from a manifold M onto itself. For our purposes we shall think of M simply as a smooth surface which, in a sufficiently small neighborhood of each point, "looks like" ordinary n-dimensional Euclidean space. That is, at each point x in M, a Euclidean tangent space T_x is defined whose elements are the tangent vectors to M at x. Two diffeomorphisms of M (i.e., from M to itself) can always be applied successively to yield a third. This operation of composition allows the definition of <u>groups</u> of diffeomorphisms of M. Such diffeomorphism groups --and the algebras of vector fields associated with them--are examples of infinite-dimensional Lie groups and Lie algebras, whose study has recently attracted great interest.

Several things in our view make unitary representations of diffeomorphism groups particularly attractive as a way of approaching quantum physics. First there is the fact that when a physical space has some global symmetry properties such as rotation or translation invariance, these can be expressed mathematically by means of a finite-dimensional Lie group of transformations. Quantization of classical motion in such a space can then be approached by studying the unitary representations of the Lie group; the self-adjoint generators in a representation correspond to observables whose values are conserved by the motion. But what of the case where there is no global symmetry, or where there is insufficient symmetry to recover a complete set of observables from the group representation? Diffeomorphism groups embody the <u>local</u> symmetry of the manifold in a certain sense, so that quantization can be accomplished through a procedure that is independent of the particular manifold in which the motion occurs.

Second, an infinite-dimensional Lie group such as a diffeomorphism group potentially describes infinitely many physical degrees of freedom, even when the underlying manifold is finite dimensional. The much richer structure of the unitary representations of such groups permits a unifying description of a wide variety of systems. For example the quantum theories of N identical particles for arbitrary N satisfying Bose statistics, Fermi statistics, or parastatistics, supermultiplets of particles with integer or half-integer spin, and systems of infinitely many particles in the thermodynamic limit are among the mutually inequivalent unitary representations of a group of diffeomorphisms of R^3.

Third, the study of the unitary representations of diffeomorphism groups is a natural way to analyze the consequences of nontrivial topology in quantum configuration spaces--as occurs, for example, in the Aharonov-Bohm effect and unusual particle statistics.

Finally, if we start with a manifold M such as R^3 at a fixed time, and the assumption that there are self-adjoint operators in a Hilbert space H that describe observations made in local regions of M, it is natural to ask what kinds of groups and algebras have subgroups and subalgebras which are associated with such regions. One answer is to introduce a Lie group G, and to consider the continuous mappings from M to G under pointwise multiplication. Then if B is a bounded region of M, the subgroup of mappings that give the identity outside B is available to us. But an even more natural answer is the group of diffeomorphisms of M, where we have the subgroup of diffeomorphisms that are trivial outside B. This structure is in a sense "given" to us along with the geometry of M, while introducing a new Lie group G puts in extra local symmetry "by hand."

Finite-dimensional Lie groups such as the 3-dimensional rotation group SO(3), its universal covering group SU(2), the Lorentz group, the Poincaré group, and higher internal symmetry groups such as SU(3) or SU(6), have long been familiar to physicists. The study of their unitary representations was motivated by the need to express mathematically the symmetries which arise in quantum physics. The desire to understand further the properties of physical systems from a group-theoretic standpoint has now led to the study of much larger groups--the infinite-dimensional Lie groups including the diffeomorphism groups discussed here.

First we develop the background, defining Lie groups of diffeomorphisms and Lie algebras of vector fields, discussing unitary group representations and self-adjoint operators in Hilbert space, and introducing some related groups and algebras. Then we review some methods of quantization, which leads to a rationale for taking unitary representations of diffeomorphism groups as central to quantum theory. The next section discusses specific unitary representations, obtained as induced representations on particle configuration spaces. In this framework, the nontrivial topology of configuration space leads to particle statistics as well as to topological quantum effects. We discuss representations which correspond to spin systems, quantum dipoles, and systems of infinitely many particles. Next we review the role of diffeomorphism groups in general relativity and quantum gravity. The paper concludes with some applications to the quantization of fluids, using the method of coadjoint orbits.

GROUPS OF DIFFEOMORPHISMS AND ALGEBRAS OF VECTOR FIELDS AS OPERATORS IN HILBERT SPACE [8,13,14-16,45,47,54-55,58]

This section provides a general introduction based on simple, intuitive descriptions. We omit many of the mathematical technicalities which underlie the development. More rigorous treatments are in the references.

Perhaps the simplest way to think of a diffeomorphism is to consider a classical fluid confined to a region of space. Each element of the fluid in its initial configuration A is labeled by its spatial coordinates x. A diffeomorphism is a smooth, invertible mapping φ which describes the displacement of all of the fluid elements to new positions $x' = \varphi(x)$, giving a new configuration B. If a second diffeomorphism ψ displaces the fluid from B to C, i.e. $x'' = \psi(x')$, then the displacement of fluid elements from configuration A to configuration C can be described by a single diffeomorphism which is the composition of the two mappings, $x'' = (\psi \circ \varphi)(x)$. The set of diffeomorphisms of a region forms a group under the operation of composition. The diffeomorphism e defined by $e(x) = x$ for all x, is the identity element in the group. The inverse of φ, denoted φ^{-1} (and also required to be smooth), satisfies $\varphi \circ \varphi^{-1} = \varphi^{-1} \circ \varphi = e$.

More generally, consider an infinitely differentiable manifold M. A C^k-diffeomorphism of M is a mapping from M to M which is one-to-one, onto, and together with its inverse is k times continuously differentiable. The C^k-diffeomorphisms of M form a group under composition, denoted $\mathrm{Diff}^k(M)$. Their mathematical importance is that they supply a certain notion of equivalence among differentiable structures. We shall mainly be concerned with the more restrictive case of infinitely differentiable or C^∞-diffeomorphisms of M, which also form a group under composition, denoted Diff(M). It is of course a subgroup of $\mathrm{Diff}^k(M)$. When the notion of volume is defined on M, the set of volume-preserving C^∞-diffeomorphisms is a natural subgroup of Diff(M), denoted $\mathrm{Diff}_v(M)$ or sDiff(M), where s stands for "special." To be specific, take $M = R^3$. Familiar transformations from R^3 to itself such as rigid rotations, translations, and uniform dilations or contractions, are all diffeomorphisms. The rigid motions are volume-preserving, while dilations and contractions are not. More interesting, though, are the infinitely many independent transformations which distort regions of R^3 in a nonuniform, but smoothly varying, manner. Even volume-preserving diffeomorphisms can do this in infinitely many independent ways (describing the possible configurations of an incompressible fluid). [58]

The smallest closed set <u>outside</u> of which φ acts trivially, in the sense that $\varphi(x) = x$, is called the support of φ, or supp φ. When supp φ is bounded, so that φ becomes trivial at infinity in all directions, we can think of it as describing a purely localized spatial disturbance. If M is finite-dimensional, a subset which is both closed and bounded is compact; we thus talk about the diffeomorphisms with compact support, which form a subgroup of Diff(M). We shall call this subgroup $\mathrm{Diff}_c(M)$; it is equal to Diff(M) if M is a compact manifold. For a compact region B in M, let $\mathrm{Diff}_B(M)$ be the subgroup of diffeomorphisms whose support is contained in B. Thus $\mathrm{Diff}_c(M) = \bigcup_{B \text{ compact}} \mathrm{Diff}_B(M)$.

It is also useful to consider diffeomorphisms of R^n, or of manifolds which behave asymptotically like R^n, which while they may not have compact support rapidly approach the identity mapping $e(x) = x$ (together with their derivatives of all orders), as $|x|$ approaches infinity. Such diffeomorphisms also form a group, denoted by K(M), or simply K. Evidently $\mathrm{Diff}_c(M)$ is a subgroup of K, which is in turn a subgroup of Diff(M); they are all the same if M is compact. If D is any of the preceding groups, and $x \in M$, the set D_x of diffeomorphisms in D for which x is a fixed point is also a group, the "stability group" of x.

Most of the groups we have mentioned can be given topologies that are mathematically compatible with the group operations, making it possible to think of one diffeomorphism as being "close" to another within the group. For example when M is compact, there is the topology of uniform convergence in all derivatives, where two diffeomorphisms are "near" if the maximum difference in their values and the values of their derivatives is small. Endowed with such topologies, certain of the groups become infinite-dimensional Lie groups, where to say that we have a Lie group means in this context that in the topology adopted, a neighborhood of the identity in the group "looks like" a suitable (infinite-dimensional) topological vector space--which, in analogy with the theory of finite-dimensional Lie groups, becomes a Lie algebra under the appropriate bracket operation (see below).

Once a diffeomorphism group has a topology, we have the concept of a closed subgroup, for which the limit of a convergent sequence in the subgroup also lies in the subgroup. For example in the above topology for Diff(M), M compact, the subgroups $Diff_x(M)$, $Diff_B(M)$, and $Diff_v(M)$ are all closed. We also have the possibility of defining continuous one-parameter subgroups. Such a (closed) subgroup is the image of a continuous map $t \rightarrow \varphi_t$ from R to the diffeomorphism group, that respects the group law (see below). Finally, we define the identity component D^e for any of the groups to be the connected subgroup containing e. For example, the inversion of the unit circle is not in $Diff^e(S^1)$, while a one-parameter subgroup is automatically contained in the identity component. We shall see that from the point of view of unitary representations, the one-parameter subgroups directly embody the most important physical information for quantum mechanics. [14-15]

Let us now move from groups to algebras. A vector field g is a smooth mapping which assigns to each point x in M a vector g(x) in the tangent space T_x to M at x. Addition of vector fields, and multiplication of a vector field by a real number, are defined in the obvious way. Denote the space of C^k vector fields on M by $Vect^k(M)$, and the space of C^∞ vector fields by Vect(M). The support of g is the smallest closed set outside of which g(x) = 0. The space of C^∞ vector fields with compact support is denoted $Vect_c(M)$.

Now a smooth vector field g defined on a compact manifold M always generates a one-parameter group φ_t of diffeomorphisms of M, by means of the differential equation,

$$\partial_t \varphi_t(x) = g(\varphi_t(x)), \qquad (1)$$

together with the initial condition $\varphi_0(x) = x$. If g is C^∞, then $\varphi_t(x)$ is also C^∞ (in both x and t). For an arbitrary (noncompact) manifold, we cannot always assert the existence of $\varphi_t(x)$ for all x and t, though we can do so for particular classes of vector fields. [8,45]

For example g(x) may be the velocity field describing the speed and direction of a particle or element of fluid located at x. Then t is the time, and $\varphi_t(x)$ is the path along which the particle or fluid element moves under the influence of g, beginning at x. More generally, $\varphi_t(x)$ can be a path in the configuration space of a many-particle system for which g describes all of the particle velocities, or it can be the trajectory in the phase space M of a classical dynamical system for which g describes the time-evolution. Particles or fluid elements are stationary outside the support of the velocity field describing their motion. When we want to emphasize the dependence of φ_t on g, we shall write the one-parameter group as φ_t^g, which we call the flow obtained by exponentiating the vector field g. We have $\varphi_{t_1}(\varphi_{t_2}(x)) = \varphi_{t_1+t_2}(x)$.

Any vector field with compact support in a manifold M will generate a one-parameter group of diffeomorphisms of M that have compact support; i.e. if $g \in Vect_c(M)$, then $\varphi_t^g \in Diff_c(M)$. The support of the diffeomorphisms will be contained in and generally coincide with the support of g. Thus there is a natural way to associate both vector fields and diffeomorphisms with bounded regions of space; $\mathrm{Vect}_B(M)$ will denote the space of C^∞ vector fields with support in B, whose elements generate diffeomorphisms in $\mathrm{Diff}_B(M)$. We can also consider C^∞ vector fields g on a manifold which behaves like R^n at infinity, whose components g^i (for i = 1, ... ,n) together with their derivatives of all orders, tend rapidly to 0 as $|x| \to \infty$; i.e., the g^i are all contained in Schwartz' space S of functions of rapid decrease for the case $M = R^n$. Such a vector field also always generates a one-parameter group, because a single point (the "point at infinity") can be adjoined to make the manifold compact (the "one-point compactification" of M); then with $g(\infty) = 0$, we again have a vector field on a compact manifold. In this case $\varphi_t^g \in K(M)$. Denote the space of such vector fields $\underline{k}(M)$.

If the vector field g on R^n satisfies div g = 0, we say it is divergenceless; then the flow φ_t^g consists of volume-preserving diffeomorphisms. If the notion of volume on a more general manifold arises from a smooth measure μ, then vector fields for which $\mathrm{div}_\mu g = 0$ generate volume-preserving flows. Denote the space of such vector fields $\mathrm{Vect}_v(M)$.

Given a C^∞ scalar function f from M to R, i.e. $f \in C^\infty(M)$, and a vector field $g \in \mathrm{Vect}(M)$, gf denotes the directional derivative (or Lie derivative) of f. In a system of local coordinates on M, we have

$$(gf)(x) = g^i(x)(\partial f/\partial x^i)(x), \tag{2}$$

where summation is understood from i = 1, ... , n. Then $gf \in C^\infty(M)$. In fact it is possible to <u>define</u> a C^∞ vector field on M by specifying its action as a derivation (i.e., as a linear mapping satisfying the product rule for derivatives) on $C^\infty(M)$. The <u>product</u> of two vector fields g and h can also be defined to act on $C^\infty(M)$ by setting (gh)f = g(hf), but this does not define a derivation and hence gh is not itself a vector field. However the <u>commutator</u> gh - hg acting on $C^\infty(M)$ does define a derivation, and thus a C^∞ vector field on M, called the Lie bracket of g and h and written [g,h]. In local coordinates we can write

$$([g,h]f)(x) = [g^i(x)(\partial h^j(x)/\partial x^i) - h^i(x)(\partial g^j(x)/\partial x^i)](\partial f(x)/\partial x^j) \tag{3}$$

or more concisely $[g,h] = (g\cdot\nabla)h - (h\cdot\nabla)g$. The Lie bracket satisfies the Jacobi identity,

$$[g_1,[g_2,g_3]] + [g_2,[g_3,g_1]] + [g_3,[g_1,g_2]] = 0, \tag{4}$$

and turns Vect(M) into a Lie algebra. It is infinite-dimensional because Vect(M) is infinite-dimensional as a linear space. [g,h] is also called the Lie derivative of h in the g-direction, and written $L_g(h)$.

To understand the meaning of the bracket operation on Vect(M) in terms of flows, let $g,h \in \mathrm{Vect}(M)$, $x \in M$, and $s,t \in R$ very small. Now let x flow by φ_s^g, i.e. under the influence of g for duration s; then by φ_t^h; then, reversing direction, by φ_s^{-g} and by φ_t^{-h}. To first order in s and t, the resulting point x' is displaced from x by a flow in the direction [g,h] for duration st.

If x is outside the support of either g or h, then [g,h](x) = 0. In particular for $g,h \in \mathrm{Vect}_c(M)$, we have $[g,h] \in \mathrm{Vect}_c(M)$, and therefore $\mathrm{Vect}_c(M)$ is also an infinite-dimensional Lie algebra. Likewise $\mathrm{Vect}_B(M)$, $\mathrm{Vect}_v(M)$, and $\underline{k}(M)$ are all infinite-dimensional Lie algebras. But note that $\mathrm{Vect}^k(M)$ is <u>not</u> a Lie algebra, because if g and h are k times continuously differentiable, [g,h] is in general only k - 1 times continuously differentiable, so that $\mathrm{Vect}^k(M)$ is not closed under the bracket operation. The requirement that the space of vector fields be a Lie algebra thus singles out the C^{∞} vector fields naturally.

We are interested in unitary representations of diffeomorphism groups, and the corresponding self-adjoint representations of algebras of vector fields. Let H be a Hilbert space; for example, the familiar space of one-particle wave functions $L^2(R^3)$ consisting of complex-valued, square-integrable functions Ψ on R^3 equipped with the usual inner product $(\Psi_1, \Psi_2) = \int \overline{\Psi_1(x)}\, \Psi_2(x)dx$. More generally, if M is a manifold and μ is a measure on M, we may consider the Hilbert space $H = L^2_{\mu}(M, C^n)$ of vector-valued wave functions on M, square-integrable in the sense that the inner product $(\Psi_1, \Psi_2) = \int_M (\Psi_1(x), \Psi_2(x))d\mu(x)$ exists, where the inner product under the integral sign is the usual one in C^n. When n = 1, we write $H = L^2_{\mu}(M)$.

A linear operator U:H→H is called unitary if $(U\Psi_1, U\Psi_2) = (\Psi_1, \Psi_2)$; i.e. U preserves the inner product in H. Equivalently, $U^* = U^{-1}$, where U^* denotes the adjoint of U. A continuous unitary representation (CUR) of a topological group G is a mapping $b \to U(b)$ which associates to every $b \in G$ a unitary operator U(b) in H, so that: (1) the group law in G is respected, i.e. $U(b_1 b_2) = U(b_1)U(b_2)$; and (2) the topology of G is respected, i.e. $(\Psi_1, U(b)\Psi_2)$ is a continuous function of b for all $\Psi_1, \Psi_2 \in H$. Since inner products of wave functions describe the outcomes of physical observations, CUR's of symmetry groups are expressions of quantum-mechanical consequences of the symmetry that is present. Two CUR's U_1 and U_2 of G in Hilbert spaces H_1 and H_2 are called unitarily equivalent if there is a unitary operator $Q:H_1 \to H_2$ (preserving the respective inner products), such that $QU_1(b)Q^{-1} = U_2(b)$ for all $b \in G$. Unitarily equivalent representations describe the same physics, at least as far as the operators represented are concerned.

A CUR of R is called a one-parameter unitary group, sometimes written U(t) for $t \in R$. The time-development of a quantum-mechanical system, in which the total probability is conserved, can be described by such a group. An important (and profound) result is Stone's theorem, which asserts a correspondence between one-parameter unitary groups and self-adjoint operators in H. Given U(t) as above there exists a unique self-adjoint operator A, defined on a domain D_A in H consisting of all vectors Ψ such that $\lim_{t \to 0}(1/it)[U(t) - 1]\Psi$ exists as a vector Ψ', and given by $A\Psi = \Psi'$ on D_A. Here D_A may or may not be the whole Hilbert space, but if it is not, it is at least dense in H; i.e., any vector in H can be approximated as closely as desired by vectors in D_A. (Self-adjointness of A means that the domain D_{A^*} of the adjoint A^* is equal to D_A, and that A and A^* take equal values on that domain.) Conversely, given a self-adjoint operator A on a domain D_A dense in H, there exists a unique one-parameter unitary group satisfying the above. We write $A = (1/i)dU(t)/dt|_{t=0}$, and $U(t) = \exp(itA)$. A is called the self-adjoint generator of U(t). For example, the time-evolution operators U(t) in quantum mechanics are generated by the Hamiltonian operator, which is self-adjoint. [54-55]

Self-adjoint operators in Hilbert space correspond in general to quantum mechanical observables. Thus Stone's theorem associates such observables with one-parameter unitary groups. When the group represented describes a symmetry of the physical system, the corresponding observables are called symmetry generators. E.g., the self-adjoint generators of one-parameter unitary groups of spatial translations are operators corresponding to measurements of linear momentum; the generators of groups of rotations correspond to measurements of angular momentum; etc. When the group represented is a gauge group for a field theory, the self-adjoint generators are called gauge currents. Gauge groups express the invariance of physical observables under certain transformations of the fields.

In the case of a CUR $V(\varphi)$ of a group of C^∞ diffeomorphisms of M, it is natural to consider the one-parameter unitary subgroups $V(\varphi_t{}^g)$ discussed above. Define the self-adjoint operator

$$J(g) = \lim_{t\to 0} (1/it)[V(\varphi_t{}^g) - 1] \tag{5}$$

on the domain where the limit exists. Under certain conditions we then obtain not only a correspondence $g \to J(g)$ between vector fields and self-adjoint operators, but a representation by self-adjoint operators of the Lie algebra of vector fields; i.e. for all g_1, g_2,

$$[J(g_1), J(g_2)] = iJ([g_1, g_2]). \tag{6}$$

There is today no general theory for the unitary representations of infinite-dimensional Lie groups. The method of induced representations, developed by Mackey for finite-dimensional groups, proved immensely valuable in elementary particle physics. The method of coadjoint orbits, motivated by the desire to carry out the geometric quantization of classical systems, provides a still more general approach in the finite-dimensional case. Both methods can be partially generalized to infinite-dimensional groups. One major obstacle to the generalization is the fact that Haar measure--a measure on the group itself that is invariant under (left) multiplication by group elements--exists in the finite- but not in the infinite-dimensional case. [3,43,47]

Unitary representations of several other kinds of infinite-dimensional groups and algebras, related to diffeomorphism groups, have been studied. We close this section by mentioning some of these briefly. The local current group Map(M,G) is the set of C^∞ mappings T from M to G under pointwise multiplication: $(T_1T_2)(x) = T_1(x)T_2(x)$ for all $x \in M$. The support of T is the smallest closed set outside of which $T(x) = e$, where e is the identity in G. In analogy with our definitions for diffeomorphism groups, we have the subgroup $\mathrm{Map}_c(M,G)$ whose elements have compact support; the subgroups $\mathrm{Map}_B(M,G)$ whose elements have support in B; and the group $S(M,G)$ whose elements T(x) rapidly approach e as $|x| \to \infty$ in M. Similarly, we have the local current algebra of C^∞ mappings from M to the Lie algebra of G, with the bracket operation taken pointwise. Given M, local current groups and algebras provide another important way to identify operators naturally with bounded regions of M. While they differ importantly from diffeomorphism groups and algebras of vector fields, there is a natural semidirect product of the two which proves helpful in constructing and interpreting certain unitary representations of the diffeomorphism groups. The group law for the semidirect product $\mathrm{Map}(M,G) \wedge \mathrm{Diff}(M)$, and its various subgroups, is given by

$$\langle T_1, \varphi_1\rangle\langle T_2, \varphi_2\rangle = \langle T_1(T_2 \circ \varphi_1), \varphi_2 \circ \varphi_1\rangle. \tag{7}$$

An important special case is the choice G = R, from which we obtain the semidirect products $C^\infty(M) \wedge \mathrm{Diff}(M)$, $S(M) \wedge K(M)$, $C_c^\infty(M) \wedge \mathrm{Diff}_c(M)$, etc.

A loop group is a special case of a local current group Map(M,G), where $M = S^1$. The corresponding loop algebra is $\mathrm{map}(S^1, \mathfrak{g})$ where $\mathfrak{g}$ is the Lie algebra of G. For $\underline{f} \in \mathrm{map}(S^1, \mathfrak{g})$ let us write $\underline{f}(s) = \sum f_i(s)Q_i$, where the f_i are real-valued functions on S^1, and the Q_i form a basis for $\mathfrak{g}$ satisfying $[Q_i, Q_j] = c_{ijk}Q_k$. Then $[\underline{f}, \underline{g}] = \underline{h}$, where $h_k = c_{ijk}f_ig_j$. Local currents indexed by the integers can be defined from the local current algebra by taking its Fourier transform. With $M_n{}^i(s) = \exp(-ins)Q_i$, the loop algebra can be written $[M_n{}^i, M_m{}^j] = c_{ijk}M^k{}_{n+m}$. The Kac-Moody algebra is a <u>central extension</u> of the loop algebra. Its bracket operation is $[M_n{}^i, M_m{}^j] = c_{ijk}M^k{}_{n+m} + cn\,\delta_{n,-m}\,\delta_{ij}P$, where P commutes with all the $M_n{}^i$

and thus is said to be in the "center" of the algebra. In a manner similar to the definition of the Kac-Moody algebra as a central extension of a loop algebra, it is possible to construct a central extension of $\mathrm{Vect}(S^1)$ called the Virasoro algebra. These groups and algebras have application to field theories in two-dimensional spacetime, and to strings. [13]

DIFF(M) AND QUANTIZATION [2,9,10,14-15,41]

Several lines of argument lead to the construction of Diff(M) or one of its subgroups in quantum theory. One line, proceeding from the study of models, was initiated by Landau who formulated his theory of superfluidity using the quantum mechanical operators for mass density $\rho(x)$ and momentum density $J(x)$, expressed in first-quantized form. The equal-time commutation relations satisfied by these operators give the semidirect product Lie algebra $C^\infty(R^3) \wedge \mathrm{Vect}(R^3)$. Subsequently Dashen, Sharp, and Sugawara explored how various model field theories could be given a complete formulation in terms of observables such as local currents. Dashen and Sharp began with second quantized Bose or Fermi fields for nonrelativistic quantum theory, and arrived at the same semidirect product algebra. This work established the completeness of the observables $\rho(x)$ and $J(x)$, assuming the completeness of the underlying fields. It was also emphasized that the same Lie algebra of observables described fermions as well as bosons. A mathematically more rigorous approach to this model was initiated by Goldin, who proposed consideration of the semidirect product group $S(R^3) \wedge K(R^3)$ obtained by exponentiating the algebra of $\rho(x)$ and $J(x)$ after averaging these operators with scalar functions and vector fields, and interpreted some of its CUR's. [9,14-15,42,57]

It is also possible to motivate the construction of Diff(M) by means of simpler, model-independent arguments. One such argument stems from consideration of observables on manifolds. To describe a single particle moving in R, one represents the Heisenberg algebra by self-adjoint position and momentum operators x_{op} and p_{op}, satisfying $[x_{op}, p_{op}] = i\hbar I$. In the Hilbert space $H = L^2(R)$, x_{op} acts on $\Psi(x)$ by multiplication, and p_{op} by $(\hbar/i)d/dx$. (We now set $\hbar = 1$). A well-known result of Stone and von Neumann asserts the uniqueness of this representation up to unitary equivalence, as an irreducible self-adjoint representation of the algebra. In R^n we choose self-adjoint vector operators x_{op} and p_{op}, satisfying $[x^i_{op}, p^j_{op}] = i\delta^{ij}I$ $(i,j = 1, \ldots, n)$, with $[x_{op}, x_{op}] = [p_{op}, p_{op}] = 0$; they act in $L^2(R^n)$ by $x_{op}\Psi(x) = x\Psi(x)$, $p_{op}\Psi(x) = (1/i)\nabla\Psi(x)$. But the operators x_{op} and p_{op} are unbounded. Consequently they are defined not on every wave function, but only on certain dense domains. To be in the domain of p_{op}, for example, the wave function must be differentiable on a set of sufficient size, and its derivative must be square integrable. Now the self-adjointness of an unbounded operator is sensitive to its domain of definition--it is required that $D_{A*} = D_A$. This is necessary for the spectral theorem to hold, so that A can be interpreted physically as a measurement, and for the exponentiation of A to a one-parameter group. It can present a problem when the region B in which the quantum particle moves is a manifold, or manifold with boundary, other than R^n. For instance if $B = [0,1] \subset R$, we may initially consider the differential operator $A = (1/i)d/dx$ on a domain D_o of wave functions such that $\Psi(0) = \Psi(1) = 0$. Then the domain of its adjoint includes wave functions taking arbitary values at 0 and 1, and A is not self-adjoint.+ One solves this problem by finding a self-adjoint extension of A; that is, widening its domain (and

+The domain D_{A*} is the set of all $\Phi \in H$ such that for some $\Phi' \in H$, $(\Phi, A\Psi) = (\Phi', \Psi)$ for all $\Psi \in D_A$. Then $\Phi' = A^*\Phi$. See Reed and Simon [54-55] for additional details.

at the same time narrowing D_{A*}) until $D_A = D_{A*}$. There exists in fact a one-parameter family of such self-adjoint extensions for $(1/i)d/dx$ in $L^2([0,1])$, associated with domains D_θ (θ fixed) whose elements satisfy $\Psi(1) = \exp(i\theta)\Psi(0)$. But the choice of a value for the relative phase θ inserts additional dynamical information into the quantum theory. The kinematics is no longer described by the statement that there is a self-adjoint representation of the Heisenberg algebra; indeed, D_θ is not invariant under x_{op}, so the bracket $[x_{op}, p_{op}] = iI$ holds only on the smaller domain D_o. The situation is quite similar when $M = S^1$. The problem is not only that the choice of self-adjoint extension is non-unique and embodies additional dynamical information: if one takes the particle domain B to be the half-line, and defines $A = (1/i)d/dx$ initially on a domain of differentiable wave functions vanishing at 0, then A has no self-adjoint extensions. Thus it is unsatisfactory to use the Heisenberg algebra for quantization on general manifolds or manifolds with boundary. On the half-line one can introduce new operators $x^2_{op}/2$ and $(x_{op}p_{op} + p_{op}x_{op})/2$ which satisfy a different Lie algebra, and are again self-adjoint; but the goal of describing quantum theory through a single prescription of a Lie algebra or Lie group is lost.

To circumvent the difficulty, localize the operator for measurement of momentum away from the boundaries, the infinities, or any singularities in B that seem to be causing the problem. In $L^2(B)$, where $B = [0,1]$, $[0,\infty)$, etc., consider the operator $g(x)p_{op} = (1/i)g(x)d/dx$, where $g(x)$ is a smooth "regularizing function" on B vanishing sufficiently rapidly at the boundary or at ∞. To make this differential operator Hermitian, we symmetrize it by defining $J(g)\Psi(x) = (1/2i)[g(x)d/dx + (d/dx)g(x)]\Psi(x)$, where the derivative is understood to act on everything to its right. For $B \subset R^n$, we similarly consider $J(g) = (1/2i)[g\cdot\nabla + \nabla\cdot g]$ where g is now a smooth vector field on B vanishing where appropriate. Such operators, defined in $L^2(B)$ initially on the domain of square integrable C^∞ functions on B, have unique, self-adjoint extensions. Furthermore the commutator satisfies Eq. (6), so that we have a self-adjoint representation of the Lie algebra Vect_B in $L^2(B)$. This will be called the 1-particle representation.

The Heisenberg algebra is thus straightforwardly replaced by an algebra of vector fields to describe a single particle in quantum mechanics. But it is clear that the 1-particle system has been substantially overdetermined. Operators x_{op} and p_{op}, representing just 2n phase space coordinates, have been replaced by an infinite family of independent momentum density operators J(g). Where there was an essentially unique irreducible representation of the Heisenberg algebra, it will develop that there are infinitely many inequivalent irreducible representations of the algebra of vector fields. These may seem like disadvantages, and have been cited as such by Isham. However the infinite dimensionality of the Lie algebra allows a substantial unification of diverse quantum theories, with the various inequivalent representations describing kinematically different physical systems. Many of these are reviewed below. We also note that in the 1-particle case the family of operators $J(g) = (1/2i)(g\cdot\nabla + \nabla\cdot g)$ is complete. Although we constructed them in order to regularize p_{op}, the introduction of the vector fields as regularizing functions eliminates the need for x_{op}; the representation is already irreducible. Nevertheless it is convenient to have a family of self-adjoint operators that act directly to localize the particle. These are provided by defining the multiplication operators $\rho(f)\Psi(x) = f(x)\Psi(x)$ for all f which are C^∞ having support in B, vanishing smoothly at the boundary and rapidly at ∞ if B is unbounded. Then $\rho(f)$ and $J(g)$ satisfy the semidirect product algebra:

$$[\rho(f_1), \rho(f_2)] = 0, \qquad [\rho(f), J(g)] = i\rho(gf),$$
$$[J(g_1), J(g_2)] = iJ([g_1, g_2]). \tag{8}$$

The group associated with Eqs. (8) is $C_B^\infty(R^n) \wedge \mathrm{Diff}_B(R^n)$, or $S(M) \wedge K(M)$, or $C_c^\infty(M) \wedge \mathrm{Diff}_c(M)$, depending on what is assumed about the supports of f and g. It is not difficult to verify that the 1-particle representation of the group is given in $L^2(R^n)$ by:

$$U(f)\Psi(x) = e^{if(x)}\Psi(x),$$
$$V(\varphi)\Psi(x) = \Psi(\varphi(x))\sqrt{\mathcal{J}_\varphi(x)}, \tag{9}$$

where $U(tf) = \exp(it\rho(f))$ and $V(\varphi_t{}^g) = \exp(itJ(g))$, with $\mathcal{J}_\varphi(x) = \det(\partial\varphi^i(x)/\partial x^j)$ being the Jacobian of φ evaluated at x.

Returning briefly to the example B = [0,1], we see that we have eliminated from the Lie algebra representation the alternative dynamical descriptions associated with the different choices of boundary condition $\Psi(1) = \exp(i\theta)\Psi(0)$. The representation of the vector fields, which vanish at 0 and 1, describes only kinematical information. But the dynamical possibilities are not lost--they are now described as they should be by the choice of a self-adjoint Hamiltonian operator, whose kinetic piece will be $-\frac{1}{2}\, d^2/dx^2$ defined on an appropriate domain in $L^2([0,1])$.

To write the 1-particle representation for $C^\infty(M) \wedge \mathrm{Diff}(M)$ where M is more general than R^n, we need to generalize from Lebesgue measure on R^n. Measures μ_1 and μ_2 on M are <u>equivalent</u> when they have the same sets of measure 0. Then the derivative of either measure with respect to the other exists and is a positive measurable function on M. We write this <u>Radon-Nikodym derivative</u> as $d\mu_1/d\mu_2$. In particular if dx refers to Lebesgue measure on R^n, and μ is any measure equivalent to dx, we have $d\mu = (d\mu/dx)dx$. Now if μ is a measure on M, and φ acts measurably on M, we denote by μ_φ the transformed measure given by $\mu_\varphi(X) = \mu(\varphi(X))$ for any Borel set X. We call μ an invariant measure if $\mu_\varphi = \mu$; we call it quasi-invariant if μ_φ and μ are equivalent. Lebesgue measure on R^n is quasi-invariant for all diffeomorphisms; and the Radon-Nikodym derivative $d\varphi(x)/dx$ is just $\mathcal{J}_\varphi(x)$. Thus any Borel measure μ on an n-dimensional manifold M, locally equivalent to Lebesgue measure, is also quasi-invariant for diffeomorphisms. In $L^2_\mu(M)$, the 1-particle representation of Eqs. (9) becomes for $V(\varphi)$,

$$V(\varphi)\Psi(x) = \Psi(\varphi(x))\sqrt{\frac{d\mu_\varphi}{d\mu}(x)}, \tag{10}$$

and the expression for J(g) becomes $J(g)\Psi = (1/i)g\Psi + (1/2i)\mathrm{div}_\mu g$; where $g\Psi$ is the directional derivative and $\mathrm{div}_\mu g$ is the μ-divergence of g. These now give us the 1-particle representation on a general manifold M. Equivalent measures on M lead to unitarily equivalent 1-particle representations.

In the past two decades, efforts have been made to bring to bear the methods of differential geometry and fiber bundles directly on the problem of quantization on manifolds. One of the outcomes of these efforts, called <u>quantum Borel kinematics</u>, provides another construction for Diff(M) in quantum theory. As a starting point, M is taken to be the "external" configuration space of the system. Internal dynamical degrees of freedom are described by constructing a fiber bundle over M. This is a space $\mathcal{A}$ (the total space), together with a projection mapping $\pi: \mathcal{A} \to M$. For $x \in M$, $\mathcal{F} = \pi^{-1}(x)$ is called a fiber. The important property of bundles is that locally $\mathcal{A}$ "looks like" the Cartesian product of M with $\mathcal{F}$. The triple $(\mathcal{A}, \pi, M)$ is called a fiber bundle when a topological group acts suitably on the fibers. Depending on the topology of M, which may have nontrivial global properties such as "holes," there may be nontrivial global properties of $\mathcal{Q}$ such as "twists," which do not show up locally. [2,10,41]

With M as configuration space and $\mathcal{Q}$ as generalized configuration space, it is assumed that the observable or physical quantities are those which can be measured in M. For example let M be the configuration space of two identical particles in R^3 which are forbidden to occupy the same point. An element of M is a non-ordered pair $\gamma = \{x_1, x_2\}$; $x_1 \neq x_2$. Then M is diffeomorphic to a projective space. It has a two-sheeted covering space $\tilde{M}$, consisting of the ordered pairs $\tilde{\gamma} = (x_1, x_2)$; $x_1 \neq x_2$. The projection $\pi: (x_1, x_2) \rightarrow \{x_1, x_2\}$. Each fiber consists of two discrete elements: (x_1, x_2) and (x_2, x_1). A covering space such as $\tilde{M}$ is a special case of a fiber bundle, in which the fiber is discrete. To describe a system having M as its (external) configuration space, introduce regions of localization X (which are Borel sets in M), with the classical motions being described by flows of the points in X under diffeomorphisms $\varphi_t{}^g$, where $g \in \mathrm{Vect}(M)$. Motivated by work of Segal, Doebner and Tolar proposed to quantize such a system directly by introducing two families of operators in $L^2_\mu(M)$, where μ is a smooth measure on M equivalent to Lebesgue measure. First, for each Borel set X there should be a projection operator E(X) acting by multiplication by indicator functions; i.e., if $I_X(x) = 1$ when $x \in X$ and 0 when $x \notin X$, then $E(X)\Psi(x) = I_X(x)\Psi(x)$. Thus E(X) localizes the system in configuration space. Second, the $\varphi_t{}^g$ should be represented in such a way that the one-parameter unitary groups $V(\varphi_t{}^g)$ are smooth and respect the local projections--that is, if $\Psi \in L^2_\mu(M)$ has support in X, $V(\varphi_t{}^g)$ must have support in the inverse image of X under the flow. (The direction is conventional; $V^*(\varphi_t{}^g)\Psi$ has support in the image of X.) This condition restricts V to be of the form

$$V(\varphi_t{}^g)\Psi(x) = \chi_t{}^g(x)\Psi(\varphi_t{}^g(x)), \tag{11}$$

where $\chi_t{}^g(x)$ is a multiplication operator depending on g and t. With $J(g) = \lim_{t \to 0} (1/it)[V(\varphi_t{}^g) - I]$, the most general form of J(g) is constrained by Eq. (11) to be $J(g)\Psi = (1/i)g\Psi + \alpha(g)\Psi$, where $g\Psi$ is again the directional derivative, and where $\alpha(g)$ is a complex linear functional of g which gives a smooth function of x for each g (a 1-form on M). The condition that J(g) give a self-adjoint representation of the algebra of vector fields now constrains the imaginary part of $\alpha(g)$ to be of the form $(1/2i)(\mathrm{div}_\mu g)(x)$, which we already saw in the 1-particle representation. But when Re $\alpha(g) \neq 0$, additional representations are possible. Even when M is topologically trivial, the choice $\alpha(g) = \lambda(\mathrm{div}_\mu g)(x)$ gives a class of unitarily inequivalent representations, for $\lambda \in R$, that have been discussed by Doebner, Tolar, Goldin, Menikoff, and Sharp. [10,19,24]

Now the ideas of quantum Borel kinematics extend to the generalized configuration spaces, with topological groups G as fibers. Inequivalent quantizations of the kinematics are given by unitarily inequivalent representations of G, which provide additional possibilities for $\chi_t{}^g$ and $\alpha(g)$. In the example of two identical particles, the group G is the symmetric group S_2 of permutations of two objects, and the inequivalent quantizations describe Bose and Fermi statistics.

In the approach just discussed, the configuration space or generalized configuration space is the starting point, and the CUR's of the semidirect product group emerge from the quantization procedure. In the remainder of the paper, we explore CUR's of Diff(M) and its subgroups, where M is just the 1-particle space. Many-particle theories, and theories with internal degrees of freedom, are obtained as inequivalent CUR's of a single group.

UNITARY REPRESENTATIONS OF DIFF(M) AND THEIR INTERPRETATION IN QUANTUM MECHANICS [14-27,32-34, 49-51]

In this section we consider various CUR's of Diff(M) or its subgroups and semidirect products that have been interpreted quantum-mechanically. Consider first a second quantized nonrelativistic field $\psi(x,t)$ satisfying either canonical commutation (-) or anticommutation (+) relations at equal times:

$$[\psi(x), \psi^*(y)]_{\pm} = \delta(x - y). \tag{12}$$

Let the Fock Hilbert space $H^{s,a} = \bigoplus_{N=0}^{\infty} H_N^{s,a}$, where $H_N^{s,a}$ is the Hilbert space of symmetric (s) or antisymmetric (a) L^2 functions of N variables in R^n. A vector $\Psi \in H^{s,a}$ may be written $(\Psi_0, \Psi_1, \Psi_2, \dots)$ with $(\Psi, \Psi) = \sum_{N=0}^{\infty} (\Psi_N, \Psi_N) < \infty$. For the canonical commutation relations, the Fock representation is given in H^s by

$$\begin{aligned} (\psi(x)\Psi)_N(x_1, \dots, x_N) &= (N+1)^{\frac{1}{2}} \Psi_{N+1}(x_1, \dots, x_N, x), \\ (\psi^*(x)\Psi)_N(x_1, \dots, x_N) &= N^{-\frac{1}{2}} \sum_{j=1}^{N} \delta(x - x_j) \Psi_{N-1}(x_1, \dots, \hat{x}_j, \dots, x_N), \end{aligned} \tag{13}$$

where $\hat{x}_j$ means that x_j is omitted. Eqs. (13) define ψ and ψ^* as operator-valued distributions. Likewise for the canonical anticommutation relations, the Fock representation is given in H^a; the equation for $\psi(x)$ is formally identical to the first of Eqs. (13), and the equation for $\psi^*(x)$ has an overall factor of $(-1)^{N+1}$ and a factor of $(-1)^{j+1}$ under the summation sign.

Define the mass density $\rho(x) = m\psi^*(x)\psi(x)$ and the momentum density $J(x) = (1/2i)\{\psi^*(x)[\nabla\psi(x)] - [\nabla\psi^*(x)]\psi(x)\}$. These are also operator-valued distributions; they become operators when averaged with smooth test functions on R^n: $\rho(f) = \int \rho(x)f(x)dx$, and $J(g) = \int J(x)\cdot g(x)dx$, where f is a C^{∞} scalar function and g a C^{∞} vector field. From these definitions, direct calculations in either of the Fock representations yield:

$$\begin{aligned} (\rho(f)\Psi)_N &= m\sum_{j=1}^{N} f(x_j)\Psi_N, \\ (J(g)\Psi)_N &= (1/2i)\sum_{j=1}^{N}[g(x_j)\cdot\nabla_j + \nabla_j\cdot g(x_j)]\Psi_N. \end{aligned} \tag{14}$$

It can be verified directly that these expressions satisfy the Lie algebra of Eqs. (8). From the point of view of field theory, it is noteworthy that the same algebra is represented whether one starts with commuting or anti-commuting fields. The distinction between Bose and Fermi particles is no longer to be made through a choice of algebra, but through the choice of representation. The subspaces $H_N^{s,a}$ of $H^{s,a}$ are invariant for the representation of Eqs. (14). Restricted to these subspaces, we obtain irreducible self-adjoint representations of the algebra, called the N-particle Bose (s) and Fermi (a) representations. [14-15,32, 35-36]

From Eqs. (14), we can now construct the N-particle Bose and Fermi representations of the group $S(R^n) \wedge K(R^n)$, or $C_c^{\infty}(R^n) \wedge \mathrm{Diff}_c(R^n)$ in $H_N^{s,a}$. They are just the symmetric and antisymmetric tensor products of N copies of the representation of Eqs. (9). This method of obtaining the quantum mechanics of N identical particles leads to a different perspective from that of quantum Borel kinematics, in that the N-particle representations for different N, as well as the different statistics, correspond to CUR's of the one group $S(R^n) \wedge K(R^n)$, without the need to consider diffeomorphisms of the N-particle configuration space manifold. We easily generalize to N Bose or Fermi particles in a manifold M (with dim M $>$ 1) having a smooth measure μ locally equivalent to Lebesgue measure, obtaining the symmetric or antisymmetric tensor product of N copies of Eqs. (10). As in the case $N = 1$, a complete set of observables is obtained from the representation V alone, of $\mathrm{Diff}_c(M)$ or $K(M)$; having the semidirect product is a convenience.

In the case of N distinguishable particles, however, the semidirect product is more than a convenience. Suppose they have different masses; then we have $U(f) = \exp[i\rho(f)]$ given by

$$U(f)\Psi_N = \exp[i\sum_{j=1}^{N} m_j f(x_j)]\Psi_N , \tag{15}$$

replacing the tensor product of identical copies of Eq. (9) with $m = 1$; and the Hilbert space for $U(f)$ is $L^2(R^{3N})$, since Eq. (15) no longer preserves the symmetry type. However the expression for $V(\varphi)$ remains formally unchanged, as the mass does not appear in $J(g)$ or $V(\varphi)$. We thus have now a reducible representation of the diffeomorphism group, with the reducing subspaces being those of fixed symmetry type, but an irreducible CUR of the semidirect product. This is the small price to be paid for deciding to represent only the diffeomorphisms of the 1-particle manifold, rather than the diffeomorphisms of the full configuration space.

Next we show how the infinite free Bose gas at zero temperature (in R^3) is described by a CUR of $\mathrm{Diff}_c(R^3)$ or $K(R^3)$, in a Fock representation of underlying canonical fields. In Eqs. (13), define $\psi'(x) = \psi(x) + \sqrt{\bar\rho}$ and $\psi'^*(x) = \psi^*(x) + \sqrt{\bar\rho}$, where $\bar\rho \in R$ will have the physical interpretation of the average density. Then ψ' and ψ'^* also satisfy the canonical commutation relations. With $\rho'(x)$ and $J'(x)$ defined in terms of them, we again obtain a self-adjoint representation of Eqs. (8), and a corresponding CUR $U'(f)V'(\varphi)$ of $S(R^3)\wedge K(R^3)$. It is easy to see that the $\rho'(f)$ and $J'(g)$ no longer leave the N-particle Fock subspaces H_N^S invariant, so that these are no longer reducing subspaces for the representation $U'(f)V'(\varphi)$. To interpret the representation physically, consider N identical free Bose particles moving in $M = T^3$ (the 3-torus), regarded as a cube of volume $\mathcal{V}$ in R^3 with corresponding points on opposite boundaries identified. Next write the N-particle Bose representation $U_{N,\mathcal{V}}(f)V_{N,\mathcal{V}}(\varphi)$ as the symmetric tensor product of N copies of Eqs. (9), where the Jacobian is with respect to the (local) Lebesgue measure on T^3 obtained by identifying it with the cube in R^3. The normalized symmetric ground state wave function for the usual free Hamiltonian is $\Omega_{N,\mathcal{V}}(x_1, \ldots, x_N) = (1/\mathcal{V})^{N/2}$. Then define the ground state expectation functional $L_{N,\mathcal{V}}(f) = (\Omega_{N,\mathcal{V}}, U_{N,\mathcal{V}}(f)\Omega_{N,\mathcal{V}}) = \{(1/\mathcal{V})\int \exp[if(x)]dx\}^N$. In the limit as $N,\mathcal{V} \to \infty$, with $N/\mathcal{V} \to \bar\rho$, we obtain:

$$\begin{aligned} L(f) = \lim_{\substack{N,\mathcal{V}\to\infty \\ N/\mathcal{V}\to\bar\rho}} L_{N,\mathcal{V}}(f) &= \lim_{N\to\infty} \left(1 + \frac{\bar\rho}{N}\int [e^{if(x)} - 1]dx\right)^N \\ &= \exp\left\{\bar\rho\int [e^{if(x)} - 1]dx\right\}. \end{aligned} \tag{16}$$

But one can verify directly that the ground state expectation functional $(\Psi_0, U'(f)\Psi_0)$ is equal to this $L(f)$, where Ψ_0 is the Fock vacuum. Similarly, it can be demonstrated that the functional $E(f,\varphi)$ obtained by taking the limit as $N,\mathcal{V} \to \infty$ with $N/\mathcal{V} \to \bar\rho$ of $(\Omega_{N,\mathcal{V}}, U_{N,\mathcal{V}}(f)V_{N,\mathcal{V}}(\varphi)\Omega_{N,\mathcal{V}})$, is equal to $\exp\{\bar\rho\int [e^{if(x)}\sqrt{\mathcal{J}_\varphi(x)} - 1]dx\}$ which is the same as $(\Psi_0, U'(f)V'(\varphi)\Psi_0)$. Thus these representations describe the infinite free Bose gas at fixed average density. [20,32,59]

In obtaining these representations, it should be remarked that we have characterized the dynamics as well as the kinematics of the quantum systems that they describe. This is very much in the spirit of Haag's theorem, which asserted (for the Weyl group) that Bose field theories with distinct dynamics (i.e., distinct scattering matrices) must be described by inequivalent representations. Thus other CUR's of $\mathrm{Diff}_c(R^3)$ or $K(R^3)$ which make use of an actual infinity of degrees of freedom, are expected to describe alternate dynamics.

The zero-temperature free Fermi gas has been studied by Menikoff from the standpoint of generating functionals for unitary representations of $S(R^3) \wedge K(R^3)$. Infinite free systems at finite temperature have also been constructed from this point of view. [12,49-50]

Another physical situation that can be set up in the Fock representation of ψ and ψ^* is the quantum theory of N particles interacting with an external, non-quantized electromagnetic field. As usual (with $m = \hbar = 1$), set $\rho(x) = \psi^*(x)\psi(x)$ and $J(x) = (1/2i)\{\psi^*(x)[\nabla - (iq/c)A(x)]\psi(x)\}$ + (hermitian conjugate), where A is the magnetic vector potential, $\nabla \times A = B$. Then the Lie algebra of Eqs. (8) is modified; the first two equations are unaltered, while the third becomes

$$[J(g_1), J(g_2)] = iJ([g_1, g_2]) + i(q/c)\,\rho(B\cdot(g_1 \times g_2)). \tag{17}$$

From the perspective of Lie group and Lie algebra extensions, the above commutation relation gives a trivial extension of the Lie algebra of vector fields, since it is obtained from Eqs. (8) by the linear transformation $J(g) \to J(g) - (q/c)\rho(A\cdot g)$. Nevertheless some nontrivial physics emerges from the representation theory. In the situation of the Aharonov-Bohm effect, for example, where the B-field is confined to a region of space, Eqs. (8) are recovered in the field-free region--which, however, is no longer simply connected. Then we must represent $S(M) \wedge K(M)$, where M is topologically nontrivial. It is interesting to note the formal resemblance between the transformation of J(g) induced by an external field, and the transformation $J(g) \to J(g) - \lambda\rho(\mathrm{div}\, g)$, which leaves Eqs. (8) invariant and leads to the class of unitarily inequivalent quantizations mentioned above. [51]

In the preceding, unitary representations of diffeomorphism groups and their semidirect products are constructed from second-quantized non-relativistic fields. This does not necessarily mean, however, that the theories described by CUR's of Diff(M) can only be non-relativistic. The same commutator algebra of Eqs. (8) is obtained if we replace $\psi(x,t)$ in the definition of $\rho(x)$ and $J(x)$ with

$$\phi_1(x,t) = \frac{1}{(2\pi)^{3/2}} \int_{k_o > 0} (k_o)^{\frac{1}{2}} e^{i(k_o t - k\cdot x)} (dk/k_o) a_k \tag{18}$$

where a_k annihilates a particle with 3-momentum k in the <u>relativistic</u> Fock representation of a Bose field ϕ; ϕ_1 and ϕ_1^* satisfy the same commutator as ψ and ψ^*. But it should be noted that $\rho(f)$ and $J(g)$ can no longer be interpreted as <u>local</u> observables in such a relativistic theory. When extended (even infinitesimally) off the fixed-time surface in R^4, they no longer commute at spacelike separations, and the assumption that they are local violates causality. J(x) can also be taken to be the fixed-time momentum density in a relativistic field theory in two-dimensional space-time. The resulting current algebra is the Virasoro algebra, a nontrivial central extension of the algebra of vector fields on M, with dim M = 1.

Next we consider <u>induced</u> representations of Diff(M), $\mathrm{Diff}_c(M)$, or K(M) over many-particle configuration spaces. The N-particle Bose and Fermi representations are recovered in this framework, as well as CUR's for particles obeying parastatistics, and unusual statistics when dim M = 2. The topological origin of particle statistics becomes especially transparent when quantum mechanics is described in this way. [22-23,27,44]

For a manifold M, define the N-particle configuration space $\Delta^{(N)}$ to be the set whose elements are subsets of M containing exactly N points. We write Δ (suppressing N) when there is no possibility of confusion. The fact that M is a smooth manifold implies that Δ is smooth. A diffeomorphism $\varphi \in \mathrm{Diff}(M)$ acts naturally and transitively on Δ: for $\gamma = \{x_1, \ldots, x_N\}$, $\gamma' = \varphi(\gamma) = \{\varphi(x_1), \ldots, \varphi(x_N)\}$. It is important here that $x_i \neq x_j$ implies $\varphi(x_i) \neq \varphi(x_j)$, since diffeomorphisms are smooth one-to-one transformations. The topology of Δ is described in part by its fundamental group or first homotopy group $\pi_1(\Delta, \gamma)$, where γ is fixed. An element of $\pi_1(\Delta, \gamma)$ is an equivalence class of loops in Δ that begin and end at γ, where two paths in Δ are equivalent (or homotopic) if one can be continuously deformed into the other. Group multiplication in $\pi_1(\Delta, \gamma)$ is defined by traversing two loops successively, an operation which respects the homotopy classes. For any $\gamma_1, \gamma_2 \in \Delta$, the groups $\pi_1(\Delta, \gamma_1)$ and $\pi_1(\Delta, \gamma_2)$ are isomorphic; thus the development is independent of the actual choice of γ. Even when M is simply connected, Δ can be multiply connected.

If $M = R^3$, the fundamental group $\pi_1(\Delta^{(N)}, \gamma)$ is isomorphic to S_N, the symmetric group whose elements are all possible permutations of N objects. This is because a path along which the points x_i exchange positions is a loop in Δ based at γ, that cannot be contracted to a point; while any path along which the points x_i return to their original positions can be untangled in R^3. The universal covering space $\tilde{\Delta}$ of Δ is the set of ordered N-tuples in R^3, with $x_i \neq x_j$ for $i \neq j$, and the projection $\pi : \tilde{\Delta} \to \Delta$ maps $\tilde{\gamma} = (x_1, \ldots, x_N)$ to γ. A loop based at γ in Δ lifts to a loop based at $\tilde{\gamma}$ in $\tilde{\Delta}$, if and only if it is homotopic to the identity in $\pi_1(\Delta, \gamma)$; otherwise it lifts to an open path whose terminal point is distinct from its initial point. Thus, if we choose a fixed element $\tilde{\gamma} \in \tilde{\Delta}$ with $\pi\tilde{\gamma} = \gamma$, it is possible to identify sheets of $\tilde{\Delta}$ with elements of S_N. The action of $\varphi \in \mathrm{Diff}(M)$ on Δ lifts to a well-defined action on $\tilde{\Delta}$ in a way compatible with the group operation; in the example at hand, $\varphi : (x_1, \ldots, x_N) \to (\varphi(x_1), \ldots, \varphi(x_N))$.

For $M = R^2$, the set of ordered N-tuples is a covering space of Δ, but not the universal covering space. The fundamental group $\pi_1(\Delta, \gamma)$ is no longer just S_N; it is the braid group B_N. B_N is an infinite, discrete group whose elements can be visualized as consisting of N braided strands of yarn connecting N points on the (2-dimensional) floor of a room with N corresponding points on the room's ceiling. Braids are combined in analogy with composition. Thus B_2 is isomorphic to the additive group of integers, while B_N is non-Abelian for $N \geq 3$. The physical meaning of the fact that the braid group is the fundamental group for $\Delta^{(N)}$ in R^2 is that as the positions of two particles are exchanged physically, it is possible to keep track consistently of the number of times they have circled each other in, say, a counterclockwise direction.

Important to the idea of induced representations is the notion of a stability group, or "little group." For the action of Diff(M) or K(M) on Δ, the little group $\mathrm{Diff}_\gamma(M)$ or $K_\gamma(M)$ is the (closed) subgroup containing diffeomorphisms φ such that $\varphi(\gamma) = \gamma$. For the case of $\Delta^{(N)}$ in R^3, there is a natural homomorphism from the little group onto S_N; in fact, to say that $\varphi(\gamma) = \gamma$ means that φ implements a permutation of the values of the points in γ. The correspondence $K_\gamma(R^3) \to S_N$ is a continuous group homomorphism, and thus a unitary representation of S_N automatically defines a CUR of $K_\gamma(R^3)$. For the case of $\Delta^{(N)}$ in R^2, the correspondence is from the little group to B_N. To establish it, we use the fact that diffeomorphisms in $K_\gamma(R^2)$ become trivial at infinity; bringing a configuration in from infinity to γ while $\varphi \in K_\gamma(R^2)$ acts on it, establishes a braid. Now a unitary representation of B_N pulls back via this homomorphism to a CUR of $K_\gamma(R^2)$.

Inducing allows us to construct a CUR of K(M) from a CUR of $K_\gamma(M)$ obtained from a unitary representation of one of the fundamental groups above. We first need a measure on Δ quasi-invariant for diffeomorphisms; this is provided by the usual volume element on $\Delta^{(N)}$. Let L be a CUR of $K_\gamma(M)$, $M = R^2$ or R^3, defined by pulling back a unitary representation $\hat{L}$ of B_N or S_N, taking values in a vector space W. The Hilbert space for the induced representation is $L^{2,\text{equiv}}(\tilde{\Delta},W)$, the space of square integrable functions $\tilde{\Psi}$ on $\tilde{\Delta}$ taking values in W, which are equivariant in that they transform according to the representation L of $K_\gamma(M)$ for all γ. Equivariance means that the inner product in W $\langle \tilde{\Psi}(\tilde{\gamma}), \tilde{\Psi}(\tilde{\gamma}) \rangle_W$ actually defines a single-valued function on Δ, not merely on $\tilde{\Delta}$; and the square integrability of $\tilde{\Psi}$ means that this inner product is an integrable function with respect to μ on Δ. The CUR $V(\varphi)$ of K(M) induced by L is now given in this Hilbert space by

$$V(\varphi)\tilde{\Psi}(\tilde{\gamma}) = \tilde{\Psi}(\varphi(\tilde{\gamma}))\sqrt{\frac{d\mu_\varphi}{d\mu}(\pi\tilde{\gamma})}\ , \tag{19}$$

where $\varphi(\tilde{\gamma})$ is the lifting of φ to $\tilde{\Delta}$.

Unitarily inequivalent representations of S_N induce unitarily inequivalent CUR's of $K(R^3)$, and thus describe distinct quantum systems. There are only two characters of S_N (one-dimensional representations); $\hat{L} = 1$ (the trivial character) leads to the N-particle Bose representation, and $\hat{L} = \pm 1$ according to whether the permutation is even or odd induces the N-particle Fermi representation. But it is apparent that higher-dimensional unitary representations of S_N also induce CUR's of $K(R^3)$, associated with the nontrivial Young diagrams; the resulting diffeomorphism group representations describe paraparticles. Goldin has noted that the action of a permutation in this context is not as a label permutation but as a physical permutation of the values of the coordinates. This requires that the values in the set $\{x_1, \ldots, x_N\}$ be ordered in some fashion. Such an ordering, in effect, fixes the arbitrary boundaries of the different sheets in $\tilde{\Delta}^{(N)}$. [18,52,56]

A character of B_N represents the braid which exchanges two points by a counterclockwise rotation of 180° (leaving other points fixed) by $\exp(i\theta)$, where θ is fixed independently of the choice of points. The resulting induced representations of $K(R^2)$ include the N-particle Bose representation (for $\theta = 0$) and Fermi representation (for $\theta = \pi$); as well as "unusual statistics" or θ-statistics for other values of θ. From the picture we have constructed, it is not difficult to see that the existence of unusual statistics in R^2 does not depend on the indistinguishability of the particles, so that in a certain sense the term "statistics" is a misnomer. If the particles are distinguishable, we take the configuration space to be the coordinate space of ordered N-tuples of points in R^2; the fundamental group for this space is still nontrivial. It is just the kernel of the natural homomorphism from B_N onto S_N, whose elements include 360° circlings of each other by pairs of particles; and its unitary representations induce representations of the diffeomorphism group describing distinct particles which obey unusual statistics. Higher-dimensional representations of B_N also exist, and induce representations of $K(R^2)$ for "para-θ-statistics." [17-19,27,46]

Now we can comment on the topological origin of statistics, θ-statistics, and parastatistics. The introduction of $\Delta^{(N)}$ as configuration space excludes the diagonals (where $x_i = x_j$ for some $i \neq j$) rather arbitrarily. Since the diagonals are a μ-measure 0 set, and the nontrivial topology resulting from their exclusion has such important consequences, one might question the decision to exclude them. In the framework of Feynman paths on configuration space, where statistics was first constructed from homotopy groups by Laidlaw and DeWitt, the exclusion of the diagonals remains an arbitrary decision. But consideration of the CUR's of a diffeomorphism group of the one-particle space <u>forces</u> their exclusion. Indeed, if we insert them "by hand" into $\Delta^{(N)}$, the resulting space simply splits into mutually disjoint orbits under the action of diffeomorphisms. Any CUR modeled on such a space can be written as a direct sum of irreducible components, one of which will be the original $\Delta^{(N)}$ with its nontrivial fundamental group. Conversely, it is not possible to obtain spurious topological effects by introducing new "holes" in $\Delta^{(N)}$ by removing measure zero sets, since $\mathrm{Diff}_c(M)$ or K(M) acts transitively on $\Delta^{(N)}$.

It is interesting that topologically induced representations of groups of diffeomorphisms of M arise even when M is topologically trivial, e.g. $M = R^n$. (The fundamental group for $\Delta^{(N)}$ when $n > 3$ is S_N, just as in the case $n = 3$.) When M itself has a nontrivial fundamental group, there is the possibility of topological quantum mechanics even in the one-particle case. The simplest such situation is that of the Aharonov-Bohm effect. Here a charged particle moves in a region outside a tightly wound infinite (or toroidal) solenoid; a potential barrier prevents the particle from entering the region inside the solenoid. The magnetic field B(x) vanishes outside the solenoid, but is nonvanishing within it due to the electric current. From Eq. (17), we recover the usual Lie algebra of vector fields with support outside the potential barrier; i.e., $\mathrm{Vect}_c(R^3 - Y)$, where Y denotes the forbidden cylindrical region. The corresponding Lie group is $\mathrm{Diff}_c(R^3 - Y)$. This group has a one-parameter family of CUR's induced from the unitary representations of $\pi_1(R^3 - Y)$, which is isomorphic to the additive group of integers. The representations are parameterized by an angle θ, the Aharonov-Bohm phase shift, which is proportional to the magnetic flux through the solenoid. We do not propose to review the long-standing physical and philosophical disagreements that have arisen concerning the Aharonov-Bohm effect, except to say that its consistent mathematical description in terms of gauge-invariant currents (which follows from the representation theory) implies in our view that no "extra" fundamental physical meaning needs to be attributed to the magnetic vector potential. [1,23,34]

The preceding discussion of induced representations bypasses the problem of the non-existence of Haar measure. It does so by making use of a natural class of measures on configuration space that is quasi-invariant for diffeomorphisms; this is what enables us to carry out the inducing construction. Such a quasi-invariant measure exists in general for finite-dimensional configuration spaces, but its existence in the infinite-dimensional case is an open question. The CUR describing an infinite free Bose gas (discussed above) can be modeled on a configuration space of infinite (but locally finite) discrete subsets of R^3 with average density $\bar{\rho}$. Then the necessary quasi-invariant measure is a Poisson measure. It would be interesting to be able to construct measures on other infinite-dimensional configuration spaces, such as spaces of strings and loops, that are quasi-invariant for diffeomorphisms (see the discussion of quantized vortex filaments below). [59]

We close this section by mentioning briefly two additional classes of CUR's of $\mathrm{Diff}_c(R^3)$ or $K(R^3)$. The first class describes tightly bound composite particles. A CUR of $S(R^3) \wedge K(R^3)$ describing a point quantum dipole is given in the Hilbert space of square-integrable functions $\Psi(x, \underset{\sim}{\lambda})$, where x is the position coordinate and $\underset{\sim}{\lambda} \neq 0$ is the (variable) dipole moment coordinate, by:

$$U(f)\Psi(x,\lambda) = \exp[i\lambda\cdot\text{grad } f(x)]\,\Psi(x,\lambda),$$
$$V(\varphi)\Psi(x,\lambda) = \Psi(\varphi(x), D_\varphi(x)\lambda)\,\mathcal{J}_\varphi(x), \tag{20}$$

where $D_\varphi(x)$ is the matrix of derivatives of φ at x. We see that mathematically, λ can be interpreted as an element of the tangent space T_x, and $D_\varphi(x): T_x \to T_{\varphi(x)}$. Physically, we can compare this representation to a two-particle representation of $S(R^3) \wedge K(R^3)$ for ordinary particles of equal and opposite charge. When the particle separations are small, so that f and φ vary slowly between x_1 and x_2, Eqs. (20) are recovered with $\lambda \approx q(x_1 - x_2)$ and $x \approx \frac{1}{2}(x_1 + x_2)$. But the correspondence holds in a weak limit: the CUR of Eqs. (20) is inequivalent to a two-particle representation. Similarly, quadrupole and higher multipole particles are described by CUR's of the same Lie group. The configuration spaces of particle coordinate and dipole and higher multipole moments partition into orbits under the action of the diffeomorphism group, and each orbit can be understood as corresponding to a different way of constructing the tightly bound composite particle from fundamental constituents. [21,25]

The second class of CUR's still to be mentioned describes particles with spin. For $x \in R^3$, define $h_x: \text{Diff}_c(R^3)$ or $K(R^3) \to SL(3,R)$, the special linear group of real 3x3 matrices having determinant one, by:

$$h_x(\varphi) = \mathcal{J}_\varphi(x)^{-1/3}\, D_\varphi(x). \tag{21}$$

If Γ is a continuous path from infinity to x, then $h_y(\varphi)$ for $y \in \Gamma$ defines a continuous path in SL(3,R) from the identity to $h_x(\varphi)$, giving us an element of the universal covering group $\widetilde{SL(3,R)}$. We obtain a well-defined map $\tilde{h}_x$ from the diffeomorphism group to $\widetilde{SL(3,R)}$ which, when restricted to the stability group at x, is a continuous homomorphism. Note that the triviality of the diffeomorphisms at infinity is necessary for this construction. The map $\tilde{h}_x$ can now be used to induce representations of $\text{Diff}_c(R^3)$ or $K(R^3)$ from CUR's of $\widetilde{SL(3,R)}$. Let $\hat{L}$ be a CUR of $\widetilde{SL(3,R)}$ in W; then

$$V(\varphi)\Psi(x) = \hat{L}(\tilde{h}_x(\varphi))\,\Psi(\varphi(x))\sqrt{\mathcal{J}_\varphi(x)} \tag{22}$$

in the Hilbert space $L^2(R^3,W)$. This CUR can be decomposed with respect to SU(2), the maximal compact subgroup of $\widetilde{SL(3,R)}$. Let Σ be the usual three generators of SU(2) acting in W, and let T_μ for $\mu = -2,-1,0,1,2$ be the quadrupole generators of $\widetilde{SL(3,R)}$ in W, obeying $T^*_\mu = (-1)^\mu T_\mu$; the operators Σ and T_μ are determined by the representation $\hat{L}$. For a vector field g, define the functionals $G_\mu(g)$ by

$$G_0(g) = \sqrt{\tfrac{1}{6}}\,(\partial_1 g^1 + \partial_2 g^2 - 2\partial_3 g^3),$$
$$G_1(g) = -\overline{G_{-1}(g)} = -\tfrac{1}{2}\,[(\partial_1 g^3 + \partial_3 g^1) + i(\partial_2 g^3 + \partial_3 g^2)], \tag{23}$$
$$G_2(g) = \overline{G_{-2}(g)} = -\tfrac{1}{2}\,[(\partial_1 g^1 - \partial_2 g^2) + i(\partial_1 g^2 + \partial_2 g^1)].$$

Then the self-adjoint generators of Eq. (22) can be written

$$J(g)\Psi(x) = J_0(g)I\Psi(x) + \tfrac{1}{2}(\nabla\times g(x)\cdot\Sigma)\Psi(x) + \tfrac{1}{2}\sum_{\mu=-2}^{2}[G_{-\mu}(g)](x)T_\mu\Psi(x), \tag{24}$$

where $J_0(g) = (2i)^{-1}[g\cdot\nabla + \nabla\cdot g]$ is the 1-particle orbital momentum density already described; the second term is the momentum density due to the spin; and the third term is a spin-changing current which could occur for excited states of nuclei or for supermultiplets of strongly interacting particles lying on Regge trajectories. Subspaces of $L^2(R^3,W)$ having a fixed spin are invariant under the spin density operators, but are connected by spin-changing currents which can be thought of as the infinitesimal generators of non-uniform spatial dilations.

The T_{μ} terms cannot simply be set equal to zero, because we would then no longer have a representation of the algebra of vector fields. But one can construct from the representation of Eq. (22) a class of operators modeled on "local rigid rotations." With respect to these, the Hilbert space decomposes into fixed spin subspaces, and on each of these there is a CUR of $[S(R^3)\otimes S(R^3,SU(2))]\wedge K(R^3)$, where $S(R^3,SU(2))$ is a local current group as in Eq. (7), and the CUR is obtained through a limiting procedure from Eq. (22). The self-adjoint generators of $S(R^3,SU(2))$ are spin density operators corresponding to a certain limit of Eq. (24).[4-5,25-26,33]

In the above, $SL(3,R)$ can be replaced by $GL(3,R)$, the general linear group, with $h_x(\varphi) = D_{\varphi}(x)$ in Eq. (21). Then an additional term λ div $g(x)I\Psi(x)$ for $\lambda \in R$ fixed, occurs in Eq. (24), which can be thought of as associated with uniform spatial dilations. We have seen this term before as a consequence of Eq. (11); it has been interpreted as a diffusion current. [19,26]

QUANTUM GRAVITY, AND QUANTIZATION OF FLUIDS [11,28-31,38-41,48]

In classical general relativity, gravitation is described by the metric tensor $g^{\mu\nu}$ of a Riemannian 4-manifold $\mathcal{M}$, which is a solution of Einstein's equations. Since there are six independent equations, the ten components of $g^{\mu\nu}$ cannot be determined uniquely by them. The four undetermined functions represent the gauge degrees of freedom in the theory of gravitation, associated with the covariance of the theory under general coordinate transformations. Mathematically, two solutions of Einstein's equations are physically equivalent if there is a diffeomorphism of $\mathcal{M}$ which takes one metric tensor into the other. Thus diffeomorphisms are the gauge transformations for the classical theory. In the case of quantum gravity, the canonical variables can be chosen as the 3-metric g^{ij} of a space-like 3-dimensional manifold M, together with a suitably defined conjugate set of variables π^{ij}. The state functional Ψ is a functional of the g^{ij}. But Einsteinian gravity is a constrained system. For i = 1,2,3, the initial value equations are equivalent to $\mathcal{H}^i(x) = -\pi^{ij}{}_{1,j} = 0$, where the subscript 1 denotes the covariant derivative taken with respect to the g^{ij}, and the $\mathcal{H}^i(x)$ are interpreted as the infinitesimal generators of the group of general coordinate transformations acting on the g^{ij}. In the quantum theory the allowed state functionals are thus restricted to obey $\mathcal{H}^i(x)\Psi = 0$, which is equivalent to the statement that Ψ is invariant under the action of any diffeomorphism generated by an element of $\mathrm{Vect}_c(M)$. Hence, Ψ may be regarded as depending only on the intrinsic 3-geometry ${}^{(3)}\mathcal{G}$ of M; i.e., $\Psi = \Psi({}^{(3)}\mathcal{G})$; and the identity component $\mathrm{Diff}_c^e(M)$ is singled out as the gauge group for quantum gravity. [38]

There can, however, exist diffeomorphisms of M that become trivial at infinity but are not connected to the identity. This can happen when M has a nontrivial topology within a bounded region B, and is asymptotically Euclidean outside B. Such a manifold is called spinorial if there exist diffeomorphisms that can be obtained by rotating a neighborhood of infinity by 360°, which nevertheless are not in $\mathrm{Diff}_c^e(M)$. Such diffeomorphisms can be represented nontrivially on states $\Psi({}^{(3)}\mathfrak{U})$; correspondingly, the angular momentum operators can have half-integral eigenvalues. Friedman and Sorkin interpret the symptotically flat metric as describing an isolated system embedded in a larger universe, and the spin as arising from rotation of the system relative to its environment. [11]

In analogy with θ-vacua in Yang-Mills theory, topological θ-sectors occur in the canonical theory of quantum gravity with the diffeomorphism group as the gauge group. In the usual canonical Yang-Mills theory, the gauge group is Map(S^3,G), where the 3-sphere S^3 is the 1-point compactification of R^3 obtained by adjoining the point at infinity, and the Lie group G is an internal symmetry group. The picture extends to the case where S^3 is replaced by a more general compact connected orientable 3-dimensional manifold M. The n-sectors are labeled by homotopy classes of maps from M to G, and the θ-sectors by the one-dimensional unitary representations of the group of such classes (for $M = S^3$, the third homotopy group $\pi_3(G)$). If $\mathbf{C}$ is the space of connections A_i on M, $\Omega \in$ Map(M,G) acts on $\mathbf{C}$ by $A_i \to \Omega A_i \Omega^{-1} + \Omega \partial_i \Omega^{-1}$. The physical configuration space Δ can be identified with $\mathbf{C}/\text{Map}_*(M,G)$, where Map_* is the subgroup consisting of gauge transformations which are the identity at the base point in M. Now we have already had occasion to see the relationship between unitary representations of the fundamental group $\pi_1(\Delta)$, and quantum theories. Here $\pi_1(\Delta)$ can be identified with the disconnected components of $\text{Map}_*(M,G)$. In the analogous development for quantum gravity, the group Diff(M) acts on the space Riem(M) of Riemannian metrics on M. The physical configuration space is $\text{Riem}(M)/\text{Diff}_*(M)$, where $\text{Diff}_*(M)$ consists of diffeomorphisms which not only map a fixed base point of M to itself, but also leave the tangent space at the base point invariant. (Thus if the base point is the point at infinity adjoined to a non-compact 3-dimensional manifold, elements of Diff_* do not asymptotically translate or rotate the non-compact manifold, and Diff_* is like the group previously labeled K.) Then $\pi_1(\Delta)$ is identified with the connected components of $\text{Diff}_*(M)$, the characters of π_1 correspond to CUR's of $\text{Diff}_*(M)$, and these describe the θ-sectors for quantum gravity. The structure of $\pi_1(\Delta)$, of course, depends on the topology of M. [39-41]

We conclude with the application of CUR's of diffeomorphism groups to the quantization of vorticity in superfluids. The simplest classical vortices are point vortices in an ideal, incompressible superfluid in R^2. When N such vortices are present, each moves convectively in the velocity field of the others. Consider two such vortices of equal circulation; with $x = x_1 - x_2$ and $y = y_1 - y_2$, the classical equations of motion can be written in Hamiltonian form as $dx/dt = \{x,H\} = \partial H/\partial y$ and $dy/dt = \{y,H\} = -\partial H/\partial x$, where $H = -(1/\pi)\ln(x^2 + y^2)$ after subtracting the individual vortex self-energies, and the Poisson bracket $\{x,y\} = 1$. One might therefore hope to quantize the system by replacing Poisson brackets with commutators. Following this procedure, it has been conjectured by Chiao, Hansen, and Moulthrop that quantum point vortices obey θ-statistics, with (for N = 2) $\theta = \pi/2$ or $3\pi/2$; or (for arbitrary N) $\theta = \pi/N$ or $\pi + \pi/N$. However, we have argued that statistics requires a multiply connected configuration space. Here the phase space (for N = 2) is $R^2 - \{0\}$, but it is not possible to choose a configuration space that is multiply connected, consistent with the interpretation of θ as arising from physical rotations. The polar angle ϕ in this model is not an observable, and as a result θ is physically indeterminate. Perhaps, though, the difficulty is due to the overidealized model, in which collective fluid motion has been described by too few coordinates. [6-7,28-29,37]

We therefore seek to quantize a more realistic theory based on the underlying momentum density field that gives rise to the vortices as disturbances. To do this, we take the configuration space for the fluid to be $K_v(R^2)$, the group of all area-preserving diffeomorphisms of the plane which tend rapidly to the identity mapping at infinity. The indistinguishability of fluid elements is a symmetry of the system under the action of diffeomorphisms. The method of coadjoint orbits permits the construction of CUR's of this group, and gives us their physical interpretation. [48]

A brief sketch of the method is as follows. The adjoint representation of a Lie algebra $\mathfrak{g}$ is given by $(\mathrm{Ad}\, Q)Q' = [Q,Q']$ for $Q,Q' \in \mathfrak{g}$. This exponentiates to the adjoint representation of the Lie group G in $\mathfrak{g}$. The coadjoint representation of $\mathfrak{g}$ in its dual space $\mathfrak{g}'$ is defined by $\langle Q\eta, Q'\rangle = \langle \eta, (\mathrm{Ad}\, Q)Q'\rangle$, for $Q \in \mathfrak{g}$ where $\eta \in \mathfrak{g}'$, and similarly for G. A coadjoint orbit is an orbit in $\mathfrak{g}'$ under the action of G. Let $\mathcal{O}$ be such an orbit, and $\eta_o \in \mathcal{O}$. The stability group K_o is the subgroup of G leaving η_o fixed. If $\mathfrak{k}_o$ is the Lie algebra of K_o, then η_o vanishes when evaluated on the bracket of an element of $\mathfrak{g}$ and an element of $\mathfrak{k}_o$; i.e., $\langle \eta_o, [\mathfrak{g},\mathfrak{k}_o]\rangle = 0$. Physically G is the symmetry group and $\mathcal{O}$ is a kind of reduced phase space for the classical system respecting the constants of the motion. $\mathcal{O}$ has 2n dimensions ($\dim G/K_o$). Next one chooses a polarization: a subgroup H of G with Lie algebra $\mathfrak{h}$, such that $\langle \eta_o, [\mathfrak{h},\mathfrak{h}]\rangle = 0$. This step selects half the phase-space coordinates as simultaneously observable ("position-like"); thus the codimension of K_o in H should equal the codimension of H in G. For appropriate orbits, η_o now exponentiates to give a character ξ of the identity component of H. Configuration space for the reduced quantum system is G/H, the leaves in a foliation of $\mathcal{O}$; and ξ together with a character of $\pi_1(G/H)$ induces a CUR of G in $L^2(G/H)$. This is the quantized theory. [3,43]

For diffeomorphism groups, coadjoint orbits can be finite- or infinite-dimensional, and we shall see examples of both types. The Lie algebra of $K_v(R^2)$ consists of divergenceless vector fields with components in Schwartz' space; think of such a vector field v(x) as the fluid velocity field. Now $\nabla \cdot v = 0$ implies $v = \nabla \times \chi$, where the stream function χ is regarded as normal to the plane. The condition $\chi(\infty) = 0$ selects a unique stream function χ_v for each v; a useful vector identity is then $\chi_{[v_1,v_2]} = v_1 \times v_2$. With this notation, the adjoint representation of $K_v(R^2)$ becomes $(\mathrm{Ad}\,\varphi)v = (D_\varphi v)\circ\varphi^{-1}$, or alternatively $(\mathrm{Ad}\,\varphi)\chi = \chi\circ\varphi^{-1}$. Now the dual to the Lie algebra of vector fields is the space of <u>generalized</u> vector fields A, whose components are distributions (generalized functions). We have $\langle A,v\rangle = \int A(x)\cdot v(x)d^2x = \int (\nabla\times A)\cdot\chi_v d^2x$; thus the value of A on an element of the Lie algebra depends only on its curl. A(x) is to be interpreted as the momentum density, and $B(x) = \nabla\times A(x)$ as the vorticity density. The coadjoint representation of $K_v(R^2)$ is given by $\varphi : B \rightarrow B\circ\varphi$.

The coadjoint orbit of a point vortex contains elements $B_a(x) = c\delta^{(2)}(x - a)$, where c is the (fixed) vorticity and $a \in R^2$ parameterizes the orbit. The stability group for B_a is $\{\varphi \mid \varphi(a) = a\}$, which is maximal in $K_v(R^2)$. Thus no polarization is possible for the system. For N point vortices, the coadjoint orbit contains elements $B_{a_1} + \ldots + B_{a_N}$, and the stability group is $\{\varphi \mid \{\varphi(a_i)\} = \{a_i\}\}$. Although it is no longer maximal, it is still impossible to define a polarization group H; the condition $\langle B_{\{a_i\}}, [\mathfrak{h},\mathfrak{h}]\rangle = 0$ fails if $\mathfrak{h}$ contains vector fields which are non-zero at any of the a_i. In short, the idealization of pure point vortices is incompatible with the quantization. However, other coadjoint orbits do admit polarizations. With $B_{(\underset{\sim}{\lambda},a)}(x) = \underset{\sim}{\lambda}\cdot\nabla\delta^{(2)}(x - a)$, the action of φ is given by $(\underset{\sim}{\lambda},a) \longrightarrow (D_{\varphi^{-1}}(a)\underset{\sim}{\lambda}, \varphi^{-1}(a))$; the stability group consists of diffeomorphisms such that $\varphi(a) = a$ <u>and</u> $D_\varphi(a)\underset{\sim}{\lambda} = \underset{\sim}{\lambda}$, and a polarization is obtained by letting $H = \{\varphi \mid \varphi(a) = a\}$ relaxing the second set of constraints. This coadjoint orbit is that of a vortex dipole. The phase space is 4-dimensional, with the plane $\lambda = 0$ excluded, and is thus multiply connected; however, the configuration space is simply connected, and there does not exist a polarization in which it is multiply connected. The resulting character ξ is trivial, and the induced representation is particle-like: $V(\varphi)\Psi(a) = \Psi(\varphi(a))$ for $\Psi \in L^2(R^2)$. Similarly the coadjoint orbit for N vortex dipoles leads to an N-particle representation of $K_v(R^2)$, for which the fundamental group of configuration space is the braid group and θ-statistics are possible. A more interesting coadjoint orbit is that of the rotating vortex dipole, with

$$B^c_{(\lambda,a)}(x) = c\,\delta^{(2)}(x-a) + \lambda\cdot\nabla\delta^{(2)}(x-a), \tag{25}$$

obtained by "gluing" a point vortex to the vortex dipole at a. $K_v(R^2)$ acts as before, so that c is fixed and (λ, a) vary with $\lambda \neq 0$; the stability group and the polarization group are as in the c = 0 case. To write the character determined by $B^c_{(\lambda,a)}$, let ω be a path from ∞ to a in R^2, and let $\varphi\omega$ be its image under φ. Then $\xi(\varphi) = \exp[icm(\varphi)]$ for $\varphi \in H$, where $m(\varphi)$ is the area between ω and $\varphi\omega$. Because φ is area-preserving, it is easy to verify that $m(\varphi)$ is independent of ω, and $m(\varphi_2 \circ \varphi_1) = m(\varphi_1) + m(\varphi_2)$. Now ξ induces a CUR of $K_v(R^2)$.

An infinite-dimensional coadjoint orbit having a polarization is given by a family of parameterized curves $C(\alpha)$ in R^2, where $0 \leq \alpha \leq 2\pi$. Such a curve is equivalently described by the unparameterized curve Γ equal to the image of C, together with $\gamma = d\alpha/ds$, where s = arc length. Now define

$$\langle B_C, \chi_v\rangle = \int_0^{2\pi} d\alpha\; \chi_v(C(\alpha)) = \int_\Gamma ds\, \gamma(s)\chi_v(C(\alpha(s))). \tag{26}$$

Thus $\gamma\delta_\Gamma$ is the vorticity density along the filament Γ. In the coadjoint representation, $\Gamma \rightarrow \varphi\Gamma$, while $\gamma \rightarrow \|\hat{s}\cdot\nabla\varphi\|^{-1}\gamma\circ\varphi^{-1}$. Suppose $C_o(\alpha) = (\cos\alpha, \sin\alpha)$ is the circle Γ_o together with the vorticity density $\gamma_o = 1$. The stability group is then the set of diffeomorphisms of R^2 which rotate the circle rigidly, while the polarization group H contains diffeomorphisms which map the circle to itself; i.e., $H \longrightarrow \mathrm{Diff}(S^1)$. The character determined by B_C is $\exp im(\varphi)$ for $\varphi \in H$, where this time $m(\varphi)$ is the area of the region bounded by ω, $\varphi\omega$, and Γ_o. To complete the construction of a CUR of $K_v(R^2)$ by inducing, one needs a measure on the space of loops which is quasi-invariant for the action of the diffeomorphism group, and for infinite-dimensional coadjoint orbits, this is still an unresolved issue. It is unlikely that such a measure will be concentrated on a single orbit; we are more likely to have a "strictly ergodic" representation, in which the measure is concentrated on uncountably many such orbits. But it is apparent from the examples discussed that we <u>must</u> sacrifice the idealization of pure point vortices. When we do so, the uncertainty principle permits us to retain all of the information about the spatial coordinates of the vorticity, while giving up information about the internal structure. [30-31]

When we seek to represent $K_v(R^3)$, there will be no point vortices. Particle-like disturbances can be recovered from vortex loops of very small diameter, but the above analysis in R^2 suggests that loops in R^3 cannot be idealized as one-dimensional filaments. Their cross-sections should be loops of vorticity looking like Eq. (26) above. It is an interesting speculation that in quantum gravity, the identity component $\mathrm{Diff}^e_c(M)$ might <u>not</u> act trivially, in which case gravitational vortex loops could occur. [53]

REFERENCES

[1] Y. Aharonov and D. Bohm (1959), Phys. Rev. 115, 485.

[2] B. Angermann, H.D. Doebner, and J. Tolar (1983), "Quantum Kinematics on Smooth Manifolds," in S.I. Andersson and H.D. Doebner (eds.), Nonlinear Partial Differential Operators and Quantization Procedures, Springer Lecture Notes in Mathematics 1037, 171-208.

[3] L. Auslander and B. Kostant (1971), Inventiones Math. 14, 255.

[4] A. B. Borisov (1978), J. Phys. A: Math. Gen. 11, 1957.

[5] A. B Borisov (1979), J. Phys. A: Math. Gen. 12, 1625.

[6] R. Y. Chiao, A. Hansen, and A. A. Moulthrop (1985), Phys. Rev. Lett. 54, 1339.

[7] R. Y. Chiao, A. Hansen, and A. A. Moulthrop (1987), Phys. Rev. Lett. 58, 175.

[8] Y. Choquet-Bruhat, C. De Witt-Morette, and M. Dillard-Bleick (1977), Analysis, Manifolds and Physics (Amsterdam: North Holland).

[9] R. F. Dashen and D. H. Sharp (1968), Phys. Rev. 165, 1857.

[10] H. D. Doebner and J. Tolar (1980), "On Global Properties of Quantum Systems," in B. Gruber and R. S. Millman (eds.), Symposium on Symmetries in Science, Carbondale, Illinois 1979 (New York: Plenum), 475-486.

[11] J. L. Friedman and R. D. Sorkin (1980), Phys. Rev. Lett. 44, 1100-1103.

[12] A. Girard (1973), J. Math. Phys. 14, 353-365.

[13] P. Goddard and D. Olive (1986), Int'l. J. Mod. Phys. A 1, 303.

[14] G. A. Goldin (1969), "Current Algebras as Unitary Representations of Groups," unpubl. Ph.D. thesis, Princeton University.

[15] G. A. Goldin (1971), J. Math. Phys. 12, 462-487.

[16] G. A. Goldin (1984), "Diffeomorphism Groups, Semidirect Products, and Quantum Theory," in J. E. Marsden (ed.), Fluids and Plasmas: Geometry and Dynamics, Contemp. Math. 28 (Am. Math. Soc.), 189-207.

[17] G. A. Goldin (1984), "Representations of Semidirect Products of Diffeomorphism Groups in Quantum Theory," in W. W. Zachary (ed.), XIII International Colloquium on Group Theoretical Methods in Physics (Singapore: World Scientific), 261-264.

[18] G. A. Goldin (1988), "Parastatistics, θ-Statistics, and Topological Quantum Mechanics from Unitary Representations of Diffeomorphism Groups," in H. D. Doebner and J. D. Hennig (eds.), Procs. of the XV. Int'l. Conference on Differential Geometric Methods in Theoretical Physics (Teaneck, NJ: World Scientific), 197-207.

[19] G. A. Goldin (in press), "On the Distinguishability of Particles Described by Unitary Representations of Diff(M)." Paper presented at the XVI Int'l. Colloquium on Group Theoretical Methods in Physics, Varna, Bulgaria, June 1987.

[20] G. A. Goldin, J. Grodnik, R. T. Powers, and D. H. Sharp (1974), J. Math. Phys. 15, 88-100.

[21] G. A. Goldin and R. Menikoff (1985), J. Math. Phys. 26, 1880-1884.

[22] G. A. Goldin, R. Menikoff, and D. H. Sharp (1980), J. Math. Phys. 21, 650-664.

[23] G. A. Goldin, R. Menikoff, and D. H. Sharp (1981), J. Math. Phys. 22, 1664-1668.

[24] G. A. Goldin, R. Menikoff, and D. H. Sharp (1981), "Induced Representations of Diffeomorphism Groups Described by Cylindrical Measures," in G. A. Goldin and R. F. Wheeler (eds.), Measure Theory and its Applications: Procs. of the 1980 Conference (DeKalb, IL: Northern Illinois Univ. Dept. of Mathematical Sciences), 207-218.

[25] G. A. Goldin, R. Menikoff, and D. H. Sharp (1983), Phys. Rev. Lett. 51, 2246-2249.

[26] G. A. Goldin, R. Menikoff, and D. H. Sharp (1983), J. Phys. A: Math. Gen. 16, 1827-1833.

[27] G. A. Goldin, R. Menikoff, and D. H. Sharp (1985), Phys. Rev. Lett. 54, 603.

[28] G. A. Goldin, R. Menikoff, and D. H. Sharp (1987), Phys. Rev. Lett. 58, 174.

[29] G. A. Goldin, R. Menikoff, and D. H. Sharp (1987), J. Math. Phys. 28, 744-746.

[30] G. A. Goldin, R. Menikoff, and D. H. Sharp (1987), Phys. Rev. Lett. 58, 2162-2164.

[31] G. A. Goldin, R. Menikoff, and D. H. Sharp (1987), "Diffeomorphism Groups, Coadjoint Orbits, and the Quantization of Classical Fluids," and "Quantized Vortex Filaments in Incompressible Fluids," in Y. S. Kim and W. W. Zachary (eds.), Procs. of the First Int'l. Conference on the Physics of Phase Space, Springer Lecture Notes in Physics 278, 360-362 and 363-365.

[32] G. A. Goldin and D. H. Sharp (1970), "Lie Algebras of Local Currents and their Representations," in V. Bargmann (ed.), Group Representations in Mathematics and Physics: Battelle Seattle 1969 Rencontres, Springer Lecture Notes in Physics 6, 300-311.

[33] G. A. Goldin and D. H. Sharp (1983), Commun. Math. Phys. 92, 217-228.

[34] G. A. Goldin and D. H. Sharp (1983), Phys. Rev. D 28, 830-832.

[35] J. Grodnik and D. H. Sharp (1970), Phys. Rev. D 1, 1531.

[36] J. Grodnik and D. H. Sharp (1970), Phys. Rev. D 1, 1546.

[37] A. Hansen, A. A. Moulthrop, and R. Y. Chiao (1985), Phys. Rev. Lett. 55, 1431.

[38] S. W. Hawking and G. F. R. Ellis (1973), The Large Scale Structure of Space-Time (London: Cambridge Univ. Press).

[39] C. J. Isham (1981), Phys. Letts. 106B, 188-192.

[40] C. J. Isham (1983), "Quantum Field Theory and Spatial Topology," in G. Denardo and H. D. Doebner (eds.), Conference on Differential Geometric Methods in Theoretical Physics: Trieste 1981 (Singapore: World Scientific), 171-185.

[41] C. J. Isham (1984), "Topological and Global Aspects of Quantum Theory," in B. DeWitt and R. Stora (eds.), Relativity, Groups and Topology II, Procs. 1983 Les Houches Summer School (Amsterdam: North Holland), 1059-1290.

[42] I. M. Khalatnikov (1965), Introduction to the Theory of Superfluidity (New York: Benjamin).

[43] A. A. Kirillov (1981), Ser. Math. Sov. 1, 351.

[44] M. G. G. Laidlaw and C. M. DeWitt (1971), Phys. Rev. D 3, 1375.

[45] S. Lang (1962), Introduction to Differentiable Manifolds (New York: Interscience).

[46] M. Leinaas and J. Myrheim (1977), Nuovo Cimento 37B, 1.

[47] G. W. Mackey (1976), The Theory of Unitary Group Representations (Chicago: Univ. of Chicago Press).

[48] J. E. Marsden and A. Weinstein (1983), "Coadjoint Orbits, Vortices, and Clebsch Variables for Incompressible Fluids," in D. K. Campbell, H. A. Rose, and A. C. Scott (eds.), Procs. of the Los Alamos Conference "Order in Chaos", Physica 7D, 305.

[49] R. Menikoff (1974), J. Math. Phys. 15, 1138.

[50] R. Menikoff (1974), J. Math. Phys. 15, 1394.

[51] R. Menikoff and D. H. Sharp (1977), J. Math. Phys. 18, 471.

[52] A. M. L. Messiah and O. W. Greenberg (1964), Phys. Rev. 136, B248.

[53] M. Rasetti and T. Regge (1975), Physica 80A, 217.

[54] M. Reed and B. Simon (1972), Methods of Modern Mathematical Physics, Vol. I: Functional Analysis (New York: Academic Press).

[55] M. Reed and B. Simon (1975), Methods of Modern Mathematical Physics, Vol. II: Fourier Analysis, Self-Adjointness (New York: Academic Press).

[56] R. H. Stolt and J. R. Taylor (1970), Nuclear Physics B 19, 1.

[57] H. Sugawara (1968), Phys. Rev. 170, 1659.

[58] W. Thirring (1978), Classical Dynamical Systems, transl. by E. Harrell (New York: Springer).

[59] A. M. Vershik, I. Gel'fand, and M. I. Graev (1975), Usp. Mat. Nauk 30, 3.

QUASI-PARTICLE GROUPS IN ATOMIC SHELL THEORY

B. R. Judd

Department of Physics and Astronomy
The Johns Hopkins University
Baltimore, Maryland 21218

INTRODUCTION

In 1968 some bizarre features of atomic shell theory were noticed. One such appeared when efforts were made to generalize the use of G_2 for atomic electrons. Racah had demonstrated the great utility of this exceptional Lie group in his classic paper on the Coulomb energies of f electrons,[1] but it was known that no analogous group existed for electrons with higher ℓ. However, the G_2 classification of the states of the f shell could be accomplished by diagonalizing a two-electron operator belonging to the irreducible representation (irrep) (111) of SO(7). It can be shown that such an operator is necessarily a scalar with respect to G_2 too. For g electrons we can form an operator that is an SO(3) scalar and which also belongs to (1111) of SO(9). As such, it is an analog of the G_2 scalar although no actual analog of the group itself exists. When the operator was diagonalized for the states of maximum multiplicity (maximum S) of g^N, many eigenvalue repetitions were noticed.[2] Moreover, several fractional parentage coefficients vanished for no obvious reason. Armstrong found that corresponding simplifications occurred in the h shell.[3]

It took some time to unravel these puzzles. A detailed account has been given elsewhere.[4] It turns out that an atomic ℓ shell is susceptible of a four-fold factorization, provided we include two parity labels p and p′ for the oddness or evenness of the numbers of electrons in the spin-up and spin-down spaces. Thus a state of ℓ^N can be written as

$$|(\ell_\lambda \ell_\mu)_p L_A,\ (\ell_\nu \ell_\xi)_{p'} L_B,\ LM_L\rangle, \tag{1}$$

where L_A and L_B are the orbital angular momenta of the spin-up and spin-down electrons, respectively. Their coupling yields the familiar total orbital angular momentum quantum number L. The angular momenta ℓ_θ (where $\theta = \lambda,\ \mu,\ \nu$ or ξ) are constrained to a limited set. For s, p, d, f and g electrons, it turns out that ℓ_θ = 0, 1/2, 3/2, 0 and 3, 2 and 5, respectively. To see how things work out, consider the g shell and take p = g (even). The terms of maximum multiplicity (for which all spin projections m_s are +1/2) are S for g^0; P, F, H and K for g^2; S, D (twice), F, G (twice), H, I (twice), K, L and N for g^4; P, F (twice), G, H, I, K and M for g^6; and G for g^8. These angular momenta can be obtained from $(d + h)^2$ provided no statistical constraints are placed on the coupling of the pairs of angular momenta.

The origin of the four-fold factorization lies in the fact that, when we consider the four sets of linear combinations of creation and annihilation operators

$$\begin{aligned}
\lambda_q^\dagger &= (1/2)^{1/2}[a_{1/2,q}^\dagger + (-1)^{\ell-q}a_{1/2,-q}],\\
\mu_q^\dagger &= (1/2)^{1/2}[a_{1/2,q}^\dagger - (-1)^{\ell-q}a_{1/2,-q}],\\
\nu_q^\dagger &= (1/2)^{1/2}[a_{-1/2,q}^\dagger + (-1)^{\ell-q}a_{-1/2,-q}],\\
\xi_q^\dagger &= (1/2)^{1/2}[a_{-1/2,q}^\dagger - (-1)^{\ell-q}a_{-1/2,-q}],
\end{aligned} \qquad (2)$$

we find that any pair of operators drawn from different sets anticommute. The group $SO(8\ell + 5)$, whose elementary spinor irrep (1/2 1/2...1/2) provides all $2^{4\ell+2}$ states of the ℓ shell,[5] decomposes according to the scheme

$$SO(8\ell + 5) \rightarrow SO_\lambda(2\ell + 1) \times SO_\mu(2\ell + 1) \times SO_\nu(2\ell + 1) \times SO_\xi(2\ell + 1), \quad (3)$$

where the generators of $SO_\theta(2\ell + 1)$ are $(\theta^\dagger\ \theta)^{(k)}$ with k odd.[6] By adding the corresponding generators of all four groups, the generators of Racah's $SO(2\ell + 1)$ group are obtained. However, the states (1) imply that the reduction $SO_\theta(2\ell + 1) \rightarrow SO(3)$ is carried out first for each θ, thus yielding the acceptable ℓ_θ. Only one irrep of $SO_\theta(2\ell + 1)$ occurs: it is the elementary spinor (1/2 1/2 ... 1/2) of dimension 2^ℓ. For $\ell = 1, 2, 3$ and 4 we find $(1/2) \rightarrow 1/2$, $(1/2\ 1/2) \rightarrow 3/2$, $(1/2\ 1/2\ 1/2) \rightarrow s + f$, $(1/2\ 1/2\ 1/2\ 1/2) \rightarrow d + h$. This reduction process provides the possible values of ℓ_θ. No duplication appears until $\ell = 9$, so the states (1) provide unambiguous descriptions of not only the f and g shells (for which the standard schemes of Racah fail), but also for the h, i, k, and l shells. Provided we can express the physical operators of interest in terms of the basic quasi-particle operators $\theta^\dagger$, we can rely on angular-momentum algebra alone and entirely avoid the use of fractional parentage coefficients. It was shown some years later by Gruber and Thomas[7] for $\ell \leq 6$ that the reduction scheme (3), taken with the more familiar reduction schemes, exhausts all the possible group chains that begin with $SO(8\ell + 5)$ and end with the physical SO(3) whose irreps are labelled by L.

EXTENSIONS TO THE NUCLEAR CASE

Elliott and Evans[8] realized that the results for the atomic ℓ shell can be generalized to the nuclear configurations j^N provided both neutrons and protons are considered simultaneously. Ranks are defined with respect to J, the total angular momentum, and T, the isospin. The coupled products

$$(a^\dagger a)^{(kr)}, \quad (a^\dagger a^\dagger)^{(kr)}, \quad (aa)^{(kr)},$$

where $0 \le k \le 2j$ and $0 \le r \le 1$, form the generators of SO(8j + 4), as was already implicit in the work of Parikh.[9] Instead, however, of taking the subgroup U(4j + 2) of Flowers[10] or SO(5) × Sp(4j + 2) of Parikh,[9] we form as a subgroup the direct product $SO_\lambda(4j+2) \times SO_\mu(4j+2)$ corresponding to the mutually commuting sets of generators $(\lambda^\dagger \lambda)^{(kr)}$ and $(\mu^\dagger \mu)^{(kr)}$, where k + r is odd, and where, in analogy to equations (2), we have

$$\lambda^\dagger_{ms} = (1/2)^{1/2}[a^\dagger_{ms} + (-1)^{j-m-s+1/2}\, a_{-m-s}],$$
$$\mu^\dagger_{ms} = (1/2)^{1/2}[a^\dagger_{ms} - (-1)^{j-m-s+1/2}\, a_{-m-s}]. \tag{4}$$

In these expressions, m and s are the magnetic substates in the J and T spaces, with $-j \le m \le j$ and $s = \pm 1/2$. The irreps of $SO_\theta(4j+2)$ are either (1/2 1/2...1/2 1/2) or (1/2 1/2...1/2 -1/2), according to whether N is even or odd.

By making the separation corresponding to (kr) = (01) and (kr) = (k0), each group $SO_\theta(4j+2)$ breaks up into $SU_\theta(2) \times Sp_\theta(2j+1)$, with the subgroup $SU_\theta(2) \times SO_\theta(3)$. The irreps (1/2 1/2...1/2 ±1/2) each yield an isospin T_θ for $SU_\theta(2)$ and the associated irrep $\langle 1^{j-x+1/2}\, 0^x \rangle$ of $Sp_\theta(2j+1)$, where $x = 2T_\theta$. For example, the irreps (1/2 1/2 1/2 1/2 1/2 ±1/2) of $SO_\theta(10)$, corresponding to j = 5/2, decompose according to the schemes

$$(1/2\ 1/2\ 1/2\ 1/2\ 1/2\ 1/2) \rightarrow {}^1\langle 111\rangle + {}^3\langle 100\rangle, \tag{5}$$

$$(1/2\ 1/2\ 1/2\ 1/2\ 1/2\ -1/2) \rightarrow {}^2\langle 110\rangle + {}^4\langle 000\rangle, \tag{6}$$

where the superscripted multiplicities specify values of $2T_\theta + 1$.

The combined direct product

$$SU_\lambda(2) \times SO_\lambda(3) \times SU_\mu(2) \times SO_\mu(3) \tag{7}$$

possesses the familiar group SU(2) × SO(3) as a subgroup, where SU(2) has **T** as its generators, and SO(3) has **J**. The outcome of the quasi-particle factorization is that the set of states

$$|J_\lambda\ T_\lambda,\ J_\mu\ T_\mu;\ J\ T\rangle \tag{8}$$

obtained by vector coupling the λ and μ spaces in both J and T cover the entire shell. These states lead to an unambiguous classification scheme for all j less than 9/2, whereas the standard scheme of Flowers[10] fails at j = 5/2.

The nuclear quasi-particle model was further developed by Hecht and Szpikowski,[11] who showed how to calculate reduced matrix elements in the λ and μ spaces. They recognized that the quantum numbers T_λ and T_μ are essentially quasi-spins for the quasi-particles: for example, the irrep <100> appearing in the decomposition (4) corresponds to $T_\theta = 1$, and its three components (-1, 0 and +1) indicate that <100> appears in the quasi-particle configurations $(5/2)$, $(5/2)^3$, and $(5/2)^5$. By embedding $\mathbf{T}_\lambda$ and $\mathbf{T}_\mu$ in the quasispin group SO(5), Hecht and Szpikowski were able to simplify the construction of states of well-defined particle number N. In terms of λ and μ, the generators of this SO(5) group are those for $SU_\lambda(2)$ and $SU_\mu(2)$, namely, the three components of both $(\lambda^\dagger\,\lambda)^{(01)}$ and $(\mu^\dagger\,\mu)^{(01)}$, together with the three components of $(\lambda^\dagger\,\mu)^{(01)}$ (or its equivalent $(\mu^\dagger\,\lambda)^{(01)}$) and the number operator $(\lambda^\dagger\,\mu)^{(00)}$. As a result of this embedding, the states (7) can be replaced by those of the form

$$|(w_1w_2)\ N\ \beta\ T;\ \alpha\ J>, \tag{9}$$

where (w_1w_2) denotes an irrep of SO(5), and α and β are multiplicity labels. Hecht and Szpikowski gave numerical values for some of the more elementary transformation coefficients relating (8) to (9). They also showed how the quasi-particle factorization could be extended to the L, S, and T spaces. The group $SO(16\ell + 8)$ stands at the head of the group-subgroup chain, and the states, in analogy to (7), can be written as

$$|(L_\lambda L_\mu)LM_L,\ (S_\lambda S_\mu)SM_S,\ (T_\lambda T_\mu)TM_T>. \tag{10}$$

This provides a complete classification for nuclear shells with $\ell \leq 2$. However, neither the particle number N nor quantum numbers such as the Wigner supermultiplet quantum numbers are preserved in this new scheme. Hecht and Szpikowski pointed out that this defect must severely limit the practical value of the scheme for nuclear ℓ^N configurations.

N AND S AS GOOD QUANTUM NUMBERS

The fact that some projection procedure would be necessary to convert the quasi-particle states to those more closely related to the physical ones was quickly realized by all who worked in the field. It was certainly appreciated by Feneuille,[12] who extended the method to cope with the atomic configurations $(s + d)^N$. Cunningham and Wybourne,[13] in a similar treatment of the general mixed configuration (that is, one involving more than a single azimuthal quantum number) remarked "The quasiparticle scheme has the weakness of yielding eigenfunctions associated with neither a well-defined number of particles nor spin quantum numbers S and M_S." This is true (in general) for the states (1), but, on the face of it, conflicts with the states (10), where both S and L appear. The resolution of this paradox lies in the phases that appear in equation (4). Both m and s are necessarily half-integral, so the reversal in signs of both of them produces no phase change. Thus the adjoint of $\lambda^\dagger_{ms}$ is proportional to $\lambda^\dagger_{-m-s}$ and not to a component of μ. (This is why Elliott and Evans had to treat both protons and neutrons together rather than separately.) If we casually replace j with ℓ in equations (4), the component m becomes integral. On taking adjoints we now find that the λ and μ spaces are no longer distinct, and the group separation cannot be made. The additional presence of isospin in the states (10) of Hecht and Szpikowski provides an additional half-integral quantum number that corrects the situation.

There is a way of providing more useful quasi-particle states for atomic shell theory. It has been realized only recently that we need not couple the angular momenta ℓ_θ; instead, we can couple the irreps of the groups $SO_\theta(2\ell + 1)$. Thus instead of states of the type (1), we propose to use the states

$$|(W_\lambda W_\mu)_p W_A, (W_\nu W_\xi)_{p'} W_B, W\tau \; LM_L>, \quad (11)$$

where τ is a multiplicity label to allow for the fact that a given irrep W may contain repeating L values. The advantage of the states (11) is, first, that there is only one irrep W_θ, namely the elementary spinor (1/2 1/2...1/2), and the acceptable irreps W_A and W_B are of the type

$$(00...0), \; (10...0), \; (110...0),..., \; (11...1), \quad (12)$$

each of which occurs just once in the decomposition of the Kronecker square $(1/2\ 1/2...1/2)^2$. Secondly, each irrep in the sequence (12) is associated with just two possible particle numbers, namely 0 or $2\ell+1$ for (00...0), 1 or 2ℓ for (10...0), etc. The parity labels p and p′ make the association unique. Thus the number of electrons N_A and N_B in the spin-up and spin-down spaces is defined, and hence the total particle number, given by $N = N_A + N_B$, is too. The coupling of W_A and W_B to W yields the identical irrep W that Racah used in his work.[1] He showed that a given W occurs with just two possible values of the total spin S. The choice to be made depends on whether N is even or odd. Since we have already specified N it follows that S must be uniquely determined too. Thus the state (11) implies a unique N and a unique S.

As an example from the f shell, we take

$$|((1/2\ 1/2\ 1/2)(1/2\ 1/2\ 1/2))_g(111),$$

$$((1/2\ 1/2\ 1/2)(1/2\ 1/2\ 1/2))_u(110), \; (210)(21)D>, \quad (13)$$

which corresponds to a D term of f^9 for which $S = 1/2$. The multiplicity label τ of (11) has been replaced by the irrep (21) of G_2 in this example.

MATRIX ELEMENTS

To use such states as (13) we have to recast angular-momentum theory, which is based on the group SO(3), in terms of the appropriate higher-order analog -- for example, SO(7) in the case of f electrons. Clebsch-Gordan coefficients, 6-j and 9-j symbols, as well as concomitant factors, all have to be replaced by corresponding expressions. The absence of any multiplicity difficulties, which is often the major bugbear of working in spaces of greater dimension than 4, makes it rather easy to do this. This work is being carried out at Johns Hopkins in collaboration with Shaozhong Li. We have only started to face the problem of converting operators of physical interest to forms in which the tensorial properties with respect to the four θ spaces are exhibited. In the case of the Coulomb interaction between f electrons, things work out very much more elegantly than might have been anticipated. Racah's four operators e_0, e_1, e_2 and e_3, in terms of which the Coulomb interaction can be expressed, present no difficulties. The first two are scalars in SO(7) and explicit expressions for their matrix elements were given by Racah.[1] The last pair correspond to the irreps (400) and (220) respectively. For e_2 we find

$$e_2 = (1081080)^{1/2}((\lambda^\dagger\mu^\dagger)^{(200)}(\nu^\dagger\xi^\dagger)^{(200)})_0{}^{(400)}, \tag{14}$$

where the subscripted zero indicates that the component of the coupled form corresponding to an SO(3) scalar must be selected. The operator e_3 requires three terms rather than just one: but, even so, it is remarkably simple. As soon as our generalized angular-momentum theory is applied to calculate the matrix elements of e_2 or e_3, new forms of the 9-j symbols arise. Since they contain nine irreps W, we propose to call them 9-W symbols. Thus, if we want to calculate the matrix element of e_2 between the ket (13) and its adjoint bra, the 9-W symbol

$$\{(111)\ (111)\ (200);\ (110)\ (110)\ (200);\ (210)\ (210)\ (400)\}$$

enters, where, for typographical convenience, the three rows of the 9-W symbol are set side by side rather than under one another. It is not as difficult to calculate such quantities as might be thought. Considerable progress has been made lately in extending angular-momentum theory to spaces of higher dimension. For a discussion, the reader is referred to a review.[14]

The Coulomb interaction does not involve electron spin. Hence all matrix elements of e_2 and e_3 must be diagonal with respect to S. Since our kets do not explicitly give S, the responsibility for giving the correct selection rules shifts to the 9-W coefficients. In the case of e_2 this is easily accomplished because of a generalized triangular condition on the third column of the 9-W symbol. For e_3 things are more interesting. Its vanishing between the states (220) and (211) of f^6, for example, which must happen because their associated spins are 0 and 1, must depend on the vanishing of the 9-W symbol

$$\{(111)\ (111)\ (200);\ (111)\ (111)\ (200);\ (220)\ (211)\ (220)\}.$$

This possesses two identical columns. We know that an ordinary 9-j symbol vanishes if two rows are identical and if the sum of the nine j values is odd. What is the analog? We have simply to examine the coupled irreps and ask whether the residues of the squares appear in the symmetric or antisymmetric parts. The relevant decompositions have been given some years ago in connection with a different problem.[15] Thus we find (220) appears in the symmetric parts of both $(111)^2$ and $(200)^2$, while (211) appears in the antisymmetric part of $(111)^2$. There is an overall antisymmetry, and the 9-W symbol vanishes. A detailed analysis of such phases is under way in collaboration with Shaozhong Li.

GROUPS WITH GENERATORS INVOLVING MORE THAN PAIRS OF ANNIHILATION AND CREATION OPERATORS

Because of the remarkable way that the quasi-particle formalism breaks up the structures of atomic shells, we might ask whether we can take advantage of the four-fold factorization to construct new groups that lie outside the group chains of Gruber and Thomas [7]. A start in that direction has been made by Labarthe [16], who recognized that the 2^ℓ states in each θ space can be transformed among themselves by operators that can be interpreted as the generators for $U(2^\ell)$. Several subgroups are easy to find. Thus, for g electrons, we get

$$U_\theta(16) \rightarrow SO_\theta(16) \rightarrow SO_\theta(10) \rightarrow SO_\theta(9)$$

for the group chain, and

$$[1] \longrightarrow (10000000) \rightarrow (1/2\ 1/2\ 1/2\ 1/2\ \pm 1/2) \rightarrow (1/2\ 1/2\ 1/2\ 1/2)$$

for the irreps. For f electrons, we have $2^\ell = 2\ell + 2$, and the coalescence of the two SO(8) groups forces the irreps (1000) and (1/2 1/2 1/2 ±1/2) to be equivalent to each other. This potential contradiction is prevented from taking place because of the well-known automorphism of SO(8).

Since we can add the generators of the four θ spaces, such classical groups as Racah's SO(7) can be embedded in SO(8). We can thus find how the states of the f shell defined by irreps W of SO(7) are distributed among the irreps of SO(8). For example, we can show that the states labelled by (221) of SO(7) must derive from the single irrep (3100) of SO(8), while (210) occurs in the decomposition of both (2200) and (2110). Interesting though such connections are, they do not immediately give rise to new selection rules. Further work in this area is required.

ACKNOWLEDGEMENTS

As mentioned in the text, some of the work described above is being carried out in collaboration with Mr. Shaozhong Li. Partial support by the United States National Science Foundation is also acknowledged.

REFERENCES

1. G. Racah, Theory of complex spectra. IV, Phys. Rev., 76: 1352 (1949).
2. B. R. Judd, Atomic g electrons, Phys. Rev., 173: 40 (1968).
3. L. Armstrong, private communication (1968).
4. B. R. Judd and J. P. Elliott, "Topics in Atomic and Nuclear Theory," Caxton Press, Christchurch, New Zealand (1970), pp. 47-50.
5. B. R. Judd, Group theory in atomic spectroscopy, in "Group Theory and Its Applications," E. M. Loebl, ed., Academic Press, New York (1968).
6. L. Armstrong and B. R. Judd, Quasi-particles in atomic shell theory, Proc. Roy. Soc. London, A315: 27 (1970); Atomic structure calculations in a factorized shell, Proc. Roy. Soc. London, A315: 39 (1970).
7. B. Gruber and M. Samuel Thomas, Symmetry chains for the atomic shell model. I. Classification of symmetry chains for atomic configurations, Kinam A2: 133 (1980).
8. J. P. Elliott and J. A. Evans, A new classification for the j^n configuration, Phys. Lett., 31B: 157 (1970).
9. J. C. Parikh, The role of isospin in pair correlations for configurations of the type $(j)^N$, Nucl. Phys., 63: 214 (1965).
10. B. H. Flowers, Studies in jj-coupling. I. Classification of nuclear and atomic states, Proc. Roy. Soc. London A212: 248 (1952).
11. K. T. Hecht and S. Szpikowski, On the new quasiparticle factorization of the j-shell, Nucl. Phys., A158: 449 (1970).
12. S. Feneuille, Traitment des configurations $(d + s)^N$ dans le formalism des quasi-particules, J. Physique, 30: 923 (1969).
13. M. J. Cunningham and B. G. Wybourne, Quasiparticle formalism and atomic shell theory II. Mixed configurations, J. Math. Phys., 11: 1288 (1970).

14. B. R. Judd, Algebraic expressions for classes of generalized 6-j and 9-j symbols for certain Lie groups, in Proceedings of the XVI International Colloquium on Group Theoretical Methods in Physics, Varna, Bulgaria (1987).
15. B. R. Judd and H. T. Wadzinski, A class of null spectroscopic coefficients, J. Math. Phys., 8: 2125 (1967).
16. J.-J. Labarthe, Atomic shell operators in the quasiparticle formalism, J. Phys. B: At. Mol. Phys., 13: 2149 (1980).

PROPERTIES OF ORTHOGONAL OPERATORS

B.R. Judd*, D.J. Newman** and Betty Ng**

*Department of Physics and Astronomy
The Johns Hopkins University, Baltimore, Md.21218, U.S.A.
**Department of Physics, University of Hong Kong
Pokfulam Road, Hong Kong

INTRODUCTION

The problem of fitting spectroscopic energies to parameters arises in many contexts, including atomic theory, crystal field theory, nuclear theory and molecular vibration/rotation theory. Both historical accident and variations in the form of the available data have led to rather different approaches being adopted in each case. The aim of this article is to show that a unified approach exists based on the concept of orthogonal operators and that group theoretical considerations can be used to determine useful parametrizations. An exhaustive list of references is provided[1-28] and the problem of using orthogonal operators for restricted sets of data is discussed in some detail.

1. THE VALUE OF ORTHOGONAL OPERATORS

In fitting spectroscopically determined energies to parameters it is necessary to expand the effective Hamiltonian, for a given basis set, in the form

$$H = \sum_{\alpha} p_{\alpha} H_{\alpha}\,, \tag{1}$$

where the H_α are a standard set of operators, each associated with a parameter p_α. Traditional methods of choosing the operators H_α are based on an examination of the mechanisms involved in producing the spectra. Racah[29] showed, in the case of atomic spectra, that group theoretical methods could be used to determine different, but nonetheless very convenient parameter sets. In the present work we survey the use of orthogonal operators[1-4] in the fitting procedure. As we shall see, the group-theoretical approach has several advantages over alternative schemes.

The set of operators H_α is chosen to form an "orthogonal" set such that

$$\mathrm{Tr}(\underset{\sim}{H}_{\alpha}) = \sum_{i} \langle i|H_{\alpha}|i\rangle = 0 \tag{2}$$

for all H_α but one, and

$$\mathrm{Tr}(\underset{\sim}{H}_{\alpha}\underset{\sim}{H}_{\beta}) = \sum_{ij} \langle i|H_{\alpha}|j\rangle\langle j|H_{\beta}|i\rangle = 0 \qquad \text{for } \alpha \neq \beta. \tag{3}$$

In these expressions the traces are evaluated over the complete basis set used in formulating the fitting procedure. This basis set is usually defined over a many-particle configuration of bosons or electrons. It is clear from the form of (2) and (3) that the orthogonality condition is independent of the choice of basis; that is, it remains unchanged if the components of the basis set are subjected to a unitary transformation.

The use of orthogonal operators has some obvious technical advantages. First of these is that the definition ensures that no linear dependences can occur between the operators. The second is that equations (2) and (3) provide simple checks on the actual construction of matrices $\underset{\sim}{H}_\alpha = [\langle i|H_\alpha|j\rangle]$ for many-particle states.

Fits of the parameters p_α to experimental data are usually accomplished by an iterative procedure which alternates linear least squares fits to H keeping the basis fixed and the diagonalization of H keeping the parameters p_α fixed. Methods of fitting are described in detail by Cowan[30], Sections 16-3 to 16-5, and by MacKeown and Newman[31], Chapter 2. In order to relate this fitting procedure to the use of orthogonal operators, we outline the salient features below.

At each iteration we determine a unitary matrix $\underset{\sim}{U}$ which diagonalises H. Hence, at the r-th iteration the residual matrix $\underset{\sim}{\varepsilon}$, defined by

$$\varepsilon_{ij} = y_i \delta_{ij} - \sum_{\substack{\alpha \\ k\ell}} p_\alpha^{(r)} U_{ik}^{\dagger(r)} H_{k\ell}^{\alpha} U_{\ell j}^{(r)}, \tag{4}$$

where y_i are the experimentally determined energies, will be diagonal. The next step is then to determine the parameters $p_\alpha^{(r+1)}$ by linear least squares fitting, keeping $\underset{\sim}{U}$ constant. Conventionally, this is done by minimizing $\sum_i \varepsilon_{ii}^2$ as described by Cowan[30]. It has been suggested[2,4,31] however, that the alternative expression $\mathrm{Tr}(\underset{\sim}{\varepsilon}^2) = \sum_{ij} \varepsilon_{ij}\varepsilon_{ji}$ could be minimised. These two procedures are not quite equivalent, for the off-diagonal elements of $\underset{\sim}{\varepsilon}$ will become non-zero as the values of p_α are varied from their initial values $p_\alpha^{(r)}$. They should, nevertheless, ultimately converge to the same minima and determine the same set of parameters p_α.

While we do not wish to discuss the relative numerical advantages of these two procedures here, it is worth pointing out that, from a mathematical point of view, $\mathrm{Tr}(\underset{\sim}{\varepsilon}^2)$ minimisation is more natural when orthogonal operators are used. If the data (y_i) available spans the whole of the fitting space we obtain the following very simple expression

$$p_\alpha^{(r+1)} = \mathrm{Tr}(\underset{\sim}{H}^\alpha \underset{\sim}{U}^{(r)} \underset{\sim}{y}\, \underset{\sim}{U}^{\dagger(r)}) \tag{5}$$

for the parameters and the variance/covariance matrix[31]

$$\begin{aligned} W_{\alpha\beta} &= \mathrm{Tr}[(\underset{\sim}{U}^{\dagger(r)} \underset{\sim}{H}^\alpha \underset{\sim}{U}^{(r)})(\underset{\sim}{U}^{\dagger(r)} \underset{\sim}{H}^\beta \underset{\sim}{U}^{(r)})] \\ &= \mathrm{Tr}(\underset{\sim}{H}^\alpha \underset{\sim}{H}^\beta) = \delta_{\alpha\beta} \end{aligned} \tag{6}$$

(providing that the operators are normalized). This latter result shows that the fitting errors of the parameters should be equal and uncorrelated[30] if $\mathrm{Tr}(\underset{\sim}{\varepsilon}^2)$ is minimized. This rigorous result can be compared to the strings of similar (but not always identical) fitting errors in the parameters that have been found for the $3d^N$ configurations when $\sum_i \varepsilon_{ii}^2$ is minimized[26].

In practice, of course, we do not always have sufficient spectroscopic data to carry out a fit over all components of the basis spanned by the state labels i,j,k,ℓ. The trace in the expression for $W_{\alpha\beta}$ will (unlike the trace in the orthogonality definition) then be evaluated only over the fitting subspace and will not reduce to $\delta_{\alpha\beta}$. A standard expression for the least squares fit parameters p_α will then have to be used, rather than equation (5).

It has been shown[4] that errors in the parameters are minimised when operators are chosen to be orthogonal. Some specific examples of this result also appear in the literature[8].

A related property is parameter independence. Clearly, parameters that can be expressed explicitly as linear combinations of the data are independent in the sense that the removal, or addition, of parameters will not affect the values of the others. We note from equation (5) that this property holds only if the removed or added parameters do not affect the matrix $\underset{\sim}{U}$.

2. INDUCED ORTHOGONALITY ALONG A SHELL

As is discussed in Section 3, the group-theoretical description of orthogonal operators ensures that orthogonality for a configuration C is automatically carried forward to a configuration C' comprising a greater number of particles. It is somewhat discomforting to us that we have not been able to formulate this property in terms of coefficients of fractional parentage. An argument[3] presented some years ago is flawed because it incorrectly treated the particle number of the products $H_\alpha H_\beta$. However, for fermion systems we can demonstrate some simple results of this type directly by using matrix algebra.

We write the one-electron operators as α, β, γ, etc. so that a (symmetrized) two-electron operator takes the form of a sum of direct products, e.g. $\frac{1}{2}[\alpha \times \beta + \beta \times \alpha]$. The symbol $\times$ separates the parts of a product that refer to different electrons. The one-electron states are represented by vectors, which we write in Dirac notation as $\langle m|$. A single-electron operator α is completely defined by the matrix $\langle m|\alpha|m'\rangle$. A two-particle state may be written $\frac{1}{\sqrt{2}}(\langle m| \times \langle n| + \varepsilon\langle n| \times \langle m|)$, where $\varepsilon = -1$ for electrons and $\varepsilon = +1$ for bosons. This state is normalised for electrons, but it is normalised for bosons only when $n \neq m$. This feature makes it more difficult to work with bosons, and we therefore restrict attention to fermion systems for the remainder of this section. It is straightforward to express traces of products of operators over many-electron states in terms of one-electron traces. Denoting the n-particle traces as Tr_n, and using the expressions

$$\mathrm{tr}(\alpha) = \sum_i \langle i|\alpha|i\rangle$$

$$\mathrm{tr}(\alpha\beta) = \sum_{i,j} \langle i|\alpha|j\rangle\langle j|\beta|i\rangle \tag{7}$$

we obtain

$$\mathrm{Tr}_2[\tfrac{1}{2}(\alpha \times 1 + 1 \times \alpha)\tfrac{1}{2}(\mu \times 1 + 1 \times \mu)]$$

$$\sim [(\frac{N}{2} - 1)\mathrm{tr}(\alpha\mu) + \tfrac{1}{2}\mathrm{tr}(\alpha)\ \mathrm{tr}(\mu)], \tag{8}$$

where N is the dimension of the one-particle basis set. This shows that, given the one-particle orthogonality conditions, namely $\mathrm{tr}(\alpha\beta) = 0$ and $\mathrm{tr}(\alpha) = 0$ for all operators but one, then the one-particle operators will be orthogonal over two-particle electron states.

In order to simplify the algebra further we now write symmetrized products of operators

$$\frac{1}{2}(\alpha \times \beta + \beta \times \alpha) = (\alpha \times \beta)_s$$

Two-particle operator orthogonality (for a two-electron system) can then be expressed as

$$\mathrm{Tr}_2[(\alpha \times \beta)_s(\mu \times \nu)_s] \sim [\mathrm{tr}(\alpha\mu)\mathrm{tr}(\beta\nu) + \mathrm{tr}(\alpha\nu)\mathrm{tr}(\beta\mu) - \mathrm{tr}(\alpha\mu\beta\nu) - \mathrm{tr}(\mu\alpha\nu\beta)] \tag{9}$$

In this case orthogonality is obtained if the component (one-particle) operators are chosen, like tensor operators, to be orthogonal and the condition

$$\mathrm{tr}(\alpha\mu\beta\nu) + \mathrm{tr}(\mu\alpha\nu\beta) = 0 \tag{10}$$

is satisfied.

The corresponding orthogonality conditions for three-electron states are as follows. For one-particle operators,

$$\mathrm{Tr}_3[(\alpha \times 1 \times 1)_s(\mu \times 1 \times 1)_s] \sim [(N-3)(N-2)\mathrm{tr}(\alpha\mu) + 2(N-2)\mathrm{tr}(\alpha)\mathrm{tr}(\mu)]. \tag{11}$$

For two-particle operators,

$$\begin{aligned}\mathrm{Tr}_3[(\beta \times \gamma \times 1)_s(\nu \times \tau \times 1)_s] \sim \{&(N-4)[\mathrm{tr}(\beta\nu)\mathrm{tr}(\gamma\tau) + \mathrm{tr}(\beta\tau)\mathrm{tr}(\gamma\nu) - \mathrm{tr}(\beta\nu\gamma\tau)\\ &- \mathrm{tr}(\nu\beta\tau\gamma)] + \mathrm{tr}(\nu)\mathrm{tr}(\beta)\mathrm{tr}(\gamma\tau) + \mathrm{tr}(\nu)\mathrm{tr}(\gamma)\mathrm{tr}(\beta\tau)\\ &+ \mathrm{tr}(\tau)\mathrm{tr}(\gamma)\mathrm{tr}(\beta\nu) + \mathrm{tr}(\tau)\mathrm{tr}(\beta)\mathrm{tr}(\gamma\nu)\\ &- \mathrm{tr}(\nu)[\mathrm{tr}(\beta\gamma\tau) + \mathrm{tr}(\beta\tau\gamma)] - \mathrm{tr}(\tau)[\mathrm{tr}(\beta\nu\gamma)\\ &+ \mathrm{tr}(\beta\gamma\nu)] - \mathrm{tr}(\beta)[\mathrm{tr}(\nu\gamma\tau) + \mathrm{tr}(\tau\gamma\nu)]\\ &- \mathrm{tr}(\gamma)[\mathrm{tr}(\nu\beta\tau) + \mathrm{tr}(\tau\beta\nu)] + \mathrm{tr}(\beta\tau\nu\gamma) + \mathrm{tr}(\beta\nu\tau\gamma)\\ &+ \mathrm{tr}(\nu\beta\gamma\tau) + \mathrm{tr}(\tau\beta\gamma\nu)\}.\end{aligned} \tag{12}$$

The right hand side of equation (11) is clearly zero if the conditions for making equation (8) zero hold, so that one-electron operator orthogonality is induced at least up to three-electron states. The first square bracket in equation (12) has the same form as the expression in equation (9), so that two-electron orthogonality requires that the rest of this expression also vanishes. It can be seen that as all other terms, except the last four, contain traces of single matrices, they all vanish identically. Hence the induction of orthogonality requires that the last four terms of this equation vanish, so that additional restrictions on the operator matrices of the form

$$\mathrm{tr}[(\gamma\beta + \beta\gamma)(\tau\nu + \nu\tau)] = 0 \tag{13}$$

must be imposed in addition to equation (10), in order to ensure the induction of two-electron operator orthogonality along a shell. We have also examined the induction of the orthogonality between one-particle and two-particle operators, and find that no additional conditions are necessary to ensure this. The need to introduce additional conditions such as equation (13) suggests that direct algebraic methods provide an inadequate characterization of orthogonality, and that a group theoretical approach might be more fruitful.

3. GROUP THEORETICAL CHARACTERIZATION

We consider systems of n-electrons or n-bosons in N possible one-particle states, so that the n-particle states span single irreducible representations on U(N): $[1^n]$ for electrons and $[n]$ for bosons. The Hamiltonian is now written in terms of orthogonal operators as

$$H = \sum_{\alpha} p_{\alpha} H_{\alpha}^{(r_{\alpha})},$$

where the superscript r_{α} distinguishes the r_{α}-particle operators. It has been shown (e.g. see refs. 17, 20 and 24) that all the r_{α}-particle operators also transform under distinct representations of U(N).

Given that the expression $\sum_{i} |i\rangle\langle i|$ is invariant if the sum spans a single irreducible representation (IR) of U(N), it follows that the expression

$$I_{\alpha\beta} = \sum_{i} \langle i| H_{\alpha}^{(r_{\alpha})} H_{\beta}^{(r_{\beta})} |i\rangle$$

vanishes if the operator product has no part which transforms as the identity representation of U(N). Hence $H_{\alpha}^{(r_{\alpha})}$ is orthogonal to $H_{\beta}^{(r_{\beta})}$ with respect to any n-particle basis if the two operators transform as components of different self-adjoint IR's or as different components of the same IR. This orthogonality criterion is clearly independent of n, so that operators chosen to have different IR labels will certainly satisfy the induced orthogonality theorem discussed in the previous section. In addition, we note that only the identity representation has non-zero trace, so that the characterization of the $H_{\alpha}^{(r_{\alpha})}$ in terms of IR's ensures that equation (2) is satisfied as well as equation (3).

The self-adjoint condition mentioned in the above discussion is satisfied automatically if the operators are described in terms of IR's of U(N) that conserve particle number. This condition will thus also be met for the subsets of operators defined using group chains, which are discussed in the following sections.

4. SUBGROUP CHAINS AND SUBSPACES

Operator orthogonality has so far been discussed in terms of <u>all</u> the basis states which span a given configuration of fermions or bosons. However, only a subset of these energy levels may be experimentally observable. It would be very useful if operator orthogonality could defined over the subspaces which are spanned by the experimentally accessible states. In other words, we should like to be able to restrict the states $|i\rangle$ and $|j\rangle$ in equation (3) to those which belong to a given subspace.

A group chain is usually selected to provide distinct group theoretical labels for the operators which are orthogonal over the entire space. Such a group chain can also be used to determine operator orthogonality over subspaces. This is particularly useful if the branching subspaces down the groups correspond to the relative magnitudes of interactions corresponding to subgroup invariants and the observed states span a single IR of a group in the chain. Recently an investigation[25] has been made of the use of a subspace tree for f^n configurations using the group chain $U(14) \supset Sp(14) \supset SO(7) \otimes SU_S(2)$, where the subscript S indicates that the generators of SU(2) are the components of the total spin, S. Examples of operators which are orthogonal over the subspaces are given, and it is shown that accurately and poorly determinable coulomb interaction parameters can be separated by using such group theoretical definitions of the operators.

To further illustrate the use of operator orthogonality over subspaces, the following two examples are given.

Example A Use of 3-boson operators in nuclear physics

The Interacting Boson Model (IBM) has been very successful in providing a general scheme of parametrization for the energy levels of atomic nuclei. An introductory account of this theory is given in the recent book by Iachello and Arima[32]. Energy levels are usually fitted to an Hamiltonian which consists of a combination of all possible one- and two-boson operators. Group theoretical techniques are used to extract the matrix elements of these operators.

Recent work by Leavitt[24] has shown that the standard operators used in the IBM can be replaced by an equivalent set of orthogonal operators. This approach also suggests a natural extension of the parametrization to include three-boson operators (see Table 4.1). Unfortunately, this

Table 4.1 Zero-, one-, two- and three-boson operators for U(6), showing (starred) operators omitted when the basis is restricted to a single SO(6) representation. Only those IR's of SO(5) and SO(6) are listed that contain SO(3) scalars.

Name	N	U(6)	SO(6)	SO(5)
$\bar{e}_0$	0	$[0^6]$	(000)	(00)
$\bar{e}_1$	1	$[10^4{-}1]$	(200)	(00)
$\bar{f}_1$	2	$[20^4{-}2]$	(000)	(00)*
$\bar{f}_2$	2	$[20^4{-}2]$	(200)	(00)*
$\bar{f}_3$	2	$[20^4{-}2]$	(400)	(00)
$\bar{f}_4$	2	$[20^4{-}2]$	(220)	(22)
$\bar{f}_5$	2	$[20^4{-}2]$	(400)	(30)
$\bar{t}_1$	3	$[30^4{-}3]$	(200)	(00)*
$\bar{t}_2$	3	$[30^4{-}3]$	(400)	(00)*
$\bar{t}_3$	3	$[30^4{-}3]$	(600)	(00)
$\bar{t}_4$	3	$[30^4{-}3]$	(310)	(30)*
$\bar{t}_5$	3	$[30^4{-}3]$	(400)	(30)*
$\bar{t}_6$	3	$[30^4{-}3]$	(330)	(30)
$\bar{t}_7$	3	$[30^4{-}3]$	(510)	(30)
$\bar{t}_8$	3	$[30^4{-}3]$	(420)	(22)
$\bar{t}_9$	3	$[30^4{-}3]$	(420)	(42)
$\bar{t}_{10}$	3	$[30^4{-}3]$	(600)	(60)

involves the addition of ten new operators to the original set of seven operators, making the parametrization impracticable for many systems because of lack of data. For example, the system $^{118}_{78}Pt_{118}$ has only 14 observed levels. It is parametrized as an $(s+d)^6$ boson configuration using the group chain $U(6) \supset O(6) \supset O(5) \supset O(3)$. Ten of the observed levels correspond to d^6 states, so it seems worthwhile investigating whether a truncated parametrization including just some of the three-boson interactions could be used for such a system.

The form of this truncated parametrization can be determined by studying the matrix elements of the 1-, 2- and 3-boson operators in the 3-boson configuration $(d+s)^3$ (see Table 4.2). Only two of the ten 3-boson states are removed if we restrict the fitting space to states within the d^3 or (300) configuration. Inspection of the table of matrix elements determined by Leavitt[24] shows, however, that this causes the removal of

Table 4.2 Three-boson basis states specified by representations in the group chain $U(6) \supset SO(6) \supset SO(5) \supset SO(3)$. Starred states are omitted in the truncated basis.

$\bar{S}_1$	$[30^5](100)(00)S^*$
$\bar{S}_2$	$[30^5](300)(00)S$
$\bar{S}_3$	$[30^5](300)(30)S$
$\bar{D}_1$	$[30^5](100)(10)D^*$
$\bar{D}_2$	$[30^5](300)(20)D$
$\bar{D}_3$	$[30^5](300)(10)D$
$\bar{F}$	$[30^5](300)(30)F$
$\bar{G}_1$	$[30^5](300)(20)G$
$\bar{G}_2$	$[30^5](300)(30)G$
$\bar{I}$	$[30^5](300)(30)I$

six operators from the parametrization, leaving just the eleven shown in Table 4.1. The removal of five of these operators can be explained simply as being due to repetitions of similar SO(6) representations. The remaining operator $\bar{t}_4$ is found to have all zero matrix elements because it is odd with respect to d-boson interchange.

The most interesting feature of this d^3 subspace parametrization is that it removes the two-boson operators $\bar{f}_1$ and $\bar{f}_2$. This suggests that the currently adopted parametrization using all one-boson and two-boson operators may not be optimal.

Example B <u>Truncation of the parameter set for the electronic configuration d^2p using terms of the core d^2 configuration.</u>

There has been considerable interest[33] over the years in the atomic configuration d^2p. The 45 levels of FeVI $3d^24p$ have been found by Ekberg[34], and a fit to a limited set of orthogonal operators has been carried out[21]. As an example of a truncated parametrization, we can ask what is entailed in fitting the energies of the levels that derive from coupling the p electron to a given term (defined by $\bar{S}$ and $\bar{L}$) of the core d^2. This amounts to limiting the group U(10), for which d^2 spans the IR [11], to the subgroup $SO_{\bar{S}}(3) \times SO_{\bar{L}}(3)$, for which the $\bar{S}\bar{L}$ term spans the IR $D_{\bar{S}} \times D_{\bar{L}}$ of dimension $(2\bar{S}+1)(2\bar{L}+1)$. The addition of the p electron leads to the quadruple direct product

$$SO_{\bar{S}}(3) \times SO_{\bar{L}}(3) \times SO_s(3) \times SO_p(3)$$

from which the subgroup $SO_S(3) \times SO_L(3)$ is extracted. The IR's of this last group are defined by the term labels S and L.

There are 30 matrix elements (counting Hermitian pairs once only) for each of the 30 possible electrostatic orthogonal operators H_i (that is, operators scalar with respect to S and L). If we begin by restricting the states to those of the type $d^2(^3F)p$, there remain just six matrix elements, corresponding to the six terms ^2D, ^2F, ^2G, ^4D, ^4F and ^4G. We might have expected the task of finding the six appropriate linear combinations of the 30 H_α to be a daunting one: but in fact nothing more is required than selecting the entries in the columns headed e_0, e_1, e_2, e_3, e_4 and e_5 of table 6 of Dothe et al[11]. This is because interactions between the d electrons are irrelevant for a single term of d^2, and all d-p interactions occur first in the configuration 3d4p, where the operators e_i make their first appearance. The entries for the three-electron operators t_i, for the six terms deriving from d^2 ^3F, are simple multiples of those for the e_i, as can be rapidly verified from table 6 of Dothe et al[11]. They thus play no role in forming the subset of orthogonal operators, though any e_i parameters obtained from a fit to the six terms in question can only be properly compared to those found from a fit to all the terms of d^2p when allowance is made for the contributions coming from the t_i operators.

Table 4.3 Sets of orthogonal operators required for complete parametrizations of the terms of d^2p that derive from single $\bar{S}\,\bar{L}$ terms of the core d^2

Core term	Operators
^1S	e_0
^3P	e_0, e_1, e_2, e_3, e_4, e_5
^1D	e_0, e_2, e_4
^3F	e_0, e_1, e_2, e_3, e_4, e_5
^1G	e_0, e_2, e_4

The properties of the core term ^3F can be generalized. The orthogonal operators required for each core term are set out in table 4.3. These are the appropriate operators to use when attention is limited to the levels of d^2p that derive from a single core term. Such truncations should be useful when a particular atomic configuration is only partially known; however, the main point we want to make here is that the classification of orthogonal operators already worked out for d^2p is very easily adapted to a core-term truncation.

REFERENCES

1. D.J. Newman, Parameterization schemes in solid state physics, Aust. J. Phys., 31: 489 (1978).
2. D.J. Newman, Matrix mutual orthogonality and parameter independence, J. Phys. A: Math. Gen., 14: L429 (1981).
3. B.R. Judd, J.E. Hansen and A.J.J. Raassen, Parametric fits in the atomic d shell, J. Phys. B: At. Mol. Phys., 15: 1457 (1982).
4. D.J. Newman, Operator orthogonality and parameter uncertainty, Phys. Lett., 92A: 167 (1982).
5. B.R. Judd, Operator averages and orthogonalities, in "Group Theoretical Methods in Physics" (Lecture Notes in Physics, Vol.201), G. Denardo, G. Ghirardi and T. Weber, eds., Springer, Berlin (1984).
6. B.R. Judd and H. Crosswhite, Orthogonalized operators for the f shell, J. Opt. Soc. Am., B1: 255 (1984).
7. B.R. Judd and M.A. Suskin, Complete set of Orthogonal scalar operators for the configuration f^3, J. Opt. Soc. Am., B1: 261 (1984).
8. B.R. Judd, Error distributions for effective atomic operators, J. Phys. B: At. Mol. Phys., 17: L617 (1984).
9. P.H.M. Uylings, Energies of N equivalent electrons expressed in terms of two-electron energies and independent three-electron parameters: a new complete set of orthogonal operators: I. Theory, J. Phys.B:At. Mol. Phys., 17: 2375 (1984).
10. P.H.M. Uylings, A.J.J. Raassen and J.-F. Wyart, Energies of N equivalent electrons expressed in terms of two-electron energies and independent three-electron parameters: a new complete set of orthogonal operators: II. Application to $3d^N$ configurations, J. Phys. B: At. Mol. Phys., 17: 4103 (1984).
11. H. Dothe, J.E. Hansen, B.R. Judd and G.M.S. Lister, Orthogonal operators for $p^N d$ and pd^N, J. Phys. B: At. Mol. Phys., 18: 1061 (1985).
12. J.E. Hansen, B.R. Judd, G.M.S. Lister and W. Persson, Observation of four-body effects in atomic spectra, J. Phys. B: At. Mol. Phys. 18: L725 (1985).
13. J.E. Hansen and B.R. Judd, Fine-structure analyses with orthogonal operators, J. Phys. B: At. Mol. Phys., 18: 2327 (1985).
14. J.-F. Wyart, A.J.J. Raassen and P.H.M. Uylings, Parametric studies of $3d^N$ ground configurations of highly charged ions, Phys. Scr., 32: 169 (1985).
15. J.E. Hansen and B.R. Judd, New effective parameters for atomic structure, Comments At. Mol. Phys., 18: 125 (1986).
16. B.R. Judd, Classification of operators in atomic spectroscopy by Lie groups, in "Symmetries in Science II", B. Gruber and R. Lenczewski, eds., Plenum, New York 265 (1986).
17. B.R. Judd and R.C. Leavitt, Many-electron orthogonal operators in atomic shell theory, J. Phys. B: At. Mol. Phys., 19: 485 (1986).
18. W.-Ü.L. Tchang-Brillet, M.-Ch. Artru and J.-F. Wyart, The $3d^4$-$3d^3 4p$ transitions of triply ionized manganese (MnIV), Phys. Scr. 33: 390 (1986).
19. H. Dothe, Orthogonal scalar operators for mixed atomic configurations and application to term-analysis residues, Thesis, The Johns Hopkins University, Baltimore, Maryland (1986).
20. R.C. Leavitt, A complete set of f-electron scalar operators, J. Phys. A: Math. Gen., 20: 3171 (1987).
21. H. Dothe and B.R. Judd, Orthogonal operators applied to term analysis residues for FeVI $3d^2 4p$, J. Phys. B: At. Mol. Phys., 20: 1143 (1987).
22. J.E. Hansen, B.R. Judd and G.M.S. Lister, Parametric fitting to $2p^N 3d$ configurations using orthogonal operators, J. Phys. B: At. Mol. Phys., 20: 5291 (1987).

23. J.E. Hansen, B.R. Judd and G.M.S. Lister, Structural content of orthogonal operators for p^Nd configurations, J. Phys. B: At. Mol. Opt. Phys., 21: 1437 (1988).
24. R.C. Leavitt, Effective operators in atomic physics, Thesis, The Johns Hopkins University, Baltimore, Maryland (1988).
25. D.J. Newman, B. Ng and C.Y. Pang, Orthogonal operators and subspace trees, J. Phys. B: At. Mol. Opt. Phys., 21: L173 (1988).
26. J.E. Hansen, A.J. Raassen, P.H.M. Uylings and G.M.S. Lister, Parametric fitting to d^N configurations using orthogonal operators, in "Atomic Spectroscopy and Highly Ionised Atoms", H.G. Berry, R. Dunford and L. Young, eds., North-Holland, Amsterdam (1988).
27. R.C. Leavitt, Effective three-body magnetic operators in the d shell, J. Phys. B: At. Mol. Opt. Phys., in press (1988).
28. J.E. Hansen, P.H.M. Uylings and A.J.J. Raassen, Parametric fitting with orthogonal operators, Phys. Scr., 37: 664 (1988).
29. G. Racah, Theory of complex spectra. IV. Phys. Rev., 76: 1352 (1949).
30. R.D. Cowan, The theory of atomic structure and spectra, University of California Press (1981).
31. P.K. MacKeown and D.J. Newman, Computational techniques in physics, Adam Hilger (1987).
32. F. Iachello and A. Arima, The interacting boson model, Cambridge University Press (1987).
33. B.R. Judd, Complex atomic spectra, Rep. Prog. Phys., 48: 907 (1985).
34. J.O. Ekberg, Term analysis of Fe VI, Phys. Scr., 11: 23 (1975).

GROUP THEORY FOR NON-PERIODIC LONG-RANGE ORDER IN SOLIDS*

Peter Kramer
Institut für Theoretische Physik der Universität
Auf der Morgenstelle 14, D-7400 Tübingen, FRG

1 Introduction

The role of groups and their representations in the physics of periodic order is well-known. The symmetry of the atomic density under a discrete translation group T yields the concepts of the lattice and of the discrete cell structure. The inclusion of point symmetry leads to a description in terms of space groups. The representations of space groups appear in the Fourier and diffraction analysis, in the electronic band structure, and in the lattice dynamics.

Experiments on the Al-Mn alloys in 1984 have lead to the new concept of non-periodic long-range order in solids. In this survey, some tools and ideas will be discussed which allow one to describe long-range order in solids beyond periodicity. The generalization of periodic and point symmetry is briefly discussed in sections 2 and 3. The tools for this generalization are taken from the theory of almost and quasiperiodic functions and from group/ subgroup techniques. This description yields non-periodic order as a subspace restriction in an embedding Euclidean space and lattice of n dimensions. In sections 4 and 5 we describe tools from algebraic topology like Euclidean cell complexes and apply them to lattices in n dimensions. This description is modified in section 6 and adapted to a subspace restriction. In section 7 it is shown how a discrete cell structure for non-periodic long-range order can be constructed from the subspace restriction.

2 Generalization of periodic order

A dominant feature of quasicrystalline material is the appearance of diffraction amplitudes restricted to a discrete set of points in **k**-space. As the diffraction amplitudes are the absolute squares of the complex Fourier amplitudes of the atomic density, the Fourier transform of the atomic density must have corresponding properties in **k**-space.

*Work supported by the Deutsche Forschungsgemeinschaft

These properties motivate a study of classes of function and their Fourier transforms. We work in two Euclidean spaces $\mathbb{E}^n$ of dimension n, the $\mathbf{x}$-space with coordinates $\mathbf{x}$ or $\mathbf{y}$, and the $\mathbf{k}$-space.

2.1 Def. The *Fourier transform* of a function f supported on $\mathbb{E}^n$ is given by

$$F^n(f)(\mathbf{k}) := \tilde{f}(\mathbf{k}) = (2\pi)^{-n} \int d^n\mathbf{y} f(\mathbf{y}) \exp(-i\mathbf{k}\cdot\mathbf{y}),$$

and its inverse by

$$f(\mathbf{y}) = (F^n)^{-1}(\tilde{f})(\mathbf{y}) = \int d^n\mathbf{k}\ \tilde{f}(\mathbf{k}) \exp(i\mathbf{k}\cdot\mathbf{y}).$$

Recall that the vector $\mathbf{k}$ denotes an irreducible representation of the continuous translation group $\hat{T}$ acting on $\mathbb{E}^n$. The Fourier transform with this interpretation is the decomposition of f into its parts with irreducible transformation properties.

Now let

$$\mathbb{E}^n \to \mathbb{E}_1^m + \mathbb{E}_2^{n-m}$$

be an orthogonal decomposition and denote the orthogonal projections of all vectors or polytopes by the subscripts 1,2 respectively.

To describe the intersection of objects in $\mathbb{E}^n$ with $\mathbb{E}^m$ introduce for a fixed vector $\mathbf{c}_2$ the *cut function*

$$v\ :\ v(\mathbf{y}) = \delta^{n-m}(\mathbf{y}_2 - \mathbf{c}_2)$$

with the Fourier transform

$$\tilde{v}(\mathbf{k}_1, \mathbf{k}_2) = (2\pi)^{-n+m}\delta^m(\mathbf{k}_1)\ \exp(-i\mathbf{k}_2\cdot\mathbf{c}_2).$$

2.2 Prop. For the Fourier transform of a function f on the intersection with $\mathbb{E}^m$ one gets by convolution

$$F^n(fv)(\mathbf{k}_1, \mathbf{k}_2) = (2\pi)^{-n+m} \int d^{n-m}\mathbf{l}_2 \tilde{f}(\mathbf{k}_1, \mathbf{k}_2 - \mathbf{l}_2) \exp(-i\mathbf{l}_2\cdot\mathbf{c}_2).$$

The Fourier transform of the function f restricted to the subspace $\mathbb{E}^m$ has the form

$$\begin{aligned} F^m(f(,\mathbf{c}_2))(\mathbf{k}_1) &= (2\pi)^{n-m} F^n(fv)(\mathbf{k}_1, 0) \\ &= \int d^{n-m}\mathbf{l}_2 \tilde{f}(\mathbf{k}_1, -\mathbf{l}_2) \exp(-i\mathbf{l}_2\cdot\mathbf{c}_2). \end{aligned}$$

Let T denote a discrete translation group acting on $\mathbb{E}^n$. Any *orbit* on $\mathbb{E}^n$ under T is a lattice Y. A *transversal* or *fundamental domain* FD is a set of points of $\mathbb{E}^n$ which contains one and only one point from each orbit. There are many possible choices of the FD, among them are the primitive cells and the Voronoi or Wigner-Seitz cell to be discussed later. A periodic function f^P on $\mathbb{E}^n$ wrt the translation group T is completely specified once its values are given on a FD, we express this by saying that f^P is supported on a FD.

2.3 Prop. Consider a translation group T acting on $\mathbb{E}^n$, the lattice Y generated by T, a function f^P periodic with respect to T supported on the fundamental domain FD, and denote by T^R the translation group of the reciprocal lattice. Then the Fourier integral of f^P collapses to the Fourier series which can be expressed in the form

$$\tilde{f}^P(\mathbf{k}) = (vol(\mathrm{FD}))^{-1} \sum_{\mathbf{k}\in T^R} \delta^n(\mathbf{k} - \mathbf{k}^R) a(\mathbf{k}^R),$$

$$a(\mathbf{k}^R) = \int_{\mathrm{FD}} d^n\mathbf{x}\; f^P(\mathbf{x}) \exp(-i\mathbf{k}^R \cdot \mathbf{x}).$$

This result can also be obtained from the orthogonality relation for the irreducible representations of the translation group.

We can express this by saying that the Fourier transform of f^P is supported in **k**-space on the Z-module spanned by the reciprocal lattice Y^R. This Z-module consists of all integer linear combinations of the n reciprocal lattice vectors.

Recall the interpretation of the reciprocal lattice Y^R with respect to the unitary irreducible representations of the translation group T: The fundamental domain in **k**-space forms the Brillouin zone BZ, each point **k**' from the BZ labels an irreducible representation of T. A general point in **k**- space denotes an irreducible representation $D^{\mathbf{k}}$ of the continuous translation group $\hat{T}$ in $\mathbb{E}^n$. Its unique decomposition of the form

$\mathbf{k} = \mathbf{k}' + \mathbf{k}^R$

characterizes the subduction rule

$\hat{T} \downarrow T: \quad D^{\mathbf{k}} \downarrow T = D^{\mathbf{k}'}$

For the restriction of a periodic function in $\mathbb{E}^n$ to the subspace $\mathbb{E}^m$ one obtains from Props. 2.2 and 2.3

2.4 Prop. The Fourier transform of the function f^P on the subspace $\mathbb{E}^m$ is given by

$$F^m(f^P(,\mathbf{c}_2)) = (vol(\mathrm{FD}))^{-1} \sum_{\mathbf{k}_1^R, \mathbf{k}_2^R \in T^R} \delta^m(\mathbf{k}_1 - \mathbf{k}_1{}^R) \exp(i\mathbf{k}_2^R \cdot \mathbf{c}_2) a(\mathbf{k}_1^R, \mathbf{k}_2^R)$$

This Fourier transform in $\mathbf{k}_1$-space is supported on the Z-module spanned by the projection of the reciprocal lattice Y^R to a subspace $\mathbb{E}^m$. The Z-module consists of all integral linear combinations of the projections of the n reciprocal lattice vectors.

At this point we can connect the method of restriction with the theory of quasiperiodic functions.

2.5 Def. A function f in $\mathbb{E}^m$ is called quasiperiodic if its Fourier amplitudes are supported on a Z-module of n, $n > m$, basis vectors linearly independent over Z.

Clearly the method of restriction described above with the result given in Prop. 2.4 yields a quasiperiodic function on $\mathbb{E}^m$. Quasiperiodic functions were considered by H. Bohr [1]-[4] as a particular class [3] of almost periodic functions. More precisely, Bohr considered the class of functions whose amplitudes in **k**-space are supported on a countable set. He then showed that these functions in **x**-space have a relatively dense set of translation numbers such that their values are almost repeated. Note the difference between quasiperiodic and periodic functions in **x**-space:

As discussed above, a periodic function f^P is supported in **x**-space on a fundamental domain FD. For a general quasiperiodic function, there is no domain in **x**-space which supports the values. There is an approximate repetition of values which can be seen from the method of restriction . This repetition occurs when the subspace passes close to two points which are related by a lattice vector in $\mathbb{E}^n$. To describe a domain for a general quasiperiodic function in $\mathbb{E}^m$ obtained by restriction, one must turn to $\mathbb{E}^n$ and use the fundamental domain FD of the translation group in $\mathbb{E}^n$.

Example 2.1. The simplest example of a quasiperiodic function is obtained by the restriction of a function periodic on a square lattice Y in $\mathbb{E}^2$ to a line $\mathbb{E}^1$. For a slope

$\tan\alpha = \phi^{-1} = \frac{1}{2}(\sqrt{5}-1)$,

the projections of the two reciprocal basis vectors are linearly independent over Z and one obtains a quasiperiodic function. Now choose in $\mathbb{E}^2$

$$\begin{aligned} f^P(\xi_1,\xi_2) &= sin(2\pi\xi_1)sin(2\pi\xi_2) \\ &= -\frac{1}{4}[\exp 2\pi i(\xi_1+\xi_2)+\exp-2\pi i(\xi_1+\xi_2) \\ &\quad -\exp 2\pi i(\xi_1-\xi_2)-\exp-2\pi i(\xi_1-\xi_2)] \end{aligned}$$

and write $\xi_1 = x_1 \cos\alpha$, $\xi_2 = x_1 \sin\alpha$.

The restriction to the line $\tan\alpha = \phi^{-1}$ is given by the function

$f'(x_1) = f^P(x_1\cos\alpha, x_1\sin\alpha)$

shown in Fig. 1. Let a_n denote the Fibonacci numbers,

$a_{n+1} = a_n + a_{n-1}, a_1 = a_2 = 1$

The line passes, for $n >> 1$, close to the lattice points

$\xi_1 = a_n$,

$\xi_2 = a_{n-1}$,

$r_n \Rightarrow [(a_n)^2 + (a_{n-1})^2]^{1/2}$

and so the values of f'(r) are approximately repeated at the points $r_n, n >> 1$, compare Fig.1.

For quasicrystals one can now choose a quasiperiodic model based on a periodic function in $\mathbb{E}^n, n > m$ and its restriction to the position space $\mathbb{E}^m, m = 3$. Then the Fourier transform and its squared absolute value have amplitudes supported on a Z-module in **k**-space. In **x**-space, the atomic density is an almost periodic function, but there is no cell description since there is no fundamental domain. A cell description is possible in the space $\mathbb{E}^n$ and so quasiperiodic models can be formulated in terms of high-dimensional crystallography.

3 Generalized point symmetry

For periodic order in $\mathbb{E}^3$, Schoenflies [5] and Fedorov [6] in 1891 established a complete classification of the 230 possible space groups which have translations and discrete point transformations as their elements. The requirement of translational symmetry in $\mathbb{E}^3$ puts restrictions on the point groups. Point groups acting on $\mathbb{E}^3$ which cannot occur in a crystallographic space group of $\mathbb{E}^3$ are called *non-crystallographic*. These point groups can occur for example in molecules and as local site symmetry in a periodic lattice, but they cannot describe a global symmetry of the long-range order in a crystal. Among the non-crystallographic point groups there are the cyclic group C(5) of order 5 and the icosahedral group A(5). The group C(5) may act on $\mathbb{E}^3$ through a five-fold rotation axis, and the group A(5) through the 60 symmetry operations of the regular icosahedron. Both these point groups appear in the diffraction patterns of quasicrystals.

The occurence of these non-crystallographic point groups has two consequences for the theoretical description of quasicrystals: First of all it excludes periodic order in $\mathbb{E}^m$, m = 2,3, in **x** - and **k**-space and suggests a more general almost or quasiperiodic order. Secondly it restricts the non-periodic scheme to the class which allows for a given point symmetry.

The tools for the analysis of point symmetry in non-periodic long-range order are conveniently described in terms of group/subgroup techniques and corresponding subduced/induced representations. We follow Kramer und Neri [7] and Haase et al.[8] in the analysis.

3.1 Prop. Consider for a lattice Y in $\mathbb{E}^n$ the holohedry group (T,G) where T is the translation group and G is the point group. Denote by D the representation of G obtained by its action on $\mathbb{E}^n$. Let H be a subgroup $\mathrm{H} < \mathrm{G}$ with the property that $\mathrm{D} \downarrow \mathrm{H}$ subduces a real orthogonal non-crystallographic representation D^α of H of dimension $m, m < n$. Let f^P denote a periodic function in $\mathbb{E}^n$ invariant under $\mathrm{D} \downarrow \mathrm{H}$. Then the restriction of f^P to the subspace $\mathbb{E}^m$ yields a quasiperiodic function invariant under the action D^α of H on $\mathbb{E}^m$.

Proof : The explicit reduction of the subduced representation of H has the form

$$D \downarrow H = M(D^\alpha + \sum_\beta D^\beta)M^{-1}$$

where M is a real orthogonal matrix. The matrix M determines an orthogonal decomposition of $\mathbb{E}^n$

$$\mathbb{E}^n \to \mathbb{E}_1^m + \mathbb{E}_2^{n-m}$$

such that H acts on $\mathbb{E}_1^m$ through the representation D^α. Application of Prop. 2.4 to f^P yields a quasiperiodic function on $\mathbb{E}_1^m$.

This construction works provided that one knows the space groups in E^n. For n=4, all space groups have been classified by Brown, Bülow, Neubüser, Wondratschek and Zassenhaus [9]. For n>4 we refer to Schwarzenberger [10] for partial results. The procedure described in Prop. 3.1 can be inverted in part by induction techniques, compare Haase et al. [8]:

3.2 Prop. Let H be a point group with a non-crystallographic representation D^α of dimension m, and assume that H has a subgroup $K < H$ such that $D^\alpha \downarrow K$ subduces a one-dimensional real representation D^β of K. Then there is a hypercubic space group $(T, \Omega(n))$, where $\Omega(n)$ is the hyperoctahedral group, with $n = |H|/|K|$ such that the conditions of Prop.3.1 are fulfilled.

Proof : Consider the representation of H induced by the representation D^β of K. Let $c_1 = e, c_2, \ldots, c_n$ denote left coset generators of K in H. The induced representation has the explicit form

$$(D^\beta \uparrow H)_{ij}(g) = D^\beta(h)\delta(c_i^{-1}gc_j, h \in H).$$

Since $D^\beta(h) = \pm 1$, this representation has the form of a signed $n \times n$ permutation matrix with $n = |H|/|K|$. The set of all signed permutation matrices forms the hyperoctahedral point group $\Omega(n)$, and $\Omega(n)$ is the point group of the symmorphic hypercubic space group $(T, \Omega(n))$. Therefore we have by embedding of representations

$$K < H < \Omega(n) < (T, \Omega(n))$$

For the representations it follows from Frobenius reciprocity that $D^\beta \uparrow H$ contains the non-crystallographic representation D^β of H, and so all the conditions of Prop.3.1 are fulfilled.

Example 3.1: The icosahedral point group

The icosahedral group $A(5)$ has the dihedral subgroups $D(5), D(3)$ and $D(2)$ of order 10, 6 and 4 respectively. The non-crystallographic irreducible representation of dimension 3 which describes the symmetry of the icosahedron we denote by $[31^2_+]$. This representation subduces non-trivial one-dimensional representations of all three dihedral groups. Since the order of $A(5)$ is 60, the construction of Prop.3.2 yields hypercubic space groups in n=6, 10 and 15 dimensions with a restriction to $\mathbb{E}^3$ and the point group action $D^{[31^2_+]}$ of $A(5)$ on $\mathbb{E}^3$.

The embedding of $\mathbb{E}^3$ into $\mathbb{E}^6$ with the icosahedral group taken as a subgroup of $\Omega(6)$ is the favoured construction for icosahedral quasicrystals as Al_4 *Mn*. Note that $\Omega(6)$ has a natural subgroup S(6) so that there is another subgroup chain

$$\Omega(6) > S(6) > S(5) > A(5)$$

The subduction in this chain does not lead to the irreducible representation $[31^2_+]$ of $A(5)$, and the embedding described in Example 3.1 is required to assure the appearance of this non-crystallographic representation.

4 Lattices and Voronoi domains in $\mathbb{E}^n$

The study of cell structure in non-periodic quasicrystals requires various tools. In sections 4 and 5 we present these tools and apply them in sections 6 and 7 to non-periodic structures.

Let Y denote a lattice in $\mathbb{E}^n$ obtained from the action T x $\mathbb{E}^n \rightarrow \mathbb{E}^n$ of a discrete translation group T. By G we denote the point group which describes the holohedry of the lattice Y. In this section we shall not use the point symmetry of the lattice. For the theory to be developed we shall work not with the usual lattice basis but rather with a set of vectors associated with the Voronoi domain of the lattice Y. For this domain we refer to Voronoi [11,12].

4.1 Def. The Voronoi domain of the lattice Y is the set of points in $\mathbb{E}^n$ which are closer to (or at the same distance from) a fixed point $\mathbf{z}$ of Y than (as) to any other lattice point. We choose $\mathbf{z} = 0$ and denote the Voronoi domain by $h(n)$.

It is well-known that $h(n)$ is bounded by a set of hyperplanes perpendicular to translation vectors of the lattice. The minimal set of these translation vectors we call the *generating or Voronoi set*.

4.2 Prop. The Voronoi domain $h(n)$ of Y with the generating set of translation vectors $\mathbf{b}_1, \ldots, \mathbf{b}_{2l}$ is the n-dimensional convex polytope

$$h(n) = \{\mathbf{y} | \mathbf{y} \cdot \mathbf{b}_i \leq \frac{1}{2}(\mathbf{b}_i \cdot \mathbf{b}_i), i = 1, \ldots, 2l\}.$$

The number of Voronoi vectors is even since $-\mathbf{b}_i$ belongs to them if $\mathbf{b}_i$ does.

4.3 Def. Let $a : i \rightarrow a(i)$ denote a permutation of the numbers $1, \ldots, 2l$. Assume that for $j = p+1, \ldots, n, 0 \leq p \leq n-1$ the vectors $\mathbf{b}_{a(i)}$ are linearly independent. We say that $h(n)$ has a $p-$*boundary* if there exists a polytope of dimension p of the form

$$h(p; a(1), \ldots, a(n-p)) =$$
$$\{ \mathbf{y} \mid \mathbf{y} \cdot \mathbf{q}_{a(j)} = \mathbf{q}_{a(j)} \cdot \mathbf{q}_{a(j)},\ j = 1, \ldots, n-p, \quad \mathbf{y} \cdot \mathbf{q}_{a(i)} \leq \mathbf{q}_{a(i)} \cdot \mathbf{q}_{a(i)}, i \neq j.\}$$

We define $\mathbf{q}_k = \frac{1}{2}\, \mathbf{b}_k$, compare Prop.4.4. Points for which the strict inequalities hold we call interior points.

If $h(n)$ has a p-boundary, this is the intersection of at least $n-p$ hyperplanes. It proves convenient to describe general intersections of hyperplanes determined by the vectors of the generating set.

4.4 Prop. For a set of vectors as described in Def.4.3, but not necessarily defining a p-boundary, define the $(n-p) \times (n-p)$ matrix

$$g = g(a(1), \ldots, a(n-p)) : g_{a(j)a(k)} = \mathbf{q}_{a(j)} \cdot \mathbf{q}_{a(k)}$$

and the vector

$$\begin{aligned} \mathbf{q}_{a(1)\ldots a(n-p)} &= \sum_{\mathbf{k}=1}^{n-p} c_{a(k)} \mathbf{q}_{a(k)}, \\ c_{a(j)} &= \sum_{\mathbf{k}=1}^{n-p} (g^{-1})_{a(j)a(k)} g_{a(k)a(k)}, j = 1, \ldots, n-p. \end{aligned}$$

The vector $\mathbf{q}_{a(a)\ldots a(n-p)}$ is the point on the intersection of the hyperplanes determined by $\mathbf{b}_{a(1)} \ldots \mathbf{b}_{a(n-p)}$ closest to $\mathbf{y} = 0$.

Proof : One easily verifies that

$\mathbf{q}_{a(j)} \cdot \mathbf{q}_{a(j)} - \mathbf{q}_{a(1)\ldots a(n-p)} \cdot \mathbf{q}_{a(j)} = 0, j = 1, \ldots, n-p,$

so that $\mathbf{q}_{a(1)\ldots a(n-p)}$ is a point of the intersection of the $n-p$ hyperplanes. A general point on this intersection may be written as

$\mathbf{y} = \mathbf{q}_{a(1)\ldots a(n-p)} + \mathbf{x},$

$\mathbf{x} \cdot \mathbf{q}_{a(j)} = 0, j = 1, \ldots, n-p$

Then $\mathbf{x} \cdot \mathbf{q}_{a(1)\ldots a(n-p)} = 0$ and $\mathbf{y} \cdot \mathbf{y}$ takes its smallest value for $\mathbf{x} = 0$.

4.5 Prop. Any pair $\mathbf{b}_i$, $\mathbf{b}_j$ from the Voronoi set obeys

$\mathbf{b}_i \cdot \mathbf{b}_i - \mathbf{b}_i \cdot \mathbf{b}_j > 0,$

$\mathbf{b}_j \cdot \mathbf{b}_j - \mathbf{b}_i \cdot \mathbf{b}_j > 0.$

Proof : Assume that the second inequality is violated. Then it is easy to see that the translation vectors $\mathbf{b}_j$ and $\mathbf{b}_i - \mathbf{b}_j$ would eliminate the vector $\mathbf{b}_i$ from the generating set.

4.6 Prop. Any vector $\mathbf{b}_i$ from the Voronoi set determines a boundary $h(n-1; i)$. A neighbourhood of the point $\mathbf{q}_i$ belongs to this boundary.

4.7 Prop. Assume the existence of a boundary $h(n-2; ij)$ of $h(n)$. Then

(1)$g_{ij} \geq 0$.

(2) If $g_{ij} > 0$, then $\pm\, (\mathbf{b}_i - \mathbf{b}_j)$ belong to the Voronoi set,

(3) if $g_{ij} = 0$, then the vectors $\pm\, (\mathbf{b}_i - \mathbf{b}_j)$ do not belong to the Voronoi set.

In case (2) there are three, in case (3) there are four Voronoi vectors associated with the boundary $h(n-2; ij)$.

Proof : (1): If $g_{ij} < 0$ then the vector $\mathbf{b}_i + \mathbf{b}_j$ would remove the assumed boundary.

(2): Consider the shifted Voronoi domains $h(n) + \mathbf{b}_j, h(n) + \mathbf{b}_i$. These two Voronoi domains must share with $h(n)$ the boundary $h(n-2; ij)$. The relations

$(-\mathbf{q}_j) \cdot (-\mathbf{q}_j) - (\mathbf{q}_i - \mathbf{q}_j) \cdot (-\mathbf{q}_j) = g_{ij} > 0,$

$(-\mathbf{q}_i) \cdot (-\mathbf{q}_i) - (\mathbf{q}_j - \mathbf{q}_i) \cdot (-\mathbf{q}_j) = g_{ij} > 0$

show that the pairs $-\mathbf{b}_j$, $\mathbf{b}_i - \mathbf{b}_j$ and $-\mathbf{b}_i$, $\mathbf{b}_j - \mathbf{b}_i$ are allowed pairs according to Prop.4.5. There are three vectors associated with the boundary $h(n-2; ij)$.

(3): For $g_{ij} = 0$ the vectors $\pm$ ($\mathbf{b}_i$ - $\mathbf{b}_j$) cannot belong to the Voronoi set. The four vectors associated with the boundary are now $\pm \mathbf{b}_i, \pm \mathbf{b}_j$.

The vectors associated with a boundary $h(n-2; ij)$ form closed polygons in 2-dimensional subspaces which we call *cycles*.

So far we dealt with properties of the Voronoi domain located at the point $\mathbf{y} = \mathbf{0}$. Now we return to the action of the translation group on this domain.

4.8 Prop. The interior points of the Voronoi domain form part of a *fundamental domain* FD or *transversal* wrt the action of T on $\mathbb{E}^n$.

Proof : We have to show that no pair of interior points of $h(n)$ can be related by a translation vector from Y. It suffices to consider a translation vector $\mathbf{b}_j, j \leq l$ from the generating set. Let $\mathbf{y}$ be an interior point,

$\mathbf{y} : \mathbf{y} \cdot \mathbf{b}_i < \frac{1}{2}\, \mathbf{b}_i \cdot \mathbf{b}_i, i = 1, \ldots, 2l.$

Then

$$(\mathbf{y} - \mathbf{b}_j) \cdot (-\mathbf{b}_j) = (-\mathbf{y}) \cdot \mathbf{b}_j + \mathbf{b}_{j+l} \cdot \mathbf{b}_{j+l}$$
$$> \tfrac{1}{2}\, \mathbf{b}_{j+l} \cdot \mathbf{b}_{j+l}$$

and so $\mathbf{y} - \mathbf{b}_j$ cannot be a point of $h(n)$.

4.9 Prop. Let $\mathbf{e}$ denote a fixed unit vector in $\mathbb{E}^n$ and choose a subset of l Voronoi vectors $\mathbf{b}_{a(1)} \ldots \mathbf{b}_{a(l)}$ with the property

$\mathbf{b}_{a(i)}$:

$\mathbf{b}_{a(i)} \cdot \mathbf{e} > 0$ or

$\mathbf{b}_{a(i)} \cdot \mathbf{e} = \mathbf{b}_{a(i)+l} \cdot \mathbf{e} = 0.$

The interior points of the subset of boundaries $h(n-1; a(i))$ form part of a fundamental domain FD with respect to the action T on $\mathbb{E}^n$.

Proof : The interior points of the boundaries $h(n-1; a(i))$ and $h(n-1; a(i)+l)$ are transformed into one another by the translation vector $\mathbf{b}_{a(i)}$. By the prescription given, one selects a single representative from each pair.

To get a complete characterization of a fundamental domain FD, this analysis must be repeated for all p-boundaries with $p = n-2, \ldots, 0$.

Example 4.1: The hypercubic lattice.
The n basis vectors $\mathbf{b}_i, i = 1, \ldots, n$ of the hypercubic lattice Y form an orthogonal set. The Voronoi or generating set is the set

$\{\mathbf{b}_1, \mathbf{b}_2, \ldots \mathbf{b}_{2n}\}$,

$\mathbf{b}_{n+i} = -\mathbf{b}_i.$

The Voronoi domain is the hypercube

$$h(n) = \{\mathbf{y} | \mathbf{y} \cdot \mathbf{q}_i \leq \frac{1}{4}, i = 1, \ldots, 2n\}$$

To describe a p-boundary, choose a subset of $n-p$ Voronoi vectors which does not contain a pair $\mathbf{b}_i, \mathbf{b}_{n+i}$. The number of these subsets is

$$\nu(n-p) = \binom{n}{p} 2^{n-p}$$

A fixed subset determines a p-boundary

$$h(p; a(1) \ldots a(n-p))$$
$$= \{\mathbf{y} | \mathbf{y} \cdot \mathbf{q}_{a(j)} = \tfrac{1}{4}, j = 1, \ldots, n-p\, , \quad \mathbf{y} \cdot \mathbf{q}_{a(i)} \leq \tfrac{1}{4}, \quad i \neq j\}$$

The vector described in Prop.4.4 becomes

$$\mathbf{q}_{a(1)\dots a(n-p)} = \sum_{j=1}^{n-p} \mathbf{q}_{a(j)}$$

and the p-boundary may be rewritten as

$$h(p; a(1)\dots a(n-p)) = \{\mathbf{y}|\mathbf{y} = \sum_{j=1}^{n-p} \mathbf{q}_{a(j)} + \frac{1}{2} \sum_{i=n-p+1}^{n} \lambda_{a(i)}\mathbf{b}_{a(i)}, -1 \leq \lambda_r \leq 1\}.$$

5 Euclidean cell complexes in $\mathbb{E}^n$ and their metrical duals

It proves useful to introduce for lattices in $\mathbb{E}^n$ some notions from algebraic topology. We refer to Munkres [13] for the details.

5.1 Def. An Euclidean cell complex K of dimension n is a collection of convex polytopes of dimensions p = n,..., 0 with the properties
(1): any q-boundary of a polytope from K belongs to K.
(2): the intersection of any two polytopes of K is empty or it is a boundary of both polytopes.

The boundaries are also called faces of the polytopes, and condition (2) requires that all polytopes be in face-to-face position. The subcomplex of all boundaries of dimension $r : r \leq p < n$ of K is called the p-*skeleton* and denoted by $K^{(p)}$.

This notion may be applied to the Voronoi domains of a lattice Y in $\mathbb{E}^n$. We extend the meaning of the lattice Y and let it describe the geometric structure obtained by the action of the translation group T on the Voronoi domain and on its boundaries of dimension $p = 0, \dots, n-1$.

5.2 Prop: The set of all translated Voronoi domains in $\mathbb{E}^n$ forms an Euclidean cell complex of dimension n which we denote by Y. The set of all translated boundaries of dimension $r, r \leq p < n$ forms the p-skeleton $Y^{(p)}$ of this complex.

Now we introduce the notion of metrical duality

5.3 Prop. For any vector $\mathbf{b}_i$ from the Voronoi set define a 1- polytope by

$$h^*(1; i) = \{\mathbf{y}|\mathbf{y} = \frac{1}{2}(1+\lambda_i)\mathbf{b}_i, -1 \leq \lambda_i \leq 1\}, i = 1, \dots, 2l$$

The collection of all translated copies of these 1-polytopes forms an Euclidean cell complex of dimension 1 which we call $(Y^*)^{(1)}$.

5.4 Def. The complexes $Y^{(n-1)}$ and $(Y^*)^{(1)}$ are called metrical duals with respect to each other. Their elements form local pairs $h(n-1; i)$ and $h^*(1; i)$ which have complementary dimension, intersect in a single point, and are spanned by two sets of vectors which are orthogonal with respect to one another.

We shall extend this notion of metrical duality to the boundaries of dimension p, keeping the condition of local pairs and local orthogonality. The corresponding dual Euclidean cell complexes will be called $Y^{(p)}$ and $(Y^*)^{(n-p)}, p = 0, \dots, n-1$. By $Y^{(0)}$ we shall denote the centers of the Voronoi domains. Then Y^* will denote the metrical dual cell complex corresponding to Y.

Example 5.1: The hypercubic cell complex.

Consider the hypercubic lattice, its Voronoi domain, and its p-boundaries described in example 4.1.

Define the dual boundary

$$h^*(n-p;\, a(1)\ldots a(n-p)) = \{\mathbf{y} | \mathbf{y} = \sum_{j=1}^{n-p} \mathbf{q}_{a(j)} + \frac{1}{2}\sum_{j=1}^{n-p} \lambda_{a(j)}\mathbf{b}_{a(j)}\}$$

It is easy to see that $h(p; a(1)\ldots a(n-p))$ and $h^*(n-p; a(1)\ldots a(n-p))$ intersect in the single point

$$\mathbf{y} = \sum_{j=1}^{n-p} \mathbf{q}_{a(j)}$$

and are spanned by mutually orthogonal sets of vectors. Extend this construction first to all boundaries of dimension p, then to all dimensions, and generate the geometric structure in $\mathbb{E}^n$ obtained by the action of the translation group. Then one finds in $\mathbb{E}^n$ two Euclidean cell complexes Y and Y^* which are the metrical duals of one another. The dual complex Y^* has a simple geometric form: it has the same intrinsic geometry as Y, but is shifted in $\mathbb{E}^n$ by the vector which connects the midpoint of the hypercube to one of the vertices.

Example 5.2: The cubic cell complexes in $\mathbb{E}^3$

In $\mathbb{E}^3$ there are three lattices with the holohedry of the octahedral group: the primitive lattice P, the face-centered lattice F, and the body-centered lattice I. The metrical dual to P is obtained as in the general hypercubic case. The metrical duals F^* and I^* are decribed in [14] and illustrated in Figs. 2 and 3.

The lattices F and I are related by reciprocity, we have the equivalences

$P^R \sim P,\ F^R \sim I,\ I^R \sim F$

It turns out that the metrical duals of F and I, taken as Euclidean cell complexes are **not** related in a similar fashion, and so for non-hypercubic lattices there is a conceptual difference between reciprocity and metrical duality. For example the cells of the metrical dual I^* are tetrahedra and hence differ from the Voronoi domains of F.

6 Lattices and Klötze in $\mathbb{E}^n$

In section 4 and 5, a lattice in $\mathbb{E}^n$ was associated with a Euclidean cell complex Y and its metrical dual Y^*. In the present section, the subspace decomposition

$$\mathbb{E}^n \rightarrow \mathbb{E}_1^m + \mathbb{E}_2^{n-m}$$

introduced in sections 2 and 3 is taken up. The complexes Y and Y^* are used to construct n-dimensional polytopes, called klötze, which are adapted to the subspace decomposition. The construction is based on the skeletons

$$Y^{(m)},\ (Y^*)^{(n-m)} \text{ and } (Y^*)^{(m)},\ Y^{(n-m)}$$

respectively.

6.1 Def. Let $Y^{(m)}$ and $(Y^*)^{(n-m)}$ denote the m-skeleton of Y and the $(n-m)$-skeleton of Y^* respectively, and consider the dual pairs of boundaries of the Voronoi domain

$h\ (m;\ a(1)\ldots a(n-m))$

$h^*(n-m;\ a(1)\ldots a(n-m))$

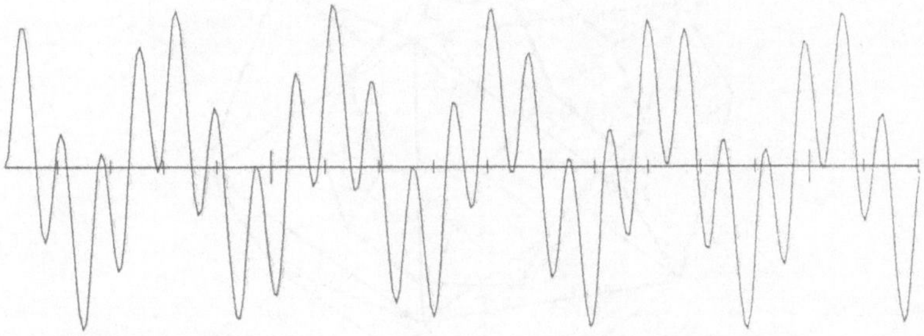

Fig.1 A quasiperiodic function with the approximate periods $r_5 = 5.83$, $r_6 = 9.43$, $r_7 = 15.26$. This function is obtained by the restriction of a periodic function on a square lattice to a line.

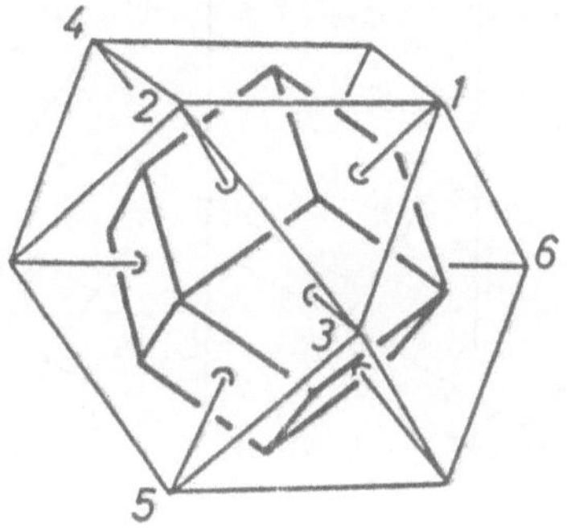

Fig.2 The Voronoi domain of the lattice F in $\mathbb{E}^3$. The faces form part of the skeleton $F^{(2)}$. The metrical dual F^* has the skeleton $(F^*)^{(1)}$ which consists of all Voronoi vectors. These vectors form the edges for the dual cells of F^*.

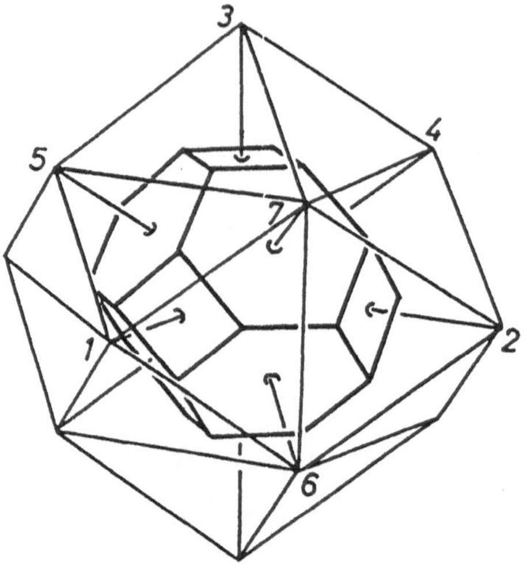

Fig.3 The Voronoi domain of the lattice I in $\mathbb{E}^3$. The faces form part of the skeleton $I^{(2)}$. The metrical dual I^* has the skeleton $(I^*)^{(1)}$ which consists of Voronoi vectors. They form the edges of the dual cells of I^*.

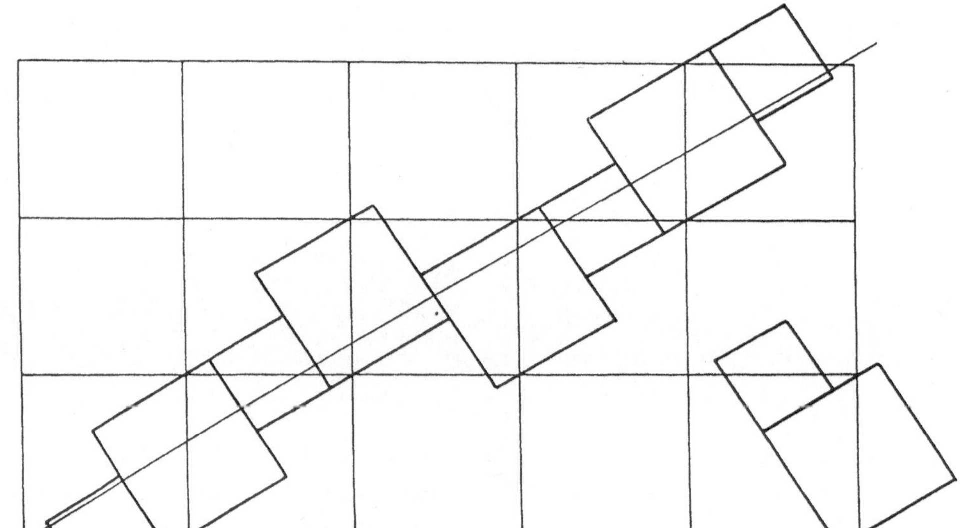

Fig.4 The klötze of the square lattice are two types of squares. One large and one small square form a fundamental domain of the lattice.

Denote by the subscripts 1,2 the projections of these boundaries to $\mathbb{E}_1^m$ and $\mathbb{E}_2^{n-m}$ respectively. The klotz associated with the pair of boundaries is the n-dimensional polytope

$$\begin{aligned} & kl(m+(n-m);\ a(1)\dots a(n-m)) \\ = \{\mathbf{y}|\mathbf{y} &= \mathbf{y}_1+\mathbf{y}_2, \\ \mathbf{y}_1 &\in h_1(m;\ a(1)\dots a(n-m)),\ \mathbf{y}_2 \in h_2^*(n-m;\ a(1)\dots a(n-m))\} \end{aligned}$$

We call the projected boundaries the *1*-chart and the *2*-chart of the klotz.

6.2 Def. Let $(Y^*)^{(m)}$ and $Y^{(n-m)}$ denote the m-skeleton of Y^* and the $(n-m)$-skeleton of Y respectively, and consider the dual pairs of boundaries of the Voronoi domain

$h^*(m;\ a(1)\dots a(m))$,

$h\ (n-m;\ a(1)\dots a(m))$.

The klotz associated with the pair of boundaries is the n-dimensional polytope

$$\begin{aligned} & kl((n-m)+m;\ a(1)\dots a(m)) \\ = \{\mathbf{y}|\mathbf{y} &= \mathbf{y}_1+\mathbf{y}_2, \\ \mathbf{y}_1 &\in h_1^*(m;\ a(1)\dots a(m)),\ \ \mathbf{y}_2 \in h_2(n-m;\ a(1)\dots a(m))\}. \end{aligned}$$

Again we call the projected boundaries the *1*-chart and the *2*-chart of the klotz.

The following properties of the klötze have been shown for a number of complexes Y, Y^* and pairs n/m:

(KL1): Any klotz has its boundaries of dimension $n-1, n-2, \dots, 1$ always parallel or perpendicular to the subspace $\mathbb{E}^m$.

(KL2): The set of all translated klötze forms a periodic space-filling of $\mathbb{E}^n$.

(KL3): Neighbouring klötze are in general not in face-to-face position with respect to their boundaries and hence do not form part of a Euclidean cell complex.

(KL4): There exists a representative set of l klötze associated with the first Voronoi domain which forms a new fundamental domain FD with respect to the translation group T of Y.

Proofs of these properties can be found in [15] for the square lattice and $n/m = 2/1$, in [16] for the hypercubic lattice and $n/m = 6/3$ with the subspace $\mathbb{E}^3$ corresponding to icosahedral point symmetry, in [17] for all 5 Bravais lattices and $n/m = 2/1$, in [14] for all lattices and $n/m = n/(n-1)$. We shall illustrate the properties (KL1) - (KL4) by various examples.

Example 6.1: Klötze for the square lattice in $\mathbb{E}^2$.

For the square lattice in $\mathbb{E}^2$, an intersecting subspace $\mathbb{E}^1$ is a line with slope $\tan\alpha$ with respect to one of the basis vectors. There are two representative klötze which are squares of edge length $\cos\alpha$ and $\sin\alpha$ repectively whose translated copies fill $\mathbb{E}^2$ periodically, compare Fig.4.

Example 6.2: The hexagonal lattice in $\mathbb{E}^2$.

There are three representative klötze which are rectangles of different size. The two constructions of Def. 6.1 and 6.2 may be obtained from one another by the interchange of the subspaces $\mathbb{E}_1^1$ and $\mathbb{E}_2^1$, compare Fig.5.

Example 6.3: The centered cubic lattices F and I in $\mathbb{E}^3$ with a subspace $\mathbb{E}_1^2$.

In the construction of Def. 6.1, the *1*-charts of the klötze are the projections of the faces of the Voronoi domain to an intersecting plane, the *2*-charts are the projections

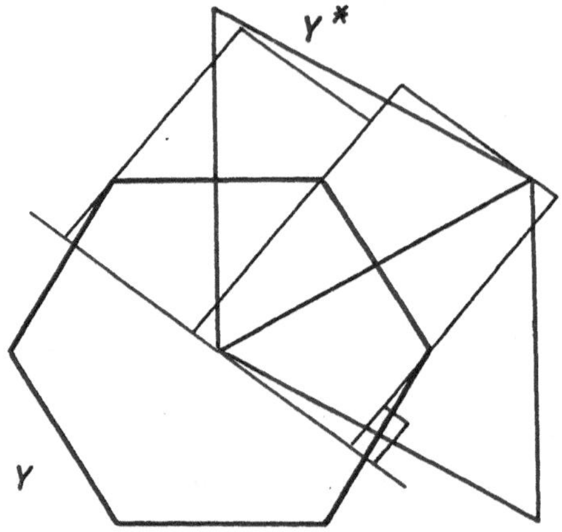

Fig.5 The Voronoi domain of the hexagonal lattice Y is a hexagon. Its dual Y^* has two types of triangular cells bounded by Voronoi vectors. For the intersection with a line, there are three rectangular klötze which form a fundamental domain of Y.

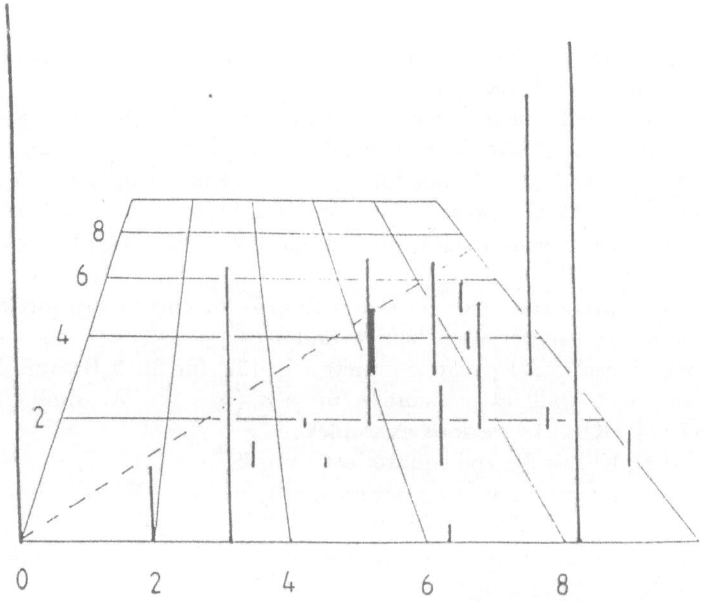

Fig.6 A diffraction pattern for an icosahedral quasicrystal model. Shown is the intensity of diffraction spots indicated by bars in a plane perpendicular to a 5-fold axis. Computation by D. Zeidler.

of Voronoi vectors to a line perpendicular to this intersecting plane. The klötze become prism whose number equals l, where $2l$ is the number of Voronoi vectors.

Example 6.4: The hypercubic lattice in $\mathbb{E}^6$ and the intersection with the subspace $\mathbb{E}_1^3$ of icosahedral point symmetry.

Both Y and Y^* correspond to hypercubic lattices, as pointed out in example 5.1, and so there is no difference between the constructions of Def.6.1 and 6.2. The *1*-chart of a klotz $kl(3+3;\ a(1)\ a(2)\ a(3))$ is a rhombohedral cell which can occur in two possible shapes, a thick and a thin rhombohedron. Each one of them can appear in 10 possible orientations. The FD is formed from 20 klötze, but the icosahedral point symmetry restricts them to two representative rhombohedra with the point symmetry $D(3)$.

7 Klötze and quasicrystal models

In section 2 it was shown that *quasiperiodicity* could provide a new paradigm for non-periodic long-range order. It was pointed out that for quasiperiodic functions there is in general no discrete cell structure in **x**-space. A completely different paradigm for non-periodic long-range order results from a *non-periodic tiling*, interpreted as a non-periodic cell structure.

Example 7.1: The non-periodic Fibonacci sequence. Consider an *alphabet* with two letters a, b and the Fibonacci *words* f_n formed according to the Fibonacci recursion relation

$$f_{n+1} = f_n\, f_{n-1},\ f_1 = b,\ f_2 = a.$$

The first Fibonacci words are

n	f_n
1	b
2	a
3	ab
4	aba
5	$abaab$
6	$abaababa$

The frequency of the letters a and b respectively in f_n are easily seen to be successive Fibonacci numbers. Since the ratio of successive Fibonacci numbers has the irrational number Φ^{-1} as its limit, it follows that the limit of f_n for $n \to \infty$ cannot be periodic. For this and many other general properties of words we refer to Lothaire [18].

This example demonstrates the combinatorial construction of a discrete non-periodic cell structure. From the concepts developed in sections 4-6, it is now possible to determine a concrete cell structure for a non-periodic object. Consider first the intersection of a fixed subspace $\mathbb{E}_1^m$ with a klotz described in section 6. The intersection of this klotz with $\mathbb{E}_1^m$ will appear as its *1*- chart, irrespective of a shift in the perpendicular direction of $\mathbb{E}_2^{n-m}$. Since the klötze form a space filling of $\mathbb{E}^n$, their intersections must have corresponding properties in $\mathbb{E}^m$, and one has

7.1 Prop. Given a periodic klotz tiling of $\mathbb{E}^n$ with the properties (KL1)- (KL4), the intersection with a subspace $\mathbb{E}^m$ yields an in general non-periodic tiling of $\mathbb{E}^m$ whose tiles are the *1*-charts of the klötze.

To determine a model for a quasicrystal, one has to specify a density f^P on the lattice Y in $\mathbb{E}^n$. Since there is a representative set of klötze according to (KL4), this density f^P may be specified on this representative set. But since the intersection of $\mathbb{E}^m$ hits translated copies of a given representative klotz in varying perpendicular positions, the different occurences of the *1*-chart for this klotz will carry different values of the density f^P, and so the density on the *1*-charts is not stable in the subspace $\mathbb{E}^m$.

7.2 Prop. Define a quasicrystal model by restricting the density f^P on the representative klotz kl_i, $i = 1, \dots, l$ by

$f_i^P(x_1, x_2) = f_i^P(x_1)$.

Then the density seen on a quasicrystal cell, defined as the *1*-chart of the klotz, remains fixed for any non-periodic occurence of that cell.

In other words, the klotz construction permits the concept of a discrete non-periodic cell structure for quasicrystal cells. The quasicrystal model described in Prop.7.2 is a submodel of the general quasiperiodic model discussed in section 2, and so the Fourier transform properties of Prop.2.4 apply to it. Due to the special properties of the quasicrystal density defined in Prop.7.2, the integral in Prop.2.3 which defines the Fourier coefficient can be carried out with respect to the perpendicular coordinate x_2 and contributes a kinematical factor. Therefore one gets

7.3 Prop. In the quasicrystal model described in Prop. 7.2, the Fourier transform of the quasiperiodic density is completely determined by integrals over the density on the l representative quasicrystal cells.

Example 7.2: The two-cell Fibonacci quasicrystal model. The klötze from the square lattice described in example 6.1 are two squares of edge length $s = sin\alpha$ and $c = cos\alpha$. On an intersecting line of slope $tan\alpha = \Phi^{-1}$, the sequence of quasicrystal cells c and s becomes identical to the Fibonacci sequence of the letters a and b considered in example 7.1. Compare Fig.4 for this construction.

An example of a diffraction pattern computed from a quasicrystal model with icosahedral point symmetry is given in Fig.6.

8 Conclusion

We conclude by giving some additional references on non-periodic long-range order. The first experimental work on quasicrystals was reported in 1984 by Shechtman, Blech, Gratias and Cahn [19]. A bibliography of experimental and theoretical work is given by Mackay [20]. Early theoretical work on discrete non-periodic tilings is due to Penrose in 1974 [21], de Bruijn [22], Mackay [23], Kramer and Neri [7]. The theoretical progress in the field of quasicrystals is presented in the reprint volume by Steinhardt and Ostlund [24]. A recent account of both experimental and theoretical work may be found in a workshop proceedings volume edited by Janot and Dubois [25].

References

[1] H. Bohr, Fastperiodische Funktionen, Berlin (1932)

[2] H. Bohr, Acta Math. 45:29 (1925)

[3] H. Bohr, Acta Math. 46:101 (1925)

[4] H. Bohr, Acta Math. 47:237 (1926)
[5] A. Schoenflies, Krystallsysteme und Krystallstruktur, Teubner, Leipzig (1891)
[6] E.S. Fedorov, The symmetry of regular systems of figures, Not. of the Imp. St. Petersburg Min. Soc. 28:1 (1891)
[7] P. Kramer and R. Neri, Acta Cryst. A40:580 (1984)
[8] R.W. Haase, P. Kramer, L. Kramer and H. Lalvani, Acta Cryst. A43: 574 (1987)
[9] H. Brown, R. Bülow, J. Neubüser, H. Wondratschek and H. Zassenhaus, Crystallographic Groups of Four-dimensional Space, Wiley, New York (1978)
[10] R.L.E. Schwarzenberger, N-dimensional Crystallography, Pitman, San Francisco (1980)
[11] G. Voronoi, Crelles J. f. d. reine und ang. Math. 134: 198(1908)
[12] G. Voronoi, Crelles J. f. d. reine und ang. Math. 136: 67 (1909)
[13] J.R. Munkres, Elements of Algebraic Topology, Addison-Wesley, Menlo Park (1984)
[14] P. Kramer, submitted for publication
[15] P. Kramer, Mod. Phys. Lett. B 1:7 (1987)
[16] P. Kramer, Int. J. Mod. Phys. B 1:145 (1987)
[17] P. Kramer, Mod. Phys. Lett. B 2:605 (1988)
[18] M. Lothaire, Combinatorics on Words, Addison-Wesley, Reading (1983)
[19] D. Shechtman, I. Blech, D. Gratias, and J.W. Cahn, Phys. Rev. Lett. 53:2477 (1984)
[20] A.L. Mackay, Int. J. of Rapid Solidification 2:S1 (1987)
[21] R. Penrose, Bull. Inst. Math. Appl. 10:266 (1974)
[22] N.G. de Bruijn, Proc. Konig. Ned. Akad. Weten. A 84:39(1981)
[23] A.L. Mackay, Physica 114 A:609 (1982)
[24] P.J. Steinhardt and St. Ostlund, The Physics of Quasicrystals, World Scientific, Singapore (1987)
[25] Ch. Janot and J.M. Dubois, Editors, Quasicrystalline Materials, World Scientific, Singapore (1988)

SYMMETRIES IN THE LAGRANGIAN AND HAMILTONIAN FORMALISM: THE EQUIVARIANT INVERSE PROBLEM

G.Marmo*, G.Morandi**, and C.Rubano*

* Dipartimento di Scienze Fisiche and INFN
Università di Napoli
Mostra d'Oltremare pad.19, 80125 Napoli – Italy
** Dipartimento di Fisica, GNSM-CISM and INFN
Università di Ferrara
Via Paradiso 12, 44100 Ferrara – Italy

ABSTRACT

In point-particle Mechanics, Helmoltz's *Inverse Problem in the Calculus of variations* consists in finding, and classifying, all the possible Lagrangian descriptions of a given second-order (i.e. Newtonian) dynamical system. In this paper we address ourselves to the *Equivariant Inverse Problem*, which arises when one wants the Lagrangian to be strictly invariant, or only quasi invariant, under the action of a Lie group of point transformations which are also symmetries for the original dynamics. The occurrence of quasi invariance will lead us to the discussion of the Lie algebra cohomology associated with the infinitesimal action of the group of symmetries on the Lagrangian.

In the solution of the Inverse Problem, one can meet obstructions associated with the De Rham cohomology of the manifold, which acts as a carrier space. Additional obstructions can arise if one looks for an invariant cohomology, i.e. one in which symmetries are required to 'pass through' the integration of forms. We discuss a general procedure for resolving the obstructions by lifting the description of the dynamical system to a suitably defined enlarged space. This will lead us to a discussion of the central extensions of the relevant Lie algebras and groups of symmetries which act then on the enlarged carrier space. The possibility of defining an equivariant momentum map in the Lagrangian context is also discussed.

We also comment on the rather similar mathematical structures which are encountered when one tries to implement classical dynamical symmetries in Quantum Field Theory, and which give rise to Wess-Zumino terms and to anomalies.

CONTENTS

1 – INTRODUCTION – EQUATIONS OF MOTION AND INVARIANCE GROUPS

A *classical dynamical system* (with a finite number n of degrees of freedom), obeying Newton's type equations of motion, is described by equations of the form:

$$\frac{dq^i}{dt} = u^i \quad ; \quad \frac{du^i}{dt} = f^i \tag{1.1}$$

with $i = 1, \ldots, n$ and $f^i = f^i(q, u)$. The q's are (generalized) coordinates, and it is generally assumed that the *configuration space* Q, locally coordinatized by the q's, has some 'nice' feature, like that of being a (connected) differentiable manifold. How one can 'build up' Q, starting from observational data, is discussed, for instance, in the first chapter of [58]. More generally, this procedure will lead eventually to the construction of a (minimal) *carrier space* M, on which the dynamical evolution is given by a set of first order equations. In the above case, $M = TQ$ (the 'velocity phase space', or the tangent bundle over TQ) , and the dynamics is called a *second order* dynamics, for obvious reasons.

Within the above general scheme, one can also fit classical descriptions of particles (or systems thereof) endowed with internal structures, like spin (see [75,10,77]).

The infinite-dimensional generalization of the scheme outlined above leads to *classical field theory*, where now the 'configuration space' is the set of fields over space-time, and the (ordinary) equations of motions (1.1) are replaced by (typically second order) partial differential equations (PDE) for the field variables [72,41].

The transition to Quantum Mechanichs (QM) requires some additional piece of structure to be added to the formalism, namely a Lagrangian, from which the equations of motion can be derived as the associated Euler-Lagrange equations. Then, either directly, via path-integrals, or passing to the Hamiltonian formalism and then applying the canonical procedure, one can build up the quantized version of a given classical theory [27,40,30,74,83].

By a *symmetry* of a dynamical system (represented by a given set of evolution equations) we mean any transformation which maps (bijectively) the set of solutions onto itself. Although not strictly required, we will add the further request that symmetries should preserve the parametrization of the solutions. In the specific case of Eq.(1.1), we may or may not require the relation between the q,s and the u's (expressed by the first equation) to be preserved. In the former case, symmetry transformations will be further qualified as *point-simmetries* (or 'Q-symmetries')[58]. Looking at many of the classical books on Classical Mechanics, written before the advent of QM [82,53], it is apparent that invariance principles associated with symmetries, important as thay may have been in Galilei's and Newton's thinking, were not considered to play a really important rôle in the framework of Mechanics. Rather, they were considered as an additional bonus through which one could add some more elegance to the exposition of the theory, but not much more than that. In a classical context, it was perhaps the formulation of Einstein's theories of relativity which began to emphasize the rôle of invariance principles, but it was really only the advent of QM which displayed their full powerfullness [25].

Quite often, in trying to build up theories of interacting fields, invariance principles[1] are used as guiding principles to select the possible (and hence admissible) interaction Lagrangians [44,33]. Somehow, therefore, the renewed interest in the study of the rôle of symmetries in a classical framework can be seen as a feedback from QM.

As already mentioned, what we will be interested in is the description of classical dynamical systems at the Lagrangian or Hamiltonian level. The main problem we shall address ourselves to in the present paper will then be that of how much of the symmetries of the original dynamical system can be made to survive as symmetries of its Lagrangian (or Hamiltonian) description.

As a simple example, we may consider the case $Q = \mathbf{R}^n$ and $f^i = -\omega^2 q^i$ in (1.1), i.e. an n-dimensional isotropic harmonic oscillator (HO) of unit mass and frequency ω.[2] It can be shown [55] that the full linear group $GL(n, \mathbf{R})$ is a symmetry group (actually a group of Q-symmetries) for the HO, but, of course, no Lagrangian exists which is invariant under the full general linear group. This example shows that, in general, passing from the description at the level of the equations of motion to the Lagrangian (or Hamiltonian) description, not all of the symmetries of the original dynamical system can be implemented as symmetries of the Lagrangian, or as canonical transformations in the Hamiltonian framework. Some of them will be lost in the passage, and one is faced with a certain amount of 'symmetry breaking' from the full group of symmetries of the original dynamical system to some residual subgroup.

We would like to choose a Lagrangian (and a Hamiltonian) in such a way that this subgroup:

i) Contains as many as possible of the symmetry transformations which are thought of as being physically relevant and
ii) it is a 'maximal' subgroup, subject to i).

[1] or quasi-invariance, in a sense which will be specified shortly below.

[2] For $\omega = 0$ we recover the free particle.

This is of course because only those symmetries which appear as symmetries[3] of the Lagrangian (or as canonical symmetries in the Hamiltonian framework) are candidate to be implemented as unitary (or antiunitary) operators in the Hilbert space description of QM.

In pursuing this program, one is faced, in a preliminary way, with the problem of the very *existence* of a Lagrangian description. The investigation of the existence and uniqueness[4] of a Lagrangian description for a given dynamical system constitutes what is known as the "Inverse Problem in the Calculus of Variations" (IP). It was formulated by Helmoltz [36] more than one century ago, and it has been since then the subject of extensive investigation. We refer to a recent review of ours [66] for a discussion of the IP and for a comprehensive list of references to it. There (but see also [62,28,45-48,23]), we also review and discuss the ambiguities in the quantization procedure, which are brought up by the (possible) existence of alternative Lagrangians for the same dynamical system.

Leaving the IP aside for a moment, let us assume that (at least) one Lagrangian exists, i.e. that a function $\mathcal{L} = \mathcal{L}(q,u)$ can be found, such that the Euler-Lagrange equations:

$$\frac{dq^i}{dt} = u^i \quad ; \quad \frac{d}{dt}\left(\frac{\partial \mathcal{L}}{\partial u^i}\right) - \frac{\partial \mathcal{L}}{\partial q^i} = 0 \tag{1.2}$$

are equivalent to (1.1). This implies that $\mathcal{L}$ should be such that the 'forces' f^i in (1.1) satisfy the set of linear equations:

$$\frac{\partial^2 \mathcal{L}}{\partial u^i \partial u^j} f^i = \frac{\partial \mathcal{L}}{\partial q^i} - \frac{\partial^2 \mathcal{L}}{\partial u^i \partial q^j} u^j \ . \tag{1.3}$$

If the Lagrangian is *regular*, i.e. if the Hessian matrix:

$$H \stackrel{\text{def}}{=} \left\| \frac{\partial^2 \mathcal{L}}{\partial u^i \partial u^j} \right\| \ , \tag{1.4}$$

is nonsingular[5] , then (1.3) has a unique solution for the f^i's. Stated otherwise, there is a unique second order dynamics associated with a given regular Lagrangian.

Conversely, if the f^i's are given, Eqns.(1.3) can be viewed as a set of linear PDE's for the unknown $\mathcal{L}$. Spelling out the integrability conditions yields the so called Helmoltz conditions [70,19,76] for the existence of $\mathcal{L}$. We are now assuming that at least one solution has been found for the Lagrangian.

As in the sequel we will be mainly concerned with Lie groups of point symmetries, let us assume that we have one such group G. At the infinitesimal level, we will have thus the transformations:

$$\delta_k q^i = X^i_k(q) \quad ; \quad \delta_k u^i = \frac{d}{dt}\delta_k q^i \ , \tag{1.5}$$

---

[3] Nöther symmetries, in the case of point transformation.

[4] When speaking of uniqueness, we mean of course uniqueness up to the addition of a total time derivative, the familiar 'gauge freedom' in the definition of the Lagrangian.

[5] Let's recall that the regularity condition $\det H \neq 0$ is invariant under point transformations.

$k = 1, \ldots, \mu$, μ being the dimension of the group **G**. Asking for **G** to be a symmetry group for Eqns.(1.1) amounts to requiring that:

$$\frac{d}{dt}\delta_k u^i = \delta_k f^i \ . \tag{1.6}$$

(All this will be reformulated in more geometrical terms in next Section).

If now a Lagrangian $\mathcal{L} = \mathcal{L}(q, u)$ is given, we may inquire whether or not **G** is a group of symmetries for $\mathcal{L}$ as well, i.e. whether or not

$$\delta_k \mathcal{L} = 0 \quad ; \quad k = 1, \ldots, \mu \tag{1.7}$$

A simple example shows that it is not so, in general. Let us consider, in $Q = \mathbf{R}^3$, the motion of a charged particle in constant and uniform electric and magnetic fields $\vec{E}$ and $\vec{B}$, at right angles to each other (i.e. $\vec{E} \cdot \vec{B} = 0$). The equations of motion can be deduced from the standard Lagrangian:

$$\mathcal{L} = \frac{1}{2}|\vec{u}|^2 + \vec{E} \cdot \vec{r} + \frac{1}{2}\vec{u} \cdot (\vec{r} \times \vec{B}) \ . \tag{1.8}$$

where the so called 'symmetric' gauge

$$\vec{A} = \frac{1}{2}\vec{r} \times \vec{B} \tag{1.9}$$

has been chosen for the vector potential. The equations of motion

$$\frac{d\vec{r}}{dt} = \vec{u} \quad ; \quad \frac{d\vec{u}}{dt} = \vec{E} + \vec{u} \times \vec{B} \tag{1.10}$$

are of course translationally invariant and the infinitesimal action of the group of translations is:

$$\delta\vec{r} = \vec{X} \quad ; \quad \delta\vec{u} = 0 \ , \tag{1.11}$$

with $\vec{X} \in \mathbf{R}^3$. However, under (1.11):

$$\delta\mathcal{L} = \vec{E} \cdot \vec{X} + \frac{1}{2}(\vec{X} \times \vec{B}) \cdot \vec{u} \ . \tag{1.12}$$

Hence, $\delta\mathcal{L} = 0$ if $\vec{X}$ is parallel to $\vec{B}$. But if $\vec{X} \perp \vec{B}$, $\mathcal{L}$ changes by:

i) a constant $\vec{E} \cdot \vec{X}$ and
ii) a total time derivative:

$$\frac{1}{2}(\vec{X} \times \vec{B}) \cdot \vec{u} = \frac{d}{dt}\left[\frac{1}{2}(\vec{X} \times \vec{B}) \cdot \vec{r}\right] \ . \tag{1.13}$$

This fact cannot be dealt with by elementary means (like changing the Lagrangian by gauge transformation), but is deeply rooted in the structure of the group of space translations. In particular, for $E = 0$, the fact that the Lagrangian changes by (1.13) under translations is at the basis of the construction, in QM, of the so called 'magnetic translation groups' [85-87].

In this paper, we will try and address ourselves to all of the problems listed above, trying to be at the same time as exhaustive (as far as our personal knowledge and expertise goes) and as pedagogical as we could. To the best of our knowledge, the reference list provided at the end is quite complete.

What has been said up to now concerning the Inverse Problem and the Inverse Problem with symmetries will be recasted, in Sect.2, in the language of modern Differential Geometry. The latter is perhaps not as familiar to physicists as the language of local coordinates we have been employing up to now. However, it is definitely more powerful as a language, and, moreover, it offers a natural setting for the problems which can arise from the global topology of the carrier space [7,1,51,58] of a given dynamical system. In the same Section, we will show how the nonexistence of strictly invariant Lagrangians (in the sense of Eq.(1.7)) can be connected to the existence of nontrivial cocycles in the Lie algebra of the symmetry group that we may have eventually selected as a 'physically reasonable' reduced symmetry group for the Lagrangian. Necessary and sufficient conditions for obtaining a strictly invariant Lagrangian if the cocycle is trivial, obtained by one of us in collaboration with other authors [59], will also be discussed.

Obstructions in the way of obtaining an invariant description (under a reduced symmetry group) will also be shown to arise from both the De Rham cohomology associated with the carrier space and from what will be termed as the *invariant* cohomology.[6] A systematic procedure for the resolution of (ordinary, i.e. not necessarly invariant) cohomological obstructions will be described in Sect.3, again borrowing from previous work of one of us in colloaboration with others [84,10], as well as from more recent work [63].

In Sect.4, we essentially discuss a general procedure for obtaining invariant descriptions of a dynamical system by lifting its description to an enlarged space, obtained by adding either a single, non dynamical, degree of freedom to the entire carrier space, or in the specific case of Lagrangian dynamics and if one does not want to loose track of the tangent bundle structure of the carrier space, by applying the same procedure to the configuration space and then by going to the tangent bundle over the enlarged space. This will lead us to discuss to some extent the problem of central extensions of Lie algebras, and also, though in a more marginal way, the related problem of central extensions of Lie groups.

In Sect.5 we shall discuss (again borrowing from previous work by two of us and others [63,64]) a problem specific to Lagrangian dynamics, namely how one can define a momentum map [35,75] in a Lagrangian context, and how the equivariance of the latter with respect to the coadjoint representation of the given Lie group of symmetries can be related to the existence of invariant Lagrangians.

The problem of whether or not symmetries of a given dynamics are implementable at the level of the Lagrangian description, and can subsequently be carried over to be symmetries of the quantum description is at the heart of the modern discussion concerning anomalies in Quantum Field Theory [89,80,14]. This may be seen as a partial (albeit very important) motivation for our discussing the same problem at the classical level of point particle dynamics.

Comments about this connection, as well as some general conclusions about what has been proved in the paper, are contained in Sect.6.

The Appendices contain some technical material, which is needed and/or referred to in the main text, concerning central extensions of Lie algebras and groups as well as the cohomology theory of Lie algebras.

[6] Loosely speaking, when the 'integration' of a form is accompanied by the requirement that its 'primitive' obeys the same invariance conditions as the original form

2 – THE INVERSE AND EQUIVARIANT INVERSE PROBLEM

We begin by recasting what has been said in Sect.1 in the intrinsic language of Differential Geometry, the one we will be mostly employing henceforth. Our notational convenctions will follow as closely as possible those of [58]. Let then Q be a smooth manifold, TQ its tangent bundle. Local coordinates in a bundle chart for TQ will be denoted by (q, u). We recall that TQ is endowed with a natural operation, the *vertical endomorphism* S, a (1-1) tensor field, whose local expression is given by:[1]

(2.1) $$S = \frac{\partial}{\partial u^i} \otimes dq^i \ .$$

Another object which is naturally defined on TQ is the dilation operator along the fibers. In local terms, it is represented by the vector field:

(2.2) $$\Delta = u^i \frac{\partial}{\partial u^i} \ .$$

A vector field $\Gamma \in \mathcal{X}(TQ)$ is called *second order* iff it satisfies

(2.3) $$S(\Gamma) = \Delta \ .$$

In local coordinates, (2.3) is easily seen to lead to the following local expression for Γ:

(2.4) $$\Gamma = u^i \frac{\partial}{\partial q^i} + f^i \frac{\partial}{\partial u^i} \quad ; \quad f^i \in \mathcal{F}(TQ) \ .$$

Hence, a system of ODE's of the form (1.1) is represented, in intrinsic terms, by a second order vector field on TQ (and viceversa).

Let now $\mathbf{G}$ be a Lie group, with a given action ϕ on Q:

(2.5) $$\phi : \mathbf{G} \times Q \to Q \ .$$

Defining:

(2.6) $$\phi_g \overset{\text{def}}{=} \phi(g, \cdot) \quad , \quad g \in \mathbf{G} \ ,$$

then $\phi_g \in \mathrm{Diff}(Q)$, and

(2.7) $$\phi_e = \mathrm{Id}_Q \quad ; \quad \phi_g \circ \phi_{g'} = \phi_{gg'} \quad ; \quad \forall g, g' \in \mathbf{G}$$

e being the identity in $\mathbf{G}$.

The tangent map $T\phi_g$, $\forall g \in \mathbf{G}$, will then define the canonical lifting of the action of $\mathbf{G}$ to an action on TQ, which is of course linear on fibers, and locally given by:

(2.8) $$T\phi_g : (q^i, u^i) \mapsto \left(\phi_g^i(q), \frac{\partial \phi_g^i}{\partial q^j} u^j \right) \ .$$

$\mathbf{G}$ will then act on Q as a group of point transformations, in the sense specified in Sect.1. More general actions can be defined, but we will be mainly concerned here with point transformations (and hence with lifted actions).

[1] For a definition in intrinsic terms see [33,20,16,42]

Let now $X_i \in \mathcal{X}(Q)$, $i = 1, \ldots, k = \dim \mathbf{G}$, be the infinitesimal generators of the action of $\mathbf{G}$ on Q:

$$X_i = X_i^j(q) \frac{\partial}{\partial q^j} , \tag{2.9}$$

locally. The generators of the lifted action on TQ will be given by the *complete lifts* X_i^c of the X_i's, locally given by:

$$X_i^c = X_i^j(q) \frac{\partial}{\partial q^j} + (L_{\Gamma_0} X_i^j) \frac{\partial}{\partial u^j} , \tag{2.10}$$

with Γ_0 any second order vector field [58].[2] The identity component of the group $\mathbf{G}$ will be a symmetry group for a given second order dinamics Γ iff

$$[X_i^c, \Gamma] = 0 . \tag{2.11}$$

Indeed, in the notation of Sect.1, considering the infinitesimal transformation associated with X_i, one can easily prove that:

$$[X_i^c, \Gamma] = \left[\delta_i f^j - L_\Gamma \delta_i u^j\right] \frac{\partial}{\partial u^j} , \tag{2.12}$$

and hence (2.11) holds iff (1.6) does. A Lagrangian $\mathcal{L}$ for Γ will be invariant under the identity component of $\mathbf{G}$ iff

$$L_{X_i} \mathcal{L} = 0 . \tag{2.13}$$

After these preliminaries, let's briefly discuss the Inverse Problem. Although we will be mainly interested in time independent dynamics, in view of future discussion of groups of transformations (like the Poincaré or the Galilei groups), which explicitely involve the time, it will be useful to cast the discussion of the problem in the context of the time dependent formalism [1,75]. On $TQ \times \mathbf{R}$, locally coordinatized by (q, u, t), we extend the definition of Γ to the vector field:

$$\tilde{\Gamma} = \Gamma + \frac{\partial}{\partial t} . \tag{2.14}$$

This is easily seen to yield the equations of motion (1.1), plus the additional equation:

$$dt/dt = 1 . \tag{2.15}$$

which fixes the parametrization. A basis of one-forms on $TQ \times \mathbf{R}$ is given by dt and by the $2n$ one-forms:

$$\alpha^i = dq^i - u^i dt \quad ; \quad \beta^i = du^i - f^i dt . \tag{2.16}$$

It is then easily seen that $\tilde{\Gamma}$ is the unique vector field satisfying:

$$i_{\tilde{\Gamma}} \alpha^i = i_{\tilde{\Gamma}} \beta^i = 0 \quad ; \quad i_{\tilde{\Gamma}} dt = 1 \tag{2.17}$$

Let now $\mathcal{L} \in \mathcal{F}(TQ)$ be given.[3] With a given $\mathcal{L}$ we can associate the 'energy function':

$$E_\mathcal{L} \stackrel{\text{def}}{=} (L_\Delta - 1)\mathcal{L} , \tag{2.18}$$

and the *Cartan one-form*[17,18]:

$$\tilde{\theta}_\mathcal{L} \stackrel{\text{def}}{=} \frac{\partial \mathcal{L}}{\partial u^i} dq^i - E_\mathcal{L} dt \tag{2.18}$$

In order to write both $\tilde{\theta}_\mathcal{L}$ and the Euler-Lagrange equations in a coordinate free setting, we need to comment briefly on the definiton of a few more operations on forms.

---

[2] With slight abuse of notation, we write here X_i^j for what should be $\pi^* X_i^j$, with $\pi : TQ \to Q$ the canonical projection.

[3] In a genuine time dependent formalism, one should consider $\mathcal{L} \in \mathcal{F}(TQ \times \mathbf{R})$, but we will not do that here.

Coordinate-free expression of the Euler-Lagrange equations

It is known [42,31] that one can associate with every (1-1) tensor field T on any manifold M a derivation of degree zero on the algebra of forms, δ_T, defined as:

$$(\delta_T \omega)(X_1, \ldots, X_k) \overset{\text{def}}{=} \omega(T(X_1), \ldots, X_k) + \cdots + \omega(X_1, \ldots, T(X_k)) , \tag{2.20}$$

$\forall \omega \in \Lambda^k(M)$. Through δ_T, one can define an antiderivation d_T as:

$$d_T \overset{\text{def}}{=} d\,\delta_T - \delta_T d , \tag{2.21}$$

whose main properties are:

$$d_T d + d\, d_T = 0 , \tag{2.22}$$

and

$$d_T d_T = 0 , \tag{2.23}$$

the latter holding iff the Nijenhuis tensor N_T [66] associated with T vanishes. In particular, on functions:

$$(d_T f)(X) = df(T(X)) \quad ; \quad \forall X \in \mathcal{X}(M) \tag{2.24}$$

If $M = TQ$, and we consider the vertical endomorphism, it is then known that $N_S = 0$ [66]. Hence, d_S satisfies the homotopy identity (2.23) and, in coordinates:

$$d_S f = \frac{\partial f}{\partial u^i} dq^i \tag{2.25}$$

In intrinsic terms, we can then also write:

$$\tilde{\theta}_{\mathcal{L}} = d_S \mathcal{L} - E_{\mathcal{L}}\, dt . \tag{2.26}$$

An easy calculation in local coordinates shows that, for a second order vector field, in the extended form (2.14),

$$L_{\tilde{\Gamma}} \tilde{\theta}_{\mathcal{L}} - d\mathcal{L} = \left[L_{\tilde{\Gamma}} \left(\frac{\partial \mathcal{L}}{\partial u^i} \right) - \frac{\partial \mathcal{L}}{\partial q^i} \right] dq^i - \left(L_{\tilde{\Gamma}} E_{\mathcal{L}} + \frac{\partial \mathcal{L}}{\partial t} \right) dt . \tag{2.27}$$

Therefore, the equation:

$$L_{\tilde{\Gamma}} \tilde{\theta}_{\mathcal{L}} - d\mathcal{L} = 0 \tag{2.28}$$

yields both the Euler-Lagrange equations:

$$L_{\tilde{\Gamma}} \left(\frac{\partial \mathcal{L}}{\partial u^i} \right) - \frac{\partial \mathcal{L}}{\partial q^i} = 0 , \tag{2.29}$$

and, if $\mathcal{L}$ is time independent, the additional equation[4] :

$$L_{\tilde{\Gamma}} E_{\mathcal{L}} = 0 \tag{2.30}$$

---

[4] The additional term in $\tilde{\Gamma}$, i.e. $\partial/\partial t$, is ineffective here. Whether or not this is the case, $L_{\tilde{\Gamma}}$ acts always as a total time derivative.

expressing the fact that the 'energy' $E_{\mathcal{L}}$ is a constant of the motion. For a time dependent $\mathcal{L}$, one would instead obtain:

$$L_{\tilde{\Gamma}} E_{\mathcal{L}} + \frac{\partial \mathcal{L}}{\partial t} = 0 \; . \tag{2.30}$$

In both cases, however, the additional equation, as well known, is a consequence of the equations of motion (2.29). Eq.(2.28) also implies:

$$i_{\tilde{\Gamma}} \tilde{\theta}_{\mathcal{L}} = \mathcal{L} \; . \tag{2.31}$$

Starting from the Cartan form, we can define the exact two-form

$$\tilde{\omega}_{\mathcal{L}} \overset{\text{def}}{=} d\tilde{\theta}_{\mathcal{L}} \equiv d\, d_{s} \mathcal{L} - dE_{\mathcal{L}} \wedge dt \tag{2.32}$$

and (2.28) becomes equivalent to:

$$i_{\tilde{\Gamma}} \tilde{\omega}_{\mathcal{L}} = 0 \; . \tag{2.33}$$

Remark: Defining:

$$\omega_{\mathcal{L}} \overset{\text{def}}{=} d_S \, d\mathcal{L} = -d\theta_{\mathcal{L}} \quad ; \quad \theta_{\mathcal{L}} \overset{\text{def}}{=} d_S \, \mathcal{L} \; , \tag{2.34}$$

we obtain, for a time independent $\mathcal{L}$, using (2.30), the equivalent expression:

$$i_{\Gamma} \omega_{\mathcal{L}} = dE_{\mathcal{L}} \; , \tag{2.35}$$

which is the usual 'canonical' form [58] of the Euler-Lagrange equations on TQ in the time independent formalism.

If $\mathcal{L}$ is regular (see Sect.1), $\tilde{\omega}_{\mathcal{L}}$ is an exact contact form on $TQ \times \mathbf{R}$. Its kernel being one dimensional, Eq.(2.33), together with the additional condition $i_{\tilde{\Gamma}}\, dt = 1$, uniquely fixes $\tilde{\Gamma}$, which turns out also to be second order. Also, $\tilde{\omega}_{\mathcal{L}}$ is invariant under the flow generated by $\tilde{\Gamma}$:

$$L_{\tilde{\Gamma}} \tilde{\omega}_{\mathcal{L}} = 0 \; , \tag{2.36}$$

and, by construction, vanishes when contracted with pairs of vertical (with respect to the canonical projection) fields:

$$\tilde{\omega}_{\mathcal{L}}(S(X), S(Y)) = 0 \quad ; \quad \forall X, Y \in \mathcal{X}(TQ \times \mathbf{R}) \; . \tag{2.37}$$

Having clarified our general setting, we shall go now to the construction of Lagrangian descriptions, if they exist.

Looking for Lagrangian descriptions

Let's now turn to the converse problem, i.e., eventually, to that of finding the Lagrangian descriptions (if they exist at all) for a given second order system. Let a second order field $\Gamma \in \mathcal{X}(TQ)$ be given, together with its extension $\tilde{\Gamma} \in \mathcal{X}(TQ \times \mathbf{R})$,and let's build up the one-forms α^i and β^i defined in Eq.(2.16). We can then construct the two-form:

$$\tilde{\omega}_{AB} = A_{ij} \alpha^i \wedge \beta^j + B_{ij} \alpha^i \wedge \alpha^j \; , \tag{2.38}$$

with $A_{ij}, B_{ij} \in \mathcal{F}(TQ)$. Then, by construction, (2.33) and (2.37) are satisfied. if now A and B are chosen (if possible) in such a way that $\tilde{\omega}_{AB}$ is also closed:

$$d\tilde{\omega}_{AB} = 0 \ , \tag{2.39}$$

then (2.36) will be satisfied as well. If we now split off explicitly the terms in (2.38), which contain dt, $\tilde{\omega}_{AB}$ can be written as:

$$\tilde{\omega}_{AB} = -\omega - \alpha \wedge dt \ , \tag{2.40}$$

where the r.h.s. defines both the two-form ω and the one-form α, which are both pull-backs of forms on TQ[5] . Of course, ω alone satisfies already (2.37). Moreover, (2.33) and (2.39) split into separate conditions on ω and α, namely:

$$d\omega = 0 \quad ; \quad i_\Gamma \omega = \alpha \ , \tag{2.41}$$

for ω, and

$$d\alpha = 0 \quad ; \quad i_\Gamma \alpha = 0 \ , \tag{2.42}$$

for α. As a consequence of (2.41-42)

$$L_\Gamma \omega = 0 \tag{2.43}$$

holds as well. Now, a theorem, first proved by Balachandran et al. [8] (but see also [19,42,69,71]), states that given a closed two-form on TQ, vanishing on any pair of vertical bivectors and satisfying (2.43), for a given second order field Γ, a Lagrangian $\mathcal{L}$ can always be found, locally at least, such that:

$$\omega = \omega_{\mathcal{L}} = d_S d\mathcal{L} \ , \tag{2.44}$$

(cfr.(2.34)) and that, moreover, $\mathcal{L}$ can always be modified, by the addition of a function of the q's alone, in such a way that

$$\alpha = dE_{\mathcal{L}} \ , \tag{2.45}$$

where the 'energy function' $E_{\mathcal{L}}$ might be defined only locally. The $\mathcal{L}$ found in this way is then an admissible Lagrangian for Γ [54].

We would like to comment here on various aspects of the theorem. The procedure outlined here for going from the equations of motion to the Lagrangian can be summarized in the following sequence:

$$\Gamma \to \tilde{\omega} \to \tilde{\theta} \to \mathcal{L} \ , \tag{2.46}$$

i.e.:

i) We first build up $\tilde{\omega}$ (as in (2.38) say), from the datum of the forces in Γ, and then try to adjust the integrating factors, i.e. the ($\mathcal{F}(TQ)$-valued) matrices A_{ij} and B_{ij}, in order to ensure the closure of $\tilde{\omega}$. Whether or not this is possible

[5] On $TQ \times \mathbf{R}$ there exist two obvious projections: $\pi_1 : TQ \times \mathbf{R} \to TQ$ and $\pi_2 : TQ \times \mathbf{R} \to \mathbf{R}$. Pull-back of forms from TQ is then via π_1

depends on the structure of the forces. Examples of second order dynamical systems which do not allow for any such two-form are given, e.g., in [62,66].

ii) Assuming that a closed two-form $\tilde{\omega}$ has been found, whether or not we can 'integrate' it to obtain a globally defined one-form $\tilde{\theta}$, s.t. $\tilde{\omega} = d\tilde{\theta}$, will depend on the cohomology class of $\tilde{\omega}$ in $H^2(TQ \times \mathbf{R}) \simeq H^2(Q)$. As α, in (2.40), is closed, the problem is actually entirely a problem in the cohomology class of ω, as we can always write:

$$\tilde{\omega}_{AB} = -\omega + d(t\alpha) \,, \tag{2.47}$$

and $t\alpha$ is a perfectly well defined one-form on $TQ \times \mathbf{R}$.

iii) On going from the one-form $\tilde{\theta}$ to the Lagrangian appropriate for Γ, we have further to integrate the one-form α, in order to eventually obtain (2.45), so that finally $\tilde{\theta}$ can be represented as in Eq.(2.26). A further obstruction to the complete global solution of the problem comes then from $H^1(TQ \times \mathbf{R}) \simeq H^1(Q)$.

All the other steps in the proof of the theorem involve only integrations along the fibers of TQ, which cannot give rise to further topological obstructions. It is only when $H^1(Q) = H^2(Q) = 0$ that we can be sure that no topological obstructions will arise and that a global Lagrangian can be defined (if any, i.e. if step i) above presents no problems). As will be seen in a short while, this is precisely the reason why in the charge monopole system no global Lagrangian (actually, no global Cartan form) can be defined.

Once a Lagrangian $\mathcal{L} \in \mathcal{F}(TQ)$ is given, the steps leading from $\mathcal{L}$ to the associated two-form $\omega_{\mathcal{L}}$ can be given entirely in terms of the exterior derivatives d_S and d, as

$$\begin{array}{ccc} \mathcal{L} & \xrightarrow{d_S} & \theta_{\mathcal{L}} \\ {\scriptstyle d}\downarrow & & \downarrow{\scriptstyle -d} \\ d\mathcal{L} & \xrightarrow{d_S} & \omega_{\mathcal{L}} \end{array} \tag{2.48}$$

and the diagram commutes, in view of Eq.(2.22).

It follows at once, from $d_S^2 = 0$, that

$$d_S\,\theta_{\mathcal{L}} = d_S\,\omega_{\mathcal{L}} = 0 \,. \tag{2.49}$$

One might wonder whether or not a two-form ω s.t. $d_S\,\omega = 0$, can be written as $d_S\,\theta$, for some one-form θ, and similarly in the case of one-forms, i.e. whether Poincaré Lemma can be established for d_S. We now discuss briefly this point, before coming again back to the additional problems connected with groups of symmetries.

A Poincaré Lemma for d_S

We recall that $d_S^2 = 0$, as the Nijenhuis tensor of S vanishes. In general, if T is a (1-1) tensor and $N_T = 0$, then

$$\alpha = d_T\beta \Rightarrow d_T\alpha = 0 \ , \tag{2.50}$$

because of $d_T^2 = 0$, but the converse fails to be true (even at the local level), unless T is invertible [66], which is not the case for the vertical endomorphism, as S is nilpotent:

$$S^2 = 0 \ , \tag{2.51}$$

a result which follows at once from the definition (2.1). For instance, on $T\mathbf{R}^2$, the two-form $\alpha = (1/2)\alpha_{ij}du^i \wedge du^j$, $\alpha_{ij} \in \mathbf{R}$, can be easily proved to be 'd_S-closed' ($d_S\,\alpha = 0$), but cannot be 'd_S-exact', i.e. there is no one-form θ s.t. $\alpha = d_S\,\theta$ [66]. However, it turns out that a (complete) Poincaré Lemma can be established for S as well, by imposing suitable restrictions on the algebra of forms on which it operates, in order to get rid of the previous counterexample. It will be enough for our purposes to establish it here for one and two-forms. Then:

i)

Let θ be a semi-basic one-form, i.e.:

$$\theta(S(X)) = 0 \quad ; \quad \forall X \in \mathcal{X}(TQ) \ , \tag{2.52}$$

and let

$$d_S\,\theta = 0 \ . \tag{2.53}$$

One can easily prove [66] that, in general,

$$d_S\,\theta(X,Y) = (L_{S(X)}\theta)(Y) - (L_{S(Y)}\theta)(X) + \theta(S([X,Y])) \ . \tag{2.54}$$

Hence, (2.52-53) together imply:

$$(L_{S(X)}\theta)(Y) = (L_{S(Y)}\theta)(X) \quad ; \quad \forall X,Y \in \mathcal{X}(TQ) \ . \tag{2.55}$$

In particular, if $X = \partial/\partial q^i$, $Y = \partial/\partial q^j$, we obtain:

$$(L_{\frac{\partial}{\partial u^i}}\theta)\left(\frac{\partial}{\partial q^j}\right) = (L_{\frac{\partial}{\partial u^j}}\theta)\left(\frac{\partial}{\partial q^i}\right) \ . \tag{2.56}$$

If θ is semi-basic, it can be written, in local coordinates, as

$$\theta = A_i\,dq^i \quad ; \quad A_i \in \mathcal{F}(TQ) \ , \tag{2.57}$$

then (2.56) implies:

$$\frac{\partial A_j}{\partial u^i} = \frac{\partial A_i}{\partial u^j} \ . \tag{2.58}$$

The latter is an integrability condition along the fibers, implying (at least locally on the base manifold) the existence of a function $\mathcal{L} \in \mathcal{F}(TQ)$ s.t.:

$$A_i = \frac{\partial \mathcal{L}}{\partial u^i} \ , \tag{2.59}$$

and hence:

$$\theta = d_S \mathcal{L} \ . \quad \blacksquare \tag{2.60}$$

The above result has the following, obvious,

Corollary:

A one-form θ is a Cartan form[6] iff it is semi-basic and satisfies $d_S \theta = 0$. ∎

ii)

Let now ω be a two-form satisfying $d_S \omega = 0$ *and* the additional condition:

$$\omega(S(X), S(Y)) = 0 \quad ; \quad \forall X, Y \in \mathcal{X}(TQ) \ . \tag{2.61}$$

We shall proceed directly in local coordinates. Eq.(2.61) implies that ω is of the form:

$$\begin{gathered} \omega = A_{ij} dq^i \wedge du^j + \frac{1}{2} B_{ij} dq^i \wedge dq^j \\ A_{ij}, B_{ij} \in \mathcal{F}(TQ) \ , \ B_{ij} = -B_{ji} \ . \end{gathered} \tag{2.62}$$

We then find:

$$\begin{aligned} d_S \omega &= \\ &= \frac{1}{2} \left[\frac{\partial A_{ij}}{\partial u^k} - \frac{\partial A_{kj}}{\partial u^i} \right] dq^k \wedge dq^i \wedge du^j + \\ &\quad \frac{1}{6} \left[\frac{\partial B_{ij}}{\partial u^k} + \text{cycl. perm. of } (i, j, k) \right] dq^i \wedge dq^j \wedge dq^k \ . \end{aligned} \tag{2.63}$$

If $d_S \omega = 0$, the two terms on the r.h.s. must vanish separately. This implies the existence of functions D_j, $C_j \in \mathcal{F}(TQ)$, s.t.:

$$A_{ij} = \frac{\partial D_j}{\partial u^i} \quad ; \quad B_{ij} = \frac{\partial C_j}{\partial u^i} - \frac{\partial C_i}{\partial u^j} \ , \tag{2.64}$$

whence:

$$\omega = d_S \theta^* \quad ; \quad \theta^* = C_i dq^i + D_j du^j \ . \quad \blacksquare \tag{2.65}$$

Poincaré Lemma has then be proved to hold true for one and two-forms which vanish when contracted with the maximal possible number of vertical fields. Let's observe that, for one-forms, condition (2.52) is equivalent to $\delta_S \theta = 0$. In the case of two-forms:

$$(\delta_S \omega)(X, Y) = \omega(S(X), Y) + \omega(X, S(Y)) \ , \tag{2.66}$$

and the condition

$$\delta_S \omega = 0 \tag{2.67}$$

[6] We recall that, in the time independent formalism, $\theta_{\mathcal{L}} = d_S \mathcal{L}$ is also called the Cartan form associated with $\mathcal{L}$.

is stronger than (2.61). Let's investigate its consequences. It can be seen at once (and it is also well known) that it implies symmetry of the matrix $\|A_{ij}\|$. But this, in conjunction with the first of (2.64), yields a further integrability contition on the D_j's, namely:

$$\frac{\partial D_j}{\partial u^i} = \frac{\partial D_i}{\partial u^j} , \tag{2.68}$$

whence:

$$D_i = \frac{\partial \tilde{\mathcal{L}}}{\partial u^i} , \tag{2.69}$$

for some $\tilde{\mathcal{L}} \in \mathcal{F}(TQ)$, and hence:

$$\theta^* = d\tilde{\mathcal{L}} + \left(C_i - \frac{\partial \tilde{\mathcal{L}}}{\partial q^i} \right) dq^i . \tag{2.70}$$

Let's now assume ω to be also closed[7] , i.e.:

$$d\omega = 0 . \tag{2.71}$$

With ω in the form (2.62), Eq.(2.71) is readily seen to imply:

$$\omega = -d\theta \quad ; \quad \theta = A_i\, dq^i , \tag{2.72}$$

i.e. that θ is semi-basic. Comparing (2.72) with (2.65) and (2.70), we find:

$$\frac{\partial A_k}{\partial u^i} = \frac{\partial^2 \tilde{\mathcal{L}}}{\partial u^i \partial u^k} , \tag{2.72}$$

and

$$\frac{\partial A_k}{\partial q^i} - \frac{\partial A_i}{\partial q^k} = \frac{\partial C_i}{\partial u^k} - \frac{\partial C_k}{\partial u^i} . \tag{2.73}$$

The first set of equations integrates to:

$$A_i = \frac{\partial \tilde{\mathcal{L}}}{\partial u^i} + g_i(q) , \tag{2.74}$$

where the g_i's are integration 'constants'. Eq.(2.74) can also be rewritten as:

$$A_i = \frac{\partial \mathcal{L}}{\partial u^i} \quad ; \quad \mathcal{L} \stackrel{\text{def}}{=} \tilde{\mathcal{L}} + u^i g_i , \tag{2.75}$$

and, apart from terms which vanish under the action of d_S, the second set yields:

$$C_i = \frac{\partial \mathcal{L}}{\partial q^i} . \tag{2.76}$$

We have thus proved the

[7] Using (2.21), we see that $d_S\,\omega = \delta_S\,\omega = 0$ only imply $\delta_S\,(d\omega) = 0$, but that $d\omega = \delta_S\,\omega = 0$ directly imply $d_S\,\omega = 0$.

Theorem:

Let ω be a *closed* two form, satisfying

(2.77) $$\delta_S \omega = 0 .$$

Then a function $\mathcal{L}$ can be found locally s.t.:

(2.78) $$\omega = \omega_{\mathcal{L}} = -d\theta_{\mathcal{L}} \quad ; \quad \theta_{\mathcal{L}} = d_S \mathcal{L} ,$$

i.e. such that ω is the Lagrangian two-form associated with $\mathcal{L}$. ∎

Compared with the theorem of Balachandran et al. [8], one sees that the single condition (2.77) incorporates both the condition (2.61) and that of the existence of a second order field which annihilates ω.

Looking for invariant Lagrangian descriptions

After this rather long digression, let's now try and state what is the main object of the present paper. We take the sequence (2.46) as summarizing the essential content of the IP. Given now a Lie group G, acting (via its lifted action) on TQ as a group of symmetries of Γ, we can state what might be called the *Equivariant Inverse Problem* (EIP) [4,5] as the requirement that *as many as possible* (in the sense discussed in Sect.1) *of the original symmetries be retained at each step in the procedure of solution, if any, of the IP.*

Of course, as, for any complete lift X^c,

(2.79) $$L_{X^c}\theta_{\mathcal{L}} = \theta_{(L_{X^c}\mathcal{L})} \Rightarrow L_{X^c}\omega_{\mathcal{L}} = \omega_{(L_{X^c}\mathcal{L})} ,$$

the residual symmetries found at each step will also be symmetries for the previous ones:

$$L_{X^c}\mathcal{L} = 0 \Rightarrow L_{X^c}\theta_{\mathcal{L}} = 0 \Rightarrow L_{X^c}\omega_{\mathcal{L}} = 0 ,$$

but not viceversa, i.e. we must in general expect some amount of 'symmetry breaking' at each step of the procedure. Therefore, the 'if any' part of the definition of the EIP given above can only be strenghtened by the further requirements we are imposing here, and one can exhibit plenty of examples (some of them have already been given in Sect.1) in which no final solution exists for the EIP, while the standard IP is perfecly well soluble.

As a simple example in which the EIP has no solution under the full group of symmetrics for the dynamics, already at the first step (from Γ to $\tilde{\omega}$), we may consider the case in which the 'forces' f^i in (2.4) are homogeneous of degree one in both the positions and the velocities. This includes the free particle as well as the simple and the damped harmonic oscillators. Then Γ is invariant under the group of simultaneous dilations in positions and velocities, generated by the vector field:

(2.80) $$\Delta^c = q^i \frac{\partial}{\partial q^i} + u^i \frac{\partial}{\partial u^i} ,$$

but no two-form of the type (2.38) can be invariant under the action of Δ^c, as it would require the A_{ij}'s and B_{ij}'s to be homogeneous functions of degree -2, which they cannot be, being assumed everywhere smooth on TQ. In view of the discussion in the previous Subsection, in which each step of the sequence (2.46) has been seen to be connected to the Poincaré Lemma (for either d or d_S), we may also say that what we are looking for, in the presence of symmetries, is what might be called an *Equivariant Poincaré Lemma* for d and/or for d_S.

We may now reexamine the example discussed in Sect.1 on the light of the previous considerations. Restricting for simplicity to $Q = \mathbf{R}^2$, with Cartesian coordinates (q^1, q^2), the motion in a uniform electric field $\vec{E} \equiv (E^1, E^2)$ is described (in the same units as in Sect.1) by the vector field:

$$\Gamma = u^i \frac{\partial}{\partial q^i} + E^i \frac{\partial}{\partial u^i} , \tag{2.81}$$

which is translationally invariant. A possible choice for the forms ω and α (cfr. Eq.(2.40)) is:

$$\omega = \delta_{ij} dq^i \wedge du^j \quad ; \quad \alpha = \delta_{ij}(u^i du^j - E^i dq^j) . \tag{2.82}$$

With this choice, both ω and α are again translationally invariant. The invariance is however broken at the next step of the integration procedure, as we can recast $\tilde{\omega} = -\omega - \alpha \wedge dt$ in the form (2.32), but with:

$$\begin{aligned} \tilde{\theta}_{\mathcal{L}} &= \delta_{ij} u^i dq^j - \left(\frac{1}{2}|\vec{q}|^2 - \vec{E} \cdot \vec{u}\right) dt \\ \mathcal{L} &= \frac{1}{2}|\vec{u}|^2 + \vec{E} \cdot \vec{q} . \end{aligned} \tag{2.83}$$

Hence, neither $\tilde{\theta}_{\mathcal{L}}$ nor the Lagrangian are translationally invariant. Note that, in the time independent formalism, we should consider only the two-form $\omega = d\theta$, with $\theta = \delta_{ij} u^i dq^j$, again translationally invariant, and the symmetry breaking would show up only at the last step of the integration procedure, as the Lagrangian would turn out to be the same as in (2.83), of course.

A similar situation shows up in the case of motion of a charged particle in a plane perpendicular to a constant, uniform magnetic field, described, on $T\mathbf{R}^2$, by:

$$\Gamma = u^i \frac{\partial}{\partial q^i} + Bu^2 \frac{\partial}{\partial u^1} - Bu^1 \frac{\partial}{\partial u^2} , \tag{2.84}$$

Let's work directly in the time independent formalism. Then the two-form can be chosen as:

$$\omega = \delta_{ij} dq^i \wedge du^j - B dq^1 \wedge dq^2 , \tag{2.85}$$

In force of the previously quoted theorem by Balachandran et al., ω is a Lagrangian two-form, i.e.:

$$\omega = -d\theta_{\mathcal{L}} \quad ; \quad \theta_{\mathcal{L}} = d_S \mathcal{L} , \tag{2.86}$$

with:

$$\theta_{\mathcal{L}} = \delta_{ij} u^i dq^j - Bq^1 dq^2 \quad ; \quad \mathcal{L} = \frac{1}{2}|\vec{u}|^2 - Bq^1 u^2 . \tag{2.87}$$

Moreover, with $E = 1/2|\vec{u}|^2$:

$$i_\Gamma \omega = dE . \tag{2.88}$$

Therefore, translational symmetry is broken already at the level of the Lagrangian one-form. One may easily convince oneself that this pathology cannot be taken care of (for $B \neq 0$) by adding to $\theta_{\mathcal{L}}$ an arbitrary closed (and hence basic) one-form, for this would simply amount to performing a gauge transformation.

As a last (and less elementary) example, let's consider the motion of a charged particle in the field of a magnetic monopole. In this case, $Q = \mathbf{R}^3 - \{0\} \simeq S^2 \times \mathbf{R}$ and the dynamical vector field is given by:

$$\Gamma = u^i \frac{\partial}{\partial q^i} + n\epsilon^i{}_{jk} u^j \frac{q^k}{r^3} \frac{\partial}{\partial u^i} , \tag{2.89}$$

where $n = 2eg$, g being the monopole charge, and $r^2 = (q^1)^2 + (q^2)^2 + (q^3)^2$.

Working again in the time independent formalism, the two-form ω can be chosen as:

$$\omega = \delta_{ij} dq^i \wedge du^j + \frac{n}{2r^3} \epsilon_{ijk} q^i dq^j \wedge dq^k , \tag{2.90}$$

and ω satisfies all the requirements for being a candidate for a Lagrangian two-form. Moreover, with $E = 1/2|\vec{u}|^2$, again:

$$i_\Gamma \omega = dE . \tag{2.91}$$

Both Γ and the two-form ω are invariant under $SO(3)$, so, if we take $\mathbf{G} = SO(3)$, the symmetry is unbroken on going from Γ to ω. However, it is quite obvious that ω cannot be exact. Indeed, the second term in (2.89) is proportional to the element of solid angle on the two-sphere, and hence it is closed but not exact. What we are facing here is therefore an example of a topological obstruction coming from $H^2(Q) = H^2(S^2 \times \mathbf{R}) \simeq H^2(S^2)$, which is nontrivial. It is only at a strictly local level that we can find a one-form θ such that $\omega = d\theta$, and θ cannot be rotationally invariant for, if it were, we could, by averaging over the group, obtain a globally defined one-form. Therefore, the breaking of the symmetry is originated by the nontrivial global topological properties of the configuration space. To be specific, a one-form A such that:

$$dA = \frac{n}{2r^3} \epsilon_{ijk} q^i dq^j \wedge dq^k \tag{2.92}$$

can only be found in each one of the components of a contractible open covering of Q. For instance, let

$$Q = U_+ \cup U_- \quad ; \quad U_\pm = \{(S^2 - (q^3 = \mp 1)) \times \mathbf{R}\} . \tag{2.93}$$

On $U_\pm$ we have (modulo a closed form):

$$A_\pm = \frac{\pm n}{r(r \pm q^3)} (q^1 dq^2 - q^2 dq^1) , \tag{2.94}$$

or, in polar coordinates, with the polar axis along the q^3-axis:

$$A_\pm = \pm n(1 \mp \cos\theta) d\phi . \tag{2.95}$$

A_+ is singular along a 'Dirac string', consisting in the negative q^3-axis (which however doesn't belong to U_+). Similarly A_- is singular along the positive q^3-axis. On the overlap region $U_+ \cap U_-$:

$$A_+ - A_- = 2n \frac{q^1 dq^2 - q^2 dq^1}{(q^1)^2 + (q^2)^2} = 2n\, d\phi , \tag{2.96}$$

and hence the two one-forms differ by a closed one (i.e., the corresponding vector potentials are related by a gauge transformation).

The Lagrangian can likewise be defined only locally, i.e. as:

$$\mathcal{L}_{\pm} = \frac{1}{2}|\vec{u}|^2 + i_\Gamma A_\pm \quad ; \quad \mathcal{L}_\pm \in \mathcal{F}(TU_\pm) \; . \tag{2.97}$$

The residual rotational freedom which is left is that of $SO(2)$ (rotations around the direction of the 'string'). As for other rotations, consider say, those around the q^1-axis, which are generated by the vector field:

$$X_1 = -\sin\phi \frac{\partial}{\partial\theta} - \cos\phi \cot\theta \frac{\partial}{\partial\phi} \; . \tag{2.98}$$

With some algebra, we find:

$$L_{X_1} \mathcal{L}_+ = \dot{f}_1 \overset{\text{def}}{=} i_\Gamma \pi^* df_1 \; , \tag{2.99}$$

or, equivalently:

$$L_{X_1} A_+ = \pi^* df_1 \; , \tag{2.100}$$

where $f_1 \in \mathcal{F}(U_+)$ is given by:

$$f_1 = \frac{\sin\theta\cos\phi}{1+\cos\theta} = \frac{q^1}{r+q^3} \; , \tag{2.101}$$

with similar results for $\mathcal{L}_-$ and for other rotations.

Under the action of $SO(3)$, the Cartan one-form changes then by a closed (actually exact) one-form, and the Lagrangian by a total time derivative.

What is relevant about this last example is that, as has been proved by two of us [60,61], the Lagrangians (2.96), in which the charged particle is 'minimally coupled' to the magnetic field, are actually *unique*, up to gauge transformations.

In all of the cases considered here, we have reached a final situation which can be summarized as follows:

The original, second order, dynamical vector field Γ is invariant under the lifted action of a Lie group G of Q-transformations, whose action on TQ is generated by the vector fields (2.10). In going, if possible, from Γ to its Lagrangian description, which, as already said, we take as the starting point for the quantization procedure, we are naturally led to pose an Equivariant Inverse Problem for Γ. Most often, in this process not only do we find obstructions, generated by the de Rham cohomology of the configuration space (attached either to d or, with the proper restrictions on forms, to d_S), but also additional obstructions, originated by the requirement of equivariance.

Remark: In gauge field theories, one often starts with a total space $\mathcal{A}$, which is a space of connections, equipped with the action of a given gauge group G. If $M = \mathcal{A}/\mathrm{G}$ has a manifold structure, $\mathcal{A}$ can be viewed as a principal G-bundle over M, which is then taken as the physical configuration space of the theory. Obstructions in the construction of a gauge invariant action, which give rise to Wess-Zumino terms and anomalies, are then viewed as due to a nontrivial topology of M, though they ultimately arise from the requirement of gauge invariance (i.e. from the equivariant Poincaré Lemma). This situation prevails every time we are considering the equivariant Poincaré Lemma on a contractible space.

After having considered few examples, we try to give here conditions to be imposed on A_{ij} and B_{ij} (appearing in (2.38)) for ω_{AB} to be G-invariant.

We denote by X_a, $a \in \{1, \ldots, n\}$, the infinitesimal generators of the action of G on the configuration space Q. By X_a^c we denote the complete lift to TQ and by X_a^v the vector field $S \cdot X_a^c$. These vector fields have the following commutation relations:

$$[X_r^c, X_s^c] = c_{rs}^t X_t^c \quad ; \quad [X_r^c, X_s^v] = c_{rs}^t X_t^v \quad ; \quad [X_s^v, X_r^v] = 0 \; .$$

In a canonical chart (q^i, u^i) for TQ, the local expression of these fields is provided by:

$$X_r^c = X_r{}^i \frac{\partial}{\partial q^i} + \frac{\partial X_r{}^i}{\partial q^j} u^j \frac{\partial}{\partial u^i} \quad ; \quad X_r^v = X_r{}^i \frac{\partial}{\partial u^i} \; .$$

When we write (see 2.38)

$$\omega_{AB} = A_{ij} \alpha^i \wedge \beta^j + B_{ij} \alpha^i \wedge \alpha^j \; ,$$

asociated with a dynamical field Γ, to look for invariant ω_{AB} we have to impose that $L_{X_r^c} \omega_{AB} = 0$ for any $r \in \{1, \ldots, n\}$ in addition to $d\omega_{AB} = 0$.

Let us define two matrices associated with the group G and ω_{AB}, namely:

$$\omega_{rs} = \omega_{AB}(X_r^c, X_s^v)$$

and

$$\alpha_{rs} = \omega_{AB}(X_r^c, X_s^c) \; .$$

These matrices are related to A_{ij} and B_{ij} by the following relations:

$$\begin{aligned} \omega_{rs} &= A_{ij} X_r^i X_s^j \\ \alpha_{rs} &= B_{ij}(X_r^i X_s^j - X_r^j X_s^i) + A_{ij}(X_r^i L_\Gamma X_s^j - X_s^j L_\Gamma X_r^i) \; . \end{aligned}$$

It is clear that A_{ij} can be derived from ω_{rs} if X_r^i are invertible matrices, or the group G is locally homeomorphic to Q. When Q is just a homogeneous space, we can try the procedure we shall consider around formula (2.120). In the general case, only the 'projection' of ω_{AB} along the tangent bundle of the orbits of G on Q will be restricted. We find:

$$L_{X_k^c} \omega_{rs} = c_{kr}^m \omega_{ms} + c_{ks}^m \omega_{rm} \qquad L_{X_k^c} \alpha_{rs} = c_{kr}^m \alpha_{ms} + c_{ks}^m \alpha_{rm}$$

and, because of $\delta_S \omega_{AB} = 0$, we get that ω_{rs} transforms as a symmetric 2-tensor and α_{rs} transforms as a skew-symmetric 2-tensor, under the adjoint representation of the Lie algebra of G.

In many practical situations, it is possible to find solutions of these equations by using an educated guess. For instance, when the rotation group acts on $\mathbf{R}^3$ in the standard way, the most general invariant two-form can be derived by noticing that both $\Delta = \frac{x^i}{r} \frac{\partial}{\partial x^i}$ and $\Omega = dx^1 \wedge dx^2 \wedge dx^3$ are invariant under rotations (for the connected component $SO(3)$). Thus the most general invariant two-form has the expression:

$$\omega_\lambda = \lambda(r) i_\Delta \Omega \; .$$

Similarly, invariant under X_a^c on $T(\mathbf{R}^3 - \{0\})$ would be the two-form:

$$\omega_F = F(r^2, \vec{r} \cdot \vec{u}, u^2) i_\Delta \Omega \; .$$

Under the action of $SO(3)$ it is actually possible to write more general invariant two-forms as a combinations of $dr^2 \wedge (d\vec{r} \cdot \vec{u})$, $dr^2 \wedge du^2$, $(d\vec{r} \cdot \vec{u}) \wedge du^2$, with coefficients functions $F(r^2, \vec{r} \cdot \vec{u}, u^2)$.

The general situation that emerges when Γ and $\omega_{\mathcal{L}}$ are **G**-invariant is that the Lagrangian is not left invariant by the **G**-action, but changes either by a constant:

$$L_{X_i^c} \mathcal{L} = c_i \quad ; \quad c_i \in \mathbf{R} \,, \tag{2.102}$$

or by a total time derivative:

$$L_{X_i^c} \mathcal{L} = \dot{f}_i \quad ; \quad \dot{f}_i = L_\Gamma \pi^* f_i \quad ; \quad f_i \in \mathcal{F}(Q) \,. \tag{2.103}$$

When this happens (and mainly when the case (2.103) occurs), we will say that the Lagrangian is *quasi invariant* under the lifted **G**-action. If, instead, $L_{X_i^c} \mathcal{L} = 0$, $\forall i$, we will say that $\mathcal{L}$ is *strictly invariant.*

If we consider the action of **G** on the Cartan form $\theta_{\mathcal{L}}$, we find that $\theta_{\mathcal{L}}$ is left unchanged, if $\mathcal{L}$ changes as in (2.102), while, if (2.103) holds:

$$L_{X_i^c} \theta_{\mathcal{L}} = df_i \tag{2.104}$$

Cocycles associated with quasi invariant Lagrangians

Let g be the Lie algebra of **G**. The infinitesimal generators X_i and X_i^c close on Lie algebras isomorphic to g, i.e., for example:

$$[X_i^c, X_j^c] = c_{ij}^k X_k^c \,, \tag{2.105}$$

the c_{ij}^k's being the structure constants of g. Let e_i be the unique element in g corresponding to X_i, and hence to X_i^c. The left hand sides of (2.102-103) can be viewed as maps:

$$\begin{aligned} c : g \to \mathbf{R} \quad &\text{by} \quad e_i \mapsto c(e_i) \stackrel{\text{def}}{=} c_i \\ \dot{f} : g \to \dot{\mathcal{F}}(TQ) \quad &\text{by} \quad e_i \mapsto \dot{f}(e_i) \stackrel{\text{def}}{=} \dot{f}_i \,, \end{aligned} \tag{2.106}$$

where $\dot{\mathcal{F}}(TQ)$ is the subset of $\mathcal{F}(TQ)$ consisting in those functions, which are linear in the fiber coordinates ('total time derivatives' in the language, more familiar to physicists, that we have been using before). The cohomology of Lie algebra and groups has been analyzed, e.g. in the classical paper by Chevalley and Eilenberg [22]. The basic results concerning the cohomology of Lie algebras will be reviewed in an Appendix to the present paper. On their basis, one can prove that both c and $\dot{f}$ are one-cocycles, in $H^1(g, \mathbf{R})$ and $H^1(g, \dot{\mathcal{F}})$ (the first cohomology groups of g with values in $\mathbf{R}$ and $\dot{\mathcal{F}}$, respectively), and that one can associate with $\dot{f}$ a unique (up to coboundaries) two-cocycle in $H^2(g, \mathbf{R})$.

In the case of (2.102), the one-cocycle condition reads:

$$c(e_i, e_j) = L_{X_i^c} c_j - L_{X_j^c} c_i \equiv 0 \,. \tag{2.107}$$

Therefore, this case cannot occur (or the cocycle is trivial) if $g = [g, g]$, which is however not always the case. In the case of (2.103), the one-cocycle condition becomes, upon using (2.105):

$$L_{X_i^c} \dot{f}_j - L_{X_j^c} \dot{f}_i - c_{ij}^k \dot{f}_k = 0 \,, \tag{2.108}$$

In this context, $\dot{f}$ will be a coboundary iff a function $\dot{F} \in \dot{\mathcal{F}}(TQ)$ exists s.t.:

$$\dot{f}_i = L_{X_i^\circ} \dot{F} \ . \tag{2.109}$$

Note that, although (2.103) looks superficially similar to (2.109), it is not a coboundary condition, as in general $\mathcal{L} \notin \dot{\mathcal{F}}(TQ)$.

As $\dot{f}_i = L_\Gamma \pi^* f_i$, (2.108) also entails:

$$L_{X_i} f_j - L_{X_j} f_i - c_{ij}^k f_k = \text{const.} = A_{ij} \ , \tag{2.110}$$

and A will be precisely the two-cocycle, $A \in H^2(g, \mathbf{R})$, mentioned above.

We can now state a necessary condition for the existence of F.

To state necessary conditions, we assume the existence of F and derive conditions on the cocycle A_{ij}. If $\mathcal{L} - \dot{F}$ is strictly invariant, we find:

$$L_{X_i^\circ} \mathcal{L} = L_{X_i^\circ} \dot{F} = \dot{f}_i \ , \tag{2.111}$$

i.e. $L_{X_i^\circ} F = f_i + a_i$, $a_i \in \mathbf{R}$, and moreover:

$$L_{X_j^\circ} L_{X_i^\circ} F = L_{X_j^\circ} f_i \quad ; \quad L_{X_i^\circ} L_{X_j^\circ} F = L_{X_i^\circ} f_j \tag{2.112}$$

implies

$$L_{[X_j, X_i]^\circ} F = c_{ji}^k L_{X_k^\circ} F = c_{ji}^k (f_k + a_k) = L_{X_j^\circ} f_i - L_{X_i^\circ} f_j \ . \tag{2.113}$$

Thus, the cocycle A_{ij} has the form:

$$A_{ij} = c_{ij}^k a_k \ , \tag{2.114}$$

i.e. it is a coboundary.

Now we are also able to show how the 'exactness' of A is a sufficient condition when the group $\mathbf{G}$ acts transitively on Q. We shall study this problem in two steps.

1) $Q \equiv \mathbf{G}$ (or $\mathbf{G}$ by a discrete subgroup, as we are working at the Lie algebra level).
2) $Q = \mathbf{G}/H$ for H a closed subgroup.

If $Q = \mathbf{G}$, the action of $\mathbf{G}$ becomes an action of $\mathbf{G}$ on itself. Assume we consider the left action:

$$\phi : \mathbf{G} \times \mathbf{G} \to \mathbf{G} \quad ; \quad \phi_g : h \mapsto gh \ . \tag{2.115}$$

Then vector fields $X_j \in \mathcal{X}(\mathbf{G})$ are right-invariant and are a basis for $\mathcal{X}(\mathbf{G})$. We can consider dual one-forms θ^j such that $\langle X_k \,|\, \theta^j \rangle = \delta_k^j$, forming a basis for $\mathcal{X}^*(\mathbf{G})$. Let us consider a one-form $\alpha = (f_k + a_k)\theta^k$, where the $f_k \in \mathcal{F}(Q)$ are as before the infinitesimal 'gauge functions' of Eq.(2.102).

By considering $d\alpha$, we find:

$$d\alpha = df_k \wedge \theta^k + (f_k + a_k) d\theta^k = df_k \wedge \theta^k - \frac{1}{2}(f_k + a_k) c_{jm}^k \theta^j \wedge \theta^m \ , \tag{2.116}$$

where we have used the formula of Maurer-Cartan. Because the vector fields X_j are a basis, we can write:

$$df_k = (L_{X_j} f_k)\theta^j \ , \tag{2.117}$$

so that:

$$
(2.118) \qquad \begin{aligned} d\alpha &= (L_{X_j} f_m)\theta^j \wedge \theta^m - \frac{1}{2}(f_k + a_k)c^k_{jm}\,\theta^j \wedge \theta^m = \\ &= \frac{1}{2}(L_{X_j} f_m - L_{X_m} f_j)\theta^j \wedge \theta^m - \frac{1}{2}(f_k + a_k)c^k_{jm}\,\theta^j \wedge \theta^m \ . \end{aligned}
$$

Thus:

$$
(2.119) \qquad L_{X_j} f_m - L_{X_m} f_j = (f_k + a_k)c^k_{jm}
$$

implies that α is closed.

From the closure of α we find that, at least locally, there exists an $F \in \mathcal{F}(\mathbf{G})$, such that $\alpha = dF$ (with F unique up to an additive constant). If, in addition, $\mathbf{G}$ is simply connected (or $H^1(\mathbf{G}) = 0$), F can be found globally and provides a solution for our problem.

Let us go now to the case $Q = \mathbf{G}/H$, with H some closed subgroup of $\mathbf{G}$. Here the additional difficulty is due to the fact that vector fields X_j's on $\mathbf{G}/H$ are not a basis[8] , as they are a redundant set (We will call them W_j and denote by X_j the corresponding ones on $\mathbf{G}$). Thus we cannot define a dual basis of one-forms satisfying $\langle W_k \mid \theta^j \rangle = \delta^j_k$. To avoid this problem, we decide to take the pull-back of functions f_j from $\mathbf{G}/H$ to $\mathbf{G}$ and try to solve the equation $L_{X_j} F = \pi^* f_j$ on $\mathbf{G}$.

After the equation has been solved on $\mathbf{G}$, we impose the condition that F should be projectable onto $\mathbf{G}/H$. That the existence of F on $\mathbf{G}$, solving $L_{X_j} F = \pi^* f_i$, is a necessary condition, is quite trivial, because the solution $\tilde{F}$ of:

$$
(2.120) \qquad L_{W_j} \tilde{F} = \pi^* f_j
$$

on $\mathbf{G}/H$ defines a solution on $\mathbf{G}$ by taking the pull-back $\pi^* \tilde{F}$.

Now let us impose that a solution of $L_{X_j} F = \pi^* f_j$ projects onto $\mathbf{G}/H$. If we call Z_a the vector fields generating the action of H, a necessary and sufficient condition for F to be projectable is:

$$
(2.121) \qquad L_{Z_a} F = 0 \ .
$$

In terms of the vector fields X_j's, as they are a basis on $\mathbf{G}$, we have:

$$
(2.122) \qquad Z_j(g) = [\mathrm{Ad}(g^{-1})]^k_j X_k(g) \ .
$$

(See [58] p.326). Then the projectability condition becomes:

$$
(2.123) \qquad [\mathrm{Ad}(g^{-1})]^k_\alpha L_{X_k} F = 0 \ ,
$$

or, equivalently:

$$
(2.124) \qquad [\mathrm{Ad}(g^{-1})]^k_\alpha (\pi^* f_k) = 0 \ .
$$

This is the additional condition to be satisfied by the infinitesimal gauge functions if a solution $\tilde{F}$ is to be obtainable from a solution F on $\mathbf{G}$.

[8] Here the X_j's are infinitesimal generators of the action of $\mathbf{G}$ on $\mathbf{G}/H$, which commutes with the action of H, used to go to the quotient.

If (and only if) a function F exists with the properties expressed by Eq.(2.109)., then it is possible to redefine the Lagrangian as:

$$\mathcal{L} \mapsto \tilde{\mathcal{L}} = \mathcal{L} - \dot{F} \ . \tag{2.125}$$

and $\tilde{\mathcal{L}}$ will be strictly invariant. ∎

Let's also remind the reader that a strictly invariant Lagrangian can always be found, starting from a quasi invariant one, if **G** is compact [59].

If the cocycles associated with the action of **G** on the Lagrangian are not trivial, there is no hope of finding a strictly invariant Lagrangian on TQ. A systematic way of dealing with this problem has been proposed in [63]. It consists in enlarging the configuration space by adding to it an extra, non dynamical, degree of freedom, and in suitably redefining the action of a new Lie algebra on the enlarged space. This new Lie algebra turns out to be a one dimensional central extension of the old one. In this way one can always recover an invariant description.

Details of this procedure, together with its interconnections with other instances of central extensions occourring in other branches of Theoretical Physics, will be discussed in the next Sections. Before doing this, we will discuss a general procedure for removing the obstructions of cohomological origin that can arise in going from the two-form ω satisfying the conditions of the theorem of Balachandran et al. to a globally defined one-form.

3 – RESOLVING THE OBSTRUCTIONS

Let us start again from the sequence analyzed in Sect.2, namely:

$$\Gamma \to \omega_{\mathcal{L}} \to \theta_{\mathcal{L}} \to \mathcal{L} \ . \tag{3.1}$$

As already discussed there, obstructions can arise in the process of going from the dynamical vector field to the Lagrangian, due to the cohomology associated with the operators d and/or d_S, and, moreover, new obstructions can arise if we insist on an *equivariant* cohomology (or on what has been called before an *Equivariant Poincaré Lemma*). Let's also recall that, if the general scenario has to be of the type outlined towards the end of Sect.2, then both Γ *and* the associated closed two-form ω (if it exists at all) are, by assumption, **G**-invariant. This is an immediate consequence of Eqns.(2.102-103). Taken this for granted, the first obstruction to be resolved is that of 'integrating' ω to a global one-form θ, s.t.:

$$\omega = d\theta \ . \tag{3.2}$$

We have already seen cases (e.g. the charge-monopole system) in which this is not possible (globally). We will try, in this and in the following Sections, to outline a general procedure for the resolution of the cohomological obstructions first, and, if possible, for rendering the solution an equivariant one.

Resolution of cohomological obstructions

Let's begin by recalling that the closed two-form ω, if found, is such that:

$$i_\Gamma \omega = \alpha \ , \tag{3.3}$$

for a closed one-form α. If no further obstructions arise, then, both ω and α are exact: $\omega = \omega_{\mathcal{L}}$, $\alpha = dE_{\mathcal{L}}$.

In the framework considered up to now, $\Gamma \in \mathcal{X}(TQ)$ is a second order field. More generally, we can consider a vector field $\Gamma \in \mathcal{X}(M)$ and a two-form $\omega \in \Lambda^2(M)$, obeying Eq.(3.3), with M a more general carrier space and $H^2(M) \neq 0$. An integration problem for ω can be posed in this more general context as well. The general scheme we are proposing here, which has already been discussed in the literature [84,62], consists in:

i) Finding a 'covering' of M by a higher dimensional manifold P, i.e. a manifold P and a smooth projection:

$$\pi : P \to M \, , \tag{3.4}$$

such that the pull-back $\pi^*\omega$ of ω on P is exact

and

ii) a vector field $\hat{\Gamma} \in \mathcal{X}(P)$, π-related [57,58] to Γ[1] such that

$$i_{\hat{\Gamma}} \pi^* \omega = \pi^* \alpha \, . \tag{3.5}$$

Whether or not ω is nondegenerate on M, the pull-back form $\pi^*\omega$ will be a degenerate form, and:

$$\ker(T\pi) \subseteq \ker(\pi^*\omega) \, . \tag{3.6}$$

Correspondingly, $\hat{\Gamma}$ will be ambiguous by the addition of fields in ker $(\pi^*\omega)$ and, in particular, of fields in $\ker(T\pi)$. The latter is closed under the Lie bracket operation [58] and can be viewed , loosely speaking, as an (infinite dimensional) Lie algebra, a sort of 'gauge' algebra of the extended dynamics. One will have therefore to add to the procedure of going from M to P some 'gauge fixing' prescription, in order to completely define $\hat{\Gamma}$.

As a specific example, let $\mathbf{G}$ be a Lie group and θ a left-invariant one-form on $\mathbf{G}$. Then:

$$\mathcal{D}_\theta \stackrel{\text{def}}{=} \ker(d\theta) \tag{3.7}$$

is closed under the Lie bracket operation and, being θ (and hence $d\theta$) left-invariant, it is a vector space of constant dimension at every point of $\mathbf{G}$, hence an involutive distribution. Going to the quotient w.r.t. the associated foliation, one can show [35] that:

$$M \stackrel{\text{def}}{=} \mathbf{G}/\mathcal{D}_\theta \tag{3.8}$$

is a smooth manifold. Moreover, $d\theta$ is, by construction, projectable w.r.t. the foliation (while θ need not be), and projects down to a closed and nondegenerate two-form ω on M, which is thus also a symplectic manifold. ω pulls back of course to an exact two-form on $\mathbf{G}$, i.e. to $d\theta$. If, in addition, $\mathbf{G}$ is semisimple, then $H^2(\mathbf{G}, R) = 0$, by Whithead's Lemma [35] , and any other closed two-form on M will pull back to an exact one on $\mathbf{G}$. Let's remark that $\mathcal{D}_\theta$ is also a left-invariant distribution. All the vector fields in $\mathcal{D}_\theta$ are thus complete and a theorem by Palais [67] assures that $\mathcal{D}_\theta$

[1] i.e. such that $T\pi(\hat{\Gamma}) = \Gamma$.

integrates to a closed Lie subgroup H of $\mathbf{G}$. We can thus also identify M as:

$$M = \mathbf{G}/H \ . \tag{3.9}$$

Remark: The above construction can also be viewed in a different, but equivalent, way. To every invariant form θ we can associate a unique element $v_\theta \in g^*$, where g^* is the dual of the Lie algebra g of $\mathbf{G}$. If θ is, say, left-invariant, and $X \in \mathcal{X}(\mathbf{G})$ is a left-invariant vector field, then $L_X \theta = i_X d\theta$. But it can be shown [51] that the element in g^* associated with $L_X \theta$ is precisely $\mathrm{ad}^*_X v_\theta$, with ad^* the coadjoint representation of g on g^*. Therefore, the subgroup $H \subset \mathbf{G}$ can also be identified with the isotropy group of v_θ in the coadjoint representation, and M becomes then an orbit in the coadjoint representation [51].

If the conditions of the above example are fulfilled, on going from M to $\mathbf{G}$ one is able to resolve *all* of the cohomological obstructions associated with $H^2(M, \mathbf{R})$, i.e. every closed two-form on M will pull back to an exact one on $\mathbf{G}$. In this sense, $\mathbf{G}$ acquires a sort of universal character, which is somehow much more than what we want, as we are interested only in the resolution of the obstructions connected with a specific closed two-form on M, and not with a generic one.

We will then be interested in finding a 'covering' P (in the sense of Eq.(3.4)), which is minimal w.r.t. the request of resolving the obstructions connected with the $\mathbf{G}$-invariant two-form ω satisfying Eq.(3.3).

There is a case, well known in Geometric Quantization [83,43], in which this 'minimal' program can be achieved, and namely when ω has integral cohomology, i.e. when:

$$\omega \in H^2(M, \mathbf{Z}) \subset H^2(M, \mathbf{R}) \ . \tag{3.10}$$

In such a case, Weil's theorem [81] gives a proof of the existence of a $U(1)$-bundle P over M and of a connection on P, whose curvature is precisely $\pi^* \omega$, which has then no more obstructions.

If there is only one nontrivial two-cycle in M for ω, it is of course always possible to normalize ω by a numerical factor in such a way as to make its cohomology integral. This amounts to normalizing the one-form on the r.h.s. of Eq.(3.3) by the same factor and, ultimately, if a Lagrangian has been found, to employing an equivalent Lagrangian description of the same dynamical system, with a Lagrangian proportional (again by the same factor) to the previous one. As we shall see in a short while, this is precisely what happens in the charge-monopole system.

Let us now briefly discuss a general procedure for going from M to the 'minimal' covering discussed above. Let's assume M to be connected and path-connected. We can build up a *path space* PM over M as follows:

Fixing a fiducial point $m_0 \in M$, we can associate to every other point $m \in M$ a fiducial path $\beta_0(m)$, i.e. a fixed path going from m_0 to m. In this way we build what some authors [24] call a *homotopy mesh* over M.

The path space PM is thus the space of all the possible paths going from m_0 to any other point $m \in M$. By a judicious use of the homotopy mesh, with any path $\beta(m)$, from m_0 to m, we can associate a loop, based at m, defined as $\gamma = \beta^{-1}(m) \circ \beta_0(m)$ and, viceversa, with any loop γ, based at m, we can associate the path $\beta(m) = \beta_o(m) \circ \gamma$, from m_0 to m. In this way, PM can also be identified with the space of all the loops based at the various points of M. The natural projection $\pi : PM \to M$, which sends every path into its end point (or every loop into its base point), gives PM a sort of 'fiber space' structure, with the 'fiber' over $m \in M$ (i.e. the inverse image $\pi^{-1}(m)$) consisting of the set of all loops based at m.

The advantage of introducing the path space PM is that it is a *contractible* space [3]. Therefore, any topological obstructions that may be present in M are *formally* resolved in PM. In particular, the pull-back $\pi^*\omega$ of ω from M to P is surely exact.

Once again, PM is a sort of 'universal space', in which all the possible cohomological obstructions can be resolved. Moreover, PM is an infinite dimensional space and hence a rather difficult object to deal with. What we need is a procedure of reduction of PM to a more manageable manifold, without introducing in this way new obstructions to the resolution of the cohomology class to which ω belongs, nor loosing physical information.

Let's begin by assuming (again this is the case of the charge monopole system) that the first homology group of M vanishes: $H_1(M, \mathbf{R}) = 0$.[2] If this is the case, every loop is a boundary. We can then set up an equivalence relation between paths from m_0 to $m \in M$ (or equivalently, in view of what was said before, between loops based at m) , according to the following definition [84]:

Any two paths β, β' from m_0 to m will be equivalent iff:

$$A(\beta, \beta') \stackrel{\text{def}}{=} \frac{1}{2\pi} \int_{\Sigma(\beta,\beta')} \omega = \text{integer} \, , \tag{3.11}$$

with $\Sigma(\beta, \beta')$ any two-surface bounded by the loop $(\beta')^{-1} \circ \beta$.

$A(\beta, \beta')$ is easily seen to enjoy the properties:

$$A(\beta, \beta') + A(\beta', \beta) = 0 \, , \tag{3.12}$$

and, for any three paths β, β', β'', ending at the same point m

$$A(\beta, \beta'') = A(\beta, \beta') + A(\beta', \beta'') \, . \tag{3.13}$$

Coming now to loops, any two loops γ, γ', based at m, will be equivalent iff:

$$A(\gamma) - A(\gamma') = \text{integer} \quad ; \quad A(\gamma) \stackrel{\text{def}}{=} \int_{\Sigma(\gamma)} \frac{\omega}{2\pi} \, . \tag{3.14}$$

where, again, $\Sigma(\gamma)$ is any two-surface, bounded by γ.

One can easily see that both (3.11) and (3.14) imply, as a consistency condition, that:

$$\int_S \frac{\omega}{2\pi} = \text{integer} \, , \tag{3.15}$$

for any closed two-surface (i.e. for any two-cycle) in M. It is then clear that we are here classifying loops according to the $U(1)$ factor:

$$I(\gamma) \stackrel{\text{def}}{=} \exp\left(i \int_{\Sigma(\gamma)} \omega \right) , \tag{3.16}$$

---

[2] Let's recall that $H_1(M, \mathbf{R}) = \pi_1/[\pi_1, \pi_1]$, where π_1 is the fundamental homotopy group. Thus, M need not be simply connected in order that $H_1 = 0$.

and that two loops are equivalent iff the same $U(1)$ factor is attached to both of them. Calling "$\sim$" the equivalence relation (3.11)(or (3.14)), it is quite clear that:

$$(\pi^{-1}(m)/\sim) \quad \simeq \quad U(1) \; . \tag{3.17}$$

It has been shown in [84] how, given a contractible open covering of m by open sets U_i, one can 'affix' smoothly the $U(1)$-fiber to each $m \in U_i$ in such a way as to provide the local trivialization:

$$\pi^{-1}(U_i) \simeq U_i \times U(1) \quad ; \quad \forall i \; , \tag{3.18}$$

and, with the help of the integrality condition (3.15), how the local bundles (3.18) can be 'patched together' in order to give to

$$\tilde{P}M \stackrel{\text{def}}{=} \bigcup_{m \in M} (\pi^{-1}(m)/\sim) \tag{3.19}$$

the structure of a nontrivial principal $U(1)$-bundle over M, with ω pulling back to an exact form on $\tilde{P}M$. The procedure of reduction of the path space to a $U(1)$-bundle, via Eqns.(3.11-16), yields therefore a constructive proof of Weil's theorem.

When $M = TQ$, which is the case of main interest for us, the topological obstructions could be removed, in principle, by adding just *one* degree of freedom. This, however, would give us a space which is not a tangent bundle. Since the topological obstructions originate in Q only, we can go to a covering $\tilde{P}Q$, with the procedure illustrated above, and then go to $T(\tilde{P}Q)$.

Remark: In the above discussion, we have assumed that $H_1(M, \mathbf{R}) = 0$. One might wonder what happens if this is not the case. As an example, we might consider $M = TQ$, where $Q = \mathbf{R}^3 - \{z\text{-axis}\}$. This case would correspond to the presence of an impenetrable, infinitely thin Aharonov-Bohm solenoid [65], located along the z axis. Being there one-cycles (loops) which are not boundaries, one has to decide what value to attribute to the 'phase factor' $A(\gamma)$ (or $A(\beta, \beta')$) associated with them. Subject to the conditions (3.12-13), but not to (3.14), which makes sense only for two-cycles, one can argue that there will be as many inequivalent $U(1)$ bundles over M as the dimension of $\mathrm{Hom}(H_1, U(1))$. In a quantum mechanical context, different $U(1)$ bundles will give rise to inequivalent quantizations of the same classical dynamical system [65,43,75,39].

We now discuss in some detail how the construction of the $U(1)$-bundle $\tilde{P}M$ works on a concrete example, namely the charge-monopole system, essentially following the scheme of [84]. In this case $M = TQ$, $Q = \mathbf{R}^3 - \{0\} \simeq S^2 \times \mathbf{R}$, and ω is given by Eq.(2.90):

$$\omega = \delta_{ij} dq^i \wedge dq^j + B \; , \tag{3.20}$$

where, in polar coordinates:

$$B = n \sin\theta \, d\theta \wedge d\phi \; . \tag{3.21}$$

From now on, we shall restrict ourselves to $Q = S^2$, since (both classically and quantum-mechanically) the radial motion decouples from the angular part and causes no problems.

According to the previous remark, we proceed directly to outline the construction of a $U(1)$ bundle over S^2 (with the two-form B playing the rôle of ω). Having removed the obstructions in P, the final carrier space of the dynamics will be TP. Being there only one nontrivial cocycle, things can always be adjusted in such a way that the integrality condition (3.15) is satisfied, so we shall assume $2n \in \mathbf{Z}$ from now on. This looks apparently related to the celebrated Dirac conditions for the product of electric and magnetic charges [25]. Let us stress, however, that in the present *classical* context, it only comes out from an essentially trivial rescaling of the two-form (3.20).

We will now exhibit explicitely the bundle P as the Hopf fibration [76] of $S^3 \simeq SU(2)$ over S^2:

$$U(1) \to S^3 \to S^2 \ , \tag{3.22}$$

referring to the literature for details of the proofs that its tangent bundle is equivalent (in the sense of equivalence of bundles [76]) to the bundle $\tilde{P}S^2$ constructed with the procedure outlined above. An element $g \in SU(2)$ can be parametrized as:

$$g = \begin{pmatrix} \alpha & -\bar{\beta} \\ \beta & \alpha \end{pmatrix} \quad ; \quad \alpha, \beta \in \mathbf{C} \quad ; \quad |\alpha|^2 + |\beta|^2 = 1 \ , \tag{3.23}$$

and the map from $SU(2)$ to S^2, which also defines the covering projection, is:

$$SU(2) \ni g \mapsto g\sigma^3 g^{-1} \stackrel{\text{def}}{=} \vec{x} \cdot \vec{\sigma} \ , \tag{3.24}$$

where $\vec{\sigma} \equiv (\sigma^1, \sigma^2, \sigma^3)$ are the Pauli matrices and $\vec{x} \in S^2$ is a normalized three vector. Any two elements: $g, g' \in \pi^{-1}(\vec{x})$ are related by:

$$g' = g \exp\left(\frac{i\theta\sigma^3}{2}\right) \ , \tag{3.25}$$

for some $\theta \in [0, 4\pi)$. This exhibits explicitly the $U(1)$ character of the fiber over $\vec{x}$, and also the fact that S^2 is a coset space.

Let us now come back to the covering of S^2 given in (2.93). $SU(2)$ is then covered by the corresponding sets:

$$\begin{aligned} V_+ &= \{g \in SU(2) \ \mid \ \alpha \neq 0\} \\ V_- &= \{g \in SU(2) \ \mid \ \beta \neq 0\} \ , \end{aligned} \tag{3.26}$$

which project down to $U_\pm$ under the covering projection (3.24). Introducing stereographic coordinates $\xi_\pm$, $\eta_\pm$ on $U_\pm$, one can coordinatize them as:

$$\vec{x} \equiv (q^1, q^2, q^3) \in U_\pm \mapsto \begin{cases} q^1 = \dfrac{2\xi_\pm}{\xi_\pm^2 + \eta_\pm^2 + 1} \\ q^2 = \dfrac{2\eta_\pm}{\xi_\pm^2 + \eta_\pm^2 + 1} \\ q^3 = \dfrac{\pm 1 \mp (\xi_\pm^2 + \eta_\pm^2)}{\xi_\pm^2 + \eta_\pm^2 + 1} \end{cases} \tag{3.27}$$

and, in $V_\pm$, the elements of $SU(2)$ are coordinatized by:

$$g \in V_+ \mapsto \begin{cases} \alpha = (\xi_+^2 + \eta_+^2 + 1)^{-1/2} e^{i\theta_+} \\ \beta = (\xi_+ + i\eta_+)\alpha \end{cases} \qquad (0 \le \theta_+ < 2\pi) \tag{3.28a}$$

$$g \in V_- \mapsto \begin{cases} \beta = (\xi_-^2 + \eta_-^2 + 1)^{-1/2} e^{i\theta_-} \\ \alpha = (\xi_- - i\eta_-)\beta \end{cases} \qquad (0 \le \theta_- < 2\pi) \tag{3.28b}$$

With these coordinatization, the two form B pulls back to:

$$\pi^* B = dA \quad ; \quad A = i\,n\,\mathrm{tr}(\sigma^3 g^{-1} dg) \ . \tag{3.29}$$

The one-form dA is thus globally defined. Explicitely, in coordinates:

$$A = \begin{cases} 2n \left(\dfrac{\xi_+ d\eta_+ - \eta_+ d\xi_+}{\xi_+^2 + \eta_+^2 + 1} + d\theta \right) & \text{on } V_+ \\ 2n \left(\dfrac{\eta_- d\xi_- - \xi_- d\eta_-}{\xi_-^2 + \eta_-^2 + 1} + d\theta \right) & \text{on } V_- \end{cases} \tag{3.30}$$

In view of the term $2nd\theta$, A turns out not to be projectable on S^2, as expected. The (global) Lagrangian on $T(SU(2))$ associated with A is now given by:

$$\mathcal{L} = \frac{1}{2}|\vec{u}|^2 + i\,n\,\mathrm{tr}(\sigma^3 g^{-1} \dot{g}) \ . \tag{3.31}$$

It is quite clear from (3.29) and (3.30) that the pull-back form $\pi^* B$ does not depend on either θ or $d\theta$. By explicit calculations, one can show that:

$$\ker(\pi^* B) \equiv \ker(T\pi) = \left\{ \frac{\partial}{\partial\theta} \right\} \ . \tag{3.32}$$

The extended dynamics $\hat{\Gamma}$ on TP is then ambiguous (see the discussion following Eq.(3.6)) exactly by a vector field along the fibers. A *global* 'gauge fixing' procedure is of course impossible, as this would amount to finding a global section of P, which is however not a trivial principal bundle.

Note also that $\partial/\partial\theta$ is a left-invariant vector field on $SU(2)$. It integrates to the right action:

$$g \mapsto gh \quad ; \quad h = e^{i\theta\sigma^3} \in U(1) \ . \tag{3.33}$$

Under this action, when $\theta = \theta(t)$, to make it into a gauge type transformation, the Lagrangian (3.31) is only quasi invariant. Indeed, under (3.33):

$$\mathcal{L} \mapsto \mathcal{L} - 2n\dot{\theta} \ . \tag{3.34}$$

To conclude this Section, let us show with an example how one can be faced with situations more general than those considered up to now (wich are essentially encompassed by Weil's theorem).

Suppose we have $k > 1$ magnetic monopoles of strenghts n_i, located at $\vec{x}_i \in \mathbf{R}^3$, $i = 1, \dots, k$, $\vec{x}_i \neq \vec{x}_j$ for $i \neq j$. A natural generalization of the Lagrangian (3.31) for a particle of unit charge and mass, moving in the field of the k monopoles is then:

$$\mathcal{L} = \frac{1}{2}|\vec{u}|^2 + i \sum_{j=1}^{k} n_j \,\mathrm{tr}(\sigma^3 g_j^{-1} \dot{g}_j) \,, \tag{3.35}$$

with $g_j \in SU(2)$. The g_j's are however not independent. Indeeed, Eq.(3.24) generalizes, in the present case, to:

$$g_j \sigma^3 g_j^{-1} = \hat{x}^{(j)} \cdot \vec{\sigma} \quad ; \quad \hat{x}^{(j)} = \frac{\vec{x} - v x_j}{|\vec{x} - v x_j|} \,, \tag{3.36}$$

with $\vec{x} \equiv (q^1, q^2, q^3)$, $\vec{x} \neq \vec{x}_j \quad \forall j$. Eq.(3.36) can be inverted to yield:

$$\hat{x}^{(j)} = \frac{1}{2} \mathrm{tr}(\vec{\sigma} g_j \sigma^3 g_j^{-1}) \,. \tag{3.37}$$

The fact that the $\hat{x}^{(j)}$'s are all uniquely determined by the simple vector $\vec{x}$ will then yield the required constraints among the g_j's. For the sake of clarity, let's work out in some detail the case $k = 2$, making reference to the geometry depicted in Fig.1.

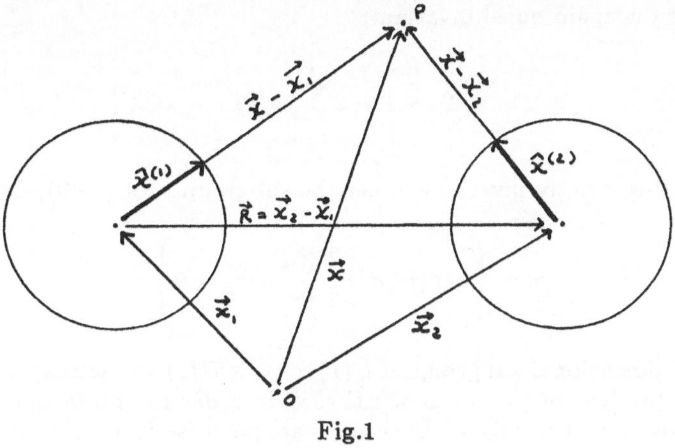

Fig.1

The geometrical constraint among the $\hat{x}^{(j)}$'s is now given, for each value of $\vec{x}$, by:

$$|\vec{x} - \vec{x}_1| \hat{x}^{(1)} - |\vec{x} - \vec{x}_2| \hat{x}^{(2)} = \vec{R} \equiv \vec{x}_2 - \vec{x}_1 \,, \tag{3.38}$$

and, using (3.37), we find the following constraints among the g's:

$$|\vec{x} - \vec{x}_1| \,\mathrm{tr}(\vec{\sigma} g_1 \sigma^3 g_1^{-1}) - |\vec{x} - \vec{x}_2| \,\mathrm{tr}(\vec{\sigma} g_2 \sigma^3 g_2^{-1}) = 2\vec{R} \,. \tag{3.39}$$

Note that Eq.(3.38) is invariant under rotations around the direction of $\vec{R}$. Eq.(3.39) yields thus only two independent constraints, thereby reducing the number of independent parameters on which the g_j's can depend from six to four. In this as well as in the general case, Eqns.(3.36-37) are invariant under the $U(1)$ right action

(3.25). Each one of the g_j's has then a $U(1)$ degree of freedom which is unconstrained, while the geometrical constraints act to reduce the number of the remaining independent parameters to the number appropriate for the description of the dynamics of the particle in $\mathbf{R}^3$ (with the exclusion of the $\vec{x}_i$'s) under the influence of the monopole fields.

It seems therefore that we should end up with a k-dimensional torus bundle, i.e. a multiple $U(1)$ bundle over the configuration space, with a $U(1)$ factor attached to each monopole. On the other hand, if, e.g., all the n_i's are pairwise rational, then we can rescale the two-form:

$$B = \sum_{i=1}^{k} B_i \ , \tag{3.40}$$

(where B_i is of the form (3.21), with the angles referred to the i-th monopole as the origin) by a factor, in order to make its cohomology integral, thereby ending up with a *single* $U(1)$ bundle, as before.

The same result can also be achieved in the following way under the generalization of the action of Eq.(3.33)

$$g_j \mapsto g_j \, e^{i\theta_j \sigma^3} \ . \tag{3.41}$$

The Lagrangian is again quasi invariant:

$$\mathcal{L} \mapsto \mathcal{L} - 2\sum_{j=1}^{k} n_j \dot{\theta}_j \ . \tag{3.42}$$

In particular, $\mathcal{L}$ is strictly invariant under the subgroup S of (3.40), defined by:

$$S = \left\{ \exp(i\theta_j \sigma^3) \;\middle|\; \sum_{j=1}^{k} n_j \dot{\theta}_j = 0 \right\} \ . \tag{3.43}$$

S is a $(k-1)$-dimensional subgroup of $U(1) \times \cdots \times U(1)$ (k factors), and we can try and go to the quotient of the latter w.r.t. to S in order to obtain the residual gauge group of $\mathcal{L}$. However, it is only when the n_i's are pairwise rational that we can obtain a smooth quotient. For, otherwise, the orbits of S wind densely on the k-torus and no smooth quotient exists.

An intermediate situation obtains if we can divide the monopoles into, say, q classes ($1 < q \leq k$), each class being composed of monopoles with mutually rational strenghts. Then according to the previous analysis, a partial reduction can be performed from the k-torus bundle to a q-torus bundle over confiquration space.

What this example shows is that we can devise a more general situation than that described by Eq.(3.15), namely one in which all the two-cycles in M can be partitioned into classes, with any two members of a class having mutually rational periods determined by ω. If we end up with q such mutually irrational distinct classes, it seems that we can substitute the single $U(1)$ bundle $\tilde{P}M$, considered up to now, with a q-torus bundle, which again plays the rôle of a minimal bundle in which the topological obstructions can be resolved. In this sense, the procedure outlined here yields also a generalization of Weil's theorem.

4 – CENTRAL EXTENSIONS AND THE LIFTING OF SYMMETRIES

Up to now, we have been interested in removing the topological obstructions which appear in the passage from a closed two-form ω (related to a dynamical vector field Γ in the manner explained in Sect.2) to its 'potential' (the one-form θ). This has been achieved by lifting the dynamics from the carrier space M ($M = TQ$ in many cases of interest) to a bundle $\tilde{P}M$ over M, where $\tilde{P}M$ is typically a $U(1)$-bundle, but, on cases, could also turn out to be a higher dimensional bundle. We shall stick from now on to the case in which ω has integral cohomology, and hence to $U(1)$-bundles over M.

Remark: When $M = TQ$, topological obstructions, if any, all come from the base manifold, i.e. form the configuration space Q. As already discussed in Sect.3, it may then be convenient to build up a $U(1)$-bundle P over Q alone, and then to consider TP as the extended carrier space. In this way we preserve the tangent bundle structure of the carrier space. In more general cases, (M, ω) will be a symplectic manifold, and $(\tilde{P}M, \theta)$, with $\pi^*\omega = d\theta$, will turn out to be an exact contact manifold.

We now address ourselves to the following, related problems:

i) Under which conditions can the G-action over M be lifted to the action of an enlarged group on $\tilde{P}M$, and how is the latter related to G.

ii) Under which conditions can the one-form θ be made invariant.

Problem i) is very similar to a problem in Quantum Mechanichs, namely that of lifting ray representations of groups to vector representations. By Wigner's theorem [82], symmetries are implemented at the quantum level by unitary or antiunitary operators, and, in general, a Lie group G of symmetries will be represented as a unitary ray group on the projective Hilbert space PH. Bargmann [12] has devised a systematic way of lifting this representation to a unitary representation on H, by first performing a central extension of the Lie algebra of G by the one dimensional algebra of $U(1)$, and by subsequently exponentiating the result to the universal covering group. This procedure is summarized in the following diagram:

$$\begin{array}{ccccc} U(1) & \longrightarrow & U(H) & \longrightarrow & U(PH) \\ & & & \nwarrow\!\!\dashrightarrow & \uparrow \\ & & & & \mathrm{G} \end{array} \tag{4.1}$$

where $U(H)$ is the set of unitary operators on H, and $U(PH)$ that of unitary operator rays [73] on the projective Hilbert space PH.

It is apparent that we have to do here with a similar problem, which might be called the construction of a 'Bargmann group' in a classical context. A systematic procedure for constructing such a group has been recently proposed by Balachandran, Gromm and Sorkin [11]. We shall only sketch here the basic ingredients, referring the intersted reader to [11] for details.

The action of G on M is defined through the map:

$$\Phi : \mathrm{G} \times M \to M \quad ; \quad (g, m) \mapsto \Phi(g, m)\Phi(g, \Phi(h, m)) = \Phi(gh, m) \ , \tag{4.2}$$

$\forall g, h, m$, where the last equality holds for a left action. We shall assume this to be the case here. Taking any point $m_0 \in M$, Φ induces the map:

$$\phi : \mathrm{G} \to M \qquad \text{by} \qquad g \mapsto \Phi(g, m_0) \ , \tag{4.3}$$

which is a submersion. The map ϕ may then be used to pull-back ω from M to G. $\phi^*\omega$ will be a closed, though not necessarily non degenerate, two-form on G, and its cohomology will be integral if that of ω is such on M. One can now proceed to the construction of the loop space over G and to its subsequent reduction to a $U(1)$-bundle $\tilde{P}$G over G. It turns out that $\tilde{P}$G is a $U(1)$ central extension of G and that there is a natural way to define an action, which we will call $\tilde{\phi}$, of $\tilde{P}$G on $\tilde{P}M$. The situation is summarized in the following commutative diagram:

$$(4.4) \qquad \begin{array}{ccc} \tilde{P}\mathrm{G} & \xrightarrow{\tilde{\phi}} & \tilde{P}M \\ {\scriptstyle \pi_G}\downarrow & & \downarrow{\scriptstyle \pi} \\ (\mathrm{G}, \phi^*\omega) & \xrightarrow{\phi} & (M, \omega) \end{array}$$

where $\pi_G : \tilde{P}\mathrm{G} \to \mathrm{G}$ is the canonical projection and $\pi : \tilde{P}M \to M$ has already been defined.

The action Φ of Eq.(4.2) induces also the family of maps:

$$(4.5) \qquad g \ni \mathrm{G} \to \phi_g : M \to M \quad \text{by} \quad m \mapsto \Phi(g, m) \ ,$$

satisfying $\phi_g \circ \phi_{g'} = \phi_{gg'}, \forall g, g'$. $\tilde{\phi}$ will induce similar maps $\tilde{\phi}_{\tilde{g}}$, with $\pi_G \tilde{g} = g$, in such a way that the diagram:

$$(4.6) \qquad \begin{array}{ccc} \tilde{P}M & \xrightarrow{\tilde{\phi}_{\tilde{g}}} & \tilde{P}M \\ {\scriptstyle \pi}\downarrow & & \downarrow{\scriptstyle \pi} \\ M & \xrightarrow{\phi_g} & M \end{array}$$

commutes. Then:

$$(4.7) \qquad \tilde{\phi}^*_{\tilde{g}} \pi^* \omega = (\pi \tilde{\phi}_{\tilde{g}})^* \omega = (\phi_g \pi)^* \omega = \pi^* (\phi^*_g \omega) \ ,$$

and hence:

$$(4.8) \qquad \tilde{\phi}^*_{\tilde{g}} (\pi^* \omega) = \pi^* \omega \ ,$$

i.e. $\tilde{P}$G will also act as an invariance group for the pull-back form $\pi^*\omega$.

From what has been said, it appears that, if there are topological obstructions coming from $H^2(M)$ (or from $H^2(Q)$, if $M = TQ$), the natural framework in which to discuss both the removal of the obstructions *and* the lifting of the action of a symmetry group *is that of an associated $U(1)$-bundle and of a $U(1)$ central extension of the original symmetry group* G.

A similar construction has been carried over in the interesting paper by Tuynman and Wiegerinck [79].

Remarks:

i) It may not be possible to realize (or to embed) G as a subgroup of $\tilde{P}$G, in which case we will be forced to employ the full group $\tilde{P}$G as a symmetry group. One case in which G is in fact a subgroup occours when the central extension is a trivial one: $\tilde{P}\mathrm{G} = \mathrm{G} \times U(1)$.

ii) If $M = TQ$, in setting up the Lagrangian formalism on $T(\tilde{P}Q)$, we should not forget that $\pi^* \omega$ is, by construction, a degenerate two-form. The discussion of the dynamics has thus to be done in the framework of the Lagrangian version of Dirac's theory of constraints [27,13] (See for instance [56]).

iii) The $U(1)$ central extension $\tilde{P}\mathbf{G}$ will be compact if $\mathbf{G}$ is. If this is the case, by averaging (if necessary) over the group, which has an invariant Haar mesure [38], we can always recover an invariant one-form θ s.t. $\pi^* \omega = d\theta$, and hence an invariant description of the dynamics.

What we have eventually ended up with is a situation of the type already encountered in Sect.2, namely a manifold, which we shall continue to call M for simplicity, on which a one-form θ is given[1] and a Lie algebra $\mathcal{X}_\theta \subseteq \mathcal{X}(M)$, composed of vector fields $X \in \mathcal{X}(M)$ such that:

$$L_X \theta = df_X \quad ; \quad f_X \in \mathcal{F}(M) . \tag{4.9}$$

Note that (4.9) implies that $\mathcal{X}_\theta$ is a Lie subalgebra of $\mathcal{X}(M)$, i.e. $X, Y \in \mathcal{X}_\theta \Rightarrow [X,Y] \in \mathcal{X}_\theta$, and that:

$$i_X d\theta = dF_X \quad ; \quad F_X \stackrel{\text{def}}{=} f_X - i_X \theta , \tag{4.10}$$

and, if θ is changed by the addition of an exact form, $\theta \mapsto \theta + dV$, then $f_X \mapsto f_X + L_X V$, but F_X will remain unchanged.

We now consider various extensions of the Lie algebra $\mathcal{X}_\theta$. First of all, $\mathcal{X}_\theta$ can be extended to a Lie algebra on $M \times \mathbf{R}$, by defining the extended vector fields:

$$\tilde{X} = X + f_X \frac{\partial}{\partial s} \quad ; \quad X \in \mathcal{X}_\theta, \ s \in \mathbf{R} . \tag{4.11}$$

Extending the definition of θ into:

$$\tilde{\theta} = \theta - ds , \tag{4.12}$$

one sees at once that:

$$L_{\tilde{X}} \tilde{\theta} = 0 \qquad \forall X \in \mathcal{X}_\theta , \tag{4.13}$$

i.e. $\tilde{\theta}$ is invariant under the action of the extended vector fields.

Remark: If $M = TQ$ and we are working in the context of Lagrangian dynamics, the above extension procedure needs modifications if we want the enlarged space to to be a tangent bundle. To be specific, let $\mathbf{G}$ be a symmetry group of Q-transformations for a second order vector field Γ, with infinitesimal generators $X_1 \dots X_\mu$, $\mu = \dim \mathbf{G}$, $X_i \in \mathcal{X}(Q)$, $i = 1 \dots \mu$. Let $\mathcal{L}$ be an admissible Lagrangian for Γ, and let it be only quasi invariant under the lifted $\mathbf{G}$-action ($L_{X_i^c} \mathcal{L}\ \dot{f}_i \in \mathcal{F}(TQ)$).

We now extend Q into $Q \times \mathbf{R}$ and the Lagrangian into:

$$\mathcal{L} \mapsto \tilde{\mathcal{L}} \stackrel{\text{def}}{=} \mathcal{L} - \dot{s} \ \in \mathcal{F}(TQ \times \mathbf{R}) , \tag{4.14}$$

which gives rise to an extended $\tilde{\theta}$ still given by Eq.(4.12).

[1] Modulo an exact form, as what finally matters is $d\theta$.

We next extend the X_i's into:

$$\tilde{X}_i = X_i + f_i \frac{\partial}{\partial s} , \tag{4.15}$$

and then consider their complete lifts:

$$\tilde{X}_i^c = X_i^c + f_i \frac{\partial}{\partial s} + \dot{f}_i \frac{\partial}{\partial \dot{s}} . \tag{4.16}$$

It is then quite clear that:

$$L_{\tilde{X}_i^c} \tilde{\theta}_{\mathcal{L}} = 0 \qquad \text{and} \qquad L_{\tilde{X}_i^c} \mathcal{L} = 0 , \tag{4.17}$$

i.e. that we have obtained in this way an invariant Lagrangian description on $TQ \times TR$. Of course, as:

$$\frac{\partial \tilde{\mathcal{L}}}{\partial \dot{s}} = -1 \tag{4.18}$$

we are dealing here with a constrained theory on the enlarged space.

Coming back now to the more general case considered above, and computing Lie brackets, we find, for $X, Y \in \mathcal{X}_\theta$:

$$[\tilde{X}, \tilde{Y}] = [X, Y] + (L_X f_Y - L_Y f_X) \frac{\partial}{\partial s} . \tag{4.19}$$

On the other hand:

$$L_{[X,Y]} \theta = L_X L_Y \theta - L_Y L_X \theta = d(L_X f_y - L_y f_X) \stackrel{\text{def}}{=} df_{[X,Y]} , \tag{4.20}$$

whence:

$$(L_X f_Y - L_Y f_X) - f_{[X,Y]} \stackrel{\text{def}}{=} c_{XY} \in \mathbf{R} , \tag{4.21}$$

and

$$[\tilde{X}, \tilde{Y}] - \widetilde{[X, Y]} = c_{XY} \frac{\partial}{\partial s} . \tag{4.22}$$

It can be easily proved that:

$$c_{[X,Y]Z} + c_{[Z,X]Y} + c_{[Y,Z]X} = 0 \quad ; \quad X, Y, Z \in \mathcal{X}_\theta , \tag{4.23}$$

which proves (see App.A) that $c : (X, Y) \mapsto c_{XY}$ is a two-cocycle in the Lie algebra, with values in $\mathbf{R}$. What (4.11) defines is therefore a central extension $\tilde{\mathcal{X}}_\theta$ of $\mathcal{X}_\theta$ by $\mathcal{F}(M)$:

$$0 \to \mathcal{F}(M) \to \tilde{\mathcal{X}}_\theta \to \mathcal{X}_\theta \to 0 , \tag{4.24}$$

where $\tilde{\mathcal{X}}_\theta$ is the algebra defined by (4.11), and the map $\mathcal{F}(M) \to \tilde{\mathcal{X}}_\theta$ is realized by $\mathcal{F}(M) \ni f \mapsto f \, \partial/\partial s$.

A different extension, namely to an algebra over $M \times U(1)$, obtains if we extend our vector fields as:

$$X \mapsto \tilde{X} = X + i\, f_X\, z \frac{\partial}{\partial z} , \tag{4.25}$$

where z is a coordinate on $U(1)$ ($z \in \mathbf{C}$, $|z| = 1$). We can likewise extend θ as:

$$\theta \mapsto \tilde{\theta} = \theta + i d(\ln z) , \tag{4.26}$$

thereby obtaining again:

$$L_{\tilde{X}} \tilde{\theta} = 0 . \tag{4.27}$$

Again, it has to be remarked that 'regauging' θ by an exact form dV leads again to $f_X \mapsto f_X + L_X V$, as before.

Lie algebra extensions and prequantization.

Let $(M, d\theta)$be an exact symplectic manifold. On such manifolds, there is a standard procedure, known as 'prequantization', to associate symmetric operators with classical observables. In short, the procedure goes as follows: with $\omega = d\theta$, ω^n is a symplectic volume. By Riesz' representation theorem [2], we can define a differentiable measure μ_θ on M, and then build up the complex Hilbert space $L_2(M, d\mu_\theta)$ of square integrable, complex-valued functions on M, with inner product:

$$\langle \psi \mid \phi \rangle = \int_M \bar{\psi} \phi \, d\mu_\theta , \tag{4.28}$$

for any $\psi, \phi \in L_2(M, d\mu_\theta)$, with the condition:

$$\|\psi\| \stackrel{\text{def}}{=} \langle \psi \mid \psi \rangle^{1/2} < +\infty \quad ; \quad \forall \psi \in L_2 . \tag{4.29}$$

Let then F be a classical observable, i.e. let F be a real and smooth function on M and let X_F be the Hamiltonian vector field associated with F by the symplectic structure:

$$i_{X_F} d\theta = dF . \tag{4.30}$$

Setting everywhere $\hbar = 1$, the prequantization procedure [83] assigns then to F the operator $\hat{F}$, defined as:

$$\hat{F}(\phi) \stackrel{\text{def}}{=} -i \left[L_{X_F} \phi + i (i_{X_F} \theta) \phi \right] + F \phi . \tag{4.31}$$

$\hat{F}$ turns out to be a symmetric operator w.r.t. the scalar product defined by Eq.(4.28). According to (4.31), constants act by multiplication and, moreover, the correspondence $F \mapsto \hat{F}$ maps Poisson Brackets into commutators, i.e. it realizes the so-called 'Dirac mapping' [26].

By lifting X_F to the vector field $\tilde{X}_F$, defined by Eq.(4.25), we obtain:

$$\tilde{X} = X_F + i(F + i_{X_F} \theta) z \frac{\partial}{\partial z} . \tag{4.32}$$

With any function $\phi : M \to \mathbf{C}$ we can associate the function $\tilde{\phi} : M \times U(1) \to \mathbf{C}$, defined by:

$$\tilde{\phi}(m,z) = z\phi(m) \; . \tag{4.33}$$

It is then easily proved that:

$$L_{\tilde{X}_F}(\tilde{\phi}) = i[\widetilde{\hat{F}(\phi)}] \; . \tag{4.34}$$

But for the factor of i, our lifted vector field $\tilde{X}_F$ acts on $\tilde{\phi}$ exactly as the operator $\hat{F}$ acts on ϕ. The prequatiztion procedure is then essentially equivalent to the lifting procedure on vector fields which has been discussed before.

The one-form (4.26) acts now as a connection one-form on the $U(1)$-bundle over M. we can then define the *horizontal lift* of any vector field $X \in \mathcal{X}(M)$ as the unique vector field $X^h \in \mathcal{X}(M \times U(1))$, which is π-related to X (with $\pi : M \times U(1) \to M$, the projection onto the first factor) and is in the kernel of the connection form:

$$i_{X^h}\tilde{\theta} = 0 \; . \tag{4.35}$$

Direct computation shows that the horizontal lift is given by:

$$X^h = X + i(i_X\theta)z\frac{\partial}{\partial z} \; , \tag{4.36}$$

and, computing Lie brackets, we find, for any two vector fields $X, Y \in \mathcal{X}(M)$:

$$[X^h, Y^h] - [X,Y]^h = id\theta(X,Y)z\frac{\partial}{\partial z} \; , \tag{4.37}$$

i.e. $d\theta$ defines, as expected, the curvature of the connection. In particular, if X is Hamiltonian, i.e. if $X = X_F$ (cfr. Eq. (4.30)) for some $F \in \mathcal{F}(M)$, then, comparing (4.36) and (4.25). we find:

$$\tilde{X}_F - X_F^h = iFz\frac{\partial}{\partial z} \; . \tag{4.38}$$

Extension of relevant infinite dimensional Lie algebras

We turn now to a more general situation, namely that in which we have a principal $U(1)$ bundle P, not necessarly trivial, over a symplectic manifold (M,ω), with a connection one-form $\tilde{\theta}$ and a curvature two-form $\Omega = d\tilde{\theta}$. In the case discussed at the beginning of this section, $P = \tilde{P}M$, the cuvature form is $\Omega = \pi^*\omega$ and, if ω is nondegenerate on M:

$$\ker\Omega \equiv \ker(T\pi) \; . \tag{4.39}$$

Otherwise stated, Ω is nondegenerate when restricted to horizontal subspaces. This can also be seen in a different way. As $\Omega^n \wedge \tilde{\theta}$ $(2n = \dim M)$ is a volume form on P, and hence defines an isomorphism between vector fields and $(2n-1)$-forms, one has:

$$0 \neq X^h \Rightarrow i_{X^h}(\Omega^n \wedge \tilde{\theta}) \equiv n(i_{X^h}\Omega)(\Omega^{n-1} \wedge \tilde{\theta}) \neq 0 \; , \tag{4.40}$$

implying $i_{X^h}\Omega \neq 0$, and viceversa.

A vertical fundamental vector field V can be defined by the condition:

$$(4.41) \qquad i_V\,\tilde{\theta} = 1 \; .$$

(In the previously examined case $V = -iz\partial/\partial z$). We may now define, for $X \in \mathcal{X}(M)$, the horizontal lift X^h exactly as in (4.35), while, if X is Hamiltonian for some $F \in \mathcal{F}(M)$ (cfr. Eq.(4.30)), its lift $\tilde{X}_F$ can also be defined starting with the condition:

$$(4.42) \qquad i_{\tilde{X}_F}\, d\tilde{\theta} = \pi^* dF \; .$$

(Note that, as $d\tilde{\theta} = \Omega = \pi^*\omega$, $i_{\tilde{X}}\,d\tilde{\theta} \equiv \pi^* i_X\Omega$). Eq.(4.42) defines of course $\tilde{X}$ modulo a vector field in the kernel of Ω, which can be fixed by requiring again[2] :

$$(4.43) \qquad L_{\tilde{X}_F}\tilde{\theta} = 0 \; .$$

Using (4.42-43), we find then:

$$(4.44) \qquad \tilde{X} = X^h - (\pi^* F)V \; .$$

We have thus seen that the $U(1)$-bundle P over M, which resolves the obstructions coming from $H^2(M)$ supports the action of a central extension of the Lie algebra of symmetries, which acts on M with Hamiltonian vector fields. The action can be defined in such a way as to preserve the connection one-form $\tilde{\theta}$ on the $U(1)$-bundle.

Up to now, the analysis has been conducted at the Lie algebra level. At the group level, we may ask ourselves the following question: suppose we had started with a Lie group G, acting on M with a Hamiltonian action. Can we conclude that a central extension of G exists which acts on $\tilde{P}M$, while preserving the one-form $\tilde{\theta}$? The affirmative answer is again provided by the theorem of Palais [67] (cfr. Sect.3). The infinitesimal generators of the G-action on M are complete. As the extension is performed by adding to them vector fields along the fibers, which are compact, the extended Lie algebra will be made again of complete vector fields, and will then integrate to a Lie group acting on $\tilde{P}M$. It is obvious that such a Lie group will cover the action of G on M, thus providing an extension of the latter.

Remarks:

i) If the action of G on M we started with was only locally Hamiltonian, we might be obliged to work on the universal covering of M and, after everything has been carried out, we should inquire about the projectability of our construction w.r.t. the fundamental group of M.

ii) The construction discussed above has an interesting application in the context of Lagrangian dynamics. Let $M = TQ$, and consider, as in (4.11), an extended vector fields of the form:

$$(4.45) \qquad \tilde{X} = X + f\frac{\partial}{\partial s},$$

[2] Loosely speaking, i.e. employing a rather sloppy notation, we can write:

$$X^h = X - (i_X\theta)V \quad ; \quad \tilde{X} = X - f_X V \; ,$$

where $L_X\theta \equiv df_X$, as before.

where we are not necessarly assuming (4.9) to hold. Guided by the form of the Euler-Lagrange equations:

$$L_\Gamma \theta_{\mathcal{L}} = d\mathcal{L} \ , \tag{4.46}$$

we may also extend the dynamical vector field Γ into:

$$\tilde{\Gamma} = \Gamma + \mathcal{L}\frac{\partial}{\partial s} \ , \tag{4.47}$$

$\mathcal{L}$ being the Lagrangian. Direct computation shows then that:

$$[\tilde{\Gamma}, \tilde{X}] = [\Gamma, X] + (L_\Gamma f - L_X \mathcal{L})\frac{\partial}{\partial s} \ . \tag{4.48}$$

Therefore:

$$[\tilde{\Gamma}, \tilde{X}] = 0 \Leftrightarrow [\Gamma, X] = 0 \quad \text{and} \quad L_X \mathcal{L} = L_\Gamma f = \dot{f} \ , \tag{4.49}$$

i.e. the fact that $\tilde{X}$ is a symmetry for the extended dynamics $\tilde{\Gamma}$ yields at once Nöther's theorem (and viceversa) as well as (as a consequence of $L_X \mathcal{L} = \dot{f}$) Eq.(4.9). Note that the Euler-Lagrange equations can be rewritten in a compact form as:

$$L_{\tilde{\Gamma}} \tilde{\theta}_{\mathcal{L}} = 0 \ . \tag{4.50}$$

where $\tilde{\theta}_L$ is the extended Cartan one-form on the extended space.

Moreover, as:

$$i_{[\tilde{\Gamma}, \tilde{X}]} \tilde{\theta}_{\mathcal{L}} = L_\Gamma (i_{\tilde{X}} \tilde{\theta}_{\mathcal{L}}) - i_{\tilde{X}} (L_{\tilde{\Gamma}} \tilde{\theta}_{\mathcal{L}}) \ , \tag{4.51}$$

we obtain, as a consequence of (4.50):

$$[\tilde{X}, \tilde{\Gamma}] = 0 \Rightarrow L_\Gamma (i_{\tilde{X}} \tilde{\theta}_{\mathcal{L}}) = 0 \ , \tag{4.52}$$

and hence the Nöther constant of the motion is expressed, in compact form, as:

$$i_{\tilde{X}} \tilde{\theta}_{\mathcal{L}} \equiv F_X + \text{const.} \ , \tag{4.53}$$

(cfr.Eq.(4.10)).

As a final application of the previous extensions, let's consider now the case where (M, ω) is a compact symplectic manifold, with ω giving rise again to a principal $U(1)$ bundle $\pi : P \to M$ and with a connection $\tilde{\theta}$, whose curvature form is precisely $\pi^* \omega$.

Let $\mathcal{X}_h(M)$ be the algebra of Hamiltonian vector fields on M. According to what has been said up to now, every vector field $X \in \mathcal{X}_h(M)$ admits a lift $\tilde{X} \in \mathcal{X}(P)$, which preserves $\tilde{\theta}$ and is π-related to X. Let's call $\tilde{\mathcal{X}}_h$ the algebra of such vector fields.

Another algebra, which we shall call $\mathcal{X}^v$, is generated by the fundamental vector field V associated with the action of $U(1)$ (cfr.Eq.(4.41)) with coefficients in $\pi^* \mathcal{F}(M)$. We obtain then the following short exact sequence of infinite dimensional Lie algebras:

$$0 \to \mathcal{X}^v \to \tilde{\mathcal{X}}_h \to \mathcal{X}_h \to 0 \ . \tag{4.54}$$

There is a rather interesting result that we shall now discuss. If M is compact, the sequence (4.54) *splits*, i.e. $\tilde{\mathcal{X}}_h$ contains a subalgebra, which is isomorphic to $\mathcal{X}_h$ (not a central extension of it). This fact is expressed by the following modification of (4.54):

$$0 \longrightarrow \mathcal{X}^v \longrightarrow \tilde{\mathcal{X}}_h \underset{\leftarrow \cdots}{\longrightarrow} \mathcal{X}_h \longrightarrow 0 . \tag{4.55}$$

Proof:

Recalling our previous construction of the $\tilde{X}$'s, we may proceed first by extending $X \in \mathcal{X}(M)$ to a horizontal field $X^h \in \mathcal{X}(P)$, which satisfies Eq.(4.35) and is π-related to X. Moreover, if X is Hamiltonian w.r.t. a function F_X, i.e.:

$$i_X \omega = dF_X , \tag{4.56}$$

X^h satisfies also:

$$L_{X^h} \tilde{\theta} = \pi^* dF_X .$$

We next define the lift $\tilde{X}$ as:

$$\tilde{X} = X^h - F_X V , \tag{4.57}$$

where V is the vertical fundamental field defined by Eq.(4.36).

Note that requiring only $\tilde{X}$ to be π-related to X and to leave $\tilde{\theta}$ unchanged is not enough to fix it uniquely, as it could be still ambiguous by addition of a constant multiple of V. So, Eq.(4.53) is actually a choice we make as to which vector should be taken as the lift of X.

Eq.(4.51) actually defines F_X only up to an additive constant, and these constants (cfr., in a different context, Eq.(4.22)) give rise in general to a central extension of the Lie algebra $\mathcal{X}_h$. If M is compact, the additive constant in F_X can always be fixed in such a way that:

$$\int_M F_X \Omega = 0 . \tag{4.59}$$

Indeed, if (4.59) is not fulfilled, and hence:

$$\int_M F_X \Omega = c_X \neq 0 , \tag{4.60}$$

we simply redefine F_X as:

$$\tilde{F}_X = F_X - c_X \left(\int_M \Omega \right)^{-1} , \tag{4.61}$$

and the definition of $\tilde{F}_X$ makes sense for a compact manifold.

Note that:

$$\int_M (L_X F_Y) \Omega \equiv \int_M L_X (F_Y \Omega) \equiv \int_M d(F_Y i_X \Omega) = 0 , \tag{4.62}$$

as $L_X \Omega = 0$. Therefore, whether or not F_Y satisfyies (4.59), $L_X F_Y$ averages always to zero over M.

The Hamiltonian function associated with $[X,Y]$, for $X,Y \in \mathcal{X}_h$ is easily found:

$$i_{[X,Y]}\omega = d(L_X F_Y) = -d(L_Y F_X) \,, \tag{4.62}$$

and hence, using (4.59) and (4.62):

$$F_{[X,Y]} = L_X F_Y = -L_Y F_X \,, \tag{4.63}$$

without ambiguities on constants. Of course, the r.h.s. of (4.64) is nothing but the Poisson Bracket of X and Y:

$$F_{[X,Y]} = L_X F_Y = -L_Y F_X = \omega(Y,X) \stackrel{\text{def}}{=} \{F_X , F_Y\} \,. \tag{4.65}$$

Horizontal fields associated with fields in $\mathcal{X}_h$ commute with the fundamental vertical field V. Indeed, for any $g \in \mathcal{F}(M)$ and any X^h, we have:

$$L_{[X^h,V]}\tilde{\theta} \equiv L_V (L_{X^h}\pi^* g) = L_V (\pi^* (L_X g)) = 0 \,, \tag{4.66}$$

which proves that $[X^h, V]$ is a vertical field. Contracting then with $\tilde{\theta}$, we find:

$$\begin{aligned} i_{[X^h,V]}\tilde{\theta} &= L_{X^h} (i_V \tilde{\theta}) - i_V L_{X^h}\tilde{\theta} = \\ &= -i_V (\pi^* dF) \equiv= L_V \pi * F = 0 \,, \end{aligned} \tag{4.67}$$

which proves that:

$$[X^h, V] = 0 \quad ; \quad \forall X \in \mathcal{X}_h(M) \,. \tag{4.68}$$

Also, we can prove, just as in the case of (4.37), that:

$$[X^h, Y^h] - [X,Y]^h = -\omega(X,Y)V \,. \tag{4.69}$$

Computing now the Lie bracket of $\tilde{X}, \tilde{Y}$, for $X,Y \in \mathcal{X}_h$, we find, using (4.66) and (4.69):

$$\begin{aligned} [\tilde{X},\tilde{Y}] &= [X^h - F_X V, Y^h - F_Y V] = \\ &= [X^h, Y^h] - (L_X F_Y - L_Y F_X)V = \\ &= [X^h, Y^h] + 2\omega(X,Y)V = [X,Y]^h - \omega(Y,X)V = \\ &= [X,Y]^h - F_{[X,Y]}V \stackrel{\text{def}}{=} \\ &\stackrel{\text{def}}{=} \widetilde{[X,Y]} \,, \end{aligned} \tag{4.70}$$

i.e. we have proved that the map $\mathcal{X}_h \ni X \mapsto \tilde{X} \in \tilde{\mathcal{X}}_h$ is a Lie algebra isomorphism onto its image, and hence that the sequence (4.54) splits. This splitting provides us with a universal setting, in the sense that any connected and simply connected Lie group, acting with a Hamiltonian action on M, can be lifted to act on P and preserve the connection one-form. ∎

Remark: In all our construction, if we define X^h to be horizontal and to project onto the vector field X such that $i_X \omega = dF_X$, no requirement is made for ω to be symplectic, i.e. ω can be the curvature form of any $U(1)$-principal bundle. Our construction is still true if we can find a volume element on P invariant under $\tilde{\mathcal{X}}_h$.

5 – LAGRANGIAN MOMENTUM MAPS

In this Section we work in a Lagrangian context, hence $M = TQ$. If then a dynamical system, i.e. a second order vector field Γ and a Lagrangian $\mathcal{L}$ are given, the lifted action of a symmetry group need not in general preserve neither the Cartan one-form $\theta_{\mathcal{L}}$ nor the two-form $\omega_{\mathcal{L}}$, associated with $\mathcal{L}$. Indeed, we have already seen that, for any complete lift X^c (*cfr.*(2.79)):

$$L_{X^c}\theta_{\mathcal{L}} = \theta_{L_{X^c}\mathcal{L}} \; . \tag{5.1}$$

Let's consider the case in which Eq.(2.103) holds. Then, we also know that (cfr. (2.104)):

$$L_{X^c}\theta_{\mathcal{L}} = \theta_{f_i} = df_i \; . \tag{5.2}$$

We recall that the situation is markedly different when one considers the canonical lift of vector fields on the cotangent bundle T^*Q. Indeed, T^*Q is endowed with a characteristic geometrical structure, the *canonical one-form*:

$$\theta_0 \overset{\text{def}}{=} p_i dq^i \; , \tag{5.3}$$

and canonical lifts are defined precisely in such a way as to preserve θ_0 [1,51,58], i.e. $X \in \mathcal{X}(Q)$ lifts to the unique $X^* \in \mathcal{X}(T^*Q)$, which is both π-related to X (i.e. $(T\pi X^*)(q) = X(q)$, $\forall q$, with $\pi : T^*Q \to Q$ (the canonical projection) and satisfies:

$$L_{X^*}\theta_0 = 0 \; . \tag{5.4}$$

The G-action on Q is then canonically lifted to a strongly Hamiltonian action [1,51,58] on T^*Q, and this is the starting point for the usual treatment of the momentum map [1,51,58]. It is also known [1,51,58] that, by virtue of Eq.(5.4), the momentum map is *equivariant* with respect to the coadjoint action of g on g^*, the dual of the Lie algebra g of G (briefly: coadjoint equivariant).

We now proceed to build up a momentum map, but in a Lagrangian context. Let's start by defining the following set of functions:

$$P_k \overset{\text{def}}{=} i_{X_k^c}\theta_{\mathcal{L}} \quad ; \quad k = 1 \ldots \mu \; . \tag{5.5}$$

We shall assume the Lagrangian $\mathcal{L}$ to be regular. The two form $\omega_{\mathcal{L}} = -d\theta_{\mathcal{L}}$ will then be a symplectic form on TQ. The equation:

$$i_{\tilde{X}_k}\omega_{\mathcal{L}} = dP_k \tag{5.6}$$

defines then uniquely the vector fields $\tilde{X}_k \in \mathcal{X}(TQ)$. Note that (5.6) and (5.5) together prove that the $\tilde{X}_k$'s satisfy::

$$L_{\tilde{X}_k}\theta_{\mathcal{L}} = 0 \; . \tag{5.7}$$

The correspondence $X_k \mapsto \tilde{X}_k$ is both a lift and a Lie algebra isomorphism i.e.:

We will only sketch here the main steps of the proof. Contracting $\tilde{X}_i$ with the Lagrangian two-form, one can prove that the following relations hold, for $i = 1 \dots \mu$:

$$i_{(\tilde{X}_i - X_i^c)}\omega_{\mathcal{L}} = \theta_{(L_{X_i^c}\mathcal{L})} \; . \tag{5.9}$$

Whether or not $L_{X_i^c}\mathcal{L} = df_i$ holds for some $f_i = f_i(q)$, the one-form on the r.h.s. of Eq.(5.9), being a Cartan one-form, is semi-basic (i.e. it vanishes on vertical vectors). Therefore, due to the structure of $\omega_{\mathcal{L}}$, the vector field on the l.h.s. must be vertical, i.e. $\tilde{X}_i$ must differ from X_i^c by a vertical field, and hence:

$$T\pi(\tilde{X}_i - X_i^c) = 0 \; , \tag{5.10}$$

proving the first of Eqns.(5.8) (as $T\pi X_i^c = X_i$).

Let's remark that, in view of this:

$$i_{X_k^c}\theta_{\mathcal{L}} \equiv i_{\tilde{X}_k}\theta_{\mathcal{L}} \; , \tag{5.11}$$

i.e. the P_k's can equally well be re-expressed in terms of the $\tilde{X}_k$'s.

The proof of the second of Eqns.(5.8) is a direct consequence of the following result:

$$L_{\tilde{X}_j} P_k = c_{jk}^r P_r \; , \tag{5.12}$$

which is obtained by contracting the two-form $\omega_{\mathcal{L}}$ with the Lie bracket of $\tilde{X}_j$ and $\tilde{X}_k$ (for details see again [64]). In conjunction with the previously obtained results, this constitutes the proof of the second of Eqns.(5.8). ∎

The original G-action on Q has been lifted to a *Hamiltonian* action on the (exact) symplectic manifold $(\omega_{\mathcal{L}}, TQ)$, with the $\tilde{X}_k$ having the same relationship with $\omega_{\mathcal{L}}$ as the canonical lifts X_k^* have with θ_0 and $\omega_0 = -d\theta_0$ on T^*Q. It is then not surprising that this *Lagrangian momentum map* turns out to be coadjoint equivariant as well, and one can actually show [64] that Eq.(5.12) constitutes precisely the proof of this fact.

Remark: It is interesting to observe that the Lagrangian momentum map defined by Eq.(5.5) yields also a quick and compact proof of Nöther's theorem. Indeed, let th X_i^c's be symmetries for the dynamics:

$$[X_i^c, \Gamma] = 0 \quad ; \quad k = 1 \dots \mu \; , \tag{5.13}$$

and satisfy:

$$L_{X_i^c}\mathcal{L} = \dot{f}_k \quad ; \quad f_k = f_k(q) \; . \tag{5.14}$$

Then:

$$0 = i_{[\Gamma, X_k^c]}\theta_{\mathcal{L}} = L_\Gamma P_k - i_{X_k^c} L_\Gamma \theta_{\mathcal{L}} = L_\Gamma P_k - i_{X_k^c} d\mathcal{L} \equiv L_\Gamma (P_j - f_k) \; . \tag{5.15}$$

Hence:

and $P_k - f_k$ is the (Nöther) constant of the motion associated with the infinitesimal symmetry X_k.

The lifting procedure leading to the vector fields $\tilde{X}_k$ has however the drawback that the (local, at least) diffeomorphisms associated with the $\tilde{X}_k$'s need not preserve the second order nature of vector fields[1]. It is then of obvious interest to inquire under which conditions the two lifting procedures can be made to coincide, i.e. under which conditions:

$$\tilde{X}_k = X_i^c \quad ; \quad k = 1 \ldots \mu . \tag{5.17}$$

If the X_k's satisfy Eq.(5.13), then Eq.(5.9) becomes:

$$i_{([\tilde{X}_k - X_k^c])}\omega_{\mathcal{L}} = df_k . \tag{5.18}$$

Coincidence of the two liftings requires then the f_k's to reduce to constants (and viceversa) and we have the result that:

If the X_k are allowed to alter the Lagrangian at most by a total time derivative, the Lagrangian momentum map they define will be coadjoint equivariant iff the Lagrangian is strictly invariant.

According to the analysis of Sect.2, this will happen if the two-cocycle (2.109) in the Lie algebra of **G** belongs to the trivial chomology class in $H^2(g, \mathbf{R})$ (for the discussion of equivalent conditions, see App.A).

We now reconsider the momentum map associated associated with the (lifted) action of **G** as a symmetry group. Altogether, such an action defines intrinsically the momentum map μ as the mapping:

$$\mu : TQ \to g^* \text{ by } \langle \mu(q,u) \mid X_i^c \rangle = P_i = i_{X_i^c}\theta_{\mathcal{L}} . \tag{5.19}$$

Employing the momentum map, we can define coordinates ξ_i, $i = 1 \ldots \mu$, on g^* via the relation:

$$\mu^*(\xi^i) \stackrel{\text{def}}{=} P_i . \tag{5.20}$$

Let's also recall that, as $\xi_i : g^* \to \mathbf{R}$, the coordinates ξ^i are actually elements in the Lie algebra g. Through them, one can define a set of fundamental Poisson Brackets on g^* as:

$$\{\xi^i, \xi^j\} \stackrel{\text{def}}{=} [\xi_i, \xi_j] = c_{ij}^k \xi_k , \tag{5.21}$$

and hence any two smooth functions $f, g \in \mathcal{F}(g^*)$ will have the P.B.:

$$\{f, g\} \stackrel{\text{def}}{=} \frac{\partial f}{\partial \xi^i}\frac{\partial g}{\partial \xi^j}\{\xi^i, \xi^j\} . \tag{5.22}$$

The momentum map defines thus a linear Poisson structure [51] on g^*.

We now inquire about the projectability on g^* of the lifted vector fields $\tilde{X}_i$, defined by Eq.(5.6). Eq.(5.12) is also a projectability condition, proving that $\tilde{X}_i$, $i = 1 \ldots \mu$ projects down to:

$$X_j = c_{jk}^i \xi^k \frac{\partial}{\partial \xi^i} . \tag{5.23}$$

Indeed:

$$L_{\tilde{X}_i}(\mu^*(\xi^j)) \equiv L_{\tilde{X}_i} P_j = c_{ij}^k P_k \equiv \mu^*(L_{X_i} \xi^j) \ . \tag{5.24}$$

The dynamics can be also defined by the equation:

$$i_\Gamma \omega_{\mathcal{L}} = dE_{\mathcal{L}} \ . \tag{5.25}$$

It will not, in general, project unless some additional assumptions are met. A necessary and sufficient projectability condition for Γ is:

$$dE_{\mathcal{L}} \wedge dP_1 \wedge \ldots \wedge dP_\mu = 0 \ , \tag{5.26}$$

if the P_i's are functionally independent ($dP_1 \wedge \ldots \wedge dP_\mu \neq 0$). Γ will then project iff $E_{\mathcal{L}}$ is a function of the generalized momenta P_i, associated with the $\tilde{X}_i$'s (see Eq.(5.6)).

Remark: g^* does not have a tangent bundle structure (not even locally). Therefore, even if Γ is projectable w.r.t. the momentum map onto some $\tilde{\Gamma} \in \mathcal{X}(g^*)$, it does not make sense to require $\tilde{\Gamma}$ to be second order. From the point of view of the momentum map, the *Hamiltonian* description of the dynamics on TQ plays then a more fundamental rôle than the Lagrangian one.

If the action of G on TQ, generated by the X_i^c's, was Hamiltonian w.r.t. $\omega_{\mathcal{L}}$, i.e. if:

$$i_{X_i^c} \omega_{\mathcal{L}} = d\mathcal{P}_i \quad ; \quad i = 1 \ldots \mu \ , \tag{5.27}$$

the $\mathcal{P}_i$'s, each one of which is however undefined by the addition of a constant, can be employed to define a new momentum map:

$$\mu' : TQ \longrightarrow g^* \ , \tag{5.28}$$

and new coordinates η_i via:

$$\mu'^*(\eta_i) = \mathcal{P}_i \ . \tag{5.29}$$

The analog of the fields (5.23) will now be the fields $\mathcal{N}_i$, defined by:

$$\mathcal{N}_i \stackrel{\text{def}}{=} c_{ij}^k \eta^j \frac{\partial}{\partial \eta^k} \ . \tag{5.30}$$

However, the X_i^c will not in general project down to the $\mathcal{N}_i$'s, and that because the arbitrary additive constants entering the definition of the $\mathcal{P}_i$'s, which will give rise to additional 'affine' terms in the projection. In other words, we will have in this case to consider vector fields on g^* of the form:

$$\mathcal{N}_i' = (c_{ij}^k \eta^i + c_k) \frac{\partial}{\partial \eta^k} \ , \tag{5.31}$$

and the new momentum map will turn out to be only 'affine equivariant' [75] (see also [51,35]).

With respect to this momentum map, the dynamical vector field Γ might either project onto the null vector field if $L_{X_j^c} E_{\mathcal{L}} = 0$ or onto a 'constant' vector field if $L_{X_j^c} E_{\mathcal{L}} = c_j$. In the older literature [80] this latter was called a 'non invariance' group.

The main point of this Section was to emphasize the rôle of $\omega_{\mathcal{L}}$ on TQ, i.e. many aspects of the formalism on TQ are genuinely dealing with the second order nature of the dynamics along with the tangent bundle structure of TQ. Other aspects, as those emerging with the reduction associated with an invariance or non invariance group, have really to do with the symplectic structure $\omega_{\mathcal{L}}$ on TQ, considered as a symplectic manifold and forgetting the tangent bundle structure (at the level of lifted vector fields from the base manifold, still a fibered nature of TQ survives, but fibers do not preserve their nature of tangent vector spaces to Q).

6 – CONCLUSIONS AND FINAL COMMENTS

We have discussed in this paper various aspects of the Equivariant Inverse Problem, which we will now summarize and rediscuss briefly, in order to give the reader a unified picture of the difficulties and the possible outcomes of the procedures discussed in the previous Sections.

First of all, we have seen that, if the dynamics admits a large group of symmetries, it may well happen that we cannot even find a two-form which is invariant under the full group, and that the search for it will already select a subgroup of the full symmetry group. Also, this reduction in symmetry may only increase as we proceed further from the two-form to the one-form, and eventually to the Lagrangian, and that even if no cohomological obstructions appear on the way. Stated otherwise, there may arise obstructions in the *invariant* cohomology, that are not present at the level of the ordinary cohomology.

In Sects.3 and 4, we have sketched a general procedure for resolving the obstructions. Going, via the construction of the path space over the carrier space (or the configuration space , if the carrier space is a tangent bundle), to a $U(1)$-bundle over the same space resolves the cohomological obstructions. The price to be paid is that we have to go from the original symmetry group G to the 'Bargmann group' $B(\mathrm{G})$ and, if it is not possible to find a natural embedding of G as a subgroup of the Bargmann group, one might be obliged to work with the full $B(\mathrm{G})$ as a new symmetry group.

Again, in the case in which an invariant two-form has been found to exist, a systematic way to recover both an invariant one-form *and* an invariant Lagrangian has been seen to consist in an enlargment of the configuration space by the addition of one extra, non dynamical, degree of freedom. This leads again to a problem of going to a central extension of the original algebra of symmetries, and, when the latter can be exponentiated, of the symmetry group.

Another price has to be paid in pursuing this program, namely the fact that we eventually end up with degenerate Lagrangians, and the dynamical problem has to be treated within the framework of Dirac-Bergmann theory of constraints [27,13]. Similar problems have been seen to arise in the discussion of the Lagrangian momentum map.

Remark: All of the above story concerning the transition from the dynamics to the Lagrangian seems also to suggest a word of caution about the procedure of guessing interactions by appropriate selection of terms in the Lagrangian. Such a procedure is perhaps more common in Field Theory than it is in point particle mechanics, but also in the latter context it leads, e.g. when one insists on Poincaré invariance, to the No-Interaction Theorem [9].

The procedure could be too restrictive if the 'experimental evidence' for symmetries is at the level of trajectories in some configuration space. Neverthless, we agree that, when looking for systems having some degree of symmetry, it is much simpler to look for an invariant function (i.e. the Lagrangian) rather than for an invariant tensor (be it the one or the two-form, or the vector field describing the dynamics).

The relevant question which arises at this point is whether the Equivariant Inverse Problem becomes actually a void problem. In other words, one might suspect that, at the end of the procedure, the only ambiguities left in the Lagrangian description amount to the usual gauge freedom, i.e. to the addition of a total time derivative, or to the multiplication by a number.

To show that this is not the case, we will now exhibit examples of alternative Lagrangians for a few dynamical systems, all admitting only a selected symmetry group.

The simplest example is that of a nonrelativistic free particle in $\mathbf{R}^3$. The dynamics is described by the vector field:

$$\Gamma = u^i \frac{\partial}{\partial q^i} \ . \tag{6.1}$$

As already remarked, an invariance group of point symmetries for Γ is provided by the inhomogeneous general linear group in three dimensions, whose lifted action is generated by the vector fields:

$$X_{A,\vec{b}} = A_i{}^j q^i \frac{\partial}{\partial q^j} + b^k \frac{\partial}{\partial q^k} + A_i{}^j u^i \frac{\partial}{\partial u^j} \ , \tag{6.2}$$

where $A = \|A_i{}^j\| \in \mathrm{Lin}(\mathbf{R}^3, \mathbf{R}^3)$ and $\vec{b} = (b^1, b^2, b^3) \in \mathbf{R}^3$. This group will not even admit of an invariant Lagrangian two-form, not to speak of a Lagrangian.

Having to select a subgroup, physical reasons seem to compel us to select the Euclidean group[1] $E(3)$, which is again generated by vector fields of the form (6.2), but ones in which A is a skew-symmetric matrix (i.e. the A's will span the Lie algebra of $SO(3)$), while $\vec{b}$ is again an arbitrary vector. As $E(3)$ does not have nontrivial central extensions, we are permitted to search for invariant Lagrangians directly on $T\mathbf{R}^3$, and it is easily seen that any Lagrangian of the form:

$$\mathcal{L} = \frac{1}{2} f(|\vec{u}|^2) \quad ; \quad \vec{u} \equiv (u^1, u^2, u^3) \quad ; \quad f \in \mathcal{F}(\mathbf{R}) \ , \tag{6.3}$$

will be an invariant Lagrangian and a nondegenrate one iff $\det \|f'\delta_{ij} + 2f'' u^i u^j\| \neq 0$. The situation changes if we want invariance not under $E(3)$, but under the Galilei group. The boost generators are given by:

$$K^i = t \frac{\partial}{\partial q^i} + \frac{\partial}{\partial u^i} \ , \tag{6.4}$$

and we have:

$$L_{K_j} f(|\vec{u}|^2) = 2f'(|\vec{u}|^2) u^j \ . \tag{6.5}$$

[1] Inclusion or not of discrete operations (parity) will not be relevant for our purposes.

We can thus achieve at most quasi invariance, and that if and only if $f' = \text{const.}$, which brings us back to the usual class of Lagrangians: $\mathcal{L} \propto |\vec{u}|^2$. This fact comes to no surprise. Indeed, the Galilei group has a nontrivial cocycle connected to the mass [75] and we will then be forced to go to a central extension [75] in order to recover an invariant description.

As a next example, let's consider a two-dimensional isotropic harmonic oscillator of unit mass and frequency. In this case:

$$\Gamma = u^i \frac{\partial}{\partial q^i} - q^i \frac{\partial}{\partial u^i} \; . \tag{6.6}$$

The most general invariance group for Γ is now the homogeneous general linear group in two dimensions. Again the requirement of invariance imposed on the two and one-form and on the Lagrangian will force us to a drastic reduction of the group, and physical reasons lead us to consider $SO(2)$ (or $O(2)$, if we want to include parity) as a candidate for the ultimate invariance group. The standard Lagrangian:

$$\mathcal{L} = \frac{1}{2}\left[(u^1)^2 + (u^2)^2 - (q^1)^2 - (q^2)^2\right] \; , \tag{6.7}$$

is of course manifestly invariant under $SO(2)$. It is well known [55] that for the harmonic oscillator one can exhibit quite a few alternative Lagrangians. For example, an admissible Lagrangian will be:

$$\mathcal{L}_1 = \frac{1}{2}\left[(u^1)^2 - (u^2)^2 - (q^1)^2 + (q^2)^2\right] \; ,$$

which is invariant under $SO(1,1)$. Other alternative Lagrangians have been found, which are quartic functions of the position and velocities. One of them is [6]:

$$\mathcal{L}_2 = \frac{1}{3}\left[(u^1)^4 + (u^2)^4\right] + 2\left[(q^1)^2(u^1)^2 + (q^2)^2(u^2)^2\right] - (q^1)^4 - (q^2)^4 \; . \tag{6.9}$$

Now, as $SO(2)$ is compact, we may try to apply the procedure of averaging over the group already discussed in Sect.2. In doing so, we find, indicating the average with a bar:

$$\bar{\mathcal{L}} = \mathcal{L} \quad ; \quad \bar{\mathcal{L}}_1 = 0 \; , \tag{6.10}$$

but:

$$\bar{\mathcal{L}}_2 = \frac{3}{4}\left\{|\vec{u}|^4 - |\vec{q}|^4\right\} + 3(\vec{q}\cdot\vec{u})^2 + \frac{1}{2}|\vec{q}\times\vec{u}|^2 - 3q^1q^2u^1u^2 \; , \tag{6.11}$$

where $\vec{q} = (q^1, q^2)$, $\vec{u} = (u^1, u^2)$.

$\bar{\mathcal{L}}_2$ is therefore an alternative, nontrivial Lagrangian satisfying the requirement of invariance.

As a final example, we may consider the Kepler problem, described, on $T(\mathbf{R}^3 - \{0\})$, by:

$$\Gamma = u^i \frac{\partial}{\partial q^i} - g\frac{q^i}{r^3}\frac{\partial}{\partial u^i} , \tag{6.12}$$

with $r^2 = (q^1)^2 + (q^2)^2 + (q^3)^2$. Selecting $SO(3)$ as the invariance group, the standard Lagrangian:

$$\mathcal{L} = \frac{1}{2}|\vec{u}|^2 - \frac{g}{r} \tag{6.13}$$

is manifestly invariant. It is again not unique, though. Indeed, in a well known paper, Henneaux and Shepley [37] have exhibited a one parameter family of $SO(3)$ invariant Lagrangians in the form:

$$\mathcal{L}_\lambda = \mathcal{L} + \lambda\frac{J}{r^2} \quad ; \quad \lambda \in \mathbf{R} , \tag{6.14}$$

where:

$$J = \quad |\vec{u}|^2|\vec{q}|^2 - (\vec{q}\cdot\vec{u})^2 \quad {}^{1/2} \tag{6.15}$$

is the angular momentum. Crampin and Prince [21,22] have further shown that the family (6.14) exhausts all possible $SO(3)$ invariant Lagrangians for the Kepler problem.

The EIP is thus not void of content, as the above examples show. It is then a program worth pursuing, with an interest of its own in point particle mechanics. As alredy mentioned, it is also of great interest in the context of field theory, where the lack of fullfillment of the program may be at the origin of anomalies and Wess-Zumino terms in the quantized version of the theory [88,89,32,29]. We hope to come back in a systematic way on this last problem in the near future.

We would like to close this paper by mentioning a possible pedagogical use of it. Cohomology and extensions were studied by physicists for the purpose of unifying space-time symmetries and internal symmetries. Later on, cohomology classes appeared in connection with the geometric quantization program. Around the same period Larry Biedenharn argued that anomalies could be interpreted as cohomology classes [14,88,32]. This interpretation was actually carried over, and it is by now a familiar characterization of anomalies, local and global, with physicists.

More recently, string theories and loop algebras have made extensive use of central extensions, cocycles and cohomology classes [68]. Recently, Lie algebra extensions have been used to discuss a possible algebraic approach to gauge theories and general relativity [46-50]. It seems to us a good opportunity to have simple examples from particle mechanics to illustrate the abstract mathematical construction that shows up in a variety of physical situations.

APPENDIX A – ORDINARY AND INVARIANT COHOMOLOGIES OF LIE ALGEBRAS

We try and give here a systematic, albeit sketchy account of the cohomology theory of Lie algebras, which has been quoted and used in various Sections of the present paper. For a comprehensive discussion of the subject, we refer to the paper by Chevalley and Eilenberg [22].

Let then g be a Lie algebra, and let Λ be any g-module. *k-cochains* are defined as multilinear alternating maps (linear maps for $k=1$):

$$\alpha : \underbrace{g \times \cdots \times g}_{k \text{ times}} \longrightarrow \Lambda \,. \tag{A.1}$$

The set of k-cochains will be denoted by $C^k(g,\Lambda)$ and, by definition: $c^0(g,\Lambda) \equiv \Lambda$. Let $e_1 \dots e_\mu$, $\mu = \dim g$, be a basis in g. A *coboundary operator* ∂ can be defined on C^k, $0 \le k \le \mu$, as:

$$\begin{aligned} &\partial : C^k \longrightarrow C^{k+1} \\ &\partial\alpha(e_1 \dots e_{k+1}) = \\ &\quad = \sum_j (-)^{j+1}(e_j \circ \alpha)(e_1 \dots \hat{e}_j \dots e_{k+1}) + \\ &\quad + \sum_{i<j} (-)^{i+j} \alpha([e_i, e_j], e_1 \dots \hat{e}_i \dots \hat{e}_j \dots e_{k+1}) \,, \end{aligned} \tag{A.2}$$

where $(e_j \circ \alpha)(\dots) \stackrel{\text{def}}{=} e_j[\alpha(\dots)]$ stands for the action of e_j on $\alpha(\dots) \in \Lambda$. We recall [35] that, if g is the Lie algebra of a simply connected group G, then C^k is isomorphic to $\Lambda^k_{\text{r.i.}}$, the linear vector space of right-invariant forms on G. The action of g can then be identified with the Lie derivative w.r.t. right-invariant vector fields on G, and then $\partial \equiv d$, the exterior differential. As defined by eq.(A.2), ∂ satisfies:

$$\partial\partial = 0 \,. \tag{A.3}$$

A cochain $\alpha \in C^k$ is a *coboundary* if $\alpha = \partial\beta$, for some $\beta \in C^{k+1}$, a *cocycle* if $\partial\alpha = 0$. Calling Z^k the space of k-cocycles, B^k that of k-coboundaries, the *k-th cohomology group of* g, $H^k(g,\Lambda)$, is defined as:

$$H^k = Z^k / B^k \,. \tag{A.4}$$

Let us now illustrate the above general situation with a few examples taken from the main text, i.e. by rediscussing in some more detail cocycles associated with quasi invariant Lagrangians (see Sect.2). If the G-action on TQ results in the Lagrangian obeying Eq.(2.102), i.e.:

$$L_{X_i^c}\mathcal{L} = \dot{f}_i \in \dot{\mathcal{F}}(TQ) \,, \tag{A.5}$$

then, as stated in Sect.2,

$$\dot{f} : g \to \dot{\mathcal{F}}(TQ) \qquad \text{by} \qquad e_i \mapsto L_{X_i^c}\mathcal{L} = \dot{f}_i \tag{A.6}$$

is a cochain in $C^1(g, \dot{\mathcal{F}})$. Now the action of g on $\dot{\mathcal{F}}$ is given by the Lie derivative: $e_i \dot{f} \equiv L_{X_i^c}\dot{f}$, with X_i^c the infinitesimal generator of the lifted action of G associated with $e_i \in g$, and that $\dot{f}$ is a cocycle simply follows from:

$$\partial\dot{f}(e_i, e_j) = L_{X_i^c}(L_{X_j^c}\mathcal{L}) - L_{X_j^c}(L_{X_i^c}\mathcal{L}) - L_{[X_i^c, X_j^c]}\mathcal{L} \equiv 0 \,. \tag{A.7}$$

Proving that, if instead (2.101) holds, then the one-cochain:

$$c \in C^1(g, \mathbf{R}) \quad ; \quad c : e_i \mapsto c(e_i) = L_{X_i^\circ} \mathcal{L} \tag{A.8}$$

is a cocycle is even simpler (see Eq.(2.106)). Turning again to Eq.(2.102), we have already seen that one can also define a two-cochain $A \in C^2(g, \mathbf{R})$ by:

$$A : (e_i, e_j) \mapsto A_{ij} \stackrel{\text{def}}{=} L_{X_i} f_j - L_{X_j} f_i - c_{ij}^k f_k \ . \tag{A.9}$$

Now, g acts trivially on $\mathbf{R}$ (as in the case of Eq.(A.8)). Therefore:

$$\partial A(e_i, e_j, e_k) = -\left\{ A\big([e_i, e_j], e_k\big) + \text{cycl. perm. of } (i, j, k) \right\} \ , \tag{A.10}$$

and

$$\partial A = 0 \Leftrightarrow c_{ij}^m A_{mk} + c_{ki}^m A_{mj} + c_{jk}^m A_{mi} = 0 \ . \tag{A.11}$$

But:

$$c_{ij}^m A_{mk} = \\ \{L_{X_i} L_{X_j} f_k - L_{X_k} L_{X_i} f_j + L_{X_k} L_{X_j} f_i - L_{X_j} L_{X_i} f_k\} - c_{ij}^m c_{mk}^s f_s \ .$$

Summing over cyclic permutations of (i, j, k), the terms in curly brackets add up to zero, and hence (A.11) is a consequence of the Jacobi identity for the structure constants of $\mathbf{G}$.

Under the conditions of Eq.(2.102), we also have[1] :

$$L_{X_i^\circ} \theta_{\mathcal{L}} = df_i \ . \tag{A.12}$$

We can then also take, as a g-module, $\Lambda = \mathcal{X}_E^*(Q)$, the *exact* one-forms on Q, with the action of g defined again by the Lie derivative, and define:

$$\mu : g \to \mathcal{X}_E^* \quad ; \quad e_i \mapsto L_{X_i^\circ} \theta_{\mathcal{L}} \ . \tag{A.13}$$

Hence, $\mu \in C^1(g, \mathcal{X}_E^*)$, and:

$$\begin{aligned} \partial\mu(e_i, e_j) &= L_{X_i} df_j - L_{X_j} df_i - \mu([e_i, e_j]) = \\ &= L_{X_i^\circ} L_{X_j^\circ} \theta_{\mathcal{L}} - L_{X_j^\circ} L_{X_i^\circ} \theta_{\mathcal{L}} - L_{[X_i^\circ, X_j^\circ]} \theta_{\mathcal{L}} \equiv 0 \ , \end{aligned} \tag{A.14}$$

and hence μ is a one-cocycle. Defining:

$$\mu(e_i, e_j) = L_{[X_i^\circ, X_j^\circ]} \theta_{\mathcal{L}} \stackrel{\text{def}}{=} df_{ij} \ , \tag{A.15}$$

we find:

$$f_{ij} = c_{ij}^k (f_k + a_k) \ , \tag{A.16}$$

with the a_k's undetermined constants. The cocycle condition (A.17) can then also be rewritten as:

$$d(L_{X_i} f_j - L_{X_j} f_i - f_{ij}) = 0 \ , \tag{A.17}$$

[1] With abuse of notation, we write here df_i for what should be $\pi^* df_i$.

which, on account of Eq.(A.16), leads again to the two-cocycle of Eq.(A.9), with the A_{ij}'s being defined 'modulo' a term of the form $c^k_{ij} a_k$, which is however a coboundary:

$$c^k_{ij} a_k = (\partial a)(e_k) \quad ; \quad a : e_k \mapsto a_k \; . \tag{A.18}$$

As in this paper we have been mainly intersted in discussing the solubility of the Equivariant Inverse Problem, we close this appendix by introducing and discussing briefly the notion of *invariant* cohomology of Lie algebras. Let $X \stackrel{\text{def}}{=} a^j e_j \in g$, with $a_j \in \mathbf{R}$. Using the definition (A.2) of the coboundary operator, one can show by direct computation that:

$$\partial(X \circ \alpha) = X \circ (\partial\alpha) \; . \tag{A.19}$$

Invariant k-cochains will be then defined as:

$$C^k_I(g, \Lambda) = \{\alpha \in c^k(g, \Lambda) \,|\, X \circ \alpha = 0 \quad \forall X \in g\} \; . \tag{A.20}$$

Because of (A.19), the coboundary operator maps invariant cochains into invariant ones, and hence satisfies (A.3) on invariant cochains as well. We can then define without ambiguities the spaces Z^k_I of invariant k-cocycles and B^k_I of invariant coboundaries, and the ***k-th invariant cohomology group*** as:

$$H^k_I \stackrel{\text{def}}{=} B^k_I / Z^k_I \; . \tag{A.21}$$

The restriction ∂_I of ∂ to invariant cochains operates as:

$$\begin{aligned} &\partial_I : C^k_I \to C^{k+1}_I \\ &\partial_I \alpha(e_1 \ldots e_{k+1}) = \sum_{i<j} (-)^{i+j} \alpha([e_i, e_j], e_1 \ldots \hat{e}_i \ldots \hat{e}_j \ldots e_{k+1}) \; , \end{aligned} \tag{A.22}$$

i.e. the first term in the definition (A.2) is missing. It is precisely the absence of this term which makes the invariant cohomology (also called the *algebraic* cohomology) to be different from the de Rham cohomology. This was already noticed, e.g. in the discussion of the motion of a charged particle in a magnetic field in Sect.2, where we found no obstructions in the de Rham complex, but, on the contrary, no invariant one-form was found to exist.

When Λ is a trivial g-module, say the field K of numbers on which the Lie algebra is defined, we have:

$$C^1(g, K) \rightleftharpoons g^* \; . \tag{A.23}$$

It follows that:

$$Z^1(g, K) = [g, g]^0 \; , \tag{A.24}$$

with $[g, g]^0$ the anihilator of $[g, g]$:

$$[g, g]^0 = \{x^* \in g^* \,|\, x^*([e_i, e_j]) = 0 \quad \forall e_i, e_j \in g\} \; . \tag{A.25}$$

Since $B^1(g, K)$ is clearly zero (for $\partial k(x) = 0$ when $k \in K$), it follows that:

$$Z^1(g, K) \equiv H^1(g, K) = [g, g]^0 = (g/[g, g])^* \; . \tag{A.26}$$

Therefore:

$$H^1(g, K) = 0 \qquad \text{iff} \qquad g = [g, g] \; . \tag{A.27}$$

What is remarkable is that it can be shown that, if g is finite dimensional, then:

$$H^1(g, K) = 0 \Rightarrow H^1_I(g, \Lambda) = 0 \; , \tag{A.28}$$

for *any* g-module Λ. If, in addition, $H^2(g, K) = 0$, then one obtains $H^2_I(g, \Lambda) = 0$ as well. In other words, Whitehead's Lemma can be extended, under the above conditions, to the invariant cohomology with values in any g-module Λ.

APPENDIX B – CENTRAL EXTENSIONS OF LIE ALGEBRAS AND GROUPS

We shall briefly recall here a few definitions concerning the extensions of Lie algebras and groups, concentrating on central extensions.

Let g be a Lie algebra. A *central extension* of g is a Lie algebra E, together with a surjective homomorphism::

$$\pi : E \to g \, , \tag{B.1}$$

such that $A = \ker(\pi)$ is in the center of E. If:

$$i : A \hookrightarrow E \tag{B.2}$$

is the inclusion mapping, a central extension is represented by the following short exact sequence:

$$0 \longrightarrow A \xrightarrow{\;i\;} E \xrightarrow{\;\pi\;} g \longrightarrow 0 \tag{B.3}$$

We shall consider in the sequel the case in which $A = V$, a linear vector space. Two central extensions E_1 and E_2 of g by V:

$$\begin{array}{c} 0 \longrightarrow V \xrightarrow{\;i_1\;} E_1 \xrightarrow{\;\pi_1\;} g \longrightarrow 0 \\ \\ 0 \longrightarrow V \xrightarrow{\;i_2\;} E_2 \xrightarrow{\;\pi_2\;} g \longrightarrow 0 \end{array} \tag{B.4}$$

will be called *isomorphic* iff there is a homomorphism:

$$j : E_1 \to E_2 \, , \tag{B.5}$$

such that the diagram:

$$\begin{array}{ccccccccc} 0 & \longrightarrow & V & \xrightarrow{\;i_1\;} & E_1 & \xrightarrow{\;\pi_1\;} & g & \longrightarrow & 0 \\ & & \| & & \downarrow{\scriptstyle j} & & \| & & \\ 0 & \longrightarrow & V & \xrightarrow{\;i_2\;} & E_2 & \xrightarrow{\;\pi_2\;} & g & \longrightarrow & 0 \end{array} \tag{B.6}$$

commutes. It actually follows that j is an isomorphism. Therefore, an isomoprphism of central extensions is an equivalence relation on the set of extensions of the given algebra g by a vector space V. We shall denote by $\mathrm{Ext}(g, V)$ the set of equivalence classes of mutually isomorphic central extensions og g by V. If now V is considered as a trivial g-module, we can define V-valued two-cocycles on g as (cfr. App.A) skew symmetric bilinear maps:

$$\begin{array}{l} c : g \times g \to V \quad \text{s.t.} \\ \{c([e_i, e_j], e_k) + \text{cycl. perm. of } (i, j, k)\} = 0 \quad ; \quad \forall e_i, e_j, e_k \in g \, . \end{array} \tag{B.7}$$

Such bilinear maps define $Z^2(g, V)$.

Remark: Had we chosen $g = V$ as a vector space, the cocycle condition would define a new Lie algebra structure on V.

Note that to any linear map:

$$f : g \to V \ , \tag{B.8}$$

we can associate the coboundary:

$$c_f : g \ni (a,b) \mapsto c_f(a,b) \stackrel{\text{def}}{=} f([a,b]) \ , \tag{B.9}$$

and viceversa. The linear maps (B.8) define then $B^2(g,V)$, and the quotient space is the second cohomology group $H^2(g,V)$.

The main result we want to recall is that there is a natural bijection:

$$H^2(g,V) \rightleftharpoons \text{Ext}(g,V) \ , \tag{B.10}$$

defined as follows: let $\beta \in H^2(g,V)$, and pick up an element $c \in \beta$. then, define a central extension g_c of g by V, taking the direct sum $g \oplus V$ as the underlying vector space of g_c, and defining a Lie bracket on g_c by:

$$[(v,x),(v',x')] = (c(x,x'),[x,x']) \quad ; \quad v,v' \in V \quad x,x' \in g \ . \tag{B.11}$$

If $i : V \hookrightarrow g_c$ is given by:

$$i(v) \stackrel{\text{def}}{=} (v,0) \ , \tag{B.12}$$

and the projection $\pi : g_c \to g$ by:

$$\pi(v,x) = x \ . \tag{B.13}$$

Then it is clear that:

$$0 \longrightarrow V \stackrel{i}{\longrightarrow} g_c \stackrel{\pi}{\longrightarrow} g \longrightarrow 0 \tag{B.14}$$

is a central extension of g by V, and hence determines an isomorphism class in $\text{Ext}(g,V)$. We have thus proved that an element in $H^2(g,V)$ determines a class in $\text{Ext}(g,V)$. This map can also be proved to be a bijection. the study of $H^2(g,V)$ is then useful to classify the central extensions of g by V.

Behaviour of $H^2(g,V)$ under Lie algebra homomorphisms

Let:

$$\phi : h \to g \tag{B.15}$$

be an homomorphism between the Lie algebras g and h. then, there is an induced homomorphism:

$$\phi^* : H^2(g,V) \longrightarrow H^2(h,V) \ , \tag{B.16}$$

which is defined as follows:

Take $\beta \in H^2(g,V)$ and a representative two-cocycle $c \in \beta$. We have therefore a bilinear map $c : g \times g \to V$. Define $\phi^* c$ as:

$$\begin{aligned} &\phi^* c = c \circ (\phi \times \phi) \\ &\text{i.e.: } (\phi^* c)(\eta_1,\eta_2) \stackrel{\text{def}}{=} c(\phi(\eta_1),\phi(\eta_2)) \quad ; \quad \eta_1,\eta_2 \in h \ . \end{aligned} \tag{B.17}$$

Therefore $\phi^* c \in Z^2(h,V)$, and hence determines a cohomology class in $H^2(h,V)$. We set:

$$\phi^*\beta \stackrel{\text{def}}{=} [\phi^* c] \quad ; \quad c \in \beta \ . \tag{B.18}$$

If V and W are vector spaces and:

$$\rho : W \to V \tag{B.19}$$

is any linear map, we define:

$$\rho_* : H^2(g,W) \to H^2(g,V) \qquad \text{by:} \quad \rho_*[c] \stackrel{\text{def}}{=} [\rho \circ c] \ . \tag{B.20}$$

Suppose we have the following diagram, whose horizontal parts are central extensions:

$$\begin{array}{ccccccccc} 0 & \longrightarrow & W & \longrightarrow & F & \longrightarrow & h & \longrightarrow & 0 \\ & & \downarrow{\scriptstyle\rho} & & & & \downarrow{\scriptstyle\phi} & & \\ 0 & \longrightarrow & V & \longrightarrow & E & \longrightarrow & g & \longrightarrow & 0 \end{array} \tag{B.21}$$

then a homomorphism:

$$\psi : F \to E \ , \tag{B.22}$$

which renders commutative the diagram will exist iff:

$$\phi^*(\beta) = \rho_*(\nu) \ , \tag{B.23}$$

with $\beta \in H^2(g,V)$, $\nu \in H^2(h,W)$, i.e. β is the class of the lower sequence and ν that of the upper one.

In general, $H^2(g,V)$ is isomorphic to $H^2(g,\mathbf{R}) \otimes V$. This can be seen as follows: if $\beta \in H^2(g,\mathbf{R})$ and $v \in V$, let $\beta \otimes v$ be the class in $H^2(g,V)$ of the two-cocycle:

$$\tilde{c} : (a,b) \mapsto c(a,b) \otimes v \quad ; \quad a,b \in g, \quad v \in V \ . \tag{B.24}$$

This identifies $H^2(g,V)$ with $H^2(g,\mathbf{R}) \otimes V$, where the first factor is often simply indicated by $H^2(g)$. Calling then K the linear dual of $H^2(g)$, $K \stackrel{\text{def}}{=} (H^2(g))^*$, there is a natural isomorphism of vector spaces between $H^2(g) \otimes V$ and $\mathrm{Hom}(K,V)$. In particular, K is itself a vector space, so we may also consider $H^2(g,K)$, which turns out to be isomorphic to $\mathrm{Hom}(K,K)$. Let then $\nu_0 \in H^2(g,K)$ be the class corresponding to the identity map $e \in \mathrm{Hom}(K,K)$. Choosing a representative two-cocycle in ν_0, we can determine a central extension g_e of g , defined by the sequence:

$$0 \longrightarrow K \xrightarrow{i_e} g_e \xrightarrow{\pi_e} g \longrightarrow 0 \ . \tag{B.25}$$

We shall call ν_0 the *universal central extension* of g. This is because of the following proposition:

Let V be any vector space, and let ρ be any element of $H^2(g, V)$ (we are thus considering central extensions of g by V). As:

(B.26) $$H^2(g, V) \simeq \mathrm{Hom}(K, V) ,$$

there exists a linear map:

(B.27) $$\tilde{\rho} : K \rightarrow V ,$$

associated with ρ, and hence an induced map between cohomology groups:

(B.28) $$\tilde{\rho}_* : H^2(g, K) \rightarrow H^2(g, V) .$$

It follows then that:

(B.29) $$\rho = \tilde{\rho}_*(\nu_0) . \quad \blacksquare$$

What the previous proposition tells us is that every central extension of g by a vector space V can be determined from the extension by K through the choice of the map (B.27). That's why g_e is called the 'universal' extension of g.

The experienced reader will have recognized that universal central extensions play, in this algebraic context, a rôle which is quite similar to that played, in the context of differentiable manifolds, by principal bundles in their relationship with associated bundles. Also, it is quite clear that the universal extension g_e will play a special rôle whenever we will have to consider the lifting problem of the action of a given Lie algebra g of symmetries on a manifold (see Sect.4).

Let us recall that, according to Bargmann, results at the Lie algebra level can be exponentiated to the simply connected Lie group having g_e as Lie algebra. The surjection $\pi_e : g_e \rightarrow g$ integrates to a surjection $G(g_e) \rightarrow G(g)$. Since these groups are both simply connected, the kernel of the surjection is a closed subgroup, with Lie algebra K, defining a central subgroup of $G(g_e)$ and

(B.30) $$G(g_e)/K \quad \simeq \quad G(g) .$$

We thus get a central extension of $G(g)$:

(B.31) $$1 \longrightarrow K \longrightarrow G(g_e) \longrightarrow G(g) \longrightarrow 1 .$$

When the starting groups are not simply connected, one has to take into account the homotopy group $\pi_1(G(g))$ and try to quotient all our construction with respect to such subgroup. The situation can still be handled, but the analysis is much more involved.

REFERENCES

1. R. Abraham and J. Marsden, "Foundations of Mechanics," Benjamin, New York, 1978.

2. R. Abraham, J. Marsden and T. Ratiu, "Manifolds, Tensor Analysis and Applications," Addison Wesley, Readin Mass., 1983.

3. J.F. Adams, "Infinite Loop Spaces," Princeton Univ. Press, Princeton N.J., 1978.

4. I. M. Anderson, *Natural Variational Principles on Riemannian Manifolds*, Ann. Math. **120** (1984), 329.

5. I.M. Anderson, *Aspects of the Inverse Problem to the Calculus of Variations*, Preprint Utah State Univ. (1987).

6. P. Antonini, G. Marmo, C. Rubano, *Alternative Lagrangians and Complete Integrability: Some Remarks*, Il Nuovo Cimento **86B** (1985), 17.

7. V.I. Arnol'd, "Mathematical Methods of Classical Mechanics," Springer, New York, 1978.

8. A.P. Balachandran, G. Marmo, B.S. Skagerstam and A. Stern, *Supersymmetric Monopoles without Strings*, Nucl. Phys. **164B** (1980), 427.

9. A.P. Balachandran, G. Marmo and A. Stern, *A Lagrangian Approach to the No-interaction Theorem*, Il Nuovo Cimento **69A** (1982), 175.

10. A. P. Balachandran, G. Marmo, B.S. Skagerstam and A. Stern, "Gauge Symmeries and Fiber bundles," LNP **188**, Springer Verlag, Berlin, 1983.

11. A.P. Balachandran, H. Gromm and R.D. Sorkin, *Quantum Symmetries from Quantum Phases. Fermions from Bosons, a* $\mathbf{Z}_2$ *Anomaly and Galileian Invariance*, Nucl. Phys. **281B** (1987), 573.

12. V. Bargmann, *On Unitary Ray Representations of Continous groups*, Ann. of Math. **59** (1954), 1.

13. P. Bergmann, I. Goldberg, *Dirac Bracket Transformations in Phase Space*, Phys. Rev. **98** (1955), 531.

14. L.C. Biedenharn, *The Structure of the Phenomenological Approach to the Adler Anomaly*, in "Colloquium on Group Theoretical Methods in Physycs," (H. Bacry Ed.), Centre de Physique Theorique, Marseille, 1972.

15. C. Chevalley and S. Eilenberg, *Cohomology Theory of Lie Groups and Lie Algebras*, Trans. Am. Math. Soc. **63** (1948), 85.

16. J.F. Cariñena and L.A. Ibort, *Geometric Theory of the Equivalence of Lagrangians for Constrained Systems*, J.Phys A: Math: Gen. **18** (1985), 3355.

17. E. Cartan, "Leçon sur les Invariants Integraux," Hermann, Paris, 1922.

18. E. Cartan, "Les Systèmes Differentielles Extérieurs et leur Application Geometriques," Hermann, Paris, 1945.

19. M. J. Crampin, *On the Differential Geometry of the Euler-Lagrange Equations, and the Inverse Problem of Lagrangian Dynamics*, J. Phys. **A14** (1981), 2567.

20. M. Crampin, *Tangent Bundle Geometry for Lagrangian Dynamics*, J. Phys. A: Math. Gen. **16** (1983), 3755.

21. M. Crampin and G. Prince, *Generalizing Gauge Variance for Spherically Symmetric Potentials*, J. Phys. A: Math. Gen. **18** (2167), 1985.

22. M. Crampin and G. Prince, *Alternative Lagrangians for Spherically Symmetric Potentials*, Preprint Open University, London (1987).

23. P. Dedecker and W.M. Tulczyjew, "Spectral Sequence on the Inverse Problem in the Calculus of Variations," LNM **836**, Springer Verlag, Berlin, 1980.

24. C. De Witt–Morette, A. Maheshwari and B. Nelson, *Path Integration in Non-relativistic Quantum Mechanics*, Phys. Rep. **50** (1979), 255.

25. P.A.M. Dirac, *Quantized Singularities in the Electromagnetic Field*, Proc. Roy. Soc. London **A133** (1931), 60.

26. P.A.M. Dirac, "The Principles of Quantum Mechanics," Clarendon Press, Oxford, 1947.

27. P.A.M. Dirac, "Lectures in Quantum Mechanics," Belfer Graduate School of Science, Yeshiva University, New York, 1964.

28. V.V. Dodonov, V.I. Man'ko, V. Skarzhinsky, *The Inverse Problem of the Variational Calculus and Non-uniqueness of the Quantization of Classical Systems*, Hadron. J. **4** (1981), 1734.

29. L.D. Faddeev, S.L. Shatashvili, *Algebraic and Hamiltonian Methods in the Theory of Non-abelian Anomalies*, Theoret. Math. Phys. **50** (1985), 770.

30. R.P. Feynmann and A.R. Hibbs, "Quantum Mechanics and Path Integral," Mc Graw-Hill, New York, 1965.

31. A. Frölicher and A. Nijenhuis, *Theory of Vector Valued Differential Forms*, Nederl. Acad. Wetensch. Proc. Ser. **A59**, Indag. Math. **18** (1956), 338.

32. M. Göckeler and T.Schücker, "Differential Geometry, Gauge Theory and Gravity," Cambridge Univ. Press, Cambridge, 1987.

33. C. Godbillon, "Géometrie Differentielle et Mécanique Analitique," Hermann, Paris, 1969.

34. H. Goldstein, "Classical Mechanics," Addison-Wesley, Reading, Mass., 1950.

35. V. Guillemin and S. Sternberg, "Symplectic Techniques in Physics," Cambridge University Press, Cambridge, 1984.

36. H. Helmoltz, *Über die Physikalische Bedeutung des Princips der Kleisten Wirkung*, Jour. f. d. Reine v. Angew. Math. **100** (1887), 137.

37. M. Henneaux, L.C. Shepley, *Lagrangians for Spherically Symmetric Potentials*, J. Math. Phys. **23** (1982).

38. G.Hochschild, "The Structure of Lie Groups," Holden–Day, San Francisco, 1978.

39. P.A. Horvathy, *Quantization in Multiply Connected Spaces*, Phys. Lett. **76A** (1980), 11.

40. C. Itzykson, J.B. Zuber, "Quantum Field Theory," Mc Graw-Hill, New York, 1980.

41. J. Kijovski and W.M. Tulczyjew, "A Symplectic Framework for Field Theories," LNP **107**, Springer Verlag, Berlin, 1979.

42. J. Klein, *Les Systèmes Dynamiques Abstraits*, Ann. Inst. Fourier, Grenoble **13** (1963), 191.

43. B.Kostant, *Quantization and Unitary Representations*, in "Lectures in Modern Analysis and Applications III," (C.T. Taom Ed.), Springer Verlag, Berlin, 1970.

44. L.D. Landau and E.M. Lifshitz, "Mechanics," Pergamon Press, Oxford, 1980.

45. G. Landi and G. Marmo, *Graded Chern-Simons Terms*, Phys. Lett. **192B** (1987), 81.

46. G. Landi and G. Marmo, *Extensions of Lie Superalgebras and Supersymmetric Abelian Gauge Fields*, Phys. Lett. **193B** (1987), 61.

47. G. Landi and G. Marmo, *Lie Algebra Extensions and Abelian Monopoles*, Phys. Lett. **193B** (1987), 429.

48. G. Landi and G. Marmo, *Algebric Reduction of the T'Hooft-Polyakov Monopole to the Dirac Monopole*, Phis. Lett. **201B** (1988), 101.

49. G. Landi and G. Marmo, *Einstein Algebras and Algebraic Kaluza-Klein Monopole*, Phys. Lett. B. (To appear)

50. B. Lawruk and W.M. Tulczyjew, *Criteria for Partial Diff. Eq. to be Euler-Lagrange Equations*, J. Diff. Eq. **24** (1977), 211.

51. P. Libermann, C.M. Marle, "Symplectic Geometry and Analytical Mechanichs," D. Reidel, Dordrecht, 1987.

52. A. Lichnerowicz, *Algèbre de Lie des Champs de Vecteurs: Cohomologie 1-differentiable et Deformations*, C. R. Acad. Sc. Paris **281A** (1975), 507.

53. T. Levi-Civita and U. Amaldi, "Lezioni di Meccanica Razionale," (3 voll.), Zanichelli, Bologna, 1922.

54. G. Marmo, *Equivalent Lagrangians and Quasi-canonical Transformations*, in LNP **50**, Springer Verlag, Berlin, 1976.

55. G. Marmo, E.J. Saletan, *Ambiguities in the Lagrangian and Hamiltonian Formalism: Transformation Properties*, Il Nuovo Cimento **40B** (1977), 67.

56. G. Marmo, N. Mukunda and J. Samuel, *Dynamics and Symmetry for Constrained Systems: a Geometrical Analysis*, Riv. Nuovo Cimento **6** (1983), 1.

57. G. Marmo, E.J. Saletan, A. Simoni, *Reduction of Dynamical Systems*, Proc. Int. Meet. on Geom. and Phys., Florence 1982, Bologna.

58. G. Marmo, E.J. Saletan, A. Simoni, B. Vitale, "Dynamical Systems, a Differential Geometric Approach to Symmetry and Reduction," J. Wiley, New York, 1985.

59. G. Marmo, E.J. Saletan and A. Simoni, *On Obtaining Strictly Invariant Lagrangians from Gauge Invariant Lagrangians*, Il Nuovo cimento **96B** (1986), 159.

60. G. Marmo, C. Rubano, *On the Uniqueness of the Lagrangian Description for a Charged Particle in External Magnetic Field*, Il Nuovo Cimento **98A** (1987), 387.

61. G. Marmo, C. Rubano, *Alternative Lagrangians for a Charged Particle in a Magnetic Field*, Phys. Lett. **119A** (1987), 321.

62. G. Marmo and C. Rubano, "Particle Dynamics on Fiber Bundles," Bibliopolis, Naples, 1988.

63. G. Marmo, G. Morandi, A. Simoni, E.C.G. Sudarshan, *Quasi-invariance and Central Extensions*, Phys. Rev. **D37** (1988), 2196.

64. G. Marmo and G. Morandi, *Inverse problem with Symmetries and the Appearance of Cohomologies in Classical Lagrangian Dynamics*, Rep. Math. Phys.. (to appear)

65. G. Morandi and E.Menossi, *Path-Integrals in Multiply Connected Spaces and the Aharonov-Bohm Effect*, Eur. J. Phys. **5** (1984), 49.

66. G. Morandi, C. Ferrario, G. Lo Vecchio, G. Marmo, C. Rubano, "The Inverse Problem in the Calculus of the Variations and the Geometry of the Tangent Bundle," (to be published).

67. R.S. Palais, *A Global Formulation of the Theory of Transformation Groups*, Mem. Am. Math. Soc. **22** (1957). Providence, R.I

68. A. Pressley and G. Segal, "Loop Groups," Clarendon Press, Oxford, 1986.

69. R. de Ritis, G. Marmo, G. Platania and P. Scudellaro, *Inverse Problem in Classical Mechanics: Dissipative Systems*, Int. J. Th. Phys. **22** (1983), 931.

70. R.M. Santilli, "Foundations of Theoretical Mechanics," Springer, New York, 1983.

71. W. Sarlet, *The Helmoltz Condition Revisited. A New Approach to the Inverse Problem of Lagrangian Dynamics*, J. Phys. A: Math. Gen. **15** (1982), 1503.

72. R.Schmid, "Infinite Dimensional Hamiltonian Systems," Bibliopolis, Naples, 1987.

73. D.J. Simms, "Lie Groups and Quantum Mechanichs," LNM **52**, Springer Verlag, Berlin, 1968.

74. D.J. Simms and N. Woodhouse, "Lectures on Geometric Quantization," LNP **53**, Springer Verlag, Berlin, 1977.

75. J.M. Souriau, "Structure des Systèmes Dynamiques," Dunod, Paris, 1969.

76. N. Stenrod, "The Topology of Fiber Bundles," Princeton Univ. Press, New Jersey, 1951.

77. W.M. Tulczyjew, *Classical and Quantum Mechanics of particles in External Gauge Fields*, Rend. Sem. Mat. Univ. Torino **39** (1981), 111.

78. W.M. Tulczyjew, *Cohomology of the Lagrange Complex*, (to appear), Ann. Scuola Norm. Sup. Pisa.

79. G.M. Tuynman and W.A.J.J. Wiegerincht, *Central Extensions and Physics*, Jour. of Geom. and Phys. **4** (1987), 207.

80. B. Vitale, *Invariance and Non Invariance Dynamical Groups*, in "Selected Topics in Solid State and Theor. Phys.," (M. Bemporad & E. Ferreira Eds.), Gordon and Breech, New York, 1967.

81. A. Weil, "Variété Kählériennes," Hermann, Paris, 1971.

82. E.T. Whittaker, "A Treatise on the Analytical Dynamics of particles and Rigid Bodies," Cambridge Univ. Press, London, 1904.

83. N. Woodhouse, "Geometric Quantization," Clarendon Press, Oxford, 1980.

84. F. Zaccaria, E.C.G. Sudarshan, J.S. Nillsson, N. Mukunda, G. Marmo and A.P. Balachandran, *Universal Unfolding of Hamiltonian Systems: from Symplectic Structure to Fiber Bundles*, Phys. Rev. **27D** (1983), 2327.

85. J. Zak, *Magnetic Translation Groups*, Phys. Rev. **A134** (1964), 1602.

86. J. Zak, *Magnetic Translation Group II. Irreducible Representations*, Phys. Rev. **A134** (1964), 1607.

87. J. Zak, *Dynamics of Electrons in Solids in External Fields*, Phys. Rev. **168** (1968), 686.

88. B. Zumino, *Chiral Anomalies and Differential Geometry*, in "Relativity, Groups and Topology II," (Ed. by R.Stora and B.S. De Witt), Les Houches 1983, North Holland, Amsterdam, 1984.

89. B. Zumino, *Cohomology of Gauge Groups: Cocycles and Schwinger Terms*, Nucl. Phys. **B253** (1985), 477.

FURTHER READING

1. I.J.R. Aitchison, *Berry Phases, Magnetic Monopoles and Wess-Zumino terms or How the Skyrmyon got its Spin*, Acta Phys. Pol. **16B** (1987), 207.

2. V. Aldaya and J.A. de Azcárraga, *Cohomology, Central extensions and (Dynamical) Groups*, Int. J. Th. Phys. **24** (1985), 141.

3. O. Alvarez, *Topological Quantization and Cohomology*, Comm. Math. Phys. **100** (1985), 279.

4. M. Asorey and P.K. Mitter, *Cohomology of the Yang-Mills Gauge Orbit Space and Dimensional Reduction*, Ann. Inst. Poincaré **45** (1986), 61.

5. M.F. Atiyah and R.Bott, *The Moment Map and Equivariant Cohomology*, Topology **23** (1984), 1.

6. H. Bacry, Ph. Combe and J.L. Richard, *Group-theoretical Analysis of Elementary Particles in an External Electromagnetic Field, I & II*, Il Nuovo Cimento **67A** (1970), 267; Il Nuovo Cimento **70A** (1970), 289.

7. G. Bandelloni, *Non Polynomial Yang-Mills Local Cohomology*, Preprint Università di Genova.

8. A.P. Balachandran, T.R. Govindarajan, B. Vijayalakshmi, *Particles of Half-integral Helicity by Quantization of a Non-relativistic free Particle, and Related topics*, Phys. Rev. **18D** (1978), 1950.

9. A.P. Balachandran, *Wess-Zumino Terms and Quantum Symmetries*, 1st Asia Pacific Workshop on High Energy Phys., Singapore, June 1987, World Scientific (to appear 1988).

10. A.P. Balachandran, *Classical Topology and Quantum Phases: Quantum Mechanics*, in "Geometrical and Algebraic Aspects of Nonlinear Field Theories," Proc. Amalfi 1988 (to appear).

11. S. Benenti and W.M. Tulczyjew, *A Geometrical Interpretation of the One-cocycles of a Lie Group*, in "Geometrodynamics," (A. Prastaro Ed.), World Sc., Singapore, 1985.

12. S. Benenti and W.M. Tulczyjew, *Cocycles of the Coadjoint Representation of a Lie Group Interpreted as Differential Forms*, (to appear), Mem. Accad. Sci. Torinp.

13. L. Bonora, P. Cotta-Ramusino, *Some Remarks on BRS Transformations, Anomalies and the Cohomology of the Lie Algebra of the Group of Gauge Transformations*, Comm. Math. Phys. **87** (1983), 589.

14. Bon-Yao Chu, *Symplectic Homogeneous spaces*, Trans. Am. Math Soc. **197** (1974), 145.

15. L. Calabi, *Sur les Extensions des Groupes Topologiques*, Ann. Mat. Pure et appl. **32** (1951), 295.

16. U. Cattaneo, *Irreducible Lie Algebra Extensions of The Poincaré Algebra*, Commun. Math. Phys. **13** (1969), 226.

17. U. Cattaneo, *Invariance Relativiste, Symetries Internes et Extensions d'Algèbre de Lie*, Thesis Université Catholique de Louvain (1970).

18. J.F. Cariñena and M. Santander, *On the Projective Unitary Representation of Connected Lie groups*, J.Math. Phys. **16** (1975), 1416.

19. J.F. Cariñena and M. Santander, *Projective Covering Group versus Representation Groups*, J. Math. Phys. **21** (1980), 440.

20. J.F. Cariñena and M. Santander, *Galilean Relativistic Wave Equations*, J. Phys. A: Math. Gen. **15** (1982), 343.

21. J.F. Cariñena, M.A. del Olmo and M. Santander, *Local Representation Groups*, J. Phys. A: Math. Gen. **17** (1984), 3091.

22. J.F. Cariñena, M.A. del Olmo and M.A. Rodriguez, *Cohomology and Locally Hamiltonian Dynamical Systems*, in "The Physics of Phase Space," (Ed. by Y.S. Kim and W.W. Zachary) LNP **278**, Springer Verlag, Berlin, 1987.

23. J.F. Cariñena, L.A. Ibort, *Noncanonical Groups of Transformations, Anomalies and Cohomology*, J. Math. Phys **29** (1988), 541.

24. J.F. Cariñena, *Canonical Group Actions*, Preprint ICTP, IC/88/37.

25. M. Crampin, P. J. Mc Carthy, *A Lifting Theorem for Compact Symplectic Manifold*, J. Phys. A: Math. Gen. **16** (1983), 3949.

26. M. Crampin , F. Pirani, "Applicable Differential Geometry," Cambridge Univ. Press, 1986.

27. B.S. De Witt, "Dynamical Theory of Groups and Fields," Blackie & Son, London & Glasgow, 1965.

28. J.S. Dowker, *Homology and Wess-Zumino Terms*, J. Phys. A: Math. Gen. **19** (1986), L415.

29. Ch. Duval, P. Horvathy, *Particles with Internal Structure: the Geometry of Classical Motions and Conservation Laws*, Ann. Phys. New York **142** (1981), 10.

30. J. Fisch, M. Henneaux, J. Stasheff, C. Teitelboim, *Existence, Uniqueness and Cohomology of the Classical BRST Charge with Ghosts of Ghosts*, Preprint Université Libre de Bruxelles.

31. M. Francaviglia, "Elements of Differential and Riemannian Geometry," Bibliopolis, Naples, 1988.

32. J.L. Friedman, R. Sorkin, *Dyon Spin and Statistics: a Fiber Bundle Theory of Interacting Magnetic and Electric Charge*, Phys. Rev. **D20** (1979), 2511.

33. A. Galindo, *Lie Algebra Extensions of the Poincaré Algebra*, J. Math. Phys **8** (1967), 768.

34. I.M. Gelfand, *The Cohomology of Infinite Dimensional Lie Algebras; Some Question of Integral Geometry*, Actes Congrès Intern. Math. **1** (1970), 95.

35. R. Giachetti, R. Ricci, E. Sorace, *Hamiltonian Formulation for the Gauge Theory of the Gravitational Coupling*, J. Math. Phys. **22** (1981), 1703.

36. P. Goddard and D. Olive, *Kaç-Moody and Virasoro Algebras in Relation to Quantum Physics*, Int. Jour. Mod. Phys. **1A** (1986), 303.

37. W. Greub, S. Halperin and R. Vanstone, "Connections, Curvature and Cohomology II," Academic Press, New York, 1973.

38. H. Hanno, *Algebraic Cohomology versus Topological Cohomology in Constructing the Wess-Zumino Term of the σ-model*, Phys. Lett. **206B** (1988), 81.

39. M. Henneaux, *Equations of Motion, Commutation Relations and Ambiguities in the Lagrangian Formalism*, Ann. Phys. **140** (1982), 45.

40. M. Henneaux, *BRST Extensions of a Lie Group*, Preprint Université Libre de Bruxelles.

41. R. Hermann, *Infinite Dimensional Lie Algebras and Current Algebra in Group Representations in Mathematics and Physics*, Battelle Rencontres (V. Bargmann ed.), Springer-Verlag, Berlin.

42. G.Hochschild, *Group Extensions of Lie Groups*, Ann. Math. **54** (1951), 96.

43. G. Hochshild, J.P. Serre, *Cohomology of Lie Algebras*, Ann. Math. **57** (1953), 591.

44. P.A. Horvathy, *Quantum Ambiguities*, Preprint Dep. of Math., University of Metz.

45. J.C. Houard, *On Invariance Groups and Lagrangian Theories*, J. Math. Phys. **18** (1977), 502.

46. J.C. Houard, *An Integral Formula for Cocycles of Lie Groups*, Ann. Inst. Poincaré **32A** (1980), 221.

47. R. Jackiw, *Three-Cocycles in Math. and Phys.*, Phys. Rev. Lett. **54** (1985), 159.

48. T. Kaluza, *Zum Unitätsproblem der Physik*, Sitz. Preuss. Akad. Wiss. (1921), 966.

49. A.A. Kirillov, "Elements of the Theory of Representations," Springer Verlag, Berlin, 1976.

50. O. Klein, *Quanten Theorie und Fünf Dimensionale Relativitätstheorie*, Phys. Zeitsch. **37** (1926), 895.

51. B. Kostant, *Lie Algebra Cohomology and the Generalized Borel-Weil Theorem*, Ann. Math. **74** (1961), 329.

52. J.L. Koszul, *Homologie et Cohomologie des Algèbres de Lie*, Bull. Soc. Math. de France **78** (1950), 65.

53. J.M. Levy Leblond, *Group-theoretical Foundations of Classical Mechanics: the Lagrangian Gauge Problem*, Comm. Math. Phys. **12** (1969), 64.

54. J.M. Levy Leblond, *Conservation Laws for Gauge-Variant Lagrangians in Classical Mechanics*, A. J. P. **39** (1971), 502.

55. G.W. Mackey, *Unitary Representations of Group Extensions*, I. Acta Math. **99** (1958), 265.

56. S. Mac Lane, *Origins of the Cohomology of Groups*, Enseign. Math. (2) **24** (1978), 1.

57. G. Marmo, E.J. Saletan, *Q-equivalent Particle Hamiltonians. The Two-dimensional Harmonic Oscillator*, Hadronic J. **1** (1978), 955.

58. G. Marmo, *Function Groups and Reduction of Hamiltonian Systems*, Proc. IUTAM-ISIMM Symp. on Modern Devel. in Anal. Mech., Acad. Scientiarum Taurinensis, Torino.

59. G. Marmo, C. Rubano, *Equivalent Lagrangians and Lax Representation*, Il Nuovo Cimento **78B** (1983), 70.

60. J.E. Marsden, A. Weinstein, *Reduction of Symplectic Manifolds with Symmetry*, Rep. Math. Phys. **5** (1974), 121.

61. L. Michel, *Sur les Extensions Centrales du Group de Lorentz Inhomogène Connexe*, Nucl. Phys. **57** (1964), 356.

62. L.Michel, *Invariance in Quantum Mechanichs and Group Extensions*, in "Group Theor. Concepts and Methods in Elem. Part. Phys.," (F. Gürsey Ed.), Gordon and Breach, New York, 1965.

63. C.C. Moore, *Extensions and Low Dimensional Cohomology Theory of Locally Compact Groups*, I. Trans. Am. Math. Soc. **113** (1964), 40.

64. N. Mukunda et al., *Evolution, Symmetry and Canonical Structure in Dynamics*, Phys Rev. **D23** (1981), 2189.

65. R.S. Palais, *The Classifications of G-spaces*, Mem. Am. Math. Soc. **36** (1960). Providence, R.I

66. R. Palais and T.E. Stewart, *The Cohomology of Differentiable Transformation Groups*, Am. J. Math. **83** (1961), 623.

67. H. Poincaré, *Remarques sur une Experience de M. Birkeland*, C.R. Acad. Sci. **123** (1896), 530.

68. J.H. Rawnsley, *Some Applications of Quantization*, Thesis Mathematical Institute, Oxford.

69. E.J. Saletan, A.H. Cromer, "Theoretical Mechanics," J. Wiley, New York, 1971.

70. W. Sarlet and F. Cantrijn, *Symmetries, First Integrals, and the Inverse problem of Lagrangian Mechanics II*, J. Phys. A: Math. Gen. **16** (1983), 1383.

71. G. Segal, *Unitary Representations of Some Infinite Dimensional Groups*, Comm. Math. Phys. **80** (1981), 301.

72. D.J. Simms, "Projective Representations, Symplectic Manifolds and Extensions of Lie Algebras," Lecture Notes at CPT/CNRS, Marseille, 1969.

73. J.D. Stasheff, *Continous Cohomology of Groups and Classifying Spaces*, Bull. Am. Math. Soc. **84** (1978), 513.

74. S. Sternberg, *Minimal Coupling and the Symplectic Mechanics of a Classical Particle in the Presence of a Yang-Mills Field*, Proc. Nat. Acad. Sci. **74** (1977), 5253.

75. T.E. Stewart, *Lifting Group Action in Fibre Bundles*, Ann. Math. **74** (1961), 192.

76. H. Tilgner, *Extensions of Lie-graded Algebras*, J. Math. Phys. **18** (1977), 1987.

77. A. Trautman, "Differential Geometry for Physicists," Bibliopolis, Naples, 1984.

78. W.M. Tulczyjew, *The Lagrange Complex*, Bull. Soc. Math. de France **105** (1977), 419.

79. A. Weinstein, *A Universal Phase Space for Particles in a Yang-Mills Field*, Letters in Math. Phys. **2** (1978), 417.

80. G.S. Whiston, *On the Gauge Variance of Action Functions Under Transformations on Space-time*, Int. Jour. of Theor. Phys. **5** (1972), 391.

81. E.P. Wigner, *Do the Equations of Motion Determine the Quantum Mechanical Commutation relations?*, Phys. Rev. **77** (1950), 711.

82. S.R. Wong, *Field Particles Equations for The Classical Yang-Mills Field and Particles with Isotopic Spin*, Il Nuovo Cimento **65A** (1970), 689.

83. T.T. Wu and C.N. Yang, *Dirac's Monopole without Strings: Classical Lagrangian Theory*, Phys. Rev. **D14** (1976), 347.

84. B. Yu-Hou and B. Yuan-Hou and P. Wang, *How to Eliminate the Dilemma in Three-cocycles*, Annals of Phys. **171** (1986), 172.

ACCIDENTAL DEGENERACY AND STRUCTURE OF MATTER IN STRONG MAGNETIC FIELDS

M. Moshinsky,* G. Loyola and A. Szczepaniak+

Instituto de Física, UNAM
Apdo. Postal 20-364
México, D.F. 01000, México

1. INTRODUCTION

One of the interesting problems in present day physics concerns the structure of matter in strong magnetic fields. The word strong is related with the comparison of the magnetic energy with the binding energy of the particle. The former is given by

$$H_{mag} = (eg/2Mc)\underline{B}\cdot\underline{J} = (eg\hbar B/2Mc)(J_z/\hbar) \qquad (1.1)$$

where M is the mass, e the change g the gyromagnetic ratio and $\underline{J}$ the total angular momentum of the particle, while $\underline{B}$ is the intensity of the magnetic field.

At the right hand side of (1.1) we have taken the z axis in the direction of the magnetic field and expressed the angular momentum in dimensionless units $(J_z/\hbar)$. The coefficient of $(J_z/\hbar)$ gives then a measure of the magnetic energy.

In the following table we mention several systems, their binding energies and the intensity of the magnetic field in gauss that would be required by formula (1.1) to give an energy the same order of magnitude. We also indicate where intensities of the order mentioned could be obsered.

TABLE 1. SYSTEMS IN STRONG MAGNETIC FIELDS

System	Binding Energy	Field Intensity in Gauss
Solid State		10^4 (Laboratory)
Giant atoms, $n \simeq 100$	10^{-3} eV	10^5 (Laboratory)
Ground state atoms, n=1	10 eV	10^9 (neutron stars)
Nuclei	10 MeV	10^{18} (vicinity of cosmic strings)
Elementary Particles	1 GeV	10^{21} (vicinity of cosmic strings)

*Member of El Colegio Nacional

+On leave of absence from the University of Warsaw

In table 1 the term of Solid State means studies such as the quantized Hall effect[1] where the electrons are assumed essentially free, but the measurements at the laboratory are usually made at magnetic intensities of the order of Teslas i.e. 10^4 gauss.

For giant atoms, where the total quantum number n is of the order of 100, the binding energy of 10^{-3} eV requires field intensities of 10^5 gauss, also available in the laboratory. For ground state of atoms i.e. n=1, or 13.5 eV in hydrogen, the fields required, of 10^9 to 10^{12} gauss, have been observed in the vicinity of neutron stars.

Finally for nuclei and elementary particles, where the binding energies go from 10 MeV to 1 GeV, the magnetic field required in (1.1) to match these binding energies go from 10^{18} to 10^{21} gauss. They may be available in the vicinity of superconducting cosmic strings as, according to Witten,[2] the latter could have magnetic fields of intensity.

$$B(r)\sim 2\times 10^{16} r^{-1} \text{ gauss,} \tag{1.2}$$

where r is tbe distance to the string in centimeters. Thus at atomic distances $r=10^{-8}$ cm, the field is of the order of 10^{24} gauss and at nuclear ones of 10^{-13} cm, we get 10^{29} gauss.

The authors have been interested in the problems of the structure of matter in strong magnetic fields[3,4,5] and furthermore are preparing an extensive new study on this subject. In the proceedings of a conference on "Symmetry in Science", the interest should center on the group theoretical aspects of the problem. Thus in the present publication we shall stress the accidental degeneracy that appears when dealing with single and composite particles in a magnetic field, and in the Lie algebra responsible for this degeneracy.

2. THE SINGLE PARTICLE IN A MAGNETIC FIELD

To be specific we shall consider our single particle to be an electron whose vectors of position and momentum in ordinary units (cgs) will be denoted by $\underline{r}''$, $\underline{p}''$ while its mass and charge are m_o,e. The vector potential $\underline{A}''$ associated with a constant magnetic field $\underline{B}$ is given by

$$\underline{A}''=(1/2)(\underline{B}\times\underline{r}'') \tag{2.1}$$

The Hamiltonian for the electron is then

$$H''=(2m_o)^{-1}[\underline{p}''+(e/2c)(\underline{B}\times\underline{r}'')]^2 \tag{2.2}$$

In atomic units $\hbar=m_o=e=1$ we have the dimensionless coordinates and momenta

$$x_i'=(m_oe^2/\hbar^2)x_i'' \ , \ p_i'=(\hbar/m_oe^2)p_i'' \ , \ i=1,2,3 \tag{2.3}$$

and choosing x_3' parallel to the field $\underline{B}$ we obtain that

$$H'\equiv(\hbar^2/m_oe^4)H''=(1/2)[(p_1'-b^2x_2')^2+(p_2'+b^2x_1')^2+p_3'^2]=H_\perp'+H_\parallel' \tag{2.4}$$

where

$$b^2=(\hbar^3B/2m_o^2ce^3) \tag{2.5}$$

is now a dimensionless parameter.

The $H'_{\perp}$, $H'_{\parallel}$ in (2.4) are respectively the Hamiltonians associated with the coordinates and momenta perpendicular and parallel to the magnetic field. As $H'_{\parallel}$ is just that of a one dimensional free particle we concentrate on H' . Carrying out the canonical transformation of dilation

$$x_i=bx'_i \ , \ p_i=b^{-1}p'_i \ , \ i=1,2 \tag{2.6}$$

we get

$$H_{\perp}=b^{-2}H'_{\perp}=(1/2)(p_1^2+p_2^2+x_1^2+x_2^2)+(x_1p_2-x_2p_1)=H_o+L_3 \tag{2.7}$$

which is the one of a two dimensional oscillator H_o plus the projection of the angular momentum in the x_3 direction i.e. L_3.

The eigenfunctions of $H_{\perp}$ are clearly the ones of the two dimensional oscillator H_o in polar coordinates given by the kets[3]

$$|\nu m\rangle=R_{\nu|m|}(\rho)(2\pi)^{-1/2}\exp(im\phi) \tag{2.8}$$

where

$$\rho=(x_1^2+x_2^2)^{1/2} \ , \ \phi=\arctan\,(x_2/x_1) \tag{2.9}$$

and

$$R_{\nu|m|}(\rho)=(-1)^{\nu}\left[\frac{2(\nu!)}{(\nu+|m|)!}\right]^{1/2}\rho^{|m|}L_{\nu}^{|m|}(\rho^2)\exp(-\tfrac{1}{2}\rho^2) \tag{2.10}$$

with L being a Laguerre polynomial. The values of ν, m are restricted to

$$\nu=0,1,2,\ldots \ , \ m=0,\pm1,\pm2,\ldots \tag{2.11}$$

The eigenvalues of H_o and $H_{\perp}$ are given respectively by

$$(2\nu+|m|+1) \text{ and } [(2\nu+|m|+1)+m] \tag{2.12a,b}$$

and they imply quite different properties of accidental degeneracy. For H_o this degeneracy is restricted to the finite set of non-negative $\nu,|m|$ such that $2\nu+|m|+1$ is fixed. For $H_{\perp}$ we see that for $m=-|m|$ we have an infinite degeneracy as for all values $m=0,-1,-2,$ the energy is $(2\nu+1)$ while for $m=|m|$ the degeneracy is finite and restricted to the set of $\nu,|m|$ for which $2(\nu+|m|)+1$ is fixed.

We shall proceed to discuss, within the same framework, the symmetry Lie algebras responsible for the accidental degeneracy of H_o and $H_{\perp}$, to stress the differences between the oscillator Hamiltonian and that of a particle in a magnetic field, despite the fact that they have the same eigenfunctions.

To achieve our purpose we start by introducing the creation and annihilation operators in cartesian components

$$\eta_i=\frac{1}{\sqrt{2}}(x_i-ip_i) \ , \ \xi_i=\frac{1}{\sqrt{2}}(x_i+ip_i) \ , \ i=1,2, \tag{2.13}$$

as well as in spherical ones

$$\eta_{\pm}=\frac{1}{\sqrt{2}}(\eta_1\pm i\eta_2) \ , \ \xi^{\pm}=\frac{1}{\sqrt{2}}(\xi_1\mp i\xi_2) \ . \tag{2.14}$$

We note that

$$[\eta_\lambda,\eta_\mu]=[\xi^\lambda,\xi^\mu]=0 \;,\; [\xi^\lambda,\eta_\mu]=\delta^\lambda_\mu, \lambda,\mu=\pm \tag{2.15}$$

and that furthermore

$$H_o=\eta_+\xi^+ +\eta_-\xi^- +1 \;,\; L_3=\eta_+\xi^+ -\eta_-\xi^- \tag{2.16a,b}$$

so we obtain

$$H_\perp =2\eta_+\xi^+ +1 \tag{2.17}$$

Turning now our attention to eigenfunctions (2.8) of both H_o and $H_\perp$, we have shown elsewhere[3] that they can be written as

$$|\nu m\rangle = \frac{\eta_+^{n_+}\eta_-^{n_-}}{\sqrt{n_+!n_-!}}|0\rangle \tag{2.18a}$$

where

$$n_+ + n_- = 2\nu+|m| \;,\; n_+ - n_- = m \;,\; n_+,n_-=0,1,2,... \tag{2.19}$$

and $|0\rangle$ is the ground state

$$|0\rangle = \pi^{-1/2}\exp(-\tfrac{1}{2}\rho^2) \tag{2.20}$$

This can be corroborated by applying (2.16) and (2.17) to (2.18) and making use of (2.15).

The eigenvalues of H_o, $H_\perp$ are given by

$$n_+ + n_- +1 \quad \text{and} \quad 2n_+ +1 \quad . \tag{2.12a,b}$$

We see clearly the finite degeneracy in H_o as the pair $[(n_+\pm k), (n_-\mp k)]$ where k=1,2,..., corresponds to the same energy as (n_+,n_-), up to the value of k in which one member of the pair vanishes. On the other hand for $H_\perp$ any values $n_-=0,1,...$ correspond to the same energy $2n_+ +1$.

The question now is which are the symmetry Lie algebras responsible for the accidental degeneracy in H_o and $H_\perp$. For H_o it is well known that it is SU(2), whose generators are

$$T_+=\eta_+\xi^- \;,\; T_o=\frac{1}{2}(\eta_+\xi^+ -\eta_-\xi^-) \;,\; T_-=\eta_-\xi^+ \;, \tag{2.22a,b,c}$$

as from (2.15) they not only commute with the H_o of (2.16a) but also close under commutation among themselves.

For H the answer is even simpler as it is clear that

$$\eta_-,\ \xi^- \quad \text{and} \quad [\xi^-,\eta_-]=1 \tag{2.23}$$

are the generators of a one dimensional Weyl Lie algebra that commute with $H_\perp$ of (2.17). In fact by applying η_- or ξ^- we can transform a state (2.18) with (n_+,n_-) into another one with $(n_+,n_-\pm 1)$ and so reach all the states with a fixed n_+. Thus the Weyl algebra is the symmetry one for the problem of a single charged particle in a magnetic field.

One can also have an SU(1,1) symmetry Lie algebra for $H_\perp$ by defining its generators as

$$S_+=\eta_-^2 \ , \ S_o=\frac{1}{2}(\eta_-\xi^-+\xi^-\eta_-) \ , \ S_-=\xi_-^2 \tag{2.24}$$

as they commute with $H_\perp$ and, from (2.23), close under commutation among themselves. We note though that now we need two states, for example

$$(n_+!)^{-1/2}\eta_+^{n_+}|0\rangle \quad \text{and} \quad (n_+!)^{-1/2}\eta_+^{n_+}\eta_-|0\rangle \tag{2.25}$$

to be able to reach all the states (2.18) with the generators of SU(1,1).

Finally we wish to remark that in reference 5 we showed, using coherent states, that the accidental degeneracy can be correlated with the arbitrariness of the center of the classical circular orbit of a particle in a magnetic field.

3. THE CHARGED COMPOSITE PARTICLE IN A MAGNETIC FIELD

In the previous section we discussed the accidental degeneracy present in the spectrum of the Hamiltonian associated with a charged single particle in a magnetic field, and obtained the Lie algebra responsible for this degeneracy. In the present section we wish to show that a similar degeneracy and symmetry algebra appears for a charged composite particle i.e. one formed from more elementary ones interacting through two body forces.

We shall first discuss the specific problem of a composite particle formed from two others of equal mass M, interacting through an harmonic oscillator force of frequency ω, and in the presence of an external magnetic field of intensity $\underline{B}$. We shall assume that one of the particles has charge e and the other is neutral to provide a reasonable model for the deuteron.[4] Once this problem is understood, we shall easily see its extension to the general case mentioned at the end of the previous paragraph.

Our Hamiltonian, again in ordinary (cgs) units where the coordinates and momenta are designated by $\underline{r}''_s$, $\underline{p}''_s$, s=1,2, is given by

$$H''=(2M)^{-1}\left[\underline{p}''_1-(e/2c)(\underline{B}\times\underline{r}''_1)\right]^2+(2M)^{-1}\underline{p}''^2_2+\frac{1}{2}M\omega^2(\underline{r}''_1-\underline{r}''_2)^2 \tag{3.1}$$

Introducing the dimensionless observables

$$\underline{r}'_s=(M\omega/\hbar)^{1/2}\underline{r}''_2 \ , \ \underline{p}'_s=(M\hbar\omega)^{-1}\underline{p}''_s \ ; \ s=1,2, \tag{3.2}$$

and writing $\underline{B}=B\underline{u}$ where $\underline{u}$ is a unit vector in the direction of the magnetic field we obtain

$$H\equiv(\hbar\omega)^{-1}H''=\frac{1}{2}\left[\underline{p}'_1-q(\underline{u}\times\underline{r}'_1)\right]^2+\frac{1}{2}\underline{p}'^2_2+\frac{1}{2}(\underline{r}'_1-\underline{r}'_2)^2 \tag{3.3a}$$

where[6]

$$q=(eB/2Mc\omega)=b^2(m_o/M)(m_oe^4/\hbar^2)(\hbar\omega)^{-1} \tag{3.3b}$$

and the dimensionless b^2 is given by (2.5).

Writing (3.2) in terms of the components x'_{is}, p'_{is} i=1,2,3, s=1,2 of coordinates and momenta, and taking x'_{3s} in the direction $\underline{u}$ of the magnetic field we obtain

$$H=H_\perp+H_{\parallel} \tag{3.4}$$

where, as in (2.4), $H_{\parallel}$, $H_{\perp}$ are the parts of the Hamiltonian associated with the observables parallel and perpendicular to the magnetic field, which are given by

$$H_{\perp} = \frac{1}{2}\left[(p'_{11}+qx'_{21})^2+(p'_{21}-qx'_{11})^2\right]+\frac{1}{2}(p'^2_{12}+p'^2_{22})+$$

$$+\frac{1}{2}\left[(x'_{11}-x'_{12})^2+(x'_{21}-x'_{22})^2\right], \tag{3.5}$$

$$H_{\parallel} = \frac{1}{2}(p'^2_{31}+p'^2_{32})+\frac{1}{2}(x'_{31}-x'_{32})^2 . \tag{3.6}$$

The $H_{\parallel}$ is the trivial problem of two particles in one dimension interacting through an harmonic oscillator potential, so the following discussion will concern only $H_{\perp}$.

While the Hamiltonian $H_{\perp}$ of (3.5) looks simple, it is not trivial. In fact, it is a member of the family of quadratic Hamiltonians in phase space discussed some time ago by Moshinsky and Winternitz.[7] What is required for its solution is a linear canonical transformation that reduces it to a form where one can apply standard techniques. To find this transformation, we start by noting that $H_{\perp}$ admits as integrals of motion two components of pseudo momentum[8,9]

$$K_1=p'_{11}+p'_{12}-qx'_{21}, \quad K_2=p'_{21}+p'_{22}+qx'_{11} . \tag{3.7}$$

The Poisson brackets $\{K_1,H_{\perp}\}$, $\{K_2,H_{\perp}\}$ will then vanish as is easily checked. On the other hand, we note that

$$\{K_1,K_2\}=-2q \tag{3.8}$$

and this suggests that we could use $K_1, -(K_2/2q)$ as part of the set of new coordinates and momenta. A similar discussion applies to the components of total momentum in the plane perpendicular to the field, and a bit of juggling on them and some other observables suggests the canonical transformation we need. Designating the new coordinates and momenta by $\bar{x}_\alpha, \bar{p}_\alpha, \alpha=1,2,3,4$ we have

$$\bar{x}_1=p'_{21}+p'_{22}-qx'_{11}; \quad \bar{p}_1=-(1/2q)(p'_{11}+p'_{12}+qx'_{21}), \tag{3.9a,b}$$

$$\bar{x}_2=x'_{11}-x'_{12}; \quad \bar{p}_2=-p'_{12}, \tag{3.9c,d}$$

$$\bar{x}_3=-p'_{22}; \quad \bar{p}_3=-(x'_{21}-x'_{22}), \tag{3.9e,f}$$

$$\bar{x}_4=p'_{11}+p'_{12}-qx'_{21}; \quad \bar{p}_4=-(1/2q)(p'_{21}+p'_{22}+qx'_{11}), \tag{3.9g,h}$$

and we easily check that $\{\bar{x}_\alpha,\bar{p}_\beta\}=\delta_{\alpha\beta}, \{\bar{x}_\alpha,\bar{x}_\beta\}=\{\bar{p}_\alpha,\bar{p}_\beta\}=0$.

In terms of $\bar{x}_\alpha,\bar{p}_\alpha,\alpha=1,2,3,4$ the Hamiltonian $H_{\perp}$ takes the form

$$H_{\perp}=(1/2)\{(2q\bar{p}_1-\bar{p}_2)^2+(\bar{x}_1+\bar{x}_3)^2+\bar{p}_2^2+\bar{x}_3^2+\bar{x}_2^2+\bar{p}_3^2\} . \tag{3.10}$$

We note that $H_{\perp}$ does not contain the integrals of motion $\bar{x}_4=K_1, \bar{p}_4=-(K_2/2q)$ and that, furthermore, no mixed terms in coordinates and momenta of the type $\bar{x}_\alpha\bar{p}_\beta$ appear. We then apply standard techniques to this Hamiltonian by first considering another canonical transformation to coordinates and momenta $\bar{x}'_\alpha, \bar{p}'_\alpha$, but now affecting them separately, and having the form

$$\bar{p}_1' = 2q\bar{p}_1 - \bar{p}_2\ , \quad \bar{p}_2' = \bar{p}_2\ , \quad \bar{p}_3' = \bar{p}_3\ , \quad \bar{p}_4' = \bar{p}_4\ , \tag{3.11a,b,c,d}$$

$$\bar{x}_1' = (\bar{x}_1/2q)\ , \quad \bar{x}_2' = (\bar{x}_1/2q) + \bar{x}_2\ , \quad \bar{x}_3' = \bar{x}_3\ , \quad \bar{x}_4' = \bar{x}_4\ . \tag{3.11e,f,g,h}$$

The Hamiltonian $H_\perp$ then becomes

$$H_\perp = \frac{1}{2}(\bar{p}_1'^2 + \bar{p}_2'^2 + \bar{p}_3'^2) + \frac{1}{2}[\bar{x}_1' \bar{x}_2' \bar{x}_3'] \begin{bmatrix} 4q^2+1 & -1 & 2q \\ -1 & 1 & 0 \\ 2q & 0 & 2 \end{bmatrix} \begin{bmatrix} \bar{x}_1' \\ \bar{x}_2' \\ \bar{x}_3' \end{bmatrix}\ , \tag{3.12}$$

where we have written the potential energy in matrix and vector notation.

We have reduced the problem to a standard one in the theory of small vibrations. If we denote by M the matrix appearing in (3.12) we need first to solve the secular equation

$$\det(M - \Lambda I) = 0\ , \tag{3.13}$$

which gives rise to the cubic equation

$$\Lambda^3 - (4q^2+4)\Lambda^2 + (8q^2+4)\Lambda - 4q^2 = 0\ . \tag{3.14}$$

Clearly this equation has no negative roots, and in fact the three of them are real and positive so we could designate them by $\Lambda = \lambda_i^2$, $i=1,2,3$. If we furthermore denote by $U = ||u_{ij}||$ the orthogonal matrix that diagonalizes M, we can then carry a last canonical transformation to coordinates and momenta x_i, p_i, $i=1,2,3$ given by

$$x_i = \lambda_i^{1/2} \sum_{j=1}^{3} (\bar{x}_j' u_{ji})\ , \quad p_i = \lambda_i^{-1/2} \sum_{j=1}^{3} (\bar{p}_j' u_{ji})\ , \tag{3.15}$$

to reduce the Hamiltonian to the form

$$H_\perp = \lambda_1 \frac{1}{2}(p_1^2 + x_1^2) + \lambda_2 \frac{1}{2}(p_2^2 + x_2^2) + \lambda_3 \frac{1}{2}(p_3^2 + x_3^2)\ . \tag{3.16}$$

We note again that $\bar{x}_4$, $\bar{p}_4$ do not appear in (3.16) so its eigenstates can be multiplied by an arbitrary function of, say, $\bar{x}_4$ and still correspond to the same eigenvalue. It is useful then to characterize further the eigenstates of (3.16) by the eigenvalues of some function of the integrals of motion K_1, K_2 and a convenient one is[8,9]

$$K^2 \equiv K_1^2 + K_2^2 = (2q\bar{p}_4)^2 + \bar{x}_4^2 = 2q(p_4^2 + x_4^2)\ , \tag{3.17}$$

where the last result was achieved by the canonical transformation

$$(2q)^{-1/2}\bar{x}_4 = x_4\ , \quad (2q)^{1/2}\bar{p}_4 = p_4\ . \tag{3.18}$$

The eigenvalues of H and K^2 are then respectively

$$E_{n_1 n_2 n_3} = \lambda_1 (n_1 + \tfrac{1}{2}) + \lambda_2 (n_2 + \tfrac{1}{2}) + \lambda_3 (n_3 + \tfrac{1}{2})\ , \tag{3.19}$$

$$\Omega_{n_4} = 4q(n_4 + \tfrac{1}{2})\ , \tag{3.20}$$

where n_α, $\alpha=1,2,3,4$ takes the values $n_\alpha = 0,1,2,\ldots$.

Introducing creation and annihilation operators by the usual definitions

$$\eta_\alpha=\frac{1}{\sqrt{2}}(x_\alpha-ip_\alpha)\ ,\ \xi_\alpha=\frac{1}{\sqrt{2}}(x_\alpha+ip_\alpha)\ ;\ \alpha=1,2,3,4 \tag{3.21}$$

we see that the eigenfunctions of $H_\perp$, K^2 associated with the eigenvalues $E_{n_1n_2n_3}$, Ω_{n_4} can be written as

$$|n_1n_2n_3n_4\rangle=(n_1!n_2!n_3!n_4!)^{-1/2}\eta_1^{n_1}\eta_2^{n_2}\eta_3^{n_3}\eta_4^{n_4}|0\rangle\ , \tag{3.22}$$

where the ground state $|0\rangle$ is given by

$$|0\rangle=\pi^{-1}\exp\left[-\frac{1}{2}(x_1^2+x_2^2+x_3^2+x_4^2)\right]\ . \tag{3.23}$$

The states (3.22) can be expressed in terms of a product of Hermite polynomials in the x_α, $\alpha=1,2,3,4$ and, at first sight, it seems that now we have to obtain the corresponding kets in the original coordinates x'_{is}; $i=1,2$; $s=1,2$. This is feasible with the help of the unitary representation[10] of linear canonical transformations, but will not be necessary for the present problem in which we only wish to discuss the accidental degeneracy and the Lie algebra associated with it.

The eigenstates (3.22) for the composite particle look very similar to those of (2.18) for the single particle problem. In both cases the corresponding eigenvalues of $H_\perp$ are independent of one of the quantum numbers, n_- in (2.18) and n_4 in (3.22). Thus for the special composite particle whose Hamiltonian is (3.5) (function only of the observables whose components are in the plane perpendicular to the magnetic field) there is an accidental degeneracy which again is associated with a one dimensional Weyl Lie algebra whose generators are

$$\eta_4=\frac{1}{2}q^{-1/2}\left[(p'_{11}+ip'_{21})+(p'_{12}+ip'_{22})+iq(x'_{11}+ix'_{21})\right]\ , \tag{3.24a}$$

$$\xi_4=\frac{1}{2}q^{-1/2}\left[(p'_{11}-ip'_{21})+(p'_{12}-ip'_{22})-iq(x'_{11}-ix'_{21})\right]\ , \tag{3.24b}$$

$$[\xi_4,\ \eta_4]=1\ , \tag{3.24c}$$

as follows from (3.21), (3.18) and (3.9g,h). The operators η_4, ξ_4 then relate all the states (3.22) of the same energy i.e. of fixed n_1,n_2,n_3 and arbitrary n_4.

Another viewpoint on the origin of the accidental degeneracy for a composite charged particle in a magnetic field is obtained through the analysis of its coherent states. We shall denote the latter by a round ket depending on the quantum numbers n_1,n_2,n_3 and a complex parameter z, and it has the form

$$|n_1n_2n_3z)=\exp(-\frac{1}{2}|z|^2)\sum_{k=0}^{\infty}(k!)^{-1}(z\eta_4)^k|n_1n_2n_30\rangle\ , \tag{3.25}$$

where the angular ket on the right hand side is given by (3.22) when $n_4=0$. The $\exp(-\frac{1}{2}|z|^2)$, where $|z|$ is the absolute value of z, is introduced for normalization.

From (3.24c) we see that the round ket $|n_1n_2n_3z)$ satisfies the equation

$$\xi_4|n_1n_2n_3z)=z|n_1n_2n_3z) \ , \tag{3.26}$$

which is the one usually associated with coherent states. Furthermore it is an eigenstate of the H in (3.5) with the eigenvalue (3.19).

When acting on the ground state in the variable x_4, the η_4^k gives rise to $2^{-k/2}H_k(x_4)$ where H_k is an Hermite polynomial.[4] Thus the coherent state (3.25) becomes

$$|n_1n_2n_3z)=\exp(-\tfrac{1}{2}|z|^2)\Big[\sum_{k=0}^{\infty}(k!)^{-1}(z/\sqrt{2})^k H_k(x_4)\Big]|n_1n_2n_30>$$

$$=\{\exp(-\tfrac{1}{2}|z|^2)\exp(-\tfrac{1}{2}z^2+\sqrt{2}\,zx_4)\pi^{-1/4}\exp(-\tfrac{1}{2}x_4^2)\}$$

$$\{\pi^{-3/4}\Big[\prod_{i=1}^{3}(2^{n_i}n_i!)^{-1/2}H_{n_i}(x_i)\Big]\exp[-\tfrac{1}{2}(x_1^2+x_2^2+x_3^2)]\} \tag{3.27}$$

where the second exponential on the right hand side is associated with the generating function of Hermite polynomials[11] and the last curly bracket is a function of only x_1,x_2,x_3.

We wish now to calculate the expectation value of the center of mass coordinate of our two body system

$$X_i'=\frac{1}{2}(x_{i1}'+x_{i2}') \tag{3.28}$$

with respect to the coherent state (3.27). For this purpose we note from (3.18) and (3.9) that

$$X_1'=-(2q)^{-1/2}p_4-(2q)^{-1}\bar{x}_1-\frac{1}{2}\bar{x}_2 \ , \tag{3.29a}$$

$$X_2'=-(2q)^{-1/2}x_4-\bar{p}_1+\frac{1}{2}\bar{p}_3 \ . \tag{3.29b}$$

Now in the expectation value

$$(n_1n_2n_3z|X_i'|n_1n_2n_3z) \ , \tag{3.30}$$

x_4, p_4 appearing in (3.29) <u>do</u> contribute as they are calculated with respect to the first curly bracket in the coherent state (3.27). On the other hand the expectation values of $\bar{x}_1$, $\bar{x}_2$, $\bar{p}_1$, $\bar{p}_3$, also appearing in (3.29), vanish as they are calculated with respect to the second curly bracket in (3.27). We thus obtain

$$(n_1n_2n_3z|X_1'|n_1n_2n_3z)=(i/2)q^{-1/2}(z-z^*)\equiv X_{01}' \ ; \tag{3.31a}$$

$$(n_1n_2n_3z|X_2'|n_1n_2n_3z)=-(1/2)q^{-1/2}(z+z^*)\equiv X_{02}' \ , \tag{3.31b}$$

where z^* is the conjugate of z and we also denote the expectation values of X_i' by the constants X_{0i}', i=1,2.

From (3.31) we then get

$$z=-i\,q^{1/2}(X_{01}'-iX_{02}') \ , \tag{3.32}$$

and thus the complex parameter z in the coherent state allows us to select the expectation value X_{0i}' of the center of mass coordinate X_i', i=1,2 at any point in the plane perpendicular to the magnetic field.

As the energy of the ket $|n_1n_2n_3z)$ is given by (3.19), independently of the value of z, we see that, from the viewpoint of coherent states,[5] the resulting accidental degeneracy is just a reflection of the fact that the corresponding classical circular orbits, of a charged composite particle in a magnetic field, can be centered at arbitrary points in the plane perpendicular to the field.

We turn now our attention to the general non-relativistic Hamiltonian for a composite particle in a magnetic field. We assume that it is composed of n more elementary particles interacting through translationally invariant two body central forces. Furthermore we shall employ throughout the atomic units (2.3) so that the masses M_s, s=1,2,...n of the particles are given in terms of the electron mass m_o and the charges Q_s in terms of e. The dimensionless coordinates x'_{is} and momenta p'_{is}; i=1,2,3; s=1,2,..n are defined as in (2.3), and taking x'_{3s} in the direction of the magnetic field $\underline{B}$ we obtain the Hamiltonian

$$H=\sum_{s=1}^{n}(2M_s)^{-1}\left[(p'_{1s}-Q_sb^2x'_{2s})^2+(p'_{2s}+Q_sb^2x'_{1s})^2+p'^2_{3s}\right]+$$

$$+\sum_{s<t=1}^{n}V\{\left[(x'_{1s}-x'_{1t})^2+(x'_{2s}-x'_{2t})^2+(x'_{3s}-x'_{3t})^2\right]^{1/2}\}, \tag{3.33}$$

where V is an arbitrary function of the variable indicated inside curly bracket and b^2 is given by (2.5).

As is well known[7,8] the Hamiltonian H has as integrals of motion the components of the pseudo momentum vector $\underline{K}$ which are given by

$$K_1=\sum_{s=1}^{n}(p'_{1s}+Q_sb^2x'_{2s}),\quad K_2=\sum_{s=1}^{n}(p'_{2s}-Q_sb^2x'_{1s}),\quad K_3=\sum_{s=1}^{n}p'_{3s}\ . \tag{3.34}$$

We easily check that the commutators $[K_i,H]$ vanish, as the total momentum $\sum_{s=1}^{n}p'_{is}$ appearing in (3.34) commutes with the translationally invariant potential V in (3.33). Furthermore this also happens for the commutator of K_i with the rest of the H. Finally we also see from (3.34) that

$$[K_1,\ K_2]=2i(\sum_{s=1}^{n}Q_s)b^2\equiv 2iQb^2\ ,\quad [K_1,\ K_3]=[K_2,\ K_3]=0\ , \tag{3.35a,b}$$

where $Q=\sum_{s=1}^{n}Q_s$ is the total charge of the composite particle which we assume different from zero.

From (3.35a) we see that K_1 and $(2Qb^2)^{-1}\ K_2$ can be interpreted respectively as coordinates and momenta as their commutator is i. Thus if we define

$$\eta=(1/\sqrt{2})\left[K_1-i(2Qb^2)^{-1}K_2\right]\ , \tag{3.36a}$$

$$\xi=(1/\sqrt{2})\left[K_1+i(2Qb^2)^{-1}K_2\right]\ , \tag{3.36b}$$

they behave as creation and annihilation operators as is corroborated by the fact that from (3.35a) their commutator is

$$[\xi,\ \eta]=1 \tag{3.36c}$$

We can now make use of the fact that η, ξ commute with H, K_3, though not among themselves, to obtain families of infinitely degenerate states associated with definite eigenvalues E of H and k of K_3. These families will be of the type $|n_1n_2n_3n_4>$ of (3.22) or $|n_1n_2n_3z)$ of (3.25) in the two body oscillator problem discussed above, where in them the degeneracy was associated either with $n_4=0,1,2,...$ or the arbitrary complex number z.

To achieve our objective we assume we can determine the ket $|Ek0>$ which is the eigenstate of the following operators

$$H|Ek0> = E|Ek0> \ , \quad K_3|Ek0> = k|Ek0> \ , \tag{3.37a,b}$$

$$\xi|Ek0> = 0 \ . \tag{3.37c}$$

Clearly then the family of states

$$|Ekn> = (n!)^{-1/2} \eta^n |Ek0> \ , \tag{3.38}$$

will correspond to the same eigenvalues E, k of H, K_3 and the $|Ekn>$ can be related to each other through successive applications of the operators η, ξ. Thus the one dimensional Weyl Lie algebra, whose generators are given by (3.36), will be the symmetry algebra associated with the states $|Ekn>$, n=0,1,2,... .

We can also introduce the coherent states for the general Hamiltonian (3.33) through the round kets defined by

$$|Ekz) = \exp(-\tfrac{1}{2}|z|^2) \sum_{k=0}^{\infty} (k!)^{-1} (z\eta)^k |Ek0> \tag{3.39}$$

where z is an arbitrary complex parameter. Again we have the same eigenvalue E, k of H, K_3 for the states $|Ekz)$ independently of the value of z. In view of the discussion carried above for the $|n_1n_2n_3z)$ of (3.25) we surmise that $|Ekz)$ reflects the freedom we have in choosing the expectation value of the center of mass

$$x'_i = \sum_{s=1}^{n} (M_s)^{-1} \sum_{s=1}^{n} (M_s x'_{is}) \ , \quad i=1,2, \tag{3.40}$$

at an arbitrary point in the plane perpendicular to the magnetic field. Note that in the direction of the magnetic field i=3 we have a precise value for the total momentum, so the position of the center of mass in this direction is completely undetermined.

We have discussed the accidental degeneracy present in the problem of a composite particle in a magnetic field, the symmetry Lie algebra responsible for this degeneracy and the physical significance of the coherent states for this problem. In the conclusion we mention some general features of the structure of matter in strong magnetic fields, which make it a problem worthwhile discussing.

4. CONCLUSION

The structure of matter in strong magnetic fields is of great current interest. The degeneracy present even in the case of the single electron, that was analyzed in section 2, and which originally was discussed by Landau,[12] is very relevant for the understanding of the quantized Hall effect.[1]

In giant atoms magnetic fields produced in the Laboratory give energy comparable to those that bind the electron to the nucleus, and the degeneracies mentioned in the present paper are quite important for the understanding the spectra observed. This also happens for atoms in their ground state for magnetic fields in the vicinity of neutron stars.

Finally, if superconducting cosmic strings exist, their effect on nuclei and elementary particles may be very interesting and would require a careful analysis, which the authors plan to present in a future publication.

REFERENCES

1. R. E. Prange and S. M. Girvin, "The Quantum Hall Effect", Springer Verlag, Berlin (1987).
2. E. Witten, Superconducting Strings, Nucl. Phys. B 249:557 (1985).
3. M. Moshinsky, N. Méndez, and E. Murow, Pseudoatoms and Atoms in Strong Magnetic Fields, Ann. Phys. 163:1 (1985).
4. M. Moshinsky and G. Loyola, Stability of Deuterons in Strong Magnetic Fields: An Exactly Solved Model, Mod. Phys. Lett. A 3:345 (1988).
5. G. Loyola, M. Moshinsky and A. Szczepaniak, Coherent States and Accidental Degeneracy for a Charged Particle in a Magnetic Field, submitted for publication in Am. J. Phys.
6. M. Moshinsky, N. Méndez, E. Murow and J. W. B. Hughes, Accidental Degeneracies in the Zeeman Effect and the Symmetry Groups, Ann. Phys. 155:231 (1984).
7. M. Moshinsky and P. Winternitz, Quadratic Hamiltonians in phase space and their eigenstates, J. Math. Phys. 21:1667 (1980).
8. D. Baye, An approximate constant of motion for the problem of an atomic ion in a homogeneous magnetic field, J. Phys. B. At.Mol. Phys. 15:L795 (1982).
9. H. Herold, H. Ruder and G. Wunner, The two-body problem in the presence of a homogeneous magnetic field, J. Phys. B. At. Mol. Phys. 14:751 (1981).
10. M. Moshinsky and C. Quesne, Linear Canonical transformations and their Unitary Representations, J. Math. Phys. 12:1772 (1971).
11. I. S. Gradshteyn and I. M. Ryzhik, "Tables of Integrals, Series and Products", Academic Press, New York (1965).
12. L. D. Landau and E. M. Lifshitz, "Quantum Mechanics", Pergamon Press, London (1965).

GALILEAN SYMMETRY

M. Omote, S. Kamefuchi

Institute of Physics
University of Tsukuba
Ibaraki 305, Japan

Y. Takahashi
Theoretical Physics Institute
Department of Physics
University of Alberta
Alberta, Canada T6G 2J1

Y. Ohnuki
Department of Physics
Nagoya University
Nagoya 464, Japan

INTRODUCTION

Since this is a symposium for symmetry, I believe that it is our duty to behave as if we all belonged to one and the same tribe living exclusively on symmetry. Thus I have chosen the theme of my lecture to be one of the oldest symmetries we, human being, have known of, that is, Galilean symmetry. It is said that the first description of this is given in one of the great books penned by Galileo Galilei:

"Dialogo sopra i due massimi sistemi del mondo tolemaico e copernicano",

published in 1632[1]. In this book he explains, by giving many examples, that any motion that takes place in a cabin of some large ship does not depend upon whether the ship is standing still or in motion. He writes, for example: "the butterflies and flies will continue their flights indifferently towards every side, nor will it ever happen that they are concentrated towards the stern...", and ends up with the statement: "the ship's motion is common to all the things contained in it, and to the air also...". At present we, of course, attribute the above phenomenón to the existence of the so-called Galilean symmetry, and understand further that this is the symmetry which should underlie any theories of nonrelativistic motion.

The present lecture consists of two parts. In the first part we present a reformulation of Galilei transformations. Usually, Galilei transformations are defined as those for 4-dimensional, space-time coordinates $\vec{x}=(x^1,x^2,x^3)$ and $t=x^4$. Here, however, we introduce a further coordinate $s=x^5$ and redefine Galilei transformations as those for the 5-dimensional coordinates x^μ $(\mu=1,2,\ldots,5)$. For distinction let us hereafter refer to the former (latter) transformations as $G_4(G_5)$. Our approach with G_5 has then the following advantages. The Lagrangians L of interesting physical systems (e.g. a system of free massive particles) become invariant under G_5: $\delta L \equiv L'-L=0$, contrary to the case of G_4 where $\delta L=df/dt$ with f being a function of dynamical variables. Accordingly, discussion of the problem is considerably simplified. Mathematically the introduction of the 5-th coordinate s corresponds to a central extension of the Galilei group G_4. Extra degrees of freedom thereby introduced are eliminated afterwards by imposing subsidiary conditions.

In the second part we are concerned with application of G_5 to nonrelativistic quantum mechanics (NQM). In formulating the theory of NQM it is customary to start with, and hence lean heavily on, the corresponding classical theory. In our case, however, the theory is completely founded on its own basis, that is, Galilean symmetry: the Schrödinger equations are obtained here as those to be satisfied by vector representations of the Galilei group G_5, and Galilei covariance is manifest throughout the formalism. This is quite contrary to the formalism with G_4 where we have to deal with projective representations of the group G_4 so that discussions are rather involved and Galilei covariance is not manifest. Another advantage of our approach is that the formulation of NQM goes in a way quite parallel to that of relativistic quantum mechanics(RQM): in fact, we have only to copy various formulae from RQM. We may say, therefore, that NQM, when formulated with G_5, can stand on its own feet, and is elevated to the same status as RQM.

REFORMULATION OF GALILEI TRANSFORMATIONS

The Fifth Coordinate

Homogeneous Galilei transformations G_4 in 4 dimensions are usually defined as[2]

$$\vec{x} \to \vec{x}' = R\vec{x} - \vec{v}t, \qquad t \to t' = t, \tag{1}$$

where R is a 3-dimensional orthogonal matrix, and $\vec{v}$ a constant vector. It is well known, however, that under G_4 Lagrangians for nonrelativistic motion in general are not invariant. For example, the Lagrangian L of a free particle with mass m, which is given by

$$L = \frac{m}{2} \dot{\vec{x}}^2 , \tag{2}$$

transforms as

$$L \to L' = L + df/dt , \tag{3}$$

with

$$f \equiv -m\vec{v}\cdot(R\vec{x}) + \frac{1}{2}m\vec{v}^2 t . \tag{3'}$$

In the corresponding quantum theory the above situation is reflected in the fact that under G_4 the wavefunction $\psi(\vec{x},t)$ is subject to the phase change $\psi(\vec{x},t) \to \psi'(\vec{x}',t')=\exp(if/\hbar)\psi(\vec{x},t)$, or $\psi(\vec{x},t)$ is taken to be a projective representation of the group G_4.

Such inconvenience will obviously be removed if we have an invariant

Lagrangian. Thus, following Marmo et al.[3] we introduce a new degree of freedom s and adopt, instead of (2), the following Lagrangian:

$$\tilde{L} = L - m\dot{s} \; . \tag{4}$$

As easily seen, $\tilde{L}$ remains invariant provided s is transformed, under (1), as

$$s \to s' = s + f/m \; . \tag{5}$$

Five-Dimensional Transformations

In view of the above let us work with a 5-dimensional space with contravariant coordinates $x^\mu (\mu=1,2,\ldots,5)$ such that $(x^1,x^2,x^3)\equiv\vec{x}$, $x^4\equiv t$ and $x^5\equiv s$, and consider a group G_5 of combined transformations of (1) and (5):

$$\begin{aligned} x^{i\prime} &= R^i{}_j x^j - v^i x^4 , \quad x^{4\prime} = x^4 , \\ x^{5\prime} &= x^5 - v_i(R^i{}_j x^j) + \tfrac{1}{2}\vec{v}^2 x^4 , \end{aligned} \tag{6}$$

where $i,j=1,2,3$ and $(v^1,v^2,v^3)=(v_1,v_2,v_3)\equiv\vec{v}$, or more compactly,

$$x^{\mu\prime} = \Lambda^\mu{}_\nu x^\nu \; . \tag{6'}$$

Group-theoretically, passing from G_4 to G_5 corresponds to a central extension. Incidentally, the transformations (6) agree with the expression which has been known for some time[4]. Another example of contravariant vectors is the 5-momentum vector $p^\mu\equiv(p,m,E_p)$ with $E_p=\vec{p}^2/2m$.

We next consider bilinear expressions $\eta_{\mu\nu}x^\mu x^\nu$ which remain invariant under (6'). As is easily seen, such a metric tensor $\eta_{\mu\nu}$ is uniquely given as $\eta_{\mu\nu}=\eta^{\mu\nu}$ with $\eta_{11}=\eta_{22}=\eta_{33}=1$, $\eta_{45}=\eta_{54}=-1$ and all other $\eta_{\mu\nu}=0$. Covariant vectors (tensors) can be constructed from contravariant vectors (tensors) by use of $\eta_{\mu\nu}$: for example, $p_\mu\equiv\eta_{\mu\nu}p^\nu=(\vec{p},-E_p,-m)$. Their transformation law can be derived from (6'): thus, for example, a covariant vector $V_\mu(x)$ is transformed, under (6'), as $V'_\mu(x')=\Lambda_\mu{}^\nu V_\nu(x)$ with $\Lambda_\mu{}^\nu=\eta_{\mu\rho}\Lambda^\rho{}_\sigma\eta^{\sigma\nu}$.

Lastly it is of interest to note that $\eta_{\mu\nu}x^\mu x^\nu=\vec{x}^2-2ts=\vec{x}^2+x_-^2-x_+^2$, where $x_\pm\equiv(x^4\pm x^5)/\sqrt{2}$. Hence, our group G_5 is a subgroup of the 5-dimensional Lorentz group such that x^4 is kept constant.

APPLICATION TO NONRELATIVISTIC QUANTUM MECHANICS

It is the paradigm of the present-day theoretical physics to base theories on symmetry: in constructing a theory we first assume a certain group of symmetry transformations and then require that basic entities be related to irreducible representations of the group and basic equations remain covariant under those symmetry transformations. In the realm of RQM we are usually faithful to this paradigm, but not at all so in NQM, at least in the conventional presentation. Thus the purpose of this second part consists in reformulating NQM on the basis of the group G_5 just as we formulate RQM on the basis of the Lorentz or Poincare group.

In the literature we find only few authors who were conscious of the paradigm and took interest in the formulation of the above-mentioned kind. Although in this connection names such as Wigner, Bargmann, Inönu,...[5] should not be forgotten, we here like to refer especially to the modern works by Lévy-Leblond[6], Hurley, Hagen[7],... However, the formulation of the latter authors, who employ the usual G_4, is not very satisfactory in that Galilei covariance is not manifestly exhibited. The main reason for this is that they have to deal with projective representations of G_4, which then make discussions rather complicated and untransparent. In what follows we show that such defects are all removed when G_5 is employed instead of G_4.

Schrödinger Equation for Spinless Particles

For this purpose we consider a 1-dimensional representation of G_5 or a scalar quantity $\chi(x)$ which remains unchanged under (6'):

$$\chi'(x') = \chi(x) \ , \tag{7}$$

and assume that it satisfies an equation of the Klein-Gordon type:

$$\eta^{\mu\nu}\partial_\mu\partial_\nu\chi(x) = 0 \ , \tag{8}$$

where $\partial_\mu \equiv \partial/\partial x^\mu$ form a covariant vector. As regards the variable x^5 we impose a subsidiary condition such that

$$(\partial_5 + im/\hbar)\chi(x) = 0 \ . \tag{9}$$

Since ∂_5 and m, being the 5-th components of covariant vectors, remain unchanged under (6'), (9) is an invariant equation. By means of (9) we extract the x^5-dependence from $\chi(x)$ and write

$$\chi(x) = \exp(-imx^5/\hbar)\psi(\vec{x},t) \ . \tag{10}$$

Then the equation for $\psi(\vec{x},t)$ immediately follow from (8):

$$i\hbar\frac{\partial}{\partial t}\psi(\vec{x},t) = -\frac{\hbar^2}{2m}\Delta\psi(\vec{x},t) \ , \tag{11}$$

in agreement with the Schrödinger equation for a particle with mass m.

The transformation property of $\psi(\vec{x},t)$ under (6') is found from (7):

$$\psi'(\vec{x}',t') = \exp[im(x^{5'}-x^5)/\hbar]\psi(\vec{x},t) = \exp(if/\hbar)\psi(\vec{x},t) \tag{12}$$

with f defined by (3'). We thus find that the role played by the usual projective representation $\psi(\vec{x},t)$ of G_4 is here taken over by a vector representation $\chi(x)$ of G_5.

The electromagnetic interaction can be introduced in a gauge- and Galilei-covariant manner. Corresponding to the usual electromagnetic scalar potential $\phi(\vec{x},t)$ and vector potential $\vec{a}(\vec{x},t)$ we define a covariant 5-vector potential such that $A_\mu(x) \equiv (\vec{a}/c,-\phi,0)$: thus $A_\mu(x)$ is assumed to be x^5-independent. The electromagnetic interaction is then introduced into (8) by making the replacement $\partial_\mu \rightarrow D_\mu \equiv \partial_\mu - (ie/\hbar)A_\mu$. On the other hand, (9) need not be modified because $A_5=0$ and $D_5=\partial_5$ in any coordinate systems.

It then turns out that $\psi(\vec{x},t)$ satisfies the Schrödinger equation with the Hamiltonian:

$$H_0 = -\frac{\hbar^2}{2m}(\vec{\nabla} - \frac{ie}{c\hbar}\vec{a})^2 + e\phi , \tag{13}$$

as expected.

Schrödinger Equation for Particles with Spin 1/2

In this case we consider a "spinor" quantity $\psi(x)$ and assume that it satisfies an equation of the Dirac type:

$$\gamma^\mu \partial_\mu \psi(x) = 0 ; \tag{14}$$

as before (14) should be supplemented by a condition of the same form as (9):

$$(\partial_5 + im/\hbar)\psi(x) = 0 . \tag{15}$$

Here, γ^μ's are matrices to satisfy

$$\gamma^\mu \gamma^\nu + \gamma^\nu \gamma^\mu = 2\eta^{\mu\nu} . \tag{16}$$

Such γ^μ's form, in fact, a 5-dimensional Clifford algebra, so that they have two inequivalent irreducible representations which are both 4-dimensional[8]. Hence, our γ^μ's are essentially the same as Dirac's. For convenience let us here adopt the representation:

$$\gamma^i = \begin{pmatrix} \sigma_i & 0 \\ 0 & -\sigma_i \end{pmatrix}, \quad \gamma^4 = \begin{pmatrix} 0 & 0 \\ -\sqrt{2} & 0 \end{pmatrix}, \quad \gamma^5 = \begin{pmatrix} 0 & \sqrt{2} \\ 0 & 0 \end{pmatrix}, \tag{17}$$

where all elements are 2×2 matrices and σ^i's are Pauli matrices. We need not consider another irreducible representation, because use of this does not lead to anything physically new. Owing to the above form of γ^μ's it is convenient to write $\psi(x)$ as

$$\psi(x) = \begin{pmatrix} \psi_1(x) \\ \psi_2(x) \end{pmatrix} \tag{18}$$

with $\psi_i(x)$ (i=1,2) having two components.

As in the usual Dirac case, the invariance of (14) under (6') is guaranteed if $\psi(x)$ simultaneously transforms according to

$$\psi(x) \to \psi'(x') = S\psi(x) \tag{19}$$

with S satisfying

$$S^{-1}\gamma^\mu S = \Lambda^\mu{}_\nu \gamma^\nu . \tag{20}$$

For the infinitesimal case of (6') where $\Lambda^\mu{}_\nu = \delta^\mu{}_\nu + \lambda^\mu{}_\nu$, S is explicitly given as

$$S = \exp(\lambda^\mu{}_\nu \Sigma^\nu{}_\mu) , \tag{21}$$

$$\Sigma^\nu{}_\mu \equiv [\gamma_\mu, \gamma^\nu]/8 . \tag{21'}$$

Defining the adjoint $\bar{\psi}(x)$ of $\psi(x)$ by $\bar{\psi}(x) \equiv \psi^*(x)\zeta$ with ζ satisfying

$$\zeta\gamma^{\mu\dagger}\zeta = -\gamma^{\mu}, \quad \zeta^2 = 1, \tag{22}$$

we can show that bilinear expressions

$$\bar{\psi}(x)\gamma^{\lambda_1}\gamma^{\lambda_2}\cdots\gamma^{\lambda_n}\psi(x) \tag{23}$$

behave under (6') as contravariant tensors of rank n. For example, the 5-current is defined by

$$j^{\mu}(x) \equiv \bar{\psi}(x)\gamma^{\mu}\psi(x)/(\sqrt{2}i), \tag{24}$$

which is conserved according to $\partial_{\mu}j^{\mu}(x)=0$.

By substituting (17) in (14) and using (15) we find that ψ_2 is expressible in terms of ψ_1, and hence a redundant component. The behaviour of ψ_1 under space rotation shows that ψ_1 rightly describes a particle with spin 1/2. Under proper Galilei transformations, on the other hand, ψ_1 remains unchanged, that is, $\psi_1'=\psi_1$. This implies that $\hat{\psi}_1(\vec{x},t)$ defined by

$$\psi_1(x) = \exp(-imx^5/\hbar)\hat{\psi}_1(\vec{x},t) \tag{25}$$

undergoes the same phase change as (12). Substituting (25) in (14) we find that $\hat{\psi}_1(\vec{x},t)$ satisfies the Schrödinger equation of the same form as (11).

The electromagnetic interaction can be introduced as before by making the replacement $\partial_{\mu} \to D_{\mu}$ in (14). The resulting equation for $\hat{\psi}_1(\vec{x},t)$ is then the Schrödinger equation with the Hamiltonian:

$$H_{1/2} = H_0 - \frac{e\hbar}{2mc}(\mathrm{rot}\,\vec{a}\cdot\vec{\sigma}), \tag{26}$$

implying that the gyromagnetic ratio g equals 2. Thus, as first noted by Lévy-Leblond[6], g=2 is a consequence not of Einsteinian but of Galilean relativity. The spin-orbit coupling can be derived in a similar manner[9].

Galilei-Covariant Formulation of NQM

Although in the above we have only sketched the manifestly Galilei-covariant formulation by referring to the simple cases of spin 0 and 1/2, it is possible to completely reformulate the general theory of NQM in the same manner. As already mentioned, the resulting formalism is quite similar, in appearance, to that of RQM; in dealing with the former, therefore, we can make full use of the knowledge and techniques of the latter, therby pushing the similarity even further.

Thus in discussing wave equations for particles with general spin, for example, we have only to adapt to NQM the Bargmann-Wigner method originally given for RQM[10]. In fact, it is an easy matter to show that a particle with spin S≡n/2 is described by a symmetric multi-spinor $\psi_{\alpha_1\alpha_2\cdots\alpha_n}(x)$ of the following properties: (1) to behave under G_5 in the same way as the direct product of n spinors ψ for the case of S=1/2 (as considered above); and (2) to satisfy a set of n equations of the Bargmann-Wigner type as well as a subsidiary condition. The resulting theory even shares a defect with that of RQM: it is not allowed in general to introduce the electromagnetic interaction by simply making the replacement $\partial_\mu \rightarrow D_\mu$ in the Bargmann-Wigner equations themselves[11].

Supplementary Remarks

Before concluding this lecture we would like to make two remarks. The first is concerned with the connection with Lorentzian symmetry. In a way similar to the above we can reformulate Lorentz transformations in a 5-dimensional form by introducing the 5-th coordinate $x^5=s$. The metric tensor $g_{\mu\nu}$ of the corresponding 5-dimensional space is such that $g_{ii}=1$, $g_{45}=g_{54}=-1$, $g_{55}=1/c^2$ and all other $g_{\mu\nu}=0$. Those transformations in such a space which keep $I_1 \equiv g_{\mu\nu} x^\mu x^\nu$ and $I_2 \equiv ct-s/c$ invariant are Lorentz transformations, and conversely. The theory of RQM can be reformulated in this space, and tends naturally to that of NQM in the Galilei limit $c \rightarrow \infty$, therby reproducing most of the results described above.

Secondly, we would like to remark that our Galilei-covariant description of NQM may be useful not only in formal discussions but also in practical applications. At the present time the only field in which NQM is still extensively used seems to be condensed-matter physics: in fact, NQM provides a principal means of research there. However, we do not think that in this field the consequence of Galilean symmetry has been fully explored and utilized[12]. It is hoped therefore that this will be done in due course, at least, to the extent that the consequence of Lorentzian or Poincaréan symmetry is explored and utilized in the field of particle physics.

Lastly, I (S.K.) would like to thank Professor Bruno Gruber for kindly inviting me to this symposium, thereby giving me ample opportunities to enjoy Symmetries in Science as well as Bregenzer Festspiele.

REFERENCES

1. Galileo Galilei, "Dialogue Concerning the Two Chief World Systems-Ptolemic and Copernican" (Univ. of California Press, Berkeley, 1967).
2. See for example, E.C.G.Sudarshan and N.Mukunda, "Classical Dynamics: A Modern Perspective" (J.Wiley, New York, 1974) chap.19.
3. G.Marmo, G.Morandi, A.Simoni and E.C.G.Sudarshan, Phys. Rev.D37: 2196 (1988).
4. See for example, D.E.Soper, "Classical Field Theory" (J.Wiley, New York, 1976) sect. 7.3.
5. E.P.Wigner, Ann. Math. 40: 149(1939); V.Bargmann, Ann. Math. 59: 1(1954); E.Inönu and E.P.Wigner, Nuovo Cimento 9: 705(1952).
6. J.M.Lévy-Leblond, J. Math. Phys. 4: 776(1963); Commun. Math. Phys. 4: 157(1967); 6: 286(1967).
7. W.J.Hurley, Phys. Rev. D3: 2339(1971); C.R.Hagen and W.J.Hurley, Phys. Rev. Letters 24: 1381(1970); C.R.Hagen, Commun. Math. Phys. 18: 97(1970). See also, J.Voisin, J. Math. Phys. 6: 1519(1965); J.J.Aghassi, P. Roman and R.M.Santilli, Phys. Rev. D1: 2753(1970); Nuovo Cimento 5A: 551(1971); P.Roman, J.J.Aghassi, R.M.Santilli and P.L.Huddleston, Nuovo Cimento 12A: 185(1972).
8. See for example, Y.Ohnuki and S.Kamefuchi, in "Progress in Quantum Field Theory" (eds. H.Ezawa and S.Kamefuchi, North-Holland, Amsterdam, 1986) p.133.
9. Y.Takahashi and C.Ropchan, Prog. Theor. Phys. (Kyoto) 76: 1187(1986).
10. V.Bargmann and E.P.Wigner, Proc. Natl. Acad. Sci. U.S. 94: 211(1948).

11. See for example, S.Kamefuchi and Y. Takahashi, Nuovo Cimento (Ser.X) 44: 1(1966).
12. Y.Takahashi and C.Ropchan, Canad. J. Phys. 65: 484(1987); Y.Takahashi, Fortsch. Phys. 36: 63, 83(1988).

SYMMETRY IN DYNAMIC ECONOMIC MODELS

Ryuzo Sato and Rama Ramachandran

Center for Japan-U.S. Business and Economic Studies
New York University Graduate School of Business Administration
New York, N.Y. 10006

I. INTRODUCTION

In their survey, Arrow and Intriligator (1981) identified three distinctive periods in the development of mathematical economics: the calculus based marginalist period (1838-1947), the set-theoretic/linear model period (1948-1960) and the current period of integration (1961 onwards). The early period was one in which economics borrowed heavily from the methodology of physical sciences, particularly the use of total and partial derivatives and the Lagrange multipliers to characterize maxima and minima. This period culminated in the classical theses of Hicks (1946) and Samuelson (1947).

Origins of the second period can be traced back to the pioneering papers on Ramsey (1927), Wald (1933-34) and Von Neumann (1937) but the new approach developed momentum only after the Second World War. Set theory, algebraic topology, particularly fixed point theorems, matrix theory, linear inequalities and game theory became standard tools of analysis. The set theoretic approach led to the definitive work of Arrow (1951) on social choice theory and Debreu (1959) on general equilibrium. A generalization by Kakutani (1941) enhanced the applicability of fixed point theorems to the existence of equilibrium. The use of input-output models was pioneered by Leontief (1941) while activity analysis was set out in a volume edited by Koopmans (1951) and by Kantorovich (1959). Game theory can be traced back to Von Neumann (1928) and became popular in economics after the publication of Von Neumann and Morgenstern (1947) and Nash (1950). Linear programming was developed during this period. These approaches were reviewed and synthesized in Dorfman, Samuelson and Solow (1958).

The wide application of mathematical techniques to virtually all areas of economics during the third period began to obliterate the distinction between economic theory and mathematical economics. The broader scope also led to an increase in the variety of mathematical tools used. Arrow and Intriligator (1981) list eleven important topics in mathematical economics but, for this survey, it is more advantageous to focus on various aspects of global analysis and dynamic economics.

In the traditional partial equilibrium analysis, the market for each commodity is considered separately. The quantities demanded and supplied are made functions of the price and, by setting excess demand equal to zero, one can, at least theoretically, calculate the equilibrium price and quantity. The stability of equilibrium was examined by Marshall (1890) and Walras (1874).

In general equilibrium analysis, the interdependence of the markets for different commodities is recognized. Walras (1874) sought to establish the existence of an equilibrium for the set of markets by showing that the number of equations equals the number of variables. The inadequacy of this approach led to the use of fixed point theorems.

In the third period of mathematical economics, calculus methods were reintroduced via differential topology. As pointed out in the surveys by Debreu (1976) and Smale (1976), three considerations motivated this approach. First, even if every agent in the economy is well behaved, there may exist in the neighborhood of equilibrium, infinitely many other equilibria. The system is then unstable and the explanation of equilibria indeterminate. Debreu used Sard's theorem to show that, for "regular economies," the set of critical values has measure zero. Second, Scarf (1973) has developed combinatorial algorithms to determine the fixed point of the system. Finally, Smale pioneered the process of convergence to the optima and even developed a Global Newton Method as the differential analog of Scarf's algorithm.

Another important development is the emergence of growth theory. Solow (1956) and Swan (1956) considered a one-sector economy whose output, Y, was a function of capital, K, and labor, L; $Y = F(K,L)$. Assuming $F(\ )$ to be linear homogeneous, output per person, $y = Y/L$, is a monotonous increasing function of $k = K/L$. Labor is assumed to grow at an exogeneous rate of $n = \dot{L}/L$ and the growth of capital is determined by the rate of savings and investment. If the savings rate, s, of an economy is given, then it can easily be shown that $\dot{k} = sF(K/L,1) - nk = sf(k) - nk$. If the production function has a number of economically desirable properties, then the model has a unique and stable equilibrium or stationary point.

These studies laid the foundation for two important extensions. First, Solow (1957) attempted an empirical estimation of the relation between y and k and concluded that less than 15% of the increases in y for the United States between 1909 and 1949 can be attributed to increase in k. The empirical conclusion and theory can be reconciled in two different ways.

The total payment to the two inputs will exhaust the output if the production function is linear homogeneous and each unit of capital and labor is paid its marginal product, $\partial Y/\partial K$ and $\partial Y/\partial L$ respectively. For this reason, Solow (1957) retained the assumption of linear homogeneity and assumed that the production function shifts due to technical progress. Now $Y = A(t)F(K,L)$. The growth in y not attributable to increases in k was allocated to $\dot{A}/A$.

A number of economists objected to this solution as an ad hoc procedure. They argued that, for economic reasons, the production function may not be linear homogeneous. If the degree of homogeneity is greater than one, the production function is considered to show "increasing returns to scale"; output will increase at a rate greater than the proportionate increase in inputs. Many empirical studies were conducted to examine the degree of homogeneity of the production function. We will examine them later when we discuss the application of Lie group.

The savings rate s determines the value of k corresponding to the singular point and this in turn determines the output per person.

Consumption per person in equilibrium is y minus the savings needed to maintain the capital-labor ratio with the given exogeneous growth in labor; $c = y - nk$. The second extension of the descriptive growth models of Solow and Swan was aimed at determining the optimality of the growth path. Phelps (1961) sought to ascertain the value of s which maximizes the steady-state c. Further, the optimal path from an economy's initial point to its stationary point or some other position preferred for policy reasons was examined in an extensive literature surveyed in Arrow and Intriligator (1981) and Hadley and Kemp (1971). We will examine some of these models in the context of conservation laws.

The central point of this section can now be made. In spite of their familiarity with many sophisticated mathematical techniques from linear inequalities to differential topology, the majority of mathematical economists, even those working on dynamic models, ignored group theory. This is in sharp contrast to the situation in natural sciences. Any discussion of the reasons for this neglect is bound to be speculative but it holds the potential of clarifying the scope and limitations of group theoretic methods in economics. The next section analyzes some general issues regarding the use of groups in economics. The third section discusses application of Lie groups to the study of the productivity controversy and brings out the underlying geometric considerations. The final section ennumerates the conservation laws discussed for various dynamic economic models.

II. GROUP THEORY IN ECONOMICS

Using Rosen (1983) as a guide, we can relate the concept of symmetry in science to the structure of economic models. Theory is an expression of casual relation in a system; it uniquely determines an output corresponding to each input. The concept of cause and effect requires some clarification. If one can think of a supersystem in which every characteristic of the elements are indexed, then the cause and effects can be thought of as subsystems. Consider an individual whose consumption pattern is to be analyzed. An exhaustive list of his characteristics will include tastes, income, the price he pays, his neighbor's tastes and the color of his skin. Modern consumer theory states that his consumption depends on his preference, his income and prices only. This is the cause subsystem. Similarly, the goods purchased can be indexed by quantity, the color of the packing, the store where it was sold and so on but the theory only considers the quantity of each commodity. The quantities, therefore, form the effect subsystem. The theory ignores all other characteristics.

Rosen (1983, p. 112) distinguishes between "ignoring in principle" and "ignoring in practice." Consumer theory argues that the consumption basket should only depend on consumers' preferences and the budget set--the combination of commodities that he can purchase given his income and the prices. Different combinations of income and prices may generate the same budget set. The concept of rational choice excludes in principle a change in the basket purchased when prices and income change without changing the budget set. In contrast, there is no theoretical reason why one individual's consumption pattern cannot depend on his neighbor's (through demonstration effect) but consumer theory ignores it. This can be treated as a case of "ignoring in practice."

If we define an equivalence relation for the state space of the system, then we can state the equivalence principle: equivalent states of a cause maps into equivalent states of its effects. A transformation of a system is mapping of a state space into itself. For a state space on which an equivalence relation is defined, some of the transformations might leave equivalence subspaces invariant. Such a transformation is called the symmetry transformation of the equivalence relation. The set of all invertible symmetry transformations of a state space of a system for an equivalence relation forms a "symmetry group"; this is a subgroup of the transformation group.

To return to economics, consider the level sets of the linear homogeneous production function, F(K,L). If we draw a ray through the input space, the level sets will show that output increases proportionately to the distance from the origin. Suppose the output of an economy increases more than proportionately, then, as noted earlier, there are two ways to modify the theory to fit the facts.

In the formative years of modern microeconomics, considerable emphasis was placed on the possibility that production function has variable returns to scale; the discussion was prompted by the study of the equilibrium of a competitive firm under different conditions. Clapham (1922) argued that economies of scale might be empirically indistinguishable from technological change. Pigou (1922), in a spirited counterattack, claimed that the distinction was vital to the foundations of microeconomics. Microeconomic theory was, in this period, evolving into a widely accepted body of doctrine and it is fair to say that most economists were happy to set aside the question, if not totally forget it, until Solow (1957).

It may be wondered why this question cannot be solved by empirical analysis. To isolate the role of returns to scale, we require a number of firms of varying sizes in equilibrium. If there is increasing returns to scale, smaller firms are at a competitive disadvantage and cannot be coexisting in equilibrium with the larger ones. If we assume that all firms are of equal efficiency at a point of time but all of them grow in size over time, then the constancy of technology cannot be assumed. Using Lie groups, Sato (1981) showed that certain types of technical progress and corresponding types of returns to scale renumber the isoquants (level sets) in an identical way and hence the two effects cannot be separated by time series analysis alone.

Two ways have been suggested to get around this problem. One can examine an economy in temporary recession; output has declined but technology can be taken as constant. But no economist will seriously suggest that firms are in equilibrium in such an economy. Another method is to consider firms across economies; firms of different sizes can be in equilibrium due to the variations in economic circumstances. Stigler (1961) used this method but Solow (1961) evoked econometric reasons to question it.

No contemporary economist will question the conceptual difference between economies of scale and technical progress. The basic problem is in identifying the source of productivity growth in a given situation. The equivalence in renumbering isoquants can be interpreted as "identity in practice." Friedman (1983) traces the interaction between the development of logical positivism and relativisitic physics. Economic theory was also influenced by positivist philosophy and counts among its achievements the expunging of concepts like cardinal utility and interpersonal comparisons that are unacceptable by positivist criterion. But due to an inability to clearly delineate the strength and limitations of its empirical analysis, the effective commitment to the doctrine of the "identity of the indiscernables" is rather limited.

Physicists have one Michelson-Morely experiment and an agreed modification of Newtonian mechanics. After two controversies--between

Clapham and Pigou, and Stigler and Solow--and many papers, economists are still unsure of its implications. Ironically, in his Nobel Prize acceptance lecture, Solow (1988) argues that economists should not confine themselves to formal econometric models using time-series data but should supplement it with the qualitative inferences by expert observers and to direct knowledge from observing economic institutions.

Economists seem to have assumed that the measurability implies the existence of a unique or distinguished coordinate system for these quantities. Existence of natural units for inputs like capital and labor was taken for granted. When confronted with quality differences and technological change, they accepted that output was a function of inputs measured in efficiency units, instead of natural units. That this argument destroyed the uniqueness of the coordinate system was not recognized; the argument in Sato (1981), regarding the equivalence of types of technical change and of returns to scale, can be interpreted as establishing an equivalence of two types of transforming the input space.

Conservation laws arise from symmetry. According to Adair (1987, p. 28), experiments on the conservation of angular momentum suggest that the mass of the universe is isotropically distributed to at least one part in 10^{27}. Assumptions of time symmetry in dynamic economic models lead to the conservation of income-wealth ratio. But any attempt to measure the constancy of the ratio is bound to run into serious conceptual and empirical problems.

The final factor that could have inhibited economists' use of group theoretic formulations is the emphasis on characterization of equilibrium rather than the process of convergence.

We have outlined an argument that the neglect of group theoretic methods in economics is due to the historical evolution of modern economic theory. Does this imply that its usefulness is going to be limited in the future also? Though economic methodology is partly borrowed from other sciences, past attempts to imitate physical or biological sciences have not been fruitful. The role of group theory in economics will be different from that in physical sciences. The community of scholars using the method is small but growing. Three areas of application seem to hold prospects. Symmetry can be used to reduce the parameters requiring estimation and it economizes information; this should be of value in a science which faces binding data limitations. Symmetry brings out the underlying geometry of the space. Finally, it permits intuitive interpretation of many economic and finance models [see Sato and Ramachandran, ed. (forthcoming)].

III. HOLOTHETICITY: RELATION BETWEEN TECHNICAL PROGRESS AND SCALE ECONOMIES

A proportionate increase in inputs can lead to a larger rate of increase in output for two reasons. First, the production process may have economies of scale. Let $\mu V_o = (\mu K_o, \mu L_o)$ be a ray in input space (taken to be two dimensional) and let the production function be $f(K,L)$. The scale elasticity is then defined as $\varepsilon(\mu) = d\log f/d\log\mu$; if $\varepsilon(\mu) > 1$, then the function indicates the prevalence of increasing returns to scale.

Second, efficiency of inputs could be increasing due to technical progress and the production function is to be written as $\overline{Y} = f(\overline{K},\overline{L})$ where the number of efficiency units $(\overline{K},\overline{L})$ are related to physical units by technical progress functions.

$$T_t: \overline{K} = \phi(K,\overline{L},t) \quad \text{and} \quad L = \psi(K,L,t) \tag{1}$$

The operator T_t is assumed to have Lie group properties. If the total effects of technical progress T is equivalent to some strictly monotone transformation of production function, f, then the production function is said to be holothetic (holo = whole, thetic = transformation) under a given T.

$$Y = f(\overline{K},\overline{L}) = F_{(t)}[f(K,L)] = F_{(t)}(Y) \tag{2}$$

The effect of a given type of technical progress has been transformed into a scale effect without changing the shape of the isoquant map. In other words, the transformation F and technical progress operator T both renumber the isoquants in an identical way.

Corresponding to the technical progress functions, we can define an infinitesimal transformation

$$U = \xi(K,L)\frac{\partial}{\partial K} + \eta(K,L)\frac{\partial}{\partial L}, \xi = (\frac{\partial\phi}{\partial t})_{t=0} \quad \text{and} \quad \eta = (\frac{\partial\psi}{\partial t})_{t=0} \tag{3}$$

The necessary and sufficient condition for holotheticity can now be stated without proof [Sato (1981, p. 30)].

Theorem 1. The necessary and sufficient condition that the production function be holothetic under a given T is that the first order measure of the impact of technical change is some function of the production function itself.

$$Uf = \xi(K,L)\frac{\partial f}{\partial K} + \eta(K,L)\frac{\partial f}{\partial L} = G(f). \tag{4}$$

Sato (1981) establishes that, for a given technical progress function, there exists one and only one holothetic technology by showing it to be equivalent to finding a general solution to the Langrangian linear partial differential equation given above.

Economic theory does not provide any logical reasons why technical progress functions should have Lie group properties. But Sato (1981, p. 52) notes that all known types of technical progress functions are special cases of the "projective" type. The one-parameter projective group is then:

$$\xi(K,L) = a_1 + a_3K + a_5L + a_7K^2 + a_8KL,$$

$$\gamma(K,L) = a_2 + a_4K + a_6L + a_7KL + a_8L^2, \tag{5}$$

where $a_1, \ldots, a_8$ are constants. Instead of a taxonomic approach, we will examine two special cases which have received considerable attention in economic literature.

First, consider Hicks-neutral technical progress with constant proportionate rate. The technical progress functions can be written as

$$\overline{K} = e^{\alpha t}K, \; L = e^{\alpha t}L \tag{6}$$

Remember that Solow (1957) used Hicks-neutral technical progress. The infinitesimal transformation is now

$$U = \alpha K\frac{\partial}{\partial K} + \alpha L\frac{\partial}{\partial L} \tag{7}$$

The production function holothetic to Hicks-neutral technical progress is the homothetic production function obtained by a monotonic increasing transformation of a linear homogeneous production function.

$$Y = g(f(K,L)), g' > 0, \lambda f(K,L) = f(\lambda K, \lambda L) \tag{8}$$

Increases in efficiency make it possible for the same output to be produced by a smaller quantity of inputs. It is as if the isoquants are uniformly reduced in the input space; for a linear homogeneous production function, all isoquants are magnification of the unit isoquant and this collapse in space is equivalent to a renumbering. But the transformation g(f) with $g' > 0$ is giving a higher number to any given level set of f and the two processes are geometically equivalent though conceptually different.

The other case is biased technical progress. Here

$$\overline{K} = e^{\alpha t}K \qquad \overline{L} = e^{\beta t}L. \tag{9}$$

The infinitesimal transformation is now

$$U = \alpha K\frac{\partial}{\partial K} + \beta L\frac{\partial}{\partial L}. \tag{10}$$

Sato (1981, pp. 33 & 398) has shown that the holothetic production function is $Y = F(K^{1/\alpha}Q(L^{\alpha}/K^{\beta}))$, where the functions Q and F must satisfy certain regularity conditions. The almost homogeneous production function proposed by Lau (1978) and the almost homothetic production proposed by Sato (1977) are special cases of this function. Lau (1988), p. 163) considers the case where one input is increased by a factor λ and the other by λ^{k_2}. When $k_2 = 0$, this degenerates to the case where one of the inputs is constant for institutional reasons. If the output increases by a factor λ^{k_1} so that

$$F(\lambda L, \lambda^{k_2}K) = \lambda^{k_1}F(K,L), \ \lambda > 0 \tag{11}$$

then the production function is taken to be almost homogeneous of degree k_1 and k_2 in inputs L and K respectively. Lau shows that this is true for a continuous differentiable function if and only if it satisfies the differential equation

$$L\frac{\partial F}{\partial L} + k_2 K\frac{\partial F}{\partial K} = k_1 F(K,L) \tag{12}$$

This is the same form as $UF = G(F)$ where U is the infinitesimal transformation of biased technical change.

One may consider that it is more realistic to assume biased technical progress but one runs into some estimation problems. Econometricians have shown that bias and the elasticity of substitution between capital and labor (relating changes in relative prices of capital and labor to the capital-labor ratio) cannot be simultaneously estimated. This is another example of the difficulties faced in testing theoretical models in economics.

The once-extended transformation of Lie has very natural economic interpretation when used in the context of technical progress functions. Going to general technical progress functions (1), the extended transformation can be written as

$$U' = \xi\frac{\partial}{\partial K} + \eta\frac{\partial}{\partial L} + \eta'(K,L,R)\frac{\partial}{\partial R} \tag{13}$$

where $R = \frac{\partial L}{\partial K}$ and

$$\eta' = \frac{\partial \eta}{\partial K} + (\frac{\partial \eta}{\partial L} - \frac{\partial \xi}{\partial K})R - \frac{\partial \xi}{\partial L}R^2 \tag{14}$$

In an economy where perfect competition prevails, R will equal the ratio of the rental of capital to wages.

A differential equation is invariant under transformation (13) if and only if U'R = 0 whenever k(K,L,R) = 0. In this case, the integral curves of this differential equation are mapped into itself by the transformation U'. To solve for holothetic production function associated with U', we define the system of ordinary differential equations

$$\frac{dK}{\xi} = \frac{dL}{\eta} = \frac{dR}{\eta'} \tag{15}$$

One first integral is obtained from the first two equations. We have characterized it as "the virtual expansion path." Given an initial amount of capital and labor, it tells us the new quantities of inputs at original efficiencies that would have the same productive capacity as the old input combination but after technical change. In economics, the path of change in inputs is called the expansion path. The other first integral determines the transformation of the marginal rate of substitution along the virtual expansion path. It is termed the differential invariant as it defines a relation involving the differential coefficient (here R) that remains constant along an integral curve.

For Hicks-neutral technical progress, the virtual expansion path can be shown to be a ray through the origin and the differential invariant is R itself. For biased technical progress, the virtual expansion path is not a straight line. The differential invariant is the ratio of shares of national income earned by the two factors.

Sato (1981, p. 30) also shows that, for any given isoquant map, there is at least one Lie type technical progress under which the production corresponding to the isoquant map is holothetic.

IV. CONSERVATION LAWS

Ramsey's (1928) paper on optimal savings presented a conservation law in an implicit form. Samuelson (1970) exploited the constancy of the Hamiltonian for Von Neumann growth model and derived the first explicit conservation law. Neither paper referred to the work of Noether. Systematic derivation of conservation laws using Noether's theorem was undertaken in Sato (1981, 1982, 1985) and Sato, Nono and Mimura (1983), Sato and Maeda (1987). Sato (1987) undertook a survey of these results.

The mathematical methods used are standard and given in Logan (1977) and Sato (1981). Except for a recapitulation to establish the basic notation and identities, we will concentrate on economic interpretation.

Let us consider a Lagrangian function $L\colon I \times R^n \times R^n \to R^1$ which is twice continuously differentiable in each of its $2n + 1$ arguments, $L \varepsilon C^2$ and $I c R^1$ is an open interval of real numbers. We then have the variational integral

$$J(x) = \int_a^b L(t, x(t), \dot{x}(t))\, dt \tag{16}$$

where $[a,b] c I$ and $x \varepsilon C_n^2[a,b]$ is the set of all vector functions $x(t) = (x^1(t), \ldots x^n(t))$, $t \varepsilon C_n^2[a,b]$. The transformations of the $(n + 1)$ variables $(t, x^1, \ldots x^n)$ depend on r real essential parameters, $\varepsilon^1, \ldots \varepsilon^r$ as follows:

$$\bar{t} = \phi(t,x,\varepsilon) \qquad \varepsilon = (\varepsilon^1, \ldots \varepsilon^r)$$

$$\bar{x} = \psi^k(t,x,\varepsilon) \qquad k = (1, \ldots n) \tag{17}$$

The infinitesimal generators are

$$\tau_s(t,x) = \frac{\partial\phi}{\partial\varepsilon_s}(t,x,0), \; \xi_s^k(t,x) = \frac{\partial\Psi^k}{\partial\varepsilon_s}(t,x,0) \tag{18}$$

The curve $x = x(t)$ in the tx plane can be mapped to $\bar{x} = \bar{x}(\bar{t})$ in $\overline{tx}$ plane. Consider a functional

$$J(x) = \int_b^a L(t, x(t), \dot{x}(t)\, dt, \qquad \dot{x}(t) = \frac{dx(t)}{dt} \tag{19}$$

Then, for the $\overline{tx}$ plane, we calculate

$$\bar{J}(x) = \int_{\bar{a}}^{\bar{b}} L(\bar{t}, \bar{x}(\bar{t}), \dot{\bar{x}}(\bar{t}))\, d\bar{t}, \qquad \dot{\bar{x}}(t) = \frac{d\bar{x}(\bar{t})}{d\bar{x}}. \tag{20}$$

Definition 1 (Absolutely Invariant). The fundamental integral (19) is absolutely invariant under the r parameter family of transformations (17) if and only if, given any curve $x\colon [a,b] \to R^n$ of Class C_n^2 and $a \leq t_1 \leq t_2 \leq b$, we have

$$\int_{\bar{t}_1}^{\bar{t}_2} L\left(\bar{t}, \bar{x}(\bar{t}), \frac{d\bar{x}(\bar{t})}{d\bar{t}}\right) - \int_{t_1}^{t_2} L\left(t, x(t), \frac{dx(t)}{dt}\right) dt = 0\ (\varepsilon) \tag{21}$$

for all $\varepsilon c U$ with $|\varepsilon| < d$ and where $\bar{t}_i = \phi(t_i, x(t_i), \varepsilon)$, $i = 1, 2$.

This is true if and only if

$$L\left(\bar{t}, \bar{x}(\bar{t}), \frac{d\bar{x}(\bar{t})}{d\bar{t}}\right)\frac{d\bar{t}}{dt} - L\left(t, x(t), \frac{d\bar{x}(t)}{dt}\right) = 0(\varepsilon) \tag{22}$$

Definition 2 (Divergence-Invariant). The fundamental integral is divergence-invariant if there exists r functions $\Phi_s\colon I \times R^n \to R^1$, $s = 1, \ldots r$ of Class C^1 such that

$$L\left(\bar{t}, \bar{x}(t), \frac{d\bar{x}(\bar{t})}{d\bar{t}}\right)\frac{d\bar{t}}{dt} - L(t, x(t), \dot{x}(t)) = \varepsilon^s \frac{d\Phi_s(t, x(t)}{dt} + 0(\varepsilon) \tag{23}$$

By totally differentiating (23) with respect to ε^s and setting $\varepsilon = 0$, we get after some manipulation

$$\frac{\partial L}{\partial t}\tau_s + \frac{\partial L}{\partial x^s}\xi_s^k + \frac{\partial L}{\partial \dot{x}^k}\left\{\frac{\partial \xi_s^k}{\partial t} + \frac{\partial \xi_s^k}{\partial x^j}\dot{x}^j - \dot{x}^k\left(\frac{\partial \tau_s}{\partial t} + \dot{x}^j\frac{\partial \tau_s}{\partial x^j}\right)\right\}$$

$$+ L\left(\frac{\partial \tau_s}{\partial t} + \frac{\partial \tau_s}{\partial x^j}\dot{x}^j\right) = \frac{d\Phi s}{dt} \qquad (s = 1, \ldots r) \tag{24}$$

Equation (24) will be seen as a set of identities in (t,x^k) for arbitrary directional arguments $\dot{x}^k$ and the coefficients of powers $\dot{x}^k$ can be equated to zero. This leads to a system of first order partial differential equations in τ_s and ξ_s^k and if this system has a solution, then we have determined a group under which the fundamental integral is invariant. Then Noether Theorem states that

$$\Omega_s \equiv \left(L - \dot{x}^k \frac{\partial L}{\partial \dot{x}^k}\right)\tau_s + \frac{\partial L}{\partial \dot{x}^k}\xi_s^k - \Phi_s$$

$$\equiv - H\tau_s + \frac{\partial L}{\partial \dot{x}^k}\xi_s^k - \Phi_s = \text{a constant.} \qquad (s = 1, \ldots r) \tag{25}$$

We thus obtain the r conservation laws corresponding to the r parameter group.

The economic models for which conservation laws are derived as above will be classified into five groups.

1. Models with Zero Discount Rate

The two famous papers of Ramsey (1928) and Samuelson (1970) fall in this group.

Ramsey's model can be related to the Solow-Swan growth model discussed earlier. Ramsey made the assumption that $n = 0$ and normalized L to be one. So $k = K$. He considered optimization over an infinite time interval but assumed that there is an upper bounds to the utility, U(C), from consumption. The upper bound was called bliss B. The objective of the society was to minimize

$$J(C) = \int_0^\infty (B - U(C))dt \tag{26}$$

subject to $C = F(K) - \dot{K}$. Multiplying (26) by minus one and converting it into a control problem in C using the Hamiltonian,

$$H = - [B - U(C)] + p(t)[F(K) - \dot{K}] \tag{27}$$

we get the value of C that maximizes H to be U'(C). Substituting in (27) and noting that the conservation law is the Hamiltonian itself (as integrand does not contain t explicitly), we get

$$- [B - U(C)] + U'(C)\dot{K} = \text{a constant} \tag{28}$$

or

$$\dot{K} = \frac{B - U(C) + \text{a constant}}{U'(C)} = \text{optimal rate of savings}$$

Ramsey did not formulate the problem as a conservation law but as a rule for optimal savings. Lord Keynes proposed an economic interpretation. If capital formation is increased at time t by h for a period Δt, then the loss in utility due to reduced consumption is approximately $hU'(C)\Delta t$. The increased output at any point of time is $hF'(K)\Delta t$ and the increased stream of utility it creates over the interval (t,∞) is

$$\int_t^\infty h\Delta t.\ F'(K)U'(C)dt \tag{29}$$

The society will equate the extra cost to the incremental utility.

$$\int_t^\infty h\Delta t.\ F'(K)U'(C)dt = hU'(C)dt \tag{30}$$

Differentiating with respect to t

$$-F'(K)U'(C) = \dot{U}'(C) \tag{31}$$

which is the Euler equation of (26). Hence the savings rule equates the intertemporal benefits of savings to its cost in terms of contemporaneous reduction in consumption.

Von Neumann model is, in a sense, an antithesis of the Ramsey Model. It is an n sectoral model but with no consumption. Hence the rate of increase of the stock of any good equals its output. The vector of capital goods at time t is $K_t = (K_t^i)$ and is used to produce the vector $\dot{K}_t$, of net capital formation (output). The transformation function relating K_t to $\dot{K}_t$ is

$$F(K_t,\dot{K}_t) = 0 \tag{32}$$

We assume that F is homogeneous of degree one, concave and smoothly differentiable. The process is considered to have begun at time $t = 0$, with capital vector K_0, and terminates at time T. If the "truncated" terminal vector $K_t' = (K_t^2, \ldots K_t^n)$ is specified, then we can define a criterion for intertemporal efficiency in terms of the maximization of K_t^1 subject to K_0 and K_t'.

The variational problem in continuous time can be written as

$$\text{Max} \int_o^T K^1{}_t dt \text{ subject to } F(K_t,\dot{K}_t) = 0 \text{ and boundary conditions.} \tag{33}$$

Since the integrand does not contain t explicitly, the Hamiltonian is constant.

$$L_t - \sum_j \dot{K}_t^j \frac{\partial Lt}{\partial \dot{K}} = \dot{K}_t^1 + \lambda_t F - \sum_j \dot{K}_t^j \frac{\partial(\dot{K}_t^1 + \lambda_t F)}{\partial \dot{K}_t^j} = -\lambda_t \Sigma \dot{K}_t^j \frac{\partial F}{\partial \dot{K}_t^j} \tag{34}$$

Samuelson (1970) points out that $-\partial F_t/\partial \dot{K}_t^j$ can be interpreted as the price of Jth good and, since the instantaneous output of Jth good is $\dot{K}_t^j$, it follows immediately that $\Sigma \dot{K}_j\ \partial F/\partial \dot{K}_j = Y_t$, the national income (the value of all goods and services produced in an economy) at time t. The first conservation law is

$$\lambda_t Y_t = \text{a constant} \tag{35}$$

Y_t is measurable but λ_t is not. Samuelson after a lengthy discussion derived another conservation law involving wealth, W_t

$$\lambda_T W_t = \text{a constant} \tag{36}$$

He then took the ratio of (35) and (36) to eliminate λ and obtain a conservation law in terms of capital-output ratio.

$$\frac{W_t}{Y_t} = \text{a constant} \tag{37}$$

Samuelson's derivation of the second conservation law (36) was not based on a general principle. Sato (1981) showed that it can be derived, after some lengthy manipulations, by using divergence invariance (25). It could also be shown that (35) and (36) are the only two conservation laws possible for the Von Neumann model.

2. Models with Fixed Discount Rate

Even though Ramsey objected to discounting on philosophical grounds, it is widely held that income or utility in future should not be added directly to income or utility in the present. Future income or utility is discounted, most commonly using a negative exponential weight. Extensions of Ramsey's analysis recently in Leviathan and Samuelson (1969) and Canton and Shell (1971) are examples of such models. Neoclassical investment by Jorgenson (1967) and Lucas (1967) and the discussion of welfare implications of national income by Samuelson (1961), Weitzman (1976) and Kemp and Long (1982) fall within this group. We will illustrate the case with the national income discussion, as it can be easily related to the Ramsey and Von Neumann models analyzed earlier.

In a modern economy, commodities are produced both for consumption and investment. The value of these goods and services is national income. In theoretical literature and popular imagination, national income is taken as an index of the welfare of the community. Investment is not an end in itself but should be valued for the future consumption that it facilitates. Notice that only consumption entered the integrand of Ramsey model.

Assume the existence of a social welfare function whose argument is a consumption vector of many goods which in turn depends on a vector of capital goods $k = (k^1, \ldots, k^n)$ and a vector of investments, $\dot{k} = (\dot{k}^1, \ldots, \dot{k}_n)$. Hence,

$$U = U(\dot{k}, k) \tag{38}$$

The society seeks to maximize the welfare functional

$$J = \int_0^\infty e^{-\rho t} U[\dot{k}(t), k(t)]\,dt \qquad \rho > 0 \tag{39}$$

subject to appropriate initial conditions. Now assume that t and k are subject to the transformations (taking r = 1).

$$\bar{t} = \phi(t, k, \varepsilon)$$

$$\bar{k}^i = \Psi^i(t, k, \varepsilon) \qquad i = 1, \ldots, n \tag{40}$$

Using divergence-invariance, Sato (1985) derived the relation

$$\frac{d}{dt}\left(U - \Sigma_i \dot{k}^i \frac{\partial U}{\partial \dot{k}^i}\right) = -\rho \Sigma_k k^i \frac{\partial U}{\partial \dot{k}^i} \tag{41}$$

The expression

$$- \Sigma \dot{k}^i \frac{\partial U}{\partial \dot{k}^i} = \sum_i p^i \dot{k}^i \tag{42}$$

is the utility-value-of-investment. The expression in parenthesis (41) is, therefore, the utility of consumption plus the utility-value-of-investment. The equation states that ($\rho > 0$)

$$\frac{\text{The rate of change of income at time t}}{\text{Utility-value-of-investment at time t}} = \text{a constant} \tag{43}$$

Now assume that the generators of (40) take the special form

$$\tau = 1, \ \xi^i = 0 \qquad i = 1, \ldots n \tag{44}$$

Then Sato (1985) shows that (for $\rho > 0$)

$$\frac{\text{Income}}{\text{Wealth}} = \rho$$

We began with a model whose objective function is similar to that of Ramsey and derived a conservation law parallel to that of Von Neumann model.

Model 3. Variable Discount Rate

Samuelson (1982) wondered whether the income-wealth conservation law held when the discount rate varies. Sato (1985) showed that the infinitesimal generators

$$\tau = \frac{1}{\rho'(t)}, \ \xi^i = o, \ \rho'(t) = d\rho(t)/dt \tag{45}$$

leads to a conservation law

$$\text{income} = \rho'(t) \times \text{general wealth.} \tag{46}$$

The generalized wealth now includes capital gains and losses.

Model 4. Technical and Taste Changes

Assume that the optimal control problem has the form

$$\text{Max} \int_o^\infty e^{-\rho t} U[x(t), \dot{x}(t), t] dt \tag{47}$$

U is directly affected by t indicating technical or taste changes. Sato, Nono and Mimura (1984) shows that for

$$\tau = e^{\rho t}, \ \xi = b(t) \exp \left[-\int_o^t \frac{U_k}{U_{\dot{k}}} ds\right] \tag{48}$$

we have the conservation law

Income + 'Value of taste (technical) change' = ρ x wealth.

Model 5. "Local" Conservation Laws

We look for conservation laws that operate in models characterizing the neighborhood of a stationary point. We begin with a local Lagrangian suggested by Samuelson (1972)

$$L = e^{-st} [- \tfrac{1}{2}\dot{x}^2 - ax\dot{x} - \tfrac{1}{2}x^2] \tag{49}$$

Sato (1981) shows that, for infinitesimal generators

$$\tau = \text{a constant}, \quad \xi = \frac{\rho\tau}{2} x \tag{50}$$

there is a conservation law that can be written as

Income + ρ times the value of capital = a constant.

V. CONCLUSION

Most of the conservation laws are derived by applying the Noether theorems to economic models originally formulated without regard to such laws. These derivations serve two purposes.

They can enhance our understanding of the models. "Income-wealth" ratios, suitably interpreted, is conserved in models with very different mathematical structures. Symmetry brings out the uniformity in the permissible transformations of space. Most of the conservation laws arise from the absence of a "distinguished origin" for the time axis. Economists would consider this symmetry self-evident and would reject any models that do not satisfy it. The second conservation law of Samuelson and Models 4 and 5 require a simultaneous transformation of the quantity axes also. No obvious economic interpretation for the specific form of these infinitesimal generators can be offered at this time. Applications of symmetry groups to a wider range of models in economics and finance will be presented in Sato and Ramachandran (forthcoming).

Though the ratio of income to wealth is theoretically measurable, an attempt to do so will run into many measurement problems.

Symmetry also provides a simple geometric interpretation of holotheticity. An effort is being made to measure technical progress using group concepts.

VI. ACKNOWLEDGMENTS

The authors thank Professor F. A. Matsen of the University of Texas for helpful discussions. Comments offered by Mr. Raghu Ramachandran on an earlier working paper were useful in preparing this survey.

VII. REFERENCES

Adair, R. K., 1987, "The Great Design," Oxford Un. Press, New York.
Arrow, K. J., 1951, "Social Choice And Individual Values," Wiley, New York.
Arrow, K. J., and Intriligator, M. D., 1981, "Handbook Of Mathematical Economics, Volume I," North-Holland, Amsterdam.
Canton, C., and Shell, K., 1971, An exercise in the theory of heterogeneous capital accumulation, Rev. Econ. Stud., 38:13.
Clapham, S. H., 1922, Of empty economic boxes, Econ. J., 32:305.
Debreu, G., 1959, "Theory Of Value," Wiley, New York.
Debreu, G., 1976, Regular differentiable economies, Amer. Econ. Rev., 66:280.

Dorfman, R., Samuelson, P. A., and Solow, R. M., 1958, "Linear Programming And Economic Analysis," McGraw-Hill, New York.
Friedman, M., 1983, "Foundations Of Space-Time Theories," Princeton Un. Press, Princeton.
Hadley, G. and Kemp, M. C., 1971, "Variational Methods In Economics," North-Holland, Amsterdam.
Hicks, J. R., 1946, "Value And Capital, 2nd Ed.," Oxford Un. Press, New York.
Jorgenson, D. W., 1967, Theory of investment behavior, in R. Ferber, ed., "Determinants of Investment Behaviour," NBER, New York.
Kakutani, S., 1941, "A generalization of Brouwer's fixed point theorem," Duke Math. J., 8:451.
Kantorovich, L. V., 1942, "The Best Use Of Economic Resources," Pergamon Press, Oxford.
Kemp, M. C. and Long, N. V., 1982, On the evaluation of social income in a dynamic economy, in G. R. Feiwel, ed., "Samuelson And Neoclassical Economics," Kluwer-Nijhoff, Boston.
Koopmans, T. C., Ed., 1951, "Activity Analysis of Production and Allocation," Wiley, New York.
Lau, L. J., 1978, Applications of profit functions, in N. Fuss and D. MacFadden, ed., "Production Economics: A Dual Approach To Theory and Applications," North-Holland, Amsterdam.
Leontief, W. W., 1941, "The Structure Of The American Economy, 1919-1939," Oxford Un. Press, New York.
Leviathan, N. and Samuelson, P. A., 1969, Notes on turnpikes: stable and unstable, J. Econ. Theory, 1:1454.
Logan, J. D., 1977, "Invariant Variational Principles," Academic Press, New York.
Lucus, R. E., 1967, Optimal investment policy and the flexible accelerator, Int. Econ. Rev., 8:78.
Marshall, A., 1890, "Principles Of Economics," MacMillan, London.
Nash, J. F., Jr., 1950, Equilibrium in n-person games, Proc. Nat. Acad. Sc., 36:48.
Phelps, E. S., 1961, Golden rule of accumulation, Amer. Econ. Rev., 55:793.
Pigou, A. C., 1922, Empty economic boxes: a reply, Econ. J., 32:458.
Ramsey, F. P., 1927, A contribution to the theory of taxation, Econ. J., 37:47.
Ramsey, F. P., 1928, A mathematical theory of saving, Econ. J., 38:543.
Rosen, J., 1983, "A Symmetry Primer For Scientists," Wiley, New York.
Samuelson, P. A., 1947, "Foundations Of Economic Analysis," Harvard Un. Press, Cambridge.
Samuelson, P. A., 1961, The evaluation of 'social income': capital formation and wealth, in F. A. Lutz and D. C. Hague, "The Theory Of Capital," St. Martin's Press, New York.
Samuelson, P. A., 1970a, Laws of Conservation of the capital-output ratio, Proc. Nat. Acad. Sc., Applied Math. Sc., 67:1477.
Samuelson, P. A., 1970b, Two conservation laws in theoretical economics, Working Paper, M.I.T.
Sato, R., 1977, Homothetic and non-homothetic functions, Am. Econ. Rev., 67:559.
Sato, R., 1981, "Theory Of Technical Change And Economic Invariance," Academic Press, New York.
Sato, R., 1982, Invariant principle and capital/output conservation laws, Working Paper, Brown University.
Sato, R., 1985, The invariance principle and income-wealth conservation laws: application of Lie groups and related transformations, J. Econometrics, 30:365.
Sato, R., 1987, Conservation laws in continuous and discrete models: a survey, Working Paper, New York University.
Sato, R., and Maeda, S., 1987, Local conservation laws of the discrete optimal growth models," Mimeo, New York University.

Sato, R., Nono, T., and Mimura, F., 1982, "Hidden Symmetries, Lie Groups And Economic Conservation Laws," Springer-Verlag, New York.

Sato, R., and Ramachandran, R., Ed., forthcoming, "Conservation Laws And Symmetry: Applications To Economics And Finance," Kluwer Academic Publishers, Boston.

Scarf, H. E., 1973, "The Computation Of Economic Equilibria," Yale University Press, New Haven, CT.

Smale, S., 1976, Dynamics of general equilibrium, Amer. Econ. Rev., 66:288.

Solow, R. M., 1956, A contribution to the theory of economic growth, Quart. J. Econ., 70:65.

Solow, R. M., 1957, Technical change and aggregate production function, Rev. Econ. Statist., 39:312.

Solow, R. M., 1961, Comment on Stigler, in "Output, Input And Productivity Measurement," Income and Wealth Series, Vol. 25, Princeton Un. Press, Princeton.

Solow, R. M., 1988, Growth theory and after, Amer. Econ. Rev., 78:307.

Stigler, G. J., 1961, Economic problems in measuring changes in productivity, in "Output, Input And Productivity Measurement," Income and Wealth Series, Vol. 25, Princeton Un. Press, Princeton.

Swan, T. W., 1956, Economic growth and capital accumulation, Econ. Record, 32:334.

Von Neumann, J., 1928, Zur Theorie der Gessellschaftsspiele, Math. Annalen, 32:47.

Von Neumann, J., 1937, Uber ein okonomisches Gleichungssystem und eine Verallgemeinerung des Brouwerischen Fixpunktsatzes, Erg. eines Math. Kolloquims, 8:73.

Von Neumann, J. and Morgenstern, O., 1947, "The Theory of Games And Economic Behavior, 2nd ed.," Princeton Un. Press, Princeton.

Wald, A., 1934-35, "Uber der Produktionsgleichungen der okonomische Wertlehre," Erg. eines Math. Kolloquiums, 7:1.

Walras, L., 1874, "Elements d'economie politique pure," L. Corbaz, Lausanne.

Weitzman, M. L., 1976, On the welfare significance of national product in a dynamic economy, Quart. J. Econ., 90:156.

HARMONIC OSCILLATOR REPRESENTATION IN THE THEORY OF SCATTERING AND REACTIONS

Yu. F. Smirnov

Institute of Nuclear Physics, Moscow State University
Moscow 117234, USSR

ABSTRACT

The problem is discussed to solve the Schrödinger equation for a particle in a potential well V in terms of the harmonic oscillator representation. The matrix of the free-motion Hamiltonian in the oscillator basis is of a simple three-diagonal form. Explicit expressions for the matrix eigenvectors corresponding to both regular and irregular solutions for free motion at an arbitrary energy E are discussed. The formulae obtained make it possible to calculate the energies of the bound and resonant states, the scattering phase and S-matrix elements. This approach is extended to "true" many body scattering. As examples of application of this method the $^{16}0(\gamma, n)$ reaction and 0^+-states in $^{12}C = 3\alpha$ system are considered.

1. INTRODUCTION

In recent years a new approach to the scattering problem was suggested [1-11] which is based on the using of square integrable functions as basic functions. The new technique also uses the fact that, in some particular bases (the oscillator and Laguerre bases), the matrix of the kinetic energy operator T is of three-diagonal form and that the eigenvectors of such Jacobi matrix (J-matrix) can be obtained in an explicit analytical form. It has been shown [2,3] that eigenvectors in the Laguerre basis may also be found for the operator matrix $T + A/r$, i.e. for the Coulomb problem. In connectioon with this, the J-matrix method proposed in [2,3] was mainly applied until recently to atomic systems, and specific calculations were made in the Laguerre basis [2-6]. At the same time, this approach proves to be very effective when applied to nuclear theory where the use of the harmonic oscillator basis to solve the problems of continuum is of main interest. The studies relevant to the application of the oscillator representation to the scattering problem are being made by G. F. Fillippov's group in Kiev [7-10] and by our group [11]. Discussed below will be the main results of these studies. The paper [7] was aimed at calculating the $\alpha - \alpha$ scattering phases. The method was later applied successfully to various cluster systems (the $^3H + p$, $^3He + n$, $d + d$, etc.) [9,10]. The general equations of the method and the problem of solving them were discussed in [8]. Essentially in these papers the approach was developed in [9] which proved to be fully equivalent to the resonating group method (RGM) [12], but is more convenient because this pure algebraic version of RGM deals with algebraic rather than integrodifferential equations. Besides that, various ways of improving RGM can readily be analyzed in such an approach. In order to illustrate the essence of the approach to the scattering problem in the harmonic oscillator basis, we shall consider at first the simplest problem of the scattering of a single particle by potential [1-3,11]. Thus, we come to the Schrödinger equation

$$\left(\frac{P^2}{2m}+V(r)\right)\psi_{\ell m}(\vec{r})=\epsilon\psi_{\ell m}(\vec{r}). \tag{1}$$

Its solution $\psi_{lm}(\vec{r}) = R_\ell(r)Y_{lm}(\Omega)$ will be sought in the form of an expansion in the eigenfunctions of the harmonic oscillator

$$R_\ell(r)=\sum_{n=0}^{\infty}C_{n\ell}R_{n\ell}(r) \tag{2}$$

where

$$R_{n\ell}(r)=(-1)^n\left(\frac{2n!}{\Gamma(n+\ell+\frac{3}{2})}\right)^{1/2}r^\ell L_n^{\ell+1/2}(r^2)e^{-r^2/2} \tag{3}$$

is the radial wave function of three-dimensional harmonic oscillator. This wave function corresponds to the eigenvalue of the oscillator energy $E_n^{osc} = (2n+\ell+3/2)\hbar\omega$ [13]. The value $r_0 = (\hbar/m\omega)^{1/2}$ is selected as length scale in relations (1) and (2). Here ω is the oscillator frequency; the energy $\epsilon = q^2/2$ is measured in units $\hbar\omega$; the wave vector k is expressed in units r_0^{-1}, $q = kr_0$ – is the dimensionless momentum. Substituting expansion (2) in (1) and multiplying (1) scalarly by $R_{nl}Y_{lm}$, we obtain the following set of equations for the coefficients $C_{n\ell}$ or, in other words, for the wave function ψ_{lm} in the harmonic oscillator or n-representations:

$$\sum_{n'}\left(H_{nn'}-\delta_{nn'}\epsilon\right)C_{n'\ell}=0, \quad n=0,1,2,\ldots. \tag{4}$$

Here, $H = T+V$ and only the following matrix elements of the kinetic energy operator $T = P^2/2$ are nonvanishing:

$$\begin{aligned}
T_{nn-1}&=-\frac{1}{2}\left[n\left(n+\ell+\frac{1}{2}\right)\right]^{1/2},\\
T_{nn}&=\frac{1}{2}\left(2n+\ell+\frac{3}{2}\right),\\
T_{nn+1}&=-\frac{1}{2}\left[(n+1)\left(n+\ell+\frac{3}{2}\right)\right]^{1/2}.
\end{aligned} \tag{5}$$

The set (4) is usually solved under the boundary condition

$$C_{-1\ell}=0. \tag{6}$$

If some states $\psi_{nlm}(\vec{r})$ with $n \le n_0$ are forbidden in virtue of Pauly's exclusion principle, as it is the case in the RGM problem [12,14], equations (4) must be solved under the boundary condition $C_{nl} = 0$ for $n \le n_0$. As to the behaviour of the coefficient

C_{nl} for $n \geq n_0$, their asymptotics are similar to the asymptotic of the wave function in the coordinate representation [8] if r is substituted by $2n^{1/2}r_0$:

$$C_{n\ell} \sim n^{1/4}\psi_{\ell m}(2\sqrt{n}r_0), \quad n \to \infty. \tag{7}$$

This result can be obtained if the WKB expression for the oscillator function $R_{nl}(r)$ is substituted in the expression for the coefficients

$$C_{n\ell} = \langle \psi_{n\ell m}(\vec{r}) | \psi_{\ell m}(\vec{r}) \rangle \tag{8}$$

and the integral (8) is calculated by the stationary phase method. The result (7) follows also from the fact that the finite-difference equation

$$\begin{aligned} -\Bigg\{ & \left[n\left(n+\ell+\frac{1}{2}\right)\right]^{1/2} C_{n-1\ell} - \left(2n+\ell+\frac{3}{2}-q^2\right) C_{n\ell} \\ & + \left[(n+1)\left(n+\ell+\frac{3}{2}\right)\right]^{1/2} C_{n+1\ell} \Bigg\} + 2\sum_{n'} \langle n\ell | V | n'\ell \rangle C_{n'\ell} = 0 \end{aligned} \tag{9}$$

in the limit $n \gg \nu = \ell/2 + 3/4$ can be replaced by the following second-order differential equation [11]:

$$X_\ell'' - \frac{\ell(\ell+1)}{x^2} X_\ell - \int_0^\infty V(x,x')\sqrt{xx'}X_\ell(x')dx' + q^2 X_\ell = 0. \tag{10}$$

Here $X = 2(n+\nu)^{1/2}$, $X_\ell(x) = X^{1/2}C_{nl}$. The boundary condition (6) takes on the form $X_\ell(2\sqrt{\nu - 1}) = 0$.

Thus, in the asymptotic limit for large n, the wave function of our system X_l for the partial wave with angular momentum l in the harmonic oscillator or n-representation obeys the conventional Schrödinger equation with nonlocal potential

$$V(x,x')\sqrt{xx'} \simeq 2\,\langle n\ell | V | n'\ell \rangle\, [(n+\nu)(n'+\nu)]^{1/4} \tag{11}$$

where the value $2(n+\nu)^{1/2}r_0$ plays the role of "coordinate". In actual calculations, the potential matrix has to be cut off by the condition $V_{nn'} = 0$, if n or (and) $n' > N$.

$$a) \quad n \leq N, \quad \sum_{n'=0}^{N} (H_{nn'} - \epsilon\delta_{nn'})C_{n'l} = -\delta_{nN}T_{NN+1}C_{N+1l}. \tag{12a}$$

$$b) \quad n \geq N+1, \quad T_{nn-1}C_{n-1l} + (T_{nn} - \epsilon)C_{nl} + T_{nn+1}C_{n+1l} = 0. \tag{12b}$$

Thus, the coefficients C_{nl} with $n > N$ obey the equation of free motion (12b) or, in the asymptotic limit of continuous n, the Schrödinger equation of free motion

$$X_\ell'' - \frac{\ell(\ell+1)}{x^2} X_\ell + q^2 X_\ell = 0.$$

It means that the condition

$$C_{n\ell} \sim 2^{-1/2} n^{-1/4} e^{-2\sqrt{n}\kappa} \tag{13a}$$

(where $\epsilon = -\kappa^2/2$ is the binding energy) must be satisfied for the bound states. The coefficients C_{nl} for the scattering problem have the following asymptotic behaviour:

$$C_{n\ell} \sim 2^{-1/2} n^{-1/4} sin \left(2q\sqrt{n} - \ell\pi/2 + \delta_\ell\right) \tag{13b}$$

where $\epsilon = q^2/2$. According to Eq. (7) the phase shift δ_l in Eq. (13b) coincides with the standard phase shift of the wave function in coordinate space. For the decaying resonance states, we get (see in [8]):

$$C_{n\ell} \sim 2^{-1/2} n^{-1/4} e^{2iq\sqrt{n}}. \tag{13c}$$

If the calculations are made up to sufficiently high values of $N \gg 1$ it is possible to use the asymptotic expressions (13) [7-10]. At modest N it is necessary to use the exact, rather than approximate, solution for the equation of free motion (12b) which was found in Refs. [3,11] in order to calculate the binding energy, the scattering phases etc. Before considering the solution for the equation of free motion, we shall note that the solution for the set (12) is equivalent to the solution for the Schrödinger equation with Hamiltonian $H = T + V^N$ containing the many-term separable potential

$$V^N = \sum_{n,n'}^{N} \langle n\ell|\ V\ |n'\ell\rangle\ |n\ell\rangle\ \langle n'\ell| \tag{14}$$

with harmonic oscillator formfactors. The technique of solving such an equation in the frame of the momentum representation was described in [15]. Here we shall describe an alternative method for solving the same problem in the n-representation.

2. SOLUTION FOR THE EQUATION OF FREE MOTION IN THE HARMONIC OSCILLATOR REPRESENTATION

Consider first the case of positive energy $\epsilon = q^2/2 > 0$. The Schrödinger equation of free motion in the coordinate space has two linear-independent solutions (regular and irregular) [16]:

$$\begin{aligned} R_\ell^{reg} &= j_\ell(kr) \sim \frac{1}{kr} sin\left(kr - \frac{\ell\pi}{2}\right), \\ R_\ell^{irreg} &= n_\ell(kr) \sim \frac{1}{kr} cos\left(kr - \frac{\ell\pi}{2}\right), \end{aligned} \tag{15}$$

In accordance with this, the finite difference equation of free motion (12b) will have also two fundamental solutions in the n-representation [3,11] namely the regular solution

$$\begin{aligned} C_{n\ell}^{reg}(q) &= \left(\frac{2\Gamma(n+\ell+\frac{3}{2})}{\Gamma(n+1)}\right)^{1/2} \frac{q^\ell}{\Gamma(\ell+\frac{3}{2})} e^{-q^2/2} M(-n, \ell+\frac{3}{2}; q^2) = \\ &= (-1)^n R_{n\ell}(q) \sim \frac{2^{3/2}}{\pi^{1/2}} n^{1/4} j_\ell(2\sqrt{n}q) \end{aligned} \tag{16}$$

satisfying the boundary condition (6) $C^{reg}_{-1l} = 0$, and the irregular solution

$$C^{irreg}_{n\ell}(q) = \left(\frac{2\Gamma(n+1)}{\Gamma(n+\ell+\frac{3}{2})}\right)^{1/2} \frac{(-1)^{\ell} q^{-\ell-1}}{\Gamma(-\ell+\frac{1}{2})} e^{-q^2/2} M(-n-\ell-\frac{1}{2}, -\ell+\frac{1}{2}; q^2) = \sim \frac{2^{\frac{3}{2}}}{\pi^{\frac{1}{2}}} n^{1/4} n_{\ell}(2\sqrt{nq}) \tag{17}$$

which is singular at the point $n = -1$.

The Kazorati determinant K_{nl} for these two solutions which plays the same role for the difference equations as the Wronskian for the differential equations [18] is of the form:

$$K_{n\ell} = \begin{vmatrix} C^{reg}_{n\ell} & C^{irreg}_{n\ell} \\ C^{reg}_{n+1\ell} & C^{irreg}_{n+1\ell} \end{vmatrix} = \frac{2}{\pi q}(-1)^{\ell+1} \left[(n+1)(n+\ell+\frac{3}{2})\right]^{-1/2} . \tag{18}$$

Since $K_{nl} \neq 0$ for any values of n and l, the expressions (16) and (17) constitute the fundamental set of solutions for equation (12). An arbitrary solution for (12b) may be presented as a linear combination of fundamental solutions. In particular, the solution for the set (12) for $n \geq N$ must be of the form

$$C_{n\ell}(q) = cos\, \delta_{\ell} C^{reg}_{n\ell}(q) + sin\, \delta_{\ell} C^{irreg}_{n\ell}(q) \tag{19}$$

whence it follows that

$$tg\ \delta_{\ell} = -\frac{C_{n\ell} C^{reg}_{n+1\ell} - C_{n+1\ell} C^{reg}_{n\ell}}{C_{n\ell} C^{irreg}_{n+1\ell} - C_{n+1\ell} C^{irreg}_{n\ell}} . \tag{20}$$

The equivalent pair of fundamental solutions for the free motion equation has the asymptotic form of the type of Hankel functions

$$C^{\pm}_{n\ell}(q) = C^{irreg}_{n\ell}(q) \pm i\, C^{reg}_{n\ell}(q) . \tag{21}$$

These solutions are useful for the calculation of the S-matrix and analyzing the decaying Gamov states. If we are interested in bound states ($\epsilon = -\kappa^2/2$, $q = i\kappa$) the solution for the equation of free motion with a corresponding asymptotic

$$C^{bound}_{n\ell}(\kappa) = i^{\ell} \left[C^{irreg}_{n\ell}(i\kappa) + i\, C^{reg}_{n\ell}(i\kappa)\right] . \tag{22}$$

must be used. The numerical values of solutions (16), (17) can be obtained by using the book [17], where the function $M(a, b; z)$ is tabulated. The dependence of $C^{reg}_{nl}(q)$ and $C^{irreg}_{nl}(q)$ on n is shown in Figs. 1 and 2 respectively for $\ell = 0$ and 1, $q = 1$ and 2. Similarly to the regular and irregular solutions of the free motion Schrödinger equation the functions C^{reg}_{nl} and C^{irreg}_{nl} are oscillating functions of n and the period of oscillations decreases with increasing energy ϵ.

3. THE SOLUTION OF THE SCATTERING PROBLEM IN THE OSCILLATOR REPRESENTATION

Consider now the solution for set (12). It follows from equations (12) that the coefficient C_{nl} for $n \geq N+1$ obey the equation of free motion with an appropriate asymptotic, i.e. $C_{nl} = C^0_{nl}$, where C^0_{nl} is the solution for the equation of free motion with asymptotic (19), (21) or (22). The coefficients C_{nl} ($n \geq N+1$) form the "external" part of the wave function in a Hamiltonian oscillator representation. The coefficients C_{nl} ($n \leq N$) belong to the "internal" part of this function. The equation (23) plays a role of "tailoring" condition of "internal" and "external" parts of the wave function. The r.h.s. of this equation has one of the form (19), (21) or (22). Into

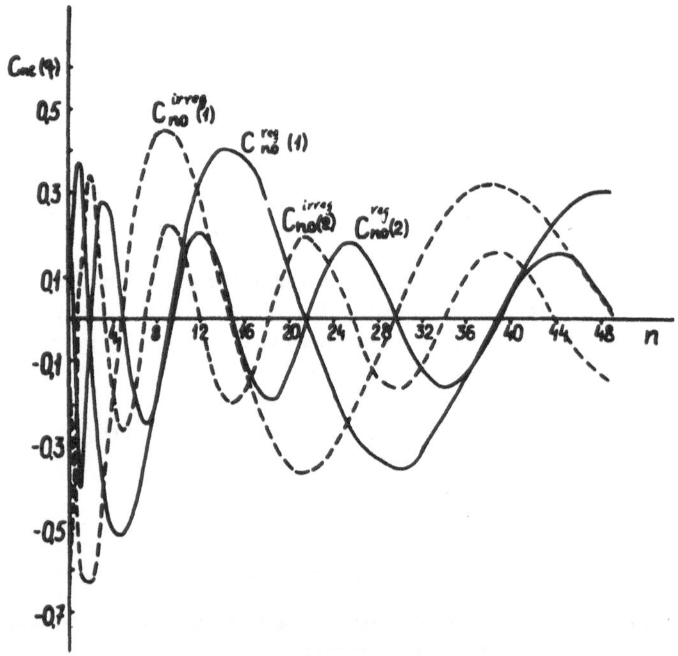

Fig. 1. n-dependence of the regular and nonregular solutions $C_{n\ell}(q)$ ($\ell = 0$, $q = 1$ and 2).

the left hand side of Eq. (23) must be substituted the solution of the set (12a). The last one can be found in the following manner [2,3]. At first we shall diagonalize the truncated Hamiltonian matrix $\|H_{nn'}\|$ using the unitary transformation Γ, i.e. turn from C_{nl} to the new factors

$$C'_{\lambda\ell} = \sum_{n=0}^{N} \Gamma_{\lambda n} C_{n\ell}, \quad \lambda = 0, 1, \ldots, N . \tag{24}$$

As a result of this transformation, equation (12a) takes the form

$$(E_\lambda - \epsilon) C'_{\lambda\ell} = -\Gamma_{\lambda N} T_{NN+1} C_{N+1\ell}, \quad \lambda = 0, 1, \ldots, N$$

i.e.

$$C'_{\lambda\ell} = -\frac{\Gamma_{\lambda N} T_{NN+1}}{E_\lambda - \epsilon} C_{N+1\ell} \tag{25a}$$

and

$$C_{n\ell} = -\sum_{\lambda=0}^{N} \frac{\Gamma^*_{\lambda n}\Gamma_{\lambda N}}{E_\lambda - \epsilon} T_{NN+1} C_{N+1\ell} \tag{25b}$$

where E_λ - is the eigenvalue of the matrix $\|H_{nn'}\|$ $(n, n' \le N)$.

Substituting the "internal" solution (25b) into Eq. (23) we can write the last one in the form

$$P\, C^0_{N+1\ell} = -C^0_{N\ell}, \quad P = \sum_{\lambda=0}^{N} \frac{|\Gamma_{\lambda N}|^2}{E_\lambda - \epsilon} T_{NN+1} \tag{26}$$

if we deal with bound states of Gamov resonances, C^0_{nl} and C^0_{N+1l} are the known functions of energy (see (21) and (22)). In such cases condition (26) is the transcendent equation which may be used to find the energies ϵ_i of the bound or resonant

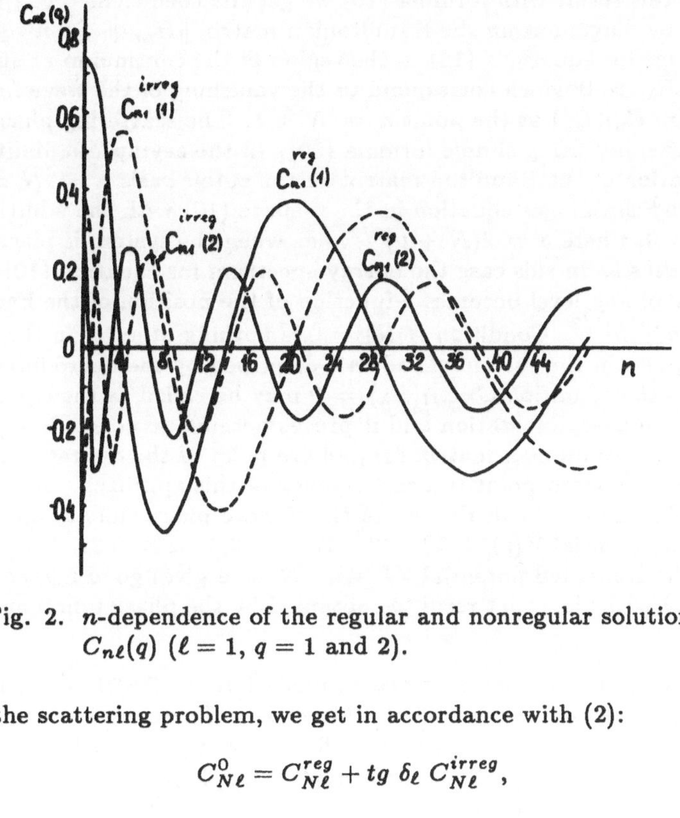

Fig. 2. n-dependence of the regular and nonregular solutions $C_{n\ell}(q)$ ($\ell = 1$, $q = 1$ and 2).

states. For the scattering problem, we get in accordance with (2):

$$C^0_{N\ell} = C^{reg}_{N\ell} + tg\ \delta_\ell\ C^{irreg}_{N\ell}, \qquad C^0_{N+1} = C^{reg}_{N+1\ell} + tg\ \delta_\ell\ C^{irreg}_{N+1\ell}\,. \tag{27}$$

Substituting these expressions in (26), we find in accordance with Refs. [2-5]:

$$tg\ \delta_\ell = -\frac{C^{reg}_{N\ell} + P\, C^{reg}_{N+1\ell}}{C^{irreg}_{N\ell} + P\, C^{irreg}_{N+1\ell}}\,. \tag{28}$$

It can be seen now that the scattering phase at an arbitrary energy ϵ will be obtained by diagonalizing the Hamiltonian matrix $\|H_{nn'}\|$ $(n, n' \le N)$ but one time.

In agreement with the Ritz variational principle, the negative eigenvalues $E_\lambda < 0$ of the Hamiltonian matrix $\|H_{nn'}\|$ $(n, n' \le N)$ may be treated as approximate values of the energies of discrete levels of a particle in the studied potential. In this case the approximation accuracy improves with increasing the size of the matrix N. The question arises, what is the sense of the matrix positive eigenvalues and of the respective wave functions? The question was answered in works [1-11,19,20] as follows. In the limit $\epsilon \to E_\lambda$ expression (28) takes on the form

$$tg\ \delta_\ell(E_\lambda) = -\frac{C^{reg}_{N+1\ell}}{C^{irreg}_{N+1\ell}}\ . \tag{29}$$

By comparing this result with formula (19) we get the coefficient $C_{N+1l}(E_\lambda) = 0$ for E_λ. Thus, by diagonalizing the Hamiltonian matrix $\|H_{nn'}\|$ $(n, n' \le N)$, we find the solutions for equations (12) in the region of the continuum at such discrete energies $E_\lambda > 0$ which correspond to the vanishing of the wave function in n-representation $C_{nl}(E_\lambda)$ at the point $n = N+1$. The scattering phase is calculated at such energies using simple formula (29). In the asymptotic limit of high n, the diagonalization of the Hamilton matrix on the cutoff basis $n \le N$ means the solution for the Schrödinger equation in the n-space (10) with the additional condition $X_l(b) = 0$, where $b = 2(N+\nu)^{1/2}$, i.e. when the system is placed within a rigid box of radius b. In this case the energy spectrum for equation (10) gets discrete and the energy of any level becomes a function of the position of the boundary point $b = 2\left[(n+\nu)\hbar/m\omega\right]^{1/2}$. Condition $X_l(b) = 0$ is nothing other than the equation for the P-matrix poles in the system of radius b described by the Schrödinger equation (10) [21]. Thus the condition $C_{N+1l}(E_\lambda) = 0$ may be called as the equation for the P-matrix poles in n-representation and it proves possible to conclude that the eigenvalues E_λ of the Hamiltonian matrix $\|H_{nn'}\|$ are poles of the discrete analogue of the P-matrix. The important point is a convergency of this approach. In order to clear this question the results of calculations of the S-wave phase shift δ_0 are shown at Fig. 3 for the Gauss potential $V(r) = V_0 e^{-\alpha r^2}$, $V_0 = -4,2$, $\alpha = 0.22$ $(r_0 = 1)$. It can be seen that the truncated potential V^N with $N = 9$ gives good agreement for δ_0^N (dashed curves) with the exact result δ_0 obtained by the phase function method (solid curve).

The good convergency was observed at modest $N \sim$ 20-30 in calculations of Kiev's group in the frame of algebraic RGM version. Therefore the using of Hamiltonian oscillator representation is a rather effective and practicable method for the study of continuum problems. Some additional examples of application of this method will be discussed in section 6.

4. MULTICHANNEL CASE

Let us consider the case of two open (binary, spinless) channel for simplicity. The wave function has the form of a column

$$\psi(r) = \begin{pmatrix} \psi_1(r) \\ \psi_2(r) \end{pmatrix} \tag{30}$$

where the wave function of the entrance channel $\psi_1(r)$ is characterized by the following asymptotic behaviour

$$\psi_1(r) \sim \left(e^{-ik_1 r} - S_{11} e^{ik_1 r}\right)/r \tag{31}$$

while in the second channel the outgoing wave is

$$\psi_2(r) \sim -\left((v_1/v_2)^{1/2} S_{21} e^{ik_2 r}\right)/r. \tag{32}$$

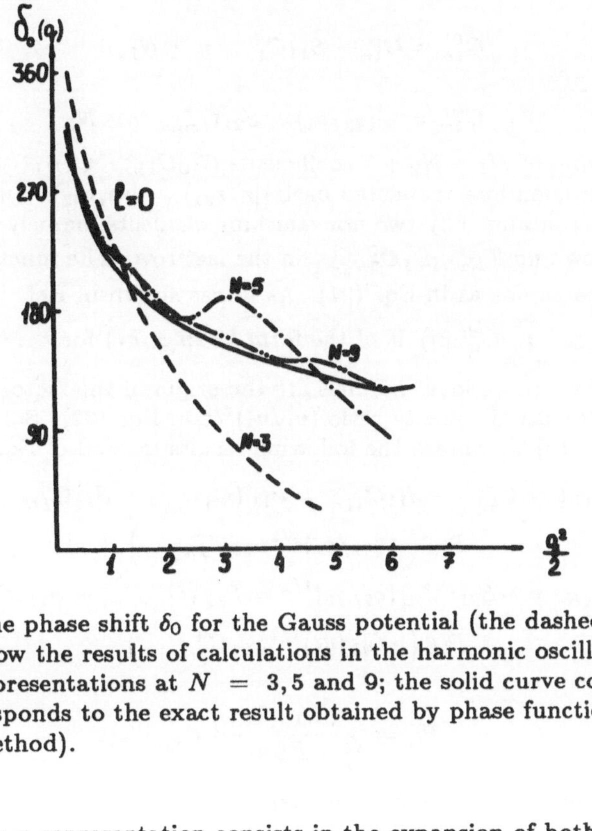

Fig. 3. The phase shift δ_0 for the Gauss potential (the dashed curves show the results of calculations in the harmonic oscillator representations at $N = 3, 5$ and 9; the solid curve corresponds to the exact result obtained by phase function method).

The transition into n-representation consists in the expansion of both channel wave functions

$$\begin{aligned} \psi_1(r) &= \sum_n C_{1n} \left|n, r_{01}\right\rangle , \\ \psi_2(r) &= \sum_m C_{2m} \left|m, r_{02}\right\rangle , \end{aligned} \tag{33}$$

in terms of Hamiltonian oscillator wave functions $|n, r_{01}\rangle$, $|m, r_{02}\rangle$ with a unique frequency $\hbar\omega$ while the linear scale parameters $r_{0i} = (h/\mu_i\omega)^{1/2}$ can be different for the channels 1 and 2 if the reduced masses μ_1 and μ_2 of the two fragments in these channels are different. Assuming that it is possible to restrict ourselves to a truncated matrix of potential energy

$$V_{1n,1n'} \quad (0 \le n,\ n' \le N_1), \quad V_{2m,2m'} \quad (0 \le m,\ m' \le N_2),$$

$$V_{1n,2m},\ V_{2m,1n} \quad (0 \le n \le N_1,\ \ 0 \le m \le N_2)$$

(generally speaking $N_1 \neq N_2$) we obtain the following set of equations for C_{1n}, C_{2m} coefficients instead of Eqs., (12), (23), (27)

$$\begin{aligned}
&(H-E)C = -TC^0, \\
&C_{1N_1} = C^0_{1N_1}, \quad C_{2N_2} = C^0_{2N_2}, \\
&C^0_{1n} = C^-_{1n} - S_{11}C^+_{1n}, \quad n \ge N_1, \\
&C^0_{2m} = -(q_2/q_1)^{1/2} S_{21} C^+_{2m}, \quad m \ge N_2.
\end{aligned} \tag{34}$$

Here C is a column of $N_1 + N_2 + 2$ coefficients $C_{10}C_{11}\ldots C_{1N1}C_{20}\ldots C_{2N2}$, H is the matrix of Hamiltonian in a truncated basis $|n, r_{01}\rangle$, $|m, r_{02}\rangle$ $(n \le N_1,\ m \le N_2)$. The column TC contains only two nonvanishing elements, namely $T_{N,N+1}C^0_{1N+1}$ in the N_1 + 1-th row and $T_{N_2N_2+1}C^0_{2N_2+1}$ in the last row. The functions $C^{\pm}_{ik} = C^{reg}_{ik} \pm iC^{irreg}_{ik}$ are the same one as in Eq. (21). As it was shown in Ref. [3] the asymptotic of the function $\psi^{\pm} = \sum_{n=0}^{\infty} C^{\pm}_n |n\rangle$ is of the form $k\,exp(\pm ikr)$ for $r \to \infty$. This fact and the difference of r_{0i} in various channels are the origin of the factor $(q_2/q_1)^{1/2}$ in Eq. (34) instead of the usual velocity ratio $(v_1/v_2)^{1/2}$ in Eq. (32). Solving the Eq. (34) similarly to Eq. (12) we obtain the following results instead of Eq. (26)

$$\begin{aligned}
C_{1N_1} &= C^-_{1N_1} - S_{11}C^+_{1N_1} = P_{11}\left(C^-_{1N_1+1} - S_{11}C^+_{1N_1+1}\right) \\
&\quad + P_{12}\left(-(q_2/q_1)^{1/2} S_{21} C^+_{2N_2+1}\right), \\
C_{2N_2} &= -S_{21}C^+_{2N_2}(q_2/q_1)^{1/2} = P_{21}\left(C^-_{1N_1+1} - S_{11}C^+_{1N_1+1}\right) \\
&\quad + P_{22}\left(-(q_2/q_1)^{1/2} S_{21} C^+_{2N_2+1}\right)
\end{aligned} \tag{35}$$

where

$$P_{ij} = \sum_{\lambda} \frac{\Gamma_{\lambda N_i}\Gamma_{\lambda N_j}}{E_\lambda - E} T_{N_jN_j+1}.$$

E_λ is the eigenvalue of the truncated matrix H, $(\Gamma_{\lambda 0}\ldots\Gamma_{\lambda N_1}\ldots\Gamma_{\lambda N_2})$ is the corresponding eigenvector of this matrix.

The relations (35) should be considered as the equations for elements of the S-matrix. The solutions of these equations are of the form

$$\begin{aligned}
S_{11} &= \frac{1}{D}\left[\left(C^-_{1N_1} + P_{11}C^-_{1N_1+1}\right)\left(C^-_{2N_2} + P_{22}C^-_{2N_2+1}\right) - P_{21}P_{12}C^-_{1N_1+1}C^-_{2N_2+1}\right] \\
S_{21} &= \frac{1}{D}\frac{2iP_{12}P_{21}}{\sqrt{\pi q_1 q_2}}, \\
D &= \left[\left(C^+_{1N_1} + P_{11}C^+_{1N_1+1}\right)\left(C^+_{2N_2} + P_{22}C^+_{2N_2+1}\right) - P_{12}P_{21}C^+_{1N_1+1}C^+_{2N_2+1}\right].
\end{aligned} \tag{36}$$

Here the property of the Kazorati determinant

$$T_{NN+1}\begin{vmatrix} C_N^- & C_N^+ \\ C_{N+1}^- & C_{N+1}^+ \end{vmatrix} = \frac{2i}{\pi q}$$

was used. Obviously the S-matrix is symmetrical in accordance with the time reflection symmetry of the Hamiltonian. As in the previous section, the eigenvalues E_λ are the poles of the discrete analogue of the P-matrix.

The eigenfunctions $\psi_\lambda = \sum_n \Gamma_{\lambda n} |n\rangle$ are the wave functions of "primitives" (in terms of the paper [21]).

The expressions (36) allow us to find the numerical values of the S-matrix elements and then to calculate the cross-sections of elastic scattering and reactions

$$\sigma_{el} = \frac{\pi}{k_1^2}|S_{11} - 1|^2,$$

$$\sigma_r = \frac{\pi}{k_1^2}|S_{21}|^2,$$

the differential cross-sections, various polarization characteristics (taking into account the spin degree of freedom) etc. If we want to describe the reaction with three-four fragments in the final states it is necessary to extend the above developed formalism which is valid only for two body (binary) channels to three, four body collisions.

5. THE DESCRIPTION OF "TRUE" MANY BODY SCATTERING IN HYPERSPHERICAL OSCILLATOR REPRESENTATION

We restrict ourselves by the case of the so called "true" many body scattering (TMBS) when the wave function of a body system is in the asymptotic region of the form $\rho \to \infty$, $\rho^2 = \sum_{i=1}^{A} (\vec{r}_i - \vec{R})^2$ is a global radius in $3(A-1)$ dimensional space, the angles Ω are hyperspherical coordinates in this space. $\vec{R}$ is the center-of-mass of the system. The approximation taking into account only the contribution of TMBS is valid if there is a "democracy" in the body system i.e. there is no pair of particles with dominating interaction between them in comparison with the rest of the interactions. The TMBS - approximation is applicable to a lot of processes of three, four body decay of light nuclei and hypernuclei [22] (for example disintegration $^{12}C \to 3\alpha$ etc.).

For the description of TMBS we shall use the expansion of the A-body wave function $\psi(\vec{r}_1 \ldots \vec{r}_A)$ in terms of A-body oscillator wave function

$$|nK\gamma\rangle = R_{nK}(\rho) Y_{K\gamma}(\Omega)$$

depending on hyperspherical coordinates ρ, Ω :

$$\psi = \sum_{nK\gamma} \langle nK\gamma|\,\psi\rangle\,|nK\gamma\rangle\,.$$

Here $K = 0,1,2,\ldots$ is a global momentum, γ substitutes the rest of the quantum numbers which are necessary for unambiguously labelling the hyperspherical harmonics $Y_{K\gamma}(\Omega)$. Further consideration is totally parallel to sections 1-4 and we represent the result in very short form. Instead of Eq. (3) we have for the many body case

$$R_{NK}(\rho) = \rho^{-(3A-4)/2}\phi_n^{\mathcal{L}}(\rho),$$
$$\phi_n^{\mathcal{L}}(\rho) = (-1)^n\sqrt{\frac{2n!}{\Gamma(n+\mathcal{L}+\frac{3}{2})}}\ \rho^{\mathcal{L}+1}e^{-\rho^2/2}L_n^{\mathcal{L}+1/2}(\rho^2)$$

$\mathcal{L} = K + (3A-6)/2$, ρ is taken in units of r_0. The Eq. (4) takes on the form

$$\sum_{n'K'\gamma'} \langle nK\gamma|\ H - E\ |n'K'\gamma'\rangle\ \langle n'K'\gamma'|\ \psi\rangle = 0.$$

The kinetic term T in the Hamiltonian $H = T + V$ is diagonal in the quantum numbers K and γ. As for the main quantum number, n, the matrix T is three diagonal with respect to n and its matrix elements coincide with Eq. (5) except for substitution of 1 by $\mathcal{L}$. We also truncate the matrix of potential energy $V = \sum\limits_{i<j}^{A} V_{ij}$ to $n \leq N$, $K \leq K_{max}$. Then for $n \geq N$ the expansion coefficients $\langle nK\gamma|\ \psi\rangle \equiv C_{n\ell}$ obey the three-term recurrent relation similar to Eq. (12b)

$$\sqrt{n(n+\mathcal{L}+\frac{1}{2})}\ \langle n-1K\gamma|\ \psi\rangle - \left(2n+\mathcal{L}+\frac{3}{2}-q^2\right)\ \langle nK\gamma|\ \psi\rangle$$
$$+\sqrt{(n+1)(n+\mathcal{L}+\frac{3}{2})}\ \langle n+1K\gamma|\ \psi\rangle = 0, \quad q = \overline{E}.$$

This difference equation has two fundamental solutions

$$C_{n\mathcal{L}}^{reg} = \sqrt{\frac{2n!}{\Gamma(n+\mathcal{L}+\frac{3}{2})}}\ q^{\mathcal{L}+1}e^{-q^2/2}L_n^{\mathcal{L}+1/2}(q^2)$$

and

$$C_{n\mathcal{L}}^{irreg} = -\frac{2q}{\pi C_{01}^{reg}(q)}v.p.\int_0^\infty \frac{C_{0\mathcal{L}}^{reg}(q')C_{n\mathcal{L}}^{reg}(q')}{q^2-q'^2}dq'$$

or the equivalent pair of solutions

$$C_{n\mathcal{L}}^{\pm} = -\frac{2q}{\pi C_{0\mathcal{L}}^{reg}(q)}\int_0^\infty \frac{C_{0\mathcal{L}}^{reg}(q')C_{n\ell}^{reg}(q')}{q^2-q'^2\pm i0}dq'.$$

The problem of TMBS is similar to the multichannel problem described in section 4. Thus the wave function with ingoing wave in some channel $K_0\gamma_0$ and outgoing waves in each channels $K'\gamma'$ under consideration takes on the form for $n > N$ (in principle the truncation boundary N may be different in the various channels $K\gamma$):

$$\langle nK\gamma|\psi\rangle = \delta_{K\gamma,k_0\gamma_0}C_{n\mathcal{L}}^{-}(q) - \sum_{K'\gamma'} S_{K\gamma,K'\gamma'}C_{n\mathcal{L}'}^{+}(q).$$

In analogy with Eq. (36) we can obtain

$$S = A^{-1}B$$

where

$$(A)_{K'\gamma',K\gamma} = P_{K'\gamma',K\gamma}C^{+}_{N+1\mathcal{L}}(q) - \delta_{K\gamma,K'\gamma'}C^{+}_{N\mathcal{L}}(q),$$

$$(B)_{K'\gamma',K\gamma} = P_{K'\gamma',K\gamma}C^{-}_{N+1\mathcal{L}}(q) - \delta_{K\gamma,K'\gamma'}C^{-}_{N\mathcal{L}}(q),$$

$$P_{K\gamma,K'\gamma'} = \sum_{\lambda} \frac{\langle NK\gamma|\lambda\rangle\langle\lambda|NK'\gamma'\rangle}{E - E_\lambda} T_{NK',N+1K'}.$$

E_λ and $\langle nK\gamma|\lambda\rangle$ are eigenvalue and eigenvector components of the truncated Hamiltonian matrix $\langle nK\gamma|H|n'K'\gamma'\rangle$ $(n, n' \leq N)$. The poles of the S-matrix (i.e. bound states and Gamov resonance states) can be found from the equation

$$\det A = 0.$$

Thus we have all expressions that are necessary for the construction of the wave function for few body states belonging to continuum or discrete spectrum in the frame of TMBS approximation.

6. EXAMPLES

In order to prove the effectiveness of the harmonic oscillator representation in an analysis of concrete nuclear processes the calculation of the cross-section of $^{16}0(\gamma, n)$ reaction was done. We restrict ourselves to the energy region of the giant dipole resonance $E_\gamma \leq 30$ MeV. For the ground state the usual shell model wave function $|0\rangle \equiv |S^4p^{12}\rangle$ was used. It was assumed that the final states have the $1p - 1h$ structure. Five $1p - 1h$ channels with total angular momentum $J^\pi = 1^-$ and isospin $T = 1$ were considered: $P^{-1}_{1/2}2S_{1/2}$, $P^{-1}_{1/2}d_{3/2}$, $P^{-1}_{3/2}S_{1/2}$, $P^{-1}_{3/2}d_{5/2}$, $P^{-1}_{3/2}d_{3/2}$. The total Hamiltonian for the $1p - 1h$ excitations is of the form

$$H = H_0 + H_p + H_h + V_{ph}$$

where H_0 is the energy of the $^{16}0$ core in the ground state. The energies H_h of the hole orbitals $P^{-1}_{1/2}$ and $P^{-1}_{3/2}$ were taken from the experiment. The wave functions of these orbitals were approximated by a single oscillator eigenfunction $|0pj\rangle$. The particle-core interaction was represented by a single local potential

$$H_p(h) = T + V(r)$$

of Wood-Saxon form with spin-orbit term. An excited nucleon can occupy S, d-orbitals in this potential as well as belong to the continuum. As usually the hole-particle interaction V_{ph} was taken in zero-range form. All parameters were chosen the same as in the paper [23].

We have used the harmonic oscillator representation and have sought the wave functions of final state in the form of the expansion

$$\psi^{j_1 \ell j_2}_{1^- T=1} = \sum_{n,\ell'=0,2,j_1'j_2'} C_n^{j_1\ell j_2, j_1'\ell' j_2'} \left|0P^{-1}j_1'n\ell'j_2' \; : \; 1^- T = 1\right\rangle .$$

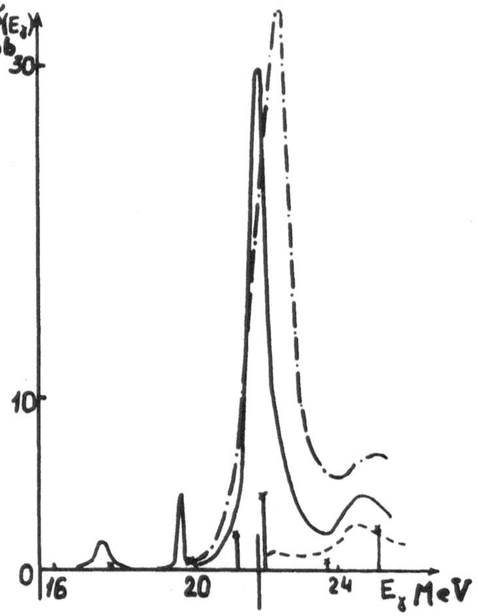

Fig. 4. The cross section of the photonuclear reaction $^{16}O(\gamma, n)$. The solid curve represents the result of our calculations, the results of the continuum shell model calculations [24] are given by dashed-dotted curve. The dashed curve represents the contribution of the channel $^{16}O(\gamma, n)^{15}O^*(3/2^-)$.

Such a function is the solution of the coupled channel problem whose asymptotic contains the ingoing wave in the entrance channel $j_1^{-1}(lj_2)$ and outgoing waves in all five $1p - 1h$ channels. The C_n were found by using the method discussed above with $N = 9$ (the matrix of order 50×50 was diagonalized). Then the matrix elements of dipole momentum operator D between the ground and excited states

$$\left\langle \psi_{1^- T=1}^{j_1 l j_2} \right| \widehat{D} \left| \psi_{ground} \right\rangle$$

were calculated. Finally the total photo absorption $\sigma_\gamma(E)$ was found. The results are shown in Fig. 4. The solid curve was obtained by us, the results of the continuum shell model calculations [24] are given by the dashed curve. The agreement of results for these two methods is rather satisfactory if some small difference in the parameters of Refs. [23] and [24] is taken into account. Besides the Coulomb *pp*-population was considered in Ref. [24] and it was neglected by us.

The results of a simple $1p-1h$ approximation without continuum effects are given by vertical arrows.

The next example is the isoscalar monopole 0^+ excitations in ^{12}C nucleus considered as 3α cluster system. The $\alpha\alpha$-interaction was taken in a form of the Ali-Bodmer [25] potential with soft repulsive core, the Coulomb $\alpha\alpha$-repulsion was neglected too. The coupling of 0^+ excitations with the continuum was considered in the frame of the "true" three-particle scattering. The truncation boundary of the potential energy matrix was taken as $N = 18$. The coupled channel problem for three-body channels with a global momentum $K \leq 10$ was solved in a manner described in the section 5 [26].

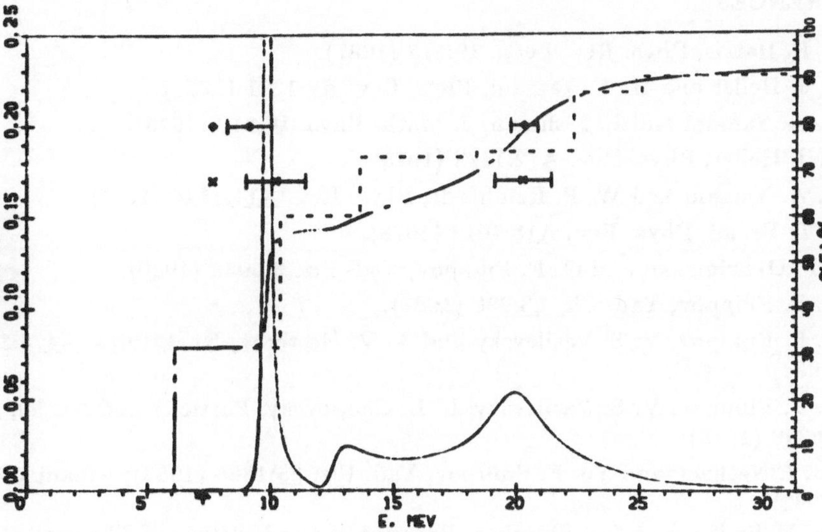

Fig. 5. The isoscalar monopole strength $F(E)$ as a function of the excitation energy E_x in the ^{12}C nucleus (solid curve). The percentage $S(E)$ of the energy weighted monopole sum rule corresponding to the energy interval O-E MeV is presented by the dashed-dotted curve. The same quantity $S(E)$ calculated by neglecting continuum effects is shown as a hystogram. The experimental data concerning the positions and widths of the EO resonances are indicated by straight line sections (Ref. [27]) and (Ref. [28]).

The binding energy of the ^{12}C nucleus with respect to 3α-channel is 7,27 MeV. The first excited 0_2^+-state is bound in our approximation (without Coulomb forces) and its excitation energy is equal to 6,11 MeV. The distribution of monopole strength in the energy region 7-25 MeV is represented in Fig. 5. It is clear that except for the strong O_2^+ level that is not shown in Fig. 5 there are two maxima of intensive monopole excitation: near 10 MeV and 20 MeV.

These results are in correspondence with the recent experiment: [27,28] $^{12}C(^6Li,^6Li')^{12}C$ where the 0^+ excitations in the ^{12}C nucleus were investigated by inelastic forward scattering of ^6Li ions. It should be noted that a considerable part of the monopole strength corresponds to the nonresonant region between 12-18 MeV. In the 3α-model the 99% of the energy weighted sum rule is exhausted up to 25 MeV of the excitation energy E_x. These examples and the calculations [7-10] in the frame of the algebraic version of RGM show that the harmonic oscillator representation is rather effective and promising method for the solution of continuum problems. In our opinion this new version of the unified theory of reactions [29] may be applied with success for the description of wide circle of nuclear and hypernuclear processes.

In conclusion the author wants to express his gratitude to Profs. G. F. Filippov and V. G. Neudatchin for illuminating discussions and to Drs. R. M. Asherova, A. M. Shirokov, T. Ya. Mihelashvily, Yu. I. Nechaev, V. A. Knyr, A. I. Mazur for valuable collaboration.

REFERENCES

1. F. E. Harris, Phys. Rev. Lett., 19:173 (1967).
2. E. J. Heller and H. A. Yamani, Phys. Rev. A9:1201 (1974).
3. H. A. Yamani and L. Fishman, J. Math. Phys. 16:410 (1975).
4. E. J. Heller, Phys. Rev. A12:1222 (1975).
5. H. A. Yamani and W. P. Reinhardt, Phys. Rev. A11:1144 (1975).
6. J. T. Broad, Phys. Rev. A18:1012 (1978).
7. I. P. Okhrimenko and G. F. Filippov, Yad. Fiz. 32:933 (1980).
8. G. F. Filippov, Yad. Fiz. 33:928 (1981).
9. G. F. Filippov, V. S. Vasilevsky and A. V. Nesterov, Nucl. Phys. A426:327 (1984).
10. G. F. Filippov, V. S. Vasilevsky, L. L. Chopovsky, Particles and Nuclei, 15:1984); 16:349 (1985).
11. Yu. I. Nechaev and Yu. F. Smirnov, Yad. Fiz. 35:1385 (1982); Kinam 4:445 (1982);
    Yu. F. Smirnov, A. M. Shirokov, Preprint of the Institute of Theoretical Physics, Kiev, 1988.
12. K. Wildermuth and Y. C. Tang, "Unified Theory of the Nucleus", Braunschweig, Vieweg Verlag, 1977.
13. M. Moshinsky, "The Harmonic Oscillator in Modern Physics: From Atoms to Quarks," Gordon and Breach, New York, London, Paris: Gordon and Breach, (1969).
14. V. I. Kukulin, V. G. Neudatchin and Yu. F. Smirnov, Particles and Nuclei, 10, no. 6:1236 (1979).
15. L. M. Kuznetzova, V. I. Kukulin and V. G. Neudstchin, Yad. Fiz. 13:694 (1971). J. Revai et al., J. Phys. G. 11:745 (1985).
16. A. Messiah, "Quantum Mechanics," Vol. 1, North-Holland, Amsterdam, (1965).
17. M. Abramowitz and I. A. Stegun, "Handbook of Mathematical Functions," National Bureau of Standard, Applied Mathematics Series 55, (1964).
18. A. A. Mirolyubov and M. A. Soldatov, "Linear homogeneous difference equations," Nauka Publishers, Moscow, (1981).
19. P. W. Langhoff and W. P. Reinhardt, Chem. Lett. 24:495 (1974).
20. W. P. Reinhardt, Comput. Phys. Comm. 6:303 (1973).
21. R. L. Jaffe and F. E. Low, Phys. Rev. D19:2105 (1979).
22. R. I. Jibuti, Particles and Nuclei, 14:741 (1983).
23. B. Buck, A. D. Hill, Nucl. Phys. A95:271 (1967).

24. I. Rotter, Particles and Nuclei 15:762 (1984).
25. S. Ali, A. R. Bodmer, Nucl. Phys. 80:99 (1966).
26. T. Ya. Michelashvily, Yu. F. Smirnov. A. M. Shirokov, Preprint of Institute of Physics, Tbilisy, (1986).
27. W. Eyrich et al., Phys. Rev. C36:416 (1987).
28. D. Lebrun et al., Phys. Lett. B97:358 (1980).
29. H. Feshbach, Ann. Phys. 19:287 (1962).

REPRESENTATIONS OF LIE GROUPS AND INTEGRAL TRANSFORMS

N. Ja. Vilenkin* and A. U. Klimyk**
*Mathematical Department, The Correspondence Pedagogical Institute (MGZPI), Moscow, 109004 USSR
**Institute for Theoretical Physics, Kiev-130, 252130 USSR

1. INTRODUCTION

The group-theoretical methods give a unified approach to main classes of special functions [1,2]. Special functions are connected with integral transforms. This is why the theory of group representations is closely related to integral transforms. The group-theoretical description is given for many integral transforms.

The Fourier, Laplace, and Mellin transforms are connected with group representations of commutative groups. Namely, an expansion of a function into Fourier series is a decomposition of the regular representation of the group $SO(2)$ into irreducible representations, and the Fourier integrals are related to a decomposition of the regular representation of the additive group R of real numbers. The Laplace transform is connected with representations of the additive semigroup R_+ of positive numbers, and the Mellin transform with representations of the multiplicative group R_+ (this group is isomorphic to the group R).

The Fourier transformation on a locally compact abelian group G is a direct generalization of these transforms [3]. If X is a group of characters for G, and dg an invariant measure on G, then the formula

$$\tilde{f}(\chi) = \int f(g)(\chi, g)dg$$

defines the isometric integral transform $F : L^2(G) \to L^2(X)$. The inverse transform has the form

$$f(g) = \int \tilde{f}(\chi)\overline{(\chi, g)}d\chi.$$

If $G = R$ we obtain the usual Fourier transform.

The well known Poisson summation formula

$$\sum_{n=-\infty} \tilde{f}(n) = 2\pi \sum_{m=-\infty} f(2\pi m)$$

connects values of a function on the subgroup Z of integers with values of its Fourier transformation on the annihilator of this subgroup. An analogous formula holds for any subgroup H of a locally compact abelian group G,

$$\int_H f(h)dh = \int_{\widehat{H}} \widehat{f}(\chi)d\chi.$$

Here $\widehat{H} = \{\chi | (\chi, h) = 1, h \in H\}$ is an annihilator of H.

The wide theory, dealing with convergence at a point, absolute convergence, convergence almost everywhere and so on, which was developed for Fourier series and integrals, is transferred onto Fourier transforms on some classes of locally compact abelian groups [4]. An expansion in Walsh functions, which is used in Information Theory, are connected with these groups.

In order to obtain a group-theoretical description of integral transforms of Mathematical Physics (Mehler-Fock, Kantorovich-Lebedev transforms and others), we have to use representations of noncommutative Lie groups. There are several methods of usage of group representations in the theory of integral transforms. They are

1. Usage of the relation $T(g)T(g^{-1}) = E$,
2. Expansion of functions, which are defined on a group or on a homogeneous space, in matrix elements of group representations,
3. Usage of intertwining operators of representations, realized on function spaces,
4. Expansion of functions in matrix elements, considered as functions of indices,
5. Expansions related to Clebsch-Gordan and Racah coefficients,
6. Usage of representations of Chevalley groups in order to obtain the q-analog of classical integral transforms.

Let us note, that under an integral transform we mean an operator from one function space into another (or into the same) function space which is given by a kernel $K(x, y)$. We suppose that functions are given on one of the sets: n-dimensional real space R^n, $n = 1, 2, \ldots$, or its subdomain, on an infinite discrete set, or a finite set.

We do not deal with the well known integral transforms which can be obtained with the help of Peter-Weyl theorem (see [1]).

2. MATRIX ELEMENTS OF UNITARY REPRESENTATIONS AS FUNCTIONS OF THE COLUMN INDEX AND RELATED TRANSFORMS

Let T be a unitary irreducible representation of a group G, and $\{e_j | j \in I\}$ be an orthogonal basis of the carrier space V for T. The matrix elements $D_{mn}(g) = (e_m, T(g)e_n)$ of T have the property

$$\sum_{n \in I} D_{mn}(g)\overline{D_{kn}(g)} = \delta_{mk}. \tag{1}$$

Setting

$$F_m(x; g) = D_{mx}(g), \quad x \in I, \tag{2}$$

we obtain an orthogonal system of functions on I. Due to the unitarity of the operators $T(g)$, every function f on I such that

$$\|f\|^2 = \sum_{x\in I} |f(x)|^2$$

can be expanded in functions (2),

$$f(x) = \sum_{m\in I} a_m F_m(x;g), \tag{3}$$

where

$$a_m = \sum_{x\in I} f(x)\overline{F_m(x;g)}. \tag{4}$$

We have

$$\|f\|^2 = \sum_{m\in I} |a_m|^2.$$

Sometimes, the functions (2) are connected with known polynomials or special functions. Then the formulas (3) and (4) give an expansion in these polynomials or functions. For example, the Krawtchouk polynomials

$$K_s(x;p,N) = {}_2F_1(-x,-s;-N;p^{-1}), \quad s = 0,1,2,...,N,$$

are related to matrix elements of the irreducible unitary representations T_l, $l = 0, \frac{1}{2}, 1, \ldots$, of $SU(2)$ [5]

$$K_s(x;sin^2t,N) = \frac{i^{s-x}[s!x!(N-s)!(n-x)!]^{1/2}}{(-1)^x N! cos^{N-x-s}t\, sin^{s+x}t}\, t^{N/2}_{s-N/2,x-N/2}(cos\, 2t).$$

It follows from here and from (3) that every function f on the set $\{0,1,2,\ldots,N\}$ can be expanded in Krawtchouk polynomials:

$$f(x) = \sum_{s=0}^{N} b_s K_s(x;p,N)$$

where

$$b_s = \sum_{x=0}^{N} \left(\frac{p}{1-p}\right)^s \binom{N}{s}\binom{N}{x} p^x(1-p)^{N-x} f(x) K_s(x;p,N).$$

In a similar way the Meixner polynomials

$$M_s(x;d,c) = \frac{\Gamma(d+s)}{\Gamma(d)}\, {}_2F_1(-x,-s;d;1-\frac{1}{c}), \quad 0 < c < 1, \quad d > 0,$$

are related to matrix elements $t^l_{mn}(cosh\, a)$ of the discrete series representations T_l^+ of $SL(2,R)$,

$$M_s(x;d,th^2a) = \left[\frac{s!x!(d+s-1)!}{(d+x-1)!}\right]^{1/2} sinh^{-x-s}a\, cosh^{d+x+s}a$$
$$\times\, t^{d/2}_{-s+d/2,-x+d/2}(cosh\, a).$$

It follows from here that every function f on the set $\{0,1,2,\ldots\}$ such that

$$\sum_{x=0}^{\infty} \frac{c^x(d+x-1)!}{x!}|f(x)|^2 < \infty$$

is expandable into the series $f(x) = \sum_{s} b_s M_s(x;d,c)$.

The matrix elements $t^{\chi}_{mn}(cosh\, t)$ of the representations T_{χ}, $\chi = (q,\epsilon)$, $q \in C$, $\epsilon \in \{0,1/2\}$, of $SL(2,R)$ are related to the Krawtchouk-Meixner functions $k_n(x;p,q)$ on the set $Z = \{0,\pm1,\pm2,\ldots\}$ [6],

$$k_n(x;p,q) = \frac{\Gamma(q+x+1)\Gamma(q-x+1)}{\Gamma(2q+1)p^q}\left(\frac{1+p}{p}\right)^{(x+n)/2} P^q_{xn}(2p+1), \quad p>0,$$

where $P^q_{mn}(cosh\, t) = T^{(q,0)}_{mn}(cosh\, t)$. Let $\chi = (i\rho - \frac{1}{2}, \epsilon)$, $\rho \in R$. Then, as it follows from (1), the Krawtchouk-Meixner functions satisfy the orthogonality relation

$$\sum_{x\in Z} M_\rho(x,p)k_n(x;p,i\rho-\frac{1}{2})\overline{k_m(x;p,i\rho-\frac{1}{2})} = p\left(\frac{1+p}{p}\right)^n \delta_{nm},$$

where

$$M_\rho(x,p) = \left|\frac{\Gamma(2i\rho)}{\Gamma(i\rho+x+\frac{1}{2})\Gamma(i\rho-x+\frac{1}{2})}\right|^2 \left(\frac{p}{1+p}\right)^x.$$

The set of functions $k_n(x;p,i\rho-\frac{1}{2})$, $n \in Z$, constitutes a complete system in the Hilbert space H of functions $f(x)$ on Z with the scalar product

$$(f_1,f_2) \doteq \sum_{x\in Z} M_\rho(x,p)f_1(x)\overline{f_2(x)}.$$

Therefore, every function f H can be expanded as

$$f(x) = \sum_{n\in Z} a_n k_n(x;p,i\rho-\frac{1}{2}),$$

where

$$a_n = \sum_{x\in Z} M_\rho(x,p)p^{n-1}(1-p)^{-n}f(x)\overline{k_n(x;p,i\rho-\frac{1}{2})}.$$

Another orthogonal system of Krawtchouk-Meixner functions is connected with the supplementary series representations of $SL(2,R)$.

Analogous transforms can be introduced with the help of matrix elements in mixed bases $D_{mn}(g) = (f_m, T(g)e_n)$, where $\{e_n\}$ and $\{f_m\}$ are orthonormal bases.

3. INTEGRAL TRANSFORMS CONNECTED WITH KERNELS OF REPRESENTATIONS IN CONTINUOUS AND MIXED BASES

In carrier spaces of representations one can consider continuous bases, analogous to the basis $(exp\, ipx)/\sqrt{2\pi}$ of the space $L^2(R)$. Representation operators in such bases are realized as integral transforms with kernels of the form $K(u,v;\chi,g)$ where u and v label basis elements and χ labels representations. The equation

$$\int K(u,w;\chi,g)K(w,v;\chi,g^{-1})dw = \delta(u-v) \tag{5}$$

holds. For unitary representations it takes the form

$$\int K(u,w;\chi,g)\overline{K(v,w;\chi,g)}dw = \delta(u-v). \tag{6}$$

Fixing χ, g, and u, we obtain the functions of x,

$$F_u(x;g) = K(u,x;\chi,g).$$

The equation (5) leads to the mutually reciprocal integral transforms with the kernel $F_u(x;g)$.

Instead of one continuous basis we can consider two continuous bases or continuous and discrete bases. They lead to corresponding integral transforms.

As an example, we consider the principal nonunitary series representations T_χ of $SL(2,R)$. They are defined by $\chi = (q,\epsilon)$, where $q \in C$ and $\epsilon \in \{0, 1/2\}$. The representation T_χ acts upon the space D_χ of functions f on R [1]. The operators $T_\chi(g)$ are given by the formula

$$(T_\chi(g)f)(x) = |bx+d|^{2q} sign^{2\epsilon}(bx+d) f\left(\frac{ax+c}{bx+d}\right), \quad g = \begin{pmatrix} a & b \\ c & d \end{pmatrix}.$$

Let us introduce three bases:

1. $\{F_{n\chi}(x) = (x+i)^{q-n-\epsilon}(x-i)^{q+n+\epsilon} \mid n = 0, \pm 1, \pm 2, \ldots\}$,
2. $\{x_+^u, x_-^u \mid u \in R\}$,
3. $\{e^{-iux} \mid u \in R\}$.

They diagonalize the one-parameter subgroups

$$g_3(t) = \begin{pmatrix} \cos t & \sin t \\ -\sin t & \cos t \end{pmatrix}, \quad g_2(t) = \begin{pmatrix} e^t & 0 \\ 0 & e^{-t} \end{pmatrix}, \quad g_-(t) = \begin{pmatrix} 1 & 0 \\ t & 1 \end{pmatrix},$$

which are called elliptic, hyperbolic, and parabolic, respectively.

Acting with the operator $T_\chi(g)$ onto the vth element of the jth basis and expanding the resulting function in elements of the ith basis, we obtain "matrix elements" (kernels) $K^{ij}(u,v;\chi,g)$. For example,

$$K^{31}(u,n;\chi,g) = \frac{1}{\pi}\int_{-\infty}^{\infty} [T_\chi(g)F_{n\chi}(x)]\, e^{iux} dx.$$

The kernels $K^{ii}(u,v;\chi,g)$ were found in [1], and the kernels $K^{ij}(u,v;\chi,g)$, $i \neq j$, were calcualted in [7,8] for unitary representations and in [9] for nonunitary representations of $SL(2,R)$. For example,

$$K^{33}(u,v;\chi,s) = \frac{(-1)^{2\epsilon\delta} e^{i\epsilon\pi}}{2\sin\pi(q+\epsilon+\frac{1}{2})}\left(\frac{v}{u}\right)^{q+1/2}\left[J_{-2q-1}(2\sqrt{uv}) - (-1)^{2\epsilon} J_{2q+1}(2\sqrt{uv})\right],$$

if $uv > 0$, and

$$K^{33}(u,v;\chi,s) = \frac{(-1)^{2\epsilon\delta}}{\pi}\left(-\frac{v}{u}\right)^{q+1/2}\left[(-1)^{2\epsilon}e^{-(q+1/2)\pi i} + e^{(q+1/2)\pi i}\right]K_{2q+1}(2\sqrt{-uv}),$$

if $uv < 0$. Here $J_a(x)$ is a Bessel function, while $K_a(x)$ is Macdonald function, $s = \begin{pmatrix} 0 & 1 \\ -1 & 0 \end{pmatrix}$, $\delta = (1 + sign\, u)/2$.

For $g_2(t) = diag(e^{t/2}, e^{-t/2})$ we have

$$K^{31}(u,n;\chi,g_2(t)) = \frac{2^{q+1}e^{\pi i(-n-\epsilon)}}{\Gamma(-q+\rho(n+\epsilon))}|u|^{-q-1} \times W_{-\rho(n+\epsilon),q+1/2}(2|u|e^{-t}), \quad \rho = sign\, u,$$

and for $g_+(t) = \begin{pmatrix} 1 & t \\ 0 & 1 \end{pmatrix}$

$$K^{32}(u,v;\chi,g_+(t)) = \frac{\Gamma(2q+v+1)\Gamma(1-v)}{2\pi i\Gamma(2q+2)}t^{v+q}(-iu)^{-q-1}M_{q+v,q+1/2}\left(-\frac{iu}{t}\right)$$

where $W_{uv}(x)$ and $M_{uv}(x)$ are Whittaker functions.

It follows from (5) that the integral transforms

$$c_n = \int_{-\infty}^{\infty} K^{13}(n,u;\chi,g)F(u)du \text{ and } F(u) = \sum_{n=-\infty}^{\infty} K^{31}(u,n;\chi,g)c_n$$

are mutually reciprocal. Setting $g = g_2(t)$ and using the expressions for the kernels K^{13} and K^{31} we obtain for $Re\, q < 1/2$ the mutually reciprocal transforms

$$c_n = -\frac{\sin\pi(q+n+\epsilon)}{\pi^2}\int_{-\infty}^{\infty} F(u)W_{-\rho(n+\epsilon),q+1/2}(2|u|e^t)|u|^{-1}du,$$
$$F(u) = \frac{1}{\pi}\sum_{n=-\infty}^{\infty} W_{-\rho(n+\epsilon),q+1/2}(2|u|e^t)c_n.$$

Using the kernels K^{32} and K^{23} we obtain the mutually reciprocal transforms

$$f(u) = \frac{1}{2\pi i\Gamma(2q+2)}\int_{a-i\infty}^{a+i\infty} M_{q+v,q+1/2}(iu)F(v)dv,$$
$$F(v) = \frac{\pi}{2}\left[\Gamma(-2q)\sin\pi v\sin\pi(2q+v+1)\right]^{-1} \times \int_{-\infty}^{\infty}(iu)^{-1}M_{-q-u,-q-1/2}(-iu)f(u)du.$$

More complicated integral transforms are derived from other kernels.

Let us write down the expressions for two kernels of the discrete series representations T_l^+, $l = 1, 3/2, 2, \ldots$:

$$K^{22}(u,v;l,g_3(t)) = e^{-i\pi l_2 i(u-v)}\frac{\Gamma(l+iv)\Gamma(l-iu)}{2\pi(2l-1)!}\, sin^{-2l}t \times (-ictg\, t)^{-2l+iu-iv}{}_2F_1(l-iu, l+iv; 2l; cos^{-2}t), \quad (7\ )$$

$$K^{21}(u,n;l,g_-(t)) = \left[\frac{n!}{2\pi(n+2l-1)!}\right]^{1/2} (l-iu)2^{l-iu}(1-it)^{-l+iu} P_n^{(2l-1,-n-iu)}\left(\frac{it+3}{it-1}\right). \quad (8)$$

With the help of the kernel (7) we obtain the mutually reciprocal transforms

$$F(u) = \frac{cos^{2l}t}{2\pi(2l-1)!}\int_{-\infty}^{\infty} {}_2F_1(l-iu, l+iv; 2l; cos^{-2}t) f(v)|\Gamma(l+iv)|^2 dv$$

$$f(v) = \frac{cos^{2l}t}{2\pi(2l-1)!}\int_{-\infty}^{\infty} {}_2F_1(l+iu, l-iv; 2l; cos^{-2}t) F(u)|\Gamma(l+iu)|^2 du,$$

where $t > 0$ is fixed, $2l \in Z$, $l > 0$, and

$$\int_{-\infty}^{\infty} |F(u)|^2|\Gamma(l+iu)|^2 du = \int_{-\infty}^{\infty} |f(v)|^2|\Gamma(l+iv)|^2 dv.$$

The kernel (8) leads to the transforms

$$c_n = \frac{1}{2\pi}\left(\frac{1+t^2}{4}\right)^l \int_{-\infty}^{\infty} P_n^{(2l-1,-n-iu)}\left(\frac{it+3}{it-1}\right) f(u)|\Gamma(l+iu)|^2 du,$$

$$f(u) = \sum_{n=0}^{\infty} \frac{n!}{(n+2l-1)!} P_n^{(2l-1,-n+iu)}\left(\frac{it-3}{it+1}\right) c_n,$$

where t is fixed and

$$\frac{1}{2\pi}\left(\frac{1+t^2}{4}\right)^l \int_{-\infty}^{\infty} |f(u)|^2|\Gamma(l+iu)|^2 du = \sum_{n=0}^{\infty} \frac{n!}{(n+2l-1)!}|c_n|^2.$$

Matrix elements of the irreducible representations of the group $ISO(2)$ in an $SO(2)$ basis are expressed in terms of Bessel functions $J_n(x)$ [1]. They lead to the mutually reciprocal transforms

$$f(n) = \sum_{m=-\infty}^{\infty} F(m) J_{n-m}(x),$$

$$F(m) = \sum_{n=-\infty}^{\infty} f(n) J_{n-m}(x),$$

where

$$\sum_{n=-\infty}^{\infty} |f(n)|^2 = \sum_{m=-\infty}^{\infty} |F(m)|^2.$$

Matrix elements of the irreducible representations of $ISO(1,1)$ are expressed in terms of Γ-functions, Hankel functions $H_a^{(1,2)}(x)$, and Macdonald functions [1]. They lead to the following pairs of mutually reciprocal integral transforms:

$$f(u) = \frac{1}{2\pi i} \int_{a-i\infty}^{a+i\infty} \Gamma(v-u)(ix)^{u-v} F(v)dv,$$

$$F(v) = \frac{1}{2\pi i} \int_{b-i\infty}^{b+i\infty} \Gamma(u-v)(-ix)^{v-u} f(u)du,$$

where $a > Re\, u,\ b > Re\, v,\ x > 0$,

$$f(u) = -\frac{1}{2} \int_{a-i\infty}^{a+i\infty} e^{(u-v)i\pi/2} H_{v-u}^{(2)}(x) F(v)dv,$$

$$F(v) = \frac{1}{2} \int_{a-i\infty}^{a+i\infty} e^{(u-v)\pi i/2} H_{u-v}^{(1)}(x) f(u)du,$$

where $-1 < Re(u-v) < 1,\ x > 0$, and

$$f(u) = \frac{1}{\pi i} \int_{a-i\infty}^{a+i\infty} e^{(u-v)\pi i/2} K_{v-u}(x) F(v)dv,$$

$$F(v) = \frac{1}{\pi i} \int_{a-i\infty}^{a+i\infty} e^{(u-v)\pi i/2} K_{v-u}(x) f(u)du,$$

where $-1 < Re(u-v) < 1,\ x > 0$.

4. FOURIER TRANSFORM ON $SL(2,R)$ AND INTEGRAL TRANSFORMS

The Fourier transform on $SL(2,R)$ can be written down in the form

$$f(g) = \frac{1}{4\pi^2} \sum_{l \in N_0/2} (l+\frac{1}{2}) Tr\,[T_{l+1}^{+}(f)(T_{l+1}^{+}(g))^* + T_{-l-1}^{-}(f)(T_{-l-1}^{-}(g))^*]$$

$$+ \sum_{\epsilon=0,1/2} \int_0^{\infty} Tr\, T_\chi(f)(T_\chi(g))^* y\ th\pi(y+i\epsilon)dy, \tag{9}$$

where $N_0 = \{0,1,2,\ldots\}$, $\chi = (iy - \frac{1}{2}, \epsilon)$,

$$T(f) = \int f(g)T(g)dg. \tag{10}$$

Traces of operators can be expressed in terms of kernels of representations. Let us use the kernel $K^{11}(m,n;\chi,g) = t_{mn}(g)$, present g in the form $g = g_3(t)g_2(s)g_3(t')$, and put $cosh\, s = x$. As a result we obtain the Jacobi transform

$$F(x) = \sum_{l=1+\epsilon}^{N} (1+\frac{1}{2}) c_{km}^{1} P_{km}^{1}(x) + \frac{1}{2} \int_{-\infty}^{\infty} c_{km}^{\chi} P_{km}^{\chi}(x) y\ th\pi(y+i\epsilon)dy, \tag{11}$$

where $\chi = (iy + \frac{1}{2}, \epsilon)$, $P^{\chi}_{km}(x)$ is expressed in terms of Jacobi functions and $P^{l}_{km}(x)$ in terms of Jacobi polynomials [1]. Here $\epsilon = 0$ if $k, m \in Z$, and $\epsilon = \frac{1}{2}$, if $k, m \in Z$, $N = max(|k|, |m|)$ if $km \geq 0$, and $N = 0$ (i.e. the sum over l is absent) if $km < 0$. For $m = k = 0$ the transformation (11) reduces to the Mehler-Fock transform.

If the kernel $K^{33}(u, v; \chi, s)$ for $\epsilon = 0$, $-u = v = \sqrt{2}/2$ is used in (9), then we obtain the Kantorovich-Lebedev transform

$$F(x) = \frac{2}{\pi^2} \int_0^\infty c(y) K_{iy}(x) y \, sinh\pi y \, dy.$$

If $\epsilon = 0$ and $u = v = \sqrt{2}/2$, then we have the integral transform

$$F(x) = \frac{1}{2} \int_0^\infty c(y) \left[J_{-iy}(x) - J_{iy}(x)\right] \frac{ydy}{sh\pi y} + \sum_{l=0}^{\infty} (4l+2) c_1 J_{2l+1}(x),$$

where

$$c(y) = \int_0^\infty f(x) \left[J_{iy}(x) - J_{-iy}(x)\right] \frac{dx}{x},$$

$$c_l = \int_0^\infty f(x) J_{2l+1}(x) \frac{dx}{x},$$

Moreover,

$$\int_0^\infty |f(x)|^2 \frac{dx}{x} = \frac{1}{2} \int_0^\infty |c(y)|^2 \frac{ydy}{sh\pi y} + \sum_{l=0} (4l+1) |c_l|^2.$$

If $\epsilon = \frac{1}{2}$, then we have the transform

$$F(x) = \frac{1}{2} \int_0^\infty c(y) \left[J_{iy}(x) + J_{-iy}(x)\right] \frac{ydy}{sh\pi y} \qquad (12)$$

where

$$c(y) = \int_0^\infty F(x) \left[J_{iy}(x) + J_{-iy}(x)\right] \frac{dx}{x}$$

and the corresponding Plancherel formula holds. Discrete summands vanish in (12), since summands, corresponding to the representations T_l^+ and T_{-l}^-, are cancelled.

With the help of kernel K^{31} we obtain from (9) the integral transform

$$f(x) = \frac{1}{2\pi^2} \int_0^\infty c(y) W_{-(n+\epsilon), iy}(x) \frac{y \, th \, (y + i\epsilon) dy}{|\Gamma(iy + n + \epsilon + 1/2)|^2}$$
$$+ \sum_{l=\epsilon}^{\infty} (2l+1) c_l x^{l+1} e^{-x^2} L_n^{2l+1}(x),$$

where

$$c(y) = \int_0^\infty f(x) W_{-(n+\epsilon), -iy}(x) \frac{dx}{x^2},$$

$$c_l = \frac{2n!}{\pi(2l+n+1)!} \int_0^\infty f(x) x^{l-1} e^{-x^2} L_n^{2l+1}(x) dx.$$

5. INTERTWINING OPERATORS AND INTEGRAL TRANSFORMS

Let T_1 be the quasi-regular representation of the group $G = SO_0(n,1)$ on the upper half H of the two-sheeted hyperboloid $[\underline{x},\underline{x}] = 1$ in R^{n+1}, $[\underline{x},\underline{x}] = x_0^2 - x_1^2 - \ldots - x_n^2$, and T_2 be the quasi-regular representation on the upper half of the cone $[\underline{z},\underline{z}] = 0$ in R^{n+1}. The Gel'fand-Graev transform

$$F(\underline{z}) = \int_H f(\underline{x})\delta([\underline{x},\underline{z}] - 1)d\underline{x}$$

intertwines the representations T_1 and T_2. Writing down this transform and its inverse (see [1]) in different coordinate systems on the hyperboloid and on the cone, and separating variables, we obtain different integral transforms of functions of one variable (Mehler-Fock, Kantorovich-Lebedev, Jacobi transforms, and others) (cf. [10]).

Let T_3 be the quasi-regular representation of $SO_0(n,1)$ on the one-sheeted hyperboloid $[\underline{y},\underline{y}] = -1$. Then the Radon transform

$$F(\underline{y}) = \int_H f(\underline{x})\delta([\underline{x},\underline{y}])d\underline{x}$$

intertwines the representations T_1 and T_3. The image consists of functions which transform according to the class 1 principal unitary series representations. The inverse Radon transform cancels functions, which transform according to the discrete series representations. The Radon transform is a special case of the transform

$$F_a(\underline{y}) = \int_H f(\underline{x})[\underline{x},\underline{y}]^a d\underline{x}, \quad a \in C.$$

Studying this transform and its inverse we obtain a collection of interesting relations for special functions.

6. CLEBSCH-GORDAN COEFFICIENTS (CGC's) AND TRANSFORMATION OF FUNCTIONS

Let us consider CGC's $C(l_1,l_2,l;m-j,j,m)$ for the group $SU(2)$. They form a unitary matrix, in which rows are labelled by l and columns by j. The orthogonality relation

$$\sum_j C(l_1,l_2,l;m-j,j,m)C(l_1,l_2,l';m-j,j,m) = \delta_{ll'}.$$

holds. Setting

$$P_l(x;l_1,l_2,m) = C(l_1,l_2,l;m-x,x,m) \tag{13 }$$

we obtain an orthogonal system of functions on the set J which consists of values of the index j. For function $f(x)$ on J we have

$$f(x) = \sum_l a_l p_l(x;l_1,l_2,m), \tag{14}$$

where $a_l = \sum_{x \in J} f(x)p_l(x;l_1,l_2,m)$.

CGC's are related to the Hahn polynomials [11]

$$Q_n(x;a,b,N) = M(x)C\left(\frac{N}{2}, \frac{N+a+b}{2}, n+\frac{a+b}{2}; \frac{n}{2}-x, x-\frac{N-a+b}{2}, \frac{a-b}{2}\right)$$

where $M(x)$ is a function of x, which depends on N, a, b, and n. This formula and the expansion (14) lead to the expansion of functions, given on the set $\{0,1,2,\ldots,N\}$, in the Hahn polynomials.

Analogous considerations are valid for other groups, including noncompact ones. For example, CGC's for the tensor product $T^+(l_1) \otimes T^+(l_2)$ of the discrete series representations of $SL(2,R)$ in hyperbolic basis have the form

$$\begin{aligned} C_p(\underline{l},\underline{u}) &\equiv C_p(l_1,l_2,l;u_1,u_2,u) \\ &= M\,{}_3F_2(l_1+l_2-1, l_1+l_2+l+1, l_2-iu_2; 2l_2, l_1+l_2-iu; 1). \qquad (15\ ) \end{aligned}$$

(an explicit expression for M is given in [12]). The right hand side without $\delta(u_1+u_2-u)$, for which $u_1+u_2=u$, will be denoted by $C(\underline{l},\underline{u})$. The unitarity of the integral operator, given by the kernel (15), means that

$$\int_{-\infty}^{\infty} C(\underline{l},\underline{u})\overline{C(\underline{l}',\underline{u})}du_2 = \delta_{ll'} \qquad (16)$$

where $\underline{l}=(l_1,l_2,l)$, $\underline{l}'=(l_1,l_2,l')$.

Let us introduce the polynomials

$$q_n(x;a,b) = {}_3F_2(-n, n+2a+b+\bar{b}+1, a-ix; 2a, a+b; 1), \quad n=0,1,2,\ldots, \qquad (17)$$

which depend on the parameters a and b. Setting in (15) $u_2 = x$, $l-l_1-l_2 = n$, $l_2 = a$, $l_1 - iu = b$, we obtain from (16) the orthogonality relation for $q_n(x;a,b)$,

$$\begin{aligned} &\frac{1}{2\pi}\int_{-\infty}^{\infty} q_n(x;a,b)\overline{q_m(x,a,b)}|\Gamma(-a-ix)\Gamma(b+ix)|^2dx \\ &= \frac{n!\Gamma(n+b+\bar{b})[\Gamma(2a)\Gamma(a+b)]^2}{(2n+2a+b+\bar{b}-1)\Gamma(n+2a+b+\bar{b}-1)\Gamma(n+2a)}\delta_{mn}\,. \end{aligned}$$

The polynomials (17) are called Hahn polynomials of an imaginary argument. It is easy to write down corresponding transform of functions on $\{0,1,2,\ldots,N\}$.

More general integral transforms are obtained with the help of CGC's from the tensor products of other representations of $SL(2,R)$.

REFERENCES

1. N. Ja. Vilenkin, "Special Functions and the Theory of Group Representations," Transl. Math. Monogr., Vol. 22, Amer. Math. Soc., Providence, R.I. (1968).
2. W. Miller, Jr., "Lie Theory and Special Functions," Academic Press, New York (1968).
3. L. S. Pontryagin, "Topological Groups," Gordon & Breach, New York (1966).

4. G. N. Ageev et al, "Multiplicative Systems of Functions and Harmonic Analysis on Nul-dimensional Groups," ELM, Bacu (1981).
5. T. H. Koornwinder, SIAM J. Math. Anal., 13:1011 (1982).
6. N. Ja. Vilenkin, A. U. Klimyk, Dokl. Acad. Nauk Ukrainian SSR, Ser. A, 7:16 (1988).
7. E. G. Kalnins, J. Math. Phys., 14:654 (1973).
8. D. Basu, K. B. Wolf, J. Math. Phys., 23:189 (1982).
9. N. Ja. Vilenkin, A. U. Klimyk, Prepring ITP-88-4, Institute for Theoretical Physics, Kiev (1988).
10. N. Ja. Vilenkin, Mat. Sb., 74:119 (1967).
11. T. H. Koornwinder, Nieuw Arch. Wisk., 29:140 (1981).
12. N. Mukunda, B. J. Radhakrishnan, J. Math. Phys., 15:1320 (1974).

PART II: SPECIAL SESSIONS

This work relates to Department of the Navy Grant N00014-88-J-9010 issued by the office of Naval Research. The United States Government has a royalty-free license throughout the world in all copyrightable material contained herein

SELF-ORGANIZATION AND SYMMETRY BREAKING IN LIVING MATTER

E. Del Giudice*, S. Doglia*, and G. Vitiello**

* Dipartimento di Fisica dell'Università,
Via Celoria, 16, 20133 Milano, Italia
and INFN, Sezione di Milano

** Dipartimento di Fisica dell'Università,
84100 Salerno, Italia
and INFN, Sezione di Napoli, Gruppo Collegato di Salerno

Abstract

A living system is schematized as a set of electric dipoles whose interaction is described by a rotationally invariant Lagrangian. A non-vanishing electric polarization appears in the ground state (spontaneous breakdown of symmetry).

Self-focusing propagation of the electric field in the correlated medium is studied. Josephson effect and temperature are also discussed.

Introduction

Living matter can be considered as 1) a dielectric, 2) a dynamically ordered structure whose components are located in compartments with preferred directions, shapes and sizes. Our aim is to understand this second feature as emerging from a basic symmetric interaction among the set of electric dipoles.

A living system is then described (1) by a rotationally invariant quantum Lagrangian. Spontaneous breakdown of the rotational symmetry produces a long range correlation (Goldstone mode) among the system dipoles. Here we should stress that the system components are connected by interactions,

which are basically electromagnetic. We focus firstly our attention on the long range radiative interaction; it has been shown that this interaction is responsible for the non-vanishing electric polarization field in the ground state of the system (2) (thereafter called the vacuum using the terminology of Quantum Field Theory). The electric polarization is thus the "order parameter" characterizing the asymmetric vacuum of the system. It is known from experiments (3) that all the relevant biomolecules display a long-lasting electric polarization which is related to the molecule hydration. Oscillation modes with frequency less than 1 Hz and relaxation times larger than 10^5 sec have been found. Highly polar metastable states have been theoretically predicted by H. Fröhlich (4).
Any description of living matter must then account for this remarkable tendency of water to produce a very big polarization around polar molecules. This property has been explained in ref. (2) as the consequence of the laser mechanism induced by the interaction between electric dipoles and the electromagnetic field in a superradiant way (5). We stress that this coherent mechanism in water is not the result of a Goldstone mechanism but it is its prerequisite. The high value of the static dielectric constant of water is just the consequence (6) of the fact that pure water, yet homogeneous and isotropic, is a system out of thermodynamic equilibrium. The vacuum of pure water is then yet rotationally symmetric, but it is highly unstable against very small disturbances.
As shown in ref. 2 a very high non vanishing electric polarization $\vec{P}$ develops when a small electrical stimulus is applied. The laser mechanism magnifies so much any applied signal that transforms it into a source of a significant dynamics governed by the mechanism of symmetry breaking.
According to the results of Ref. 2 described in the next section, we could then take the ground state of such a system as characterized by a quite high value of polarization $\vec{P}$

$$\langle 0|\vec{P}|0\rangle = \vec{P} \neq 0 \qquad (1)$$

The value of P in equation (1) is not uniquely defined so that we handle with many different Hilbert spaces each one emerging from a different vacuum. At infinite volume all these states are unitarily inequivalent.

In Ref. 7 we have considered the problem of the extension and validity of general theorems of Quantum Field Theory with spontaneous breakdown of symmetry to non-equilibrium systems. For sake of brevity we refer to Ref. 7 for a detailed discussion. Here we only observe that to each value of the order parameter $\vec{P}$ is associated one representation of the physical states (Hilbert space) among the infinitely many unitarily inequivalent representations of the canonical commutation relations. Since in each one of these representations the formalism of spontaneous breakdown of symmetry holds true, we can describe non-equilibrium systems with non-constant order parameter (see also ref. 8 and 9) by this plurality of Hilbert spaces.

Following the above analysis we can prove that in a medium whose ground state is given by equation (1) the electro-magnetic field propagates according to the Anderson Higgs Kibble mechanism by which it undergoes self-focusing propagation in the ordered (polarized) medium (10).

Such a kind of propagation is responsible for emergence of selective forces acting on molecules and leading to the formation of polymer-like structures which may be associated to cytoskeleton filaments in the cell.

In section 3 we analyze the possible emergence of Josephson-like phenomena (11).

The relation between the temperature and the collective modes will be then commented upon.

Section 2 - Spontaneous electric polarization around impurities in water

The starting point of our analysis is equation (1), namely the statement that a nonvanishing (quite big) electric polarization develops around polar molecules in aqueous solutions.

This result emerges from the remarkable property of water reported in section 1. Let us sketch briefly the content of Refs. 2 and 6. Suppose we take N molecules of water and let us concentrate on the dynamics between the lowest two rotational levels of the single molecule.

The energy difference between these levels is $E = h\nu = hc/\lambda$ (ν=25 cm^{-1}, E=3 meV).

The photon emitted during the transition between these levels has a wavelength $\lambda=c/\nu=420$ μm. Within a sphere of radius $R=\lambda/2$ there are in liquid water about 10^{18} molecules.

The small coupling between the electromagnetic radiative field and the electric dipoles is then magnified, as shown in ref. 2 and 6 by a factor $N^{1/2} \sim 10^9$. This enhancing factor transforms the so far "neglected interaction", between radiative field and matter into the dominant mechanism of the dynamics of condensed matter.

All the water molecules inside the sphere of radius $\lambda/2$ will then be involved in a coherent process started by the electromagnetic radiation of frequency $\nu=24$ cm^{-1} and their dynamic evolution could be described as a typical superradiance phenomenon (ref. 12).

By following the formalism of reference 5, we introduce three complex amplitudes $b_{\pm}=B_{\pm}e^{i\omega_{\pm}t}$, $a=Ae^{i\omega t}$:

$$|b|^2 = N_{\pm}/N$$
$$|a|^2 = N_{\gamma}/N \qquad (2)$$

where N_+ and N_- are the populations of the upper and lower states involved in the coherent subdynamics and N_γ is the photon number. In reference 2 it has been shown that this amplitude obeys the equation 3:

$$\dot{b}_- = -3\Omega\, a^* b_+$$
$$\dot{b}_+ = -\,\Omega a b_- \qquad (3)$$
$$\dot{a} = 2\Omega b_+ b^*_-$$

where

$$\Omega = G\omega_o \sim 2\pi\sqrt{\lambda}(17\cdot 24\ cm^{-1})$$
$$|b(0)|^2 = \lambda\cos^2\theta_o$$
$$|b(0)|^2 = \lambda/3\,\sin^2\theta_o$$
$$|a(0)|^2 = 0$$
$$\lambda = (N_+ - N_-)/N$$

System (3) admits the limit cycle:

$$B_- = \sqrt{\lambda}\,\alpha = \sqrt{\lambda}/\sqrt{3}\ \left(1+\cos^2\theta_o + \sqrt{1-1/4\,\sin^2 2\theta_o}\right)^{1/2}$$
$$B_+ = \sqrt{\lambda}/\sqrt{3}\,(1-\alpha^2)^{1/2}$$
$$A = \sqrt{2\lambda/3}\ (\alpha^2-\cos^2\theta_o)^{1/2} \qquad (4)$$

$$\dot{\omega}_+ - \dot{\omega}_- = (2\, B_+ B_- /A)\ \Omega$$

$$\omega_+ - \omega_- - \omega \quad = \pi/2$$

and the small oscillations around the limit cycle have a pulsation:

$$\omega_i = 2\sqrt{2}\ (1-1/4\ \sin^2 2\theta_o\ )^{1/4}\ \Omega \qquad (5)$$

From equation (4) it can be deduced that the population of the two states have the difference $N_- - N_+ = 5(N_- - N_+)_{eq.}$ which is very much higher than the difference expected by Boltzmann factor at thermal equilibrium. In reference 6 it has been shown that the static dielectric constant of water acquires its well known high value just because of this shift from thermal equilibrium.

The same non equilibrium explains the permanent electric polarization produced around an external electric disturbance. If we introduce in water a molecule producing an electric disturbance $V_d = E \cdot d$ where d is the water dipole moment, along a given direction, the resulting polarization is (2)

$$\vec{P} = (N/V)\ d/\sqrt{3}\ \ \vec{E}/E \sin 2\alpha\ (N_- - N_+) \qquad (6)$$

where:

$$\text{tg}\ \alpha = \frac{\nu/2\pi - ((\nu/2\pi)^2 + 4\ E^2 d^2\,)^{1/2}}{2\ Ed} \qquad (6')$$

We can see that $|\vec{P}|$ acquires the necessary high value because of the difference $(N_- - N_+)$. At thermal equilibrium we would have got quite a negligible value. Equation (6) implements the requirements of equation (1) and allows to proceed safely through the symmetry breaking approach.

Section 3 - Non Maxwellian propagation of electromagnetic field in a correlated medium

Electric dipoles in biological systems oscillate coherently under the Goldstone mechanism outlined in sect. 1. The resulting phase correlation spoils the local gauge invariance which requires just the freedom of changing independently the phase at each point. The electrodynamics valid when the matter field is coherent has been developed firstly by Anderson (13).

As discussed in ref. 10 a propagation of an electromagnetic field in a correlated medium requires not only the breaking of rotational invariance, but also the breaking of the U(1) symmetry respect to the rotations around the polarization direction. This last breaking implies the presence of a preferred winding around this direction, leading to a parity non conservation. Under the above conditions it can be shown (10) that the electromagnetic field A_μ is the solution of the equation:

$$(\Box + M^2)A_\mu = J_\mu \qquad (7)$$

where $M^2 = \text{const.}\ |\vec{P}| \qquad (8)$

This equation is relevant when the four momentum K_μ of the electromagnetic field is such that $K_\mu K^\mu = K^2 \sim M^2$. When $K^2 >> M^2$ the ordinary Maxwellian propagation is recovered.

Equation (7) with the condition (8) admits stationary solutions confined within filaments with radius

$$R = \hbar/c \cdot 1/M \qquad (9)$$

In a completely aqueous medium, in the extreme hypothesis of a totally polarized medium, we get $M \sim 13$ eV and $R \sim 150$ Å.

We point out that the mass term in equation (7) means that an observer totally immersed in the medium would measure a mass M for the photon.

An external observer would interpret this term only as the evidence that the field is trapped within a region of size R. This last observation is the only empirical feature which can be inferred from equation (7). Furthermore we point out that the AHK mechanism implies the disappearance of any coherent correlation among dipoles within the filament. We therefore expect that water within the filament must not show the properties of section 2.

A possible solution of this dynamical requirement is that no liquid water could be present within the filament, which then behaves as a true vacuum allowing a complete transparence respect to incoming electromagnetic radiation. It is interesting to compare this prediction with the phenomenon of "light piping" reported in the literature (14).

The trapping of the e.m field in the filaments produces a strong field gradient on the boundary of each filament. As a consequence the molecules surrounding the filament are acted upon by the force (10):

$$\vec{F} = \text{const grad } E^2 \sum_k \frac{\nu_k^2 - \nu^2}{(\nu_k^2 - \nu^2)^2 + \Gamma_k^2} \tag{10}$$

Here ν is the e.m. field frequency, ν_k is the frequency of the k-th molecular mode with damping Γ_k. Eq. (10) shows that for $\nu \sim \nu_k$ a strong force acts on the molecules belonging to the k-th mode. A molecular coating of the field filament becomes possible.

This mechanism can be related (10,15) with the formation and dynamics of cytoskeleton.

It should be noticed that, as mentioned above, M^2 depends on P and thus a time-dependent pattern emerges since $\vec{P}$ may change in time reflecting changes in time of the coupling system environment. Consequently a time-dependent pattern of molecular processes and of chemical reactions also emerges. In general, chemical reactions occurring along a filament release an heat output (negative or positive). This heat output does not propagate in a diffusive way but in a wave form on the correlation, producing a polarization wave. A change in the pattern of frequencies may be therefore induced with a consequent change in the selective action of the force (10).

A further consequence of the Anderson-Higgs-Kibble mechanism is the magnetic flux quantization. Indeed, in the region where the current J_μ vanishes and θ is the Goldstone field arising by the U(1) breaking, one obtaines (8)

$$h\, \partial_\mu \theta = qA_\mu \tag{11}$$

which integrated along a closed line gives

$$\phi(H) = \frac{h}{q} \cdot 2\pi n \qquad (12)$$

The magnetic flux quantization expressed by eq. (12) is a standard result in superconductivity. As a matter of fact, non-maxwellian electrodynamics implied by eq. (7) necessarily leads to currents flowing without losses on the boundaries of filaments where the electric field is trapped.

A Josephson-like effect could be the consequence from the magnetic flux quantization described by Eq. (12). A number of experiments suggests indeed such a possibility (11). Smith and coworkers (16) have performed a class of experiments where they consider dividing yeast cells and report a) emission of e.m. radiation from the system, b) step-like voltage-current characteristics and their modifications in presence of e.m. disturbances at proper frequencies.

The Goldstone bosons introduced in the above analysis must be considered massless at infinite volume. An effective mass M_{eff} appears due to the finite size R of the system. We have shown in ref. 10 that under the assumption that the Goldstone bosons behaves as an ideal gas R is connected with .temperature by the law

$$R = \frac{\pi hc}{6k} \cdot \frac{1}{T} \qquad (13)$$

The finite size of the system appears then as a consequence of the nonzero temperature.

In equation (13) h,c, and k, are the Planck constant, speed of light and Boltzmann constant respectively.

At T=300 K a size of 25 μm corresponds which is in reasonable agreement with the average size at the cell.

References

1. E. Del Giudice, S. Doglia, M. Milani and G. Vitiello, Nucl. Physics 251B (FS 13), 375 (1985).
2. E. Del Giudice, G. Preparata and G. Vitiello, Phys. Rev. Letters 61, 1085 (1988).

3. S. Celaschi and S. Mascarenhas, Biophysics J. 20 273 (1977).
J.P. Hasted, H.M. Millany and D. Rosen, J. Chem. Soc. Faraday Trans. 77, 2289 (1981).
4. H. Fröhlich, Int. J. Quantum Chem. 2, 641 (1968); in Advances in Electronic and Electronic Physics 53 85 (980).
5. G. Preparata, Phys. Rev. A38, 233 (1988).
6. E. Del Giudice and G. Preparata, Some remarks on the electrostatics of water, submitted for publication.
7. E. Del Giudice, S. Doglia, M. Milani and G. Vitiello, Physica Scripta, 38, 505 (1988).
8. E. Del Giudice, R. Manka, M. Milani and G. Vitiello, Phys. Lett. B206, 661 (1988).
9. H. Umezawa and T. Arimitsu, Progr. Theor. Phys. Suppl. 86, 243 (1986).
10. E. Del Giudice, S. Doglia, M. Milani and G. Vitiello, Nucl. Phys. B275 (FS 17), 185 (1986).
11. E. Del Giudice, S. Doglia, M. Milani, G. Vitiello and C.W. Smith, Magnetic flux quantization and Josephson behaviour in living systems, submitted for publication.
12. R.H. Dicke, Phys. Rev. 93, 99 (1954).
P.V. Anderson, Phys. Rev. 110 (1958), 827; 130 (1963), 439.
14. D.F. Mandoli and W. Briggs, Proc. Natl. Acad. Sci. USA 79 (1982) 2902
15. E. Del Giudice, S. Doglia, M. Milani and G. Vitiello, in Biological Coherence and response to external stimuli, H. Fröhlich ed., Springer, Berlin 1988, p. 49.
16. C.W. Smith, in "Energy Transfer Dynamics" T.W. Barrett and H.A. Pöhl eds., Springer Berlin 1987.
A.H.Jafary-Asl and C.W. Smith, IEEE Publ. 83 CH 1902-6, 350 (1987).

ON KIDA CLASS OF VORTEX FILAMENT MOTIONS*)

Adam Doliwa and Antoni Sym

Institute of Theoretical Physics
of Warsaw University
ul. Hoża 69, 00-681 Warszawa
Poland.

0. INTRODUCTION

The problem formulated in [1] and partially solved therein now is completely solved.

Subject. Consider the following <u>nonlinear</u> system

$$\left\{ \begin{array}{l} \mathbf{r}_{,t} = \mathbf{r}_{,x} \wedge \mathbf{r}_{,xx} \\ \mathbf{r}_{,x} \cdot \mathbf{r}_{,x} = 1 \end{array} \right. \tag{0.1}$$

where $\mathbf{r}=\mathbf{r}(x,t)$ is an $\mathbb{E}^3$-valued function of two real variables x and t, the comma means differentiation, and $\wedge$ $(\cdot)$ denotes the skew (scalar) product in $\mathbb{E}^3$.

There are a few physical applications of the system (0.1). In particular, eqs. (0.1) constitute an approximate equation of motion for a <u>single</u> (self-interaction !) and very thin vortex filament. In the vortex hydrodynamics the underlying approximate setting is known as the "Localized Induction Approximation" [2,3,4]. The usefulness of eqs. (0.1) as a mathematical model of vortex phenomena has been confirmed in experiments [5,6,7].

Hasimoto map. The eqs. (0.1) when rewritten in Cartesian component form constitute a pretty complicated non-linear system of four equations for three real functions. H. Hasimoto was the first to discover a remarkable connection between eqs. (0.1) and the Non-linear Schrödinger (Nl.S.) eq.

*)Work supported in part by Polish Ministry of Science and Higher Education, Research Problem CPBP 01. 03.

$$iq_{,t} + q_{,xx} + 2|q|^2 q = 0 \ . \tag{0.2}$$

Namely, in [8] Hasimoto constructed an one-to-many map H

$$q \xrightarrow{H} r \tag{0.3}$$

from the (smooth and global) solution space of eq. (0.2) onto the (smooth and global) solution space of eqs. (0.1). This map is called the Hasimoto map. Unfortunately, to compute the image $H(q)$ according to the original Hasimoto prescription is a highly non-trivial task ! Indeed, Hasimoto was able to compute $H(q)$ only in the case q = one-soliton solution of eq. (0.2). The resulting vortex filament motion is now identified as "Hasimoto vortex".

Spectral reformulation of Hasimoto map. The Nl.S. eq. (0.2) is a classical soliton system with all its spectral features [9]. In the framework of soliton surfaces [10] (spectral generalization of the pseudospherical geometry of the sine-Gordon eq.) one can reformulate the Hasimoto map in an entirely (just spectral) way [11]. See also [1]. This result enables one to compute $H(q)$ in many cases [12,13 and 1]. The Hasimoto approach becomes a pretty efficient method !

The detailed description of the geometric and spectral aspects of the discussed problem will be published in [14].

For the Reader's convenience the concise description of the spectral definition of the Hasimoto map is described in Sec. 1.

Stationary wave solution of Nl.S. eq. Eq. (0.2) is a non-linear (and one-dimensional) analogy of the "free" Schrödinger eq. of the Quantum Mechanics. Obviously, the corresponding non-linear "stationary states" are of the form

$$q(x,t) = g(x;E)e^{iEt} \ , \tag{0.4}$$

where g is a complex-valued function while E ("energy") is a real parameter. Any solution to eq. (0.2) of the form (0.4) is called a stationary wave solution.

In Sec.2 we present a general form of the g function. It depends on 3 real parameters: E, C and M. The solution (0.4) is denoted by $q(x,t;E,C,M)$. A similar notation holds for g.

Galileo boosts. The Nl.S. eq. (0.2) admits the following symmetry

$$q(x,t) \longrightarrow q_V(x,t) = q(x-Vt,t)e^{i(\frac{V}{2}x-\frac{V^2}{4}t)} \ , \tag{0.5}$$

where ν is an arbitrary real parameter. For obvious reasons the symmetry (0.5) is called the Galileo boost symmetry (corresponding to the value ν).

Travelling wave solution of Nl.S. eq. The action of Galileo boost transformations on stationary wave solutions produces 4-real-parameter family of solutions to Nl.S. eq. (0.2)

$$q_{\nu}(x,t;E,C,M) = g(x-\nu t;E,C,M)e^{i[\frac{\nu}{2}x+(E-\frac{\nu^2}{4})t]}. \quad (0.6)$$

Any solution of the form (0.6) is called a travelling wave solution.

Kida observation. Of course, a calculation of the image $H[q_{\nu}(x,t;E,C,M)]$ gives the corresponding 4-real-parameter family of the vortex filament motions.

S. Kida was the first to investigate the problem: how to compute the image $H[q_{\nu}(x,t;E,C,M)]$? [15]. The discussed family of the vortex filament motions is called the Kida class.

In principle, Kida calculated the image $H[q_{\nu}(x,t;E,C,M)]$. His successful trick consists in the following observation: any vortex filament of the Kida class can be described as a rigid motion of some spatial curve of the invariant shape. This is a simple consequence of the so called Fundamental Theorem of Curves in $\mathbb{E}^3$ [16]. In Kida's paper this observation is the starting point of the investigations.

Aims of the paper. Our paper is aimed at a novel and improved description of the Kida class of vortex filament motions. The image $H[q_{\nu}(x,t;E,C,M)]$ is calculated explicitly from the spectral definition of the Hasimoto map (Sec. 1). The main result of the paper is the Theorem on Kida class (Sec. 3). In particular, we point out that we employ new and very convenient parameters (ones replacing ν, E, C and M parameters).

Remarks on the symmetries. In the paper we don't apply symmetry techniques explicitly. We point out, however, we make use of some unusual properties of the Nl.S. eq. which are strictly related to its fundamental feature: the existence of the infinite-dimensional Lie-Bäcklund (generalized) symmetry group [17].

1. SPECTRAL REFORMULATION OF THE HASIMOTO MAP

In this section we present a concise description of the announced earlier spectral definition of the Hasimoto map (0.3).

Spectral features of Nl.S. eq. [9]. Consider the following linear system for an unknown 2x2 matrix Φ

$$\Phi,_x = \begin{bmatrix} i\zeta & q(x,t) \\ -q^*(x,t) & -i\zeta \end{bmatrix} \Phi \quad . \tag{1.1a}$$

$$\Phi,_t = \begin{bmatrix} -2i\zeta^2 + i|q(x,t)|^2 & -2\zeta q(x,t) + iq(x,t),_x \\ 2\zeta q^*(x,t) + iq^*(x,t),_x & 2i\zeta^2 - i|q(x,t)|^2 \end{bmatrix} \Phi \quad . \tag{1.1b}$$

where $q(x,t)$ is a given smooth and complex-valued function on $\mathbb{R}^2$, ζ is a complex parameter and asterisk stands for complex conjugation.

One can show that the eqs. (1.1) admit a smooth and global solution Φ iff q is a solution to Nl.S. eq. (0.2).

According to standard definitions of soliton theory eq. (1.1a), parameter ζ and unknown $\Phi = \Phi(x,t;\zeta)$ are called spectral problem (of eq. (0.2)), spectral parameter and the wave function, resp.

It is not difficult to show that $\Phi = \Phi(x_0,t_0;\zeta) \in SU(2)$ at some point $(x_0,t_0) \in \mathbb{R}^2$ and for any $\zeta \in \mathbb{R}$ implies $\Phi(x,t;\zeta) \in SU(2)$ everywhere ($\zeta \in \mathbb{R}$). In the sequel we assume $\zeta \in \mathbb{R}$ implies $\Phi(x,t;\zeta) \in SU(2)$.
As a result we have also

$$\Phi^{-1}(x,t;0)\cdot\Phi,_\zeta(x,t;0) \in su(2) \quad . \tag{1.2}$$

Algorithm. This is a spectral prescription to compute the image $H(q)$ under the Hasimoto map (0.3).

1) Select an arbitrary smooth and global solution q to eq. (0.2).
2) Insert q into q-dependent matrices of R.H.S. of eqs. (1.1).
3) Find $SU(2)$-valued wave function Φ ($\zeta \in \mathbb{R}$). Φ is defined modulo right ζ-dependent 2x2 matrix factor. This freedom is immaterial for the future constructions.
4) Real, global and smooth functions $X^k(x,t)$ ($k = 1,2,3$) are defined as follows (summation convention !)

$$i\Phi^{-1}(x,t;0)\cdot\Phi,_\zeta(x,t;0) = X^k(x,t)\cdot\sigma_k \quad , \tag{1.3}$$

where we have used (1.2) and σ_k are standard Pauli matrices.

5) We put

$$\mathbf{r}_q(x,t) = X^1(x,t)\mathbf{i} + X^2(x,t)\mathbf{j} + X^3(x,t)\mathbf{k} \quad , \tag{1.4}$$

where (**i**, **j**, **k**) is a fixed orthonormal reper in $\mathbb{E}^3$.

And the final statement is: $H(q) = \mathbf{r}_q$.

Remarks on soliton surfaces. As yet we have not discussed any geometric meaning of the approach. In fact, the "**Algorithm**" (discussed above) is a particular application of the setting of soliton surfaces [10].

The image $\mathbf{r}_q(\mathbb{R}^2)$ is an example of soliton surface (of Nl.S. eq. (0.2) corresponding to the solution q and to $\zeta = 0$). In general, it is not 2-manifold !

An open subset U_q of $\mathbb{R}^2$ is defined by

$$U_q = \{ (x,t) \in \mathbb{R}^2 \ : \ q(x,t) \neq 0 \} \tag{1.5}$$

It is not difficult to show that $\mathbf{r}_q(U_q)$ is an immersed surface [18] in $\mathbb{E}^3$ [14]. In this case self-intersections are allowed. In any case the instantaneous shape of the vortex filament is a smooth (without any cusps !) curve.

2.MANIFOLD OF TRAVELLING WAVE SOLUTIONS OF NL.S. EQ.

Convention: As far as elliptic functions are involved we employ terminology and notation of [19].

General stationary wave solution of Nl.S. eq. The totality of all stationary wave solutions to Nl.S. eq. of the form (0.4) is a " 3-dim. manifold ". In fact, put $g = a + ib$ and insert the ansatz (0.4) into eq. (0.2). The resulting equations (prim $= \frac{d}{dx}$) read

$$\begin{cases} a'' - Ea + 2(a^2+b^2)a = 0 \quad , & (2.1a) \\[2ex] b'' - Eb + 2(a^2+b^2)b = 0 \quad . & (2.1b) \end{cases}$$

The system (2.1) is identical with the Newton's eqs. for a unit mass point moving on (a,b) plane in the potential $U(a,b) := -\frac{1}{2}E(a^2+b^2) + \frac{1}{2}(a^2+b^2)^2$.

Make use of the (almost) polar representation

$$g = re^{i\phi} \quad , \tag{2.2}$$

where r – a real function but now r (being smooth !) can assume also negative values !

The obvious integrals of motion of eqs. (2.1) read

$$M = r^2\phi' \qquad \text{(angular momentum)} \quad , \qquad (2.3)$$

$$C = \frac{1}{2}(r'^2+r^2\phi'^2) - \frac{1}{2}Er^2 + \frac{1}{2}r^4 \qquad \text{(energy)} \quad . \qquad (2.4)$$

Put $R = r^2$. Eqs. (2.3) and (2.4) imply the following " conservation energy law "

$$R'^2 + 4M^2 - 8CR - 4ER^2 + 4R^3 = 0 \quad . \qquad (2.5)$$

It is not difficult to show that non-negative solutions of eq. (2.5) exist iff the (one-dimensional) "potential"

$$V(R) := 4M^2 - 8CR - 4ER^2 + 4R^3 \qquad (2.6)$$

admits 3 real equilibrium points : $-\alpha$, β and γ. Moreover,

$$\alpha,\beta,\gamma \geq 0 \quad \text{and} \quad \beta \leq \gamma \ . \qquad (2.7)$$

The map $(E,C,M) \longrightarrow (\alpha,\beta,\gamma)$ is almost bijection : (α,β,γ) defines M^2.

Not entering into technical details we present the most convenient set of parameters : k, u and p. These are defined as follows.

The parameter k serves as a modulus of the standard Jacobi's elliptic functions

$$0 \leq k \leq 1 \quad . \qquad (2.8a)$$

The complex parameter u is given by

$$u = K + iv \quad , \quad v \in [-K',K') \quad , \qquad (2.8b)$$

where $K = K(k)$ is the standard complete elliptic integral of the I kind, $K' = K(k')$ and $k'^2 + k^2 = 1$.

Finally,

$$p \geq 0 \quad . \qquad (2.8c)$$

To sum up, the admissible values of E, C and M can be described either by parameters (α,β,γ) (not ideal choice !) or by the parameters (k,u,p) (ideal choice !) :

$$E = p^2 \frac{1+cn^2u+dn^2u}{sn^2u} , \tag{2.9a}$$

$$C = -\frac{p^4}{2}\cdot\frac{cn^2u\cdot dn^2u+cn^2u+dn^2u}{sn^4u} , \tag{2.9b}$$

$$M = \frac{sn^3u\cdot cnu\cdot dnu}{p^3} . \tag{2.9c}$$

where snu, cnu and dnu are standard Jacobi elliptic functions. From now on we write $g = g(x;E,C,M) = g(x;k,u,p)$.

Not entering into the calculation details we present a general stationary wave solution. This is given by eq. (0.4). Moreover,

$$g(x;k,u,p) = \frac{p}{snu}\cdot\frac{\vartheta_o\cdot\vartheta_o\left(\frac{z-u}{2K}\right)}{\vartheta_o\left(\frac{z}{2K}\right)\cdot\vartheta_o\left(\frac{u}{2K}\right)}\cdot e^{zB(u)} , \tag{2.10a}$$

where $\vartheta_o(\zeta)$ is the Jacobi's theta function ($\vartheta_o = \vartheta_o(0)$),

$$B(u) = znu + \frac{cnu\cdot dnu}{snu} \quad \left(zn(2K\zeta) = \frac{1}{2K}\frac{\vartheta_o'(\zeta)}{\vartheta_o(\zeta)} \right) , \tag{2.10b}$$

$$z = px , \tag{2.10c}$$

and $E = E(k,u,p)$ is given by (2.9a).

Some particular (and important for physical applications) cases of the function (2.10) are collected in the subsequent table. Similar formulae can be found in the framework of the Inverse Method of soliton theory. See [20] and references quoted therein. Our derivation of (2.10) is direct.

General travelling wave solution of Nl.S. eq. It is given by eq. (0.6), where g is given by (2.10a). Of course, the parameters E, C and M are replaced by k, u and p. For some reasons the fourth parameter of eq. (0.6) v is replaced by a_o

$$V = -\frac{2i}{p}\left[\frac{sna_o}{snu\cdot sn(a_o+u)} + \frac{cna_o\cdot dna_o}{sna_o}\right] . \tag{2.11}$$

Eq. (2.11) describes an interesting diffeomorphism:

$$(0,2iK') \ni a_o \longrightarrow V \in \mathbb{R} . \tag{2.12}$$

Manifold of travelling wave solutions of Nl.S. Eq. It is conveniently described by four parameters k, u, p and a_o. See (2.8) and (2.12).

3.KIDA CLASS REVISITED

We turn now to the main result of the paper: the spectral description of the Kida class of vortex filament motions. It is summarized as the following

Theorem on Kida class:

The travelling wave solution of Nl.S. eq. (0.2) specified by the fixed parameters k, u, p and a_o and described in the last two points of the Sec. 2) is mapped under Hasimoto map into the following vortex filament motion of the Kida class

$$X = -\frac{p}{\omega sn(2a_o+u)}\sqrt{1-k^2sn^2(2a_o+u)sn^2z}\ \cos\gamma(z,t) , \tag{3.1a}$$

$$Y = -\frac{p}{\omega sn(2a_o+u)}\sqrt{1-k^2sn^2(2a_o+u)sn^2z}\ \sin\gamma(z,t) , \tag{3.1b}$$

$$Z = -\frac{p}{\omega}\left\{ E(z)^{*)} + \frac{z}{2}\left[\frac{1}{sn^2(a_o+u)} + \frac{1}{sn^2a_o} - 2\right] + it\cdot\frac{\partial\omega}{\partial a_o}\right\} , \tag{3.1c}$$

where

$$\omega = \omega(k,p,u,a_o) = \frac{p^2}{2}\left[\frac{1}{sn^2a_o} - \frac{1}{sn^2(a_o+u)}\right] , \tag{3.2}$$

$$\gamma(z,t) = \tag{3.3}$$

$$= \gamma(z,t;k,p,u,a_o) = i\Pi^{**)}(z,2a_o+u) - i\frac{sn(2a_o+u)}{sna_o sn(a_o+u)}\cdot z + 2\omega t ,$$

and $z = px$.

----

*) E is the elliptic integral of the II kind.
**) Π is the elliptic integral of the III kind in the Jacobi's form.

TABLE:

PARTICULAR CASES OF FUNCTION $g(x;k,u,p)$.

Nr	k, u, p	E, C, M	$g(x;k,u,p)$ ($z=px$)
1	k-arb.	$E=p^2(2-k^2)$	$p\cdot dnz$
	$u=K$	$C=\frac{p^4}{2}(k^2-1)$	
	p-arb.	$M=0$	
1'	$k=1$	$E=p^2$	$p\cdot sechz$
	$u=K$	$C=0$	soliton
	p-arb.	$M=0$	
2	k-arb.	$E=p^2(2k^2-1)$	$kp\cdot cnz$
	$u=K+iK'$	$C=\frac{p^4}{2}k^2(1-k^2)$	
	p-arb.	$M=0$	
3	$k=0$	$E=p^2\frac{2+\cos^2u}{\sin^2u}$	$\frac{p}{\sin u}\cdot e^{iz\cdot tgu}$
	u-arb.	$C=-\frac{1}{8}E^2$	harmonic plane wave
	p-arb.	$M=\frac{\sin^3u\cdot\cos u}{p^3}$	

Remarks. A few remarks are in order.

1) A proof of the Theorem consists in application of the **Algorithm** of Sec. 1. We point out, however, the points 3 and 4 of the **Algorithm** are highly non-trivial ! Many particular results of the Jacobi's theory of theta functions are required to carry out this computing program. The details in [14].

2) The presence "i" in the formulae (3.1) does not mean the Cartesian components X, Y or Z are complex ! They are real !

3) Special choices of the parameters (see Table) lead to particular vortex filament motions. For instance [1] deals with the case *1*.

4) The results of this paper can be used to calculate one-soliton excitations (Bäcklund transforms) of any vortex filament motion of the Kida class by the application of the Theorem II of [1].

Acknowledgement

One of us (A. S.) wishes to thank the Organizer of the symposium "Symmetries in Science III", Prof. Bruno Gruber, for so kind invitation to attend the symposium.

REFERENCES

[1]. A. Sym, Fluid Dynamics Research 3 (1988), p. ?.
[2]. F. R. Hama, Phys. Fluids 5 (1962), p. 1156.
[3]. R. J. Arms and F. R. Hama, Phys. Fluids 8 (1965), p. 553.
[4]. G. K. Batchelor, "Introduction to Fluid Dynamics" (1970), Cambridge University Press, Cambridge.
[5]. T. Kambe and T. Takao, Phys. Soc. Jpn. 31 (1971), p. 591.
[6]. H. Viets and P. Sforza, Phys. Fluids 15(1972), p. 230.
[7]. H. Aref and E. P. Flinchem, J. Fluid Mech. 148 (1984), p. 477.
[8]. H. Hasimoto, J. Fluid Mech. 51 (1972), p. 477.
[9]. G. L, Lamb Jr., "Elements of Soliton Theory" (1980), Wiley, New York.
[10]. A. Sym, in : "Geometric Aspects of the Einstein Equations and Integrable Systems" (ed. R. Martini) v. 239 (1985), Springer-Verlag, New York.
[11]. A. Sym, Lett. Nuovo Cim. 36 (1983), p. 307.
[12]. D. Levi, A. Sym and S. Wojciechowski, Phys. Lett. 94 A (1983), p.408. See also Y. Fukumoto and T. Miyazaki, J. Phys. Soc. Jpn. 55 (1986), p. 4152.
[13]. J. Cieśliński, P. Gragert and A. Sym, Phys. Rev. Lett. 57 (1986), p. 1507.
[14]. A. Sym, "Geometry of Solitons" v.1 (1989), Reidel Publ. Comp., Dordrecht.
[15]. S. Kida, J. Fluid Mech. 112 (1981), p. 397.
[16]. M. do Carmo , "Differential Geometry of Curves and Surfaces" (1976), Prentice-Hall, Engelwood Cliffs, NJ.
[17]. P. J. Olver, "Applications of Lie Groups to Differential Equations" (1986), Springer-Verlag, New York.
[18]. Y. Matsushima, "Differentiable Manifolds" (1972), Marcel Dekker, Inc., New York.
[19]. F. Oberhettinger and W. Magnus, "Anwendung der Elliptischen Funktionen in Physik und Technik" (1949), Springer-Verlag, Berlin.
[20]. J. Mertsching, Fortschr. Phys. 35 (1987), p. 519.

DYNAMICAL SYMMETRY OF THE KALUZA-KLEIN MONOPOLE

L. Gy. Fehér
JATE, Bolyai Intézet
H-6720 Szeged, Hungary
and
P. A. Horváthy
Dipartimento di Fisica, Università di Napoli
Mostra d'Oltremare, pad. 19
I-80125 Napoli, Italy

1. A SUMMARY OF KALUZA-KLEIN THEORY[1,2]

One of the oldest and most enduring ideas regarding the unification of gravitation and gauge theory is Kaluza's five dimensional unified theory. Kaluza's hypothesis was that the world has four spatial dimensions, but one of the dimensions has curled up to form a circle so small as to be unobservable. He showed that ordinary general relativity in five dimensions, assuming such a cylindrical ground state, contained a local U(1) gauge symmetry arising from the isometry of the hidden fifth dimension. The extra components of the metric tensor constitutes the gague fields of this symmetry and could be identified with the electromagnetic vector potential.

To be more specific, consider general relativity on a five dimensional space-time with the Einstein-Hilbert action

$$S = -\frac{1}{16\pi G_K}\int d^5x \sqrt{-g_5}\, R_5 \tag{1.1}$$

where R_5 is the five-dimensional curvature scalar of the metric g_{AB}, and G_K is the five-dimensional coupling constant. Our conventions are: upper case Latin letters A, B, C denote five-dimensional indices 0, 1, 2,

3, 5; lower case Greek indices run over for dimensions 0, 1, 2, 3, whereas lower case Latin indices run over four-dimensional spatial values 1, 2, 3, 5. The signature of g_{AB} is -,+,+,+,+, and the Riemann tensor is

$$R^K{}_{LMN} = \partial_M \Gamma^K{}_{LN} - \partial_N \Gamma^K{}_{LM} + \Gamma^K{}_{JM} \Gamma^J{}_{LN} - \Gamma^K{}_{JN} \Gamma^J{}_{LM}, \quad R_{LM} = R^K{}_{LKM}, \quad R_5 = R^L{}_L$$

and, except where indicated, $h/2\pi = c = 1$.

In the absence of other fields the equations of motion are of course $R_{AB} = 0$. The basic assumption of Kaluza and Klein was that the correct vacuum is the space $M^4 \times S^1$, the product of four dimensional Minkowski space with a circle of radius R. The radius of the circle in fifth dimension is undetermined by the classical equations of motion, since any circle is flat. If R is sufficiently small then all low-energy experiments will simply average over the fifth dimension. In fact the components of the metric, $g_{AB}(x^\mu, x^5)$, can be expanded in a Fourier series,

$$g_{AB}(x^\mu, x^5) = \sum g_{AB}(x^\mu) \exp \frac{inx_5}{R} \tag{1.2}$$

and all modes with $n \neq 0$ will have energies greater than hc/R. Thus the effective low-energy theory can be deduced by considering the metric g_{AB} to be independent of x^5. Under these assumptions the theory invariant under general coordinate transformations that are independent of x^5. In addition to ordinary four dimensional coordinate transformations $x^\mu \rightarrow x^\mu(x^\nu)$, we have a U(1) local gauge transformation $x^5 \rightarrow \Lambda(x^\mu)x^5$, under which $g_{\mu 5}$ transforms as a vector gauge field,

$$g_{\mu 5}(x) \rightarrow g_{\mu 5}(x) + g_{\mu 5}(x)\partial_\mu \Lambda \, . \tag{1.3}$$

Therefore the low-energy theory should be a theory of four dimensional gravity plus a U(1) gauge theory, i.e. electromagnetism, with the massless modes of $g_{\mu\nu}$ ($g_{\mu 5}$) corresponding to the graviton (photon). The low-energy theory is also invariant under scale transformations in the fifth dimension,

$$x^5 + \lambda x^5, \qquad g_{55} \rightarrow \lambda^{-2} g_{55}, \qquad g_{\mu 5} \rightarrow \lambda^{-1} g_{\mu 5}. \tag{1.4}$$

This global scale invariance is spontaneously broken by Kaluza-Klein vacuum (since R is fixed) thus giving rise to a Goldstone boson, the dilaton.

To exhibit the low-energy theory we write the metric as follows

$$g_{AB} = \begin{pmatrix} g_{\mu\nu} + A_\mu A_\nu V & A_\mu V \\ A_\nu V & V \end{pmatrix}$$

$$g_5 \equiv \det g_{AB} = \det g_{\mu\nu} V \equiv g_4.V,$$

$$ds^2 = V(dx^5 + A_\mu dx^\mu)^2 + g_{\mu\nu} dx^\mu dx^\nu. \qquad (1.5)$$

The five-dimensional curvature scalar can be expressed in terms of the four dimensional curvature R_4, the field strength $F_{\mu\nu} = \partial_\mu A_\nu - \partial_\mu A_\nu$, and the scalar field V,

$$R_5 = R_4 - \frac{2}{\sqrt{V}} \Box \sqrt{V} \qquad (1.6)$$

Thus the effective low energy theory is described by the four dimensional action

$$S = -\frac{1}{16\pi G} \int d^4x \sqrt{-g_4}\, V^{1/2} \left\{ R_4 + \frac{1}{4} V F_{\mu\nu} F^{\mu\nu} \right) \qquad (1.7)$$

where we have dropped the terms in R_5 involving V, since these yield (when multiplied by $\sqrt{g_5}$) a total derivative, and

$$G = G_K/2\pi R \qquad (1.8)$$

is Newton's constant, determined by $\sqrt{hG/(\sqrt{2\pi})c^3} = 1.6 \times 10^{-33}$ cm.

This theory is recognizable as a variant of the Brans-Dicke theory[3] of gravity, with $V^{1/2}$ identified as a Brans-Dicke massless scalar field, coupled to electromagnetism. V indeed sets the local scale of the gravitational coupling. In the vacuum V = 1. Also in the Brans-Dicke theory the coupling of V to matter is somewhat arbitrary, here it is totally fixed by five-dimensional covariance.

The radius R is determined by the electric charge. To see this consider a complex field Φ with action

$$S_\Phi = \int d^5x \sqrt{-g_5}\, (\partial_A \Phi)(\partial_B \Phi')g^{AB}. \qquad (1.9)$$

The Fourier component of Φ with non-trivial x^5 dependence, $\Phi^{(n)} = \exp(inx^5/R)$, will behave as a particle of charge $e = \sqrt{16\pi G}n/R$ and mass n/R, since

$$(\partial_A \Phi)(\partial_B \Phi\dagger)\, g^{AB} = \left| \left(\partial_\mu + i\frac{n}{R} A_\mu \right) \Phi \right|^2 + V^{-1} \frac{n^2}{R^2} |\Phi|^2 \qquad \text{(II.1.10)}$$

(1.10)

(Note that the properly normalized gauge field is $(16\pi G)^{-1/2} A_\mu$). Thus

$$\alpha = e^2/2h\,c = 2h\,G/\pi c^3 R^2, \quad R = (1/\pi\sqrt{\alpha})\, hG/c^3 \simeq 3.7 \times 10^{-32}\ \text{cm}. \quad (1.11)$$

Consider a classical point test particle of unit mass. It has action

$$S = \int d\tau \sqrt{g_{AB} \frac{dx^A}{d\tau} \frac{dx^B}{d\tau}}$$

(1.12)

and the motion is given by a five-dimensional *geodesic*

$$\frac{d^2x^A}{d\tau^2} + \Gamma^A_{BC} \frac{dx^B}{d\tau} \frac{dx^C}{d\tau} = 0$$

(1.12)

Now since the space-time possesses a Killing vector, namely

$$K = K^A \partial/\partial x^A = \partial/\partial x^5, \quad (1.13)$$

it is guaranteed that $K_A dx^A/d\tau$ is a constant of the motion. Indeed a first integral of eqn. (1.12) is

$$K_A dx^A/d\tau = V dx^5 + A_\mu dx^\mu/d\tau = q. \quad (1.14)$$

The remaining equations of motion then take the form

$$\frac{d^2x^\mu}{d\tau^2} + \Gamma^\mu_{\alpha\beta} \frac{dx^\alpha}{d\tau} \frac{dx^\beta}{d\tau} = q F^\mu{}_\nu \frac{dx^\nu}{d\tau} + q^2 \frac{\partial^\mu V}{2V^2}$$

(1.15)

$\Gamma^\mu{}_{\alpha\beta}$ is the four dimensional connection constructed from $g_{\mu\nu}$. We recognize on the right the Lorentz force if $q\sqrt{16\pi G}$ is identified with the *charge* of the particle, plus an interaction with the scalar field V.

2. The Kaluza-Klein monopole of Gross, Perry and Sorkin[2]

Now we are searčhing for *solitons* in the five-dimensional Kaluza-Klein theory sketched above. By a soliton we mean a non-singular solution to the classsical field equations which represent spatially localized lumps that are topologically stable. Such solutions are expected to have a large mass (of the order M/g, where M is the mass scale of the theory and g a dimensionless coupling constant), and are the starting points for the semi-classical construction of quantum mechanical particle states.

Our goal is to construct solutions of the five dimensional field equations that approach the vacuum solution: $V = 1$, $A_\mu = 0$, $g_{\mu\nu} = \eta_{\mu\nu}$ at spatial infinity. It is natural to consider static metrics with $\partial/\partial t$ as Killing vector. It is also natural to look for solutions with $g_{0A} = \delta_{0A}$. In this case the space-time is totally flat in the 'time' direction and the field equations are simply

$$R_{ij} = R_{5i} = R_{55} = 0, \quad (2.1)$$

namely the four-dimensional, wholly space-like, manifold at each fixed t has a vanishing Ricci tensor. These are simply the equations of four dimensional euclidean gravity, where we can think of x^5 as representing euclidean, periodic time. Our task is greatly simplified by the fact that the equations of four-dimensional euclidean gravity have been extensively studied. For example, the *Kaluza-Klein monopole* of Gross and Perry, and of Sorkin[2] (which is the object of our considerations here) is obtained by imbedding the Taub-NUT gravitational instanton into five dimensional Kaluza-Klein theory. Its line element is expressed as

$$ds^2 = -dt^2 + \left(1+\frac{4m}{r}\right)(dr^2 + r^2(d\theta^2 + \sin^2\theta\, d\phi))^2 + \frac{(d\psi + 4m\cos\theta\, d\phi)^2}{1+\frac{4m}{r}} \tag{2.2}$$

where $r > 0$, and the angles θ, ϕ, ψ $(0 < \theta < \pi,\ 0 \le \phi < 2\pi)$ parametrize $\mathbb{S}^3 \approx SU(2)$. The apparent singularity at the origin is unphysical[4] if ψ is periodic with period $16\pi m$. Since we want our solution to approach the vacuum for large r we must identify $16\pi m$ with $2\pi R$. Thus

$$m = R/8 = \sqrt{\pi}\, G\,/2e \tag{2.3}$$

The gauge field, A_μ, is clearly that of a monopole, $A_\phi = 4m\cos\theta$, $\mathbf{B} = \nabla \times \mathbf{A} = 4m\,\mathbf{r}/r^3$, and has a string singularity along the whole z axis. As usual, this singularity is an artifact if and only in the period of x^5 is equal to that of $16\pi m$.

The magnetic charge of our monopole is thus fixed by the radius of the Kaluza-Klein circle. If we scale the magnetic field so as to have the proper normalization, $\mathbf{B} \to (16\pi G)^{-1/2}\mathbf{B}$, we find that the magnetic charge is

$$g = 4m/\sqrt{16\pi G} = R/2\sqrt{16\pi G} = 1/2e. \tag{2.4}$$

Thus, as expected, our monopole has one unit of Dirac charge.

The mass of a static, asymptotically flat spacetime can be defined. In our case it is

$$M = -\frac{2\pi R}{16\pi G_K}\int d^3x\, \nabla^2 \frac{1}{V} = \frac{m}{G} \tag{2.5}$$

Since m is fixed by the radius if the vacuum circle, which in turn is fixed by e and G, the soliton mass is determined to be

$$M^2 = m_P^2/16\,\alpha, \tag{2.7}$$

where $m_P = \sqrt{hc/2\pi G} = 2.17 \cdot 10^{-5}$ g is the Planck mass. As is costumary for solitons, the monopole mass is 1/e times heavier than the mass scale of the theory.

Remarkably, the Kaluza-Klein monopole has re-emerged recently[5] as the asymptotic limit of the curved manifold whose geodesics describe the scattering of self-dual monopoles.

3. Classical dynamics

Let us consider the geodesic motion of a particle in the Kaluza-Klein monopole field, with Lagrangian

$$\pounds = \frac{1}{2} g_{\mu\nu}\dot{x}^{\mu}\dot{x}^{\nu} = \frac{1}{2}\left\{(1+\frac{4m}{r})v^2 + \frac{(\dot{\psi}+4m\cos\theta\,\dot{\phi})^2}{1+\frac{4m}{r}}\right\} \tag{3.1}$$

where $\mathbf{v} = \dot{\mathbf{r}}$. To the two cyclic variables ψ and t are associated the conserved quantities

$$q = (1+4m/r)^{-1}(\dot{\psi} + 4m\cos\theta\,\dot{\phi}) \tag{3.2}$$

and

$$E = (1/2)(1+4m/r)(\mathbf{v}^2 + q^2), \tag{3.3}$$

interpreted as electric charge and energy, respectively.

It is convenient to introduce the *mechanical 3-momentum*

$$\mathbf{p} = (1+4m/r)\mathbf{v}. \tag{3.4}$$

The 5-dimensional geodesics are the solutions to the Euler-Lagrange equations associated to (3.1). The projection into 3-space of the motion is governed hence by the equation,

$$\frac{dp_i}{dt} = -4mq\,\epsilon_{ijk}\frac{v_j r_k}{r^3} + 2mq^2\frac{r_i}{r^3} - 2mv^2\frac{r_i}{r^3} \tag{3.5}$$

This complicated equation contains, in addition to the Dirac-monopole plus a Coulomb terms, also a velocity-square dependent term, typical for the motion in curved space. Due to manifest spherical symmetry, the monopole angular momentum,

$$\mathbf{J} = \mathbf{r}\times\mathbf{p} + (4mq)\,\mathbf{r}/r, \tag{3.6}$$

is conserved. The presence of the velocity-square dependent force changes, when compared to the pure Dirac + Coulomb case, the situation dramatically. Indeed, energy conservation implies that

$$(\dot{r})^2 = \{(2E-q^2)\,r^2 - (E-q^2)\,r - J^2\}/(r+4m)^2. \tag{3.7}$$

For positive m the particle cannot reach the centre. Indeed, eqn. (3.3) shows that, for the $m > 0$ the energy is at least $q^2/2$, and hence

$$r \geq r_o = 4mq^2/(2E-q^2) > 0 . \tag{3.8}$$

There are no bound motions, because the coefficient of r^2 is positive.

For $m < 0$ instead, the energy may be smaller then $q^2/2$. In such a case the coefficient of r^2 is negative and the system *does* admit *bound motions.* More details are given below.

Despite the complicated form of the equations of motion, the classical motions are surprisingly simple. The clue is the observation[6] that, in addition to the angular momentum, **J** there is also conserved 'Runge-Lenz' vector, namely

$$\mathbf{K} = \mathbf{p} \times \mathbf{J} - 4m(E-q^2)\, \mathbf{r}/r . \tag{3.9}$$

This can be verified by an explicit calculation, using the equations of motion.

The knowledge of these conserved quantitites allows for a complete description of the motion[6]. Indeed, eqns. (3.5) and (3.6) allow to prove that

$$\mathbf{J}.\mathbf{r}/r = -4mq \tag{3.10}$$

and

$$[\, \mathbf{K} + \mathbf{J}\ (E-q^2)/q\,] \,.\, \mathbf{r} = \mathbf{J}^2 - (4mq)^2 . \tag{3.11}$$

The first of these equations implies that, as usual in monopole interactions, the the particle moves on a *cone* with axis **J** and opening angle $\cos\alpha = 4mq/J$. The second implies in turn that the motions lie in the *plane* perpendicular to the vector $\mathbf{N} = q\mathbf{K} + \mathbf{J}(E-q^2)$. They are therefore *conic sections*, see Fig. 1.

The form of the trajectory depends on β, the plane's inclination, being smaller or larger then the complement of the cone's opening angle, $\{\pi/2 - \alpha\}$. Now, $\cos\beta = \mathbf{N}.\mathbf{J}/NJ$. Using the relations

$$\mathbf{K}.\mathbf{J} = -\, (4m)^2 q\ (E-q^2) \tag{3.12}$$

and

$$\mathbf{K}^2 = (2E-q^2)(\mathbf{J}^2 - (4mq)^2) + 4m^2\,(E-q^2)^2, \tag{3.13}$$

a simple calculation gives that for

- $E < q^2/2$ (only possible for $m < 0$)		ellipses;
- $E > q^2/2$	the trajectories are	hyperbolae;
- $E = q^2/2$		parabolâe.

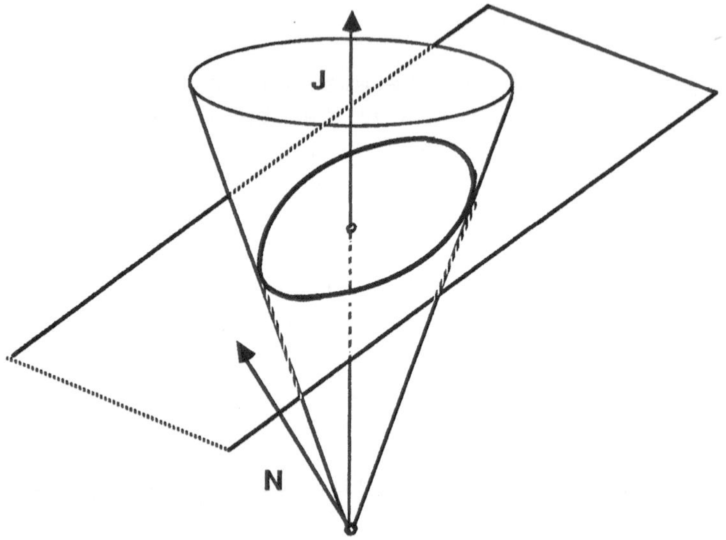

FIG. 1. The particle moves on a cone whose axis is **J**, conserved angular momentum. The velocity vector is perpendicular to **N** = q**K**+**J**(E-q^2). The trajectory is therefore a *conic section.*

Let us now study the unbound motions. Using the two conserved quantities **J** and **K**, the classical scattering can also be described. Let ω be the scattering angle, γ the twist, and $\boldsymbol{v}$ denote the velocity at large distances. Then $E = (\nu^2 + q^2)/2$, $L = |\mathbf{L}|$, the module of orbital angular momentum, is $L = y\boldsymbol{v}$, where y is the impact parameter. From the conservation of the component of **J** which is parallel to the particle's plane we get

$$\mathrm{tg}\ (\pi-\theta)/2 = -\ y\boldsymbol{v}\ \sin\gamma/\ 4qm \tag{3.14}$$

and from the conservation of the parallel component of **K** we get

$$\mathrm{tg}\ (\pi-\theta)/2 = -\ y\boldsymbol{v}\ \cos\gamma/\ 4m\ (E-q^2) \tag{3.15}$$

From these equations we deduce

$$\mathrm{tg}\ \gamma/2 = q/\boldsymbol{v} \qquad \text{and} \qquad \mathrm{tg}\ \theta/2 = -\ 2m\ \left(1 + (q/\nu)^2\right)/y. \tag{3.15}$$

The classical cross-section is therefore

$$d\sigma/d\omega = y\ |dy/d\theta|/\sin\theta = m^2\ (1 + (q/\nu)^2)^2\ /\sin^4\theta/2\ . \tag{3.16}$$

In order to clarify the structure of the conserved quantities, it is convenient switch over to the Hamiltonian formalism. Those momenta canonically conjugate to the coordinates are $\pi_i = (1+4m/r)x_i - qA_i$, i = 1,

2, 3, and $\pi_4 = q$, where $A_r = A_\theta = 0$, $A_\phi = -4m\cos\theta$. The Poisson brackets are $\{x^i, \pi_j\} = \delta^i{}_j$, $\{\psi, \pi_4\} = 1$. The mechanical momentum $\mathbf{p} = \pi + q\,\mathbf{A} = (1+4m/r)\mathbf{v}$ satisfies therefore the Poisson bracket relations

$$\{p_i\,,\,p_j\} = -(4qm)\epsilon_{ijk}x^k/r^3, \qquad \{x^i,\,p_j\} = \delta^i{}_j, \qquad \{\psi,\,p_j\} = A_j\,. \tag{3.17}$$

The Hamiltonian is

$$H = \frac{1}{2}\left(\frac{p_i p^i}{1+\frac{4m}{r}} + (1+\frac{4m}{r})\,q^2\right) \tag{3.18}$$

The Poisson brackets of angular momentum and the Runge-Lenz vector are

$$\{j_i,\,j_k\} = \varepsilon_{ikn}J^n, \quad \{j_i,\,K_k\} = \varepsilon_{ikn}K^n\,, \quad \{K_i,\,K_k\} = (q^2 - 2H)\,\varepsilon_{ikn}j^n. \tag{3.19}$$

For each fixed value E set

$$\mathbf{M} = \begin{cases} (q^2 - 2E)^{-1/2}\,\mathbf{K} & E < q^2/2 \\ \mathbf{K} \quad \text{for} & E = q^2/2\,. \\ (2E - q^2)^{-1/2}\,\mathbf{K} & E > q^2/2 \end{cases} \tag{3.20}$$

For energies lower then $q^2/2$ the angular momentum **J** and the rescaled Runge-Lenz vector **M** form an o(4) algebra . For $E > q^2/2$ the dynamical symmetry generated by **J** and **M** is rather o(3,1). In the parabolic case $E = q^2/2$ the algebra is $o(3)\otimes_s\mathbb{R}^3$.

4. Quantization

First, we find the quantum operators by DeWitt[7]'s rules. $x_i\hat{}$ is multiplication by x_i, and the canonical momentum operator is

$$\pi_i\hat{} = -i\,\{\partial_i + (1/4)\,(\partial_i \ln g)\}\,.$$

$x_i\hat{}$ and $\pi_i\hat{}$ are hermitian with respect to $L^2(\sqrt{g}\,d^4x)$, and satisfy the basic commutation relations $[x^i\hat{}, x^j\hat{}] = [\pi_i\hat{}, \pi_j\hat{}] = 0$, $[x^i\hat{}, \pi_j\hat{}] = i\delta^i{}_j$. It is, however, more useful to introduce rather the modified (with respect to the Taub-NUT volume element non-self-adjoint !) operator

$$p_k\hat{} = -i\,\partial_k + q\,A_k \tag{4.1}$$

with commutation relations

$$[p_i^\wedge, p_j^\wedge] = -i(4mq)\ \varepsilon_{ijk}\ x^k/r^3, \quad [x^{i\wedge}, p_j^\wedge] = i\,\delta^i{}_j, \quad [\psi^\wedge, p_i^\wedge] = iA_i \,. \quad (4.2)$$

The charge operator $q = -i\partial_\psi$ has eigenvalues $q = s/4m$, $s = 0, \pm 1/2, \pm 1, \ldots$ Our 4-metric is Ricci-flat, so the quantum Hamiltonian is[7]

$$H^\wedge = -(1/2)\, g^{-1/4}\, \pi_\mu \sqrt{g}\, g^{\mu\nu}\, \pi_\nu\, g^{-1/4} = -(1/2)\, g^{-1/2} \partial_\mu(\sqrt{g}\ \partial^\mu) =$$

$$-\Delta/2, \quad (4.3)$$

the covariant Laplacian on curved 4-space. A lengthy calculation yields then the three-dimensional expression

$$H^\wedge = \frac{1}{2}\left(\frac{p_i p^i}{1 + \frac{4m}{r}} + \left(1 + \frac{4m}{r}\right) q^2 \right) \quad (4.4)$$

which is formally the same as classically. Notice that the order of the operators <u>is</u> important. The charge operator q commutes with the Hamiltonian, and so is conserved. The angular momentum operator is

$$\mathbf{J}^\wedge = \mathbf{r}^\wedge \times \mathbf{p}^\wedge + 4mq\, \mathbf{r}^\wedge / r. \quad (4.5)$$

Inserting $\mathbf{J}^2$, the square of the angular momentum, into the Hamiltonian, we get

$$H^\wedge = \frac{1}{2}\left(\frac{ -\frac{1}{r^2}\partial_r(r^2 \partial_r) + \frac{J^2 - (4qm)^2}{r^2} }{1 + \frac{4m}{r}} + \left(1 + \frac{4m}{r}\right) q^2 \right) \quad (4.6)$$

Substituting here $j(j+1)$ for J^2, $s/4m$ for q, introducing the new wave function $\psi = r\Psi$, multiplying by $2r(1+4m/r)$, the eigenvalue equation $H^\wedge \Psi = E\Psi$ becomes

$$\left\{ \frac{d^2}{dr^2} - \frac{j(j+1)}{r^2} + \frac{(-s^2/2m + 8mE)}{r} - \left(\left(\frac{s}{4m}\right)^2 - 2E \right) \right\} \psi = 0 \quad (4.7)$$

By chosing $\psi \sim r^{j+1}$ as boundary condition near $r = 0$, we obtain a well-defined problem for which (4.7) is self-adjoint on $0 \leq r \leq \infty$. Continuum solutions of eqn. (4.7) bounded at $r = 0$ have the form

$$\psi(r) = r^{j+1} e^{ikr} F(i\lambda+j+1,\ 2j+2,\ -2ikr) \tag{4.8}$$

where

$$k^2 = 2E - q^2, \tag{4.9}$$

and

$$\lambda = -4m \frac{(q^2 - E)}{\sqrt{2E - q^2}} \tag{4.10}$$

and F is the confluent hypergeometric function

$$F(a,b,u) = 1 + \frac{a}{b}u + \frac{a(a+1)\, u^2}{b(b+1)\, 2\,!} + \dots \tag{4.11}$$

Square-integrable ground states correspond to values of λ such that $\lambda^2 = -(k+j+1)^2 = -n^2$, $k = 0, 1, 2, \dots$, i.e.

$$-4m \frac{(q^2 - E)}{\sqrt{-2E + q^2}} = n = |s|+1,\ |s|+2,\ \dots \tag{4.12}$$

For $q^2 > 2E$ (i.e. for $s^2 > 32Em^2$) which, by eqn. (4.4), is only possible for negative m), one gets the bound-state energy levels

$$E = \frac{1}{(4m)^2}\sqrt{n^2 - s^2}\left(\pm n - \sqrt{n^2 - s^2}\right), \quad n = |s|+1, |s|+2, \dots \tag{4.13}$$

with degeneracy $n^2 - s^2$. (Observe that $n^2 - s^2 = (n+s)(n-s)$ is always an integer, since n and s are simultaneously integers or half-integers). The two signs correspond to the lightly bound states $E > 0$ and to the tightly bound states with $E < 0$.

For $E > q^2/2$ one gets *scattering states*. The quantum cross-section can be calculated[6] by solving the problem in parabolic coordinates. It is found to be identical to the classical expression (3.16) with $q/\boldsymbol{v}$ replaced by $s/4mk$, i.e.

$$d\sigma/d\omega = m^2\,(1 + (s/4mk)^2\,)^2/\sin^4 \theta/2\,. \tag{4.14}$$

Quantizing the Runge-Lenz vector **K** is a hard task. The clue is[8] that **K** is associated with 3 *Killing tensors* $K^i{}_{\mu\nu}$, $i = 1, 2, 3$. Remember that a Killing tensor is a symmetric tensor $K_{\mu\nu} = K_{\nu\mu}$ on (curved) space, such that $K_{(\mu\nu;\alpha)} = 0$. To a Killing tensor is associated a conserved quantity which is quadratic in the velocity, namely $K = K_{\mu\nu}\dot{x}^{\mu}\dot{x}^{\nu}/2$. For example, the metric tensor $g_{\mu\nu}$ itself satisfies these conditions; the corresponding conserved quantity is the energy. According to Carter[8], the quantum operator of K is

$$K^\wedge = -(1/2)\,\nabla_\mu K^{\mu\nu}\nabla_\nu = -(1/2)\,\nabla^\mu K_{\mu\nu}\nabla^\nu \qquad (4.15)$$

(∇_μ is the covariant derivative). The classical expression (2.3) of the Runge-Lenz vector allows us to identify the components of the Killing tensor and use then Carter's prescription to get the quantum operators. A long and complicated calculation yields

$$\mathbf{K}^\wedge = (1/2)\,\{\mathbf{p}^\wedge \times \mathbf{J}^\wedge - \mathbf{J}^\wedge \times \mathbf{p}^\wedge\} - 4m(\mathbf{r}^\wedge/r)\,(H^\wedge - q^{\wedge 2})\,. \qquad (4.16)$$

Again, the order of the operators is relevant. For notational convenience we drop the 'hat' ^ from our operators in what follows.

5. Spectrum from dynamical symmetry

Using the fundamental commutation relations (3.2) one verifies that the operators **J** and **K** satisfy the quantized version of (2.7), namely

$$[J_i, J_k] = i\varepsilon_{ikn} J^n, \qquad [J_i, K_k] = i\varepsilon_{ikn} K^n,$$

$$[K_i, K_k] = i(q^2 - 2H)\,\varepsilon_{ikn} J^n. \qquad (5.1)$$

On the fixed-energy eigenspace $H\Psi = E\Psi$ define the rescaled Runge-Lenz operator **M** (cf. (2.8)) by

$$\mathbf{M} = \begin{cases} (q^2 - 2E)^{-1/2}\,\mathbf{K} & E < q^2/2 \\ \mathbf{K} \quad \text{for} & E = q^2/2 \\ (2E - q^2)^{-1/2}\,\mathbf{K} & E > q^2/2 \end{cases} \qquad (5.2)$$

M and **J** close, just like classically, to an <u>o(4) algebra</u> for $E < q^2/2$,

$$[J_i, J_k] = i\varepsilon_{ikn} J^n, \quad [J_i, M_k] = i\varepsilon_{ikn} M^n, \quad [M_i, M_k] = \varepsilon_{ikn} J^n\,; \qquad (5.3a)$$

to an <u>o(3,1) algebra</u> for $E > q^2/2$,

$$[J_i, J_k] = i\varepsilon_{ikn} J^n, \quad [J_i, M_k] = i\varepsilon_{ikn} M^n, \quad [M_i, M_k] = -\varepsilon_{ikn} J^n; \qquad (5.3b)$$

and to an $o(3)\otimes_s\mathbb{R}^3$ in the parabolic case $E = q^2/2$,

$$[J_i\,,\,J_k\,] = i\varepsilon_{ikn}\,J^n, \qquad [J_i\,,\,M_k\,] = i\varepsilon_{\,ikn}M^n, \qquad [M_i,\,M_k] = 0\;; \qquad (5.3c)$$

Following Pauli[9], this allows to calculate the energy spectrum. We have indeed the constraint equations

$$\mathbf{K.J} \;= -\,(4m)^2\,q\,(H - q^2), \qquad (5.4)$$

and

$$\mathbf{K}^2 = (2E - q^2)\,(\mathbf{J}^2 - (4qm)^2 + 1) + (4m)^2\,(E - q^2)^2\,. \qquad (5.5)$$

Observe that the quantum expression differs from its classical counterpart (3.12-13) by an important term $(h/2\pi)^2 = 1$ in our units.

Those states Ψ with constant energy $E < q^2/2$ and charge,

$$-\,4mq\,\Psi = s\,\Psi, \qquad H\,\Psi = E\,\Psi \qquad (5.6)$$

form a representation space for o(4). It is more convenient to consider the commuting operators

$$\mathbf{A} = (\mathbf{J} + \mathbf{M})/2, \qquad \mathbf{B} = (\mathbf{J} - \mathbf{M})/2 \qquad (5.7)$$

which generate two independent so(3)'s, $o(4)\approx o(3)\oplus(3)$. A common eigenvector Ψ of the commuting operators q, H, A^2, A_3, B^2, B_3 satisfies

$$A^2\,\Psi = a(a{+}1)\,\Psi,\;\; A_3\,\Psi = a_3\,\Psi,\;\; B^2\,\Psi = b(b{+}1)\,\Psi,\;\; B_3\,\Psi = b_3\,\Psi, \quad (5.8)$$

where a and b are half-integers, and $a_3 = -\,a, -a+1,\ldots, a$; $b_3 = -\,b, -b+1,\ldots, b$. Consider the (so far non-negative real) number

$$n = \frac{-4m(q^2-E)}{\sqrt{q^2-2E}} \qquad (5.9)$$

The operator identities (5.4-5) imply that

$$a(a{+}1) + b(b{+}1) = (s^2 - 1 - n), \qquad a(a{+}1) - b(b{+}1) = sn^2. \qquad (5.10)$$

Some algebra yields then

$$2a+1 = \pm\,(n+s) \;\text{ and }\; 2b+1 = \pm\,(n-s)$$

$\Rightarrow$

$$a-b = \pm\,s \;\text{ and }\; a+b+1 = \pm n = n, \qquad (5.11)$$

since a and b are non-negative. Due to the first of these relations, n is integer or half-integer, according to s being integer or half-integer. Eqn. (5.9) is thus identical to eqn. (4.12), yielding the bound-state spectrum (4.13) once more.

6. Algebraic calculation of the S-matrix [10,11]

The scattering states form rather a representation space for o(3,1). This can be used to derive algebraically the S-matrix. Let us indeed re-write the Runge-Lenz operator **K** as

$$\mathbf{K} = M\{i\mathbf{v} - \mathbf{J} \times \mathbf{v}\} - 4m\,(\mathbf{r}/r)\,(H - q^2)\,, \tag{6.1}$$

where $M = 1+4m/r$, and **v** is the velocity operator

$$\mathbf{v} = -i\,[\mathbf{r},\ H] = M^{-1}\,\mathbf{p}\,. \tag{6.2}$$

We consider incoming (respectively outgoing) wave packets which are sharply peaked around momentum **k** and have fixed charge $q = s/4m$ and energy $E = k^2/2 + q^2/2$. We argue that they satisfy the relations

$$(\mathbf{J}.\hat{\mathbf{k}})\,\big|\,\mathbf{k}\,(^{\text{in}}_{\text{out}})> \;=\; \mp\, s\,\big|\,\mathbf{k}\,(^{\text{in}}_{\text{out}})>\,, \qquad (\hat{\mathbf{k}} = \mathbf{k}/|\mathbf{k}|\,) \tag{6.3}$$

$$(\mathbf{K}.\hat{\mathbf{k}})\,\big|\,\mathbf{k}\,(^{\text{in}}_{\text{out}})> \;=\; (ik \pm 4m(E-q^2)\,\big|\,\mathbf{k}\,(^{\text{in}}_{\text{out}})>\,. \tag{6.4}$$

The states $|\mathbf{k}(\text{in})>$ and $|\mathbf{k}(\text{out})>$ are solutions of the complete time-dependent Schrödinger equation controled at $t = \mp\infty$. It is enough to check the validity of eqn. (6.3) and (6.4) at $t = \mp\infty$, since H commutes with the operators $\mathbf{J}.\hat{\mathbf{k}}$ and $\mathbf{K}.\hat{\mathbf{k}}$ which are therefore constants of motion. Let us apply the left and right hand sides of the operator identity $\mathbf{J}.\mathbf{r}/r = 4mq$ to the states $|\mathbf{k}(\text{in})>$ and $|\mathbf{k}(\text{out})>$ and take into account that $|\mathbf{k}(\text{in})>$ and $|\mathbf{k}(\text{out})>$ approach eigenstates of $\hat{\mathbf{r}}$ with eigenvalues $\mp\hat{\mathbf{k}}$ as $t \to \mp\infty$, since $|\mathbf{k}(\text{in})>$ and $|\mathbf{k}(\text{out})>$ represent wave-packets incoming from (outgoing to) the direction $\hat{\mathbf{k}}$ when $t \to \mp\infty$. This yields (6.3).

To make (6.4) plausible, apply rather the operator identity

$$\mathbf{K}.\hat{\mathbf{k}} = M\left\{i\mathbf{v}.\hat{\mathbf{k}} - \mathbf{J}.(\mathbf{v}\times\hat{\mathbf{k}}\,)\right\} - 4m(\hat{\mathbf{r}})(\hat{\mathbf{k}})\,(H - q^2) \tag{6.5}$$

to $|\mathbf{k}(in)>$ and $|\mathbf{k}(out)>$ and notice that, besides approaching eigenstates of $\hat{\mathbf{r}}$, they also approach *velocity-eigenstates* with eigenvalue $\mathbf{k}$ as $t \to \mp \infty$. Similarly, the position-dependent mass $M = 1+4m/r$ tends to unity since the incoming and outgoing wave-packets are far away from the monopole's location. The equations (4.3) and (4.4) can also be tested on the explicit solution of the Schrödinger equation, obtained by Gibbons and Manton[6] in parabolic coordinates.

Having established (6.3) and (6.4) let us now take into account that, for fixed charge and energy, the scattering states span a (non normalizable) representation of the dynamical o(3,1) algebra. This representation is characterized by the eigenvalues of the Casimir operators, fixed by the constraints (5.4) and (5.5).

Among those states with fixed energy and charge we have, in addition to those bases which consist of (distorted) plane-wave-like scattering states $|\mathbf{k}(in)>$ and $|\mathbf{k}(out)>$, also the standard angular momentum basis (spherical waves) $|j, j_3>$ $(j = |s|, |s|+1, \ldots)$,

$$\mathbf{J}^2 |j, j_3> = j(j+1) |j, j_3>, \qquad J_3 |j, j_3> = j_3 |j, j_3>. \tag{6.6}$$

The angular momentum basis is easy to handle. We would obtain the matrix elements $<\mathbf{l}(out)|\mathbf{k}(in)>$ if we could expand $|\mathbf{k}(in)>$ and $|\mathbf{k}(out)>$ in the angular momentum basis. The desired expansions can be algebraically derived, since one has the o(3,1) algebra and all the basis vectors $|\mathbf{k}(in)>$, $|\mathbf{k}(out)>$ and $|j, j_3>$ are eigenvectors of the appropriate components of the generators. The details are practically identical to those presented by Zwanziger[10]. This leads to the final expression

$$S(\mathbf{l}, \mathbf{k}) = <\mathbf{l}(out)|\,\mathbf{k}(in)> = \sum_{j \geq |s|} (2j+1) \left[\frac{(j-i\lambda)!}{(j+i\lambda)!}\right] D^j_{(-s,s)}\,(R^{-1}_{\mathbf{l}}\,R_{\mathbf{k}}), \tag{6.7}$$

with λ given in (4.10). The D are the rotation matrices and $R_{\mathbf{k}}$ is the rotation which brings the $\mathbf{z}$ direction into the direction of $\mathbf{k}$. (6.7) is consistent with eqn. (2.4) for the cross-section, and its poles yield the bound-state spectrum (4.13).

It should be emphasised that the derivation of the S-matrix given here above crucially depends on the expression (6.1) of the Runge-Lenz vector $\mathbf{K}$ in terms of the 3 dimensional variables and on the constraint equations (5.4) and (5.5), which fix the actual representation of o(3,1). These relations are in turn straightforward consequences of Carter's covariant expressions.

7. Extension to o(4,2)

Now we extend[12,13] the o(4) symmetry to the <u>conformal algebra o(4,2)</u>. Let us rearrange the time-independent Schrödinger equation H $\Psi = E\Psi$ as

$$\left[\frac{p_i p^i}{1+\frac{4m}{r}} + \left(1+\frac{4m}{r}\right)q^2 - 2E\right]\Psi = 0$$

(7.1)

Under multiplication by $r(1+4m/r) = r+4m$ from the left, the operator on the l.h.s. becomes, after rearrangement,

$$r(p_i p^i) + r(q^2 - 2E) + \frac{(4qm)^2}{r} = 8m\,(E - q^2)$$

(7.2)

Let us work first with bound motions, $q^2/2 > E$, so that we can introduce the new variables

$$\mathbf{R} = (q^2 - 2E)^{1/2}\,\mathbf{r} \quad \text{and} \quad \mathbf{P} = = (q^2 - 2E)^{-1/2}\,\mathbf{p}\,. \tag{7.3}$$

The new operators **R** and **P** satisfy the commutation relations

$$[P_j, P_k] = -i(4mq)\,\varepsilon_{jkn}\,R^n/r^3, \quad [R^j, R^k] = 0\,, \quad [R^j, P_k] = i\delta^j{}_k, \tag{7.4}$$

which show that, for fixed charge and energy, the transformation (**p**,**r**) → (**P**, **R**) is canonical. In terms of **R** and **P** eqn. (7.1) becomes

$$R(P_i P^i) + R + \frac{(4qm)^2}{R} = -8m\,\frac{q^2 - E}{\sqrt{q^2 - 2E}}$$

(7.5)

On the left hand side of (4.6) we recognize Γ_o, the generator of an <u>o(2,1) algebra</u>. Indeed, as a consequence of the commutation relations (7.5), the operators

$$\Gamma_o = (1/2)\{\, R\mathbf{P}^2 + R + (4qm)^2/R \,\} \tag{7.6a}$$

$$\Gamma_4 = (1/2)\{R\mathbf{P}^2 - R + (4qm)^2/R\} \tag{7.6b}$$

$$D = \mathbf{R}.\mathbf{P} - i \tag{7.6c}$$

satisfy the o(2,1) relations $[\Gamma_o, \Gamma_4] = iD$, $[\Gamma_4, D] = -i\Gamma_o$, $[D, \Gamma_o] = i\Gamma_4$.The energy spectrum (but not the degeneracy) is recovered from this at once: the o(2,1) generator Γ_0 has eigenvalues $n = |s|+1, |s|+2, \ldots$ Equating the r.h.s. of eqn. (7.5) with 2n, we get once more the crucial relation (4.12). Let us complete (7.6a-c) by 12 more operators, namely with

$$\mathbf{V} = \mathrm{R}.\mathbf{P} \tag{7.6d}$$

$$\mathbf{J} = \mathbf{R} \times \mathbf{P} + (4qm)\, \mathbf{R}/\mathrm{R} \tag{7.6e}$$

$$\mathbf{M} = (1/2)\, \mathbf{R}(\mathbf{P})^2 - \mathbf{P}(\mathbf{R}.\mathbf{P}) - \mathbf{R}/2 - 4qm\, \mathbf{J}/\mathrm{R} + (4qm)^2\, \mathbf{R}/2\mathrm{R}^2 \tag{7.6f}$$

$$\mathbf{U} = (1/2)\mathbf{R}(\mathbf{P})^2 - \mathbf{P}(\mathbf{R}.\mathbf{P}) + \mathbf{R}/2 - 4qm\, \mathbf{J}/\mathrm{R} + (4qm)^2\mathbf{R}/2\mathrm{R}^2 \tag{7.6g}$$

The commutation relations (7.4) imply that these operators extend the o(2,1) algebra (4.6a-c) into an o(4,2) *conformal algebra.* Those operators commuting with Γ_0 form an o(2)xo(4) generated by Γ_0 itself and by **J** and **M.** Expressed in terms of **r** and **p**, we see that **J** is just the angular momentum operator. On the other hand, **M** is also written as

$$\mathbf{M} = (1/2)(\mathbf{P}\times\mathbf{J} - \mathbf{J}\times\mathbf{P}) + (\mathbf{R}/\mathrm{R})\Gamma_0 \tag{7.7}$$

Using the relation (7.5) this becomes

$$\mathrm{M}_i = \frac{1}{2} \epsilon_{ijk}[\mathrm{P}_j, \mathrm{J}_k] - \frac{\mathrm{R}_i}{\mathrm{R}} \left(4m \frac{\mathrm{E} - q^2}{\sqrt{q^2 - 2\mathrm{E}}} \right) \tag{7.8}$$

Substituting here **r**, **p**, **M** reduces to (the restriction onto $H\Psi = E\Psi$ states) of the rescaled Runge-Lenz vector in (3.11), as anticipated by the notation. We conclude that the o(4,2) algebra (7.6) extends the original dynamical o(4) symmetry.

The scattering case $E > q^2/2$ is treated exactly the same way. Defining

$$\mathbf{R} = (2\mathrm{E} - q^2)^{1/2}\, \mathbf{r} \tag{7.9}$$

one discovers, after suitable rearrangement, the non-compact operator Γ_4 with continuous spectrum. The extension to o(4,2) proceeds along the same lines as before. Those operators commuting with Γ_4 form o(2)xo(3,1), generated by Γ_4, **J**, and **U**, this latter being now identified with the rescaled Runge-Lenz vector **M**.

Note finally, that a similar procedure works for a test particle in a self-dual monopole's field[14].

Acknowledgements

Some of the results presented here were obtained in collaboration with B. Cordani, to whom we express our indebtedness. We also thank P. Forgács, Z. Horváth, L. O'Raifeartaigh and L. Palla for discussions.

References

1 Kaluza T 1916, Sitzunbgsber. Preus. Akad. Wiss. Phys. Math. K1, 996; Klein O 1926 Z. Phys. **37**, 895

2 Gross D J and Perry M J 1983 Nucl. Phys. **B226**, 29; Sorkin R 1983 Phys. Rev. Lett. **51**, 87

3 Brans C and Dicke R H 1961 Phys. Rev. **124**, 95

4 Misner C W 1963 Journ. Math. Phys. **4**, 924

5 Manton N S 1982 Phys. Lett. **110B**, 54; Atiyah M F and Hitchin N J 1985 Phys. Lett. **107A** , 21; Manton N S 1985 Phys. Lett. **154B**, 397 (E) **157B**, 475; Atiyah M F and Hitchin N J 1988 The Geometry and Dynamics of Magnetic Monopoles, Princeton University Press; Temple-Raston M 1988 Cambridge Preprint DAMTP-88/15;

6 Gibbons G W and Manton N S 1986 Nucl. Phys. **B274**,183; Fehér L Gy and Horváthy P A 1987 Phys. Lett. **182B**, 183;

7 DeWitt B 1957 Rev. Mod. Phys. **29**, 377

8 Carter B 1977 Phys. Rev. **D16**, 3395

9 Pauli W 1926 Z. Phys. **36**, 33

10 Zwanziger D 1968 Phys. Rev. **176**, 1480;

11 Alhassid Y, Gürsey F and Iachello F 1983 Ann. Phys. **148**, 896; 1986, *ibid* **167**, 181;

12 Barut A O and Bòrnzin G L 1971 Journ. Math. Phys. **12**, 841; D'Hoker E and Vinet L 1985a Phys. Rev. Let. **55**, 1043; 1985b Nucl. Phys. **B260**, 79

13 Gibbons G W and Ruback P 1987 Phys. Lett. **188B**, 226 Gibbons G W and Ruback P 1988 Commun. Math. Phys. **115**,

Cordani B, Fehér L Gy and Horváthy P A 1988 Phys. Lett. **201B**, 481
Fehér L Gy 1988 in Proc. *2nd Hungarian Relativity Workshop, Budapest' 87*, ed. Perjés, World Scientific (to be published)

14 Schönfeld J F 1980 Journ. Math. Phys. **21** 2528;
Fehér L Gy 1986 J. Phys. **A19**, 1259 ; 1984 Acta Phys. Pol. **B15**, 919; 1985 Acta Phys. Pol. **B16**, 217;
1987 in Proc. Siófok Conference on Non-Perturbative methods in Quantum Field Theory, ed. Z. Horváth, L. Palla and A. Patkós, World Scientific;
Fehér L Gy and Horváthy P A 1988 Mod. Phys. Lett. **A**

NONLINEAR OPTICS IN QUANTUM CONFINED SEMICONDUCTORS

Paul Horan and Werner Blau
Department of Pure and Applied Physics
Trinity College, Dublin 2, Ireland

ABSTRACT

The field of nonlinear optics is dominated by the search for materials which show a large nonlinear optical response. Physical confinement of charge carriers, whether in 1, 2 or 3 dimensions, can greatly enhance the nonlinear optical behaviour. We have studied the nonlinear optical properties of small particles, in the size range 100-10 Å which provide 3-D charge confinement. The larger particles are too big to show significant quantum confinement effects. Smaller particles show distinct confinement effects and significant optical nonlinearity.

1. INTRODUCTION

The interaction of light with matter is usually characterized by several phenomena, such as light absorption, refraction, scattering and luminscence. These are regarded as properties of the material, independent of light intensity. However, for sufficiently large light intensities, such as are available with lasers, these optical characteristics become a function of intensity. This is the field of nonlinear optics.[1]

The great activity in the area of nonlinear optics arises both from the interesting basic physics and the possible applications in the fields of optical computing and signal processing. Towards this end, materials with a large nonlinear response are very desirable i.e. materials which exhibit a large change in some characteristic optical property for a low light intensity. Most work has been done in semiconductors.[2] In the search for larger nonlinearities organic polymer materials show great promise.[3] Another interesting approach is the field of quantum confined semiconductor systems. When charge carriers are physically confined within a dimension comparable with the exciton radius r_{ex}, the energy structure is radically altered. This technique allows the engineering of novel optical materials.

Structures which produce confinement within one dimension i.e. where the electron or hole is free to move only within a plane but confined perpendicular to that plane, are known as quantum wells.[4] These have shown very interesting nonlinear properties. Confinement of the charge carriers in all three dimensions should show an enhanced nonlinear response.[5] Such structures are known as "quantum dots". The purpose of this paper is to investigate the nonlinear optical properties of such quantum dots. Towards this end we have studied several systems exhibiting different degrees of confinement. We observe an enhanced nonlinear response in the well confined samples.

2. THE QUANTUM DOT SYSTEM

By a quantum dot we refer only to particles retaining the bulk lattice structure. This imposes a lower size limit, since at some point rearrangement to a closer packed "cluster" structure will occur so as to minimize the number of broken bonds. As an upper limit, particles must be small enough to provide charge confinement, typically smaller than the bulk exciton radius. This defines a range of approximately 20-200 Å.

The electronic behaviour can be simply modelled by the "particle-in-a-box" problem. Consider an electron confined within a cubic volume of semiconductor of side length a. The wave function must be zero outside and go to zero at the boundaries. The simplest solutions are sine waves; i.e. only particular "standing waves" which satisfy the boundary conditions are selected from the conduction band continuum. The resulting energies are given by:

$$E_{n1m} = E_g + \frac{\hbar^2 n^2}{2m_e a^2}(n^2 + 1^2 + m^2) \qquad (1)$$

where E_g is the bulk semiconductor bandgap, m_e is the electron effective mass and $n, 1, m = 1, 2, 3, \ldots$. These are the only allowed energies, and the absorption spectrum should show a series of sharp absorption lines. It is because of this concentration of bulk oscillator strength into single spectral lines and small volumes that an enhanced nonlinear response is to be expected.

3. EXPERIMENTAL METHOD

Linear optical spectra were taken using a Philips SP8-200 UV/VIS double beam spectrophotometer. The nonlinear response was investigated by two methods. The variation of sample transmission with intensity was investigated using a nitrogen pumped dye laser system, PRA LN1000 high pressure nitrogen laser and PRA LN107 high resolution dye laser. This system gave 50-100 μJ pulses of ~500 ps duration, with a spectral bandwidth of 5 Å. The same laser system was also used for Laser Induced Grating (L.I.G) experiments.[6] In this technique two coherent beams are made to overlap at a small angle within the sample. The resulting interference pattern modulates the materials' absorption/refractive index. The pulses self-diffract from the created grating. The energy diffracted out depends on the ease with which the materials' properties may be altered. Measurement of the diffraction efficiency allows direct calculation of the effective third order susceptibility $\chi^{(3)}$, a figure of merit of the nonlinear response.[7]

4. CdS_xS_{1-x} DOPED GLASS

Initial studies were carried out with semiconductor doped glasses. These are commercially available as sharp cut-off high pass filters. They are formed by doping a boroscilicate glass with Cd, S and Se and annealing at high temperatures ($\sim 700°C$).

The microcrystallites grow by the process of Ostwald ripening to give particles of $\sim 100 \pm 30$ Å dimension with occasional bulk islands.[8] These are not small enough to provide strong confinement, the exciton radius in CdS is $\sim$ 7Å, and a sharp bulk-like band edge is observed in absorption. See Figure 1.

We carried out wavelength resolved L.I.G. experiments on a series of glasses for a different ratio x.[7] Typical results are illustrated in Figure 1, where we observed an effective $\chi^{(3)}$ proportional to absorption. A simple band-filling model has been proposed to account for this behaviour.[9] However, the effective $\chi^{(3)}$ is enhanced compared with the bulk due to surface effects and local field effects.

5. QUANTUM CONFINED CdSe DOPED GLASS

For semiconductor doped glasses the particle size is determined by the annealing process. The commercial glasses used above were obtimised to give a sharp absorption cut-off, rather than for particle size. By careful control of the annealing process a series of glasses were produced showing clear quantum confinement effects.[10] The absorption spectrum for one of these glasses (Corning 203 AME) is illustrated in Fig. 2a, showing the first and second transitions. This spectrum corresponds to a particle size of about 25 to 30 Å.

We studied the intensity dependence of the transmission for this sample,[11] the results are shown in Fig. 2b. Fitting with a simple two level saturation model (solid curve) does not give a good fit. The change in transmission is not as large as expected, suggesting a second mechanism is at work. If we allow for an excited state absorption a much better fit to data is obtained (dashed curve). Higher levels of the quantum confinement series or conduction states in the glass could provide the necessary absorbing states. L.I.G. measurements give a calculated effective $\chi^{(3)}$ of the same order of magnitude as the nonconfined samples of Sec. 4.

6. BiI_3 COLLOID

To further investigate well confined samples we prepared colloidal BiI_3 as described by Sandroff et al.[12] This colloidal technique exploits the layered honeycomb structure of BiI_3. The colloid forms a single layer sandwich of bismuth between iodine and seems to be particularly stable for the first complete rings of 12Å and 16Å illustrated in Fig. 3a. The observed Gaussian absorption peaks are attributed to the first transition for these two particle sizes. The width of these peaks is probably due to a spread in particle sizes.

We carried out nonlinear transmission and L.I.G. measurements at the absorption maximum of 362 nm. The variation of transmission and diffraction efficiency with intensity are shown in Fig. 3b. We can see the onset of absorption saturation. At the same intensities the cubic dependence of diffraction efficiency is also seen to drop off, as expected for saturation behaviour. A large effective $\chi^{(3)}$ is also observed, of the same magnitude as the quantum-confined glass.

7. CONCLUSIONS

Quantum confined systems show great promise for nonlinear optics. Proper understanding of these systems is currently limited by the quality of samples available.

Particles of homogeneous size, exhibiting clear quantum confinement are required. This will probably require invoking some form of self-regulating mechanism such as is observed with colloids.

ACKNOWLEDGEMENTS

We would like to thank D. W. Hall of Corning Glass Company who supplied the sample of quantum confined glass. Financial support for part of this work was provided by EOLAS, the Irish science and technology agency.

REFERENCES

1. Y. R. Shen, "The Principles of Nonlinear Optics" (Springer Verlag, Berlin, (1982).
2. P. Horan and W. Blau, Contemp. Phys. 28:59 (1987).
3. J. Zyss and D. S. Chemla, "Nonlinear Optic Properties of Organic Polymers", 2 Vols., Academic Press, (1987).
4. D. S. Chemla and D. A. B. Miller, J. Opt. Soc. Am. B2:1155 (1985).
5. S. Schmitt-Rink, D. A. B. Miller and D. S. Chemla, Phys. Rev. B. 35:813 (1987).
6. H. J. Eichler, "Laser Induced Dynamic Gratings", Springer Verlag, Berlin (1986).
7. P. Horan and W. B. Blau, Semicond. Sci. Technol. 2:382 (1987).
8. J. Warnock and D. D. Awschalom, Phys. Rev. B. 32:5529 (1985).
9. P. Roussignol, D. Ricard, J. Lukasik and C. Flytzanis, J. Opt. Soc. Am. B4:5 (1987).
10. D. W. Hall and N. F. Borelli. Submitted to J. Appl. Phys.
11. P. Horan and W. Blau. To be published.
12. C. J. Sandroff and D. M. H. Wang, J. Chem. Phys. 85:5337 (1986).

SYMMETRY AND SCALE: FROM LOCAL TO GLOBAL SYMMETRIES

Richard Kerner
L.P.T.P.E., Université Pierre et Marie Curie
Tour 16, 1er étage
4, Place Jussieu, 75005 Paris, France

ABSTRACT

The explanation of the properties of macroscopic bodies and large-scale phenomena by exploring the features of the microscopic constituents is one of the oldest paradigms of physical sciences. Among the features which are perhaps the most obvious and often the most intriguing are the observed symmetries of material objects.

The purpose of this article is to review some recent developments in solid state physics which throw a new light on the way in which the global symmetries can be – or can not be – deduced from the symmetry properties of the microscopic constituents. It has to be underlined that the first attempt in this direction is due to Johannes Kepler.

We discuss some simple models of covalent or metallic networks, and present a theory of nucleation and growth. The distinction is made between the ordinary crystals, the quasi-crystals represented by the Penrose tilings, and the amorphous networks.

Finally, we show how the two ingredients are necessary in order to foresee the global symmetry of the resulting macroscopic structure. The first is the symmetry of the elementary constituent itself, and the second is the symmetry of the interactions (forces) between the constituents.

1. INTRODUCTION: KEPLER'S IDEAS ON SYMMETRY BUILDING

In a beautiful booklet entitled "De nive sexangula", published in 1613, Johannes Kepler investigates the problem of the apparent hexagonal symmetry of snowflakes.

It is amazing how far Kepler could advance in his conclusions with so few assumptions at hand.

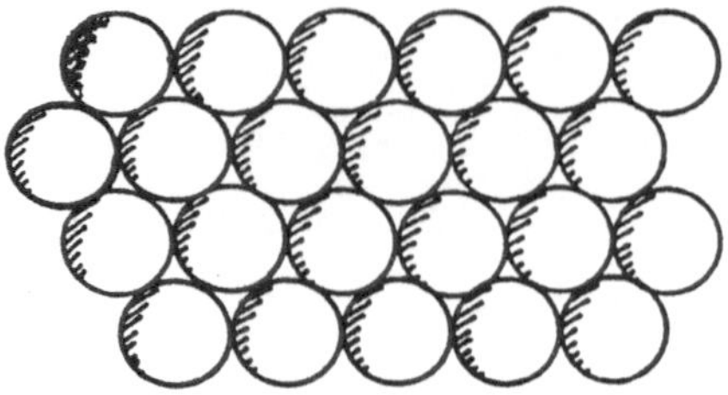

Fig. 1.1. Close packing of identical spheres on the plane

First of all, Kepler adopted the atomist point of view, trying to explain the macroscopic shapes of matter by the microscopic features of atoms and their possible arrangements in space.

As water was considered one of the elements at that time, it is no wonder that its elementary constituents ("atoms of water") were thought to have the form of perfect spheres. Then the problem arises, how the hexagonal symmetry would evolve from objects which themselves have a different (in this case spherical) symmetry.

Kepler's key argument is the following: "the shape of the resulting construction is dependent not only on the properties of the subject, but also on the properties of the agent".

It has to be remembered that the language Kepler and his scientific contemporaries used was not exactly the same that the physicists or mathematicians use nowadays.

As far as a reasonable translation can be done, the aforementioned sentence means that the macroscopic symmetry properties are not only a function of the microscopic constituent particles ("subject"), but also of the kind of forces ("agent") acting between them. In other words, no conclusion can be drawn without knowledge of interactions.

In order to explain the hexagonal shapes of the snowflakes, supposedly built out of identical spheres, Kepler has assumed the binary attractive interactions, leading to the closest packing possible. He shows then how the hypothesis that the particles of water attract each other with the strength whose intensity is independent of direction, leads naturally to the well-known six-coordinate triangular lattice on the plane (Fig. 1.1).

It had been discovered later that the molecules of water are built out of two hydrogen atoms and one oxygen atom. It has been taken for granted that the angle between the bonds H-O-H in the molecule of water is naturally equal to $2\pi/3 = 120°$.

The surprise was very big when it had been discovered that this angle was close to 104°, and that some extra assumptions concerning the intermolecular foces had to be made in order to explain the resulting hexagonal symmetry.

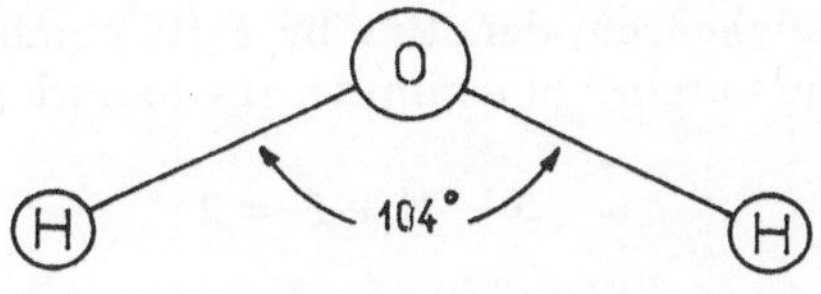

Fig. 1.2. The molecule of water, H_2O

The aim of this paper is to investigate the relationship between the simplest assumptions one can make about the elementary constituents and their interactions, and the symmetries of the resulting large-scale structures.

In doing so, we shall be led by Kepler's idea of combining the symmetry properties of the elementary building blocks with the symmetries of interactions between them.

We shall discuss first some global properties of networks and tilings, and then come to some simple models of structure building. The methods developed here will apply to metallic and covalent solids only, in which the interactions between closest neighbors are not polarised (i.e. are described by scalar and not vector variables; the water is not of that kind because of strong electric polarization of its molecules). One of the most important questions we shall also address will be the reasons for which in some circumstances, the same elementary building blocks prefer to assemble into networks of different symmetry, or without no symmetry whatsoever. We shall try to find some reasons for which quasi-crystalline lattices may be created instead of crystalline ones.

Most of the analysis will concern the simplest models possible, i.e. two-dimensional networks on the plane. The vertices can represent atoms, and the edges represent the bonds between them; also dual networks may be considered, in which the atoms are supposed to be placed inside the polygons of the network, the bonds between them being transverse to the edges shared by the polygons. The generalization to three dimensions, which is of real physical interest, is less obvious. The general covalent networks in three dimensions can not always be divided into polyhedra that one would like to see as the natural generalization of polygons in two dimensions. The tiling of three-dimensional euclidean space with polyhedra is the basis of crystallography. However, the polyhedra of crystallographic groups are built in the dual space rather than in the real space: the atoms are usually not placed only in their vertices, and the edges do not represent the bonds. In what follows, because we want to include the quasi-crystals (aperiodic lattices) and the amorphous networks, we shall adopt the real-space description, in which the atoms are placed in the vertices and the edges represent the real bonds.

2. EULER'S THEOREM: FINITE AND LARGE SCALE CONSEQUENCES

One of the simplest and the most beautiful results due to Leonhardt Euler is the theorem bearing his name. It can be stated as follows:

Given any convex polyhedron, denoting by F its number of faces, by E its number of edges and by S the number of summits, the following relation holds:

$$S - E + F = 2 \tag{2.1}$$

This is always referred to as "Euler's relation". Any convex polyhedron defines a partition of the two-dimensional sphere into spherical polygons; the number two on the right-hand side of (2.1) is called Euler's number or Euler's characteristic of a sphere.

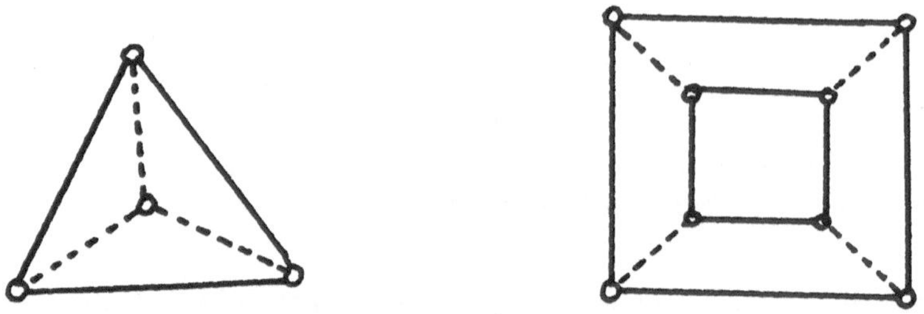

Fig. 2.1. The tetrahedron The cube
(central projection onto a plane)

The proof can be obtained by induction. First one observes that Euler's relation holds for some given convex polyhedron; next, one can prove that any convex polyhedron can be obtained by adding new summits, edges or faces to a simpler one; whatever one does, the right-hand side does not change.

We do not give an exhaustive proof here, but a sketch of the reasoning that leads to it. Any convex polyhedron can be represented on a plane, as projection, or as seen through one of its faces from a distance close enough to make all other faces appear inside this one which is closest to the eye. Fig. 2.1 displays such views of a tetrahedron and a cube.

In both cases, Euler's relation obviously holds.

We can produce more complicated polyhedra from simpler ones by one of the following elementary operations:

a) Dividing a face into two new ones, without changing S. Then one has to create an edge joining two summits, and $S \to S$, $F \to F+1$, $E \to E+1$. The Euler number

$$X = S - E + F = 2 \tag{2.2}$$

does not change.

b) Creating a new summit (vertex) without changing the number of faces. Then one of the edges is divided into new ones; and we get $F \to F$, $S \to S+1$, $E \to E+1$. Again, $X \to X$.

c) Creating a new vertex and a new edge: $S \to S+1$, $F \to F+1$, $E \to E+2$. This is just the combination of a) and b). Again, $X \to X$.

d) Slicing of a corner: if the vertex that is sliced is C-coordinate, it will create C new eges, one new face, $C-1$ new summits (because one has been sliced away). Again,

$$S \to S + C - 1, \quad F \to F + 1, \quad E \to E + C \quad \text{and} \quad X \to X.$$

Figure (2.2) a), b), c), d) displays the aforementioned situations:

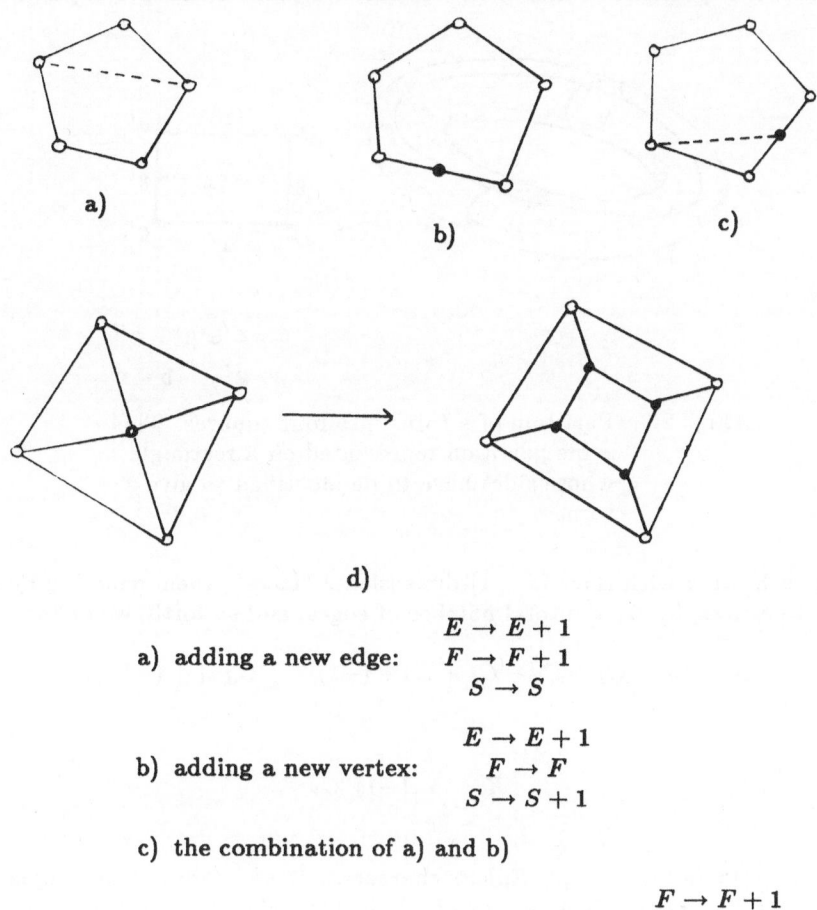

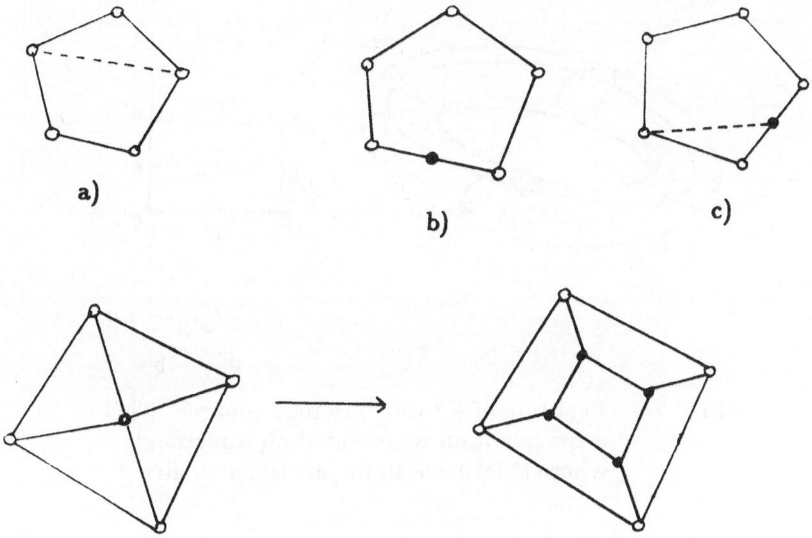

d)

a) adding a new edge: $E \to E+1$, $F \to F+1$, $S \to S$

b) adding a new vertex: $E \to E+1$, $F \to F$, $S \to S+1$

c) the combination of a) and b)

d) slicing away a four-coordinate vertex: $F \to F+1$, $E \to E+4$, $S \to S+3$

Fig. 2.2. Elementary operations on polyhedra

These elementary operations are all that is needed to create any convex polyhedron out of a simpler one.

The Euler characteristic of the sphere is equal to 2; should we analyze the partitions of the two-dimensional torus, the result would be different. It is enough to see what happens with any simple partition, e.g. the one represented on Fig. 2.3:

Here we have: $S = 4$, $E = 8$, $F = 4$, so that

$$X = 4 - 8 + 4 = 0 \tag{2.3}$$

The Euler characteristic of the 2-torus is equal to zero.

The notion of Euler characteristic can be extended to any compact d-dimensional manifold. If it is partitioned into some number of d-dimensional polytopes, which are

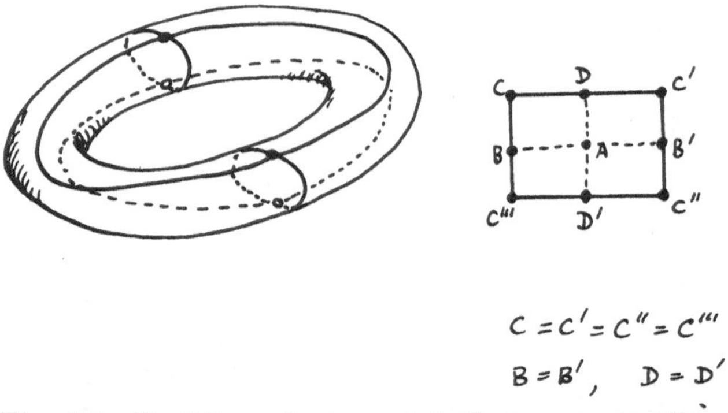

Fig. 2.3. Partition of a torus into four squares. The same partition represented on a rectangle whose sides have to be identified to give a torus

touching each other with their $(d-1)$-dimensional "faces", then, denoting by C_0 the number of vertices, by C_1 the total number of edges, and so forth, we define

$$X = C_0 - C_1 + C_2 - C_3 + \ldots + (-1)^{(d-1)} C_{d-1} + (-1)^d C_d \tag{2.4a}$$

or just

$$X = \sum_{k=0}^{d} (-1)^k C_k \tag{2.4b}$$

Let us evaluate for example Euler's characteristic of a 3-dimensional sphere S^3. It is sufficient to partition it by means of a four-dimensional generalization of the cube ("a tessaract"), whose border will provide us with the partition needed. Its projection onto 3 dimensions is shown on Fig. 2.4. It is easy to verify that $C_0 = 16$, $C_1 = 32$, $C_2 = 24$ and $C_3 = 8$, and

$$X(S^3) = 16 - 32 + 24 - 8 = 0 \tag{2.5}$$

The same result may be obtained with even simpler partition by a 3-dimensional simplex (Fig. 2.5).

We have now $C_0 = 5$, $C_1 = 10$, $C_2 = 10$ and $C_3 = 5$; again $X = 0$.

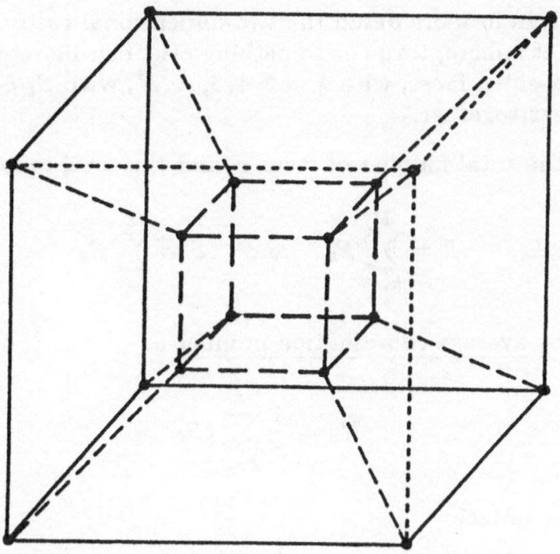

Fig. 2.4. A tessaract (four-dimensional analog of a cube) as as seen in three dimensions

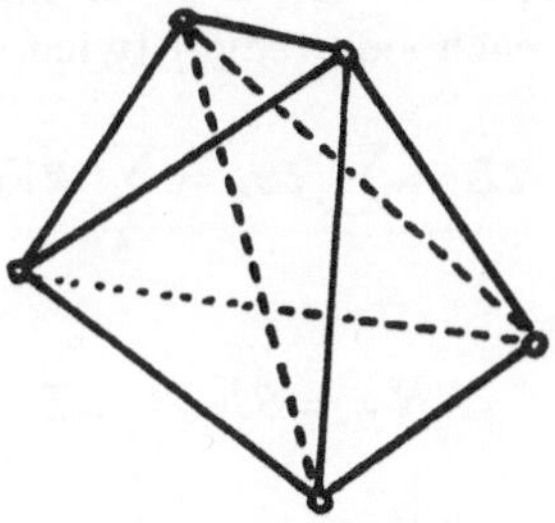

Fig. 2.5. A four-dimensional simplex as seen in three dimensions

Euler's characteristic depends only on the topological properties of the manifold. The Gauss-Bonnet theorem shows the relation that exists between the local properties of a two-dimensional compact manifold and its global topology:

$$\int_{M_2} R\sqrt{|g|}d^2x = 2\pi X(M_2) \tag{2.6}$$

The integral of the scalar curvature over the compact 2-dimensional manifold is proportional to its Euler's characteristic. That is why one can realize a flat metric on a 2-torus, but not on a 2-sphere, which has always intrinsic curvature, no matter how we squash it.

Now let us analyze in more detail the two-dimensional case. Any partition of such a manifold, be it sphere, torus or something else, can be represented by a polyhedron which has F_k k-sided faces, with $k = 3, 4, 5, \dots N$, with S_ℓ ℓ-coordinate summits, at which ℓ edges meet together.

Then we have the total number of faces F and the total number of summits S given by

$$F = \sum_{k=3}^{N} F_k \quad \text{and} \quad S = \sum_{\ell=3}^{L} S_\ell \tag{2.7}$$

Let us define the average coordination number as

$$\overline{N}_c = \frac{1}{S} \sum_{\ell=3}^{L} \ell S_\ell \tag{2.8}$$

and the average type of face

$$\overline{N}_f = \frac{1}{F} \sum_{k=3}^{N} k F_k \tag{2.9}$$

The number of edges E can be computed in two different ways: either we count ℓ times the ℓ-coordinate summits, $\sum \ell S_\ell$, and each edge is counted twice, because it belongs to two summits at once; or we can count k times the k-sided faces, $\sum kF_k$, and once again, we shall count each side (edge) twice. Therefore

$$2E = \sum_{\ell} \ell S_\ell = \sum_{k} k F_k \tag{2.10}$$

so that we can write

$$F \overline{N}_f = S \overline{N}_c = 2E \tag{2.11}$$

Let us now re-write Euler's relation expressing everything by means of one variable, say F, only:

$$S = \frac{\overline{N}_f}{\overline{N}_c} F \quad \text{and} \quad E = \frac{1}{2} \overline{N}_f F \tag{2.12}$$

so that Euler's relation becomes

$$S - E + F = \left[\frac{\overline{N}_f}{\overline{N}_c} - \frac{1}{2} \overline{N}_f + 1 \right] F = X \tag{2.13}$$

A more symmetric formula is obtained by dividing both sides by $\overline{N}_f F$:

$$\frac{1}{\overline{N}_c} + \frac{1}{\overline{N}_f} = \frac{1}{2} + \frac{X}{\overline{N}_f F} \tag{2.14}$$

or

$$\frac{1}{\overline{N}_f} + \frac{1}{\overline{N}_c} = \frac{1}{2} + \frac{X}{\overline{N}_c S} \tag{2.15}$$

The two last versions of the same formula show clearly the existence of duality between the summits and the faces: if one finds a solution of (2.14), one has at the same time the solution of (2.15) by interchanging the corresponding numbers F_k and S_ℓ, $\overline{N}_f$ and $\overline{N}_c$, and F with S.

The immediate corollary is the existence of five perfect (Platonic) polyhedra, known already in the Antiquity. They are solutions of (2.14) or (2.15) with given N_f and N_c (without averaging), all the summits and all the faces supposed to be identical. For $X = 2$ (a 2-sphere) we can have in such a case

dual	$N_c = 3$	$N_f = 4$	$F = 6$	a cube
	$N_c = 4$	$N_4 = 3$	$F = 8$	octahedron
dual	$N_c = 3$	$N_f = 5$	$F = 12$	dodecahedron
	$N_c = 5$	$N_f = 3$	$F = 20$	icosahedron
	$N_c = 3$	$N_f = 3$	$F = 4$	tetrahedron (dual to itself)

By admitting more than one kind of faces and more than one kind of summits, one can generate an infinite number of polyhedra.

There exist only three perfect polyhedra on a torus, but their number of faces is arbitrary:

dual	$N_c = 3$	$N_f = 6$	tri-coordinate hexagonal lattice
	$N_c = 6$	$N_4 = 3$	six-coordinate triangular lattice
	$N_c = 4$	$N_f = 4$	four-coordinate square lattice (dual to itself)

Note that what is true for a torus is also true for the infinite plane $\mathbb{R}^2$.

It is also interesting to investigate some less familiar situations than a two-sphere or a two-torus. If we take a genus $= 2$ surface with two holes (Fig. 2.6), whose Euler's characteristic is $X = -2$.

It is quite easy to see how this surface can be partitioned into eight pentagons, with ten summits at each of which four pentagons meet, so that the number of edges is 20:

$$S = 10, \quad E = 20, \quad F = 8, \quad N_c = 4, \quad N_f = 5$$

and

$$S - E + F = 10 - 20 + 8 = -2 \tag{2.16}$$

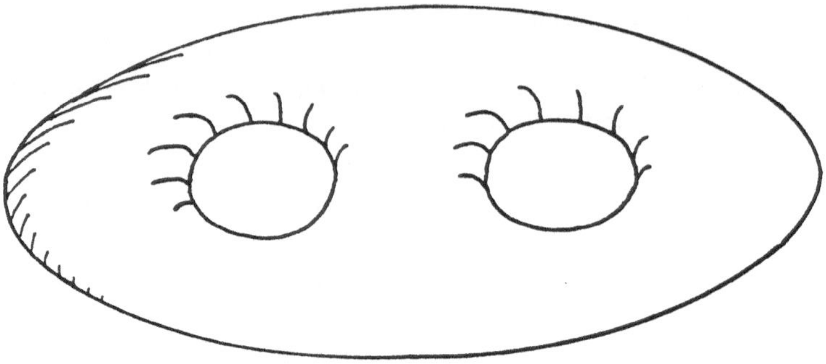

Fig. 2.6. Genus = 2, $X = -2$ surface

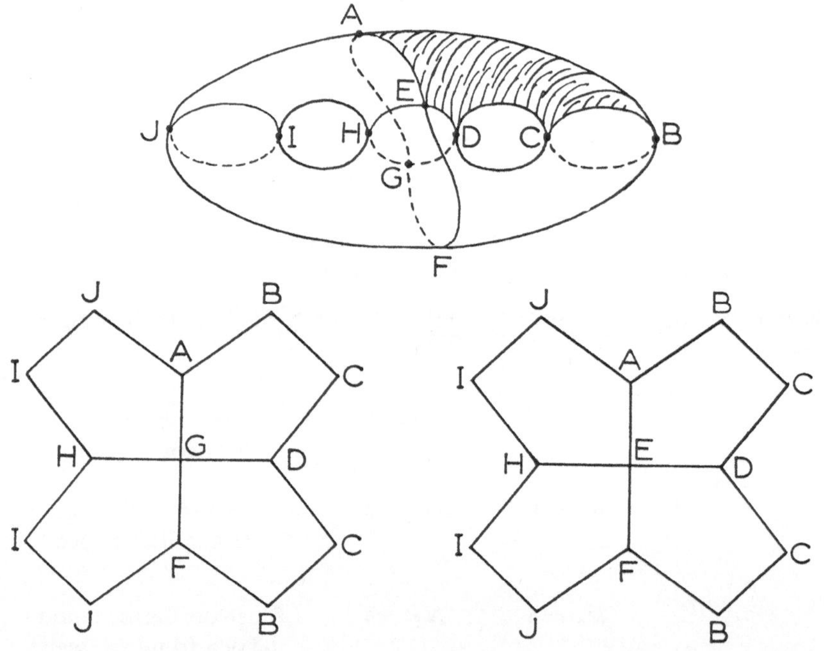

Fig. 2.7. A surface of genus = 2 partitioned into a hyperbolic octahedron. One of the eight pentagons is shaded. All the edges are visualized. If embedded on a surface of negative curvature, all the pentagons can be made equilateral and equiangular

(see Fig. 2.7).

Hyperbolic octahedron: $F = 8$, $N_f = 5$, $N_c = 4$. One of the pentagons is shaded. All the edges are visualised. If embedded on a surface of negative curvature, all the pentagons can be made equilateral and equiangular.

The negative Euler characteristic indicates via the Gauss-Bonnet theorem that the mean curvature of this space has to be negative.

there are much more "perfect" polyhedra in the space of negative curvature and $X = -2$. The dual to the aforementioned octahedron is a decahedron

$$S = 8, \quad E = 20, \quad F = 10; \quad N_c = 5, \quad N_f = 4 \tag{2.17}$$

Another mutually dual couple is:

$$S = 28, \quad E = 42, \quad F = 12, \quad N_c = 3, \quad N_f = 7 \tag{2.18}$$

a dodecahedron composed of twelve three-coordinate heptagons, and

$$S = 12, E = 42, F = 28\, N_f = 7, N_c = 3 \tag{2.19}$$

twenty-eight seven-coordinate triangles.

There exist also other 'platonic" polyhedra in the space of $X = -2$.

For very large networks, when one can assume $F \to \infty$, the result is the same as for the torus, i.e. flat space:

$$\frac{1}{N_c} + \frac{1}{N_f} = \frac{1}{2} \tag{2.20}$$

Any infinite tiling of the plane with arbitrary convex polygons and various coordination numbers has to respect Euler's constraint (2.20) on the average, but not locally, of course.

3. CREATING NETWORKS WITH GIVEN LOCAL SYMMETRY

When a network of atoms is about to be created, either during the solidification from liquid state, or crystallization or amorphous growth from a solution, there is usually a supplmentary constraint given by a fixed valence (coordination number). Let us investigate first a simplest model of structure building in two dimensions.

Suppose that the atoms are placed at the vertices of the network, that the bonds between the closest neighbors are all of the same length – which means that the resulting polygons will be equilateral – and that the coordination number is fixed, and equal to three everywhere.

Let us try to build a lattice starting with one vertex, joining three closest neighbors to it, then building three adjacent convex polygons, and so forth. If the three polygons have k_1, k_2 and k_3 faces, and if we require them to be perfect, i.e. not only equilateral, but also equiangular, then we must satisfy the following relation:

$$\frac{(k_1 - 2)\pi}{k_1} + \frac{(k_2 - 2)\pi}{k_2} + \frac{(k_3 - 2)\pi}{k_3} = 2\pi \tag{3.1}$$

The sum of the three angles around any vertex is equal to 2π (the assumption being that our space is flat $\mathbb{R}^2$ and the curvature is vanishing everywhere!)

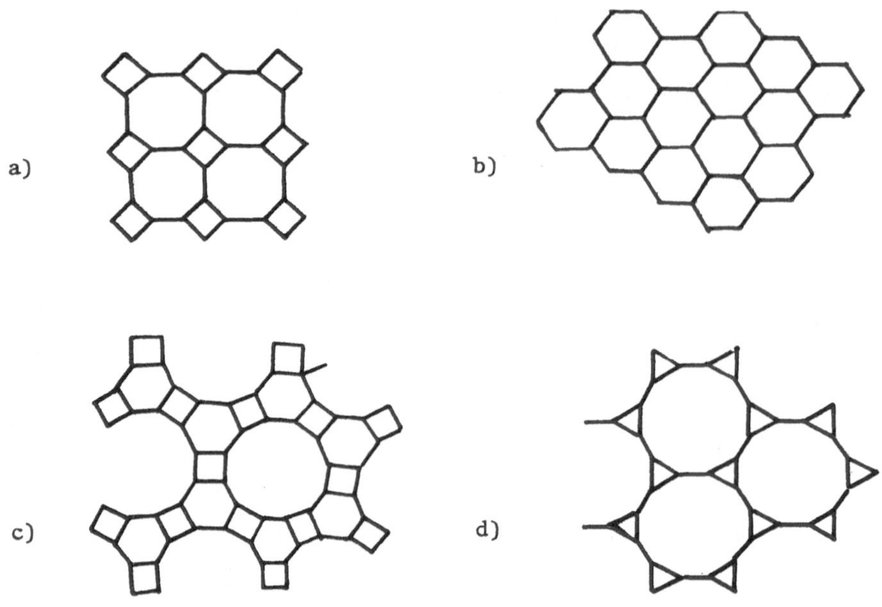

Fig. 3.1. The four perfect homogeneous 3-coordinate tilings of $\mathbf{R}^2$.
a) (6,6,6) b) (4,8,8) c) (3,12,12) d) (4,6,12)

If our lattice is suppose to be homogeneous, i.e. all the vertices are identical, then only four solutions of (3.1) exist: the triplets (k_1, k_2, k_3) are the following:

$$(6,6,6); \quad (4,8,8); \quad (3,12,12) \quad \textit{and} \quad (4,6,12) \tag{3.2}$$

The corresponding perfect homogeneous tri-coordinate tilings of the plane are displayed on the Fig. 3.1. By fixing $N_c = 3$, we obtain from Euler's relation $\overline{N}_f = 6$. It is quite easy to verify that all the four tilings (3.2) satisfy this relation. As the lattices displayed have translational symmetry, it is enough to take out an elementary subset that is repeated ad infinitum by translations, covering the whole plane. The relative proportion of each kind of polygons in such an elementary crystallographic cell is the same in the infinite network.

If our network is four-coordinate, homogeneous and perfect (all the polygons equiangular), then there are three solutions of the corresponding equation

$$\sum_{i=1}^{4} \frac{(k_i - 2)\pi}{k_i} = 2\pi \tag{3.2}$$

These are (6,3,6,3) or (6,6,3,3), (4,4,4,4) and (3,4,6,4). The corresponding tilings are displayed on Fig. 3.2.

It is also interesting to note that the overall crystallographic symmetry is independent of the coordination number: for example, the tilings (6,6,6) or (4,6,12) have a rhombohedral symmetry, while the tiling (4,8,8) has a cubic symmetry.

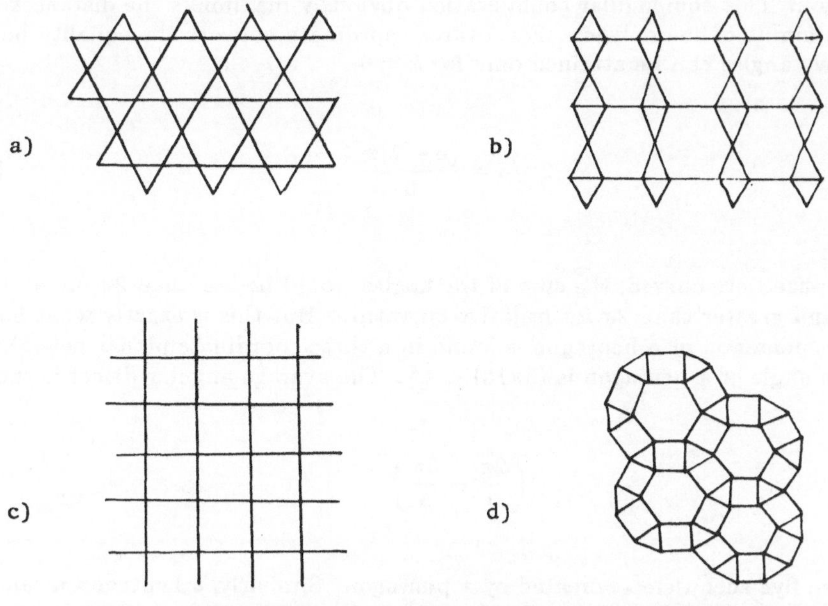

Fig. 3.2. The three perfect homogeneous 4-coordinate tilings of $\mathbf{R}^2$. The tiling b) is not really homogeneous: there are two distinct types of vertices a) (6,3,6,3) b) (6,6,3,3) c) (4,4,4,4) d) (3,4,6,4)

Let us now make some simple (but quite realistic) assumptions concerning the nature of the interactions between the atoms placed in the vertices. The simplest idea is that once the bonds between the N_c closest neighbors are formed, and the network starts to grow, the atoms caught in the network try to get as far from each other as possible. This hypothesis is realistic for the so-called covalent substances, like Si, SiO_2, B_2O_3, and many others. The distances between the atoms are in these networks at least as big as the average diameters of the atoms. The metals behave differently, which leads them to closest packing.

Now, the statement "atoms try to get as far as possible from each other", i.e. the assumption that the residual forces are repulsive and central, yields different results depending on what kind of subset of the network we take under consideration. If a vertex (i.e. just any atom with N_c closest neighbors linked to it via N_c bonds) could be cut out of the network for a while, the bonds would put themselves in a position in which the N_c angles between them would be equal to $2\pi/N_c$. Then, in a 3-coordinate network, if the elementary tripods thus defined were very stiff, only haxagonal lattice would be possible.

If a polygon could be cut out of the network, a similar reasoning would tell us that all its angles would become equal to $(k-2)\pi/k$, where k is the number of sides of the polygon. This equiangular configuration obviously maximizes the distances between the summits of the polygon. For a three-coordinate network the equality between the two angles can be attained only for $k = 6$:

$$2\pi/3 = \frac{(6-2)\pi}{6} \tag{3.3}$$

If the space were curved, the sum of the angles would be less than 2π for positive curvature, and greater than 2π for negative curvature. But this is exactly what happens when a pentagon or a heptagon is found in a three-coordinate planar network. The average angle of a pentagon is $(3\pi/5) < \frac{2\pi}{3}$. The average angular defect is then

$$\left(\frac{2\pi}{3} - \frac{3\pi}{5}\right)$$

and there are five such defects created by a pentagon. Similarly, a heptagon mean angle has an excess of $\left(\frac{5\pi}{7} - \frac{2\pi}{3}\right)$, and there are seven such positive defects. It is easy to verify that the total effects of these defects or excesses cancel each other:

$$5 \times \left(\frac{2\pi}{3} - \frac{3\pi}{5}\right) + 7 \times \left(\frac{2\pi}{3} - \frac{5\pi}{7}\right) = 0 \tag{3.4}$$

That is why on the surfaces of Euler's characteristic +2 and −2 there were two dodecahedrons, one made of pentagons, another made of heptagons: the corresponding defects in angles being the same, the number of these polygons was also equal in both cases. By joining a sphere ($X = 2$) with a "pretzel" ($X = -2$) one can produce a torus, which is flat ($X = 0$).

The effects of creating local curvature must have an impact on the potential energy stored in parts of the network. The simplest approximation consists in assuming that when in a perfect tri-coordinate hexagonal lattice a pentagon is put instead of a hexagon, it costs an energy ΔE; the same will be true for a heptagon creation, but if we put a pentagon and a heptagon together, the elastic strain is falling almost to zero because the corresponding defects cancel each other. This energy measure, which is proportional to the departure from flatness in a network of given coordination number, can be expressed by the following association matrix: the energy cost of assembling together of two polygons whose numbers of sides are i, j is propoortional, in a network with the coordination number N_c, to the quantity

$$\Delta E_{ij} = \Delta E \times \left(\frac{4N_c}{N_c - 2} - i - j\right) \tag{3.5}$$

(here $\frac{4N_c}{N_c-2} = 2\overline{N}_f$ according to Euler's formula for the flat space $X = 0$. In the case when one considers a three-coordinate network on the plane, $\overline{N}_f = 6$, and ΔE_{ij} can be represented as follows:

$j\backslash i$	4	5	6	7	8
4	4	3	2	1	0
5	3	2	1	0	1
6	2	1	0	1	2
7	1	0	1	2	3
8	0	1	2	3	4

in units ΔE

ΔE is some common measure of energy, which has to be derived from the physico-chemical properties of atoms under consideration. In known examples (SiO_2 etc.) ΔE is of the order 0.1 eV.

Let us also introduce the variables that will describe the networks. We shall speak of the proportion of i-sided polygons, or the probability of finding an i-sided polygon in the network, and shall denote it by P_i. If the polygons that can be encountered are $3, 4, \ldots$ up to N-sided, then

$$\sum_{i=3}^{N} P_i = 1 \tag{3.6}$$

so that we are left with $N-1$ independent variables, neither of which can be greater than 1, and all taking positive values only.

In an infinite network realized on a plane, for a given N_c, there is another constraint coming from Euler's equation:

$$\sum_{k=3}^{N} kP_k = \overline{N}_f = \frac{2N_c}{N_c - 2} \tag{3.7}$$

One can easily check this relation on the networks on Fig. 3.1: for example, in (4,8,8) we have $P_4 = P_8 = \frac{1}{2}$ and $4P_4 + 8P_8 = 4 \cdot \frac{1}{2} + 8 \cdot \frac{1}{2} = 6$, whereas for the tiling (4,6,12) we have $P_4 = \frac{1}{2}$, $P_6 = \frac{1}{3}$ and $P_{12} = \frac{1}{6}$.

Although the variables P_k do not give full information about the network, we shall use them in order to have some more insight into possible ways of growth of these structures. If we call P_{ij} the probability of finding a couple of two adjacent polygons (i,j), these numbers give much more information about the network, and so forth.

Let us try now to analyze a simple model of growth of the network on a plane, constructed out of 5, 6 and 7-sided polygons only, and 3-coordinate.

4. STATISTICAL MODEL OF GROWTH

Having established a simplified relation between the local symmetry imposed by the coordination number, the curvature effects coming from polygons whose number of sides deviate from $\overline{N}_f$ required by Euler's condition, and the energy, we can now describe – although in a very simplistic manner – the process of nucleation and growth, and investigate the properties of resulting networks.

In the simplest model the atoms (vertices) are 3-coordinated, the lowest energy polygon is a hexagon, and we shall suppose that at given physical conditions some number of defects, i.e. 5 and 7-sided polygons, is created at constant rate. The relative frequencies of getting 5, 6 or 7-sided polygons will be denoted by

$$\overset{0}{P}_5 + \overset{0}{P}_6 + \overset{0}{P}_7 = 1 \tag{4.1}$$

Only two variables are independent; let us choose $\overset{0}{P}_5$ and $\overset{0}{P}_7$, then $\overset{0}{P}_6 = 1 - \overset{0}{P}_5 - \overset{0}{P}_7$.

As the density of polygons becomes greater, at some point doublets will be observed, i.e. polygons which grow up on another polygon already in place. According to our analysis, the association of two hexagons or of a pentagon with a heptagon do not change local curvature, therefore the energy cost can be neglected. In other cases some energy has to be paid, and the correspondent probabilities should contain the appropriate Boltzmann factor. We shall write then:

$$\overset{(1)}{P}_{ij} = \frac{1}{Q}(2 - \delta_{ij})\overset{0}{P}_i\overset{0}{P}_j e^{-\Delta E_{ij}/kT} \tag{4.2}$$

with

$$\Delta E_{ij} = \Delta E(12 - i - j) \tag{4.3}$$

$$Q = \sum_{i,j}(2 - \delta_{ij})\overset{0}{P}_i\overset{0}{P}_j e^{-\Delta E_{ij}/kT} \tag{4.4}$$

is the normalizing factor, so that

$$\sum_{i,j} \overset{(1)}{P}_{ij} = 1 \tag{4.5}$$

There are six different doublets, (55), (56), ... (77), whose probabilities are denoted by $\overset{(1)}{P}_{ij}$.

Now we can ask ourselves, what is the probability of observing a k-sided polygon among all the doublets? This can be readily evaluated as

$$\overset{1}{P}_k = \frac{1}{2}\left(2\overset{(1)}{P}_{kk} + \sum_{j\neq k} P_{kj}\right) \tag{4.6}$$

This is because there are two k-sided polygons in the (k)-doublet, and only one in (kj)-doublet. As the right-hand side can be expressed as function of $\overset{0}{P}_5$, $\overset{0}{P}_7$ only, we obtain:

$$\overset{1}{P}_t = \frac{1}{Q}\left[\overset{0}{P}_5{}^2 e^{-2\alpha} + \overset{0}{P}_5 \overset{0}{P}_7 + \overset{0}{P}_5(1 - \overset{0}{P}_5 - \overset{0}{P}_7)e^{-\alpha}\right] \tag{4.7}$$

The difference between $\overset{1}{P}_k$ and $\overset{0}{P}_k$ can be interpreted as the derivative with respect to some parameter s (measuring the number os steps in agglomeration of polygons):

$$\frac{dP_k}{ds} = \overset{1}{P}_k - \overset{0}{P}_k \tag{4.8}$$

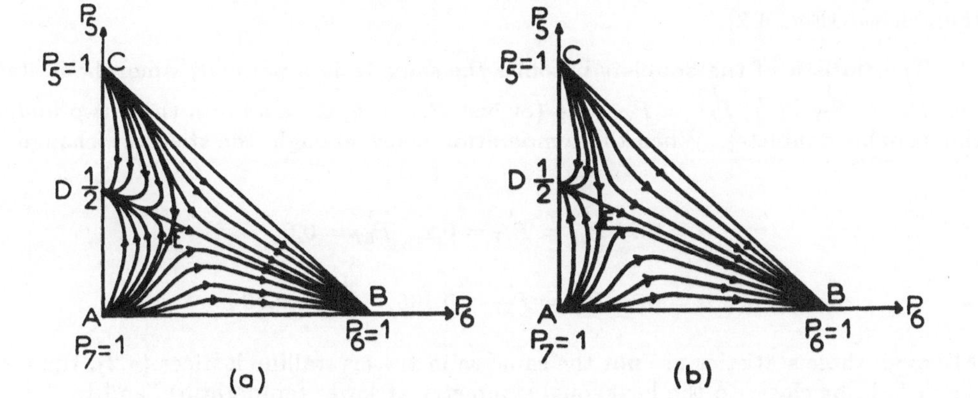

Fig. 4.1. The trajectories of the differential system (4.8). a) $\Delta E/kT = 2$ b) $\Delta E/kT = 0.3$

Replacing $\overset{0}{P}_k$ by a common variable P_k can be done without danger if we are in vicinity of the singular points of the non-linear system thus obtained, where $\frac{dP_k}{ds} \to 0$. Here are the two non-linear equations; expressed in the variables P_5 and P_6:

$$\begin{aligned}\frac{DP_5}{ds} = \frac{P_5}{Q}\big[&(1 - e^{-2\alpha}) + 3P_5(e^{-2\alpha} - 1) - P_6(1 - e^{-\alpha})^2 + \\ &+ 2P_5P_6(1 - e^{-2\alpha}) + 2P_5^2(1 - e^{-2\alpha}) - P_6^2(1 - e^{-\alpha})^2\big]\end{aligned} \tag{4.8}$$

where $\alpha = (\Delta E/kT)$. $\frac{dP_7}{ds}$ can be obtained from (4.8) just by replacing P_7 by P_5 and vice versa, because of the supposed symmetry of the enerrgy table.

The singular points (solutions of the systems in the form of constants) are the following:

$$\begin{aligned} &P_6 = 1; \quad P_5 = P_7 = \tfrac{1}{2}; \quad P_5 = 1; \quad P_7 = 1 \\ \text{and} \quad &P_5 = P_7 = (3 - e^{-\alpha})^{-1}, \quad P_6 = (1 - e^{-\alpha})/(3 - e^{-\alpha}) \end{aligned} \tag{4.9}$$

By linearizing our system in the neighborhood of these points, we can see if the corresponding points are stable (negative eigenvalues) or unstable (positive or positive and negative eigenvalues). The trajectories of our system look like shown on Fig. 4.1 a) b), with different values of $\Delta E/kT$:

We display the trajectories on the plane of barotropic coordinates, $P_5 + P_6 + P_7 = 1$.

All the singular points except the central saddle point do not depend on the temperature.

There are two stable points, to which the trajectories converge: $\dot{P}_6 = 1$, representing the pure hexagonal lattice, and $P_5 = P_7 = \frac{1}{2}$, representing the mixture of heptagons and pentagons. The saddle point which respects Euler's constraint $P_5 = P_7$, moves with the temperature. It approaches the stable point $P_5 = P_7 = \frac{1}{2}$ when the temperature grows up ($\alpha \to 0$). The repartition between the different doublets varies with temperature, too. With the same overall statistics $P_5 = P_7 = \frac{1}{2}$, one can realize different tilings; some of them have even crystallographic symmetry, if the pentagons and hexagons are specially deformed from their equiangular shape in order to adapt themselves. (Fig. 4.2)

The statistic of the doublets becomes the same as in a perfectly amorphous state, i.e. $\overset{1}{P}_{55} = \overset{1}{P}_{77} = \frac{1}{4}$, $\overset{1}{P}_{57} = \overset{1}{P}_{75} = \frac{1}{4}$ (or just $\overset{1}{P}_{57} = \frac{1}{2}$, if we add up these two undistinguishable doublets). When the temperature is low enough, the statistics change, e.g.

$$\alpha = 0.2 \quad \overset{1}{P}_{55} = \overset{1}{P}_{77} = 0.2, \quad \overset{1}{P}_{57} = 0.6$$

$$\alpha = 0.5 \quad \overset{1}{P}_{55} = \overset{1}{P}_{77} = 0.135, \quad \overset{1}{P}_{57} = 0.73$$

Although these statistics are not the same as in the crystalline lattices (4.2), the tendency is to be closer to the hexagonal symmetry at lower temperature, and to the rhombohedric at higher one.

In order to investigate the growth process closer, one has to include the next step, when the triplets are formed: (Fig. 4.3)

If we have a doublet of polygons (i, j), a third one of type k may grow on it in two ways: as a chain, with the statistical factor $\left(\frac{i+j-6}{i+j-4}\right)$, and as a cell, with three polygons meeting at one vertex, with the statiscal factor $\left(\frac{2}{i+j-4}\right)$. These factors give us the feeling of what kind of configurations give rise to more entropy. Then, the new probabilities of triplets, $\overset{2}{P}_{ijk}$, may be computed from the agglomeration processes which lead to them, i.e. $(iu) + (k)$, or $(ki) + (j)$, or $(jk) + (i)$ if all are different, $(ii) + (k)$ or $(ik) + (i)$ if only two are different, etc. For the chains, which amount to creation of new doublets, the new probability shall be computed as proportional to

$$\overset{1}{P}_{ij}\overset{1}{P}_k = \overset{1}{P}_{ij}\left(\overset{0}{P}_k + \frac{dP_k}{ds}\right) \tag{4.10}$$

For the cells, which bring in a new interaction, we put

$$\overset{1}{P}_{ij}\overset{0}{P}_k e^{c_{ij,k}\Delta E/kT} \tag{4.11}$$

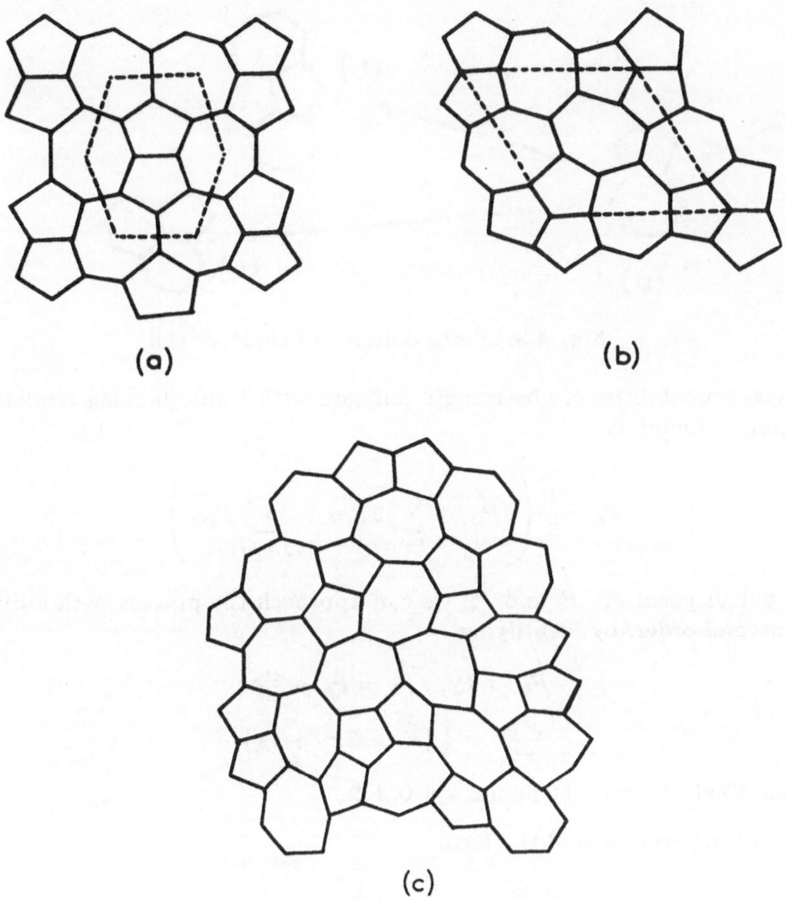

Fig. 4.2. Different tilings with 5- and 7-polygons For all the three, $P_5 = P_2 = 1/2$, but a), b) are crystalline (with different types of symmetry), whereas c) is amorphous.

where $C_{ij,k}$ is equal to 0 or ± 1 depending on the resulting "curvature" of the triplet. The computations can be found elsewhere ( ), here we give just two examples out of ten:

$$\overset{2}{P}_{555} = \frac{1}{Z}\left[\frac{2}{3}\overset{1}{P}_{55}\left(\overset{0}{P}_5 + \frac{dP_5}{ds}\right) + \frac{1}{3Q}\overset{0}{P}_5{}^3 e^{-3\alpha}\right] \tag{4.12}$$

$$\overset{2}{P}_{556} = \frac{1}{Z}\left[\frac{2}{3}\overset{1}{P}_{55}\overset{1}{P}_6 + \frac{5}{7} + \overset{1}{P}_{56}\overset{1}{P}_5 + \frac{19}{21Q}\overset{0}{P}_5{}^2\overset{0}{P}_6 e^{-2\alpha}\right] \tag{4.13}$$

and so on. Z is an overall normalizing factor which assures that $\Sigma\overset{2}{P}_{ijk} = 2$.

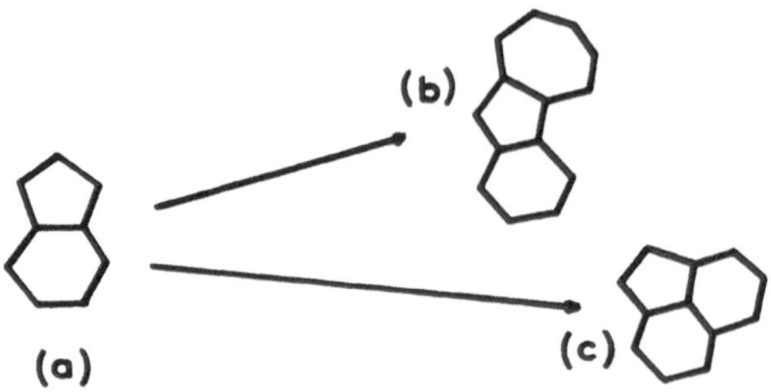

Fig. 4.3. From doublet to chain or cell

The new probabilities of observing a polygon with k-sides among triplets of adjacent polygons is found as

$$\overset{2}{P}_i = \frac{1}{3}\left(3\overset{2}{P}_{iii} + \sum_{k\neq i} 2\overset{2}{P}_{iik} + \sum_{j,k\neq i} \overset{2}{P}_{ijk}\right) \tag{4.14}$$

Having at our disposal $\overset{2}{P}_i$, $\overset{1}{P}_i$ and $\overset{0}{P}_i$, we can approach the process with a differential system of second order, by identifying

$$\begin{aligned} &\overset{0}{P}_k \to P_k, \quad \overset{1}{P}_k = P_k + \tfrac{dP_k}{ds}, \\ &\tfrac{d^2 P_k}{ds^2} = \tfrac{1}{2}(\overset{2}{P}_k - 2\overset{1}{P}_k + P_k) \end{aligned} \tag{4.15}$$

like a formal Taylor's series at points $s = 0, 1, 2$.

The resulting system is of the form

$$\frac{d^2 P_k}{ds^2} = D(P_i)\frac{dP_k}{ds} + W_k(P_i) \tag{4.16}$$

Its singular solutions are similar to the singular points of the first-order system, i.e. we have still two repulsive points $P_5 = 1$, $P_7 = 1$, two attractive ones: $P_6 = 1$ *or* $P_5 = P_7 = \frac{1}{2}$, and one saddle point in the middle:

Only the isoclines are shown on Fig. (4.4). The saddle point corresponding to a meta-stable amorphous configuration disappears at finite temperature, which is to be identified with critical temperature; at the same moment, the point $P_5 = P_7 = \frac{1}{2}$ becomes unstable.

Also the regime of growth changes with the temperature: as usual in second-order systems, we may obtain real or complex eigenvalues when we linearize the system around its singular point. It happens here that at the stable point $P_6 = 1$, corresponding to a pure hexagonal configuration, has all real negative eigenvalues at $\alpha < 0.72$, and two complex conjugate eigenvalues with negative real part at $\alpha > 0.72$. The corresponding oscillating behaviour will result in a layer structure of the resulting network, i.e. a layer with an excess of pentagons, followed by a layer with the excess of heptagons, and so on.

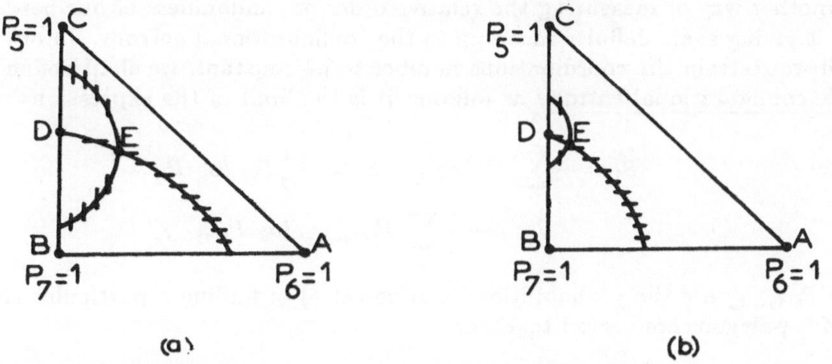

Fig. 4.4. Isoclines and singular points of the system (4.16). a) $\Delta E/kT = 2$ b) $\Delta E/kT = 1$

5. DIGRESSION ON CONFIGURATIONAL ENTROPY AND LOCAL ORDER

In the previous section we have discussed a model of growth which, under different circumstances (i.e. temperature, the energy of defects and the initial conditions on the probability of formation of elementary constituents such as rings of atoms in our example), can lead to the formation of a perfect crystalline lattice, or to the formation of amorphous networks as well.

The work "amorphous" is much more ambiguous than the work "crystalline". As we have seen in the previous example, the pure configuration $P_6 = 1$ determines uniquely the tri-coordinate hexagonal lattice; in contrast to it, the stable configuration $P_5 = P_7 = \frac{1}{2}$ can correspond to many different networks (Fig. 4.2). In terms of our model, the natural way of distinguishing between these networks is to compare the corresponding probabilities of the doublets of polygons sharing an edge, $\overset{1}{P}_{ij}$, then the probabilities of chains and cells, $\overset{1}{P}_{ijk}$, etc. For example, the corresponding doublet statistics of the configurations presented on Fig. 4.2 are

$$\begin{aligned} Fig.4.2a) &: P_{55} = \frac{1}{12}, \; P_{57} = \frac{2}{3}, \; P_{77} = \frac{1}{4} \\ Fig.4.2b) &: P_{55} = \frac{1}{6}, \; P_{57} = \frac{1}{2}, \; P_{77} = \frac{1}{3} \\ Fig.4.2c) &: P_{55} \approx 0.1, \; P_{57} \approx 0.64, \; P_{77} \approx 0.26 \end{aligned} \tag{5.1}$$

whereas the singlet's statistic is the same for the three configurations, i.e. $P_6 = 0$, $P_5 = P_7 = \frac{1}{2}$.

It is quite obvious that when we investigate in this manner a crystalline network, which has a well-defined elementary cell whose translations reproduce the whole lattice, then everything that can be said about this configuration will be contained in the statistics of the doublets, triplets, etc. <u>contained</u> in the elementary cell. On the contrary, for an amorphous network these statistics oscillate around some average values, but converge only when we consider the infinitely big agglomerates.

Another way of measuring the relative order or randomness of our networks is giving some definite meaning to the configurational entropy. In our case, if we restrain the coordinations number to be constant, we shall define the specific configurational entropy as follows: it is the limit of the expressions

$$S_1 = -\sum P_i \, log \, P_i, \;\; S_2 = -\frac{1}{2} P_{ij} \, log \, P_{ij}, \qquad (5.2)$$
$$S_n = -\frac{1}{n} \sum P_{i_1 i_2 \ldots i_n} \, log \, P_{i_1 i_2 \ldots i_n}$$

where $P_{i_1 i_2 \ldots i_n}$ are the probabilities (relative rates) of finding a particular configuration of n polygons connected together.

Let us give some examples of computation of this expression. Consider one of the crystalline hoomogeneous 3-coordinate lattices (4,8,8), displayed oon Fig. 3.1 (b). It is obvious that the number of squares is the same as the number of octagons; therefore $P_4 = \frac{1}{2}$, $P_8 = \frac{1}{2}$; so that

$$S_1 = -\frac{1}{2} \, log \, \frac{1}{2} - \frac{1}{2} \, log \, \frac{1}{2} = log \, 2 \cong 0.69315 \qquad (5.3)$$

If there were no correlation whatsoever, then the probabilities of finding (ij)-doublets of polygons sharing one edge would be just products of P_i, P_j; the same is true for any kind of multiplets. Then, if we put $P_{i_1 i_2 \ldots i_n} = P_{i_1} P_{i_2} \cdots P_{i_n}$, it is easy to show that

$$S_n = -\frac{1}{n} \sum P_{i_1} P_{i_2} P_{i_3} \cdots P_{i_n} \, log \, (P_{i_1} P_{i_2} \cdots P_{i_n}) \qquad (5.4)$$
$$= -\frac{1}{n} \left(n \sum P_{i_k} \, log \, P_{i_k} \right) = S_1$$

In other words, in a completely disordered amorphous network the series S_n is stationary. The same is true for the other "extreme": the perfect hexagonal lattice, for which $S_n \equiv 0$, because only one of the probabilities is equal to 1, no other possibilities being available, and $log \, 1 \equiv 0$.

In our case there is precise order in the network: this is seen in the fact that the entropy per polygon falls down when we consider bigger agglomertes. Looking at the doublets in the network (4,8,8), we find:

$$P_{88} = \frac{1}{5}, \;\; P_{48} = \frac{2}{5}, \;\; P_{84} = \frac{2}{5}, \;\; P_{44} = 0,$$

which gives

$$S_2 = \frac{1}{2} \left[-\frac{1}{5} \, log \, \frac{1}{5} - \frac{2}{5} \, log \, \frac{2}{5} - \frac{2}{5} \, log \, \frac{2}{5} \right] \qquad (5.5)$$
$$= \frac{1}{2} \, log \, 5 - \frac{2}{5} \, log \, 2 = 0.52746$$

The fact that the specific entropy has fallen down means that we have recovered some new information about the network by considering the correlated doublets.

For the correlated triplets in the same network one finds:

$$P_{444} = P_{448} = P_{844} = 0, \;\; P_{484} = \frac{3}{11},$$

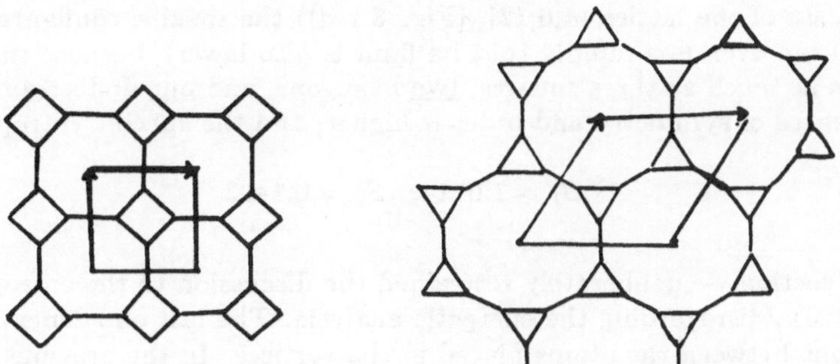

Fig. 5.1. Crystalline lattices (4,8,8) and (3.12,12) with their elementary cells and translations that generate the whole lattice.

$$P_{488} = P_{884} = \frac{7}{33}, \; P_{848} = \frac{1}{6}, \; P_{888} = \frac{3}{22} \tag{5.6}$$

The computation of S_3 yields:

$$S_3 = \frac{1}{3}\left[\frac{10}{33} \log 2 + \frac{6}{33} \log 3 + \frac{5}{6} \log 11 - \frac{14}{33} \log 7\right] = 0.52750 \tag{5.7}$$

The fact that S_3 is up to 10^{-4} equal to S_2 shows that almost all the information needed in order to produce this particular network can be found in the arrangement of the doublets.

Similar analysis of the network (3,12,12) displayed on Fig. 3.1 leads to the following evaluations:

$$\text{Singlets: } P_3 = \frac{2}{3}, \; P_{12} = \frac{1}{3}$$

$$S_1 = -\frac{2}{3} \log \frac{2}{3} - \frac{1}{3} \log \frac{1}{3} = 0.636514$$

$$\text{Doublets: } P_{33} = 0, \; P_{3,12} = P_{12,3} = 0.3, \; P_{12,12} = 0.4,$$

$$S_2 = 0.54445;$$

Finally,

$$S_3 = 0.5037.$$

The convergence to the limit is less rapid than in the previous case. This can be put into correspondence with the symmetry properties of the respective lattices. If we look at the elementary minimal cell which generates the whole network by discrete translations, we see that such a minimal cell for the network (4,8,8) contains only two polygons (one square and one octagon), whereas the minimal cell of the network (3,12,12) contains three polygons (two triangles and one dodecagon). (Fig. 5.1)

This is an adequate mathematical measure of the intuitive feeling one gets while looking on these tilings, that the lattice (4,8,8) is more primitive, or less "organized", than the lattice (3,12,12), which seems a little bit more sophisticated, although both are composed of two different kinds of polygons only.

In the case of the lattice (4,6,12), (Fig. 3.1 d)) the specific configurational entropy falls down even less rapidly (but its limit is also lower), because the elementary cell contains as much as three squares, two hexagons, and one dodecagon. The corresponding degree of symmetry and order is higher, and the specific entropy is lower:

$$S_1 = 1.0114, \quad S_2 = 0.1493$$

In this section we deliberately restrained the discussion to the entropy of the resulting network, disregarding the energetic analysis. The last one depends on the type of interactions between the atoms placed in the vertices. In the previous section, the entropy effects were manifested via the statistical weights attributed to the particular doublets, triplets, etc., while the most important driving force that decided which configurations were to be preferred has been the energy entering via Boltzmann factors. We think that in the real world the outcome is due to the intimate interplay of both phenomena.

The equilibrium configurations should correspond to the minimum of the free energy

$$F = U - TS \tag{5.8}$$

In principle, this expression has to be computed for all possible infinitely big networks, and then minimized. Such a program is obviously too difficult. Our approach enables us to reformulate the problem and to try to solve it by successive approximations. We have seen how to evaluate the configurational entropy for small subsets of the network; the potential part of the energy (i.e. U at $T = 0$) can be also evaluated by assuming that there are two main contributions to it: the bond-binding potential

$$\begin{aligned} U_B &= \sum_i \alpha_{1i}^2 + \alpha_{2i}^2 + \alpha_{3i}^2, \\ with \alpha_3 &= 2\pi - \alpha_1 - \alpha_2 \end{aligned} \tag{5.9}$$

where $\alpha_{1i}, \alpha_{2i}, \alpha_{3i}$ are the angles between the bonds surrounding the i-th vertex, and the surface contribution

$$U_P \cong -\lambda \sum_k S_k \tag{5.10}$$

where S_k is the surface of k-th polygon, λ a constant depending on the chemistry. Minimizing U_P is equivalent to maximize polygons' surfaces. Only in the case of pure hexagonal lattice U_B and U_P happen to have the same minimum at the value of all the angles equal to $2\pi/3$; in other cases two distinct minima can appear.

Let us now discuss the possibility of using these ideas to describe the aperiodic structures.

6. APPLICATION TO APERIODIC STRUCTURES: THE EXAMPLE OF PENROSE TILINGS

The starting point in any discussion of crystalline symmetries is the observation that not all discrete rotations are compatible with the action of a discrete subgroup of translation group, that would serve to generate the whole infinite lattice from an elementary cell. The classical argument goes as follows: consider a crystalline lattice,

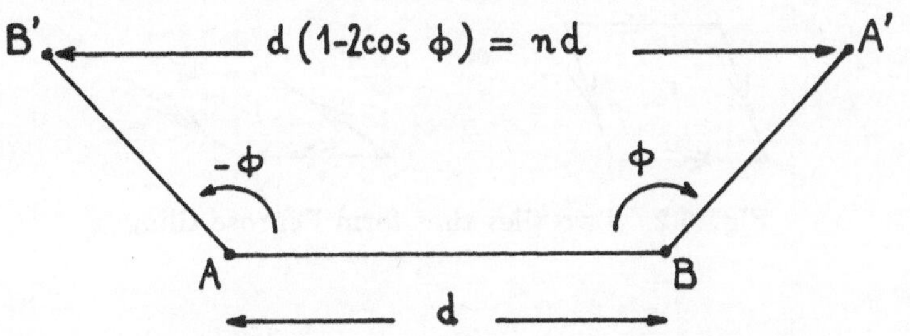

Fig. 6.1. Closest neighbors in a crystalline lattice with rotational symmetry ϕ

and take two atoms which are separated by a minimal distaince d; let us call them A and B. Now, if there is a discrete symmetry group of rotations with elements of the form $e^{in\phi}$, then, if we rotate the vector $\overrightarrow{AB}$ by the angle $-\phi$ around A, and rotate the vector $\overrightarrow{BA}$ by the angle ϕ around B (Fig. 6.1), the positions of new points A', B' thus obtained should correspond to the positions of atoms in the lattice; if the translation group acts on the lattice, too, this means that A' and B' have to be separated by an integer multiple of the minimal distance d. Therefore,

$$\|\overrightarrow{A'B'}\| = nd = d(1 - 2\cos\phi) \tag{6.1}$$

and

$$\cos\phi = \frac{1}{2}(1-n) \tag{6.2}$$

As $|\cos\phi| \leq 1$, the only possible solutions are $n = 1, 2, 3$ which correspond to the values of

$$\phi = \frac{2\pi}{m} \quad \text{with} \quad m = 1, 2, 3, 4 \ or \ 6. \tag{6.3}$$

Any other angle should produce a disordered and dense set of atoms on plane or in the 3-space.

The discovery by Schechtman et al, in 1984, of the materials (a particular, rapidly quenched AlMn alloy) whose diffraction patterns displayed local 5-fold and 10-fold symmetries, has created quite a surprise among the solid state physicists, and quite a lot of activity. It has been noted that the diffraction peaks were placed exactly at the vertices of the so-called Penrose tilings, which have been invented by Roger Penrose already before. The mathematical theory of Penrose tilings has been developed by de Bruijn in the early 1980's (see ref. (1981)). The generalisation to three-dimensional tilings of this type has been produced by Kramer and Neri. The Penrose tiling in two dimensions is a pattern made out of two tyupes of rhombi, a fat one, with angles $(\frac{2\pi}{5}, \frac{3\pi}{5})$, and a thin one, with angles $(\frac{\pi}{5}, \frac{4\pi}{5})$, which are assembled with some incidence rules, i.e. sticking always the right sides, with double or single arrows, together. (Fig. 6.2)

The resulting lattice, called the Penrose tiling, is represented on Figure (6.3). The eight types of vertices encountered are represented on Fig. 6.4.

The fact that all the bond angles are multiples of $(\frac{\pi}{5})$ exclude the translational symmetry. That is why the Penrose tilings are called aperiodic. In other words,

Fig. 6.2. Two tiles that form Penrose tilings

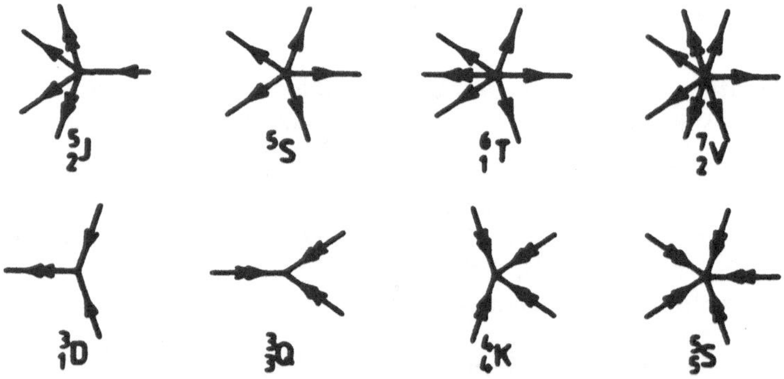

Fig. 6.3. A Penrose tiling ('regular')

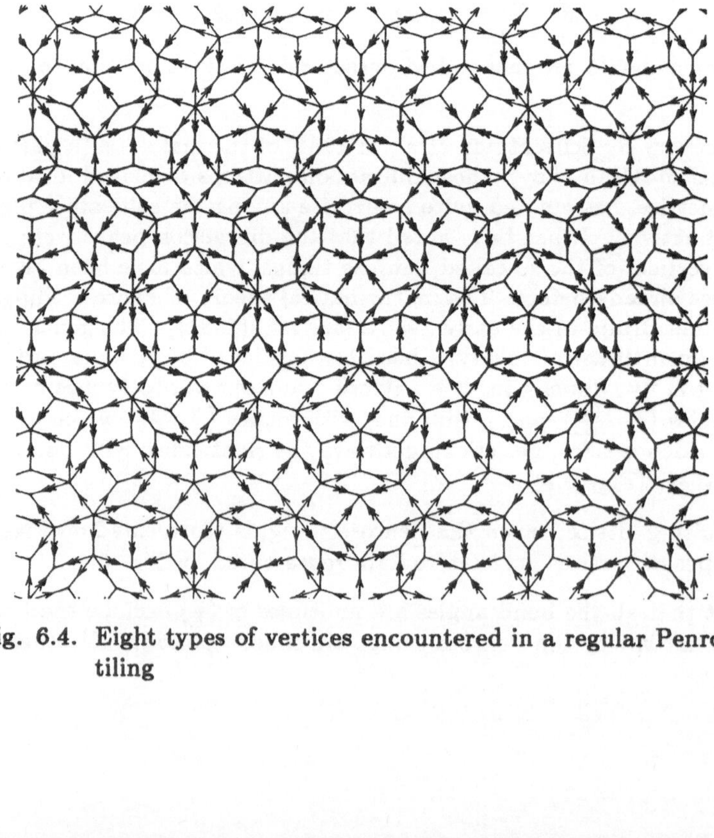

Fig. 6.4. Eight types of vertices encountered in a regular Penrose tiling

a) There is no translation which would leave such a pattern invariant.

b) An infinite pattern can not be determined by any finite region. Starting from any finite subset, there is an uncountable infinity of ways to complete the pattern.

c) Different infinite patterns are equivalent in the following sense: although they are not superposable globally, any finite region, however large it might be, can be found in any other infinite tiling.

Finally, these tilings are self-similar in the following sense: it is possible to associate to any Penrose tiling a different one whose tiles are smaller and in the length ratio $\tau = \frac{1+\sqrt{5}}{2}$ with the former, and which includes all the vertices of the original tiling.

The mathematical theory of Penrose tilings has been developed by de Bruijn, and then by Kramer and Neri. We shall just remind briefly that two equivalent constructions may serve to produce a Penrose tiling: one is called the pentagrid, another the projection from five dimensions.

The pentagrid is a kind of dual pattern for the Penrose tiling. It is composed of five families of parallel straight lines, which are equidistant and perpendicular to five pentagonal directions

$$\vec{n}_k = \left(cos\,\frac{2k\pi}{5},\ sin\,\frac{2k\pi}{5} \right) \qquad (k = 0, 1, 2, 3, 4). \tag{6.3}$$

Any point belonging to a pentagrid obeys the equation

$$\overrightarrow{OM} \cdot \vec{n}_k + \gamma_k = integer, \tag{6.4}$$

with five numbers γ_k satisfying

$$\sum_{0}^{4} \gamma_k = integer \tag{6.5}$$

There are as many different Penrose tilings as there are combinations of γ_k satisfying (6.5).

The pentagrid can be also viewed as a projection of the five families of 4-dimensional hyperplanes orthogonal to the five basis unit vectors $\vec{e}_k$ in a five-dimensional Euclidean space, on a two-dimensional plane. Consider the following five vectors:

$$\vec{d} = \vec{u}_0 = \frac{1}{\sqrt{5}} \sum_{k=0}^{4} \vec{e}_k$$

$$\vec{u}_1 = \sqrt{\frac{2}{5}} \sum_{k=1}^{4} cos\left(\frac{2k\pi}{5}\right) \vec{e}_k, \qquad \vec{u}_2 = \sqrt{\frac{2}{5}} \sum_{k=1}^{4} sin\left(\frac{2k\pi}{5}\right) \vec{e}_k \tag{6.6}$$

$$\vec{u}_3 = \sqrt{\frac{2}{5}} \sum_{k=1}^{4} cos\left(\frac{4k\pi}{5}\right) \vec{e}_k,\ \vec{u}_4 = \sqrt{\frac{2}{5}} \sum_{k=1}^{4} sin\left(\frac{4k\pi}{5}\right) \vec{e}_k,$$

Let us call D the 1-dimensional subspace spanned by $\vec{d}$, P the 2-dimensional space spanned by $(\vec{u}_1, \vec{u}_2)$ and P' the 2-dimensional space spanned by $(\vec{u}_3, \vec{u}_4)$.

The five families of 4-dimensional hyperplanes, when projected into P, give a pentagrid. To each mesh of the pentragrid corresponds a vertex of the Penrose tiling.

An equivalent way of obtaining the same result is the method of projection: whatever the set of five numbers γ_k such that $\Sigma_{k=0}^{4}\gamma_k = 0$, it is sufficient to project on the plane P the vertices of the five-dimensional cubic lattice contained in a strip parallel to P whose thickness is equal to $\frac{1}{\sqrt{5}}$. (For more details see de Bruijn and Kramer).

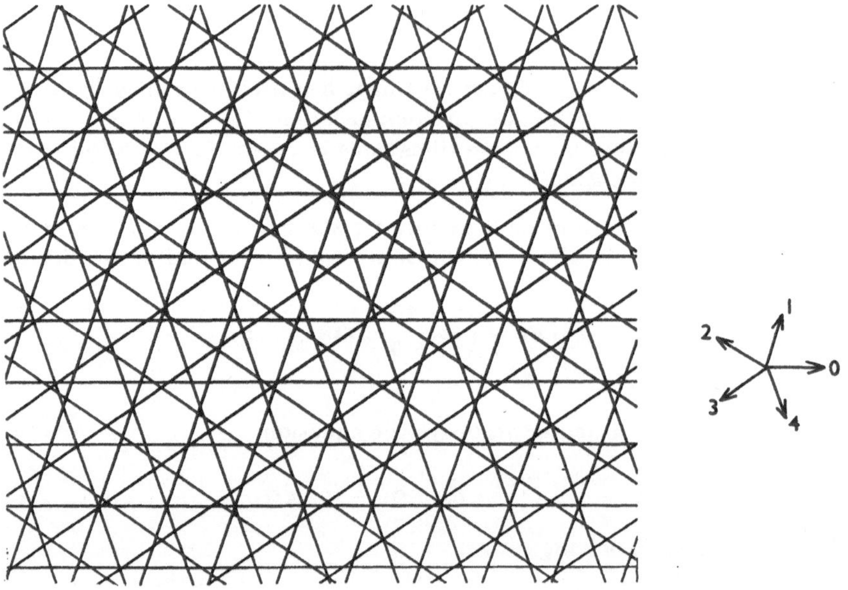

Fig. 6.5. A pentagrid with five vectors $\overrightarrow{n_k}$

Now, when looking on the Penrose tilings, one is tempted to try the statistical approach that has been exposed above. Instead of assembling polygons, it is natural to assemble vertices. As the tiling is composed with rhombi, $\overline{N}_f = N_f = 4$, so that $\overline{N}_c$, the average coordination number, is strictly four. All other vertices are deviations from the ideal four-coordinate vertex.

The assembling of vertices into doublets and triplets is shown on Fig. (6.6).

The statistics of the vertices in Penrose tilings is the following:

$$\begin{array}{lcl} 3-\text{vertices}: & P_3 = \tau^{-4} + \tau^{-2} & \approx 52.8\% \\ 4-\text{vertices}: & P_4 = \tau^{-5} & \approx 9.02\% \\ 5-\text{vertices}: & P_5 = \tau^{-3} + \frac{(\tau^{-5}+\tau^{-7}}{2\tau-1} & \approx 29.18\% \\ 6-\text{vertices}: & P_6 = \tau^{-7} & \approx 3.45\% \\ 7-\text{vertices}: & P_7 = \tau^{-6} & \approx 5.57\% \end{array} \tag{6.7}$$

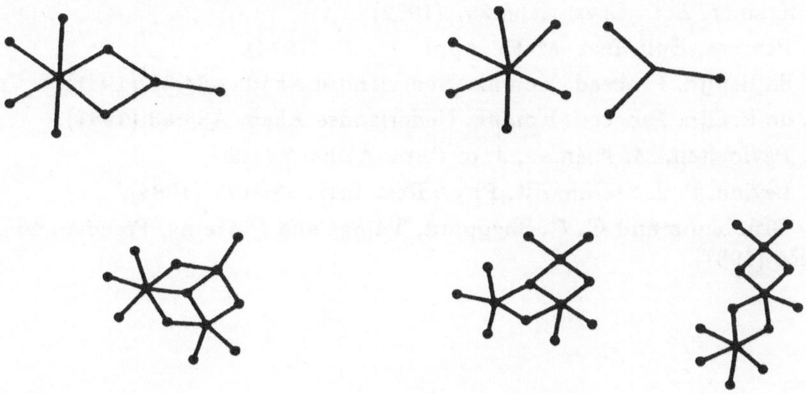

Fig. 6.6. Assembling of the vertices into doublets and triplets: chains and cells.

In a recent work, we have reproduced the statistical model of growth to the vertices with the coordination numbers from 3 to 7. The energy table has been chosen in the simplest manner, i.e.

$$\Delta E_{ij} = |8 - i - j| \Delta E \tag{6.8}$$

Although it is a very crude approximation, and although our analysis stopped at triplets only, we have been able to find, at quite a wide range of temperatures, several singular points of the system of second order, one of which being a saddle point with the following coordinates:

$$\begin{aligned} P_3 &\approx 52\% \\ P_4 &\approx 10\% \\ P_5 &\approx 28\% \\ P_6 &\approx 10\% \end{aligned} \tag{6.8}$$

Such an agreement with the real statistics (6.7) seems to be encouraging for the further investigations in this direction.

REFERENCES

1. Johannes Kepler, De Nive Sexangula, (1611).
2. M. Gardner, Scientific American, January (1977), p. 110.
3. N.H. Christ, R. Friedberg, T.D. Lee, Nucl. Phys. B202:89 (1982).
4. A. Jauner, T. Janssen, Phys. Rev. B15:643 (1977).
5. A. Mackay, Sov. Phys. Crystallogr. 26:517 (1981).
6. M. Kléman, L. Michel, G. Toulouse, J. de Physique Lett. 38:L195 (1977).
7. J.F. Sadoc, R. Mosseri, Journ. of Non-Crystalline Solids, 61-62:487 (1984).
8. R. Kerner, Phys. Rev. B28:756 (1983).
9. R. Kerner, D. M. dos Santos, Phys. Rev. B37:37 (1988).

10. P. Kramer, Acta Cryst. A38:257 (1982)
11. R. Penrose, Bull. Inst. Math. Appl., 7/8:10 (1974).
12. N. de Bruijn, Proceed. Konink. Nederlandse Akad. A84:39 (1981).
N. de Bruijn, Proceed. Konink. Nederlandse Akad. A84:53 (1981).
13. A. Pavlovitch, M. Kléman, J. of Phys. A20:687 (1987).
14. D. Levine, P. J. Steinhardt, Phys. Rev. Lett. 53:2477 (1984).
15. G. Grünbaum and G. C. Sheppard, Tilings and Patterns, Freeman ed. San Francisco (1987).

THE MAPPING CLASS GROUP IN STATISTICAL MECHANICS : A CONCISE REVIEW

Mario Rasetti*

School of Natural Sciences
The Institute for Advanced Study
Princeton, NJ 08540, USA

1. Introduction

The group of mapping classes, *i.e.* the group $\mathcal{M}_g = \pi_0 \, Diff^{(+)}(\Sigma_g)$ of path components of isotopy classes of orientation preserving diffeomorphisms of surfaces Σ_g of *genus* g, has recently entered the collection of mathematical tools relevant to physics, appearing first in the statistical mechanics of the 3-dimensional *Ising Model* [1], then in *String Theory*[2].

In this paper we shall briefly review the former case, discussing how the mapping class group plays a role in dealing with the Ising model, in particular when this is defined over a lattice homogeneous under some finitely presented group.

We simply mention here, for the sake of completeness, how it enters string theory. At critical dimensions, closed string amplitudes are straightforwardly obtained from vertex-operator correlation functions on a Riemann surface by summing first over all inequivalent surfaces of given topological genus, then over all genera (*i.e.* numbers of handles) [3]. Thus string amplitudes can be analyzed in terms of the complex analytic structure of the moduli space of inequivalent

* Permanent address : Dipartimento di Fisica and Unitá CISM del Politecnico, Torino, ITALY.

Riemann surfaces of given genus (for example one need an invariant measure over such space) : the moduli space of smooth curves of genus g, however, is but the quotient of the action of $\mathcal{M}_g$ over the *Teichmüller space*, and its construction requires a faithful representation of $\mathcal{M}_g$.

In sect.2. we review the basic notion of mapping class group, introduce its minimal presentation in terms of *Dehn-Lickorish twists*, and discuss how this bears on the group representation. In sect.3. we discuss its application to the Ising model, and discuss how the information about presentations and representations bears on the possibility of evaluating rigorously partition function and correlation functions.

2. The Mapping Class Group

2.1 Definitions

The mapping class group $\mathcal{M}_g$ of an orientable 2-*manifold* (briefly a *surface*) Σ_g of genus g is the group of path components (*i.e.* modulo isotopy) of the group of all orientation preserving homeomorphisms of Σ_g .

More precisely : consider first a surface Σ_g of *finite type*† , namely one that can be enbedded into a compact surface Σ'_g in such a way that $\Sigma'_g \setminus \Sigma_g$ consists at most of a finite number of disks or points, and whose first homotopy and homology groups, $\pi_1(\Sigma_g)$ and $H_1(\Sigma_g)$, are finitely generated. Allow also the surface to have m *holes*. The *Van Kampen-Seifert theorem* states that the fundamental group of Σ_g has $(2g+m)$ generators, and presentation

$$\pi_1(\Sigma_g) \approx < h_1, \ldots, h_m; t_1, u_1, \ldots, t_g, u_g \,|\, h_1 \ldots h_m \prod_{i=1}^{g} [t_i, u_i] > \quad ; \tag{2.1}$$

where $\{t_i, u_i \,|\, i = 1, \ldots, g\}$ are a canonical basis for the first homology group $H_1(\Sigma_g)$; whereas the *Nielsen theorem* says that if $\alpha : \pi_1(\Sigma_g) \to \pi_1(\Sigma_g)$ is an automorphism such that

$$\alpha(h_j) = p_j \circ h_{i_j}^{\sigma_j} \circ p_j^{-1} \quad , j = 1, \ldots, m \quad , \quad p_j \in \pi_1(\Sigma_g) \quad , \quad \sigma \in \mathbf{Z}_2 = \{1, -1\} \, , \tag{2.2}$$

i.e. such that homotopy classes corresponding to boundary curves are mapped into classes of the same type, then there is a homeomorphism $\mu : \Sigma_g \to \Sigma_g$ such that $\mu|_{\#} \equiv \alpha$ (we shall define $\#$ rigorously a little further, its meaning here is

† Only in the concluding part of the paper we shall discuss a tentative extension of the notions presented here to the case in which Σ_g is *not* of finite type; *i.e.* – roughly – to the limit $g \to \infty$.

intuitive). [Notice that $\sigma_j \cdot \omega(p_j)$, where $\omega : \pi_1(\Sigma_g) \rightarrow \mathbf{Z}_2$ is the homomorphism describing the *orientation* behaviour, is independent on j].

We denote by $Aut_\star\, \pi_1(\Sigma_g)$ the group af all automorphisms α of $\pi_1(\Sigma_g)$ which have the property (2.2), and define $Out_\star\, \pi_1(\Sigma_g) := Aut_\star\, \pi_1(\Sigma_g) \,/\, Inn\, \pi_1(\Sigma_g)$, where Inn is the set of all the inner automorphisms. A mapping $\mu : \Sigma_g \rightarrow \Sigma_g$ induces, in general, an endomorphism on $\pi_1(\Sigma)$ only after the base point $(\star)$ has been moved back : this can be done by an isotopy which is determined obviously only up to any other isotopy of the identity $\mathbf{1}$ into a mapping which moves the base point back. Thus the effect on $\pi_1(\Sigma_g)$ is in turn determined only up to an inner automorphism, and to a homeomorphism $\mu : \Sigma_g \rightarrow \Sigma_g$ corresponds indeed an element of $Out_\star\, \pi_1(\Sigma_g)$. We denote by $\Omega : Homeo\, \Sigma_g \rightarrow Out_\star\, \pi_1(\Sigma_g)$ the homomorphism naturally defined in this way.

The *Baer theorem* states that the mapping

$$\mathcal{M}_g \sim \Omega : Homeo\, \Sigma_g / Isot\, \Sigma_g \longrightarrow Aut_\star\, \pi_1(\Sigma_g) \,/\, Inn\, \pi_1(\Sigma_g) \tag{2.3}$$

is an *isomorphism*. There is also another isomorphism, namely the one we used above without definition, $\# : Homeo\,(\Sigma_g, \star) \,/\, Isot\,(\Sigma_g, \star) \rightarrow Aut_\star\, \pi_1(\Sigma_g)$. Since an isotopy class of homeomorphisms is called a *mapping class*, the homeotopy group $\mathcal{M}_g \equiv Homeo\, \Sigma_g \,/\, Isot\, \Sigma_g$ is the **group of mapping classes** [4] of Σ_g. Calling $Out\, \pi_1(\Sigma_g)$ the group of *outer automorphisms* of the surface, a concise formulation of the Baer-Nielsen theorems gives us the equivalent definiton :

the mapping class group of a surface is isomorphic to the outer automorphism group of its fundamental group.

2.2 Presentation and Representations

We shall restrict henceforth our attention to the case in which the surface has no holes (*i.e.* m, number of boundary components, equals zero). In this case it is possible to give two different finite presentations of $\mathcal{M}_g$ that we shall discuss briefly; one which is minimal[5] ($(2g+1)$ generators), but gives very few clues about the representations one can possibly derive from it‡ , the other which holds also in the general case[6],[7] ($m \neq 0$), has features which are very suggestive in terms of representations, but has so many relations it is difficult to handle it in complex situations. Both have as generators *Dehn-Lickorish twists*, which are defined in the following way.

‡ Since $\mathcal{M}_g$ is residually finite, one can in principle derive from a suitable finite presentation for it its representation as a group of matrices in some field

Let $\gamma \in \Sigma_g$ be a simple closed curve, and let $\mathcal{N}_\gamma$ be a neighbourhood of γ in Σ_g. $\mathcal{N}_\gamma$ is homeomorphic with a cylinder, and one can parametrize it by the coordinates (z, ϑ) , with $0 \leq \vartheta \leq 2\pi$ the polar angle and $-1 \leq z \leq 1$ along the axis of the cylinder. Assume also that γ is defined by $z = 0$. The map $\mathcal{D}_\gamma : \Sigma_g \to \Sigma_g$ which is the identity on $\Sigma_g \setminus \mathcal{N}_\gamma$ and on $\mathcal{N}_\gamma$ is given by

$$\mathcal{D}_\gamma(z, \vartheta) = (z, \vartheta + \pi(z+1)) \tag{2.4}$$

is known as a Dehn twist about γ. It has two properties which are relevant :

(i) even though $\mathcal{D}_\gamma$ does depend on the choice (*i.e.* the orientation) of the coordinate system on $\mathcal{N}$, it does *not* depend on the orientation of γ ;

(ii) if λ and μ are simple closed curves in Σ_g , and for $x \in I$, where I is the unit interval, $\iota_x : \Sigma_g \to \Sigma_g$ is an isotopy, with $\iota_0 = \mathbf{1}$ and $\iota_1(\lambda) = \mu$, then there is an isotopy connecting $\mathcal{D}_\lambda$ and $\mathcal{D}_\mu$.

In order to give explicitly the presentations mentioned above, we need to define the Dehn twists with respect to a particular set of curves, referred to as *Humphries generators* [8]. These curves are :

[a] the simple closed curves (for simplicity, since we don't distinguish between isotopic curves, we shall call them *circles*) $\{\alpha_i \,|\, i = 1, \ldots, g\}$: for $2 \leq i \leq g$, α_i goes once around the handle separating the $(i-1)$-th from the i-th holes ; α_1 goes once around the first handle;

[b] the circles $\{\beta_i \,|\, i = 1, \ldots, g\}$: β_i goes once around the throat of the i-th hole ; for $i \leq g-1$, β_i intersects transversally once both the circle α_i and the circle α_{i+1} ; β_g intersects α_g and the circle δ_g. The latter, which corresponds to the auxiliary element $\mathcal{D}_g$ – entering the fourth relation below – , goes once around the g-th handle [notice that $\mathcal{D}_g$ is *not* a Humphries generator];

[c] the circle δ, which intersects β_2 transversally, going once around the second handle.

We denote by $\mathcal{A}_i$, $\mathcal{B}_i$, $i = 1, \ldots, g$; $\mathcal{D}$ the Dehn twists with respect to the circles α_i , β_i , $i = 1, \ldots, g$; δ respectively : they are the Humphries generators .

The Wajnryb Presentation [5]

Using the convention of writing composition of homeomorphisms in $\mathcal{M}$ from left to right (we also omit the composition symbol $\circ$), let us define now the auxiliary elements :

$$\mathcal{T}_i := \mathcal{B}_i \mathcal{A}_i \mathcal{A}_{i+1} \mathcal{B}_i \quad ; \quad i = 1, \ldots, g-1 \, ; \tag{2.5}$$

$$\mathcal{V}_0 := \mathcal{A}_1\mathcal{B}_1\mathcal{A}_2\mathcal{B}_2\mathcal{D}\,[\mathcal{A}_1\mathcal{B}_1\mathcal{A}_2\mathcal{B}_2]^{-1} \quad ; \quad \mathcal{V}_1 := \mathcal{D} \quad ;$$

$$\mathcal{V}_i := \mathcal{T}_{i-1}\mathcal{T}_i\mathcal{V}_{i-1}\,[\mathcal{T}_{i-1}\mathcal{T}_i]^{-1} \quad , \quad i = 2, \ldots, g-1 \; ; \tag{2.6}$$

$$\mathcal{U}_0 := \mathcal{A}_3\mathcal{B}_3\mathcal{T}_2\mathcal{D}\,[\mathcal{A}_3\mathcal{B}_3\mathcal{T}_2]^{-1} \quad ;$$

$$\mathcal{U}_i := \mathcal{A}_i\mathcal{B}_i\mathcal{A}_{i+1}\mathcal{B}_{i+1}\mathcal{V}_i\,[\mathcal{B}_{i+1}\mathcal{A}_{i+1}\mathcal{B}_i\mathcal{A}_i]^{-1} \quad , \quad i = 1, \ldots, g-1 \; ; \tag{2.7}$$

$$\mathcal{D}_g := [\mathcal{U}_1\mathcal{U}_2 \ldots \mathcal{U}_{g-1}]^{-1}\mathcal{A}_1\mathcal{U}_1\mathcal{U}_2 \ldots \mathcal{U}_{g-1} \, . \tag{2.8}$$

Then the *Wajnryb theorem* states that if Σ_g has no boundary components, $\mathcal{M}_g$ has the following *minimal* presentation :

$$\mathcal{M}_g \approx < \mathcal{A}_1, \mathcal{B}_1, \ldots, \mathcal{A}_g, \mathcal{B}_g, \mathcal{D} \,|\, \{\mathcal{R}_i\}_{i=1}^4 > \quad ; \tag{2.9}$$

where the four relations $\{\mathcal{R}_i\}_{i=1}^4$ are given by :

$$\mathcal{R}_1 \quad : \quad \mathcal{A}_i\mathcal{B}_i\mathcal{A}_i = \mathcal{B}_i\mathcal{A}_i\mathcal{B}_i \;\; , \;\; \mathcal{A}_{i+1}\mathcal{B}_i\mathcal{A}_{i+1} = \mathcal{B}_i\mathcal{A}_{i+1}\mathcal{B}_i \;\; , \;\; \mathcal{B}_2\mathcal{D}\mathcal{B}_2 = \mathcal{D}\mathcal{B}_2\mathcal{D} \; ;$$

$$\forall \; \textit{allowed } i\text{'s} \; \& \; \textit{any other pair of generators commute} \, . \tag{2.10}$$

$$\mathcal{R}_2 \quad : \quad [\mathcal{A}_1\mathcal{B}_1\mathcal{A}_2]^4 = \mathcal{D}\,[\mathcal{B}_2\mathcal{A}_2\mathcal{B}_1\mathcal{A}_1\mathcal{A}_1\mathcal{B}_1\mathcal{A}_2\mathcal{B}_2]^{-1}\mathcal{D}\,\mathcal{B}_2\mathcal{A}_2\mathcal{B}_1\mathcal{A}_1\mathcal{A}_1\mathcal{B}_1\mathcal{A}_2\mathcal{B}_2 \; ; \tag{2.11}$$

$$\mathcal{R}_3 \quad : \quad \mathcal{D}\,\mathcal{T}_2\mathcal{D}\mathcal{T}_2{}^{-1}\mathcal{T}_1\mathcal{T}_2\mathcal{D}[\mathcal{A}_1\mathcal{A}_2\mathcal{A}_3\mathcal{T}_1\mathcal{T}_2]^{-1} =$$

$$= [\mathcal{U}_0\mathcal{B}_1\mathcal{A}_2\mathcal{B}_2\mathcal{A}_3\mathcal{B}_3]^{-1}\,\mathcal{V}_0\mathcal{U}_0\mathcal{B}_1\mathcal{A}_2\mathcal{B}_2\mathcal{A}_3\mathcal{B}_3 \; ; \tag{2.12}$$

$$\mathcal{R}_4 \quad : \quad \mathcal{D}_g \;\; \textit{commutes with} \;\; \mathcal{B}_g\mathcal{A}_g \cdots \mathcal{B}_1\mathcal{A}_1\mathcal{A}_1\mathcal{B}_1 \cdots \mathcal{A}_g\mathcal{B}_g \, . \tag{2.13}$$

The Hatcher-Thurston-Harer Presentation [6],[7]

The Wajnryb relations explicitly realize, in terms of Humphries generators, the presentation for $\mathcal{M}_g$ derived in quite different form by Hatcher and Thurston (and successively simplified by Harer). We review here the basic results, in order to extract from them the relevant information on how to possibly attack the construction of a faithful representation of $\mathcal{M}_g$ itself. The fundamental notion is that of *cut system*. A cut system C is an isotopy class of a collection $\{\chi_1, \ldots, \chi_g\}$ of disjoint circles on Σ_g , such that $\Sigma_g \setminus [\bigcup_{i=1}^g \chi_i]$ is connected (*i.e.* is a punctured sphere with $2g$ holes). Replacing χ_i by another circle, say χ'_i which intersects χ_i transversally at one point and doesn't intersect any other χ_k, $k \neq i$, one obtains a new cut system $C' \neq C$. The operation $\mu : C \to C'$; $\{<\chi_i> \Rightarrow <\chi'_i>\}$ is referred to as a *simple move*. If now $\mathcal{G}_1^{(g)}$ is an abstract graph whose vertices are labelled by cut systems on Σ_g and two vertices are joined by an edge (*i.e.* a 1-dimensional cell) if they are connected by a simple move , and one transforms it into a cellular complex $\mathcal{X}_2^{(g)}$ by attaching a plaquette (*i.e.* a 2-dimensional cell) to each cycle of simple moves of one of the following types :

(i) $<\chi_i> \Rightarrow <\chi'_i> \Rightarrow <\chi''_i> \Rightarrow <\chi_i>$
($\chi_i ; \chi'_i ; \chi''_i$ all different) ;

(ii) $<\chi_i, \chi_j> \Rightarrow <\chi'_i, \chi_j> \Rightarrow <\chi'_i, \chi'_j> \Rightarrow <\chi_i, \chi'_j> \Rightarrow <\chi_i, \chi_j>$
($\chi_i ; \chi'_i ; \chi_j ; \chi'_j$ all different) ;

(iii) $<\chi_i, \chi_j> \Rightarrow <\chi_i, \chi'_j> \Rightarrow <\chi'_i, \chi'_j> \Rightarrow <\chi'_i, \chi''_j> \Rightarrow <\chi_j, \chi''_j> \Rightarrow <\chi_j, \chi_i>$
($\chi_i ; \chi'_i ; \chi_j ; \chi'_j ; \chi''_j$ all different) ;

or

$(iii)'$ $<\chi_i, \chi_j> \Rightarrow <\chi_i, \chi'_j> \Rightarrow <\chi_i, \chi''_j> \Rightarrow <\chi'_i, \chi''_j> \Rightarrow$
$\Rightarrow <\chi_j, \chi''_j> \Rightarrow <\chi_j, \chi''_i> \Rightarrow <\chi_j, \chi_i>$
($\chi_i ; \chi'_i ; \chi''_i ; \chi_j ; \chi'_j ; \chi''_j$ all different) .

when they exist; the first main result of ref. [6] asserts that $\mathcal{X}_2^{(g)}$ is *connected* and *simply connected.*

The presentation of Hatcher and Thurston is now obtained as follows. First, fix the cut system $C_0 \equiv \{\alpha_1, \ldots, \alpha_g\}$ on Σ_g . Define then the new family of closed simple curves on Σ_g, $\{\omega_{i,j} ; 1 \leq i < j \leq 2g\}$: $\omega_{i,j}$ interlaces handles i and j [more precisely, $\omega_{i,j}$ enters hole i , goes around handle i , comes out of hole $(i-1)$, enters hole j , goes around handle j , comes out of hole $(j-1)$ and closes]. Denote by $\mathcal{W}_{i,j}$ the Dehn's twist with respect to $\omega_{i,j}$. Define moreover the following new homeomorphisms of Σ_g : $\mathcal{P} := \mathcal{A}_g \mathcal{B}_g \mathcal{A}_g$, which is a simple move permuting α_g and β_g; $\mathcal{L} := \mathcal{B}_g \mathcal{A}_g \mathcal{A}_g \mathcal{B}_g$, which reverses the orientation of α_g . Notice finally that each $\mathcal{T}_i$, $i = 1, \ldots, g-1$ defined in (2.5) permutes the circles α_i and α_{i+1}.

The mapping class group $\mathcal{M}_g$ is *generated* by $\{\mathcal{L} ; \mathcal{P} ; \mathcal{A}_i, i = 1, \ldots, g ; \mathcal{T}_j, j = 1, \ldots, g-1 ; \mathcal{W}_{i,j}, 1 \leq i < j \leq g\}$.

Let now $\mathcal{H}_0$ be the subgroup, generated by $\{\mathcal{A}_i ; \mathcal{W}_{i,j}\}$, of elements of $\mathcal{M}_g$ which leave the circles $\{\alpha_i\}$ fixed ; and $\mathcal{H}$ the subgroup, generated by $\{\mathcal{H}_0 ; \mathcal{L} ; \mathcal{T}_i\}$, of elements which leave the cut system C_0 invariant . $\mathcal{H}$ is defined by the exact sequences :

$$1 \longrightarrow \mathcal{H}_0 \longrightarrow \mathcal{H} \xrightarrow{\vartheta} \pm S_g \longrightarrow 1 \quad ; \tag{2.14}$$

$$1 \longrightarrow [\mathbf{Z}/2\mathbf{Z}]^g \longrightarrow \pm S_g \longrightarrow S_g \longrightarrow 1 \quad ; \tag{2.15}$$

where $\vartheta(\mathcal{L}) \in [\mathbf{Z}/2\mathbf{Z}]^g$ and $\vartheta(\mathcal{T}_i)$ is the transposition $(i, i+1)$ in the symmetric group S_g .

The relations of $\mathcal{M}_g$ are generated by $\{\mathcal{H}, \mathcal{P}\}$. Instead of reporting them explicitly (they are numerous and cumbersome) we simply *describe* them here :

(*I*) $\mathcal{P}$ commutes with $\mathcal{H}_g$ (the subgroup of elements of $\mathcal{H}$ which leave α_g and β_g invariant) ;

(*II*) $\mathcal{P}^2 \equiv \mathcal{A}_g \mathcal{L} \mathcal{A}_g \in \mathcal{H}$;

(*III*) $\mathcal{P}\mathcal{F}\mathcal{P}\mathcal{F}\mathcal{P} \in \mathcal{H}$ whenever $\exists$: *(1).* a circle γ on Σ_g which intersects once transversally both α_g and β_g and does not intersect any other $\alpha_i, i \neq g$, and *(2).* a map $\mathcal{F} \in \mathcal{H}$ such that $[\mathcal{P}\mathcal{F}]^{-1} \gamma \mathcal{P}\mathcal{F} = \beta_g$; $[\mathcal{P}\mathcal{F}]^{-1} \beta_g \mathcal{P}\mathcal{F} = \alpha_g$; $[\mathcal{P}\mathcal{F}]^{-1} \alpha_g \mathcal{P}\mathcal{F} = \gamma$. All the *triangular* plaquettes of type *(i)* above are generated from these relations.

(*IV*) $\mathcal{P}$ commutes with $\tilde{\mathcal{F}}\mathcal{P}\tilde{\mathcal{F}}^{-1}$ where $\tilde{\mathcal{F}} \in \mathcal{H}$ maps the simple closed curve $\tilde{\beta}$ encircling holes $(g-1)$ and g onto β_g . All the *square* plaquettes of type *(ii)* are produced by these relations.

(*V*) $\mathcal{P}\mathcal{F}_1\mathcal{P}\mathcal{F}_2\mathcal{P}\mathcal{F}_3\mathcal{P}\mathcal{F}_4\mathcal{P} \in \mathcal{H}$ whenever $\exists$: *(1).* a circle δ on Σ_g which intersects once transversally both α_{g-1} and β_g and does not intersect β_{g-1} nor any other $\alpha_i, i \neq g-1$, and *(2).* the maps $\mathcal{F}_j \in \mathcal{H}; j = 1, \ldots, 4$ satisfy – upon defining $\mathcal{E}_{(1)} := \mathcal{P}\mathcal{F}_1$; $\mathcal{E}_{(2)} := \mathcal{E}_{(1)}\mathcal{P}\mathcal{F}_2$; $\mathcal{E}_{(3)} := \mathcal{E}_{(2)}\mathcal{P}\mathcal{F}_3$; $\mathcal{E}_{(4)} := \mathcal{E}_{(3)}\mathcal{P}\mathcal{F}_4$ (in terms of which the element of $\mathcal{H}$ we are considering reads $\mathcal{E}_{(4)}\mathcal{P}$) – the following relations : $\mathcal{E}_{(1)}\beta_{g-1}\mathcal{E}_{(1)}^{-1} = \beta_g$; $\mathcal{E}_{(2)}\delta\mathcal{E}_{(2)}^{-1} = \beta_g$; $\mathcal{E}_{(3)}\alpha_g\mathcal{E}_{(3)}^{-1} = \beta_g$; $\mathcal{E}_{(4)}\alpha_{g-1}\mathcal{E}_{(4)}^{-1} = \beta_g$. All the *pentagonal* plaquettes of type *(iii)* are generated by these relations.

The explicit construction of the relations of $\mathcal{M}_g$ consists in writing the relations of $\mathcal{H}_0$ and of $\mathcal{H}$ and the presentation of $\mathcal{H}_g$, and in finding the elements of $\mathcal{H}$ corresponding to all possible independent choices of γ and δ . It was proved by Harer that four choices of γ and one of δ suffice.

The second main result of ref.[6] is that in the case when Σ_g has no punctures the isotropy subgroup $\mathcal{H}$ of $\mathcal{X}_2^{(g)}$ in included in the exact sequence

$$\mathbf{Z} \longrightarrow \mathbf{Z}^g \oplus B_{2g-1} \longrightarrow \mathcal{H} \longrightarrow \pm S_g \longrightarrow 1 \; ; \qquad (2.16)$$

where B_{2g-1} is the *Artin* coloured *braid group* over $(2g-1)$ strings. It is worth recalling here that $\pm S_g$, the group of signed permutations of g objects , is isomorphic with the group of $g \times g$ matrices having just one non-zero entry, equal to ± 1, in each row and column.

Representations

Deriving a faithful representation of $\mathcal{M}_g$ from the above presentations is certainly not an easy task. We summarize here the present state of the problem (see refs.[9],[10] for more extended accounts) . For $g = 1$, $\mathcal{M}_g \sim SL(2, \mathbf{Z})$, the *classical* (as opposed to the *Teichmüller* or *many-handled*) **Modular Group**. The related moduli space $Mod(1)$ is a space whose points correspond to conformal isomorphism classes of tori. For arbitrary $g > 1$, denote by $\mathcal{I}(\Sigma_g)$ the set of isotopy classes of all the closed (non oriented) curves enbedded in Σ_g , and let Φ_g be a foliation whose leaves are geodesics for for some metric on Σ_g (since Σ_g has negative Euler characteristics, the metric is hyperbolic), with a transverse measure $\mu_\perp$. The latter is given by a positive real function $\mu_\perp(\bullet)$ assigning to each arc $\sigma \in \Sigma_g$ transverse to the leaves of Φ_g and with extremal points in $\Sigma_g \setminus \Phi_g$ an invariant weight , defined by the following conditions :

(*a*). $\mu_\perp(\sigma) = \mu_\perp(\sigma')$ if σ is homotopic to σ' through arcs transverse to Φ_g and with endpoints in $\Sigma_g \setminus \Phi_g$;

(*b*). if $\sigma = \bigcup_i \sigma_i$; with $\sigma_i \cap \sigma_j \subset \partial\sigma_i \cap \partial\sigma_j$; then $\mu_\perp(\sigma) = \sum_i \mu_\perp(\sigma_i)$;

(*c*). $\mu_\perp(\sigma) \neq 0$ if $\sigma \cap \Phi_g \neq \emptyset$.

The collection of all these measured geodesic foliations constitutes a space Ξ_g on which $\mathcal{M}_g$ acts in a natural way. In particular the elements $m \in \mathcal{M}_g$ are classified according to the following scheme : m is said to be

periodic , if it is of finite order in $\mathcal{M}_g$;

reducible , if there is a point in $\mathcal{I}(\Sigma_g)$ which is invariant with respect to the element m itself ;

pseudo-Anosov , if $\exists$ mutually transverse geodesic foliations $\Phi_g^{(s)}$, $\Phi_g^{(u)} \in \Xi_g$ (s stands for *stable* , u for *unstable*) , such that $m(\Phi_g^{(s)}) = \frac{1}{\varepsilon}\Phi_g^{(s)}$ and $m(\Phi_g^{(u)}) = \varepsilon\Phi_g^{(u)}$ for some real $\varepsilon > 1$.

In order to derive a faithful representation from a finite presentation, one should first prove that no normal subgroup $\mathcal{N}\mathcal{M}_g$ of $\mathcal{M}_g$ can have all of its elements $\neq \mathbf{1}$ which are pseudo-Anosov, because only then[♯] one can identify an homeomorphism $m_o \in \mathcal{N}\mathcal{M}_g$ fixing some $\iota \in \mathcal{I}(\Sigma_g)$ and then proceed in the construction of an *induced* faithful representation of $\mathcal{M}_g$ as a group of matrices (possibly with entries in a field of characteristics $\neq 0$ or of anticommuting variables) .

We discuss this on a special case (*no* global proof is available at present). Let π be a path on Σ_g which crosses the curve α_i at a finite number ℓ of points $\{p_1^{(i)}, \ldots, p_\ell^{(i)}\}$. When we act on Σ_g with $\mathcal{A}_i$, the effect on π is that it is broken at each point $p_k^{(i)}$ and a copy of α_i is inserted at the discontinuity in such a way as to coalesce (also in orientation) with the adjacent fragments of π. Resorting to the property that on any compact surface such as Σ_g there exists at least a pair of essential simple closed curves, say γ , γ' , which *fill* the surface but are such that one can find another essential closed curve $\tilde{\gamma}'$, disjoint from γ' , such that $\gamma \cup \tilde{\gamma}'$ does *not* fill the surface ; it was shown in ref.[9] that – denoting by $\mathcal{D}_\bullet$ the corresponding Dehn twists – $\mathcal{D}_\gamma \mathcal{D}_{\gamma'}^{-1}$ is isotopic to a pseudo-Anosov map [11]. Then $\gamma'' \equiv \mathcal{D}_\gamma \mathcal{D}_{\gamma'}^{-1} \circ \tilde{\gamma}'$ is a curve disjoint from any essential simple curve $\tilde{\gamma}$ having no intersections with $\gamma \cup \tilde{\gamma}'$. Thus there exists a map

$$\mathcal{D}_{\tilde{\gamma}'}^{-1} \mathcal{D}_\gamma \mathcal{D}_{\gamma'}^{-1} \mathcal{D}_{\tilde{\gamma}'} \mathcal{D}_{\gamma'} \mathcal{D}_\gamma^{-1} \equiv \mathcal{D}_{\tilde{\gamma}'}^{-1} \mathcal{D}_{\gamma''} \tag{2.17}$$

which fixes $\tilde{\gamma}$ and hence is *not* pseudo-Anosov.

Considering the action of $\mathcal{M}_g$ on the projective space Ξ_g of measured geodesic foliations , Dehn's twists should be treated as maps with parabolic action, since they are locally conjugate to the element $\begin{pmatrix} 1 & 1 \\ 0 & 1 \end{pmatrix} \in PSL(2, \mathbf{Z})$. Moreover, recalling the presentation (2.1) of the fundamental group – which in present notation should be rewritten identifying the t_i's and u_i's with $\mathcal{A}_i$'s and $\mathcal{B}_i$'s respectively and getting rid of the h_j's (since there are no boundary components) – and noticing that its elements which act parabolically on the hyperbolic projective space are only those which may be freely homotoped into cusps and that just these elements are non-Anosov, all that remains to be done is to check – by using the presentations given above – whether $\mathcal{M}_g$ has a geometrically finite subgroup $\mathcal{S}\mathcal{M}_g$ on which it acts by conjugation. Then, *unless* the normal closure in $\pi_1(\Sigma_g)$ of the elements of the action of $\mathcal{M}_g$ on $\mathcal{S}\mathcal{M}_g$ excludes all the cusp generators[♮]

♯ This is so because the only possible overlap in the classification given above of the elements of the mapping class group is between periodic and reducible mapping classes.

♮ Because of Jørgensen's inequality [12] the elements of a discrete group can never get too close to 1.

not all of its elements $\neq \mathbf{1}$ are pseudo-Anosov.

It is worth pointing out that this conclusion does not hold for $\mathcal{M}_1$ but only for $g \geq 2$, when $\mathcal{M}_g$ has – as one can verify from the presentations – a set of elementary homeomorphisms equivalent to *global braids*. The corresponding matrix representation , when it exists , is but – as (2.16) implies – that induced from the *monodromy* representation associated with the *Lefschetz fibration* [13] of Σ_g .

3. The Ising Model

The Ising model is one of the most studied models in statistical mechanics. It describes magnetic as well as lattice gas systems. It is simply given by assigning:

i) a *lattice* Λ , enbedded in $\mathbf{R}^d$ (we shall be here mainly concerned with the case in which the dimension of the ambient space $d = 3$) ;

ii) a set of *dynamical variables* $\{\sigma_{\mathbf{i}}\,;\, \mathbf{i} \in \Lambda\}$ taking values in $\mathbf{Z}_2 \equiv \{-1, 1\}$ (they are referred to as *spins* , each sitting on a site of the lattice) ;

iii) a *Hamiltonian* H describing the energy associated with each possible distribution of spin values over Λ :

$$H = -\sum_{\mathbf{i},\mathbf{j} \in \Lambda} J_{\mathbf{i},\mathbf{j}}\, \sigma_{\mathbf{i}}\, \sigma_{\mathbf{j}}\ ; \tag{3.1}$$

where the functions $J_{\mathbf{i},\mathbf{j}}$ represent the coupling energy between site pairs (due to the global translational and rotational invariance of the system it is typically $J_{\mathbf{i},\mathbf{j}} \equiv J(|\mathbf{i} - \mathbf{j}|)$). The above form of the Hamiltonian does not include any coupling of the spins to an external field. We assume also – as it is done in general – that :

$$J(|\mathbf{i} - \mathbf{j}|) = \begin{cases} J_\alpha, & \text{if } \mathbf{i} = n.n.(\mathbf{j}) \text{ and } (\mathbf{i} - \mathbf{j}) \parallel \alpha \text{ in } \Lambda\ ; \\ 0, & \text{otherwise}\ ; \end{cases} \tag{3.2}$$

where $n.n.$ stands for *nearest neighbours* and α labels the *principal directions* in Λ (for simplicity one can think of Λ as simple cubic , in which case α assumes values x, y, z).

To *solve* the model in the frame of statistical mechanics one has to compute its *partition function* and *n-pont correlation functions* ; respectively given by :

$$Z = Z(\beta) = \sum_{\{\sigma_i\}} \exp\{-\beta H\} ; \tag{3.3}$$

and

$$C^{(n)}(\mathbf{i}_1, \ldots, \mathbf{i}_n) = \frac{1}{Z(\beta)} \sum_{\{\sigma_i\}} \sigma_{\mathbf{i}_1} \cdots \sigma_{\mathbf{i}_n} \exp\{-\beta H\} ; \tag{3.4}$$

where the sums are over all possible *configurations*, *i.e.* spin value distributions, on Λ ; $\mathbf{i}_1, \ldots, \mathbf{i}_n \in \Lambda$, and $\beta \equiv (\kappa_B T)^{-1}$; T denoting the temperature and κ_B the Boltzmann 's constant. We refer to the huge existing literature [14],[15] for the several methods devised to solve the Ising model for $d = 1$ and $d = 2$. At present *no* exact solution is known for $d = 3$.

The approach to the model which appears to be the most promising for extension to the $d = 3$ case is that referred to as the *Pfaffian* (or *dimer*) method, whose formulation holds – to a certain extent – for any number d of dimensions. We briefly review here the formulation of such a method that was proposed in ref. [1] (and references therein) as a possible candidate to attack some three-dimensional cases. It holds when Λ is homogeneous under some finitely presented (not necessarily finite) group G , and consists of a number of steps :

[*a*] the *decorated* lattice Λ_δ is derived from Λ following Fisher's scheme [16] ;

[*b*] the positional degrees of freedom in Λ_δ are relabelled in terms of a set of anticommuting *Grassmann* variables η_{g_ℓ} , in one-to-one correspondence with the group elements g_ℓ of G ;

[*c*] the group G is extended to the group $\tilde{\mathcal{G}}$ in such a way that all the bond orientations of Λ_δ compatible with the combinatorial constraints imposed by the global generalization of the *Kasteleyn*'s theorem[♭] to a non-planar case (*i.e.* to one in which the lattice Λ cannot be enbedded into a surface of genus zero, but can – yet preserving the lattice coordination – be enbedded into one of genus $g \geq 0$) and only those can be obtained as the invariant (under $\tilde{\mathcal{G}}$) set of configurations of the graph Γ covering Λ_δ 2^{2g} times.

[*d*] The reduced partition function of the model on Λ (*i.e.* (3.3) up to a numerical factor $\prod_{\alpha=1}^{d}\{\cosh(\beta J_\alpha)\}^{N_\alpha}$, where N_α is the number of sites in Λ per row in the direction α) is then given by

$$\tilde{Z}(\Lambda) = \mathrm{Pf}\,\tilde{\mathfrak{F}} \quad . \tag{3.5}$$

♭ See ref.[1] for a complete discussion of this delicate issue.

$\tilde{\mathfrak{S}}$ is the incidence matrix of Λ_δ , extended with respect to $\tilde{\mathcal{G}}$ and Pf denotes the Pfaffian (for a skew-symmetric matrix such as $\tilde{\mathfrak{S}}$, Pf $\equiv \sqrt{\det}$) .

[e] If both G and g are finite, then $\tilde{\mathcal{G}}$ is finite, and recalling that the regular representation $\mathcal{R}$ of a finite group $\tilde{\mathcal{G}}$ is the direct sum of its irreducible representations , labelled by an index j , each contained as many times as its dimension $dim\, j$, (3.5) reduces in a natural way to :

$$\tilde{\mathcal{Z}}(\Lambda) = \prod_{j^{(F)}} \left(\det \mathcal{R}\left[\tilde{\mathfrak{S}}^{(j^{(F)})}\right]\right)^{\frac{1}{2} dim\, j^{(F)}} \quad ; \tag{3.6}$$

where the extra-index F refers to *Fermionic* representations , as required by the generalized Kasteleyn's theorem, and $\mathfrak{S}^{(j)}$ is a matrix of rank j .

Of course the key point of the whole procedure is the choice of Λ and $\tilde{\mathcal{G}}$, both of which are limited by topological and combinatorial constraints. We already mentioned that Λ 's major constraint is that it should be enbeddable in a two-dimensional orientable compact surface Σ_g of genus g. As for $\tilde{\mathcal{G}}$, its definition requires a little more care.

Let us recall first that the group $\mathcal{G}$ is called the *extension* of the group G by the group Π *if*: having G presentation $G \approx < \Xi \mid \Omega >$, where Ξ denotes the set of *generators* and Ω the set of *relations* , and similarly having Π presentation $\Pi \approx < \Upsilon \mid \Theta >$; we have the exact sequence

$$1 \longrightarrow G \xrightarrow{\iota} \mathcal{G} \xrightarrow{\pi} \Pi \longrightarrow 1 \quad ; \tag{3.7}$$

and – upon denoting by φ a mapping which is the inverse of the inclusion ι ; $\varphi : \Pi \to \mathcal{G}$ with $\pi \circ \varphi = 1_\Pi$; and by $\Upsilon^{(\varphi)} \sim \varphi(\Upsilon)$ the restriction of the relations of Π to $\mathcal{G}$ – $\mathcal{G}$ has presentation :

$$\mathcal{G} \approx < \Xi \cup \Upsilon^{(\varphi)} \mid \Omega \cup \{\varphi^{-1}(v)\, \xi\, \varphi(v) \lambda_v^{-1}(\xi) \ : \ \xi \in \Xi \, ; \, v \in \Upsilon\} \cup$$

$$\cup \{\mathcal{W}_\vartheta(\xi)\, \vartheta(\varphi(v)) \ : \ \vartheta \in \Theta\} > \ . \tag{3.8}$$

where $\lambda_\varpi \ : \ G \to G$ is the automorphism of G induced by the action of the element $\varpi \in \Pi$ on G : $g_\ell \mapsto \varphi^{-1}(\varpi)\, g_\ell\, \varphi(\varpi)$; and $\mathcal{W}_\vartheta$ is some suitable word (one is to be selected for each $\vartheta \in \Theta$) bringing each element of $\mathcal{G}$ into the form $\mathcal{W}(\xi) \cdot \gamma(\varphi(v))$ for some $\gamma \in \iota(G)$.

Of course, each automorphism λ_ϖ can be altered by an inner automorphism of G with no essential effect. If we factor out the group of inner automorphisms we obtain a new mapping $\kappa : \Pi \rightarrow Out\, G = Aut\, G \,/\, Inn\, G$ which is a homomorphism and is basic for the extension, in that equivalent extensions define the same homomorphism. The triple $\{\Pi, G, \kappa\}$ is called an *abstract kernel*[17] , and a group $\mathcal{G}$ togehter with the exact sequence (3.7) is called an extension with respect to the abstract kernel if for $\gamma \in \pi^{-1}(\varpi)$, $\varpi \in \Pi$, the automorphism of G defined by $g_\iota \mapsto \iota^{-1}[\gamma^{-1}\iota(g_\iota)\,\gamma]$ belongs to the equivalence class of $\kappa(\varpi)$.

The cases of physical interest are those in which G is a *Fuchsian* group and Σ_g is a *factor surface* of G. The center $\aleph$ of G can therefore be considered as a Π-*module* with an operation in the equivalence class of $\kappa(\varpi)$, $\varpi \in \Pi$, if Π is identified with the fundamental group $\pi_1(\Sigma_g)$. Considering now the family of *cohomology* groups $H^n(\Pi, G)$, $n \geq 1$ of Π with coefficients in G (*i.e.* the cohomology groups of the *cochain complexes* defined by $\{C^n(\Pi, G)\, ,\, \partial^n\}_{n \in \mathbf{Z}}$, where $\partial^n : C^n(\Pi, G) \rightarrow C^{n+1}(\Pi, G)$ is the *boundary operator* and C^n is an n-dimensional cochain†), one notices that $C^3(\Pi, \aleph)$ – upon regarding $C^n(\Pi, \aleph)$ as an abelian group whose operation we write multiplicatively – is zero (one says that there is a trivial *obstruction*). The theorem of *Zieschang* states then the extension $\mathcal{G}$ of the abstract kernel $\{\Pi, G, \kappa\}$ exists, and that $\mathcal{G}$ is a proper subgroup of the mapping class group $\mathcal{M}_g$.

Thus the homeomorphism $Ext : G \rightarrow \tilde{\mathcal{G}}$ required in [c]‡ acts locally by attaching a Kasteleyn's phase to the circuits on Σ_g homotopic to zero, and globally by an extension by the fundamental group, *i.e.* mapping $\pi_1(\Sigma_g)$ to $\mathbf{Z}_2$. On the other hand, all possible surfaces in which Λ_δ can be enbedded are equivalent from the combinatorial point of view, and we can restrict to one *e.g.* by fixing a cut system on Σ_g. Moreover, the relations of the mapping class group – as one can easily check from the Hatcher-Thurston presentation – all follow from relations supported in certain subsurfaces of Σ_g finite in number and of genus at most 2. There follows – as explicitly derived in ref.[1] – that the most general choice for $\tilde{\mathcal{G}}$ is :

$$\tilde{\mathcal{G}} = \Re \bigotimes_{wr} S_{2g} \; ; \tag{3.9}$$

where $\otimes_{wr}$ denotes the *wreath*-product[18], whose elements can be taken to be all $2g \times 2g$ permutation matrices in which the non- zero elements have been replaced

---

† Recall that $C^n(\Pi, G)$, $n \geq 1$ is the group of all functions $f : \Pi^n \equiv \overbrace{\Pi \times \cdots \times \Pi}^{n \text{ times}} \rightarrow G$ such that $f(\varpi_1, \ldots, \varpi_n) = 0$ if some ϖ_i , $1 \leq i \leq n$ equals 1 .

‡ It should be kept in mind that maps and spaces are to be thought of in the *PL* (*piecewise-linear*) category.

by elements of $\mathfrak{R}$; whereas $\mathfrak{R} = \mathcal{M}_g / \mathcal{H}$, $\mathcal{H}$ being the stabilizer group of $\mathcal{M}_g$ defined in sect.2 , namely the subgroup of diffeomorphisms of Σ_g which preserve the isotopy class of a maximal, unordered, non separating system of g disjoint, smoothly enbedded cycles $\{\alpha_i \, ; \, i = 1, \ldots, g\}$ (non contractible and non isotopic). $\mathfrak{R}$ is then essentially generated by the elements representing homology exchange between any pair of circles (α_i, α_j) ; $i, j = 1, \ldots, g$.

The final step is now to return to the quantities of interest in statistical mechanics. Whereas very little can be said at this stage about correlation functions, we have – at least in principle – all the ingredients necessary to compute rigorously the thermodynamic equilibrium features of the model. Eq.(3.6) indeed allows us to write the *free energy* $\mathcal{F} \equiv -\kappa_B T \ln Z$ as

$$-\beta\mathcal{F} = \sum_{\alpha=1}^{d} N_\alpha \ln\cosh(\beta J_\alpha) + \frac{1}{2}\sum_{j^{(F)}} dim\, j^{(F)}\, \mathrm{Tr}\Big(\ln \mathcal{R}\big[\tilde{\mathfrak{S}}^{(j^{(F)})}\big]\Big) \; ; \qquad (3.10)$$

from which it appears clearly that while Z can be expanded in terms of *characters* of $\mathfrak{R}$, $\mathcal{F}$, as given by expression (3.10), could be rewritten in terms of invariant symmetric functions (polynomials) for $\mathfrak{R}$. This has been done in a number of cases concerning finite lattices enbedded in surfaces of genus 2 and 3 [1]. The coefficients of such an invariant expansion retain some of the original combinatorial flavour of the problem : they count the numbers of *words* in $\mathfrak{R}$ equivalent to the identity, *i.e.* provide a solution for the Dehn's word problem for the subgroup $\mathcal{H}$ of $\mathcal{M}_g$.

4. Conclusions

The whole theory described in this paper has been rigorously derived for physical problems leading to surfaces Σ_g of finite type. Actually the cases solved so far considered finite lattices generated by tesselations of Riemann surfaces of low genus (typically we studied tesselations and factor surfaces induced by the modular group and its congruence subgroups of prime level p , with p small [19]). These do have a physical relevance – even though typically in statistical mechanics one is interested in the *thermodynamic limit*, namely the limit in which the lattice volume , *i.e.* the number of sites of Λ , $N = \prod_{\alpha=1}^{d} N_\alpha$, becomes infinite – in several respects : they describe exactly finite *clusters*, which can be used as starting elements for a generalized mean field approach to the problem, and, above all, in cases when the system exhibits defects (*e.g.* dislocations or disclinations) or frustration provide a regular symmetric (in the Lobachvskiĭ space) description which simulates the irregular situation in the ambient (Euclidean) space.

The challenge of course remains of evaluating what happens in the thermodynamic limit. For instance, for a simple cubic lattice with periodic boundary conditions the genus $g \to \infty$ proportionally to N in the limit. However, the expression of $\mathcal{F}$ in terms of invariant polynomials for $\Re$, together with the fact that $\mathcal{F}/N$ must naturally remain finite, makes it plausible that a direct functional dependence of $Z(\beta)$ on g may be found, or at least a recursion relation connecting the result for (even) genus g to those for genus $(g-2k)$, $k = 1, \dots, \frac{1}{2}g$; the former being obtained from the latter by successive additions of pairs of handles – as suggested by the Hatcher-Thurston presentation. In both cases the evaluation of the limit would hopefully be feasible.

Acknowledgements

The author wishes to thank the Director and the Faculty of the Institute for Advanced Study, Princeton, NJ for the kind hospitality extended to him during the completion of this work.

References

[1] M. Rasetti, *Ising Model on Finitely Presented Groups*, in *"Group Theoretical Methods in Physics"*, M. Serdaroglu and E. Inönü eds.; Springer Verlag, Lecture Notes in Physics **180**; Berlin, 1983
plus references therein and in
G. Jacucci and M. Rasetti, J. Phys. Chem. **91**, 4970 (1987)

[2] L. Alvarez-Gaumé, G. Moore and C. Vafa, Commun. Math. Phys. **106**, 1 (1986)

[3] A.M. Polyakov, Phys. Lett. **103 B**, 207 and 211 (1981)

[4] J.S. Birman, *Braids, Links and Mapping Class Groups*; Annals of Mathematics Studies **82**; Princeton University Press; Princeton, 1974

[5] B. Wajnryb, Israel J. Math. **45**, 157 (1983)

[6] A. Hatcher and W. Thurston, Topology **19**, 221 (1980)

[7] J. Harer, Inventiones Math. **72**, 221 (1983)

[8] S. Humphries, *Generators for the Mapping Class Group*, in *Topology of Low-dimensional Manifolds*, R. Fenn, ed.; Springer Verlag, Lecture Notes in Methematics **722**; Berlin, 1979

[9] A. Montorsi and M. Rasetti, *The mapping Class Group : Homology and Linearity*, in *Group Theoretical Methods in Physics*, H.D. Döbner and T.D. Palev, eds.; World Scientific Publ. Co.; Singapore, 1988

[10] N.V. Ivanov, Uspekhi Mat. Nauk **42**:3, 49 (1987)

[11] A. Fathi, F. Laudenbach and V. Poenaru, Astérisque, **66- 67**, 33 (1979)

[12] T. Jørgensen, Amer. J. Math. **98**, 839 (1976)

[13] R. Mandelbaum and J.R. Harper, Can. Math. Soc. Conf. Proc. **2**, 35 (1982)

[14] See, for instance,

C. Domb and M.S. Green, (eds.), *Phase Transitions and Critical Phenomena*; Academic Press; London,

vol.**1** (C.J. Thompson, *One-dimensional Models : Short Range Forces*,

H.N.V. Temperley, *Two-dimensional Ising Model*), 1972

vol.**3** (C. Domb, *Ising Model*), 1974

[15] B. Mc Coy and T.T. Wu, *The Two-dimensional Ising Model*; Harvard University Press; Cambridge, 1973

and

R.J. Baxter, *Exactly Solved Models in Statistical Mechanics*; Academic Press; London, 1982

[16] M.E. Fisher, J. Math. Phys. **7**, 1776 (1966)

[17] H. Zieschang, *Finite Groups of Mapping Classes*; Springer Verlag, Lecture Notes in Mathematics **875**; Berlin, 1981

[18] A. Kerber, *Representations of Permutation Groups*; Springer Verlag, Lecture Notes in Mathematics **240**; Berlin, 1971

[19] F. Hirzebruch and G. van der Geer, *Lectures on Hilbert Modular Surfaces*; Les Presses de l'Université de Montréal; Montréal, 1981

and

R.A. Rankin, *The Modular Group and its Subgroups*; mimeographed notes, The Ramanujan Institute; Madras, 1969

ALGEBRAIC PROPERTIES AND SYMMETRIES OF INTEGRABLE EVOLUTION EQUATIONS

P.M. Santini

Dipartimento di Fisica, Università "La Sapienza"

00185 Roma, Italy

1. INTRODUCTION

In recent years many progresses have been made in the comprehension of the rich mathematical structure of nonlinear integrable systems, which very often turn out to be relevant in the description of natural phenomena [1]).

For example, well-known integrable systems in 1+1 dimensions, like the Korteweg-de Vries (KdV)

$$u_t = u_{xxx} + 6uu_x, \tag{1.1}$$

the nonlinear Schrodinger (NLS)

$$iu_t + \frac{1}{2}u_{xx} - u|u|^2 = 0, \tag{1.2}$$

the intermediate long wave (ILW)

$$u_t = \hat{T}u_{xx} + 2uu_x, (\hat{T}f)(x) := (\eta)^{-1}P\int_{-\infty}^{+\infty} coth((\pi/\eta)(\xi - x))f(\xi)d\xi, \tag{1.3a}$$

and its $\eta \to \infty$ limit, the Benjamin-Ono (BO)

$$u_t = \hat{H}u_{xx} + 2uu_x, \quad (\hat{H}f)(x) := \pi^{-1}P\int_{-\infty}^{+\infty}(\xi - x)^{-1}f(\xi)d\xi, \tag{1.3b}$$

the Chiral fields (CF)

$$(g^{-1}g_t)_x = \alpha(g^{-1}g_x)_t \tag{1.4}$$

equations and their 2+1 dimensional analogues: the Kadomtsev-Petviashvili (KP)

$$u_t = u_{xxx} + 6uu_x + \alpha^2\partial_x^{-1}u_{yy}, \tag{1.5}$$

the Davey-Stewartson (DS)

$$iu_t + \frac{1}{2}(u_{x_1x_1} + \alpha^2 u_{x_2x_2}) = uv \tag{1.6a}$$

$$v_{x_1x_1} - \alpha^2 v_{x_2x_2} = (|u|^2)_{x_1x_1} + \alpha^2(|u|^2)_{x_2x_2}, \tag{1.6b}$$

the 2+1 dimensional version of the ILW and BO

$$u_t = \hat{T}D^2u + 2uDu, \; D :== \alpha\partial_y, \tag{1.7a}$$

$$u_t = \hat{H}D^2u + 2uDu, \tag{1.7b}$$

and the self-dual Yang-Mills

$$(g^{-1}g_t)_x = \alpha(g^{-1}g_y)_z \tag{1.8}$$

equations play an important role in different areas of Physics and Applied Mathematics, from Fluid Dynamics to Plasma Physics, from Nonlinear Optics to Field Theory.

In this review we discuss the algebraic properties of nonlinear integrable systems, often referring to equations (1.1-8) as illustrative examples; in particular:

i) we give special emphasis to the fact that the equations (1.1-8) are nothing but *different* realizations of the *same* operator structure and then they are essentially *equivalent*, from an algebraic point of view! This unified perspective allows a more profound understanding of the basic algebraic properties underlying integrability (§2);

ii) we review the theory developed by Fokas and the author [2,3,4] associated with the recursion operators and the bi-hamiltonian structures of integrable systems; this theory incorporates very naturally the notion of Backlund transformations (BT) and leads to the existence of infinitely many symmetries and constants of motion in involution under distinct Poisson brackets (§3);

iii) we review the dimensional deformations [5], an approach used to explore the richness of the operator structure associated with a given integrable system, with the goal of generating integrable generalizations (deformations) in higher dimensions (§4).

2. A COMMON OPERATOR STRUCTURE UNDERLYING INTEGRABILITY

2.1 Bilocal Representation of Integrable Systems

It has been shown in recent years [6,2,3,4,11] that integrable systems like equations (1.1-8) are elements of hierarchies of integrable equations that can be represented in the following *bilocal* form

$$\delta(x-x')u'_t = \beta_n\delta(x-x')K^{(n)}(u,u'), \quad u = u(x,t), u' := u(x',t), \tag{2.1a}$$

$$K^{(n)}(u,u') := \Phi^n\hat{K}^0 \cdot 1, \tag{2.1b}$$

which obviously implies

$$u_t = \beta_n \int_{\Re^\nu} dx'\delta(x-x')K^{(n)}(u,u') = \beta_n K^{(n)}(u,u) =: \beta_n k^{(n)}(u). \tag{2.1c}$$

where x, x' are ν - dimensional vectors of components $x_i, x_{i'}$, i=1,..,ν,

$$\delta(x-x') = \prod_{i=1}^{\nu} \delta(x_i - x'_i)$$

(δ is the Dirac function), $K^{(n)}$ belongs to a suitable space S of polynomials in u=u(x,t), $u' := u(x',t)$ and their integrals and derivatives, and the "recursion operator" Φ and the "starting operator" $\hat{K}^0$ are operator valued functions on S. Through this paper m and n are nonnegative integers.

Integrable systems admit in general *more than one* bilocal representation (2.1), corresponding to more than one pair Φ, $\hat{K}^0$ [7,4], but in this presentation we shall focus only on a specific one, which is the operator structure *common* to all the equations (1.1-8).

Precisely it turns out that the integrable systems (1.1-8) can be viewed as *different bilocal representations* of the *same operator structure*, given by

$$\Phi = q^-(d^-)^{-1}, \tag{2.2a}$$

$$\hat{K}^0 = q^- C, \quad d^- C = 0, \tag{2.2b}$$

where the basic operators q^-, d^- are defined in the following way

$$\left(\begin{pmatrix} q^- \\ d^- \end{pmatrix} f\right)(x,x') := \int_{\Re^\nu} dx'' \left(\begin{pmatrix} q(x,x'') \\ d(x,x'') \end{pmatrix} f(x'',x') - f(x,x'') \begin{pmatrix} q(x'',x') \\ d(x'',x') \end{pmatrix} \right) \tag{2.3}$$

in terms of their kernels $q(x,x')$ and $d(x,x')$, that can be thought as NxN matrix functions.

If $d = \delta(x-x')\sigma$, where σ is a constant NxN diagonal matrix, then d^- reduces to the usual commutation operator between matrices

$$d^- f = \hat{\sigma} f := \sigma f - f\sigma, \tag{2.4}$$

and we can derive the following interesting "off-diagonal" reduction of the operator structure (2.2)

$$\Phi = (q_D{}^- + (1-\Pi)q_O{}^- - q_O{}^-(q_D{}^-)^{-1}\Pi q_O{}^-)\hat{\sigma}^{-1}, \tag{2.5a}$$

$$\hat{K}^0 = q_O{}^- C_D, \tag{2.5b}$$

where q_O and q_D are the off-diagonal and diagonal parts of q, Πf is the diagonal part of f and C_D is a diagonal matrix.

2.2 Different Representations

Different choices of the kernels q and d give rise to different bilocal representations of the nonlocal operators q^- and d^-; in particular *differential* equations like (1.1-8) are generated by kernels q and d proportional to the Dirac function and its derivatives; in the following we list the different bilocal realizations which generate equations (1.1-8).

Examples 2.1

1) If $d = \delta'(x_1 - x_1')$, $q = \delta(x_1 - x_1')u' + \delta''(x_1 - x_1')$ and u is a scalar; then $d^- = \partial_{+1}$, $q^- = u - u' + \partial_{+1}\partial_{-1}$, where

$$\partial_{i\pm}{}^j := (\partial_{x_i} \pm \partial_{x_i'})^j, \tag{2.6}$$

and (2.2) reduces to the bilocal representation of the KdV class; the KdV equation (1.1) is the fourth member (n=3,$\beta_3 = 1/2$) of the class (2.1) [7,4].

2) If $d = \delta'(x_1 - x_1'), q = \delta(x_1 - x_1')u' + \delta''(x_1 - x_1') + \alpha\delta'(x_2 - x_2')$, then $d^- = \partial_{1+}, q^- = u-u'+\partial_{1+}\partial_{1-}+\alpha\partial_{2+}$ and (2.2) reduces to the bilocal representation of the KP class (2.1); the KP equation (1.5) is its fourth member (n=3, β_3 =1/2)[7,4].

3) If $d = -(1/2)\delta(x_1 + i\eta - x_1')$, $q = \delta(x_1 - x_1')u' + i\delta'(x_1 - x_1')$, then

$$d^- = (1 - ee')(2e')^{-1}, \quad q^- = u - u' + i\partial_{1+},$$

$$(ef)(x,x') := f(x + i\eta, x'), \quad (e'f)(x,x') := f(x, x' + i\eta)$$

and (2.2) reduces to the bilocal representation of the ILW class; the ILW equation (1.3a) is its third member (n=2, $\beta_2 = 1/4i$) [18].

4) If $d = (-1/2)\delta(x_1 + i\eta - x_1')$, $q = \delta(x_1 - x_1')u' + \alpha\delta'(x_2 - x_2')$, then $d^- = (1-ee')(2e')^{-1}$, $q^- = u-u'+\alpha\partial_{2+}$ and (2.2) becomes the bilocal representation of the 2+1 dimensional ILW equation (1.7a), corresponding to n=2, $\beta_2 = 1/4i$ in (2.1) [5].

5) If $d = \delta'(x_1-x_1')$, $q = \delta(x_1-x_1')u'$, then $d^- = \partial_{1+}$, $q^-f = uf - fu'$ and we obtain the bilocal representation of the CF algebra; if $d = \delta'(x_1-x_1')\delta(x_2-x_2')$, $q = \delta u' + \delta(x_1-x_1')\delta'(x_2-x_2')$, then $d^- = \partial_{1+}$, $q^-f = uf - fu' + \alpha\partial_{2+}f$ and we obtain the bilocal representation of the SDYM algebra. The CF and SDYM equations are generated by these algebras via the mechanism discussed in §4.2 (see also [5]).

6) If N=2, $d = \delta\sigma$, $\sigma = diag(1,-1)$, $q_0 = \delta u_0'$, $q_D = I\delta'(x_1 - x_1')\delta(x_2 - x_2')$, then $d^-f = \hat{\sigma}f, q_D{}^- = \partial_{1+}$, $q_0{}^-f = uf - fu'$ and (2.5) reduces to the bilocal representation of the NLS class; the NLS equation comes from (2.1) for n=2,$\beta_2 = 1/4i, \bar{u}_{12} = u_{21} = u$ [2,5].

7) If N=2,$d = \delta\sigma$, $\sigma = diag(1,-1)$, $q_0 = \delta u_0'$, $q_D = \delta_1{}^1 I + \alpha\delta_2{}^1\sigma$, where

$$\delta_i^j := \delta^j(x_i - x_i') \prod_{s\neq i} \delta(x_s - x_s'), \tag{2.6a}$$

$$\delta^j(x_i - x_i') := \partial^j \delta(x_i - x_i')/\partial x_i{}^j; \tag{2.6b}$$

then $d^-f = \partial_{1+}f + \alpha(\sigma f_{x_2} + f_{x_2'}\sigma)$ and (2.4) reduces to the representation of the DS class; the DS equation coincides with (2.1) for n=2, $\beta_2 = 1/4i$ [2,5].

All the equations (1.1-8), generated by the structures (2.2,5), are described by the theory summarized in the next section.

3. ALGEBRAIC PROPERTIES OF INTEGRABLE SYSTEMS

In this section we review the basic results of the theory developed in [2,3,4,5] associated with recursion operators of integrable systems. We shall always refer to the explicit Examples 2.1 for concreteness, although the results presented have a much more general validity (see for example [5]).

3.1 Operator Structure and Canonical Lie Algebra

The operator structure of a given integrable system is completely described when we assign

i) the recursion operator Φ;

ii) the starting operator(s) $\hat{K}^0$;

iii) a Lie algebra $\tilde{\mathrm{H}}$ of suitable functions $H(x,x')$ endowed with the bracket

$$[H^{(1)},H^{(2)}]_I := \int_{\Re^\nu} dx''(H^{(1)}(x,x'')H^{(2)}(x'',x')-$$
$$-H^{(2)}(x,x'')H^{(1)}(x'',x')), \tag{3.1}$$

and containing an *abelian* subalgebra $\tilde{\mathrm{h}}$ of functions $h = h(x-x')$;

In the Examples 2.1 Φ and $\hat{K}^0$ depend on the basic operators q^-, d^- defined in (2.3). Definition (2.3) implies the following operation

$$q^{-\prime}[g(x,x')]f(x,x') = \int_{\Re^\nu} dx''(g(x,x'')f(x'',x')-$$
$$-f(x,x'')g(x'',x')) =: (g^- f)(x,x'), \tag{3.2}$$

where $q^{-\prime}[g]$ denotes the Frechet derivative of the operator q^- in the direction of the integral operators g^- and, consequently,

$$(L(q^-))'[g^-] = \partial_\epsilon L(q^- + \epsilon g^-)|_{\epsilon=0}. \tag{3.3}$$

Moreover we assume that d^- is functionally constant, namely that $(d^-)'[g] = 0$.The following properties hold:

Properties 3.1

i) The recursion operator Φ is a hereditary or Nijenhuis operator, namely

$$\Phi'[\Phi f]g - \Phi\Phi'[f]g \textit{ is symmetric w.r.t. } f \textit{ and } g, \tag{3.4}$$

ii) The starting operators $\hat{K}^0$, acting on elements of $\tilde{\mathrm{H}}$, satisfy the equation

$$[\hat{K}^0 H^{(1)}, \hat{K}^0 H^{(2)}]_d = c\Phi^i \hat{K}^0[H^{(1)},H^{(2)}]_I,\ H^{(1)},H^{(2)} \in \tilde{H}, \tag{3.5}$$

for some nonnegative integer i and constant c, in terms of the Lie bracket $[,]_d$ defined by

$$[X^{(1)},X^{(2)}]_d := X^{(1)\prime}[X^{(2)}] - X^{(2)\prime}[X^{(1)}]. \tag{3.6}$$

iii) Φ is a strong symmetry for $\hat{K}^0 H$ or, equivalently, the Lie derivative of Φ in the direction $\hat{K}^0 H$ is zero, namely

$$\Phi'[\hat{K}^0 H] + \Phi(\hat{K}^0 H)' - (\hat{K}^0 H)'\Phi = 0; \tag{3.7}$$

iv) Φ admits a factorization $\Phi = \Theta^{(2)}(\Theta^{(1)})^{-1}$ in terms of the two compatible hamiltonian operators $\Theta^{(1)} = d^-, \Theta^{(2)} = q^-$, namely if $\Theta := \Theta^{(1)} + \xi\Theta^{(2)}$, then for any constant ξ, we have that

a)

$$\Theta^* = -\Theta, \tag{3.8a}$$

where L^* denotes the adjoint of L with respect to the symmetric bilinear form

$$< g,f > := tr \int_{\Re^{2\nu}} dx dx' g(x',x)f(x,x') \tag{3.8b}$$

(the trace operation is dropped if g and f are scalars);

b) Θ satisfies the Jacobi identity w.r.t. the bracket

$$\{a, b, c\} :=< a, \ \Theta'[b]c >; \tag{3.9}$$

v) $\hat{\Gamma}^0 H := (\Theta^{(1)})^{-1} \hat{K}^0 H$ is a gradient, namely

$$(\hat{\Gamma}^0 H)' := (\hat{\Gamma}^0 H)'^{*}. \tag{3.10}$$

vi) The eigenfunctions of $\Phi^* = (d^-)^{-1} q^-$ are the "bilocal squared eigenfunctions" of the associated linear problem, namely if

$$\Phi^* w = \lambda w, \ \ w = w(x, x'), \tag{3.11}$$

then

$$w = v(x)\tilde{v}(x'), \tag{3.12}$$

where v and $\tilde{v}$ are the right and left eigenfunctions of the associated spectral problem

$$d_R v = \lambda q_R v, \ \ d_L \tilde{v} = \lambda q_L \tilde{v}, \tag{3.14a}$$

$$(\begin{pmatrix} q_R \\ d_R \end{pmatrix} f)(x) := \int dx'' \begin{pmatrix} q \\ d \end{pmatrix} (x, x'') f(x''), \tag{3.14a}$$

$$(\begin{pmatrix} q_L \\ d_L \end{pmatrix} f)(x) := \int dx'' f(x'') \begin{pmatrix} q \\ d \end{pmatrix} (x'', x), \tag{3.14b}$$

$$q_R - q_L = q^-; \tag{3.14c}$$

Remark 3.1

i) The factorization $\Phi = \Theta^{(2)}(\Theta^{(1)})^{-1}$ in terms of the two hamiltonian operators implies the following "well-coupling" condition $\Phi\Theta^{(1)} = \Theta^{(2)}\Phi^*$.

ii) Properties 3.1 are not independent, for example:

a) the factorization of a compatible pair of hamiltonian operators gives rise to a hereditary operator Φ;

b) there exists a one to one correspondence between spectral problems and their associated recursion operators [2,8]; in particular spectral problems yield hereditary operators Φ (see §4.E of [2]);

c) the recursion operator Φ algorithmically implies $\hat{K}^0 H$, furthermore if Φ is hereditary, it is also a strong symmetry for $\hat{K}^0 H$ (see §4 of [2]).

Properties 3.1 characterize the Lie algebra underlying integrability; namely we have the following

Theorem 3.1

Let μ be the linear hull of the operators $\hat{K}^{(m)} := \Phi^m \hat{K}^0$, namely

$$\mu := \{\hat{K} : \ \hat{K} = \sum_j c_j \hat{K}^{(m_j)}, m_j \geq 0, c_j \in C\}, \tag{3.15}$$

then Properties 2.1,i-iii) imply that the vector space

$$\kappa_{\tilde{H}} := \{X : \ X = \hat{K}H, \ H(x, x') \in \tilde{H}, \hat{K} \in \mu\} =: \mu(\tilde{H}) \tag{3.16}$$

is a Lie algebra endowed with the Lie bracket $[,]_d$ and

$$\mu(c_1 H^{(1)} + c_2 H^{(2)}) = c_1 \mu(H^{(1)}) + c_2 \mu(H^{(2)}), \tag{3.17a}$$

$$[\mu(H^{(1)}), \mu(H^{(2)})]_d \subset \mu([H^{(1)}, H^{(2)}]_I), H^{(i)} \in \tilde{H}, i = 1, 2, \tag{3.17b}$$

where $\mu(H)$ is the hull of vectors $\hat{K}H, \hat{K} \in \mu$.

Remark 3.2

i) Equations (3.17) imply that $\kappa_{\tilde{h}} := \{X : X = \hat{K}h, h = h(x - x') \in \tilde{h}, \hat{K} \in \mu\}$ is an abelian subalgebra of $\kappa_{\tilde{H}}$.

ii) While the abelian algebra $\kappa_{\tilde{h}}$ generates commuting (time independent) symmetries and constants of motion in involution, the nonabelian algebra $\kappa_{\tilde{H}}$ generates time dependent symmetries[3)] and is strictly related to the mastersymmetries theory[9,10)].

Definition 3.1

The vector space $\kappa_{\tilde{H}}$, satisfying the set of properties illustrated above, is called a *canonical* Lie algebra.

3.2 Distributional Projection and Directional Derivative

As we have seen in §2, the evolution equations (1.1-8) are obtained choosing q and d to be generalized functions; correspondly the basic operator q^- becomes a bilocal operator u^-, and Φ and $\hat{K}^0$ assume a bilocal structure involving the function u evaluated at two different points x and x' (see Examples 2.1).

Since $\kappa_{\tilde{H}}$ is canonical with respect to the Frechet derivative of the nonlocal operator q^-, *any* particular choice (like the one given in Examples 2.1) of $q(x, x')$ preserves the canonicity of the obtained bilocal algebra with respect to the so called *directional d-derivative*

$$u^-{}_d[g(x, x')] := q^{-'}[g(x, x')] = g^-. \tag{3.18}$$

of the bilocal operator u^-.For this reason any bilocal realization of an operator canonical Lie algebra $\kappa_{\tilde{H}}$ is hereafter called canonical too.

The d-derivative of the bilocal operator u^- satisfies the following *projective* formula

$$u^-{}_d[\delta(x - x')g(x, x')]f(x, x') = g(x, x)f(x, x') - f(x, x')g(x', x') =: u^-{}_f[g]f(x, x'), \tag{3.19}$$

where the subscript f denotes the usual Frechet derivative with respect to u, namely

$$L_f(u, u')[g] = \partial_\epsilon L(u + \epsilon g(x, x), u' + \epsilon g(x', x'))|_{\epsilon=0}. \tag{3.20}$$

The d-derivative and the usual Frechet derivative are the main tool of this theory.

3.3 Integrable Evolution Equations

Now we are ready to state the main theorem for integrable systems.

Theorem 3.2

Let $\Theta^{(1)} + \xi\Theta^{(2)}$ be a bilocal hamiltonian operator for all constants ξ . Define $\Phi := \Theta^{(2)}(\Theta^{(1)})^{-1}$, $\hat{\Gamma}^0 := (\Theta^{(1)})^{-1}\hat{K}^0$ and assume that

i) Φ is a strong symmetry for $\hat{K}^0 H$;

ii) the starting symmetries $\hat{K}^{\circ}H$ satisfy equation (3.5);
then
1) Φ is a hereditary or Nijenhuis operator;
2) $\Phi^m\Theta^{(1)}$ are hamiltonian operators;
3) $\kappa_{\tilde{H}}$ is a canonical Lie algebra endowed with the Lie bracket $[,]_d$.
Moreover if, in addition,
iii) $\hat{\Gamma}^{\circ}H$ is an extended gradient,
iv) the distribution $\delta(x-x')K^{(n)}(u,u')$ belongs to the abelian sub-algebra $\kappa_{\tilde{h}}$,
then
4) $\Sigma^{(m)} := \Phi^m\hat{K}^{\circ}\cdot 1$ and $\Gamma^{(m)} := (\Theta^{(1)})^{-1}\Sigma^{(m)} = (\Phi^*)^m\hat{\Gamma}^{\circ}\cdot 1$ are extended symmetries and extended gradients of the conserved quantities I_n in involution respectively, for equation (2.1), namely

$$\Sigma^{(m)}{}_f[K^{(n)}(u,u)] = (\delta K^{(n)})_d[\Sigma^{(m)}], \tag{3.21a}$$

$$\Gamma^{(m)}{}_f[K^{(n)}(u,u)] = -(\delta K^{(n)})_d{}^*[\Gamma^{(m)}], \tag{3.21b}$$

$$(\Gamma^{(m)})_d = (\Gamma^{(m)})_d{}^* \Leftrightarrow I_{md}[f] =< \Gamma^{(m)}, f >, \tag{3.21c}$$

$$\{I_m, I_n\}_i := < \delta\Gamma^{(m)}, \Theta^{(i)}\Gamma^{(n)} >, \; i = 1,2; \tag{3.21d}$$

5) equations (2.1a,b) are *extended bi-hamiltonian systems*, since they can be written in the following two extended hamiltonian forms

$$\delta(x-x')u'_t = \delta(x-x')\Theta^{(1)}\Gamma^{(n)} = \delta(x-x')\Theta^{(2)}\Gamma^{(n-1)}; \tag{3.22}$$

6) $\sigma^{(m)} = \sigma^{(m)}(u) := \Sigma^{(m)}(u,u)$ and $\gamma^{(m)} = \gamma^{(m)}(u) := \Gamma^{(m)}(u,u)$ are commuting symmetries and gradients of the conserved quantities I_m in involution respectively for equations (2.1), namely

$$\sigma^{(m)}{}_f[k^{(n)}] = k^{(n)}[\sigma^{(m)}], \tag{3.23a}$$

$$\gamma^{(m)}{}_f[k^{(n)}] = -k^{(n)}{}_f{}^+[\gamma^{(m)}], \tag{3.23b}$$

$$\gamma^{(m)}{}_f = \gamma^{(m)}{}_f{}^+ \Leftrightarrow I_{mf}[f] = (\gamma^{(m)}, f) := \int_{\Re^\nu} dx\gamma^{(m)}f; \tag{3.23c}$$

7) the equations $\Sigma^{(m)} = \Sigma^{(m)}(u,u') = 0$ are auto-BT's of equations (2.1).
Remark 3.3
i)Theorem 3.2 shows that symmetries and BT's of an integrable system originate from the same entity: the *extended symmetry*, defined in equation (2.21a). In this equation two *different* operations on bilocal structures appear: the usual Frechet derivative and the novel d-derivative, which takes account of the nonlocal origin of bilocal operators.
ii) If the hypothesis iv) of Theorem 3.2 is satisfied, namely if

$$\delta(x-x')K^{(n)}(u,u') = \sum_{l=1}^{n} b_{n,l}\Phi^{n-l}\hat{K}^{\circ}\cdot h^{(l)}, \;\; h^{(l)} \in \tilde{h}, \tag{3.24}$$

for some constants $b_{n,l}$, then the operators $(\delta K^{(n)})_d$, appearing in the equation (2.36), are well defined and read

$$(\delta K^{(n)})_d = \sum_{l=1}^{n} b_{n,l}(\Phi^{n-l}\hat{K}^{\circ}\cdot h^{(l)})_d. \tag{3.25}$$

iii) In the BT $\Sigma(u, u') = 0$, the functions u, u' are now viewed as two different solutions $u = u(x)$, $u' = u'(x)$ of $u_t = k(u)$ and, correspondly, in equations (3.21) the even derivative ∂_{i+} of the operator $(\delta K^{(n)})_d$ is replaced by ∂x_i; the odd derivative ∂_{i-} is absent (see theorem 4.1 of [2]).

Theorem 3.2 provides a nonambiguous definition of integrability.

Definition 3.2

A nonlinear evolutionary system $u_t = k(u)$ is integrable iff it is associated with a canonical Lie algebra $\kappa_{\tilde{H}}$ in the way prescribed by Theorem 3.2, namely the following *two* conditions must be simultaneously satisfied:

A) there exists an underlying canonical Lie algebra $\kappa_{\tilde{H}}$ generated by the bilocal operators Φ and $\hat{K}^0$;

B) the evolution equation takes the distributional form (2.1a), and $\delta K^{(n)}(u, u')$ belongs to the canonical algebra $\kappa_{\tilde{h}}$, namely equation (3.24) holds.

Conditions A and B are at the basis of the deformation approach presented in the following section.

4. DEFORMATIONS AND INTEGRABLE SYSTEMS

As we have seen in the previous sections, the evolution equations (1.1-8) are different bilocal realizations of the same (and elementary!) operator structures (2.2,4), corresponding to the different specifications of the basic operators q^-, d^- illustrated in the Examples 2.1.

A natural question arises at this point: does *any* choice of q and d give rise to integrable systems and, if not, what are the necessary limitations? Equivalently, given a couple of basic operators q^- and d^- which give rise to an integrable system, how can one *deforme* them in new operators $q^- \rightarrow q^- + \chi^-$, in order to generate other integrable systems, eventually in higher or lower dimensions?

These and other questions are exaustively answered by the " deformation approach summarized in this section. This approach is inspired by the theory reviewed in §3, which is based on two facts:

i) the existence of a *canonical* Lie algebra,

ii) the associated evolution equations belong to this algebra.

Consequently the dimensional deformation approach is based on the following two steps:

1) we introduce a suitable dimensional deformation of the bilocal Lie algebra of the integrable system under consideration, in order to generate a higher dimensional generalization; this can always be done in a straightforward manner, once the operator structure of the integrable system is known;

2) we check whether the obtained multidimensional Lie algebra can be used to generate integrable multidimensional systems. This check consists of computing a simple commutator relation and it is also very straightforward.

This approach allows us to derive in a natural, straightforward and unified manner known examples of integrable multidimensional systems and some novel ones which, by construction, are integrable too [5].

4.1 The Deformations Approach

A deformation of the Lie algebra $\kappa_{\tilde{H}}$ is obtained through a suitable deformation of the basic operators q^-, d^-. This is acheived replacing their kernels q and/or d by

$q+\chi$ and/or $d+\chi$, $\chi=\chi(x,x')$; then, correspondly,

$$q^- \to q^- + \chi^-. \tag{4.1}$$

Of particular interest for us are deformations of the type (4.1) which are associated with an increase of the number of dimensions of the vectors x and x' and, correspondly, give rise to higher dimensional systems; they are called *dimensional deformations.*

An example of *dimensional deformation* from, say, a one dimensional canonical algebra to a three dimensional one is given by

$$\chi^- = \alpha(\partial_{x_2} - \partial_{x_2'}) + \beta(\partial^2_{x_3} - \partial^2_{x_3'}), \tag{4.2a}$$

corresponding, via the mechanism (2.3), to the kernel

$$\chi = \alpha\delta_2^1 + \beta\delta_3^2, \tag{4.2b}$$

Remark 4.1

With respect to the d-derivative

$$(q^- + \chi^-)_d[g] := g^- \tag{4.3}$$

the obtained algebra is still canonical (see also §3.2), then deformations of a canonical Lie algebra allow to construct very easily multidimensional canonical Lie algebras satisfying condition A of §3. On the other hand it is also very easy to check whether or not the associated multidimensional system (2.1a,b) belongs to this algebra (condition B of §3). In order to check condition B or, equivalently, equation (3.24), it is indeed enough to compute the commutators $[\Phi,h]$, $[\hat{K}^0,h]$. It is perhaps convenient to illustrate this in two explicit examples.

4.2 Examples of Dimensional Deformations

1) If we consider the bilocal algebra generating the NxN NLS class, described in Examples 2.1, 4), and we look for deformations χ of $q_D = I\delta_1{}'$ of the type

$$\chi = \sum_{i=2}^{\nu} J_i\delta_i^1, \tag{4.4a}$$

then the "new" operator $q_D{}^-$ reads

$$q_D{}^- f = \sum_{i=1}^{\nu}(J_i f_{x_i} + f_{x_i'} J_i), \quad J_1 = I, \tag{4.4b}$$

and

$$[\Phi,\delta] = (\sum_{i=2}^{\nu}\delta_i^1\hat{J}_i)(\hat{a})^{-1}, \; [\hat{K}^0,\delta] = 0. \tag{4.5}$$

Condition B is satisfied if

$$\hat{J}_i = b_i\hat{a} \Leftrightarrow J_i = e_i I + b_i a, \tag{4.6}$$

since it is straightforward to check that $K^{(n)}$ satisfy equation (3.24) with $b_{n,l} = (-1)^l$ and $h^{(l)} := (\sum_{i=2}^{\nu} b_i \partial_{x_i})^l \delta(x - x')$. The N-wave interaction in $\nu + 1$ dimensions

$$u_{ij_t} = (a_i - a_j)^{-1} \sum_{l=1}^{\nu} ((e_l + b_l a_i) c_j - (e_l + b_l a_j) c_i) u_{ij_{x_l}} - \sum_{k=1}^{N} (\frac{c_i - c_k}{a_i - a_k} - \frac{c_k - c_j}{a_k - a_j}) u_{ik} u_{kj} \tag{4.7}$$

is the second member of the hierarchy (2.1) for n=1, $\beta_1 = 1$.

Remark 4.2

It is known [12] that the algebraic constraint (4.6) implies the existence of a change of variables reducing the class (2.1) to 2+1 dimensional equations; then only the $\nu = 2$ case determines a nontrivial increase of dimensions.

2) It is very easy to convince oneself that, for a generic deformation of the canonical Lie algebra, condition B is *not* satisfied. Then the associated multidimensional systems are *not* integrable, according to Definition 3.2, and *none* of the beautiful properties stated in Theorem 3.2 are enjoyed by the associated class of evolution equations.

For example, if $\nu = 3, q = \delta u'$, $d = \delta_1{}^1$, then the deformation $\chi = \alpha \delta_2{}^1 + \beta \delta_3{}^2$ of the operator algebra (2.2) implies that

$$[\Phi, h] = 2\beta h' \partial_{3+} \partial_{1+}{}^{-1}, \ [\hat{K}^0, h] = 2\beta h' \partial_{3+}; \tag{4.8}$$

consequently it is easy to see that $\delta \Phi^n \hat{K}^0 \cdot 1 \notin \kappa_{\tilde{h}}$ and the associated 3+1 dimensional equations are not integrable.

The problem of the existence of nontrivial integrable structures in more than 2+1 dimensions is open and it is, in our opinion, one of the most outstanding one. This approach, in its simplicity, allows to focus concretely the associated theoretical and technical difficulties.

Technically speaking, the almost total absence of integrable systems in multidimensions (more than 2+1 dimensions) is related to the difficulty of satisfying condition B for a generic deformation of the *available* canonical algebras.

4.3 Elementary Deformations and Integrable Systems with an Arbitrary Number of Time Variables.

The simplest possible deformation of a canonical Lie algebra is of course achieved

i) increasing the number of dimensions of the vectors x, x',

ii) keeping unchanged the structure of q and d, namely choosing $\chi = 0$ in formula (4.1).

We refer to it as an *elementary* dimensional deformation.

This apparently trivial deformation gives rise to interesting integrable multidimensional systems containing an *arbitrary* number of variables that can not be distinguished from the time variable t. Their derivation relies upon the existence of *ghost* classes (see [5]), a common feature the bilocal realizations of the algebras (2.2.5) for which

$$[\Phi, h] = [\hat{K}^0, h] = 0, \ \ \hat{K}^0 = u^-, \ u^- f = uf - fu', \tag{4.9}$$

and it consists of the following two steps:

i) we introduce an elementary deformation, making the vectors x and x' ν-dimensional;

ii) we make use of the following distributional identity

$$\delta_i^1 \Phi^{n_i} u^- \cdot 1 = -\delta(x-x')\Phi^{n_i} u'_{x'_i}, \tag{4.10}$$

which follows directly from equation (4.9). Then the distributional evolution equation

$$\delta u'_t = \delta K^{(n)}(u,u'), \tag{4.11}$$

$$K^{(n)}(u,u') = \sum_{i=1}^{\nu} c_i \Phi^{n_i} u'_{x'_i} + \beta_n \Phi^n \hat{M} \cdot 1, \tag{4.12}$$

is integrable, since the identity (4.10) implies

$$\delta K^{(n)}(u,u') = -\sum_{i=1}^{\nu} c_i \Phi^{n_i} u^- \cdot \delta_i^1 + \beta_n \Phi^n \hat{M} \cdot \delta \in \kappa_{\tilde{h}}, \tag{4.13}$$

namely condition B is satisfied.

Then the results of Theorem 3.2 apply to the class of evolution equations (4.11,12), whose projection gives

$$u_t = \sum_{i=1}^{\nu} c_i \phi^{n_i} u_{x_i} + \beta_n \phi^n \hat{m} \cdot 1 =: k^{(n)}, \tag{4.14a}$$

$$\phi := (\hat{u} + \alpha \partial_{x_2})(\partial_{x_1})^{-1}, \tag{4.14b}$$

$$\hat{m} := \hat{u} + \alpha \partial_{x_2}. \tag{4.14c}$$

In particular $\Sigma(u,u') = a_1 \Phi^{m_1} u^- \cdot 1 + a_2 \Phi^{m_2} \hat{M} \cdot 1$ and $\Gamma(u,u') = (\Theta^{(1)})^{-1} \Sigma(u,u')$ are extended symmetries (BT's) and extended gradients of conserved quantities in involution for the evolution equations (4.14).

Remark 4.3

i) The evolution equation (4.14) shows that the variables x_i, $2 \leq i \leq \nu$, introduced via the deformation approach described in this section, play the *same* role as t; for this reason they will be called *time variables*. In this sense, it is worthwhile to remark again that integrable *evolution* equations should be correctly viewed as *elementary dimensional* deformations (corresponding to the introduction of the variable $x_2 = t$) of their *stationary* analogues.

ii) Calogero was the first to derive examples of integrable equations with an arbitrary number of time variables [13,14].

Examples 4.2

i) In order to derive the SDYM equation (1.8) we notice that the bilocal algebra of Examples 2.1,5) satisfies conditions (4.9) for c=I. Then the distributional identity (4.10) holds and the class of integrable multidimensional systems (4.14) can be constructed. The SDYM equation (1.8) is equivalent, with the positions $u = g^{-1} g_{x_2}$, $u_{x_3} = (g^{-1} g_t)_{x_1}$, to the equation [15]

$$u_t = \alpha \partial_{x_1}{}^{-1} u_{x_2 x_3} + u_- \partial_{x_1}{}^{-1} u_{x_3}, \tag{4.15}$$

obtained from (4.14a) for $c_i = 0, i \neq 3, c_3 = 1,\ n_3 = 1$. The CF equation is obtained from (1.8) in the reduction $x_2 = x_1,\ x_3 = t$ [15,16].

ii) Elementary deformations can also be used to generate the following novel (and peculiar) class of multidimensional equations

$$v_t = \beta_n \prod_{i=1}^{n+1} v_{x_i} =: \beta_n k^{(n)}. \tag{4.16}$$

If we consider the canonical algebra (2.2) with $q = \delta u', d = \delta'(x_1 - x_1{}')$ and u scalar, the identity (4.10) generalizes to

$$(\prod_{i=1}^{n+1} \delta^1(x_i - x_i'))\Sigma^{(n)}(u, u') = (-1)^{n+1}\delta(x - x')(\prod_{i=1}^{n+1} \partial_{x_1'}^{-1}\partial_{x_i'}u')_{x_1'}, \tag{4.17a}$$

$$\Sigma^{(n)}(u, u') := \Phi^n u^- \cdot 1 = ((u - u')\partial_{1+}^{-1})(u - u') =$$

$$\frac{1}{(n+1)!}(\partial_{1+}(\partial_{1+}^{-1}(u - u'))^n). \tag{4.17b}$$

Then the evolution equation

$$\delta(x - x')u_t' = \beta_n \delta(x - x')(\prod_{i=1}^{n+1} \partial_{x_1'}^{-1}\partial_{x_i'}u')_{x_1'} \tag{4.18}$$

belongs to the canonical algebra $\kappa_{\tilde{h}}$ and reduces to equation (4.16) via the change of variable $v_{x_1} = u$.

Remark 4.4

i) $\Sigma^{(m)} = ((m+1)!)^{-1}(\partial_{1+}((v - v')^{m+1}))$ are extended symmetries of the evolution equations (4.18), namely equations (3.36a) are satisfied, *but* the corresponding BT's $\Sigma^{(m)}(v, v') = 0$ are trivial (they imply $v' = v$), and the corresponding commuting symmetries are zero.

ii) Analogously $\Gamma^{(m)} = \partial_{1+}^{-1}\Sigma^{(m)}$ are extended gradients of conserved quantities for equations (4.18), but the projected gradients $\gamma^{(m)}$ are zero and then the corresponding constants of motion are trivial (they do not depend on v).

The only result of Theorem 2.2 which does not trivialize is

$$k^{(n)}{}_f[k^{(m)}] = k^{(m)}{}_f[k^{(n)}], \tag{4.19}$$

which follows from Th.4. of [2]. Then $k^{(n)}$ form an infinite dimensional abelian Lie algebra or, in other words, the evolution equations (4.16) describe commuting flows.

iii) It turns out that equations (4.16) can be integrated using the methods of the characteristics [17].

REFERENCES

1. F.Calogero and A.Degasperis,Spectral Transform an Solitons I, Studies in Mathematics and its Applications, North-Holland Publishing Company, Amsterdam,1982. M.Ablowitz and H.Segur,Solitons and the Inverse Scattering Transform,in SIAM Studies in Applied Mathematics, No.4 CSIAM,Philadelphia,1981. S.Novikov, S.

V.Manakov, L.P.Pitaevskii and V.E.Zakharov, Theory of Solitons, The Inverse Scattering Method, Contemporary Soviet Mathemaics,Consultants Bureau. New York and London,1984.

2. P.M.Santini and A.S.Fokas,Recursion operators and bi-hamiltonian structures in multidimensions I, Comm.Math.Phys.**115**, 375 (1988).
3. A.S.Fokas and P.M.Santini, Recursion operators and bi-hamiltonian structures in multidimensions II, Comm.Math.Phys.**116**, 449 (1988).
4. P.M.Santini and A.S.Fokas, The bi-hamiltonian formulations of integrable evolution equations in multidimensions, in Nonlinear Evolutions, Proceedings of the IV Workshop on Nonlinear Evolution Equations and Dynamical Systems; edited by J.Leon, World Scientific Publishing Company, Singapore (1988).
5. P.M.Santini, Dimensional deformations of integrable systems, an approach to integrability in multidimensions.I, Preprint 586 Dipartimento di Fisica, Universita' di Roma I,1988; Inverse Problems (submitted to).
6. A.S.Fokas and P.M.Santini, Stud.Appl.Math.**75**, 179 (1986).
7. A.S.Fokas and P.M.Santini, J.Math.Phys.**29**, 604 (1988).
8. M.Bruschi, private communication.
9. A.S.Fokas and B.Fuchssteiner, Phys.Lett.**86A**, 341 (1981).
10. W.Oevel and B.Fuchssteiner,Phys.Lett.**88A**, 323 (1982). H.H.Chen, Y.C.Lee and J.E.Lin, Physica **9D**, 439 (1983). B.Fuchssteiner, Progr.Theoret.Phys.**70**,150 (1983).
11. B.G.Konopelchenko, Recursion and group theoretical structures of the integrable equations in 1+1 and 2+1 dimensions. Bilocal approach, Preprint INP Novosibirsk, USSR, 1987.
12. S.V.Manakov: private communication. A.S.Fokas, Phys.Rev.Lett. **57**, 159 (1986).
13. F.Calogero, Lett. Nuovo Cimento **14**, 443, (1975).
14. F.Calogero, Lett.Nuovo Cimento **14**, 537 (1975); F.Calogero and A.Degasperis, Nuovo Cimento **32B**, 201 (1976); Nuovo Cimento **39B**, 1 (1977).
15. M.Bruschi, D.Levi and O.Ragnisco, Lett.Nuovo Cimento **33**, 263 (1982).
16. M.Bruschi, D.Levi and O.Ragnisco, Phys.Lett.**88A**, 379 (1982).
17. We thank W.Craig for showing us that equation (4.16) can be integrated using the method of the characteristics.
18. P.M.Santini, Bi-hamiltonian formulations of the Intermediate Long Wave equation,Preprint INS 80,Clarkson University,1987;Inverse Problems (submitted to).

ENERGY OF SKYRMIONIC CONFIGURATIONS

E.Sorace and M.Tarlini
Istituto Nazionale di Fisica Nucleare
Largo E.Fermi 2
I-50125 Firenze (Italy)

1. The Skyrmion

At the beginning of the '60 Skyrme [1] during his research [2] of a non linear field theory of fundamental elementary particle strong interactions proposed the following Lagrangian (with massless pion fields):

$$L = \frac{F_\pi{}^2}{16} Tr(\partial_\mu U \partial_\mu U^\dagger) + \frac{1}{32e^2} Tr([U^\dagger \partial_\mu U, U^\dagger \partial_\nu U]^2) \tag{1.1}$$

where the Skyrme field $U(x)$, $x \epsilon R^{3,1}$, is a $SU(2)$ matrix:

$$U : R^{3,1} \to SU(2).$$

The quartic term in the derivatives is the simplest correction to the chiral kinetic term to escape the Derrick theorem. The F_π is the pionic coupling constant and e is the only free parameter.

The associated Hamiltonian, adding the chiral symmetry breaking term that takes into account the non zero mass of the pionic field, is:

$$H = \frac{1}{16} \int dx^3 \{F_\pi{}^2 Tr(V_0{}^2 + \vec{V}^2) + \frac{1}{e^2} Tr(\vec{C}^2 + \vec{D}^2) - 2m_\pi{}^2 F_\pi{}^2 Tr(U - 1)\} \tag{1.2}$$

with

$$V_0 = iU^\dagger \partial_0 U \ , \quad \vec{V} = iU^\dagger \vec{\partial} U \ , \quad \vec{C} = \frac{1}{i}[V_0, \vec{V}] \ , \quad \vec{D} = \frac{1}{2i}(\vec{V} \times \vec{V}).$$

At any given time the finite energy configurations which imply constant value of U at infinity are continuous mappings:

$$U : S^3 \to S^3$$

where the domain is the R^3 manifold with the infinity boundary reduced to a single point and the image is the manifold of $SU(2)$. So the fields $U(\vec{x})$ are classified by the homotopy group $\Pi_3(S^3)$. Then the mere continuity of the evolution in time makes the fields to be characterized by integer topological numbers [3]. The topological invariant interpreted as the barionic number B [4] is given by:

$$B = \frac{1}{24\pi^2}\epsilon_{ijk} \int dx^3 Tr\left[V_i V_j V_k\right] \tag{1.3}$$

An elegant geometrical interpretation of static energies of this form has been recently given by N.S.Manton [5]. The field is considered as a mapping between two riemannian manifolds of the same dimension and the energy will then represent a measure of the distortion of the metric by the mapping. This can be realized by means of a symmetric and positive defined strain tensor D whose invariants must be the constitutive terms of the energy. In the tree dimensional case they are $Tr(D)$; $\frac{1}{2}Tr(D^2)$; $Det(D)$. One can easily see that $Tr(D)$ is the kinetic energy term while the combination $\frac{1}{2}(Tr(D))^2 - \frac{1}{2}Tr(D^2)$ realizes the Skyrme quartic term. The $Det(D)$ corresponds to the topological number and this allows a simple explanation of the Bogomol'nyi type lower bound for the ground state energy.

An explicit representation of $U(x)$ in terms of real valued fields is given by:

$$U = \exp\left[i\pi_a(x)\tau_a\right]. \tag{1.4}$$

$\pi_a(x)$ is the pion field and τ_a the Pauli matrices.

The field equations are in general extremely complicated. The energy of the static configurations ($V_0 = \vec{C} = 0$) in terms of $\pi_a(\vec{x})$ is:

$$\begin{aligned} E = \int dx^3 \{ &\frac{{F_\pi}^2}{8}\left[(\partial_i|\pi|)^2 + \sin^2|\pi|\,(\partial_i n_a)^2\right] \\ &+ \frac{1}{2e^2}\sin^2|\pi|\left[(\partial_i n_a)^2\,(\partial_i|\pi|)^2 - (\partial_i n_a \partial_i|\pi|)(\partial_k n_a \partial_k|\pi|)\right] \\ &+ \frac{1}{4e^2}\sin^4|\pi|\left[(\partial_i n_a)^2\,(\partial_k n_b)^2 - (\partial_i n_a \partial_k n_a)^2\right] \\ &+ \frac{1}{4}{m_\pi}^2{F_\pi}^2(1-\cos|\pi|)\} \end{aligned} \tag{1.5}$$

with $n_a = \frac{\pi_a}{|\pi|}$ ($|\pi| = 0$ at ∞). It is obvious then the necessity of good ansatz and an adequate numerical strategy.

The $B = 1$ spherical skyrmion is obtained by solving the variational state equation derived by (1.5) using the Skyrme's original and successful ansatz:

$$\pi_a(\vec{x}) = F(r)\frac{x_a}{r} \qquad \begin{cases} F(0) = \pi \\ F(\infty) = 0. \end{cases}$$

The equation $\delta E[F] = 0$ can be solved numerically as an ordinary differential equation. Nobody knows the analytical form of $F(r)$, it is possible however to get good approximate expressions, for example the continuous fraction (for $m_\pi = 0$):

$$\overline{F}(r) = 2\arcsin\cfrac{1}{\alpha_0 r^2 + \cfrac{1}{\alpha_1 r^2 + \cfrac{1}{\cdots}}} \tag{1.6}$$

reproduces the correct asymptotic behaviours near zero and infinity at every step. At the first step one finds for the static mass:

$$\overline{M}_0 = 2\frac{F_\pi}{e}\pi^2\{(\frac{51}{8} - \frac{8}{\sqrt{2}})(\frac{7}{2} + \frac{137}{128})\}^{\frac{1}{2}} \approx \frac{F_\pi}{e}36.845 \tag{1.7}$$

At the second step the value of $\overline{M}_0$ is the one obtained with the numerical solution ($\approx \frac{F_\pi}{e}36.50$) for $\alpha_0 = .2072$, $\alpha_1 = .0897$; from the third step the value of the mass is stabilized at about $\frac{F_\pi}{e}36.47$.

2. The Spinning Skyrmion

Let us give some comments about the quantization of the field configurations with $B = 1$ and an outlook of its physical meaning. The very important point is that the Skyrme field solution U has less invariances than the Lagrangian. The rotations of U under $SU(2)$ matrices (and also translations) are solutions like U, but are different from U. This means that we have zero modes (i.e. eigenvalues zero in the linearized equation of small oscillations around U) and we can associate dynamical collective coordinates to the parameters of the transformations. About this point we remind the geodesic motion in the so called "moduli space" as in the lecture of Horvathy given in this conference. Moreover let us mention the recent global concept introduced by Manton [6] to treat collective motion of weakly interacting solitons, this motion takes place on "an unstable manifold" constituted by the union of the paths of steepest descend of the family of all the saddle points configurations of the field energy appearing when the flat valleys of the given minimum have acquired a weak negative slope.

Of course a complete semiclassical quantization of the Skyrme field theory should consider even the pion fluctuations around the skyrmion [7] taking into account the presence of zero modes, this is of course absolutely necessary to get pion–nucleon system properties.

As is well known however the quantization of the Skyrme solution (see ref. [8]) as a finite–dimensional system produced very good nucleons from spinning skyrmions. The semiclassical Hamiltonian is:

$$H = M[F] + \frac{I(I+1)}{2\Lambda[F]} \tag{2.1}$$

The spin is equal to the isospin for the "equivariance" of the static solution, F is the static solution and Λ is the momentum of inertia. The idea of varying the energy (2.1) with respect to F to find a better approximation for the rotating quantized skyrmion [9] seems not to be a good one. In fact it has been shown [10], proposing an alternative equation of motion, that the optimum one can get in this reference frame and with these collective coordinates has been reached by the procedure of ref. [8].

3. The $B = 2$ Configurations

Let us now consider the $B = 2$ sector, the main question is how to exploit the classical field theory to construct a physically sensible approximation of two interacting nucleons.

A stable (i.e. $E < 2M$) configuration with $B = 2$ was found numerically last year by two groups [11]. The form of the solution is given in cylindrical coordinates (ρ, y, φ) by two independent fields:

$$\begin{aligned} \pi_1 &= \eta_1(\rho, y)\cos 2\varphi, \\ \pi_2 &= \eta_2(\rho, y), \\ \pi_3 &= \eta_1(\rho, y)\sin 2\varphi. \end{aligned} \tag{3.1}$$

The russian authors have guessed this form, while the other group has found it as the result of numerical calculations on a possible configuration without any continuous symmetry, this last numerical work was correctly interpreted by Manton [12] who guessed the axial continuous symmetry. So it is highly probable that it is actually the exact $B = 2$ solution. This solution is compatible with the minimal energy configuration at great distance given by the product ansatz [13]:

$$U_{pa} = U(x, y, z - \frac{r_0}{2})AU(x, y, z + \frac{r_0}{2})A^\dagger \tag{3.2}$$

where A is a $SU(2)$ matrix representing a rotation by an angle π about an axis perpendicular to the z–axis.

Of course the static $B = 2$ solution can be semiclassically quantized on its zero modes, but the situation at a first glance is rather embarrassing if one wants to describe interacting nucleons. The rotations parameters allowed with arbitrary value in the $B = 1$ skyrmions are now frozen at a particular value to minimize the energy, at large skyrmion separation freezing the relative orientation means not to give the quantum numbers of the nucleons to the separated skyrmions [6]. In some sense even the peculiar toroidal shape of the iso–energy surfaces seems to stress that at that point we have a different mechanical system. So unless one finds a way out we agree that the interpretation of these quantized configurations as interacting nucleons is "questionable" [14]. However a precise analysis of this $B = 2$ solution and the phenomenological implications of its collective coordinates quantization is in order [15], a clear and exhaustive discussion can be found in the lecture of Dothan in this volume.

We studied a simple model (it is a constrained solution) which tries to avoid some difficulties and allows for stable composite systems [16]. We used a bipolar coordinate system with the two poles identified to the centers of two $B = 1$ static skyrmions and positioned at $+a$ and $-a$ on the z–axis:

$$\begin{aligned} x &= \frac{a\sin\vartheta\cos\varphi}{\cosh\eta - \cos\vartheta} & \eta &= \ln\frac{r_-}{r_+} \\ y &= \frac{a\sin\vartheta\sin\varphi}{\cosh\eta - \cos\vartheta} & \vartheta &= \arctan\frac{2a\rho}{r^2 - a^2} \\ z &= \frac{a\sinh\eta}{\cosh\eta - \cos\vartheta} & \varphi &= \arctan\frac{y}{x} \end{aligned} \tag{3.3}$$

where $r_\pm$ is the distance from $(0, 0, \pm a)$, r from the origin and ρ from the z–axis. The interesting properties of such coordinates are:

$$\eta \to \pm\infty \quad \Rightarrow \quad |\eta| \approx \ln\frac{2a}{r_\pm}$$

and

$$a \to \infty \quad \Rightarrow \quad \eta_{>0} = f(r_+), \quad \eta_{<0} = f(r_-)$$

This suggests in first approximation to search for configurations of the form:

$$U = \exp\{i f(\eta) \hat{\eta} \cdot \vec{\tau}\} \tag{3.4}$$

with boundary conditions $f(+\infty) = \pi$, $f(-\infty) = -\pi$ which imply $f(0) = 0$ and $B = 2$. We solved numerically the equation of state using the ansatz (3.4) for the function $f(\eta)$ for every value of a.

For $a \to \infty$ we then have $U(\eta) = U_{B=1}(r_+)U_{B=1}(r_-)$ and total mass $M = 2m_{B=1}$. Because $U(0) = 1$ one can factorize smoothly

$$U(\eta) = U_+(\eta)U_-(\eta) \tag{3.5}$$

with

$$U_+(\eta) = U(\eta) \quad \text{for} \quad \eta \geq 0 \qquad U_+(\eta) = 1 \quad \text{for} \quad \eta < 0$$

$$U_-(\eta) = U(\eta) \quad \text{for} \quad \eta < 0 \qquad U_+(\eta) = 1 \quad \text{for} \quad \eta \geq 0$$

The substitution $U \to AU_+A^\dagger BU_-B^\dagger$ (with independent A and B) does not change the value of the mass which can be written as the sum of two independent integrals on two separate semispaces.

One can therefore proceed as in the $B = 1$ skyrmion case and get at the end the "rotating mass" of each lobe:

$$\mathcal{M}_i(\alpha,\beta) = \frac{M(\alpha,\beta)}{2} + \frac{1}{2\mathcal{A}(\alpha,\beta)} \frac{\vec{S}_i^2 + \vec{I}_i^2}{2} + \frac{1}{2}\left(\frac{1}{\mathcal{C}(\alpha,\beta)} - \frac{1}{\mathcal{A}(\alpha,\beta)}\right) S_{3i}^2 \qquad i = 1,2 \tag{3.6}$$

where $\alpha = eF_\pi a$,$\beta = \frac{m_\pi}{eF_\pi}$, $\vec{S}_i = \vec{I}_i$ are spin and isospin of each nucleon, S_{3i} is the third component of the spin in the moving frame of each configuration, $\mathcal{A}$ and $\mathcal{B}$ are the two independent momenta of inertia. They are very complicated integrals of $f(\eta)$ and $f'(\eta)$ and depend strongly from the normalized distance α so that in spite of the repulsive character of the static mass $M(\alpha,\beta)$ we can get attractive potentials.

Given these encouraging results we tried to work with less constrained fields hoping to get classical stability maintaining the nice features of the simpler model described previously. We allowed the field to have the general axial symmetry with boundary conditions that recover the Skyrme spherical solution at infinite separation limit and give $B = 2$. Our numerical analysis used a variational direct method and gave as answer a great lowering of the static mass figure in function of the separation and a questionable minimum at short distances ($\approx 1fm$) but we did not get a spontaneous annihilation of the field on the plane $z = 0$ and so it has been impossible to repeat the independent collective coordinate semi–quantization as in the simpler model [17]. At this point we can impose the annihilation on the separation plane for the general field configuration and quantize separately the two skyrmions, getting certainly more reliable spectra than in the one dimensional configuration previously described due to the greater freedom of the fields.

However of course an important point is to exploit the true $B = 2$ minimum, trying to give a physical description of the separation of the system in two nucleons. May be it will be useful from one side to explore the configurations of minimal energy of two $B = 1$ kinks starting from a complete separation to the collapsed toroidal shape, from the other side could be useful to explore a quantization procedure which takes in account all the collective coordinates, the vibrational degree of freedom, the possible bound states and eventually their correlations [18].

References

[1] T. H. R. Skyrme, Proc.Roy.Soc.London **A260**, 127 (1961); **A262**, 237 (1961);

[2] Can be of some historical interest to recall that it was during this research that for the first time a solitonic behaviour of a solitary wave (of a Sine Gordon soliton) was numerically observed, see I. K. Perring and T. H. R. Skyrme, Nucl.Phys. **31**, 550 (1962);

[3] D. Finkelstein and Ch. W. Misner, Ann. of Phys. **6**, 230 (1959);

[4] There are many good review papers about the Skyrme model which explain its connection with low energy quantum cromodynamics and therefore the problem of the identification of B. See, *e.g.*, I. Zahed and G.E.Brown, Phys. Rep. **142**, 1 (1986);

[5] N. S. Manton, Com. Math. Phys. **111**, 469 (1987);

[6] N. S. Manton, Phys. Lett. **60**, 1916 (1988);

[7] H. J. Schnitzer, Nucl. Phys. **B261**, 546 (1985)
S. Saito, T. Otofuji and M. Yasuno, Prog. Theo. Phys. **75**, 68 (1986);

[8] G. Adkins, C. R. Nappi and E. Witten, Nucl.Phys. **B228**, 522 (1983);

[9] M. Bander and F. Hayot, Phys.Rev **D30**, 1837 (1984);
R. Rajaraman, H. M. Sommerman, J. Wambach and H. W. Wyld, Phys. Rev. **D33**, 287 (1986);

[10] B. A. Li, K. F. Liu and M. M. Zhang, Phys. Rev **D35**, 1693 (1987);

[11] V. B. Kopeliovich and B. E. Shtern, JEPT Lett **45**, 203 (1987);
J. J. M. Verbaarschot, T. S. Walhout,J. Wambach and H. W. Wyld Nucl. Phys. **A468**, 520 (1987); **A468**, 520 (1987);

[12] N. S. Manton, Phys. Lett. **B192**, 177 (1987);

[13] A. Jackson, A. D. Jackson and V. Pasquier, Nucl.Phys. **A432**, 567 (1985);

[14] J. J. M. Verbaarschot, Phys. Lett. **B195**, 235 (1987);

[15] E. Braaten and L. Carson, "The Deuteron as Toroidal Skyrmion", Preprint Univ. of Minnesota UMN–TH–645/88

[16] E Sorace and M. Tarlini, Phys. Rev. **D33**, 253 (1986);

[17] E Sorace and M. Tarlini, Nuovo Cimento **B101**, 85 (1988)

[18] R. Giachetti E. Sorace and V. Tognetti, Phys. Lett. **A128**, 256 (1988).

REMARKS ABOUT THE DYNAMICS OF THE SOLITARY WAVES

Luis Vázquez

Departamento de Física Téorica
Facultad de Ciencias Físicas
Universidad Complutense
28040 - Madrid (Spain)

The solitary waves appear as solutions of nonlinear partial differential equations which modelate three large classes of physical phenomena [1,2]: a) Propagation of waves in continuous media. b) Condensed matter physics. c) Particle physics. The characteristic properties of the solitary waves are the following:

1. Constance in time of their wave form and velocity.
2. They represent structures which do not spread: the effect of the dispersion is compensated by the effect of the nonlinearity.
3. In the framework of the classical extended particles, the solitary waves are localizations of energy, momentum, charge, magnetic moment as well as other physical quantities.

The general questions related to the nonlinear wave equations appear reflected in the simplest nonlinear relativistic equation of the mathematical physics: the scalar field with a polynomial selfinteraction:

$$\mathcal{L} = \partial_\mu \phi \, \partial^\mu \phi^* - m^2 \phi \phi^* - g(\phi \phi^*) \tag{1}$$

$$\phi_{tt} - \Delta \phi + m^2 \phi + f(\phi \phi^*) \phi = 0 \qquad (f = g') \tag{2}$$

$$j^\mu = \frac{i}{2} \left(\phi^* \phi^{,\mu} - \phi^{*,\mu} \phi \right) \tag{3}$$

$$T^{\mu\nu} = \partial^\mu \phi \, \partial^\nu \phi^* + \partial^\nu \phi \, \partial^\mu \phi^* - g^{\mu\nu} \mathcal{L} \tag{4}$$

The solitary waves of (1) are stationary solutions (without radiation) of the form

$$\phi(\vec{x}, t) = e^{-i\omega t} \phi(\vec{x}) \tag{5}$$

and localized, in the sense that the associated physical

quantities must be finite

$$Q = \int j^0 d^3x = \omega \int \phi^2(\vec{x})\, d^3\vec{x} < \infty \tag{6}$$

$$E = \int T^{00} d^3x = \int [(\omega^2 + m^2)\phi^2 + |\nabla\phi|^2 + g(\phi)]\, d^3\vec{x} < \infty \tag{7}$$

The charge Q and the energy E define the natural functional spaces to study the associated mathematical problems. In this context we can summarize the following topics to be studied:

A. The existence of solitary waves [3-5].
B. The general evolution problem [6,7] associated to the nonlinear wave equation.
C. The stability problem [8].
D. The quantum properties [9,10].
E. Integrability [1].
F. Computational problems [1].

In relation with the quantum analysis of the solitary waves we must remark the following aspects:

1. The creation and annihilation of particles, explained with the help of the second quantization, has an analogue mechanism in the integrable nonlinear wave equations, i.e. the creation of solitons.
2. The ordinary quantum mechanics deals with point particles. In this way a natural question is to consider the quantum treatment of the extended particles.
3. An important task is to understand completely the relation between the quantum and classical nonlinear wave equations.

INTEGRABILITY AND STABILITY

The dynamics of the integrable wave equations is well understood [1]. On the other hand in non-integrable systems the dynamics of the solitary waves is more complicated and there is no canonical method to find out their features. Some progress has been made to obtain a new KAM theorem for infinite degrees of freedom [11]. Concerning the dynamics of the non-integrable system we can consider three main approaches:

1. The stability of simple solitary waves [8, 12].
2. Analysis of the effect of the perturbation, either deterministic or stochastic, for the non-integrable system [13, 15].
3. Study of the interaction between the solitary waves associated to the non-integrable system. The phenomenology of such interaction is richer than in the case of an integrable system [17].

In this context the analysis of the stability of the solitary waves appears as an important tool in the study of the non-integrable dynamics.

A solitary wave is stable in a general sense if it is not destroyed under the influence of a general perturbation. Thus we have different definitions of stability, that can be classified into two large group:

A. Stability with respect to a perturbation of initial data.
   1. Dynamical stability. A solitary wave is dynamically stable when small perturbations of its amplitude do not destroy it.
   2. Liapunov stability: It happens when the initially perturbed solitary wave remains in a neighbourhood (defined in an appropiate functional space) of the unperturbed solitary wave.
   3. Stability with constraints.
   4. Energetic stability. A solitary wave is energetically stable if it is a minimum of the associate energy.
   5. Soliton stability. A solitary wave is called a soliton if preserves its shape after collision with other solitary waves.

B. Structural stability: Stability with respect to a perturbation of the evolution equation.
   1. Static stability: Analysis of the behaviour of the solitary waves under external static potentials.
   2. Stability under time-dependent potentials.
   3. Stability under stochastic perturbations.

In this context it is worth to compare general properties and the dynamics of the Sine-Gordon and ϕ^4 systems [1, 14, 17, 18].

$\varphi_{tt} - \varphi_{xx} + \sin\varphi = 0$ (Integrable)		$\varphi_{tt} - \varphi_{xx} - \varphi + \varphi^3 = 0$ (Non-Integrable)
Yes	Lax Pairs	No
Yes	Inverse Scattering Transform	No
Yes	Bäcklund Transforms	No
Infinite	Conservation laws	Finite
Yes	Breather solutions	No
Soliton stability	Behaviour under collisions	Very rich Phenomenology
Stable	Behaviour under weak stochastic perturbations	Stable

The dynamic of these two related nonlinear models offer a promising example to be located in the framework of the symmetry theory.

REFERENCES

1. R.K.Dodd,J.C.Eilbeck,J.D.Gibbon and H.C.Morris,"Solitons and Nonlinear Wave Equations", Academic Press,London,1982.
2. A.F.Rañada, "Classical nonlinear Dirac field models of extended particles",in "Quantum Theory,Groups,Fields and Particles".Ed. A.O.Barut,Reidel,Dordrecht,1983.
3. H.Berestycki and P.L.Lions, Arch.Rat.Mech.Anal. 82,313 (1983). And references therein.
4. H.Brezis and E.H.Lieb, Comm.Math.Phys. 96,97 (1984).
5. T.Cazenave and L.Vázquez, Comm.Math.Phys. 105,35 (1986).
6. A.Haraux, "Nonlinear Evolution Equations.Global Behaviour of Solutions".Lecture Notes in Mathematics 841,Springer-Verlag,Berlin,1981.
7. J.Ginibre and G.Velo, "Nonlinear evolution equations:Cauchy problem and scattering theory",BiBoS preprint 102/85, Bielefeld,1985.

8. J.Stubbe and L.Vázquez, "Stability of solitary waves: mathematical and physical aspects", in "Mathematics + Physics. Lectures on Recent Results", Vol.3, Ed. L. Streit, World Scientific, Singapore (In press).
9. R.Jackiw, Rev.Mod.Phys. 49,681 (1977).
10. L.Vázquez, Phys. Rev. D33, 2478 (1986).
11. M.Vittot, "Theorie Classique des Perturbations et Grand Nombre de Degrés de Liberté", These de Doctorat de L'Université de Provence (Aix-Marseille I) 1985.
12. J.Shatah and W.Strauss, Comm. Math.Phys. 100, 173 (1985).
13. J.P.Pascual and L.Vázquez, Phys. Rev. B32, 8305 (1985).
14. F.G.Bass, Yu. S.Kivshar,V.V.Konotop and Yu.A.Sinitsyn, Phys. Rep. 157, 63 (1988).
15. J.C.Ariyasu and A.R.Bishop, Phys. Rev. B35, 3207 (1987).
16. J.M.Ghidaglia and R.Teman, J.Math. Pures et Appl. 66, 273 (1987).
17. D.K.Campbell and M.Peyrard, Physica 18D, 47 (1986).
18. M.J.Rodriguez and L.Vázquez, "Behaviour of the ϕ^4 kinks under stochastic perturbations". To appear.

A METHOD OF STUDYING INTEGRALS OF DYNAMICAL SYSTEMS BASED ON FROBENIUS' INTEGRABILITY THEOREM

S. Wojciechowski*

Department of Mathematics
Linköping University
581 83 Linköping, Sweden

1. INTRODUCTION

Solving nonlinear ODE's is the art of guessing a solution. It relies on finding an integral of motion which enables us to reduce order of equation by one. In the case of hamiltonian systems or Euler-Lagrangé equations it is possible to reduce the order of the system by two.
Thus solving ODE's is an endless search for integrals of motion and this requires accurate guesses. It may be done by smart partial integration of equations,by a suitable ansatz for an integral of motion or by discovery of a one parameter symmetry group. Surprisingly, there are not many more tools at our disposal when studying given equations.

Here we are going to advocate yet another method[1] which is based on the Frobenius integrability theorem and applies best to 3-dimensional dynamical systems. This method is related to(but more general then) the Lie symmetry method.
We shall consider the question of finding integrals (of motion) for a given 3-dimensional dynamical system namely an autonomous system of equations

$$\dot{x}_k = f_k(x) \quad , \quad k=1,2,3 \tag{1}$$

which we shall represent by the vector field (v.fld) $X(x)=f_k\partial^k$.

Dot denotes time derivative, $\partial^k=\partial/\partial x_k$, and the summation convention for repeating indices applies. The study of such systems in the large is difficult mainly because they may exhibit very complicated (chaotic) behaviour. So we are interested in the cases when one integral of motion can be explicitly found or, at least, an invariant foliation exists.

*Research supported by the contract No: F-FU 8677-102

2. FROBENIUS' INTEGRABILITY THEOREM

Frobenius integrability theorem gives an answer to the question when a smooth set of linearly independent v.flds $X_1,...,X_k$ tangent to a smooth manifold M^n (n>k) has a k-dimensional integral manifold $N^k \subset M^n$. That is at each point $x \in N^k$ the tangent space $(TN^k)_x$ is spanned by these v.flds.

<u>Frobenius theorem</u>[2]

Let each point $x \in M^n$ of a smooth manifold M^n have k linearly independent v.flds $X_1,...,X_k$ belonging to the tangent space $(TM^n)_x$. Then every point x has a coordinate neighbourhood $\phi: U \to R^n$ such that the submanifold

$$\phi^{-1}(x_1,...,x_k,c_{k+1},...,c_n)$$

with the c_j's constant, has at each point x, as its tangent subspace, the subspace generated by $X_1,...,X_k$ if and only if $[X_i,X_j]=c_{ij}{}^r X_r$ (for all $1 \le i,j,r \le k$). That is, the commutator lies in the subspace spanned by $X_1,...,X_k$.

Such a set of v.flds is often called integrable but we shall use the term <u>compatible</u>. It yields a local foliation of M^n into k-dimensional leaves N^k.

We shall consider the case of two independent v.flds X,Y in R^3. Then the integrability condition of Frobenius' theorem for v.flds reads

$$[X,Y] = \alpha(x)X + \beta(x)Y \tag{2}$$

($\alpha(x)$, $\beta(x)$ are some functions) or

$$\det(X,Y,[X,Y]) = 0 \tag{3}$$

for vectors X,Y,[X,Y]. Two vectors which satisfy (2) or (3) are called <u>compatible</u>.

If the condition (2) is fulfilled then in a neighbourhood of each nonsingular point $x \in R^3$, such that X and Y are linearly independent, there is a local foliation of R^3 into two dimensional leaves tangent to the planes X,Y. Each leaf of such local foliation can be extended to a maximal connected one. However the two dimensional foliation obtained is not, in general, defined by level surfaces of some function. We are interested in the case when such a function can be found.

Geometrically the condition (3) is equivalent to the condition of integrability for the Pffaf form $h_1dx_1+h_2dx_2+h_3dx_3$ where the vector $(h_1,h_1,h_3) = Z = X \times Y$ is orthogonal to the plane X,Y. The condition (3) reads then

$$Z \cdot \mathrm{rot} Z = 0$$

3. METHOD OF SEARCHING FOR INTEGRALS

The main idea is to use the Frobenius theorem for finding integrals of motion of the system (1). That is for a given v.fld $X(x)=f_k\partial^k$ we are looking for another v.fld $Y(x)=g_k\partial^k$ which is compatible with X and then integrate the resulting equations for the function (an integral of motion) defining the leaves of the foliation tangent to the planes X,Y.

The condition (2) leads to 3 linear PDE's on the functions $g_k(x)$ which are, in general, impossible to solve unless a suitable ansatz for a solution is taken. When the v.fld Y is known and u(x),v(x) are two independent integrals of Y then the problem of finding an integral of X is reduced to solving one 1-st order ODE[3].

Let u(x),v(x) be two functionally independent integrals of Y (Yu=Yv=0) then the common integral of X and Y has the form F(u,v). The integral F satisfies

$$0=XF=\frac{\partial F}{\partial u} Xu + \frac{\partial F}{\partial v} Xv=(Xv)\left(\frac{\partial F}{\partial u}\frac{Xu}{Xv} + \frac{\partial F}{\partial v}\right)=(Xv)\left(\frac{\partial F}{\partial u} G(u,v) + \frac{\partial F}{\partial v}\right) \quad (4)$$

The coefficient Xu/Xv=G(u,v) is a function of u,v, only because the compatibility condition (2) implies

$$Y(Xu/Xv)=(Xv)^{-2}[(YXu)(Xv)-(Xu)(YXv)]=(Xv)^{-2}[(aXv)(Xu)-(Xv)(aXu)]=0$$

and, therefore, Xu/Xv is also a first integral of Y. Solving the equation (4) is equivalent to finding a general integral of the equation

$$\frac{du}{dv} = G(u,v) \quad (5)$$

More precisely if F(u,v) is a general integral of (4) then F(x)=F(u(x),v(x)) is a first integral of the v.fld X.

Thus the process of finding an integral of motion is divided into two steps

(i) finding a compatible v.fld Y,
(ii) integrating the equation du/dv =G(u,v).

We shall compare this with the standard method of a direct search for an integral of motion F(x) which requires solving one 1-st order PDE

$$0 = XF = f_k\partial^k F$$

Both methods require some ansatz: either for Y or for F. An ansatz for F may work. If it doesn't work then a new ansatz has to be made and one is usually tempted to work within a very narrow class of algebraic ansatzes for F. More complicated ansatzes for F are difficult to conceive because of little experience with transcendental functions. In the Frobenius theorem method even very elementary ansatzes for Y may lead to complicated equation (5) and integral of motion may be given by a very complicated function or be implicitly defined as a general solution of (5). It is quite acceptable if we can deduce global properties of the solution on the basis of the equation (5). Obviously such integrals are impossible to find by the ansatz method for F.

On the other hand a few words of warning are necessary. In the process of using Frobenius theorem several things may go wrong. First it is nontrivial to guess a suitable Y. Next it may be difficult to find two independent integrals u(x),v(x). Finally it

may be impossible to integrate equation (5) or to say something about global properties of its solutions. And this means that usually a very simple ansatz for Y has to be made. We shall work mainly in the framework of a linear ansatz for Y because two independent solutions u(x),v(x) are explicitly known. Occasionally we shall also use a quadratic ansatz for Y. An important gain is that one, simple ansatz for Y may correspond to several algebraically unrelated ansatzes for F(x) and yields, in the process of integration, quite complicated integrals F(x). However the most important gain of our method is that it provides a new opportunity to work simultaneously on two levels: with integrals F and with compatible v.flds Y. An accurate guess on one level may support a systematic research on the other one. For instance a simple Y may lead to a complicated expression for F(x) which gives further hints how to generalise an ansatz for F. This, on its own, determines new compatible v.flds Y and instructs how to modify suitably ansatzes for Y. It will be illustrated by examples in the next sections.
In the framework of this method we encounter situations when it is impossible to integrate explicitly equation (5) and global properties of its solutons are unclear. In such cases we can not claim the existence of a global integral of motion. But certainly there exists an invariant two-dimensional foliation which excludes, in principle, the ocurrence of chaos.In this case we obtain new class of relatively regular dynamical systems which may be called "weakly integrable".
So far we have discussed only the direct problem of theory of ODE's: find integrals of motion for given equations. This is a very difficult question. But our method enables us to study, as in the case of the Lie symmetry theory, an inverse question. Given Frobenius structure of compatible v.flds - find the class of equations which admit this structure. We shall discuss this question in section 7.

4. LOTKA-VOLTERRA EQUATIONS

As the first example of the application of our method to given equations we consider Lotka-Volterra (L-V) equations[4] describing a population of three interacting species. They read

$$\begin{bmatrix} x_1 \\ x_2 \\ x_3 \end{bmatrix}^{\bullet} = \begin{bmatrix} x_1(Cx_2+x_3) \\ x_2(x_1+Ax_3) \\ x_3(Bx_1+x_2) \end{bmatrix} + \begin{bmatrix} \lambda x_1 \\ \mu x_2 \\ \nu x_3 \end{bmatrix} = K + L \qquad (6)$$

where A,B,C,λ,μ,ν are real parameters. We present results of the study of compatibility with linear v.fld which has been done "by hand" in[1].Our method selects all previously known cases and several new ones not covered by the ,so called, Painlevé analysis[5]. Systematic, computer study of compatibility with general, homogeneous, linear $Y=a_i{}^j x_j \partial^i$ has been done recently in collaboration with B.Grammaticos, J.Moulin-Ollagnier, A.Ramani, J-M.Strelcyn and will be published soon. Further new cases have been found. The case of a compatible quadratic v.fld presented in section 4.2 is also new in comparison with the content of[1].

4.1 Linear ansatz for Y

We consider first compatibility with a diagonal, linear v.fld $Y=Mx_1\partial^1+Nx_2\partial^2+Px_3\partial^3$ because in this case we know explicitly two independent integrals $u(x),v(x)$ which satisfy $Yu=Yv=0$ and, therefore, the equation $du/dv = G(u,v)$ can be found explicitly. The compatibility condition reads

$$0 = \det(K+L,Y,[K,Y]) = \det(K,Y,[K,Y]) + \det(L,Y,[K,Y]) \qquad (7)$$

and polynomial terms of different order give two sets of equations . The 4-th order term $\det(L,Y,[K,Y])=0$ gives

$$\begin{aligned} M[B(\lambda N-\mu M) + (\nu M-\lambda P)] &= 0 \\ N[C(\mu P-\nu N) + (\lambda N-\mu M)] &= 0 \\ P[A(\nu M-\lambda P) + (\mu P-\nu N)] &= 0 \end{aligned} \qquad (8)$$

and the 5-th order term $\det(K,Y,[K,Y])=0$ gives

$$\begin{aligned} (M-N)(BCN+M-CP) &= 0 \\ (N-P)(N-AM+CAP) &= 0 \\ (P-M)(-BN+ABM+P) &= 0 \end{aligned} \qquad (9)$$

Equations (8) and (9) are quadratic in M,N,P but possible to handle due to factorizations.Tedious analysis leads to the first 3 cases in table 1.

Separately we have found also two nondiagonal compatible v.fld

$$Y_1=[(C+B^{-1}\alpha)x_1+(C\beta+B^{-1}\gamma)x_2+(\delta(C+B^{-1}\alpha)+\varepsilon(C\beta+B^{-1}\gamma))x_3]\partial^1+$$
$$+ [x_1+\beta x_2+(\delta+\varepsilon\beta)x_3]\partial^2 + [\alpha x_1+\gamma x_2+(\delta\alpha+\varepsilon\gamma)x_3]\partial^3$$
$$Y_2=[(-C-(C+1)(\alpha+\beta))x_1+C(C+1)\alpha x_2+C\alpha x_3]\partial^1+$$
$$+ [-(C^{-1})(C+1)(\alpha+\beta)x_1+(C+(C+1)\alpha)x_2-(\alpha+\beta)x_3]\partial^2 +$$
$$+ [(C+1)x_1-C(C+1)x_2+(C+1)\beta x_3]\partial^3$$

which correspond to the cases 4 and 5 respectively; $\alpha,\beta,\gamma,\delta,\varepsilon$ are real parameters. Simplest nontrivial cases of Y_1 and Y_2 are presented in the table 1.

As an example we integrate the case 5. The v.fld Y_2 from table 1 generates the linear equations

$$dx_1/ds = -Cx_1 \quad , \quad dx_2/ds = Cx_2 \quad , \quad dx_3/ds = (C+1)(x_1-Cx_2)$$

with two independent integrals $u=x_1x_2$, $v=C^{-1}x_1+x_2+(C+1)^{-1}x_3$. They satisfy the equations

$$\dot{u} = (K+L)u = x_1x_2\ (x_1+Cx_2+C(C+1)^{-1}x_3) = Cu$$
$$\dot{v} = (K+L)v = 2x_1x_2 = 2u$$

The equation (5) reads here $du/dv=Cv/2$ and consequently

$$F(x_1,x_2,x_3) = F(u,v) = Cv^2-4u = C^{-1}x_1^2 + Cx_2^2 + C(C+1)^{-2}x_3^2 +$$
$$-2x_1x_2+2C(C+1)^{-1}x_2x_3 + 2(C+1)^{-1}x_3x_1$$

The integral in table 1 is equal to $CF(x)$.

<u>Table1</u> Integrable cases of the Lotka-Volterra equations. Other integrable cases are obtained by simultaneous cyclic permutation of (A,B,C), (λ,μ,ν) and (x_1,x_2,x_3).

Case	Parameters		Compatible v.f. Y	Integral
1a	$\lambda=\mu=\nu$	A,B,C	$x_1\partial^1+x_2\partial^2+x_3\partial^3$	?
1b		ABC+1=0	$x_1\partial^1+x_2\partial^2+x_3\partial^3$	$\lvert x_1\rvert^{AB}\lvert x_2\rvert^{-B}\lvert x_3\rvert$ $\lvert x_1-Cx_2+B^{-1}x_3\rvert^{-AB+B-1}$
1c		B=1/(1-A) C=(A-1)/A	$x_1\partial^1+x_2\partial^2+x_3\partial^3$	$\lvert x_1\rvert^{-A}\lvert x_2\rvert\lvert x_3\rvert^{A-1}$
1d		A=C=1	$x_1\partial^1+x_2\partial^2+x_3\partial^3$	$\lvert x_2x_3\rvert\lvert x_3-Bx_1\rvert^{-1}\lvert x_2-x_1\rvert$
2	$\lambda=\mu$ $\nu=0$	A=1	$x_1\partial^1+x_2\partial^2$	$\lvert Cx_2-x_1\rvert^{BC+1}\lvert x_1\rvert^{-1}$ $\lvert x_2\rvert^{-BC}\lvert x_3\rvert^{C}$
3	λ,μ-any $\mu=\lambda A-$ $-\nu AC$	ABC+1=0	$Mx_1\partial^1+(AM-ACP)x_2\partial^2+$ $+Px_3\partial^3$	$\lvert x_1\rvert^{-A}\lvert x_2\rvert\lvert x_3\rvert^{AC}$
			$-ACx_1(x_3+1)\partial^1+$ $+x_3(x_1-\nu AC)\partial^2$	$(x_1-Cx_2+ACx_3)+AC$ $(-\nu\log\lvert x_1\rvert+\lambda\log\lvert x_3\rvert)$
4	$\lambda=\mu=$ $=\nu=0$	ABC+1=0	$Cx_1\partial^1+(x_1+B^{-1}x_2)\partial^2-$ $-Cx_2\partial^3$	$x_1-Cx_2+ACx_3$
			$Mx_1\partial^1+Nx_2\partial^2+$ $+(BN-ABM)x_3\partial^3$	$\lvert x_1\rvert^{AB}\lvert x_2\rvert^{-B}\lvert x_3\rvert$
5	$\lambda=\mu=$ $=\nu=0$	ABC-1=0 $A=-(C+1)^{-1}$ $B=-(C+1)C^{-1}$	$-Cx_1\partial^1+Cx_2\partial^2+$ $+(C+1)(x_1-Cx_2)\partial^3$	$x_1^2+C^2x_2^2+C^2(C+1)^{-2}x_3^2-$ $-2Cx_1x_2+2C^2(C+1)^{-1}x_2x_3+$ $+2C(C+1)^{-1}x_3x_1$

<u>4.2 Quadratic ansatz for Y</u>

There are many trivial quadratic compatible v.flds which follow from the known linear compatible Y by multiplying these by a linear form. So we are interested in finding nontrivial ones. We shall look for a v.fld Y in the form

$$Y=x_1(A_2x_2+A_3x_3+\lambda)\partial^1+x_2(B_1x_1+B_3x_3+\mu)\partial^2+x_3(C_1x_1+C_2x_2+\nu)\partial^3$$

which contains the L-V v.fld and, therefore, has at least, one nonzero solution. It is also justified by the fact that presence of only mixed quadratic terms x_ix_j, $i\neq j$ reduces considerably the number of algebraic conditions on A_i, B_i, C_i. One nontrivial solution of the compatibility condition (2) exists for

$$ABC+1=0 \quad \text{and} \quad A\lambda-\mu-AC\nu=0 \tag{10}$$

then

$$Y=-ACx_1(x_3+\lambda)\partial^1+x_3(x_1-AC\nu)\partial^3$$

This v.fld has an obvious integral: $u(x)=x_2$. As a v.fld in the plane x_1, x_3 it represents two-dimensional L-V equations which have the integral

$$v(x)=x_1+ACx_3-AC\nu\log|x_1|+AC\lambda\log|x_3|$$

Thus $du/dv=C^{-1}$ and hence the integral of motion is

$$F(x)=v(x)-Cu(x)=x_1-Cx_2+ACx_3-AC\nu\log|x_1|+AC\lambda\log|x_3| \tag{11}$$

This means that the case 3 in table 1 has two functionally independent integrals of motion: one for a linear v.fld and a second one related to quadratic Y.

Once the integral (11) is known it is natural to investigate closer the ansatz

$$F(x) = ax_1+bx_2+cx_3+\alpha\log x_1+\beta\log x_2 +\gamma\log x_3$$

It leads to a set of linear equations on the coefficients a,b,c and α,β,γ:

$$0 = \alpha\lambda+\beta\mu+\gamma\nu \qquad \text{at constant term} \tag{12}$$

$$\begin{aligned} 0 &= a\lambda+\beta+\gamma B \\ 0 &= b\mu+\alpha C+\gamma \\ 0 &= c\nu+\alpha+\beta A \end{aligned} \qquad \text{at linear terms} \tag{13}$$

$$\begin{aligned} 0 &= aC+b \\ 0 &= bA+c \\ 0 &= a+cB \end{aligned} \qquad \text{at quadratic terms} \tag{14}$$

Equations (14) have nonzero solution iff $ABC+1=0$ and $a\neq0$. Then: $b=-aC$, $c=aAC$. Solvability of (13) requires $A\lambda-\mu-AC\nu=0$ and then $\alpha=-\beta A-aAC\nu$, $\gamma=\beta AC+aAC\lambda$, β-parameter. Equations (12) are satisfied identically. So we get a one-parameter (β/a) family of integrals of motion:

$$F_1(x) = (x_1-Cx_2+ACx_3) + AC(-\nu\log|x_1|+\lambda\log|x_3|)+$$
$$+(\beta/a)(-A\log|x_1|+\log|x_3| +AC\log|x_3|)$$

So far we have found only one nontrivial particular solution of the compatibility condition for a quadratic v.fld Y and we found a one-parameter family of integrals. Each of these integrals determines a new solution to the compatibility condition (2) for the L-V v.fld. Namely any quadratic v.fld orthogonal to grad F_1 fulfills in R^3 the condition (2). But when integral F_1 is known it is much easier to solve for Y the othogonality condition

$Y \perp \text{grad } F_1$ then the condition (2). The question of finding other nontrivial, quadratic, compatible v.fld Y remains open.
The main results of our (limited) study of compatibility of linear and quadratic v.flds Y with the L-V v.fld X are listed in table 1. We have found several new cases of integrability of the L-V equations which have not been covered by the Painlevé analysis in[5]. In all cases, except 1, it has been possible to determine the integral of motion explicitly. These are of differing algebraic nature and would be difficult to find directly. Our results show that the unifying factor is the linear nature of one of the compatible v.flds. There are two cases (3 and 4) where we know two integrals of motion which determine trajectories up to time parametrization. The integrals in the table may be a good starting point for further study of more general ansatzes for the integral of motion.

5. MAY-LEONARD EQUATIONS

The May -Leonard equations[6] describe a population of 3 competing species. They have the form:

$$\begin{bmatrix} x_1 \\ x_2 \\ x_3 \end{bmatrix}^{\bullet} = \begin{bmatrix} -x_1(x_1+\alpha x_2+\beta x_3) \\ -x_2(\beta x_1+x_2+\alpha x_3) \\ -x_3(\alpha x_1+\beta x_2+x_3) \end{bmatrix} + \begin{bmatrix} x_1 \\ x_2 \\ x_3 \end{bmatrix} = K + E \tag{15}$$

where α, β are real parameters. They are compatible with the linear v.fld $E = x_k \partial^k$ since

$$[K+E, E] = -K \qquad \text{and} \qquad \det(K+E, E, -K) = 0$$

So there is always an invariant foliation. The v.fld E has two independent integrals $u(x) = x_2 x_1^{-1}$, $v(x) = x_3 x_1^{-1}$ and the equation (5) for the system (15) reads here

$$\frac{du}{dv} = \frac{u((1-\beta)+(\alpha-1)u+(\beta-\alpha)v)}{v((1-\alpha)+(\alpha-\beta)u+(\beta-1)v)} \tag{16}$$

So far it has been possible to solve it only in the following four cases

a) $\alpha+\beta=2$ -integral $F(u,v) = \dfrac{(1+u+v)^3}{uv} = \dfrac{(x_1+x_2+x_3)^3}{x_1x_2x_3}$

b) $\alpha=\beta$ -integrals are $F_1(u,v) = \dfrac{(u-1)v}{(v-1)u} = \dfrac{(x_2-x_1)x_3}{(x_3-x_1)x_2}$, $F_2 = \dfrac{(x_3-x_2)x_1}{(x_1-x_2)x_3}$

The integral F_2 follows from F_1 by cyclic permutation of indices.

c) $\alpha=\beta=0$ -equations (15) decouple

d) $\alpha=\beta=1$ -integrals are $F_1(u,v) = x_2x_1^{-1}$, $F_2(u,v) = x_3x_1^{-1}$ but the equations (15) may be solved by hand.

For general α, β little is known about global properties of integrals of (16).

6. LORENZ EQUATIONS

In application to the Lorenz equations

$$\dot{x}_1 = \sigma(x_2-x_1)$$
$$\dot{x}_2 = -x_1x_3 + (\rho x_1-x_2)$$
$$\dot{x}_3 = x_1x_2 - bx_3$$

Our method allows for systematic calculation of the Sigur time independent integral of motion which exists for $\sigma=1/2$, $\rho=0$ and $b=1$. Sigur[7] guessed two time dependent integrals: $x_2^2+x_3^2=a_1\exp(-2t)$ and $x_1^2-x_3=a_2\exp(-t)$, (a_1,a_2 -const) which on elimination of time yield $F(x_1,x_2,x_3)=(x_2^2+x_3^2)/(x_1^2-x_3)^2$. A diagonal ansatz for Y leads to the compatible v.fld $Y=x_1\partial^1+2x_2\partial^2+2x_3\partial^3$ with two independent integrals $u=x_2x_1^{-2}$, $v=x_3x_1^{-2}$. Then from the Lorentz v.fld we get

$$\dot{u} = Xu = x_1^{-3}(-x_1^2x_3-x_2^2)$$
$$\dot{v} = Xv = x_1^{-3}(x_1^2x_2-x_2x_3)$$

and $du/dv = (-v-u^2)/(u-vu)$ .The Pfaff form $(u-vu)du+(v+u^2)dv$ has the integrating factor $(1-v)^{-3}$. This yields again the integral of motion $F(u,v)=(u^2+v^2)/(1-v)^2 = (x_2^2+x_3^2)/(x_1^2-x_3)^2$.

7. CONSTRUCTION OF INTEGRABLE SYSTEMS IN R^3

Here we consider the system (1) to be integrable if it has one integral of motion or, at least, an invariant foliation.
The direct question of finding a compatible v.fld Y for a given v.fld X is, in general, an impossible task like the problem of finding Lie symmetries for a given ODE. Most of the classical Lie theory is concerned with the inverse problem of constructing all equations (within some class) admitting a given one-parameter symmetry group. Here we are in a quite analogous situation despite positive results for a few examples.
Dynamical systems integrable in our sense may be easily constructed from any pair of compatible v.flds $D_1=d_{k1}(x)\partial^k$, $D_2=d_{k2}(x)\partial^k$ satisfying $[D_1,D_2]=\alpha_1D_1+\alpha_2D_2$ for some functions $\alpha_1(x),\alpha_2(x)$. Then any dynamical system of the form

$$\begin{bmatrix} x_1 \\ x_2 \\ x_3 \end{bmatrix}^{\bullet} = L_1(x)\begin{bmatrix} d_{11}(x) \\ d_{21}(x) \\ d_{31}(x) \end{bmatrix} + L_2(x)\begin{bmatrix} d_{12}(x) \\ d_{22}(x) \\ d_{32}(x) \end{bmatrix} = L_1D_1+L_2D_2 \qquad (17)$$

where $L_1(x),L_2(x)$ are arbitrary functions, has as compatible v.fld D_1 (or D_2) and hence an invariant foliation. If D_1 or D_2 are so choosen that we can integrate equation (5) then we know explicitly integral of motion $F(x)$.

An important class of compatible v.flds follows from Poisson structures[8]. A Poisson structure on a set of (C^∞) differentiable functions is defined by the bilinear operation

$$\{F,G\} = d_{ij}\partial^i F \partial^j G$$

which is skew-symmetric and satisfies the Jacobi identity

$$\{\{F,G\},H\} + \{\{H,F\},G\} + \{\{G,H\},F\} = 0$$

This requires that the functions $d_{ij}(x)$ are skew-symmetric

$$d_{ij}(x) = -d_{ji}(x) \tag{18}$$

and satisfy the set of equations

$$d_{rk}(x)\partial^r d_{ij}(x) + d_{rj}(x)\partial^r d_{ki}(x) + d_{ri}(x)\partial^r d_{jk}(x) = 0 \tag{19}$$

$i,j,k,r=1,2,3$. If we consider 3 v.flds $D_k = d_{sk}\partial^s$, $k=1,2,3$ then it is easy to see that (by skew-symmetry (18)) they are linearly dependent. Each pair D_j, D_k is compatible since by (18) and (19) we get

$$[D_j, D_k] = -(\partial^r d_{jk})D_r \tag{20}$$

and from their linear dependence follows

$$[D_j, D_k] = \alpha_j(x)D_j + \alpha_k(x)D_k \tag{21}$$

for some functions α_j, α_k. The condition (21) and linear dependence of D_k is sufficient for the set of equations

$$0 = D_r C(x) \quad , \qquad k=1,2,3$$

to have a nontrivial solution for the function $C(x)$ which is called a Casimir function. This function is the required integral of motion for any equation of the form

$$\dot{x} = L_r(x)D_r \tag{22}$$

where D_r stands for column vectors and summation convention applies. Equations (22) acquire a more special structure[9] if $L_k(x)=\partial^k H(x)$ that is if they are hamiltonian. Then equations (22) can be written as

$$\dot{x}_k = d_{kr}\partial^r H = \{x_k, H\} \tag{23}$$

and have a second integral of motion: the hamiltonian $H(x)$. Indeed by skew-symmetry

$$\dot{H} = x_k\partial^k H = d_{kr}\partial^r H \partial^k H = 0$$

The integral $H(x)$ is functionally independent of $C(x)$ because the function C generates trivial hamiltonian equations of motion

$$\dot{x}_k = \{x_k, C\} = d_{kr}\partial^r C = D_k C = 0$$

Thus equations (23) admit an additional compatible v.fld related to the integral $H(x)$. For such a v.fld we can choose any v.fld which is orthogonal to grad H, e.g. $E=J\,\mathrm{grad}\,H$ where $J=-J^T$ is any skew-symmetric matrix, possibly depending on x. The construction of equations (23) raises the question whether all 3-dimensional dynamical systems posessing two integrals of motion arise in this way. That is whether they may be represented as a hamiltonian system with respect to some Poisson structure.
A general solution to the equations (18) and (19) is not known. Many particular solutions are obtained under more specific assumptions on dependence on x. When this dependence is linear

we obtain the, so called, Lie-Poisson structures and all $d_{ij}=c_{ij}{}^{k}x_k$ are given through the structure constants $c_{ij}{}^{k}$ of 3-dimensional Lie algebras. All these algebras give rise to pairs of compatible v.flds which may be used for constructing integrable quadratic v.flds by assuming $L_k(x)$ to be linear forms[9]. If $L_k(x)=\partial^k H(x)$ (where $H(x)=c^{ij}x_i x_j+b^i x_i$) then we get integrable Riccati systems of equations with two integrals of motion[10] Such systems are very interesting in connection with the examples studied before. One would like to know how large a class of Riccati equations can be generated by compatible pairs of linear v.flds and whether all such pairs are related to 3-dimensional Lie algebras.

Acknowledgement

I would like to thank P.Basarab-Horwath for reading the manuscript of this paper.

References

1. J-M.Strelcyn and S.Wojciechowski, "A method of studying integrals for 3-dimensional dynamical systems", Phys.Lett.A (in press).
2. D.W.Kahn, "Introduction to global analysis",Academic Press, New York (1980).
3. L.E.Dickson "Differential equations from the group stand point", Ann.Math.Ser A2, 25:287(1924).
4. N.Goel,S.Maitra,E.W.Montroll,"On the Volterra and other models of interacting populations", Rev.Mod.Phys. 41:231 (1971).
5. T.C.Bountis,A.Ramani,B.Grammaticos,B.Dorizzi,"On the complete and partial integrability of nonhamiltonian systems",Physica 128A:268 (1984).
6. R.M.May,W.J.Leonard,"Nonlinear aspects of competition between three species",SIAM J.Appl.Math. 29:243(1975).
7. H.Segur,"Solitons and inverse scattering transform",Lectures at International School "Enrico Fermi", Varenna,Italy (1980).
8. A.Weinstein,"The local structure of Poisson manifolds", J.Diff.Geom.18:523(1983).
9. S.Wojciechowski,"Construction of integrable Riccati systems by the use of low dimensional Lie algebras", in "Structure, coherence and chaos in dynamical systems" ed.Christiansen and Permentier,Manchester University Press,Manchester(1988).
10. S.Wojciechowski, "A new test for integrability of Riccati systems of equations",in Proceedings of the workshop on finite dimensional integrable dynamical systems,ed.P.G.Leach and W.H.Steeb,World Scientific,Singapore(1988).

PART III: SPECIAL RESEARCH PAPEERS AND REPORTS

FORM FACTORS OF MASSIVE NEUTRINOS IN EXTENDED $SU(2)_L xU(1)$ MODELS*)

M. Abak and C. Aydin

Department of Physics

Karadeniz Technical University, Trabzon, Turkey

Abstract

Assuming the standard SU(2) xU(1) model with the leptons in left-handed doublets and right-handed singlets plus a single Higgs doublet and using 't Hooft-Feynman gauge, we study the electromagnetic form factors of the massive neutrinos.

I. Introduction

Interest in $SU(2)_L xU(1)$ models proposed by Weinberg[1)] and Salam[1)] is still increasing and has received several important experimental confirmations. The main point of these theories is that they are renormalizable so that useful higher-order calculations can be made. The questions are: what are the results of such calculations, how large are they and can they be tested experimentally?

The questions of whether the neutrinos have masses and if so, whether they are Dirac or Majorana particles are two of the most important issues in both particle physics and astrophysics[2)]. Neutrino mass is also great importance for cosmology.

A crucial point in the standard SU(2)xU(1) model is played by the notion of internal, continous symmetry. The

---

*) Supported by the Scientific and Technical Research Council of Turkey (TÜBİTAK, Ankara)

concept was first introduced by W.Heisenberg to account for the similarity of the nuclear interactions of protons and neutrons. The further step which leads to the standard model is to use the symmetry to determine the interaction rather than simply restrict it.

In the standard model the neutrinos are massless,neutral particles of spin 1/2. The structure of the leptons in the standard model is given by doublets and singlets:

$$\binom{\nu_e}{e}_L , \binom{\nu_\mu}{\mu}_L , \binom{\nu_\tau}{\tau}_L ,\ldots,$$

$$e_R , \mu_R , \tau_R ,\ldots,$$

One of the consequences of this structure is that the neutrinos are massless. This is since there is no ν_R in the standard theory so that the Dirac mass term $\bar{\nu}_L\nu_R$ is absent and there is no Majorana term of the form $\nu_L^c\nu_L$.

There is many possible extensions of the standard model to give massive neutrinos[3]. Extensions of the standard model involving new SU(2)-singlet neutral fermions (for Dirac mass terms) or new Higgs representations (to generate Majorana masses) allow non-zero masses.

Neutrinos could be either Dirac particles or Majorana particles. A Dirac field ν which is different from the corresponding charge conjugate field ν^c describes left- and right-handed neutrinos with their antineutrinos.

For Majorana neutrino field $\nu=\nu^c$ which reduces the number of objects describing by the Dirac equation from four to two (the helicity states of a single self-charge-conjugate fermion).

In the case that neutrino has masses, the neutrino mass eigenstates (ν_1, ν_2,...,) and neutrino-lepton-number eigentates (ν_e, ν_μ, ν_τ,...,) are not necessarily coincident[4]. Thus there may exist neutrino mixing which is analogous to quark mixing as described by a generalized Cabibbo-like mixing,

$$\nu_\ell = \sum_\ell u_{\ell i}\nu_i \qquad \ell\equiv e,\mu,\tau,\ldots, \quad i\equiv 1,2,3,\ldots, \tag{1}$$

This could result in neutrino oscillations when mas differences are small and decays of heavy neutrinos into lighter ones.

The leptonic weak charged current in the presence of neutrino mass has the following form[3,5].

$$J_W^{\mu +} = \frac{ig}{2\sqrt{2}} (\bar{\nu}_e \; \bar{\nu}_\mu \; \bar{\nu}_\tau) \; U^+ \gamma^\mu (1-\gamma_5) \begin{pmatrix} e- \\ \mu- \\ \tau- \end{pmatrix} \tag{2}$$

where U is the analogue of the Cabibbo-Kobayashi-Maskawa mixing matrix[6] describing the relative strengths of the weak transition between various charged leptons and neutrinos of definite mass.

The relevant diagrams for the standard SU(2)xU(1) model with the leptons in left-handed doublets and right-handed singlets plus a single Higgs doublet (Dirac neutrinos ν^D) are shown in Fig.1.

The diagrams involving the unphysical G^+ cannot be ignored even though the coupling of G^+ is proportional to a lepton mass. This coupling may be written[7]

$$- \frac{ig}{2\sqrt{2}M_W} \bar{\nu}_\ell U^+ [(m_\ell - m_\nu) + (m_\ell + m_\nu)\gamma_5] \ell \; G^+ + \text{H.c} \tag{3}$$

where m_ℓ ($\ell = e, \mu, \tau, \ldots,$) is the charged-lepton mass, m_ν is the neutrino mass, $g^2/8M_W^2 = G_F/\sqrt{2}$, G_F is the Fermi constant and U is the unitary matrix relating the neutrino mass eigenstates ν_i ($i=1,2,3,\ldots,$) to the weak eigenstates ν_ℓ.

After the discovery of violation of parity and CP-conserving in weak interactions, many authors[8] studied the interaction of a charged spin 1/2 particle with a weak electromagnetic field which has four static couplings described by its charge, magnetic moment, charge radius and the anapole moment (In CP-conserving theory no electric dipole moment is possible).

The electromagnetic current of leptons ($\ell = e, \mu, \tau, \ldots,$) is given by

$$J_\mu^{e.m} = ie \sum_\ell \bar{u}_\ell \gamma_\mu u_\ell$$

The matrix element of the electromagnetic current J_μ^{em} between ν_i and ν_f states is[9]

$$<\nu_f(p_f)|J_\mu|\nu_i(p_i)>=ie\bar{u}(p_f)\{\gamma_\mu F_1(q^2)+\mu'(q^\mu/2m)\sigma_{\mu\nu}F_2(q^2)$$

$$+\frac{g^2}{16\pi^2}\cdot\frac{1}{M_W^2}a\gamma_5(q^2\gamma_\mu-\gamma\, qq_\mu)G_1(q^2)$$

$$+d\gamma_5(q^\mu/2m)\sigma_{\mu\nu}G_2(q^2)\}\ u(p_i) \qquad (4)$$

in terms of charge e, (anomalous) magnetic moment μ', anapole moment a and electric dipole moment d with the associated form factors satisfying

$$F_1(q^2) = \frac{1}{6}q^2 <r^2> + 0(q^4)\ ,\ <r^2>\ :\ \text{charge radius}$$

$$F_2(q^2) = 1+0(q^2)\ ;\ G_1(q^2)=1+0(q^2),\ G_2(q^2)=1+0(q^2) \qquad (5)$$

These form factors are physically observable quantities at zero momentum transfer ($q^2=0$). The terms $F_1(0)$ and $F_2(0)$ are well known from quantum electrodynamics while the terms $G_1(0)$ and $G_2(0)$ are axial vector terms and arise as a result of the weak interaction.

In the nonrelativistic interaction, the magnetic moment μ', the electric dipole moment d and the anapole moment a have forms of $-\mu'.\vec{\sigma}.\vec{B}$, $d.\vec{\sigma}.\vec{E}$ and $a.\vec{\sigma}.\text{curl}\vec{B}$ or $a.\vec{\sigma}.\vec{j}$ respectively. Where $\vec{\sigma}$ is the spin of the particle, $\vec{B}$ is the magnetic field, $\vec{E}$ is the electric field and $\vec{j}$ is the current density.

While a Dirac neutrino ν^D has four electromagnetic form factors (F_1,F_2,G_1 and G_2), a Majorana neutrino ν^M has only one (G_1) [9].

The form factors of the neutrinos (charge radius[10],magnetic moment[11], anapole moment[12] and the electric dipole moment[13]) has been calculated in the gauge theories.

II. Form Factors of the Neutrinos

For the calculations of the form factors we use the following vertices and propagators:

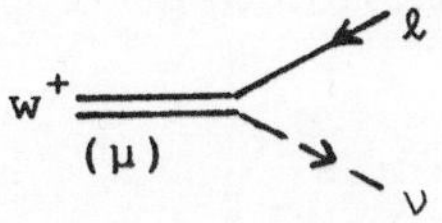

$$: \frac{ig}{2\sqrt{2}} U^{+}\gamma^{\mu}(1-\gamma_5)$$

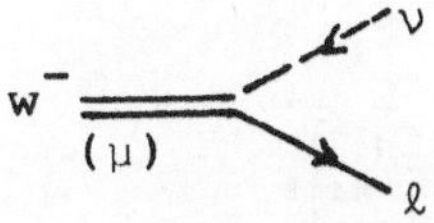

$$: \frac{ig}{2\sqrt{2}} U\gamma^{\mu}(1-\gamma_5)$$

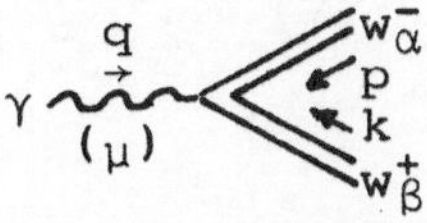

$$: \quad -ie\,\Gamma^{\alpha\beta\mu}(p,k,q)$$

$$\Gamma^{\alpha\beta\mu}(p,k,q) = g^{\alpha\beta}(p-k)^{\mu} + g^{\beta\mu}(k-p)^{\alpha} + g^{\mu\alpha}(q-p)^{\beta}$$

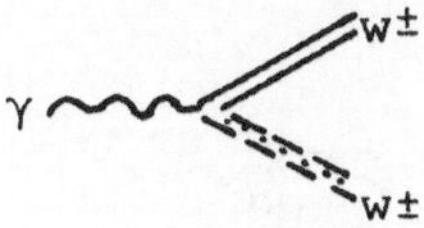

$$: -ieM_W g^{\mu\nu}$$

$$: -ie(p-q)^{\mu}$$

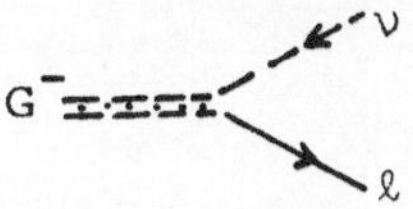

$$: \frac{-ig}{2\sqrt{2}M_W} U[(m-m_{\nu})-(m+m_{\nu})\gamma_5]$$

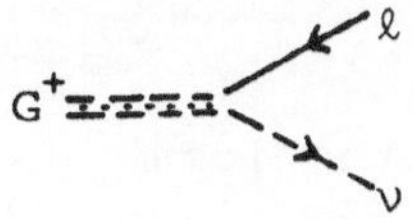

$$: \frac{-ig}{2\sqrt{2}M_W} U^{+}[(m-m_{\nu})+(m+m_{\nu})\gamma_5]$$

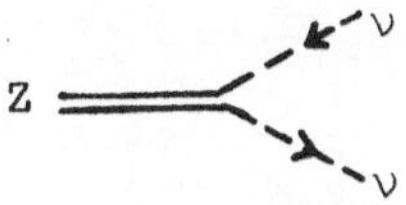

$$: \frac{ie}{2}\gamma^{\mu}(1-\gamma_5)C_A\,, \qquad C_A = \frac{M_Z^2}{2M_W(M_Z^2-M_W^2)^{1/2}}$$

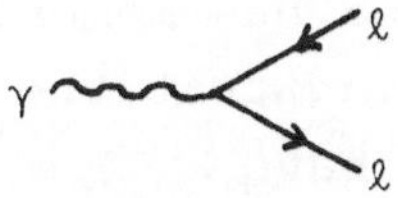

$$: ie\gamma^{\mu}$$

$$: \frac{i}{\not{p}-m_{\nu}}$$

$$: \frac{i}{\not{p}-m_{\ell}}$$

$$: \frac{-i}{p^2-M_W^2}\left[g_{\mu\nu}-\left(1-\frac{1}{\xi}\right)\frac{p_{\mu}p_{\nu}}{p^2-M_W^2/\xi}\right]$$

$$: \frac{-i}{p^2-M_Z^2}\left[g_{\mu\nu}-\left(1-\frac{1}{\xi}\right)\frac{p_{\mu}p_{\nu}}{p^2-M_Z^2/\xi}\right]$$

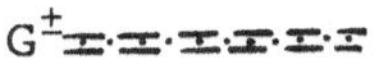

$$: \frac{i}{p^2-M_W^2/\xi}$$

Feynman diagrams which contribute to the form factors of the neutrinos are

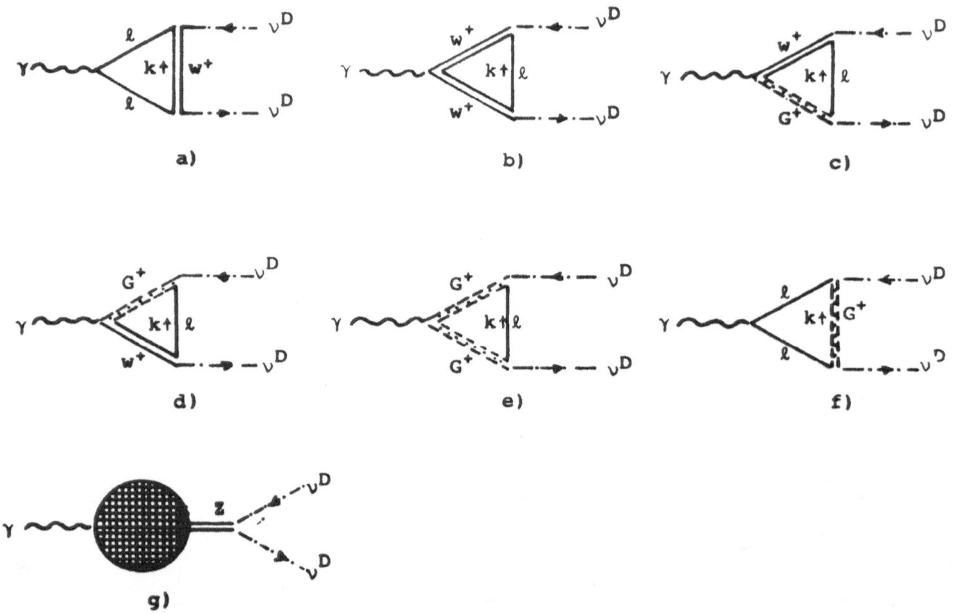

Fig.1 Feynman diagrams which contribute to the form factors of the neutrinos.

Choosing the Breit frame in which the neutrino has momentum initially $(p-\frac{q}{2})^{\mu}$ and finally $(p+\frac{q}{2})^{\mu}$ and requiring that the neutrino is physical (massive Dirac neutrino) we have

$$(p-\frac{q}{2})^2 = (p+\frac{q}{2})^2 = m_{\nu}^2 \Longrightarrow p^2 = m_{\nu}^2 - \frac{q^2}{4} \; ; \; p.q = 0$$

Using dimensional regularization and Feynman-'t Hooft gauge, we find the following matrix elements

$$M^{(a)} = \frac{eg^2}{8} U^{+}U \int \frac{d^{2\omega}k}{(2\pi)^{2\omega}} \frac{\gamma^{\alpha}(1-\gamma_5)(\not{k}+\not{p}+\frac{\not{q}}{2}+m)\gamma^{\mu}(\not{k}+\not{p}-\frac{\not{q}}{2}+m)\gamma_{\alpha}(1-\gamma_5)}{[(k+p+\frac{q}{2})^2-m^2][(k+p-\frac{q}{2})^2-m^2][k^2-M_W^2]} \tag{6}$$

$$M^{(b)} = \frac{eg^2}{8} U^{+}U \int \frac{d^{2\omega}k}{(2\pi)^{2\omega}} \frac{\gamma_{\beta}(1-\gamma_5)(\not{k}+m)\gamma_{\alpha}(1-\gamma_5)\Gamma^{\alpha\beta\mu}}{[k^2-m^2][(k+p+\frac{q}{2})^2-M_W^2][(k+p-\frac{q}{2})^2-M_W^2]} \tag{7}$$

with

$$\Gamma^{\alpha\beta\mu} = \Gamma^{\alpha\beta\mu}(k+p+\frac{q}{2}, -k-p-\frac{q}{2}, q) = 2g^{\alpha\beta}k^{\mu} + g^{\beta\mu}(-k-p-\frac{3}{2}q)^{\alpha} + g^{\mu\alpha}(\frac{3}{2}q-k-p)^{\beta} \quad (8)$$

$$M^{(c)} = \frac{eg^2}{8} U^{+}U \int \frac{d^{2\omega}k}{(2\pi)^{2\omega}} \frac{[(m-m_{\nu})+(m+m_{\nu})\gamma_5](\not{k}+m)\gamma^{\mu}(1-\gamma_5)}{(k^2-m^2)[(k+p+\frac{q}{2})^2-M_w^2][(k+p-\frac{q}{2})^2-M_w^2]} \quad (9)$$

$$M^{(d)} = \frac{eg^2}{8} U^{+}U \int \frac{d^{2\omega}k}{(2\pi)^{2\omega}} \frac{\gamma^{\mu}(1-\gamma_5)(\not{k}+m)[(m-m_{\nu})-(m+m_{\nu})\gamma_5]}{(k^2-m^2)[(k+p+\frac{q}{2})^2-M_w^2][(k+p-\frac{q}{2})^2-M_w^2]} \quad (10)$$

$$M^{(e)} = \frac{eg^2}{8} U^{+}U \int \frac{d^{2\omega}k}{(2\pi)^{2\omega}} \frac{[(m-m_{\nu})+(m+m_{\nu})\gamma_5]k^{\mu}(\not{k}+m)[(m-m_{\nu})-(m+m_{\nu})\gamma_5]}{(k^2-m^2)[(k+p+\frac{q}{2})^2-M_w^2][(k+p-\frac{q}{2})^2-M_w^2]} \quad (11)$$

$$M^{(f)} = \frac{eg^2}{8} U^{+}U \int \frac{d^{2\omega}k}{(2\pi)^{2\omega}} \frac{[(m-m_{\nu})+(m+m_{\nu})\gamma_5](\not{k}+\not{p}+\frac{\not{q}}{2}+m)\gamma^{\mu}(\not{k}+\not{p}-\frac{\not{q}}{2}+m)[(m-m_{\nu})-(m+m_{\nu})\gamma_5]}{[(k+p+\frac{q}{2})^2-m^2][(k+p-\frac{q}{2})^2-m^2](k^2-M_w^2)} \quad (12)$$

$$M^{(g)} = \frac{ie}{4} \cdot \frac{M_z^2}{M_w^2(M_z^2-M_w^2)^{1/2}} \gamma^{\mu}(1-\gamma_5) \frac{-i}{q^2-M_z^2} \, i\, \Pi^{z\gamma}_{\mu\nu}(q^2) \quad (13)$$

with[14])

$$\tilde{\Pi}^{z\gamma}_{\mu\nu} = (-\frac{2}{3}M_w\sqrt{M_z^2-M_w^2} + \delta Y_{wz}\frac{M_w}{\sqrt{M_z^2-M_w^2}} - \frac{1}{3}\frac{\sqrt{M_z^2-M_w^2}}{M_w}$$

$$\Pi^{z\gamma}_{\mu\nu}(q^2) = \frac{g^2}{16\Pi^2} q^2 g_{\mu\nu} \tilde{\Pi}^{z\gamma} \quad (14)$$

Projecting out the axial part proportional to

$$\frac{i}{16\Pi^2} eg^2\gamma_5\gamma^{\mu}\frac{q^2}{M_W^2} \quad (15)$$

and after long but straightforward calculations, we obtain for the anapole part of the Dirac neutrinos the following expressions:

$$a^{(a)} = U^{+}U \left[\frac{5}{48} - \frac{3}{16}\ln\left(\frac{m_{\nu}^2}{M_W^2}\right) + \frac{1}{12m_{\nu}^2} + \ldots\right] \quad (16)$$

$$a^{(b)} = U^+U\,[-\frac{19}{24} - \frac{1}{48m_\nu^2} - \frac{3}{4\,m_\nu^4} + \ldots] \qquad (17)$$

$$a^{(c)} = U^+U\,[-\frac{m^2}{24M_W^2} + \ldots] \qquad (18)$$

$$a^{(d)} = U^+U\,[-\frac{m^2}{24M_W^2} + \ldots] \qquad (19)$$

$$a^{(e)} = U^+U\,[\frac{m^2(m^2-m_\nu^2)}{24m_\nu^2\,M_W^2} + \ldots] \qquad (20)$$

$$a^{(f)} = U^+U\,[-\frac{5(m^2-m_\nu^2)}{96\,M_W^2}\,\ell n(m_\nu^2/M_W^2) + \frac{(m^2-m_\nu^2)}{M_W^2} + \ldots] \qquad (21)$$

$$a^{(g)} = -\frac{M_W}{4(M_Z^2-M_W^2)^{1/2}}\,\tilde{\Pi}^{Z\gamma} \qquad (22)$$

The total anapole moment is

$$a = a^{(a)} + a^{(b)} + a^{(c)} + a^{(d)} + a^{(e)} + a^{(f)} + a^{(g)} \qquad (23)$$

The magnetic moment of a Dirac particle which is proportional to the operator

$$ie\bar{u}(p_f)\sigma_{\mu\nu}q^\nu u(p_i) \qquad (24)$$

can also be calculated.

Projecting out the (tensor) part proportional to $\sigma_{\mu\nu}q^\nu$, we find the following expression:

$$\mu_\nu \approx U^+U\,\frac{g^2}{16\Pi^2M_W^2}\,mm_\nu \qquad (25)$$

(We have used the approximation that all the charged lepton masses m are much smaller than the gauge boson mass M_W).

Our results parallels results found by B.W.Lee and R.E. Shrock, W.J.Marciano and A.I.Sanda, and J. Liu[11].

III. Conclusions

For Majorana particle many authors[15] have obtained the most general expression for the electromagnetic form factor using the constraint of CPT invariance.

The Majorana neutrino will have no γ_μ or $\sigma_{\mu\nu}q^\nu$ electromagnetic form factors. It will develop no $\gamma_5\sigma_{\mu\nu}q^\nu$ form factor either, because, such a form factor would be CP violating and the original one-loop diagram common to Dirac neutrino and Majorana neutrino doesn't contain any CP violation to begin with.

It can be shown that in the massless limit there are no differences between Dirac, Majorana and Weyl neutrinos.

For massless neutrinos in a model where all weak currents are left-handed, there is no distinction between two component Dirac (i.e. Weyl) neutrino and a Majorana neutrino. That is

$$\langle\nu^M|J^{em}|\nu^M\rangle \rightarrow \langle\nu^D|J^{em}|\nu^D\rangle \tag{26}$$

The remaining form factor is proportional to $\gamma_\mu\gamma_5$ which is G_1 (anapole moment).[12]

Notice that in the second-loop approximations the electric-dipole moment G_2 appears.

In the static limit $q^2 \rightarrow 0$, in one-loop approximation the Dirac neutrino ν^D has two electromagnetic characteristics-magnetic moment (which vanish for massless neutrino) and anapole moment (which is different from zero in the massless limit).

At present, the form factors cannot be measured because of their smallness.

ACKNOWLEDGMENTS

The authors would like to thank Prof. B. Gruber and acknowledge with pleasure that value of continued correspondence with Prof. A. O. Barut. One of the authors (C.A.) is grateful to the TURAN-BARUT Physics Foundation.

REFERENCES

1. S. Weinberg, Phys. Rev. Lett. 19, 1264 (1967); A. Salam, Proceedings of the VIII Nobel Symposium, edited by N. Svartholm (Stockholm, 1968), p. 367.

2. B. Pontecorva and S. Bilenky, Neutrinos Today. El:2-87-567 (1987); J. Bernstein, CERN 84-06.

3. P. Langacker, DESY 88-022; DESY 88-023.

4. S. M. Bilenky and B. Pontecorvo, Lett. Nuovo Cim. 17:569 (1976); Phys. Rep. 41:225 (1987).

5. J-L. Vuilleumier, Rep. Prog. Phys. 49:1293 (1986); L. F. Li and F. Wilczek, Phys. Rev. D25:143 (1982).

6. M. Kobayashi and T. Maskawa, Progr. Theor. Phys. 19:652 (1973).

7. B. P. Palash and L. Wolfenstein, Phys. Rev. D25:766 (1982).

8. J. Bernstein, M. Ruderman, and G. Feinberg, Phys. Rev. 132:1227 (1963); W. K. Cheng and S. Bludman, Phys. Rev. 136:B1787 (1964); T. D. Lee and J. Bernstein, Phys. Rev. Lett. 11:512 (1963); I. Meyer and D. Schiff, Phys. Lett. 8:217 (1964); Ya. B. Zeldovich and A. M. Perelomov, Sov. Phys. JETP 12:777 (1961); M. A. Bég, W. J. Marciano and M. Ruderman, Phys. Rev. D17:1395 (1978); M. Abak, Nuovo Cim. 55A:289 (1980).

9. B. Kayser, Phys. Rev. D26:1662 (1982); J. F. Nieves, Phys. Rev. D26:3152 (1982); N. Dombey and A. Kennedy, Phys. Lett. 91B:428 (1980).

10. A. Grau and J. A. Grifols, Phys. Lett. 166B:233 (1986); J. L. L. Martinez, A. Rosado and A. Zepeda, Phys. Rev. D29:1539 (1984).

11. J. E. Kim, Phys. Rev. D14:3000 (1976); A. V. Kyuldjiev, Nucl. Phys. B243:387 (1984); J. Liu, Preprint (CMU-HEP 86-15); B. W. Lee and R. E. Shrock, Phys. Rev. D16:1444 (1977); W. J. Marciano and A. I. Sanda, Phys. Lett. 67B: 303 (1977); R. N. Mohapatra and G. Senjanovic, Phys. Lett. 44:912 (1980); Phys. Rev. D23:165 (1981); A. O. Barut, Preprint ICTP (IC-79-51). M. J. Duncan, J. A. Grifols, A. Mendez and S. U. Sankar, Preprint (UPR-0318-T/UAB-FT-163); R. N. Mohapatra, Preprint (UM pp. 88-172); M. B. Voloshin and M. I. Vysotsky, Preprint ITEP-1 (1986).

12. M. Abak and C. Aydin, Europhys. Lett. 4:881 (1987).

13. L. B. Okun, Preprint ITEP-20 (1986).

14. S. Sakakibara, PITHA 79/17; Phys. Rev. D24:1149 (1981).

15. B. Kayser and A. S. Goldhaber, Phys. Rev. D28:2341 (1983); R. Shrock, Nucl. Phys. B206:359 (1982); J. Schechter and J. Valle, Phys. Rev. D25:1883 (1981).

INCOMPATIBLE OBSERVABLES AND THE MEASUREMENT OF BERRY'S PHASES

K. Datta

Department of Physics and Astrophysics
University of Delhi
Delhi 110007, India

ABSTRACT

Berry has shown that topological obstructions in parameter space for systems described by adiabatically varying parameter dependent Hamiltonians induce non-integrable phase changes in pure states. We study the effect of such obstructions in those descriptions of quantum states which allow for the introduction of mixtures and consider the question of the general character of measurements in which the topological phase manifests itself in the final collapsed density matrix.

1. INTRODUCTION

Ever since Berry's important and interesting discovery[1] a few years ago that the quantum adiabatic theorem requires modifications for cyclically varying Hamiltonians, there has been a spate of investigations which have discovered hitherto unsuspected areas of physics where the so-called "Berry phase" manifests itself. Thus, for example, spin 1/2 particles in slowly rotating magnetic fields,[2] spin resonance experiments in a slowly modulated magnetic field,[3] nuclear quadrupole resonance spectra for slowly rotating samples,[4] photons propagating down a helically wound optical fibre,[5] atoms with an odd number of electrons in a slowly rotating electric field,[6] the quantum Hall effect,[7] the fractional quantum Hall effect[8] have all been restudied keeping Berry's discovery in view. While in most cases these studies have led to a deeper understanding of effects already known, in some instances (such as the sign change in the wave function of integer spin systems under suitable circumstances) new observable effects have been noted. It is in this spirit that we shall take a fresh look at some aspects of the quantum measurement problem which bear upon Berry's discovery.

2. THE QUANTUM ADIABATIC THEOREM

The quantum adiabatic theorem[9] says that if one considers a system (which is not isolated but interacts with an environment which alters slowly) with a Hamiltonian $H(\alpha(t))$, $\alpha(t)$ denoting collectively a set of time dependent parameters, the state of the system satisfies, in the Schrödinger picture,

$$H(\alpha(t))\ |\psi(\alpha(t))\rangle = i\frac{\partial}{\partial t}\ |\psi(\alpha(t))\rangle \tag{1}$$

at every instant of time. At every instant, there exists an instantaneous complete set $|n(\alpha(t))\rangle$

$$H(\alpha(t))\ |n(\alpha(t))\rangle = E_n(\alpha(t))\ |n(\alpha(t))\rangle \tag{2}$$

with no simple connection between basis states $|n(\alpha(t))\rangle$ and $|m(\alpha(t'))\rangle$ at different instants of time. Generally, if $H(\alpha(t))$ varies arbitrarily with the time, it will connect the states $|n(\alpha(t))\rangle$ and $|m(\alpha(t'))\rangle$ through transitions. However, if the change of $H(\alpha(t))$ is sufficiently slow (periods of change of typical matrix elements of $H(\alpha(t))$ large compared to the natural $\frac{1}{\hbar}(E_n(t) - E_m(t))$ then the adiabatically changing Hamiltonian "drags" the state along. If, say,

$$|\psi(t=0)\rangle = |n(\alpha(0))\rangle \tag{3}$$

then, at $t = T$, the system is to be found in the corresponding eigenstate $|n(\alpha(T))\rangle$. This state vector will, of course, be accompanied by the appropriate dynamically evolved phase, the analog of $exp(-i\,E_nT)$ being, in the present case, $exp\left(-i\int_0^T E_n(t')dt'\right)$; more, it is accompanied by a second phase, which, in the approximation

$$\left|\frac{1}{\omega_{mn}}\left\langle\frac{\partial H}{\partial t}\right\rangle\right| \ll |E_n - E_m| \tag{4}$$

is given by $exp(i\,\gamma_n(t))$ with

$$\gamma_n(t) \simeq i\int_0^t dt'\ \langle n(\alpha(t'))|\ \frac{d}{dt'}\ |n(\alpha(t'))\rangle \tag{5}$$

is purely real; this led, in older versions of the adiabatic theorem, to the statement that this phase was of no consequence since we are free to redefine the phase associated with the state vector at every instant of time, given that the Schrödinger equation (2) and the normalization equation

$$\langle n(\alpha(t))|n(\alpha(t))\rangle = 1 \tag{6}$$

are invariant under redefinitions of the phase. Berry[1] argued that in situations where the Hamiltonian undergoes cyclical changes $\alpha(T) = \alpha(0)$ along a circuit C in α space:

$$H(\alpha(t)) = H(\alpha(0)) \tag{7}$$

the initial state may be compared with that which arrives dragging in the phase $exp(i\,\gamma_n(C))$

$$\gamma_n(C) = i\int_0^T dt\ \langle n(\alpha(t))|\ \frac{d}{dt}\ |n(\alpha(t))\rangle \tag{8}$$

The topological character of this phase is revealed by writing it as a surface integral in parameter space. Since the states $|n(\alpha(t))\rangle$ depend on t only through the dependence implicit in $\alpha(t)$, we have

$$\gamma_n(C) = i\oint_C \langle n(\alpha(t))|\ \frac{\partial}{\partial\alpha_j}\ |n(\alpha(t))\rangle \tag{9}$$

so that

$$\mathcal{A}_j(\alpha(t)) \equiv i \langle n(\alpha(t))| \frac{\partial}{\partial \alpha_j} |n(\alpha(t))\rangle \tag{10}$$

$$\gamma_n(C) = \oint \mathcal{A}_j(\alpha(t)) d\alpha_j \tag{11}$$

is the line integral of a non-integrable "vector" function; a "Stokes" theorem allows it to be written as a surface integral

$$\gamma_n(C) = \iint_S (\epsilon_{ijk}\delta_j \mathcal{A}_k) dS_i \tag{12}$$

As long as the surface spanned in parameter space cannot be shrunk to zero, $\gamma_n(C)$ is non-vanishing; hence its essentially topological character.

3. DENSITY MATRIX DESCRIPTION & QUANTUM MEASUREMENTS

We have so far confined our attention to pure states i.e. to states describable as state vectors. Such a description is inadequate for describing the measurement process in quantum theory, for which it becomes necessary to invoke the description of states of a quantum mechanical system in terms of non-negative unit trace elements of the Banach space of bounded, Hermitian operators on the Hilbert space $\mathcal{H}$

$$\begin{aligned} &\rho \in B(\mathcal{H}, \mathcal{H}), \text{ the Banach space of bounded operators on } \mathcal{H} \\ &\rho \geq 0 \\ &tr\, \rho = 1 \end{aligned} \tag{13}$$

Pure states are extremal elements of convex sets

$$\rho^2 = \rho \quad \text{and are of the form}$$

$$\rho = |\psi\rangle\ \langle\psi|\ ;$$

this description allows, in addition, for non-extremal states

$$\rho^2 \neq \rho$$

which are not one-dimensional projections. Such states are called mixtures and figure in an essential way in the complete quantum description of a measurement process which consists of two stages which may be described as follows.

Stage I. This stage is characterized by the setting up of correlations between pure apparatus and system states

$$\rho = |A_0\rangle\ \langle A_0| \otimes \sum_{ss'} C_s C_{s'}^*\ |s\rangle\ \langle s'| \tag{14}$$

$$\rightarrow \rho = \sum_{ss'} C_s C_{s'}^*\ |A_s\rangle\ \langle A_{s'}|\ \otimes\ |s\rangle\ \langle s'| \tag{15}$$

These correlations are set up through suitable apparatus-system interactions. The state of the system-apparatus combination is pure; off-diagonal coherences between different basis (apparatus and system) states are present in ρ.

Stage II. This stage is characterized by the collapse to a mixture in which no such off-diagonal correlations persist.

$$\rho = \sum_{s} |C_s|^2 \, |A_s\rangle \, \langle A_s| \times |s\rangle \, \langle s| \tag{16}$$

The measured system operator is $\sum \lambda_s \, |s\rangle \, \langle s|$.

The problem of the theory of measurement consists in arriving at a satisfactory understanding, within the calculational scheme of quantum theory, of the transition from stage I to stage II of the measurement process. Here, we pose a restricted question:

Given this general characteristic of quantum measurements, what is the general nature of measurements in which the topological phase, if introduced in stage I, persists in stage II?

4. DENSITY OPERATOR UNDER CYCLIC ADIABATIC HAMILTONIANS

We describe the system by the density operator and work in the Heisenberg picture. Instead of attempting to solve the appropriate Heisenberg equation of motion

$$\dot{\rho} = \frac{1}{i}[\rho(t), H(t)] + \frac{\partial \rho}{\partial t} \tag{17}$$

we note that a pure state which is a superposition of instantaneous Hamiltonian eigenstates

$$\begin{aligned} \rho(0) &= |\psi(0)\rangle \, \langle \psi(0)| \\ &= \sum_{nm} a_n(0) a_m^*(0) \, |n(\alpha(0))\rangle \, \langle m(\alpha(0))| \end{aligned} \tag{18}$$

evolves into

$$\begin{aligned} \rho(T) &= |\psi(T)\rangle \, \langle \psi(T)| \\ &= \sum_{nm} a_n(T) a_m^*(T) \, |n(\alpha(T))\rangle \, \langle m(\alpha(T))| \end{aligned} \tag{19}$$

with T chosen such that $H(\alpha(t)) = H(\alpha(0))$. Further, in (13),

$$a_n(T) = exp(i\,\gamma_n(C)) exp\left(-i \int_0^T E_n(t') dt'\right) a_n(0) \tag{20}$$

Since we choose

$$|n(\alpha(t))\rangle = |n(\alpha(0))\rangle \tag{21}$$

$$\begin{aligned} \rho(T) = \sum_{nm} & a_n(0) a_m^*(0) \, |n(\alpha(0))\rangle \, \langle m(\alpha(0))| \times \\ & \times exp\Big(i(\gamma_n(C) - \gamma_m(C))\Big) exp\left(-i \int_0^T (E_n(t') - E_m(t')) dt'\right) \end{aligned} \tag{22}$$

In studying the evolution of the density operator in the Heisenberg picture, we must be careful in evaluating $\rho(t)$ in a time-fixed basis. Using the set $\{ |n(\alpha(0))\rangle \}$ as the natural one in the present case, we have

$$\rho_{nm}(T) = \rho_{nm}(0)\, exp\Big(i(\gamma_n(C) - \gamma_m(C)) \Big) \times \\ \times\, exp\left(-i \int_0^T (E_n(t') - E_m(t'))dt' \right) \tag{23}$$

We therefore conclude that a pure state density operator evolves in such a way that the populations ρ_{nn} (in an appropriately chosen time-independent basis) remain unaffected by topological phases whereas the coherences ρ_{nm}, $n \neq m$ carry their effects. This has the consequence that only in those measurements where the collapsed density operator depends on such off-diagonal coherences of the density operator in the Heisenberg representation (basis provided by $|n(\alpha(t))\rangle$ will the topological phase manifest itself.

5. PREFERRED POINTER BASIS

From the earlier statement of the completed measurement process ending in collapse (eqs. 15, 16)

$$\rho = \sum_{ss'} C_s C^*_{s'}\ |A_s\rangle\ \langle A'_s|\ \otimes\ |s\rangle\ \langle s'|$$

$$\rightarrow \rho = \sum_s |C_s|^2\ |A_s\rangle\ \langle A_s|\ \otimes\ |s\rangle\ \langle s|$$

it would appear that, generally, the off-diagonal coherences do not figure in the final collapsed state. Since the distinction between coherences and populations in the density operator is not basis independent, it needs to be emphasised that going to a different (e.g. preferred pointer)[10] basis before collapse does not alter this conclusion. For, consider a different pointer basis $\{ |a_r\rangle \}$ s.t.

$$|A_s\rangle = \sum_r \langle a_r|A_s\rangle\ |a_r\rangle \tag{24}$$

in terms of which the pre-collapsed density operator is

$$\rho = \sum_{ss'} \sum_{rr'} C_s C^*_{s'}\ \langle a_r|A_s\rangle\ |a_r\rangle\ \langle a_{r'}|\ \times\ \langle A_{s'}|a_{r'}\rangle\ \otimes\ |s\rangle\ \langle s'| \tag{25}$$

Introducing Everett's relative system states[11]

$$d_r\ |r\rangle = \sum_s C_s\ \langle a_r|A_s\rangle\ |s\rangle \tag{26}$$

which are, in general, normalized by not orthogonal,

$$d_r d^*_{r'}\ \langle r'|r\rangle = \sum_{ss'} C_s C^*_{s'}\ \langle a_r|A_s\rangle\ \langle A_{s'}|a_{r'}\rangle\ \delta_{ss'}$$

$$= \sum_s |C_s|^2\ \langle a_r|A_s\rangle\ \langle A_s|a_{r'}\rangle$$

so that

$$|d_r|^2 = \sum_s |C_s|^2 \, |\langle a_r | A_s \rangle|^2 \tag{27}$$

and we have

$$\rho = \sum_r d_r d^*_{r'} \; |a_r\rangle \; \langle a_{r'}| \otimes |r\rangle \; \langle r'| \quad ;$$

the collapsed density matrix in the relative state basis

$$\rho = \sum_r |d_r|^2 \; |a_r\rangle \; \langle a_r| \otimes |r\rangle \; \langle r| \tag{28}$$

carries no trace of the off-diagonal coherences $C_s C^*_{s'}$.

6. COLLAPSE FOLLOWING AN INCOMPLETE MEASUREMENT

The quantum measurement algorithm producing collapse can be retraced to single out cases where the off-diagonal coherences induced in stage I of the measurement process contribute to the collapsed density matrix. We work with the system density operator only; after the establishment of correspondences, the apparatus merely mimics the system. Given

$$\rho = \sum_{kj} C_{s_k} C^*_{s_j} \; |s_k\rangle \; \langle s_j| \tag{29}$$

the collapse following the measurement of the system operator $\sum_i \lambda_i \; |s_i\rangle \; \langle s_j|$ is given by the algorithm

$$\rho \to \rho' = \sum_i P_{s_i} \rho P_{s_i} \tag{30}$$

(where P_{s_i} is the projection operator associated with λ_i)

$$\begin{aligned} &= \sum_{ikj} C_{s_k} C^*_{s_j} \; |s_i\rangle \; \langle s_i | s_k\rangle \; \langle s_j | s_i\rangle \; \langle s_i| \\ &= \sum_i |c_{s_i}|^2 \; |s_i\rangle \; \langle s_i| \end{aligned} \tag{31}$$

Consider now the following situation:

A measuring arrangement is organized which sets up correspondences between eigenstates of $\sum_i |s_i\rangle \; \langle s_i|$ and apparatus states; the state of the system after the correspondence is established is

$$\rho = \sum_{kj} C_{s_k} C^*_{s_j} \; |s_k\rangle \; \langle s_j| \tag{32}$$

The measurement is now not completed through the algorithm just stated (projection through the P_{s_i}, eq. (30)); instead, a measurement of an incompatible observable B is completed. Following the stated algorithm

$$\rho \to \rho' = \sum_i P_{b_i} \sum_{kj} C_{s_k} C^*_{s_j} \; |s_k\rangle \; \langle s_j| \; P_{b_i} \tag{33}$$

$$= \sum_{i;kj} C_{s_k} C^*_{s_j} \; |b_i\rangle \; \langle b_i|s_k\rangle \; \langle s_j|b_i\rangle \; \langle b_i|$$

$$= \sum_i |\beta_i|^2 \; |b_i\rangle \; \langle b_i| \tag{34}$$

$$\beta_i \equiv \sum_k C_{s_k} \; \langle b_i|s_k\rangle \tag{35}$$

$$\beta_i^* = \sum_j C^*_{s_j} \; \langle s_j|b_i\rangle \tag{36}$$

$$\beta_i|^2 = \sum_{kj} C_{s_k} C^*_{s_j} \; \langle b_i|s_k\rangle \; \langle s_j|b_i\rangle \tag{37}$$

The off-diagonal coherences contained in the density operator in eq. (32) are now to be seen in the collapsed density operator.

7. CONCLUSION

We have investigated the nature of measurements in which Berry's topological phase, if introduced, persists in the collapsed density matrix characteristic of the final stage of every quantum measurement process. In general, even a non-global phase introduced in pure states at some intermediate stage of the measurement process will disappear from the final collapsed density matrix. This is because the collapsed density matrix contains no off-diagonal pure state coherences and, hence, is invariant under redefinitions of phases of the pure states. A different situation, however, emerges if an arrangement which sets up correspondences between eigenstates of $\sum \lambda_i \; |s_i\rangle \; \langle s_i|$ and apparatus states is followed, before the measurement process is complete, by the measurement of an incompatible observable B, where $[B, \sum \lambda_i \; |s_i\rangle \; \langle s_i|\,] \neq 0$. In this case, the final collapsed density operator is of the form $\sum |C_k|^2 \; |b_k\rangle \; \langle b_k|$ and is not invariant under phase redefinitions $|s_i\rangle \to |s_i\rangle \; e^{i\theta_i(s_i)}$ introduced under interactions with adiabatically varying time dependent Hamiltonians (as pointed out by Berry) at some intermediate stage of the measurement process.

REFERENCES

1. M. V. Berry, Proc. Roy. Soc. A392:45 (1984).
2. T. Bitters and D. Dubbers, Phys. Rev. Lett. 59:251 (1987).
3. J. Moody, A. Shapero and F. Wilczek, Phys. Rev. Lett. 56:893 (1986).
4. R. Tycko, Phys. Rev. Lett. 58:2281 (1987).
5. R. Y. Chiao and Y. S. Wu, Phys. Rev. Lett. 57:933 (1986).
6. C. A. Mead, Phys. Rev. Lett. 59:161 (1987).
7. D. Thouless, M. Kohomoto, M. Nightingale and M. den Nijs, Phys. Rev. Lett. 49:405 (1982).

   J. E. Avron, R. Seiler and B. Simon, Phys. Rev. Lett. 51:51 (1983).

B. Simon, Phys. Rev. Lett. 51:2167 (1983).

8. G. Senenoff and P. Sodano, Phys. Rev. Lett. 57:1195 (1986).
9. L. I. Schiff, "Quantum Mechanics," p. 289 ff., 3rd Edition, McGraw Hill, N.Y. (1968).

   A. Messiah, "Quantum Mechanics," p. 744 ff., North Holland, Amsterdam (1966).
10. W. H. Zurek, Phys. Rev. D24:1516 (1981).
11. H. Everett, Rev. Mod. Phys. 29:454 (1957).

A preliminary version of this paper was presented as a talk at a conference on the Philosophical Foundations of Quantum Mechanics held in the National Institute of Science, Technology and Development Studies, New Delhi 110012 in March 1988.

HIDDEN SYMMETRIES OF A SELF-DUAL MONOPOLE

L. Fehér[1,2] , P. Horváthy[2,3] , and L. O'Raifeartaigh[2]

[1] Bolyai Institute, Jate H-6720 Szeged (Hungary)
[2] Dublin Institute for Advanced Studies, Dublin (Ireland)
[3] Maths. Dept., Université, F-57 Metz (France)

A self-dual (SD) $SU(2)$ monopole is a static solution of the first order Bogomolny[1] equations

$$D_k\Phi = B_k \left(= \frac{1}{2}\epsilon_{kij}F_{ij}\right) \tag{1}$$

The monopole's field can be identified with a pure Yang-Mills configuration $A_0 = 0$, A_k, $A_4 = \Phi$ in (1+4)-dimensional flat space which does not depend on the coordinates x^0 and x^4. The Dirac equation

$$i\frac{\partial\Phi}{\partial x^0} = i\gamma^0 \not{D}_4 \Psi \tag{2}$$

plays a decisive role[2,3,4] in describing the fluctuations around a SD monopole. Here we are interested in the symmetries of the 4-dimensional, Euclidean Dirac operator

$$\not{D}_4 = \gamma^k D_k + \gamma^4 D_4 = \frac{\sqrt{2}}{i}\begin{pmatrix} 0 & Q \\ Q^+ & 0 \end{pmatrix}, \tag{3}$$

where

$$Q = \frac{1}{\sqrt{2}}\left(\Phi \cdot 1_2 - i\vec{\pi}\cdot\vec{\sigma}\right), \quad \vec{\pi} = -i\vec{D} = -i\vec{\nabla} - \vec{A}^a I_a, \quad D_4 = \frac{\partial}{\partial x^4} - i\Phi^a I_a \tag{4}$$

We use the following γ-matrices:

$$\gamma^k = \begin{pmatrix} 0 & i\sigma^k \\ -i\sigma^k & 0 \end{pmatrix}, \quad \gamma^4 = \begin{pmatrix} 0 & 1_2 \\ 1_2 & 0 \end{pmatrix}, \quad \gamma^0 = i\gamma^5 = i\begin{pmatrix} 1_2 & 0 \\ 0 & -1_2 \end{pmatrix}. \tag{5}$$

The I_a $(a = 1,2,3)$ are the standard isospin matrices in some representation and we suppose that nothing depends on the extra coordinate x^4.

Let us now consider the supersymmetric quantum mechanical system defined by $(-1)^F = \gamma^5$, and

$$H = \left(\frac{i}{\sqrt{2}}\not{D}_4\right)^2 = \begin{pmatrix} QQ^+ & 0 \\ 0 & Q^+Q \end{pmatrix} = \begin{pmatrix} H_1 & 0 \\ 0 & H_0 \cdot 1_2 \end{pmatrix}. \tag{6}$$

Here

$$H_O = \frac{1}{2}\left(\pi^2 + \Phi^2\right), \qquad H_1 = H_O \cdot \mathbf{1}_2 - \vec{\sigma} \cdot \vec{B}. \tag{7}$$

As a consequence of self-duality, the spin drops out of Q^+Q, while QQ^+ contains the term $\vec{\sigma} \cdot \vec{B}$ characteristic of a particle of "apparent" gyromagnetic ratio 4. Note that H_O is the hamiltonian of a nonrelativistic, spinless particle moving in the field of the monopole. The Pauli hamiltonian H_1 describes a nonrelativistic particle of spin $\frac{1}{2}$. For $E \neq 0$ the relation

$$H_1 = U^+\left(H_O \cdot \mathbf{1}_2\right)U, \qquad U = \frac{1}{\sqrt{H_O}}Q^+ \tag{8}$$

is satisfied. Thus H_1 and $H_O \cdot \mathbf{1}_2$ have identical spectra apart from zero modes which are counted by the index theorem.[5] Let us now consider particles of definite momenta, helicity, and electric charge

$$g = \frac{I_a \Phi^a}{\|\Phi\|} \tag{9}$$

scattering on the monopole. At large distances A_k tends to the potential of a Dirac monopole, while the Higgs field behaves as $\Phi \sim q\left(1 - \frac{1}{r}\right)$. This implies that the transformation U preserves the quantum numbers of the scattering states. Therefore H_1 and $H_O \cdot \mathbf{1}_2$ have identical S-matrices with a factorized, purely kinematical spin dependence. (This spin-charge factorization of the S-matrix is inherited[6] by the vector fluctuations scattering off the monopole.) One can understand the factorization of the S-matrix also as a consequence of the conservation of $\vec{S_0} = \frac{1}{2}\begin{pmatrix} 0 & 0 \\ 0 & \vec{\sigma} \end{pmatrix}$ for $H_O \cdot \mathbf{1}_2$ and

$$\vec{S_1} = U^+\vec{S_0}U = \frac{-1}{2H_1}\vec{\Omega}, \quad \vec{\Omega} = \frac{1}{2}(\pi^2 - \Phi^2)\vec{\sigma} + \Phi(\vec{\pi} \times \vec{\sigma}) - (\vec{\sigma} \cdot \vec{\pi})\vec{\pi} \tag{10}$$

for H_1. Now we introduce the following vector supercharges commuting with H:

$$Q_1 = \frac{i}{\sqrt{2}}\not{D}_4 = \begin{pmatrix} 0 & Q \\ Q^+ & 0 \end{pmatrix}, \quad Q_2 = -i\gamma^5 Q_1, \quad \vec{Q_\alpha} = 2i\left[\vec{S_0}, Q_\alpha\right], \quad (\alpha = 1,2) \tag{11}$$

The even generators γ^5, $\vec{S_0}$, $\vec{S_1}$ and the odd ones Q_α, $\vec{Q_\alpha}$ yields, for fixed $E \neq 0$, a closed superalgebra determined by the commutation relations:

$$\left[S^i_\alpha, S^j_\beta\right] = i\delta_{\alpha\beta}\epsilon_{ijk} \qquad (\alpha, \beta = 0, 1), \tag{12.a}$$

$$\left.\begin{aligned}
&\left[\gamma^5, Q_\alpha\right] = 2i\epsilon_{\alpha\beta}Q_\beta, && \left[\gamma^5, Q^k_\alpha\right] = 2i\epsilon_{\alpha\beta}Q^k_\alpha \qquad (\alpha, \beta = 1, 2), \\
&\left[S^k_0, Q_\alpha\right] = -\frac{1}{2}iQ^k_\alpha, && \left[S^k_0, Q^j_\alpha\right] = \frac{1}{2}i\left(\delta_{jk}Q_\alpha + \epsilon_{kjn}Q^n_\alpha\right), \\
&\left[S^k_1, Q_\alpha\right] = \frac{1}{2}iQ^k_\alpha, && \left[S^k_1, Q^j_\alpha\right] = -\frac{1}{2}i\left(\delta_{jk}Q_\alpha - \epsilon_{kjn}Q^n_\alpha\right),
\end{aligned}\right\} \tag{12.b}$$

$$\left.\begin{aligned}
&\left\{Q_\alpha, Q_\beta\right\} = 2\delta_{\alpha\beta}H, \qquad \left\{Q_\alpha, Q^k_\alpha\right\} = -4H\epsilon_{\alpha\beta}\left(S^k_0 + S^k_1\right), \\
&\left\{Q^k_\alpha, Q^j_\beta\right\} = 2H\delta_{\alpha\beta}\delta_{jk} - 4H\epsilon_{\alpha\beta}\epsilon_{kjn}\left(S^n_1 - S^n_0\right)
\end{aligned}\right\} \tag{12.c}$$

For fixed $E > 0$, this algebra is $SU(2/2) \oplus U(1)$. The $U(1)$ factor is generated by γ^5. Observe that $\vec{S} = \vec{S_0} + \vec{S_1}$ generates an $SO(3)$ invariance algebra of the Dirac operator $\not{D}_4$. The same algebraic structure is present (at least locally) in an instanton and in a SD gravitational instanton background. This could be useful for analysing the Dirac equation.

Our investigations presented so far are valid in a general SD background, in the whole space. From now on we shall restrict ourselves to the behaviour at large distances. We show, in fact, that for the asymptotic, long range field of the BPS 1-monopole there exists also a large dynamical symmetry.[7] As $r \to \infty$, the fields tends to that of a singular monopole

$$A_i^a = \epsilon_{aij}\frac{x_j}{r^2}, \qquad \Phi^a = -\frac{x^a}{r}\left(1 - \frac{1}{r}\right) \tag{13}$$

For the asymptotic system, the electric charge is conserved and the angular momentum (without spin) becomes

$$\vec{L_0} = \vec{J} - \frac{\vec{\sigma}}{2} = \vec{r} \times \vec{\pi} - q\hat{r}, \tag{14}$$

since A_i^a describes an embedded Dirac monopole. The hamiltonian H_0 is separated as

$$H_0 = -\frac{1}{2}\Delta_r + \frac{L_0^2 - q^2}{2r^2} + \frac{q^2}{2}\left(1 - \frac{2}{r} + \frac{1}{r^2}\right) = -\frac{1}{2}\Delta_r + \frac{L_0^2}{2r^2} - \frac{q^2}{r} + \frac{q^2}{2} \tag{15}$$

The $\frac{q^2}{r^2}$ terms coming from the centrifugal potential and from Φ^2 cancel and we are left with a Coulomb type system. This is the "MIC-Zwanziger" Hamiltonian[8,9] introduced originally just for the sake of its symmetries 20 years ago. It commutes with the Runge-Lenz vector

$$\vec{K_0} = \frac{1}{2}\left(\vec{\pi} \times \vec{L_0} - \vec{L_0} \times \vec{\pi}\right) - q^2\hat{r} \tag{16}$$

Correspondingly, $\vec{L_1} = U^+\vec{L_0}U = \vec{J} - \vec{S_1}$ and $\vec{K_1} = U^+\vec{K_0}U$ are constants of motion for H_1. The hamiltonian H_1 has been investigated recently by D'Hoker and Vinet[10] but without its present interpretation.[11] The spin $\frac{1}{2}$ "Runge-Lenz vector" $\vec{\Lambda_1}$ they have found is

$$\vec{\Lambda_1} = \vec{K_1} - q\vec{S_1} = K_0 \cdot 1_2 + \vec{\pi} \times \vec{\sigma} + \left(\frac{q}{r} - \frac{q}{2}\right)\vec{\sigma} - \left(\vec{\sigma} \cdot \vec{B}\right)\vec{r} \tag{17}$$

For bound states $0 < E < \frac{q^2}{2}$, and from (12), the angular momentum and the Runge-Lenz vector combine into an $[SU(2/2) \oplus U(1)] \oplus O(4)$ dynamical symmetry algebra of H. For $E = \frac{q^2}{2}$ and $E > \frac{q^2}{2}$ $O(4)$ is replaced by $E(3)$ and $O(3,1)$, respectively. The bound state spectrum (without zero modes)

$$E = \frac{q^2}{2}\left(1 - \frac{1}{n^2}\right), \qquad n = |q| + 1,\ |q| + 2, \ldots \tag{18}$$

is calculated from the $O(4)$ symmetry by Pauli's method.[12] The multiplicity (for fix $q \neq 0$) is $2\left(n^2 - q^2\right)$. The S-matrix is also easy to derive from the bosonic symmetry $O(3) \oplus O(3,1)$ (the $O(3)$ factor comes from $\vec{S}$). In fact, Zwanziger's method[8] yields

$$\begin{aligned} S\left(\underline{k}', s' \mid \underline{k}, s\right) &= \delta\left(E_{k'} - E_k\right) \cdot \mathcal{D}^{\frac{1}{2}}_{s',s}\left(R^{-1}_{\underline{k}'} R_{\underline{k}}\right) \cdot S_0\left(\underline{k}' | \underline{k}\right), \\ S_0\left(\underline{k}' | \underline{k}\right) &= \sum_{l \geq |q|} (2l+1) \left[\frac{\left(l - \frac{i}{k}\right)!}{\left(l + \frac{i}{k}\right)!}\right] \mathcal{D}^{l}_{-q,q}\left(R^{-1}_{\underline{k}'} R_{\underline{k}}\right) \end{aligned} \tag{19}$$

Here $\underline{k}$ is the velocity and s is the helicity of the incoming particles of charge q, $k = \sqrt{2E_k - q^2}$. The primed quantities refer to the outgoing particles. $R_{\underline{k}}$ is some rotation, chosen by convention, which brings the $\hat{z}$-direction into the direction of $\underline{k}$. The poles of S are at $l - \frac{i}{k} = -T$, $T = 1, 2, \ldots$ yielding the $E > 0$ bound state spectrum (18) with $n = T + l = |q| + 1, |q| + 2, \ldots$.

Finally, we show that, for the singular monopole (13), $H_0 \cdot \mathbf{1}_2$ and H_1 are essentially identical. In terms of the quantities

$$x = \vec{\sigma} \cdot \vec{J} - \frac{1}{2}, \qquad H_r = -\frac{1}{2}\Delta_r - \frac{q^2}{r} + \frac{q^2}{2}, \tag{20}$$

$H_0 \cdot \mathbf{1}_2$ is rewritten as

$$H_0 \cdot \mathbf{1}_2 = H_r + \frac{J^2 + \frac{1}{4} - x}{2r^2} \tag{21}$$

The operator x commutes with $H_0 \cdot \mathbf{1}_2$ and with $\vec{J}$ and has eigenvalues $\pm\left(j + \frac{1}{2}\right)$ because of the identity

$$x^2 = J^2 + \frac{1}{4} \tag{22}$$

Thus, by separating according to J^2, J_3 and x_1 (21) becomes

$$H_0 \cdot \mathbf{1}_2 = H_r + \begin{cases} \dfrac{\left(j - \frac{1}{2}\right)\left(j + \frac{1}{2}\right)}{2r^2}, & j = |q| + \dfrac{1}{2}, \ldots \\ \dfrac{\left(j + \frac{1}{2}\right)\left(j + \frac{3}{2}\right)}{2r^2}, & j = |q| - \dfrac{1}{2}, \ldots \end{cases} \tag{23}$$

On the other hand, the operator

$$y = \vec{\sigma}\left(\vec{J} + 2q\hat{r}\right) - \frac{1}{2} \tag{24}$$

allows us to rewrite $H_1 = H_0 \cdot \mathbf{1}_2 - \vec{\sigma} \cdot \vec{B}$ as

$$H_1 = H_r + \frac{J^2 + \frac{1}{4} - y}{2r^2} \tag{25}$$

Moreover, y commutes with H_r and $\vec{J}$, and has eigenvalues $\pm\left(j + \frac{1}{2}\right)$. Therefore, by diagonlizing the commuting set J^2, J_3, y (25) is converted into

$$H_1 = H_r + \begin{cases} \dfrac{\left(j - \frac{1}{2}\right)\left(j + \frac{1}{2}\right)}{2r^2}, & j = |q| - \dfrac{1}{2}, |q| + \dfrac{1}{2}, \ldots \\ \dfrac{\left(j + \frac{1}{2}\right)\left(j + \frac{3}{2}\right)}{2r^2}, & j = |q| - \dfrac{1}{2}, \ldots \end{cases} \tag{26}$$

This is the same equation as (23) describing $H_0 \cdot \mathbb{1}_2$! The associated eigenvalue problem can be solved in terms of hypergeometric functions. The result is consistent with (18). The only difference between $H_0 \cdot \mathbb{1}_2$ and H_1 is that for H_1, in the $y = \left(j + \frac{1}{2}\right)$ sector j starts from $|q| - \frac{1}{2}$, while for $H_0 \cdot \mathbb{1}_2$, in the $x = \left(j + \frac{1}{2}\right)$ sector j starts from $|q| + \frac{1}{2}$. This is the only effect produced by the term $\vec{\sigma} \cdot \vec{B}$. In fact, this leads to the spectral asymmetry required by the index theorem.[5] The zero modes of H_1 carry the "unpaired" quantum numbers $j = |q| - \frac{1}{2}$, $y = |q|$. For fixed $q \neq 0$, the multiplicity of the 0-energy state is $(2j_{min} + 1) = 2|q|$.

REFERENCES

1. For a review, see P. Goddard and D. Olive, Rep. Prog. Phys. 41:1357 (1978).
2. L. S. Brown, R. D. Carlitz and C. Lee, Phys. Rev. D16:417 (1977).
3. R. Jackirv and C. Rebbi, Phys. Rev. D13:3398 (1976).
4. F. A. Bais and W. Froost, Nucl. Phys. B178:125 (1981).
5. E. Weinberg, Phys. Rev. D20:936 (1979)
6. K. J. Bieble and J. Wolf, Nucl. Phys. B279:571 (1987).
7. J. F. Schönfeld, Journ. Math. Phys. 21:2528 (1980).
   L. Gy Fehér, J. Phys. A19:1259 (1986).
   G. W. Gibbons and P. Ruback, Commun. Math. Phys. 115:226 (1988).
8. D. Zwanziger, Phys. Rev. 176:1480 (1968).
9. H. V. McIntosh and A. Cisneros, Journ. Math. Phys. 11:896 (1970).
10. E. D'Hoker and L. Kinet, Phys. Rev. Lett. 55:1043 (1985);
    Lett. Math. Phys. 12:71 (1986)
11. L. Gy Fehér and P. A. Horváthy, Mod. Phys. Lett. A (to be published) (1988).
12. W. Pauli, Z. Phys. 36:33 (1926).

ON THE IMPORTANCE OF THE WEYL GROUP FOR KAC-MOODY AND STRING ALGEBRAS

N. Gorman,[1] W. McGlinn,[2] and L. O'Raifeartaigh[1]

[1] Dublin Institute for Advanced Studies

[2] Notre Dame University, USA.

ABSTRACT

The Weyl Group of Kac-Moody algebras is investigated and is shown to play a rich and varied role both for these algebras and for string theory.

For ordinary compact simple Lie[1] algebras G_0 with generators

$$[T_0^a, T_0^b] = if_c^{ab} T_0^c \,, \tag{1}$$

the Weyl group W_0 has two equivalent definitions:

(a) W_0 is the group generated by reflexions in the planes orthoganal to the roots $\vec{\alpha}$, i.e. by the reflexions

$$\vec{p} \to \vec{p} - \frac{2(\alpha,p)}{\alpha^2}\vec{\alpha} \tag{2}$$

where $\vec{p}$ is any vector in the l-dimensional root space.

(b) W_0 is the quotient group $W_0 = N_0/C_0$ where N_0 and C_0 are the normalizer and centralizer of the Cartan subalgebra respectively (in the simply-connected covering group generated by G_0).

The second definition also suggests an immediate generalization to the case when N_0 and C_0 are no longer taken with respect to the covering group generated by G_0 (which may be regarded as the group of inner automorphisms of G_0) but with respect to the group of all automorphisms of G_0. In this way one picks up not only the traditional Weyl group, but also the Weyl group $\widetilde{W}_0$ which includes the (discrete) outer automorphisms of G_0.

Let us now consider the case of Kac-Moody[2] algebras,

$$[T_m^a, T_n^b] = if_e^{ab} T_N^c + \delta^{ab} K \delta_N \,, \tag{2}$$

where m, n are integers, $N = m + n$ and K is a central element. These algebras admit the well-known Virasoro automorphism

$$[L_n, T^a_m] = -mT^a_N \,, \tag{3}$$

for which the consistency (Jacobi) conditions are just the Virasoro algebra relations

$$[L_n, L_m] = (n-m)L_N + \frac{c}{12}n(n^2-1)\delta_N \,, \tag{4}$$

where c is a central term (which is arbitrary unless the L_n are constructed as bilinears in the T^a_m, in which case it takes a fixed value depending only on G_0 and K). Equations (3) and (4) simply state that the KM-algebra (2) is conformally invariant.

The Cartan subalgebra of the KM-Virasoro system (2)-(4) is $\{L_0, \overrightarrow{H_0}, K\}$, since these operators label the generators E^α_n and L_n uniquely (K in a trivial way of course). The roots are then $(l+2)$-dimensional vectors of the form $\overrightarrow{a} = (n, \overrightarrow{\alpha}, 0)$ and arbitrary vectors in root-space are $(l+2)$-dimensional vectors of the form $\overrightarrow{P} = (\epsilon, \overrightarrow{p}, k)$. The natural invariant forms in the $(l+2)$-dimensional root space are

$$2KL_0 - H_0^2 \quad \text{and} \quad 2\epsilon k - p^2 \,. \tag{5}$$

One can see that these are actually Galilean forms, either from the analogy with the form $2mE - p^2$ of non-relativistic mechanics, or by noting that (5) are Minkowski forms with one of the two lightlike components (K or k) kept fixed.

As in the case of ordinary Lie algebras the Weyl group W has two equivalent definitions:

(a) W is the group generated by the $(l+2)$-dimensional "reflexions"

$$\overrightarrow{P} \to \overrightarrow{P} - \frac{2(P,a)}{a^2}\overrightarrow{a} \tag{6}$$

(b) W is the quotient group N/C, where N and C are the normalizer and centralizer of the Cartan $\{L_0, \overrightarrow{H_0}, K\}$ respectively (in the group of the KM-Virasoro algebra). Furthermore, W can be extended to $\widetilde{W} = \widetilde{N}/\widetilde{C}$, where $\widetilde{N}, \widetilde{C}$ are the normalizer and centralizer of the Cartan in the group of *all* automorphisms of the KM-Virasoro algebra, and in the KM-case $\widetilde{W}$ is actually the most natural group to consider.

Using the definitions (b) it turns out that $\widetilde{W}$ and W have the semi-direct product structures

$$\widetilde{W} = \widetilde{W_0} \wedge \widetilde{A} \quad \text{and} \quad W = W_0 \wedge A \tag{7}$$

where $\widetilde{W_0}, W_0$ are the Weyl groups for ordinary Lie algebras discussed earlier, and the invariant subgroups $\widetilde{A}$ and A are the new feature peculiar to KM-algebras. It then turns out that $\widetilde{A}$ and A are abelian groups, parameterized by an l-vector $\overrightarrow{v}$ (whose domain distinguishes $\widetilde{A}$ and A) and that the action of these groups on the KM-Virasoro system is

$$\begin{aligned} \overrightarrow{H_n} &\to \overrightarrow{H_n} + K\overrightarrow{v}\delta_n \,, \\ E^\alpha_n &\to E^\alpha_{n+(v.\alpha)} \,, \\ L_n &\to L_n + v.H_n + \frac{v^2}{2}K\delta_n \,. \end{aligned} \tag{8}$$

Thus the Weyl group of the KM-Virasoro system is simply an extension of the ordinary Weyl group W_0 by the transformations (8) with suitable $\vec{v}$. The equations (8) are therefore the key expressions for the action of the Weyl group in the KM case. One sees by inspection that the transformations (8) are actually Galilean accelerations, which is not so surprising in view of the invariant forms (5).

Let us now consider the properties of the groups $\widetilde{A}$ and A defined by (8).

First let us consider the possible values of the parameters $\vec{v}$. There are three interesting domains.

(i) $\vec{v} \in \Gamma_{\widetilde{\alpha}}$, where $\Gamma_{\widetilde{\alpha}}$ is the coroot lattice i.e. the lattice generated by the coroots $\widetilde{\alpha}$ defined as $\widetilde{\alpha} = \frac{2\vec{\alpha}}{\alpha^2}$. This is the domain for the inner automorphisms A and correspond to the Weyl transformations obtained by other authors[3] using the "reflexion" definition (a) of W.

(ii) $\vec{v} \in \Gamma_{\widetilde{\omega}}\ (\notin \Gamma_{\widetilde{\alpha}})$, where $\Gamma_{\widetilde{\omega}}\ (\supset \Gamma_{\widetilde{\alpha}})$ is the coweight lattice, i.e. the lattice of all l-vectors $\widetilde{\omega}$ satisfying $(\widetilde{\omega}, \alpha) \in Z$ (integers) for all roots. This is the domain for the strictly outer Weyl tranformations $\widetilde{A}/A$, and, in contrast to the case of ordinary Lie algebras where outer automorphisms cannot link non-congruent representations unless they are of the same dimension (conjugate), these $\widetilde{A}/A$ transformations link *all* the congruency classes of representations of the simply-laced KM-algebras (e.g. link all l congruency classes for $SU(n)$).

(iii) $v \notin \Gamma_{\widetilde{\omega}}$. In this case the A-transformations (8) are not automorphisms at all, since, from (8) they do not transform the E_n^α into themselves. Nevertheless, they play an important role, becuase the $E_{n+(v.\alpha)}^\alpha$ then generate the so-called pseudo-twisted KM-algebras uniquely and can be used to untwist them.

Next let us consider the other KM-algebraic properties of the $\widetilde{A}$-transformations (8). One sees that the transformations have the property that they link the Virasoro and KM-algebras together (Indeed these two algebras form a reducible but not fully reducible representations of $\widetilde{A}$). Perhaps the most important algebraic property of the $\widetilde{A}$-group, however, is that it plays a vital role in the so-called 'vertex' construction of non-abelian KM-algebras. In fact, the 'vertex' construction is nothing more than the statement that (for $k = 1$ at least) the non-abelian elements E_n^α of the KM-algebra for $n \neq 0$ can be constructed from polynomials in the ordinary Lie algebra E_0^α, (ordered) exponential functions in the abelian KM generators H_n^i, and the elements of A. In other words,

$$E_n^\alpha = f\left(E_0^\alpha, H_m^i, A\right), \quad n \neq 0, \quad k = 1 \tag{9}$$

Thus, for $k = 1$, A is a key part of the non-abelian KM-structure.

Finally, let us consider the properties of the A-transformations in the case of string theory.[4] In this case the KM-algebra is abelian, so there are no E_n^α elements to restrict the range of $\vec{v}$. Thus A takes the form

$$A = exp\, i\vec{v}.\vec{X}\,, \quad \vec{v} \text{ real}, \tag{10}$$

where the infinitesimal generator X has the action

$$[X^i, L_n] = H_n^i\,, \quad [X^i, H_n^j] = k\delta^{ij}\,, \quad [X^i, k] = 0\,, \tag{11}$$

on the string variables H_n^i and their Virasoro algebra L_n. From these equations those familiar with string theory will recognize $\vec{X}$ as the centre-of-mass of the string. Thus the Weyl automorphism (11) of the string-Virasoro algebra introduces the centre-of-mass of the string in a very natural way.

But this is by no means the end of the story. In string theory the gauge in which all the states are physical is the light-cone gauge, characterized by some arbitrary lightlike vector (n^μ say). In this gauge the theory is not manifestly covariant with respect to the full Poincafe group $\mathcal{P}$, but is manifestly covariant with respect to the maximal subgroup of $\mathcal{P}$ (little group of $\mathcal{P}$) that leaves the lightlike vector n^μ invariant. This little group is a Galilean group G and if one examines the theory in detail one sees that it is just the Weyl group ($G = W = W_0 \wedge A!$). Thus the Weyl group is not only an "internal" automorphism group of the string algebra, but is also a "space-time" group G. To sum up, in string theory the Weyl group provides both the centre-of-mass of the string and the little group of the light-cone.

Details of the above discussion can be found in the Proceedings of the 1987 Varna Summer School (World Scientific 1988) and a more up-to-date version (including the case of the twisted KM algebras) in a recent DIAS preprint.

REFERENCES

1. J. F. Cornwell, "Group Theory in Physics II," Academic Press (1984).
2. V. Kac, "Infinite-Dimensional Lie Algebras," Birkhauser (1983), Cambridge (1985).
3. P. Goddard and D. Olive, Int. Journ. Mod Phys. A1:303 (1986).
4. M. Green, J. Schwartz and E. Witten, "Superstring Theory," Cambridge (1987).

ON REPRESENTATIONS OF THE SYMPLECTIC GROUP

V. A. Groza

Mathematical Department, Kiev State University

Kiev-17, 252601 USSR

One can consider the compact group $Sp(n)$ as a subgroup of the unitary group $U(2n)$. The imbedding of $Sp(n)$ into $U(2n)$ is given by the isomorphic imbedding of the skew field of quaternions into the linear algebra of 2×2 matrices,

$$q = a + bi + cj + dk \rightarrow \begin{pmatrix} z & w \\ -\overline{w} & \overline{z} \end{pmatrix}, \quad z = a + bi, \quad w = c + di.$$

The formula $T(g)f(\underline{z}) = f(g^{-1}\underline{z})$, $g \in U(2n)$, $\underline{z} = (z_1, \ldots, z_{2n})^t$, defines the representation of $U(2n)$ in the space H of holomorphic homogeneous polynomials of degree p in the variables $z_1, \ldots, z_{2n}$. This representation is irreducible [1]. The highest weight of T is $(p, 0, \ldots, 0)$. Let T' be the restriction of T onto the subgroup $Sp(n)$. By direct calculations one can see that there is only one function f_0 in H such that $T'(X_\alpha)f_0 = 0$ for all positive weights α of the Lie algebra of $Sp(n)$. It means that T' has only one highest weight, i.e. T' is an irreducible representation of $Sp(n)$. It is easy to check that its highest weight is $(p, 0, \ldots, 0)$. Thus, the restriction of the irreducible representation T of the group $U(2n)$ with the highest weight $(p, 0, \ldots, 0)$ onto $Sp(n)$ is the irreducible representation of $Sp(n)$ with the highest weight $(p, 0, \ldots, 0)$. It follows from here that the Gel'fand-Zeitlin basis of the representation T of $U(2n)$ serves as an orthonormal basis for the representation T' of $Sp(n)$.

Matrix elements of representations of $U(n)$ are known (see, for example, [2]). Calculating matrix elements for T and considering matrices of this representation for elements of the subgroup $Sp(n)$, we obtain matrix elements of the irreducible representation of $Sp(n)$ with the highest weight $(p, 0, \ldots, 0)$. We present the result obtained. For this we take into account that any matrix $g \in Sp(n)$ can be represented in the form $g = hd_n(q)g_n(\theta)h'$, where $h, h' \in Sp(n-1)$,

$$d_n(q) = diag(1, \ldots, 1, s), \quad s = \begin{pmatrix} z & w \\ -\overline{w} & \overline{z} \end{pmatrix}, \quad q = z + wi, \quad |q| = 1. \qquad (1)$$

$$g_n(\theta) = diag(1, \ldots, 1, g(\theta)),$$

$$g(\theta) = \begin{pmatrix} \cos\theta & 0 & -\sin\theta & 0 \\ 0 & \cos\theta & 0 & -\sin\theta \\ \sin\theta & 0 & \cos\theta & 0 \\ 0 & \sin\theta & 0 & \cos\theta \end{pmatrix}.$$

Therefore, by virtue of the property $T'(g_1 g_2) = T'(g_1)T'(g_2)$ it is sufficient to define matrix elements for the operators corresponding to the elements $d_n(q)$ and $g_n(\theta)$ of the group $Sp(n)$. The matrix s from (1) belongs to $SU(2) \sim Sp(1)$. Hence, it is defined by the Euler angles ϕ, θ, ψ [3]. We have $d_n(q) = d_n(\phi, \theta, \psi)$.

A Gel'fand-Zetlin pattern of the representation T is defined by $2n$ numbers $p_0, p_1, p_2, \ldots, p_{2n-1}$, which will be denoted by P. The symbol P will denote the corresponding basis element too. We have

$$\langle P'|\, T'(d_n(\phi,\theta,\psi))\, |P\rangle = \frac{i^{-a}}{a!}\left[\frac{(p_0 - p_1')!(p_1 - p_2)!}{(p_0 - p_1)!(p_1' - p_2)!}\right]^{1/2}$$

$$\times\ e^{-i\psi(p_0+p_2-2p_1)/2 - i\phi(p_0+p_2-2p_1')/2} (tg\theta)^{-p_1-p_1'} (\cos\theta)^{-p_0-p_2}$$

$$\times\ (\sin\theta)^{2p_1}\, {}_2F_1(p_1 - p_0, p_1 - p_2 + 1; a+1; \sin^2\theta),$$

where $P' = (p_0, p_1', p_2, \ldots, p_{2n-1})$, $a = p_1 - p_1'$, and $p_1 \geq p_1'$. In the case $p_1 < p_1'$ one has to interchange p_1 and p_1' on the right and multiply the right hand side by $(-1)^{p_1+p_1'}$. For other P and P' this matrix element vanishes [4,5].

The matrix $g_n(\theta) \in Sp(n)$ is represented as

$$g_n(\theta) = s_1 g_{2n-2,2n-3}(\theta) s_1^{-1} s_2 g_{2n-1,2n-2}(\theta) s_2^{-1},$$

where

$$s_1 = diag(1,\ldots,1,s,1), \quad s_2 = diag(1,\ldots,1,s), \quad s = \begin{pmatrix} 0 & -1 \\ 1 & 0 \end{pmatrix}$$

and $g_{k,k-1}(\theta) \in U(2n)$ is an ordinary rotation in the plane $(k-1,k)$ by the angle θ. Therefore,

$$T'(g_n(\theta)) = T(s_1)T(g_{2n-2,2n-3}(\theta))T(s_1^{-1}s_2)T(g_{2n-1,2n-2}(\theta))T(s_2^{-1}). \qquad (2)$$

Using matrix elements of the representation T of $U(2n)$, we have

$$\langle P'|\, T'(g_n(\theta))\, |P\rangle = \left[\frac{(p_0 - p_1 + 1 - k)!(p_2 - p_3)!(p_1 - 1 - p_3')!(p_3 - p_4)!}{(p_0 - p_1)!(p_1 - p_2)!(p_2' - p_3')!(p_3' - p_4)!}\right]^{1/2}$$

$$\times\ \frac{tg^1\theta\,(\cos\theta)^{-p_0-2p_1+p_2-p_3-p_4}}{(1-k)!k!}\, {}_2F_1(p_1 - p_0, p_2 - p_3 + 1; l - k + 1; \sin^2\theta)$$

$$\times\ {}_2F_1(p_2 - p_1, p_3 - p_4 + 1; k+1; \sin^2\theta),$$

where $P = (p_0, p_1, \ldots, p_{2n-1})$, $P' = (p_0, p_1 - 1 + k, p_2', p_3', p_4, \ldots, p_{2n-1})$. Here $1 \geq k \geq 0$. Similar expressions are obtained from (2) for other conditions for 1 and k.

REFERENCES

1. W. Rudin, "Function Theory in the Unit Ball of C^n," Springer, Berlin (1980).
2. A. U. Klimyk, A. M. Gavrilik, J. Math. Phys. 20:1624 (1979).
3. N. Ja. Vilenkin, "Special Functions and the Theory of Group Representations," Transl. Math. Monogr. 22, Amer. Math. Soc., Providence, R.I. (1968).
4. M. K. F. Wong, Hsin-Yang Yeh, J. Math. Phys. 21:630 (1980).
5. A. U. Klimyk, B. Gruber, J. Math. Phys. 25:743 (1984).

TENSOR PRODUCT OF FINITE AND INFINITE REPRESENTATIONS IN PHYSICS

W.F.Heidenreich
Institut für Theoretische Physik A
TU Clausthal
W. Germany

1 Introduction

In field theory, particles with gauge freedom, like the photon, are usually not described as unitary irreducible representations of the spacetime symmetry group, but they appear in indecomposable representations with indefinite invariant scalar product. We call such representations Gupta-Bleuler triplets. There is not yet a general representation theory, but some techniques are available to deal with them [1,2]. One—which is very useful in physical applications—stems from the observation, that the tensor product of finite and infinite representations may contain Gupta-Bleuler triplets in the reduction.

Many free field equations for particles with spin can be brought in the form

$$(\Box - m^2)\Psi_I(x) = 0, \tag{1}$$

where I denotes some tensor- or spinor indices, which characterize a finite representation D_I of the Poicaré group with trivial action of the translations. The solutions of the Klein-Gordon equation for a scalar field carry a massive spin 0 representation $D(m,0)$ of positive energy, and another one of negative energy. If the action of the generators on the field $\Psi_I(x)$ is of the form

$$J = L + S, \tag{2}$$

where L acts only on the coordinates, and S acts only on the indices, then the positive energy solution space of the field equation (1) carries the tensor product of a finite and an infinite representation,

$$D(m,0) \otimes D_I. \tag{3}$$

For example in the case of the Proca-field $A_\mu(x)$ we consider the tensor product with the 4-dimensional vector representation, which reduces to

$$D(m,0) \otimes D_4 = D(m,1) \oplus D(m,0). \tag{4}$$

The resulting spin 0 part is described by the scalar $\partial^\mu A_\mu$. In general to get the reduction for massive fields, one restricts the finite representation of the Lorentz group to $SO(3)$, and then induces from the resulting direct sum to the Poincaré group.

For massless fields the corresponding restriction to the "little group" E(2) gives indecomposable representations. The field equations (1) with mass zero have positive energy solutions which carry indecomposable representations of the Poincaré group [3,4]. For the vector field we have the following sequence of equations which project on invariant subspaces, and the corresponding positive energy representations

$$
\begin{array}{ccccc}
\Box A_\mu = 0 & \supset & \partial^\mu A_\mu = 0 & \supset & A_\mu = \partial_\mu \phi,\ \Box\phi = 0 \\
D(0,0) & \rightarrow & [D(0,+1) \oplus D(0,-1)] & \rightarrow & D(0,0) \\
\text{scalar} & & \text{physical} & & \text{gauge.}
\end{array} \tag{5}
$$

The arrows indicate, that the invariant subspaces are *not* invariantly complemented. In Gupta-Bleuler quantization, the scalar and gauge modes are conjugate to one another; the corresponding indecomposable representation with conjugate (equivalent) irreducible representations in the smallest invariant subspace, and in the quotient space with respect to the largest invariant subspace is a Gupta-Bleuler triplet.

As another example we may consider the Weyl-spinor with its invariant subspace described by the Weyl-equation:

$$
\Box\psi = 0 \quad \supset \quad \sigma^\mu \partial_\mu \psi = 0 \tag{6}
$$

$$
D(0,-1/2) \quad \rightarrow \quad D(0,+1/2). \tag{7}
$$

We could use the non-invariant $D(0,-1/2)$, to describe a neutrino with gauge freedom (s.also [5]).

Such representations of the Poincaré group can be generalized in various ways, e.g. to the conformal group SO(4,2), or the de Sitter group SO(3,2). Here we first want to discuss as a simple model the case of SO(1,2), which contains already many features of the de Sitter and conformal theories. References to some formulations of de Sitter and conformally invariant theories which use similar techniques, will be given in the end.

2 Some properties of so(1,2)

Basis of the Lie algebra

A basis of the Lie algebra so(1,2) are the operators $M_{ij} = -M_{ji}$, $i,j = 1,4,6$ with

$$
[M_{ik}, M_{kj}] = -i\eta_{kk} M_{ij}, \text{ and } \eta_{kk} = (1,-1,-1). \tag{8}
$$

Another basis of ladder operators are

$$
M^{\pm} = M_{61} \mp i M_{14},\ M_{46}, \tag{9}
$$

with the commutation relations

$$
[M^{\pm}, M_{46}] = \mp M^{\pm},\ [M^{+}, M^{-}] = -2M_{46}. \tag{10}
$$

We will call the eigenvalues of M_{46} "energy".

Lowest weight representations

The lowest weight is a state $\mid 0\,\rangle_{E_0}$ with the properties

$$
M^{-} \mid 0\,\rangle_{E_0} = 0, \text{ and } M_{46} \mid 0\,\rangle_{E_0} = E_0 \mid 0\,\rangle_{E_0}. \tag{11}
$$

We define higher states by

$$
M^{+} \mid n\,\rangle_{E_0} = \mid n+1\,\rangle_{E_0}; \tag{12}
$$

they satisfy

$$
M_{46} \mid n\,\rangle_{E_0} = (E_0 + n) \mid n\,\rangle_{E_0}, \tag{13}
$$

$$
M_{-} \mid n\,\rangle_{E_0} = n(2E_0 + n - 1) \mid n-1\,\rangle_{E_0}. \tag{14}
$$

The invariant norm ($(M^+)^\dagger = M^-$) of the states is with ${}_{E_0}\langle 0 \mid 0 \rangle_{E_0} = 1$:

$$ {}_{E_0}\langle n \mid n \rangle_{E_0} = n(2E_0 + n - 1)\ {}_{E_0}\langle n-1 \mid n-1 \rangle_{E_0} = n! \frac{\Gamma(2E_0 + n)}{\Gamma(2E_0)}. \tag{15} $$

For $E_0 > 0$ the matrix elements $n(2E_0 + n - 1)$ are positive for all $n \geq 1$; the norm of all states $\mid n \rangle_{E_0}$ is positive, and the lowering operator acting on each state $\mid n+1 \rangle_{E_0}$ gives a finite multiple of $\mid n \rangle_{E_0}$. The representations are irreducible and unitary.

In the cases of $E_0 = -m/2$, $m = 0, 1, 2, \ldots$ the matrix elements $n(2E_0 + n - 1)$ become zero for n=1+m. This implies that there is a new lowest weight state

$$ \mid 1 - 2E_0 \rangle_{E_0} \tag{16} $$

from which the raising operators generate an invariant subspace. The other states with $n < 1 - 2E_0$ do not span an invariant subspace; we have an indecomposable representation, which we denote by

$$ D(E_0) \rightarrow D(1 - E_0). \tag{17} $$

The states in the subspace $D(1 - E_0)$ have norm zero. The finite representations act on the quotient space

$$ \frac{D(E_0) \rightarrow D(1 - E_0)}{D(1 - E_0)} = D(E_0). \tag{18} $$

We could also get them by introducing in Eq.(12) a coefficient on the right hand side which is determined by using only states normalized to ± 1.

For the other lowest weights, with $E_0 < 0$ neither integer nor half-integer, we get again irreducible representations; they are not unitarizable, as the matrix elements $n(2E_0 + n - 1)$ and therefore the invariant norm of states can become negative.

Weyl-equivalence

The only Casimir operator of so(1,2) is

$$ C_2 = \tfrac{1}{2} M_{ij} M^{ij} = M^+ M^- + M_{46}^2 - M_{46}. \tag{19} $$

Acting on a lowest weight it yields

$$ C_2 \mid 0 \rangle_{E_0} = E_0(E_0 - 1) \mid 0 \rangle_{E_0}. \tag{20} $$

The representations $D(E_0)$ and $D(1 - E_0)$ have the same Casimir-eigenvalues: such representations are called Weyl-equivalent. We get Weyl-equivalent lowest weights by reflection at the point $E_0 = 1/2$ ("Weyl-plane").

The irreducible parts of indecomposable representations have to be Weyl-equivalent [6]. In addition, the difference of the energies of the lowest weights has to be integer, so that the ladder operators can map between them. These necessary conditions give as possible inequivalent parts of indecomposable representations

$$ D(E_0) \text{ and } D(1 - E_0) \text{ with } E_0 = 0, -1/2, -1, \ldots. \tag{21} $$

One of them is always finite, the other one is infinite-dimensional, e.g. D(0) and D(1), or $D(-1/2)$ and D(3/2).

Apart from this last property, the same analysis can be performed for so(3,2) and so(4,2), and also for the superalgebras which contain them as even part.

3 Tensor product of a finite and an infinite representation

The tensor product of two finite resp. two infinite representations of so(1,2) gives a Clebsch-Gordan-series of finitely many finite resp. infinitely many infinite representations. They can be calculated easily by comparing weight diagrams. Indecomposable representations cannot occur, as there are no Weyl-equivalent lowest weights.

The situation is completely different for the tensor product of a finite and an infinite representation. Consider for example $D(-1) \otimes D(1)$: comparing the weight diagrams gives the trivial D(0), twice D(1), and D(2) (s.Fig.1). D(0) and D(1) are Weyl-equivalent. We want to show, that they form a triplet

$$D(-1) \otimes D(1) = [D(1) \to D(0) \to D(1)] \oplus D(2). \tag{22}$$

For this we consider the states in the tensor product at energy 0 and 1, that is

$$\begin{aligned} \mid 0 \rangle_p &= \mid 0 \rangle_{-1} \otimes \mid 0 \rangle_1, \\ (1)_1 &= \mid 0 \rangle_{-1} \otimes \mid 1 \rangle_1, \\ (1)_2 &= \mid 1 \rangle_{-1} \otimes \mid 0 \rangle_1. \end{aligned} \tag{23}$$

Its shown easily that

$$M^+ \mid 0 \rangle_p = (1)_1 + (1)_2 =: 2 \mid 0 \rangle_g, \text{ with } M^- \mid 0 \rangle_g = 0; \tag{24}$$

so $\mid 0 \rangle_g$ is cyclic for an irreducible representation $D(1)$, and $\mid 0 \rangle_p$ is cyclic for a $D(0) \to D(1)$. Applying M^- on any non-vanishing linear combination $a(1)_1 + b(1)_2$ with $a \neq b$ yields a multiple of $\mid 0 \rangle_p$; so it is cyclic for the full Gupta-Bleuler triplet in Eq. (22).

Figure 1: Some weights of the product $D(-1) \otimes D(1)$.

We can calculate the norms of the various cyclic states in the tensor product, with the result

$$\begin{aligned} {}_p\langle 0 \mid 0 \rangle_p &= 1, \\ {}_g\langle 0 \mid 0 \rangle_g &= 0, \\ {}_s\langle 0 \mid 0 \rangle_s &= 0, \text{ where } \mid 0 \rangle_s = \tfrac{1}{2}[(1)_1 - (1)_2], \\ {}_s\langle 0 \mid 0 \rangle_g &= 1. \end{aligned} \tag{25}$$

These norms fix $\mid 0 \rangle_s$ uniquely up to a relative factor between $\mid 0 \rangle_s$ and $\mid 0 \rangle_g$.

The Casimir operator acting on the states $\mid 0 \rangle_p$ and $\mid 0 \rangle_g$ gives the eigenvalue zero, while $\mid 0 \rangle_s$ is not a eigenstate:

$$C_2 \mid 0 \rangle_s = 4 \mid 0 \rangle_g. \tag{26}$$

So the Casimir operator can be used to find equations which are solved by the physical and gauge modes only i.e., it supplies a generalized Lorentz condition. Its square can be used to find equations for the full Gupta-Bleuler triplet.

In summary we have:

$$\begin{array}{llll} & D(1) \quad - & D(0) \quad \rightarrow & D(1) \\ \text{cyclic state:} & \mid 0\rangle_s & \mid 0\rangle_p & \mid 0\rangle_g \\ \text{norm:} & 0 & +1 & 0 \\ \text{Casimir:} & 4\mid 0\rangle_g & 0 & 0. \end{array} \tag{27}$$

The invariant two-point functions of the tensor product are simply the two-point functions of the scalar representation and the finite one. This makes free field quantization of fields with spin quite straightforward.

To get an overview of possible indecomposable representations in tensor products, we consider their weight diagrams. As it can be seen in Fig. 1, the weight diagram of the product is obtained easily: all weights of the finite representation are shifted by the lowest eigenvalue E_{02} of the infinite representation, and then they become lowest weights in the reduction. If they label a finite representation, there are also the weights of the Weyl-equivalent one. Using Eq. (21) we get a list of tensor products, whose reduction can contain indecomposable representations. The reduction of the tensor product of the finite $D(E_{01})$ with

$$E_{01} = -1/2, -1, \ldots \tag{28}$$

and the infinite $D(E_{02})$ with

$$E_{02} = 1/2, 1, \ldots, -E_{01} \tag{29}$$

contains the lowest weights of the finite $D(E_0')$ with

$$E_0' = E_{01} + E_{02}, E_{01} + E_{02} - 1, \ldots, (0 \text{ or } -1/2). \tag{30}$$

In addition there are two copies of the Weyl-equivalent representations. So there can be triplets of the form

$$D(1 - E_0') \rightarrow D(E_0') \rightarrow D(1 - E_0'). \tag{31}$$

This actually happens. We sketch a proof, which in this form only works in the present situation. Highest weights in the tensor product, which satisfy $M^+ \mid \,\rangle = 0$, must contain a term with the highest weight of the finite representation. This does not happen at the energies, where irreducible $D(E_0')$ would have there highest weight states. This proves the lower leak. The tensor product has an invariant indefinite scalar product; so the indecomposable $D(E_0') \rightarrow D(1 - E_0')$ must be part of a Gupta-Bleuler triplet (31).

4 A field-theoretic realization

Next we want to give a realization, which is used in de Sitter and conformal field theories. We will apply it here to calculate the cyclic states of some Gupta-Bleuler triplets in tensor products, using computer algebra.

Consider flat space R^3 with metric

$$dx_1^2 - dx_4^2 - dx_6^2. \tag{32}$$

The Lie algebra

$$L_{ij} = -i(x_i\partial_j - x_j\partial_i) \tag{33}$$

acts on a scalar field $\Phi(x)$. The ladder operators (9) become first order differential operators,

$$\begin{aligned} L^+ &= -2x_1\partial_+ - x_-\partial_1, \\ L^- &= 2x_1\partial_- + x_+\partial_1, \\ L_{46} &= -x_+\partial_+ + x_-\partial_-, \end{aligned} \tag{34}$$

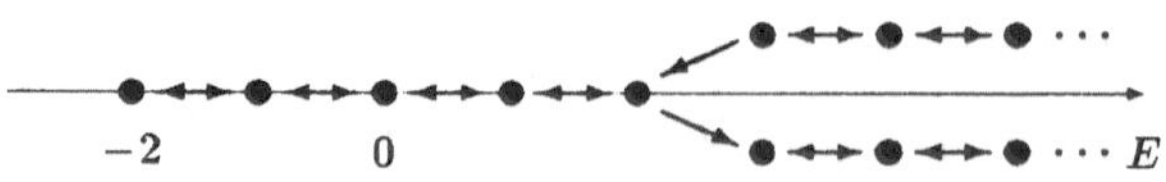

Figure 2: Action of the ladder operators on some states in $D(3) \to D(-2) \to D(3)$.

with $x_\pm = x_4 \pm i x_6$, and $\partial_\pm = \partial/\partial x_\pm$. Lowest weight states with energy E_0 are

$$x_+^{-E_0}. \tag{35}$$

Action of the raising operator L^+ gives for $E_0 > 0$ the representation space of a unitary irreducible $D(E_0)$.

The finite representations with $E_0 = 0, -1, -2, \ldots$ can be obtained in the same way. To distinguish them in notation, we replace x by z and L by S. Then the lowest weight states are

$$z_+^{-E_0}. \tag{36}$$

The Lie algebra

$$J = L + S. \tag{37}$$

acts on the tensor product $D(E_{01}) \otimes D(E_{02})$ with E_{01}, E_{02} from Eqs. (28,29). To find the various cyclic states in the Gupta-Bleuler triplets (31), we follow a scheme for which we are giving explicit expressions in the most simple case of Eq. (22). It can be used similarly in de Sitter and conformal field theory.

- Calculate the lowest state Φ_0 at energy E_0', which is unique up to normalization. To do this we look for a linear combination of all states in the tensor product at energy E_0', which satisfies $M^-\Phi_0 = 0$. In the example $\Phi_0 = z_+/x_+$.
- Calculate the "highest" state Φ_{pmax} with energy $-E_0'$ by applying the raising operator $2E_0'$ times. In the example it is Φ_0 again.
- If $D(E_0')$ is realized irreducibly, then $M^+\Phi_{pmax} = 0$ holds. Otherwise $\Phi_g := M^+\Phi_{pmax} \neq 0$ is a lowest weight and is cyclic for an invariant subspace, which carries $D(1 - E_0')$. In the example, $\Phi_g = (2x_1 z_+ - 2x_+ z_1)/x_+^2$.
- At energy $1 - E_0'$, look for a state which satisfies

  $$M^-\Phi_s = \Phi_{pmax}. \tag{38}$$

  It is fixed only up to addition of multiples of Φ_g. If such a state exists, then there is a Gupta-Bleuler triplet (31).

The action of the ladder operators is shown in Fig. 2 for an example.

This scheme can be followed automatically. A REDUCE 3.3 [7] program was written, which calculates the necessary states and performs the described tests. The most simple tensor product which can contain

$$D(1 - E_0) \to D(E_0) \to D(1 - E_0)$$

is $D(E_0 - 1) \otimes D(1)$. For this case we give some runtimes for the IBM 3090-150E of the computer centre at TU Clausthal:

E_0	time (sec.)	E_0	time (sec.)	E_0	time (sec.)
0	0.17	−4	12.4	−8	236
−1	0.78	−5	32	−9	419
−2	2.2	−6	70	−10	697
−3	5.7	−7	131		

Other cyclic states were also calculated, with comparable runtimes.

The method described above has been used to construct free quantum field theories for massless spin 1 and spin 2 particles in conformal space [1,8] and de Sitter space [9]. The tensor product of a finite and an infinite representation has also been used in quantum theories for the conformal neutrino with gauge freedom, and the spin 1/2 Dirac-singleton [10], as well as for Bargmann-Wigner equations in conformal space [11].

Acknowledgements

I am indebted to J.Wolf and W.Schmidt for a discussion. I further have the pleasure to thank Prof. H.D.Doebner for encouragement and advice.

References

[1] B.Binegar, C.Fronsdal, and W.Heidenreich, "Conformal QED", J.Math.Phys. **24**, 2828 (1983).

[2] H.Araki, "Indecomposable representations with invariant inner product: A theory of the Gupta-Bleuler triplet", Comm.Math. Phys. **97**, 149 (1985).

[3] A.O.Barut, R.Raczka "Properties of non-unitary zero mass induced representations of the Poincaré group on the space of tensor-valued functions", Ann.Inst.H.Poincaré **17**, 111 (1972).

[4] W.Heidenreich, "On solution spaces of massless field equations with arbitrary spin", Journ.Math.Phys. **27**, 2154 (1986).

[5] A.O.Barut, and J.McEwan, "The four states of the massless neutrino with Pauli coupling by spin-gauge invariance", Lett.Math.Phys. **11**, 67 (1986).

[6] I.N.Bernshtein, I.M.Gel'fand, S.J.Gel'fand, Funct. Anal. Appl., **5**, 1 (1971), G.Pinczon, J.Simon, Rep. Math. Phys. **16**, 49 (1979).

[7] A.C.Hearn, "Reduce user's manual", Santa Monica 1987.

[8] B.Binegar, C.Fronsdal, and W.Heidenreich, "Linear conformal quantum gravity", Phys.Rev. **D27**, 2249 (1983).

[9] B.Binegar, C.Fronsdal, and W.Heidenreich, "de Sitter QED", Ann.Phys.(N.Y.) **149**, 254 (1983), C.Fronsdal, and W.F.Heidenreich, "Linear de Sitter Gravity", Journ.Math.Phys. **28**, 215 (1987).

[10] W.Heidenreich, "Quantum theory of spin-1/2 fields with gauge freedom", Nuovo Cimento **A80**, 220 (1984).

[11] W.F.Heidenreich, and M.Lorente, "Quantization of conformally invariant Bargmann-Wigner equations with gauge freedom", Journ.Math.Phys. **29**, 1698 (1988).

STRING AND SUPERSTRING SPECTRA

Ronald C. King
Mathematics Department, University of Southampton
Southampton, SO9 5NH, England

Abstract

The spectrum generating functions for a variety of string and superstring models are written down. They are expanded in such a way as to make clear the transformation properties of the states with respect to the action of the appropriate transverse spacetime symmetry group. The Green-Schwarz superstring spectrum generating function is shown to be identical with that of the D=10 GSO-projected Neveu-Schwarz-Ramond superstring. The same techniques are then applied to the heterotic string with gauge group $E_8 \times E_8$.

1. Introduction

It is of some interest not only to enumerate the string and superstring states of a given mass in any particular model, but also to determine their transformation properties under the action of the groups relevant to that model **[R2]**. It is shown here that these tasks can be accomplished in a systematic way for a wide variety of models by exploiting generating function methods involving group characters **[CT]**.

Let the strings and superstrings under consideration be set up in D-dimensional spacetime with $D = 2k+2$. The corresponding transverse symmetry group is $SO(D-2) = SO(2k)$. Massless states span irreducible representations of this group, whilst massive states span irreducible representations of the group $SO(D-1) = SO(2k+1)$. It is convenient to denote the eigenvalues of an arbitrary element of this group by

$$(\mathbf{x}) = (x_1, x_2, \ldots, x_k, x_1^{-1}, x_2^{-1}, \ldots, x_k^{-1}, 1) \ , \qquad (1.1)$$

where $x_j = \exp(i\phi_j)$ and $x_{k+j} = \exp(-i\phi_j)$ for $j = 1,2,\ldots,k$. The character of the irreducible representation of SO(2k+1) having highest weight λ is then denoted by $[\lambda](\mathbf{x})_{2k+1}$.

The spectrum generating function for each string or superstring, when expressed in terms of an expansion parameter q, takes the general form:

$$\kappa(\mathbf{x},q) = q^{L_0} \sum_{L=0}^{\infty} \sum_{\lambda} n^{\lambda}_{L} \, [\lambda](\mathbf{x})_{2k+1} \, q^{L} \, . \qquad (1.2)$$

The coefficient n^{λ}_{L} gives the number of times the irreducible representation $[\lambda]_{2k+1}$ of SO(2k+1) having highest weight λ and character $[\lambda](\mathbf{x})_{2k+1}$ occurs at level L. In flat spacetime the mass squared of this level is $L_0 + L$, where L_0 is the mass squared of the vacuum state.

The plan of the paper is as follows. In Section 2 the spectrum generating functions κ_B **[CGoT]**, κ_{NS} **[CGhT]** and κ_R **[CGhT]** for the open bosonic, the Neveu-Schwarz **[NS]** and the Ramond **[R1]** strings, respectively, are written down and then expanded in the form (1.2) using S-function methods **[L,M,K2,BKW,KLDW]**. These results are used in Section 3 to write down the spectrum generating function, κ_{NSR}, for the Neveu-Schwarz-Ramond superstring involving the Gliozzi-Scherk-Olive projection **[GSO]**. The Green-Schwarz superstring **[GS]** spectrum generating function, κ_{GS}, is then established and a proof is outlined of the remarkable Theorem which states that $\kappa_{GS} = \kappa_{NSR}$ **[N]**. The proof depends upon the triality of SO(8) and a theta function identity due to Jacobi **[J]**.

Finally in Section 4 closed string models with a gauge group are discussed. The heterotic superstring **[GHMR]** is described and the full spectrum generating function is written down in the case of the gauge group $E_8 \times E_8$. The lowest lying states are enumerated.

Further details of all the material presented here are to appear elsewhere **[FKW]**.

2. The open bosonic, Neveu-Schwarz and Ramond strings

In the case of the open bosonic string the vacuum state is an SO(2k) singlet state which is tachyonic with mass squared $L_0 = -1$, whilst excited states are generated by the action of the bosonic operators a^{i}_{-n}

with $i = 1,2,\ldots,2k$ and $n = 1,2,\ldots,\infty$. These operators, for each fixed n, transform as the basis states of the defining vector 2k-dimensional irreducible representation $[1]_{2k}$ of SO(2k). Moreover each operator a^i_{-n} contributes n to the mass squared value. It follows that the required spectrum generating function (1.2) is given by[CT]

$$\kappa_B(\mathbf{x},q) = q^{-1} \prod_{n=1}^{\infty} \prod_{i=1}^{2k} (1 - x_i q^n)^{-1} . \tag{2.1}$$

The factor $x_i q^n$ is associated with the operator a^i_{-n}; x_i taking care of the SO(2k) transformation properties and q^n taking care of the mass squared contribution. The inverse power of each factor in the product allows for the fact that the operators are indeed bosonic so that multiple excitations may occur.

The lowest lying states of this bosonic string model may easily be enumerated [R2] and are given in the following table, along with the representations of SO(2k) or SO(2k+1) which they span at each level.

Open bosonic string spectrum

Level	L = 0	L = 1	L = 2	L = 3
States	$\|\psi_0\rangle$	$a^i_{-1}\|\psi_0\rangle$	$a^i_{-2}\|\psi_0\rangle$	$a^i_{-3}\|\psi_0\rangle$
			$a^i_{-1}a^j_{-1}\|\psi_0\rangle$	$a^i_{-1}a^j_{-2}\|\psi_0\rangle$
				$a^i_{-1}a^j_{-1}a^k_{-1}\|\psi_0\rangle$
Rep	$[0]_{2k}$	$[1]_{2k}$	$[2]_{2k+1}$	$[3]_{2k+1} + [1^2]_{2k+1}$

The Neveu-Schwarz string [NS] can be treated in exactly the same way. The vacuum state is once again an SO(2k) singlet state but has mass squared given by $L_0 = -k/8$. There now exist both bosonic excitation operators, a^i_{-n}, and fermionic excitation operators, $b^i_{-(n-1/2)}$. In each case $i = 1\ \ 1,2,\ldots,2k$ and $n = 1,2,\ldots,\infty$. Moreover for each fixed value of n each of these two sets of operators transforms as the basis of the vector representation $[1]_{2k}$ of SO(2k). The spectrum generating function therefore takes the form:

$$\kappa_{NS}(\mathbf{x},q) = q^{-k/8} \prod_{n=1}^{\infty} \prod_{i=1}^{2k} (1 - x_i q^n)^{-1} (1 + x_i q^{n-1/2}) , \qquad (2.2)$$

where the distinction between bosonic and fermionic operators shows itself in the fact that only the first of the factors in the product involves an inverse power. The content of the lowest lying levels is given in the following table.

Neveu-Schwarz string spectrum

Level	L = 0	L = 1/2	L = 1	L = 3/2
States	$\lvert\psi_0\rangle$	$b^i_{-1/2}\lvert\psi_0\rangle$	$a^i_{-1}\lvert\psi_0\rangle$	$b^i_{-3/2}\lvert\psi_0\rangle$
			$b^i_{-1/2}b^j_{-1/2}\lvert\psi_0\rangle$	$a^i_{-1}b^j_{-1/2}\lvert\psi_0\rangle$
				$b^i_{-1/2}b^j_{-1/2}b^k_{-1/2}\lvert\psi_0\rangle$
Rep	$[0]_{2k}$	$[1]_{2k}$	$[1^2]_{2k+1}$	$[2]_{2k+1} + [1^3]_{2k+1}$

In the case of the Ramond string **[R2]** the vacuum state has dimension 2^k and transforms as the basis of the reducible spin representation whose character may be denoted **[KLDW,BKW]** by $\Delta(\mathbf{x})_{2k} = \Delta_+(\mathbf{x})_{2k} + \Delta_-(\mathbf{x})_{2k}$ of SO(2k). It is massless, having $L_0 = 0$. There now exist bosonic excitation operators, a^i_{-n}, and fermionic excitation operators, d^i_{-n}. In each case $i = 1,2,\ldots,2k$ and $n = 1,2,\ldots,\infty$. Once more for each fixed value of n each of these two sets of operators transforms as the basis of the vector representation $[1]_{2k}$ of SO(2k). The spectrum generating function therefore takes the form:

$$\kappa_R(\mathbf{x},q) = \Delta(\mathbf{x})_{2k} \prod_{n=1}^{\infty} \prod_{i=1}^{2k} (1 - x_i q^n)^{-1} (1 + x_i q^n) . \qquad (2.3)$$

The content of the lowest lying levels is as follows.

Ramond string

Level	L = 0	L = 1	L = 2
States	$\lvert\psi_0\rangle$	$a^i_{-1}\lvert\psi_0\rangle$	$a^i_{-2}\lvert\psi_0\rangle$
		$d^i_{-1}\lvert\psi_0\rangle$	$d^i_{-2}\lvert\psi_0\rangle$
			$a^i_{-1}a^j_{-1}\lvert\psi_0\rangle$
			$a^i_{-1}d^j_{-1}\lvert\psi_0\rangle$
			$d^i_{-1}d^j_{-1}\lvert\psi_0\rangle$

Rep $\quad [\Delta]_{2k} \quad 2[\Delta;1]_{2k+1} \quad 2([\Delta;2] + [\Delta;1^2] + [\Delta;1] + [\Delta])_{2k+1}$

Whilst it is possible to continue in this way, enumerating the states, identifying their transformation properties and collecting them into sets associated with irreducible representations of SO(2k+1), the method rapidly becomes tedious. It is much better to note that the generating functions (2.1), (2.2) and (2.3) may all be expanded very easily using the formulae **[L, M]**

$$\prod_{n=1}^{N} \prod_{i=1}^{2k} (1 - x_i y_n)^{-1} = \sum_{\sigma} \{\sigma\}(\mathbf{y})_N \{\sigma\}(\mathbf{x})_{2k} \tag{2.4}$$

and

$$\prod_{n=1}^{N} \prod_{i=1}^{2k} (1 + x_i y_n) = \sum_{\tau} \{\tau'\}(\mathbf{y})_N \{\tau\}(\mathbf{x})_{2k} . \tag{2.5}$$

The summation in (2.4) is carried out over all partitions $\sigma = (\sigma_1, \sigma_2, \ldots)$. Such a partition specifies both an irreducible representation of GL(N) with character $\{\sigma\}(\mathbf{y})_N$ and an irreducible representation of GL(2k) with character $\{\sigma\}(\mathbf{x})_{2k}$. These characters are symmetric functions of their arguments known as Schur functions or S-functions **[L, M]**. The expansion (2.5) is much the same as (2.4) except that (2.5) involves the partition τ' conjugate to τ .

In applying (2.4) to the expansion of the bosonic string generating function (2.1) it is necessary to set $y_n = q^n$ and take the limit $N \longrightarrow \infty$, and then to rewrite characters of GL(2k) in terms of characters of SO(2k+1) **[K]**. This yields the formula

$$\kappa_B(\mathbf{x}, q) = q^{-1} \sum_{\mu} \{\mu\}(q)_\infty \, H(q)_\infty \, [\mu](\mathbf{x})_{2k+1} , \tag{2.6}$$

where

$$\{\mu\}(q)_\infty = \prod_{i,j}^{F^\mu} q^i (1 - q^{h_{ij}})^{-1} \tag{2.7}$$

and

$$H(q)_\infty = \prod_{1 \le m < \infty} (1 + q^m)^{-1} \cdot \prod_{1 \le m < n \le \infty} (1 + q^{m+n})^{-1}$$
$$= 1 - q + q^4 + q^6 + 2q^8 + 3q^{10} + q^{11} + 6q^{12} \ldots . \tag{2.8}$$

The notation in (2.7) is such that the product is taken over all boxes of the Young diagram F^{μ} specified by the partition μ, and h_{ij} is the hook length of the box in the ith row and jth column of F^{μ}. This hook length is given by $h_{ij} = \mu_i - j + \mu'_j - i + 1$, where μ'_j is the number of boxes in the jth column of F^{μ}, just as μ_i is the number of boxes in the ith row of F^{μ}. The factor (2.8) is a generating function for the number of scalars of SO(2k) up to the (k+1)(2k+1)th level. Thereafter modification rules **[K1,BKW]** are required. This factor (2.8) has been called a prefactor **[CGoT]**.

In expanding the Neveu-Schwarz spectrum generating function (2.2) it is necessary to use not only (2.4) but also (2.5). The subsequent exploitation of the algebra of S-functions and S-function series **[L,K1]** in connection with tensor characters of SO(2k+1) then yields the final result which can be written in the form

$$\kappa_{NS}(\mathbf{x},q) = q^{-k/8} \sum_{\mu} \{\mu\}(q/q')_{\infty}\, H(q/q')_{\infty}\, [\mu](\mathbf{x})_{2k+1}, \tag{2.9}$$

where

$$\{\mu\}(q/q')_{\infty} = \prod_{i,j}^{F^{\mu}} (q^{i} + q^{j-1/2})\,(1 - q^{h_{ij}})^{-1} \tag{2.10}$$

and

$$H(q/q')_{\infty} = \prod_{1\le m<\infty} (1 + q^{m})^{-1}(1 + q^{m-1/2})^{-1} .$$

$$\prod_{1\le m<n<\infty} (1 + q^{m+n})^{-1}\,(1 + q^{m+n-1})^{-1} \prod_{1\le m,\,n<\infty} (1 + q^{m+n-1/2}) \tag{2.11}$$

$$= 1 - q^{1/2} + q^{2} + q^{7/2} + 2q^{4} + 3q^{11/2} + 5q^{6} \;\ldots$$

In exactly the same way, but this time using spin characters **[L,K2,KLDW]** of SO(2k+1), the Ramond spectrum generating function (2.3) may be expanded to yield

$$\kappa_{R}(\mathbf{x},q) = \sum_{\mu} \{\mu\}(q/q)_{\infty}\, B(q/q)_{\infty}\, [\Delta;\mu](\mathbf{x})_{2k+1}, \tag{2.12}$$

where

$$\{\mu\}(q/q)_{\infty} = \prod_{i,j}^{F^{\mu}} (q^{i} + q^{j})\,(1 - q^{h_{ij}})^{-1} \tag{2.13}$$

and

$$B(q/q)_{\infty} = \prod_{1\le m<n<\infty} (1 + q^{m+n})\,(1 - q^{m+n})^{-1} \tag{2.14}$$

$$= 1 + 2q^{2} + 4q^{3} + 8q^{4} + 16q^{5} + 32q^{6} + \;\ldots$$

The formulae (2.6), (2.9) and (2.12) are the required spectrum generating functions of the type (1.2). Apart from the matter of modification rules **[K1, KLDW, BKW]** appropriate to particular values of k, these formulae give the required string spectra in full. The effect of the modification rules and a tabulation of the results are given elsewhere **[FKW]** along with more details of the calculations summarised here.

3. The Green-Schwarz superstring

A very significant advance was made **[GSO]** by restricting the Neveu-Schwarz string to states of even G-parity and combining it with the Ramond string modified in such a way that its vacuum state transforms as a Majorana-Weyl spinor. The corresponding GSO-projected Neveu-Schwarz-Ramond superstring has as its spectrum generating function

$$\kappa_{NSR}(\mathbf{x},q) = \kappa^{B}_{NSR}(\mathbf{x},q) + \kappa^{F}_{NSR}(\mathbf{x},q) \tag{3.1}$$

where the bosonic and fermionic parts are defined by

$$\kappa^{B}_{NSR}(\mathbf{x},q) = \frac{1}{2}\left\{ \kappa_{NS}(\mathbf{x},q) + (-1)^{k/4}\,\kappa'_{NS}(\mathbf{x},q) \right\} \tag{3.2}$$

and

$$\kappa^{F}_{NSR}(\mathbf{x},q) = \frac{1}{2}\left\{ \kappa_{R}(\mathbf{x},q) + \kappa'_{R}(\mathbf{x},q) \right\}, \tag{3.3}$$

with

$$\kappa_{NS}(\mathbf{x},q) = q^{-k/8} \prod_{n=1}^{\infty} \prod_{i=1}^{2k} (1 - x_i q^n)^{-1} (1 + x_i q^{n-1/2}), \tag{3.4}$$

$$\kappa'_{NS}(\mathbf{x},q) = q^{-k/8} \prod_{n=1}^{\infty} \prod_{i=1}^{2k} (1 - x_i q^n)^{-1} (1 - x_i q^{n-1/2}), \tag{3.5}$$

$$\kappa_{R}(\mathbf{x},q) = \prod_{i=1}^{k} (x_i^{1/2} + x_i^{-1/2}) \prod_{n=1}^{\infty} \prod_{i=1}^{2k} (1 - x_i q^n)^{-1} (1 + x_i q^n), \tag{3.6}$$

$$\kappa'_{R}(\mathbf{x},q) = \prod_{i=1}^{k} (x_i^{1/2} - x_i^{-1/2}) . \tag{3.7}$$

These definitions are appropriate to the case $D = 2k+2$.

The Green-Schwarz superstring **[GS]** is a $D = 10$ supersymmetric model. The vacuum state has dimension 16 and transforms as the basis of the reducible representation $\mathbf{8}_V + \mathbf{8}_S = [1](\mathbf{x})_8 + \Delta_+(\mathbf{x})_8$ of SO(8). It is massless having $L_0 = 0$. The bosonic excitation operators, a^i_{-n}, with $i = 1 \; 1,2,\ldots,8$ and $n = 1,2,\ldots,\infty$, transform, for each value of n, as the basis of the vector representation $\mathbf{8}_V = [1](\mathbf{x})_8$ of SO(8). The fermionic excitation operators, s^a_{-n}, with $i = 1,2,\ldots,8$ and $n = 1,2,\ldots,\infty$, transform, for each value of n, as the basis of the spin representation $\mathbf{8}_C = \Delta_-(\mathbf{x})_8$ of SO(8). It follows that the spectrum generating function of the Green-Schwarz superstring takes the form

$$\kappa_{GS}(\mathbf{x},q) = \Big[[1](\mathbf{x})_8 + \Delta_+(\mathbf{x})_8\Big] \prod_{n=1}^{\infty} \prod_{i=1}^{8} (1 - x_i q^n)^{-1} \; (1 + z_i q^n).$$

$$= \Delta(\mathbf{z})_8 \prod_{n=1}^{\infty} \prod_{i=1}^{8} (1 - x_i q^n)^{-1} \; (1 + z_i q^n) \tag{3.8}$$

where the triality of SO(8) is such that

$$\begin{aligned} \mathbf{8}_V &= [1](\mathbf{x})_8 &&= \Delta_+(\mathbf{y})_8 &&= \Delta_+(\mathbf{z})_8 \\ \mathbf{8}_S &= \Delta_+(\mathbf{x})_8 &&= [1](\mathbf{y})_8 &&= \Delta_-(\mathbf{z})_8 \\ \mathbf{8}_C &= \Delta_-(\mathbf{x})_8 &&= \Delta_-(\mathbf{y})_8 &&= [1](\mathbf{z})_8 \, . \end{aligned} \tag{3.9}$$

The relationship between $\mathbf{x}$, $\mathbf{y}$ and $\mathbf{z}$ is such that

$$\begin{aligned} y_1 &= \bar{y}_5 = x_1^{1/2} x_2^{1/2} x_3^{1/2} x_4^{1/2} & z_1 &= \bar{z}_5 = x_1^{1/2} x_2^{1/2} x_3^{1/2} \bar{x}_4^{1/2} \\ y_2 &= \bar{y}_6 = x_1^{1/2} x_2^{1/2} \bar{x}_3^{1/2} \bar{x}_4^{1/2} & z_2 &= \bar{z}_6 = x_1^{1/2} x_2^{1/2} \bar{x}_3^{1/2} x_4^{1/2} \\ y_3 &= \bar{y}_7 = x_1^{1/2} \bar{x}_2^{1/2} x_3^{1/2} \bar{x}_4^{1/2} & z_3 &= \bar{z}_7 = x_1^{1/2} \bar{x}_2^{1/2} x_3^{1/2} x_4^{1/2} \\ y_4 &= \bar{y}_8 = x_1^{1/2} \bar{x}_2^{1/2} \bar{x}_3^{1/2} x_4^{1/2} & z_4 &= \bar{z}_8 = x_1^{1/2} \bar{x}_2^{1/2} \bar{x}_3^{1/2} \bar{x}_4^{1/2} \, , \end{aligned} \tag{3.10}$$

where $\bar{x}_j = x_j^{-1}$ etc.

The way is now open to prove the following

Theorem In the case $D = 10$ the spectrum generating function of the GSO-projected Neveu-Schwarz-Ramond model coincides with that of the Green-Schwarz superstring, that is

$$\kappa_{GS}(\mathbf{x},q) = \kappa_{NSR}(\mathbf{x},q) \, . \tag{3.11}$$

The proof depends on recognising the fact that all the spectrum generating functions we have encountered may be written in terms of theta functions **[N]**. In fact if $\psi(\xi,\tau)$ is defined by

$$\kappa_B(\mathbf{x},q) = q^{-1} \, \phi(q) \, \psi(\xi,\tau) \, , \tag{3.12}$$

where

$$\phi(q) = \prod_{n=1}^{\infty} (1 - q^n) \tag{3.13}$$

with $q = \exp(i2\pi\tau)$ and $x_j = \exp(i2\pi\xi_j)$ for $j = 1,2,\dots,k$, then

$$\kappa_{NS}(\mathbf{x},q) = q^{-k/8}\ \theta_3(\xi,\tau)\ \psi(\xi,\tau) \tag{3.13a}$$

$$\kappa'_{NS}(\mathbf{x},q) = q^{-k/8}\ \theta_4(\xi,\tau)\ \psi(\xi,\tau) \tag{3.14b}$$

$$\kappa_{R}(\mathbf{x},q) = q^{-k/8}\ \theta_2(\xi,\tau)\ \psi(\xi,\tau) \tag{3.14c}$$

$$\kappa'_{R}(\mathbf{x},q) = i^k\ q^{-k/8}\ \theta_1(\xi,\tau)\ \psi(\xi,\tau)\ . \tag{3.14d}$$

Thanks to (2.1), (2.3), (3.12) and (3.14c), (3.8) can be recast in the form

$$\kappa_{GS}(\mathbf{x},q) = \kappa_R(\mathbf{z},q)\ \kappa_B(\mathbf{x},q)\ /\ \kappa_B(\mathbf{z},q) = q^{-1/2}\ \theta_2(\zeta,\tau)\ \psi(\xi,\tau) \tag{3.15}$$

where $z_j = \exp(i2\pi\zeta_j)$ for $j = 1,2,3,4$. The theorem (3.11) is then proved by noting an identity due to Jacobi **[Jp505]** which is peculiar to the $k = 4$ case, namely

$$\theta_2(\zeta,\tau) = \frac{1}{2}\left\{\theta_3(\xi,\tau) - \theta_4(\xi,\tau) + \theta_2(\xi,\tau) + \theta_1(\xi,\tau)\right\}\ . \tag{3.16}$$

Substituting this into (3.15) gives (3.11) immediately by virtue of the definitions (3.1)-(3.3) and (3.14) with, once more, $k = 4$.

4. Heterotic superstring

In the case of a closed string model the existence of left and right sectors implies that the spectrum generating function takes the form

$$\kappa(\mathbf{x},t,q,r) = \kappa^R(\mathbf{x},r)\ \kappa^L(\mathbf{x},t,q) \tag{4.1}$$

with

$$\kappa^R(\mathbf{x},r) = r^{R_0} \sum_{R=0}^{\infty} \sum_{\lambda} m^{\lambda}{}_R\ [\lambda](\mathbf{x})_{2k+1}\ r^R\ , \tag{4.2}$$

exactly as in (1.2), and

$$\kappa^L(\mathbf{x},t,q) = q^{L_0} \sum_{L=0}^{\infty} \sum_{\mu,\nu} n^{\mu,\nu}_L\ [\mu](\mathbf{x})_{2k+1}\ ch^{\nu}(\mathbf{t})_G\ q^L\ , \tag{4.3}$$

where $ch^{\nu}(t)_G$ denotes the character of an irreducible representation of the relevant gauge group G. The boundary conditions for closed string models imply that the levels of the right and left hand sectors must be matched **[R2]**. Thus in the expansion of (4.1) as the product of (4.2) and (4.3) it is only necessary to retain those terms for which

$$R_0 + R = L_0 + L \quad . \tag{4.4}$$

To take an explicit example, consider the heterotic string model with gauge group $G = E_8 \times E_8$ **[GHMR]**. In this model the right hand sector has $D = 10$, and consists indeed of nothing other than the Green-Schwarz superstring discussed in Section 3. The spectrum generating function is then given by

$$\kappa^{R}_{HET}(\mathbf{x},r) = \kappa_{NSR}(\mathbf{x},r) = \kappa_{GS}(\mathbf{x},r) \quad . \tag{4.5}$$

In contrast to this the left hand sector has $D = 26$, and consists of a $D = 10$ open bosonic string contribution, as described in Section 2, together with a $D = 16$ gauge group contribution associated with the 16-dimensional, even self-dual lattice $\Gamma = \Gamma_8 \times \Gamma_8$ generated by the roots of the gauge group $G = E_8 \times E_8$ **[R2]**. The corresponding spectrum generating function takes the form

$$\kappa^{L}_{HET}(\mathbf{x},\mathbf{t},q) = \kappa_B(\mathbf{x},q)\, \kappa_{E_8}(\mathbf{u},q)\, \kappa_{E_8}(\mathbf{v},q) \tag{4.6}$$

where $\kappa_B(\mathbf{x},q)$ is given by (2.1), $\mathbf{t} = (\mathbf{u},\mathbf{v})$ and

$$\kappa_{E_8}(\mathbf{u},q) = \prod_{n=1}^{\infty} (1 - q^n)^{-8} \sum_{\mathbf{m}\in\Gamma_8} \mathbf{u}^{\mathbf{m}}\, q^{(\mathbf{m}.\mathbf{m})/2} \quad . \tag{4.7}$$

The first factor is associated with bosonic excitation operators $\tilde{a}^{I}_{-n}$, with $I = 1,2,\dots,8$, transforming as SO(8) scalars. The second is associated with momentum operators p^I, with $I = 1,2,\dots,8$, specifying lattice points $\mathbf{m}$ of Γ_8. The notation in (4.7) is such that $\mathbf{u}^{\mathbf{m}} = u_1^{m_1} u_2^{m_2} \dots u_8^{m_8}$. The lattice is defined by

$$\Gamma_8 = \{\, \mathbf{m} \mid 2m_i \in \mathbb{Z},\ m_i - m_j \in \mathbb{Z},\ |\mathbf{m}|/2 \in \mathbb{Z} \ \text{ for } \ i,j = 1,2,\dots,8 \,\}. \tag{4.8}$$

Remarkably, (4.7) can be re-expressed successively in the forms

$$\kappa_{E_8}(\mathbf{u},q) = \prod_{n=1}^{\infty} (1 - q^n)^{-8}\, \frac{1}{2}\Big\{\theta_1(\alpha,\tau) + \theta_2(\alpha,\tau) + \theta_3(\alpha,\tau) + \theta_4(\alpha,\tau)\Big\} \tag{4.9}$$

$$= \frac{1}{2}\Big\{ \prod_{n=1}^{\infty} \prod_{I=1}^{16} (1 + u_I q^{n-1/2}) + \prod_{n=1}^{\infty} \prod_{I=1}^{16} (1 - u_I q^{n-1/2})$$

$$+ q\, \Delta(\mathbf{u})_{16} \prod_{n=1}^{\infty} \prod_{I=1}^{16} (1 + u_I q^{n}) + q\, \Delta'(\mathbf{u})_{16} \prod_{n=1}^{\infty} \prod_{I=1}^{16} (1 - u_I q^{n}) \Big\} \qquad (4.10)$$

$$= \sum_{\substack{\mu \\ |\mu| \text{ even}}} \{\mu'\}(q')_\infty\, D(q)_\infty\, [\mu](\mathbf{u})_{16}$$

$$+ q \sum_{\mu} \{\mu'\}(q')_\infty\, F(q)_\infty\, [\Delta;\mu]_{(-)^{|\mu|}}(\mathbf{u})_{16} \qquad (4.11)$$

where

$$D(q)_\infty = 1 + q^2 + q^3 + 3q^4 + 3q^5 + 7q^6 + \ldots \qquad (4.12)$$

and

$$F(q)_\infty = 1 + q + 2q^2 + 4q^3 + 7q^4 + 12q^5 + 21q^6 + \qquad (4.13)$$

The final step is to express the characters of SO(16) appearing in (4.11) in terms of characters of E_8 **[MP, W]**.

Putting all the above factors together one obtains

$$\kappa_{HET}(\mathbf{x},t,q,r) = \kappa^R_{HET}(\mathbf{x},r)\, .\, \kappa^L_{HET}(\mathbf{x},t,q)$$

$$= \kappa_{GS}(\mathbf{x},r)\, .\, \kappa_B(\mathbf{x},q).\kappa_{E_8}(\mathbf{u},q).\kappa_{E_8}(\mathbf{v},q), \qquad (4.14)$$

giving the following states after matching levels of the left and right sectors.

Heterotic string spectrum with gauge group $E_8 \times E_8$

		Right hand sector	.	Left-hand sector
Level	0	**(8 + 8)**	.	**(1.496 + 8.1)**
	1	**(128 + 128)**	.	**(1.69752 + 8.496 + 44.1)**
	2	**(1152 + 1152)**	.	**(1.2115008 + 8.69752 + 44.496 + 192.1)**

The notation is such that the factors arising from $\kappa_{GS}(\mathbf{x},r)$ are

$$(8 + 8) = [1]_8 + \Delta_{-8}$$

$$(128 + 128) = [2]_9 + [1^3]_9 + [\Delta;1]_9$$

$$(1152 + 1152) = [21^2]_9 + [1^4]_9 + [3]_9 + [21]_9 + [1^2]_9 + [1]_9$$
$$+ [\Delta;2]_9 + [\Delta;1^2]_9 + [\Delta;1]_9 + [\Delta;0]_9\, ,$$

those from $\kappa_B(\mathbf{x},q)$ are

$$\mathbf{1} = [0]_8 \qquad \mathbf{8} = [1]_8 \qquad \mathbf{44} = [2]_9 \qquad \mathbf{192} = [3]_9 + [1^2]_9 ,$$

and those from $\kappa_{E_8}(\mathbf{u},q)\, \kappa_{E_8}(\mathbf{v},q)$ are **[DS]**

$$\begin{aligned}
\mathbf{1} &= \mathbf{1.1} \\
\mathbf{496} &= \mathbf{248.1} + \mathbf{1.248} \\
\mathbf{69752} &= \mathbf{3875.1} + \mathbf{1.3875} + \mathbf{248.248} + 2(\mathbf{1.1}) \\
\mathbf{2115008} &= \mathbf{30380.1} + \mathbf{1.30380} + \mathbf{3875.248} + \mathbf{248.3875} + \mathbf{3875.1} \\
&\quad + \mathbf{1.3875} + 2(\mathbf{248.248}) + 3(\mathbf{248.1} + \mathbf{1.248}) + 2(\mathbf{1.1}) ,
\end{aligned}$$

where the representations of E_8 have been identified by their dimension **[MP]**.

All of these calculations may be extended to higher levels if required **[FKW]** since the spectrum generating functions are completely general. More important perhaps is the fact that the techniques used here apply not just to the models mentioned but can be used in the context of other string and superstring models **[T-M]**.

References

[BKW] G.R.E. Black, R.C. King and B.G. Wybourne, J. Phys. **A16** (1983) 1555

[CGhT] T.L. Curtright, G.I. Ghandour and C.B. Thorn, Phys. Lett. **182B** (1986) 45

[CGoT] T.L. Curtright, J. Goldstone and C.B. Thorn, Phys. Lett. **175B** (1986) 47

[CT] T.L. Curtright and C.B. Thorn, Nucl. Phys. **B274** (1986) 520

[DS] L. Dolan and R. Slansky, Phys. Rev. Lett. **54** (1985) 2075

[GSO] F. Gliozzi, J. Scherk and D. Olive, Nucl. Phys. **B122** (1977) 253

[GS] M.B. Green and J.H. Schwarz, Nucl. Phys. **B181** (1981) 502

[GHMR] D.J. Gross, J.A. Harvey, E. Martinec and R. Rohm, Nucl. Phys. **B256** (1985) 253

[J] C.G.J. Jacobi, "Gesammelte Werke" Vol 1. 2nd.Edition (Chelsea Pub. Co., New York, 1969)

[K1] R.C. King, J. Math. Phys. **12** (1971) 1588

[K2] R.C. King, J. Phys. **A8** (1975) 429

[KLDW] R.C. King, Luan Dehuai and B.G. Wybourne, J. Phys. **A14** (1981) 2509

[L] D.E. Littlewood, "The theory of group characters" 2nd. Edition (Clarendon Press, Oxford, 1950)

[M] I.G. Macdonald, "Symmetric functions and Hall polynomials" (Clarendon Press, Oxford, 1979)

[MP] W.G. McKay and J. Patera "Tables of dimensions, indices and branching rules for representations of simple Lie algebras" (Dekker, New York, 1981)

[N] W. Nahm, Commun. Math. Phys. **105** (1986) 1

[NS] A. Neveu and J.H. Schwarz, Nucl. Phys. **B31** (1971) 86

[R1] P. Ramond, Phys. Rev. **D3** (1971) 2415

[R2] P. Ramond, "Group theory for string states" University of Florida preprint UFTP-85-10 (1985)

[T-M] J. Thierry-Mieg, Phys. Lett. **171B** (1986) 163

[W] B. G. Wybourne, J. Phys. **A17** (1984) 1397

THE SHIFT OPERATORS AND THE WAVE FUNCTIONS IN GINOCCHIO MODEL

Yinsheng Ling, Zhengkun Chu, and Dehuang Ji

Suzhou University
People's Republic of China

INTRODUCTION

As well known, there are three subalgebra chains in IBM:

$$su(6)---so(6)---so(5)---so(3) \quad (A1)$$

$$su(6)---su(5)---so(5)---so(3) \quad (A2)$$

$$su(6)------su(3)------so(3) \quad (A3)$$

The spectra of the nucleons in the low energy region can be described successfully by IBM[1].

The realistic nucleon states should be the fermion states. If the nucleon angular momenta can be separated as

$$\vec{j} = \vec{k} + \overrightarrow{3/2} \quad . \quad (1)$$

All the even nucleons states carry the spin representation (1/2,1/2---1/2) of $so\{8(2k+1)\}$. There is a subalgebra chain in $so\{8(2k+1)\}$:

$$so\{8(2k+1)\} ------so(2k+1)+so(8).$$

The identity representation space of so(2k+1) can be constructed by the following S^+ pairs and D^+ pairs:

$$S^+=\sqrt{2k+1}\ (b^+_{k,3/2}\ b^+_{k,3/2})^{(0,0)}_{0,0} \, , \quad (2)$$

$$D^+_m=\sqrt{2k+1}\ (b^+_{k,3/2}\ b^+_{k,3/2})^{(0,2)}_{0,m} \, . \quad (3)$$

Here

$$b^+_{km_k;im_i} = \sum_{jm} \langle km_k, im_i | jm \rangle\ a^+_{jm} \, . \quad (4)$$

The a^+_{jm} is the fermion creation operator. This subspace carries the spin representation $(\Omega/4, \Omega/4, \Omega/4, \Omega/4)$ of so(8), $\Omega = 1/2\sum_j (2j+1)$. The dimension of this subspace is exactly equal to the total dimensions of the IBM states in the corresponding fermion shell. The states between these two models are one-to-one correspondence[2]. So that there are interesting relationships between Ginocchio so(8) model and IBM.

There are some things different. The boson states $s^+ \cdot s^+|0\rangle$ and $d^+ \cdot d^+|0\rangle$ are orthoganal. But the fermion states $S^+ \cdot S^+|0\rangle$ and $D^+ \cdot D^+|0\rangle$ are not orthoganal. So that the analytic expressions of the wave functions in Ginocchio model are more complicated than those in IBM.

We list the subalgebra chains in so(8) model and the correspondence quantum numbers:

$$\begin{array}{lllllll} \mathrm{so}(8) & --- & \mathrm{u}(1)+\mathrm{so}(6) & --- & \mathrm{u}(1)+\mathrm{so}(5) & --- & \mathrm{u}(1)+\mathrm{so}(3) \\ & & N \quad (\sigma,0,0) & & (\tau,0) & & L \end{array} \qquad (G1)$$

$$\begin{array}{lllllll} \mathrm{so}(8) & --- & \mathrm{so}(7) & --- & \mathrm{u}(1)+\mathrm{so}(5) & --- & \mathrm{u}(1)+\mathrm{so}(3) \\ & & \langle\Sigma,0,0\rangle & & N \quad (\tau,0) & & L \end{array} \qquad (G2)$$

$$\begin{array}{lllllll} \mathrm{so}(8) & --- & \mathrm{su}(2)+\mathrm{so}(5) & --- & \mathrm{u}(1)+\mathrm{so}(5) & --- & \mathrm{u}(1)+\mathrm{so}(3) \\ & & S' \quad (\tau,0) & & S_0 & & L \end{array} \qquad (G3)$$

The generators of these algebras are

$$\mathrm{so}(8): \{P^{(1)}_m, P^{(2)}_m, P^{(3)}_m, S^+, D^+_m, \tilde{S}, \tilde{D}_m, S_0\}$$

$$\mathrm{so}(7): \{P^{(1)}_m, P^{(3)}_m, D^+_m, \tilde{D}_m, S_0\}$$

$$\mathrm{so}(6): \{P^{(1)}_m, P^{(2)}_m, P^{(3)}_m\}$$

$$\mathrm{su}(2)+\mathrm{so}(5): \{S^+, \tilde{S}, S_0; P^{(1)}_m, P^{(3)}_m\}$$

$$\mathrm{so}(5): \{P^{(1)}_m, P^{(3)}_m\}$$

$$\mathrm{so}(3): \{P^{(1)}_m\}.$$

Here

$$P^{(r)}_m = \sqrt{2k+1}\,(b^+_{k,3/2}\,\tilde{b}_{k,3/2})^{(0,r)}_{0,m}\,, \qquad (5)$$

$$S_0 = P^{(0)} - \Omega/2\,, \qquad (6)$$

$$\tilde{S} = \sqrt{2k+1}\,(\tilde{b}_{k,3/2}\,\tilde{b}_{k,3/2})^{(0,0)}_{0,0} \,, \tag{7}$$

$$\tilde{D}_m = \sqrt{2k+1}\,(\tilde{b}_{k,3/2}\,\tilde{b}_{k,3/2})^{(0,2)}_{0,m} \,, \tag{8}$$

$$\tilde{b}_{km_k;im_i} = (-1)^{k+m_k+i+m_i}\, b_{k,-m_k;i,-m_i} \,. \tag{9}$$

The S_0 is related to the nucleon number N:

$$S_0 = 1/2(N-\Omega)\,. \tag{10}$$

Using the following shift operators we can easily find out the highest weight states of irreducible representations(irreps) of so(5). The method to find out the states in an irrep of so(5) is already solved[3].

(I) so(8)---u(1)+so(6)---u(1)+so(5) chain

The highest weight states of irreps of so(5) can be labeled as

$$|N(\sigma,0,0)(\tau,0)(\tau,0)\rangle = |N,\sigma,\tau,\tau\rangle\,. \tag{11}$$

Here

$$N = 0,\ 2,\ 4,\text{------}\ 2\Omega\,, \tag{12}$$

$$\sigma = \begin{cases} N/2,\ N/2-2,\text{------}\ 0 \text{ or } 1 & (N\leq\Omega) \\ \Omega-N+2,\ \Omega-N/2-2,\text{---}0 \text{ or } 1\,, & (N>\Omega) \end{cases} \tag{13}$$

$$\tau = 0,\ 1,\ 2,\text{------}\ \sigma\,. \tag{14}$$

Since

$$P^{(2)}_2|N,\sigma,\tau,\tau\rangle = \sqrt{\frac{(\sigma-\tau)(\sigma+\tau+4)(\tau+1)}{2\tau+5}}\,|N,\sigma,\tau+1,\tau+1\rangle\,. \tag{15}$$

We only need to consider the (0,0) reps of so(5).

Define

$$\mathcal{J}^+ = S^+\cdot S^+ - D^+\cdot D^+\,. \tag{16}$$

It is easy to show that

$$[\mathcal{J}^+,\, P^{(r)}_m] = 0\,. \qquad (r=1,2,3) \tag{17}$$

So that $\mathcal{J}^+$ only increases the nucleon number N by 4, not influences the quantum number of so(6):

$$(\mathcal{J}^+)^k|N=2\sigma,\sigma,0,0\rangle = \mathcal{N}_{\sigma,k}|N=2\sigma+4k,\sigma,0,0\rangle\,. \tag{18}$$

The wave functions $|N=2\sigma,\sigma,0,0\rangle$ can be constructed by the operator

$$\mathcal{A}^+(\sigma) = \sum_p^{[\sigma/2]} \beta_p(\sigma)(S^+)^{\sigma-2p}(\mathcal{J}^+)^p\,, \tag{19}$$

$$\mathcal{A}^+(\sigma)|0\rangle = |N=2\sigma,\sigma,0,0\rangle\,. \tag{20}$$

From

$$P^{(2)}\cdot P^{(2)}|N=2\sigma,\sigma,0,0\rangle = \sigma(\sigma+4)|N=2\sigma,\sigma,0,0\rangle \qquad (21)$$

we can get the recurrence formula for the coefficients

$$\beta_p(\sigma) = -\frac{(\sigma-2p+2)(\sigma-2p+1)}{4p(\sigma-p+2)}\beta_{p-1}(\sigma) . \qquad (22)$$

(II) so(8)---so(7)---U(1)+so(5) chain

The highest weight states of irreps of so(5) can be labeled as

$$|\langle\Sigma ,0,0\rangle\ N(\tau,0)(\tau,0)\rangle = |\langle\Sigma\rangle ,N,\tau,\tau\rangle . \qquad (23)$$

Here

$$\Sigma = 1/2|\Omega-N| + p , \qquad p=\begin{cases} 0,\ 1,\ 2,\text{------}\ N/2 & (N\leq\Omega) \\ 0,\ 1,\ 2,\text{---}\Omega-N/2 & (N>\Omega) \end{cases} \qquad (24)$$

$$\tau = p,\ p-2,\text{------}\ 0 \text{ or } 1 . \qquad (25)$$

We only consider $N\leq\Omega$ case. The shift operators are

$$(1)\ \sum_{p=0}^{[N/4]} \beta_p(N)(S^+)^{N/2-2p}(D^+\cdot D^+)^p|0\rangle = |\langle\Omega/2-N/2\rangle ,N,0,0\rangle \qquad (26)$$

From

$$D^+\cdot D^+|\langle\Omega/2-N/2\rangle ,N,0,0\rangle = 0 , \qquad (27)$$

we can get the recurrence formula for the coefficients

$$\beta_p(N) = \frac{(N/2-2p+2)(N/2-2p+1)}{2p(\Omega+2p+5-N)}\beta_{p-1}(N) . \qquad (28)$$

$$(2)\ D_2^+|\langle\Sigma\rangle ,N,\tau,\tau\rangle = \mathcal{N}_2|\langle\Sigma\rangle ,N+2,\tau+1,\tau+1\rangle . \qquad (29)$$

$$(3)\ D^+\cdot D^+|\langle\Sigma\rangle ,N,\tau,\tau\rangle = \mathcal{N}_4|\langle\Sigma\rangle ,N+4,\tau,\tau\rangle . \qquad (30)$$

(III) so(8)---su(2)+so(5)---u(1)+so(5) chain

The highest weight states of irreps of so(5) can be labeled as

$$|S,S_0;(\tau,0)(\tau,0)\rangle = |S,S_0;\ \tau,\tau\rangle . \qquad (31)$$

Here

$$S=\Omega/2 - p , \qquad p=\begin{cases} 0,\ 1,\ 2,\text{------}\ N/2 & (N\leq\Omega) \\ 0,\ 1,\ 2,\text{---}\Omega-N/2 & (N>\Omega) \end{cases} \qquad (32)$$

$$\tau = p,\ p-2,\text{------}\ 0 \text{ or } 1 . \qquad (33)$$

We can use the following shift operators to construct these states:

$$(1)\ (D_2^+)^p|0\rangle = \sqrt{\frac{p!\,\Omega!!}{(\Omega-2p)!!}}\,|\Omega/2-p,-\Omega/2+p;\ p,p\rangle . \qquad (34)$$

(2) We abbreviate

$$|S,S_0\rangle = |S,S_0;\ \tau,\tau\rangle . \qquad (35)$$

Fig. 1 The relationships of the wave functions of (0,0) irreps of so(5) between three chains in so(8) model are shown. The wave functions of so(6) chain are as basis. The wave functions are labeled as

(G1) chain: $|\sigma,0,0\rangle$,

(G2) chain: $|\langle\Sigma\rangle,0,0\rangle$,

(G3) chain: $|S;\ 0,0\rangle$.

(a) N=2, $S_0=-\Omega/2+1$

chain \ basis(G1)		$\|1,0,0\rangle$
(G2)	$\|\langle\Omega/2-1\rangle,0,0\rangle$	1
(G3)	$\|\Omega/2\ ;\ 0,0\rangle$	1

(b) N=4, $S_0=-\Omega/2+4$

chain \ basis(G1)		$\|2,0,0\rangle$	$\|0,0,0\rangle$
(G2)	$\|\langle\Omega/2-2\rangle,0,0\rangle$	$\sqrt{\frac{5(\Omega+4)}{6(\Omega+3)}}$	$\sqrt{\frac{\Omega-2}{6(\Omega+3)}}$
	$\|\langle\Omega/2\ \rangle,0,0\rangle$	$\sqrt{\frac{\Omega-2}{6(\Omega+3)}}$	$-\sqrt{\frac{5(\Omega+4)}{6(\Omega+3)}}$
(G3)	$\|\Omega/2\ ;\ 0,0\rangle$	$\sqrt{\frac{5(\Omega-2)}{6(\Omega-1)}}$	$\sqrt{\frac{\Omega+4}{6(\Omega-1)}}$
	$\|\Omega/2-2;\ 0,0\rangle$	$\sqrt{\frac{\Omega+4}{6(\Omega-1)}}$	$-\sqrt{\frac{5(\Omega-2)}{6(\Omega-1)}}$

(c) N=6, $S_0=-\Omega/2+3$

chain \ basis(G1)		$\|3,0,0\rangle$	$\|1,0,0\rangle$
(G2)	$\|\langle\Omega/2-3\rangle,0,0\rangle$	$\sqrt{\frac{5(\Omega+4)}{8(\Omega+1)}}$	$\sqrt{\frac{3(\Omega-4)}{8(\Omega+1)}}$
	$\|\langle\Omega/2-1\rangle,0,0\rangle$	$\sqrt{\frac{3(\Omega-4)}{8(\Omega+1)}}$	$-\sqrt{\frac{5(\Omega+4)}{8(\Omega+1)}}$
(G3)	$\|\Omega/2\ ;\ 0,0\rangle$	$\sqrt{\frac{5(\Omega-4)}{8(\Omega-1)}}$	$\sqrt{\frac{3(\Omega+4)}{8(\Omega-1)}}$
	$\|\Omega/2-2;\ 0,0\rangle$	$\sqrt{\frac{3(\Omega+4)}{8(\Omega-1)}}$	$-\sqrt{\frac{5(\Omega-4)}{8(\Omega-1)}}$

(d) N=8, $S_0=-\Omega/2+4$

chain \ basis(G1)		$\|4,0,0\rangle$	$\|2,0,0\rangle$	$\|0,0,0\rangle$
(G2)	$\|\langle\Omega/2-4\rangle,0,0\rangle$	$\sqrt{\frac{7(\Omega+2)(\Omega+4)}{16(\Omega-1)(\Omega+1)}}$	$\sqrt{\frac{(\Omega-6)(\Omega+2)}{2(\Omega-1)(\Omega+1)}}$	$\sqrt{\frac{(\Omega-4)(\Omega-6)}{16(\Omega-1)(\Omega+1)}}$
	$\|\langle\Omega/2-2\rangle,0,0\rangle$	$\sqrt{\frac{21(\Omega-6)(\Omega+4)}{40(\Omega-1)(\Omega+3)}}$	$-\sqrt{\frac{(2\Omega+13)^2}{15(\Omega-1)(\Omega+3)}}$	$-\sqrt{\frac{5(\Omega-4)(\Omega+2)}{24(\Omega-1)(\Omega+3)}}$
	$\|\langle\Omega/2\rangle,0,0\rangle$	$\sqrt{\frac{3(\Omega-4)(\Omega-6)}{80(\Omega+1)(\Omega+3)}}$	$-\sqrt{\frac{7(\Omega-4)(\Omega+4)}{30(\Omega+1)(\Omega+3)}}$	$\sqrt{\frac{35(\Omega+2)(\Omega+4)}{48(\Omega+1)(\Omega+3)}}$
(G3)	$\|\Omega/2;0,0\rangle$	$\sqrt{\frac{7(\Omega-4)(\Omega-6)}{16(\Omega-1)(\Omega-3)}}$	$\sqrt{\frac{(\Omega-4)(\Omega+4)}{2(\Omega-1)(\Omega-3)}}$	$\sqrt{\frac{(\Omega+2)(\Omega+4)}{16(\Omega-1)(\Omega-3)}}$
	$\|\Omega/2-2;0,0\rangle$	$\sqrt{\frac{21(\Omega+4)(\Omega-6)}{40(\Omega-1)(\Omega-5)}}$	$\sqrt{\frac{(17-2\Omega)^2}{15(\Omega-1)(\Omega-5)}}$	$-\sqrt{\frac{5(\Omega-4)(\Omega+2)}{24(\Omega-1)(\Omega-5)}}$
	$\|\Omega/2-4;0,0\rangle$	$\sqrt{\frac{3(\Omega+2)(\Omega+4)}{80(\Omega-3)(\Omega-5)}}$	$-\sqrt{\frac{7(\Omega-6)(\Omega+2)}{30(\Omega-3)(\Omega-5)}}$	$\sqrt{\frac{35(\Omega-4)(\Omega-6)}{48(\Omega-3)(\Omega-5)}}$

(e) N=10, $S_0=-\Omega/2+5$

chain \ basis(G1)		$\|5,0,0\rangle$	$\|3,0,0\rangle$	$\|1,0,0\rangle$
(G2)	$\|\langle\Omega/2-5\rangle,0,0\rangle$	$\sqrt{\frac{7(\Omega+2)(\Omega+4)}{24(\Omega-1)(\Omega-3)}}$	$\sqrt{\frac{25(\Omega-8)(\Omega+2)}{48(\Omega-1)(\Omega-3)}}$	$\sqrt{\frac{3(\Omega-6)(\Omega-8)}{16(\Omega-1)(\Omega-3)}}$
	$\|\langle\Omega/2-3\rangle,0,0\rangle$	$\sqrt{\frac{7(\Omega-8)(\Omega+4)}{12(\Omega-3)(\Omega+1)}}$	$-\sqrt{\frac{(\Omega+22)^2}{24(\Omega-3)(\Omega+1)}}$	$\sqrt{\frac{3(\Omega-6)(\Omega+2)}{8(\Omega-3)(\Omega+1)}}$
	$\|\langle\Omega/2-1\rangle,0,0\rangle$	$\sqrt{\frac{(\Omega-6)(\Omega-8)}{8(\Omega-1)(\Omega+1)}}$	$-\sqrt{\frac{7(\Omega-6)(\Omega+4)}{16(\Omega-1)(\Omega+1)}}$	$\sqrt{\frac{7(\Omega+2)(\Omega+4)}{16(\Omega-1)(\Omega+1)}}$
(G3)	$\|\Omega/2;0,0\rangle$	$\sqrt{\frac{7(\Omega-6)(\Omega-8)}{24(\Omega-1)(\Omega-3)}}$	$\sqrt{\frac{25(\Omega-6)(\Omega+4)}{48(\Omega-1)(\Omega-3)}}$	$\sqrt{\frac{3(\Omega+2)(\Omega+4)}{16(\Omega-1)(\Omega-3)}}$
	$\|\Omega/2-2;0,0\rangle$	$\sqrt{\frac{7(\Omega-8)(\Omega+4)}{12(\Omega-1)(\Omega-5)}}$	$\sqrt{\frac{(26-\Omega)^2}{24(\Omega-1)(\Omega-5)}}$	$-\sqrt{\frac{3(\Omega-6)(\Omega+4)}{8(\Omega-1)(\Omega-5)}}$
	$\|\Omega/2-4;0,0\rangle$	$\sqrt{\frac{(\Omega+2)(\Omega+4)}{8(\Omega-3)(\ -5)}}$	$-\sqrt{\frac{7(\Omega-8)(\Omega+2)}{16(\Omega-3)(\Omega-5)}}$	$\sqrt{\frac{7(\Omega-6)(\Omega-8)}{16(\Omega-3)(\Omega-5)}}$

It is easy to show that

$$D^{+}\cdot D^{+}|S,-S\rangle = c_1|S+2,-S+2\rangle + c_2|S,-S+2\rangle + c_3|S-2,-S+2\rangle \ , \quad (36)$$

$$S^{+}\tilde{S}(D^{+}\cdot D^{+})|S,-S\rangle = -4(2S+1)c_1|S+2,-S+2\rangle -2(2S-1)c_2|S,-S+2\rangle \ , \quad (37)$$

$$(S^{+})^2(\tilde{S})^2(D^{+}\cdot D^{+})|S,-S\rangle = 24(S+1)(2S+1)c_1|S+2,-S+2\rangle + 4S(2S-1)c_2|S,-S+2\rangle \ . \quad (38)$$

Define

$$\mathcal{J}^{+} = \beta_0 D^{+}\cdot D^{+} + \beta_1 S^{+}\tilde{S}(D^{+}\cdot D^{+}) + \beta_2(S^{+})^2(\tilde{S})^2(D^{+}\cdot D^{+}) \ , \quad (39)$$

here

$$\beta_1 = \frac{5S+2}{4(2S-1)(2S+1)}\beta_0 \ , \quad (40)$$

$$\beta_2 = \frac{1}{8(2S-1)(2S+1)}\beta_0 \ . \quad (41)$$

We can show that

$$\mathcal{J}^{+}|S,-S;\tau,\tau\rangle = |S-2,-S+2;\tau,\tau\rangle \ . \quad (42)$$

(3) $$S^{+}|S,S_0;\tau,\tau\rangle = \sqrt{(S-S_0)(S+S_0+1)}\,|S,S_0+1;\tau,\tau\rangle \ . \quad (43)$$

When we calculate the module, we will meet the oblique basis

$$(S^{+})^m(D^{+}\ D^{+})^n(D_2^{+})^{\tau}|0\rangle \ . \qquad (2m+4n+2\tau=N) \quad (44)$$

Define

$$|\tau\rangle = \sqrt{\frac{(\Omega-2\tau)!!}{\tau!\,\Omega!!}}\,(D_2^{+})^{\tau}|0\rangle \ , \quad (45)$$

$$I(m,n;p,q) = \langle\tau|\,S^m(\tilde{D}\cdot\tilde{D})^n(S^{+})^p(D^{+}\cdot D^{+})^q|\tau\rangle \ . \quad (46)$$

The metric matrix can be calculated by the recurrence relationships and the initial conditions[2]

$$I(m,n;p,q) = p(\Omega-2\tau-p-4q+1)I(m-1,n;p-1,q) -2q(2q+2\tau+3)I(m-1,n;p+1,q-1) \ , \quad (47)$$

$$I(0,n;0,n) = 2^n\,\frac{n!(2\tau+3+2n)!!(\Omega-2\tau)!!(\Omega+3)!!}{(2\tau+3)!!(\Omega-2\tau-2n)!!(\Omega+3-2n)!!} \ . \quad (48)$$

Some results are shown in Tabl.

This work is supported by National Natural Science Foundation of China.

REFERENCES

1. A.Arima and F.Iachello, Ann phys 253(1976)99, Ann phys 201(1978)111, Ann phys 468(1979)123

2. J.N.Ginocchio, Ann phys 126(1980)234

3. S.Szpikowski and A.Gozdz, Nucl phys A340(1980)76

HIGHEST WEIGHT REPRESENTATIONS OF THE LIE ALGEBRA gl_∞ IN A GEL'FAND-ZETLIN BASIS

Tchavdar D. Palev*

Arnold-Sommerfeld Institute for Mathematical Physics

3392 Clausthal-Zellerfeld, West Germany

In the last few years the algebra gl_∞ of infinite matrices is of increasing interest in physics. The reason for this stems from the observation that the irreducible gl_∞ modules appear as natural carrier spaces for Virasoro algebras, loop algebras and Kac-Moody algebras [1]. Therefore, the gl_∞ modules provide a natural background for study of various physical applications of the algebras, mentioned above [2].

A class of representations of gl_∞, called fundamental representations, were constructed in Ref.[3] (see also [1] and [4], where further results and references can be found). These representations and also the irreducible representations, obtained from tensor products of fundamental representations, are highest weight representations.

In the present paper we study also highest weight representations of the algebra gl_∞, which are different from the representations, mentioned above. Within each module we introduce a basis and write down relations for the transformation of the basis under the action of the algebra generators.

We define the Lie algebra gl_∞ by:

$$gl_\infty = \{(a_{ij})_{i,j\in \mathbf{N}} | \text{all but a finite number of the } a_{ij} \text{ are } 0\}, \tag{1}$$

where $\mathbf{N}$ is the set of all positive integers. The Lie bracket [,] on gl_∞ is the ordinary matrix commutator.

Let $e_{ij} \in gl_\infty$ be an infinite-dimensional Weyl matrix, i.e., e_{ij} has 1 as the (i,j) entry and all other entries 0. The $e_{ij}, i,j \in \mathbf{N}$ form a basis for gl_∞. The commutation relations of gl_∞ are linear extension of the commutation relations of the e_{ij}:

$$[\![e_{ij}, e_{kl}]\!] = \delta_{jk} e_{il} - \delta_{li} e_{kj}. \tag{2}$$

We recall that the representation (in the particular realization (1) of gl_∞) is said to be a highest weight representation, if the corresponding to it module admits a non -zero vector (M), called a highest weight vector, such that

*Permanent address:Institute for Nuclear Research and Nuclear Energy,boul.Lenin 72, 1184 Sofia, Bulgaria

$$e_{kk}(M) = M_k(M) \quad \forall\, k \in \mathbf{N}, \tag{3}$$

$$e_{ij}(M) = 0 \quad \forall\, i < j \in \mathbf{N}. \tag{4}$$

In such a case the collection of numbers

$$[M] = [M_1, \ldots, M_k, \ldots] = [M_k |\, k \in \mathbf{N}] \tag{5}$$

is the highest weight of the representation. The Weyl matrices e_{kk}, $k \in \mathbf{N}$ span a basis in the Cartan subalgebra H of gl_∞. As an ordered basis in the dual space H^* of H we take all linear functionals

$$e^1, e^2, \ldots, e^k, \ldots \quad , \tag{6}$$

where

$$e^i(e_{jj}) = \delta^i_j. \tag{7}$$

With respect to this ordering the generators

$$e_{ij} \quad \forall\, i < j \in \mathbf{N} \tag{8}$$

are the positive root vectors of the algebra and the numbers $M_k, k \in \mathbf{N}$ are the coordinates of (M) in the basis (6).

The algebra gl_∞ contains an infinite chain of subalgebras

$$gl(1) \subset gl(2) \subset \ldots \subset gl(N) \subset \ldots \subset gl_\infty, \tag{9}$$

where

$$gl(N) = \text{lin.env.}\{e_{ij} | i, j = 1, \ldots, N\}. \tag{10}$$

The basis of the irreducible modules of gl_∞, which we introduce, will be a natural generalization of the Gel'fand and Zetlin basis of the finite-dimensional irreducible modules (fidirmods) of $gl(N)$ [5].

We recall that every finite-dimensional irreducible module (fidirmod) $V([m]_N)$ of $gl(N)$ is labeled by its signature

$$[m_{1N}, m_{2N}, \ldots, m_{NN}] = [m]_N, \tag{11}$$

where $m_{1N}, m_{2N}, \ldots, m_{NN}$ are,in general, complex numbers such that

$$m_{iN} - m_{jN} \in \mathbb{Z}_+ \quad \forall i < j = 1, \ldots, N. \tag{12}$$

Throughout the paper $\mathbb{Z}$ and $\mathbb{Z}_+$ denote all integers and all nonnegative integers, respectively.The basis $\Gamma([m]_N)$ in $V([m]_N)$ can be chosen to consists of all Gel'fand-Zetlin patterns. This is the set of all patterns

$$
(m) \equiv \begin{bmatrix} [m]_N \\ \vdots \\ [m]_i \\ \vdots \\ [m]_2 \\ m_{11} \end{bmatrix} \equiv \begin{bmatrix} m_{1N}, & m_{2N}, & . & . & . & m_{NN} \\ \vdots & \vdots & \vdots & \vdots & \vdots & \\ m_{1i}, & m_{2i}, & . & m_{ii} & & \\ \vdots & \vdots & \vdots & \vdots & & \\ m_{12}, & m_{22} & & & & \\ m_{11} & & & & & \end{bmatrix}, \tag{13}
$$

which are consistent with the conditions:

$$
m_{i,j+1} - m_{ij} \in \mathbb{Z}_+, \; m_{ij} - m_{i+1,j+1} \in \mathbb{Z}_+ \quad \forall\, i \leq j = 1, \ldots, N-1. \tag{14}
$$

Let (m) be an arbitrary Gel'fand-Zetlin pattern (13).Denote by

$$
(m)_{\pm ij} \tag{15}
$$

the scheme obtained from (m) by the replacement

$$
m_{ij} \rightarrow m_{ij} \pm 1. \tag{16}
$$

Set

$$
l_{ij} = m_{ij} - i. \tag{17}
$$

The transformation of $V([m]_N)$ is completely determined from the transformations of the Gel'fand-Zetlin basis (GZ basis) (13) under the action of the Cartan generators $e_{kk}, k = 1, \ldots, N$ and the root vectors $e_{k,k-1}, e_{k-1,k}, k = 2, \ldots, N$:

$$
e_{kk}(m) = (m_{1k} + \ldots + m_{kk} - m_{1,k-1} - \ldots - m_{k-1,k-1})(m), \tag{18}
$$

$$
e_{k,k-1}(m) = \sum_{j=1}^{k-1} \left| \frac{\prod_{i=1}^{k}(l_{ik} - l_{j,k-1} + 1) \prod_{i=1}^{k-2}(l_{i,k-2} - l_{j,k-1})}{\prod_{i \neq j=1}^{k-1}(l_{i,k-1} - l_{j,k-1} + 1)(l_{i,k-1} - l_{j,k-1})} \right|^{1/2} (m)_{-j,k-1}, \tag{19}
$$

$$
e_{k-1,k}(m) = \sum_{j=1}^{k-1} \left| \frac{\prod_{i=1}^{k}(l_{ik} - l_{j,k-1}) \prod_{i=1}^{k-2}(l_{i,k-2} - l_{j,k-1} - 1)}{\prod_{i \neq j=1}^{k-1}(l_{i,k-1} - l_{j,k-1})(l_{i,k-1} - l_{j,k-1} - 1)} \right|^{1/2} (m)_{j,k-1}. \tag{20}
$$

We now proceed to define a gl_∞ module $V([M])$ with a highest weight (5).Let

$$
M_1, M_2, \ldots, M_j, \ldots \tag{21}
$$

be an infinite sequence of complex numbers with the following properties:

$$
M_i - M_j \in \mathbb{Z}_+ \quad \forall i < j \in \mathbf{N}. \tag{22}
$$

Denote

$$
\begin{aligned}
[M] &= [M_1, M_2, \ldots, M_j, \ldots], && (23)\\
[M]_j &= [M_1, M_2, \ldots, M_j], && (24)\\
[m]_j &= [m_{1j}, m_{2j}, \ldots, m_{jj}]. && (25)
\end{aligned}
$$

Let

$$
(m) \equiv (m_{ij}), \forall i < j \in \mathbf{N} \tag{26}
$$

be an infinite number scheme, similar to the Gel'fand-Zetlin schemes (GZ schemes)[see (13)], i.e.,

$$(m) = \begin{bmatrix} \vdots & \vdots & \vdots & \vdots & \vdots \\ m_{1j}, & m_{2j}, & . & m_{jj} & \\ \vdots & \vdots & \vdots & \vdots & \\ m_{12}, & m_{22} & & & \\ m_{11} & & & & \end{bmatrix} = \begin{bmatrix} \vdots \\ [m]_j \\ \vdots \\ [m]_2 \\ m_{11} \end{bmatrix} \tag{27}$$

with the following properties:

(1) There exists a positive integer $N[(m)] \in \mathbf{N}$, such that

$$\forall j > N[(m)] \ \ [m]_j = [M]_j; \tag{28}$$

(2)

$$m_{i,j+1} - m_{ij} \in \mathbf{Z}_+, \ m_{ij} - m_{i+1,j+1} \in \mathbf{Z}_+ \ \ \forall\, i \leq j \in \mathbf{N}. \tag{29}$$

The condition (28) means that whenever

$$j > N[(m)], \text{then } m_{ij} = M_i \ \ \forall\, i = 1, 2, \dots, j. \tag{30}$$

Clearly, (28) holds if and only if (30) is fulfilled.

Denote by $\Gamma([M])$ the set of all schemes (27), which are compatible with the conditions (28),(29) and let $V([M])$ be a linear space with a basis $\Gamma([M])$,

$$V([M]) = \sum_{(m)\in\Gamma([M])} \oplus \mathbf{C}(m). \tag{31}$$

We postulate that the generators $e_{ii}, i \in \mathbf{N}$ and $e_{k,k-1}, e_{k-1,k}$, $k = 2, 3, \dots$, transform each pattern $(m) \in \Gamma([M])$ according to the relations (18)- (20). The action of any other generator e_{ij} we define,setting

$$e_{ij} = [[[\dots[[e_{i,i+1}, e_{i+1,i+2}], e_{i+2,i+3}], \dots,]e_{j-2,j-1}], e_{j-1,j}]. \tag{32}$$

In the rest of the paper we shall prove that the relations (18)-(20) together with (32) turn $V([M])$ into a gl_∞ module with a highest weight (23). First we introduce some notations.Consider a vector $(m) \in \Gamma([M])$,

$$(m) = \begin{bmatrix} \vdots \\ [m]_{n+2} \\ [m]_{n+1} \\ [m]_n \\ [m]_{n-1} \\ \vdots \\ m_{11} \end{bmatrix}. \tag{33}$$

Divide the pattern (33) into two parts:one infinite pattern

$$(m)^{up(n)} = \begin{bmatrix} \vdots \\ [m]_{n+2} \\ [m]_{n+1} \end{bmatrix}, \tag{34}$$

which will be called the n-upper part of (m) and a finite pattern

$$(m)^{low(n)} = \begin{bmatrix} [m]_n \\ [m]_{n-1} \\ \vdots \\ m_{11} \end{bmatrix}, \tag{35}$$

which is said to be the n-lower part of (m).

Consider the subalgebra

$$gl(n) = \{e_{ij} | i,j = 1, \ldots, n\} \subset gl_\infty. \tag{36}$$

Observation 1 . *Let $e_{ij} \in gl(n) \subset gl_\infty$. Then $e_{ij}(m)$, $(m) \in \Gamma([M])$, is a linear combination of vectors from $\Gamma([M])$ with one and the same n-upper part $(m)^{up(n)}$, which is the n-upper part of (m). More generally, let a be an element from the universal enveloping algebra $U[gl(n)]$ of $gl(n)$. Then $a(m)$ is a linear combination of vector (27) with one and the same n-upper part $(m)^{up(n)}$.*

Let $(m) \in \Gamma([M])$. From (29) it follows that in the case $N = n$ $(m)^{low(n)}$ is one of the patterns (13), which is consistent with the conditions (14). Therefore, we shall consider $(m)^{low(n)}$ as an element from the basis $\Gamma([m]_n)$ of a $gl(n)$ module $V([m]_n)$ with a signature $[m]_n$ (see (11) in the case $N = n$), i.e., it transforms under the action of $gl(n)$ according to the relations (18)-(20).

Observation 2 . *Denote by U the universal enveloping algebra of gl_∞. Let $a \in U[gl(n)] \subset U$ and*

$$(m), (m)_1, \ldots, (m)_q \in \Gamma([M]).$$

Then

$$a(m) = \sum_{i=1}^{q} \alpha_i (m)_i \quad \textit{if and only if} \quad a(m)^{low(n)} = \sum_{i=1}^{q} \alpha_i (m)_i^{low(n)}. \tag{37}$$

Denote by (M) the following vector from the basis $\Gamma([M])$:

$$(M) = (m_{ij}), \text{ where } m_{ij} = M_i \ \forall\, i \le j \in \mathbf{N}. \tag{38}$$

Then for any set of basis vectors $(m)_1, \ldots, (m)_p \in \Gamma([M])$ there exists $N \in \mathbf{N}$ such that

$$(m)_i^{up(N)} = (M)^{up(N)}, \quad i = 1, \ldots, p. \tag{39}$$

Consider any two vectors

$$x = \sum_{i=1}^{p} \alpha_i (m)_i, \quad (m)_i \in \Gamma([M]), \ \alpha_i \in \mathbb{C}, \ i = 1, \ldots, p \tag{40}$$

and

$$y = \sum_{j=1}^{q} \beta_j (\bar{m})_j, \quad (\bar{m})_j \in \Gamma([M]), \ \beta_j \in \mathbb{C}, \ j = 1, \ldots, q. \tag{41}$$

Let N be such that for any $i = 1, \ldots, p$ and $j = 1, \ldots, q$

$$(m)_i^{up(N)} = (\bar{m})_j^{up(N)} = (M)^{up(N)}. \tag{42}$$

Set

$$x^{low(N)} = \sum_{i=1}^{p} \alpha_i (m)_i^{low(N)} \in V([M]_N), \tag{43}$$

$$y^{low(N)} = \sum_{j=1}^{q} \beta_j (\bar{m})_j^{low(N)} \in V([M]_N). \tag{44}$$

Then from Observation 2 one concludes.

Observation 3 . *For every $a \in U[gl(N)] \subset U$ and for any two vectors (40) and (41), for which (42) holds*

$$ax = y \text{ if and only if } ax^{low(N)} = y^{low(N)}. \tag{45}$$

In particular,

$$ax = 0 \text{ if and only if } ax^{low(N)} = 0. \tag{46}$$

Take any two generators $e_{ij}, e_{kl} \in gl_\infty$ and an arbitrary

$$x = \sum_{r=1}^{p} \alpha_r (m)_r, \quad (m)_r \in \Gamma([M]), \ \alpha_r \in \mathbb{C}, \ r = 1, \ldots, p. \tag{47}$$

Choose N to be such that $i, j, k, l \leq N$ and in the same time

$$(m)_r^{up(N)} = (M)^{up(N)} \ \forall r = 1, \ldots, p. \tag{48}$$

Consider

$$x^{low(N)} = \sum_{r=1}^{p} \alpha_r (m)_r^{low(N)} \tag{49}$$

as an element from $V([M]_N)$. Since $V([M]_N)$ is a $gl(N)$ module and $e_{ij}, e_{kl} \in gl_(N)$,

$$(e_{ij}e_{kl} - e_{kl}e_{ij} - \delta_{jk}e_{il} + \delta_{li}e_{kj})x^{low(N)} = 0. \tag{50}$$

Then according to (46)

$$(e_{ij}e_{kl} - e_{kl}e_{ij} - \delta_{jk}e_{il} + \delta_{li}e_{kj})x = 0 \ \forall x \in V([M]). \tag{51}$$

The latter relation indicates that any two generators $e_{ij}, e_{kl} \in gl_\infty$,considered as operators in in $V([M])$, satisfy the relation

$$[e_{ij}, e_{kl}] = \delta_{jk}e_{il} - \delta_{li}e_{kj}. \tag{52}$$

Thus, $V([M])$ is a gl_∞ module.

Consider any two vectors x, y [see (40) and (41)] and choose N to be such that also (42) holds. Since $x^{low(N)}$ and $y^{low(N)}$ [see (43) and (44)] are elements from the irreducible $gl(N)$ module $V([M]_N)$, there exists a polynomial a of the $gl(N)$ generators, $a \in U[gl(N)] \subset U$, such that

$$ax^{low(N)} = y^{low(N)}. \tag{53}$$

Then according to (45)

$$ax = y. \tag{54}$$

Therefore, $V([M])$ is an irreducible gl_∞ module. The relations (18)-(20) yield for (M) [see (38)]

$$e_{ii}(M) = M_i(M) \ \forall\, i \in \mathbf{N}, \tag{55}$$

$$e_{k-1,k}(M) = 0 \ \forall\, k \in \mathbb{Z}_+. \tag{56}$$

From the identity (32) and (56) it follows now that

$$e_{ij}(M) = 0 \ \forall\, i < j \in \mathbf{N}. \tag{57}$$

Hence, (M) is the highest weight vector of $V([M])$.

We collect the results obtained so far in a proposition.

Proposition 1 . *Let*

$$[M] = [M_1, M_2, \ldots, M_j, \ldots] \tag{58}$$

be an infinite sequence of complex numbers, such that

$$M_i - M_j \in \mathbb{Z}_+ \ \ \forall i < j \in \mathbf{N}. \tag{59}$$

Denote by $\Gamma([M])$ *be the set of all schemes*

$$(m) = \begin{bmatrix} M_1, & M_2, & . & M_j, & . \\ \vdots & \vdots & \vdots & \vdots & \vdots \\ m_{1j}, & m_{2j}, & . & m_{jj} & \\ \vdots & \vdots & \vdots & \vdots & \\ m_{12}, & m_{22} & & & \\ m_{11} & & & & \end{bmatrix} = \begin{bmatrix} [M] \\ \vdots \\ [m]_j \\ \vdots \\ [m]_2 \\ m_{11} \end{bmatrix} \tag{60}$$

with the following properties:

(1) For any (m) *there exists a positive integer* $N[(m)] \in \mathbf{N}$*, such that*

$$\forall j > N[(m)] \ \ [m]_j = [M]_j; \tag{61}$$

(2)

$$m_{i,j+1} - m_{ij} \in \mathbb{Z}_+, \ m_{ij} - m_{i+1,j+1} \in \mathbb{Z}_+ \ \ \forall\, i \leq j \in \mathbf{N}. \tag{62}$$

Denote by

$$(m)_{\pm ij} \tag{63}$$

the scheme obtained from (m) by the replacement

$$m_{ij} \to m_{ij} \pm 1. \tag{64}$$

Define

$$V([M]) = \sum_{(m)\in\Gamma([M])} \oplus \mathbb{C}(m) \tag{65}$$

to be the infinite-dimensional linear space with a basis $\Gamma([M])$*. Then the transformations* $(l_{ij} = m_{ij} - i)$

$$e_{kk}(m) = (m_{1k} + \ldots + m_{kk} - m_{1,k-1} - \ldots - m_{k-1,k-1})(m), \tag{66}$$

$$e_{k,k-1}(m) = \sum_{j=1}^{k-1} \left| \frac{\prod_{i=1}^{k}(l_{ik} - l_{j,k-1} + 1)\prod_{i=1}^{k-2}(l_{i,k-2} - l_{j,k-1})}{\prod_{i\neq j=1}^{k-1}(l_{i,k-1} - l_{j,k-1} + 1)(l_{i,k-1} - l_{j,k-1})} \right|^{1/2} (m)_{-j,k-1}, \tag{67}$$

$$e_{k-1,k}(m) = \sum_{j=1}^{k-1} \left| \frac{\prod_{i=1}^{k}(l_{ik} - l_{j,k-1})\prod_{i=1}^{k-2}(l_{i,k-2} - l_{j,k-1} - 1)}{\prod_{i\neq j=1}^{k-1}(l_{i,k-1} - l_{j,k-1})(l_{i,k-1} - l_{j,k-1} - 1)} \right|^{1/2} (m)_{j,k-1}. \tag{68}$$

turn $V([M])$ into an irreducible highest weight module of gl_∞ with a highest weight $[M]$ and a highest weight vector

$$(M) = (m_{ij}), \text{ where } m_{ij} = M_i \ \forall\, i \leq j \in \mathbf{N}. \tag{69}$$

References

[1] See V.G.Kac and A.K.Raina, Bombay Lectures on Highest Weight Representations of Infinite Dimensional Lie algebras (Advanced Series in Mathematical Physics, vol.2 ,World Scientific Publishing Co.Pte.Ltd.) and the references therein.

[2] See, for instance, P.Goddard and D.Olive, Internat.J.Mod.Phys. **A1**,303 (1986) and the references therein.

[3] V.G.Kac and D.H.Peterson, Proc.Nat.Acad.Sci. USA **78**,3308 (1981).

[4] V.G.Kac,Infinite Dimensional Lie Algebras (Birkhäuser Boston, Inc.,1983).

[5] I.M.Gel'fand and M.L.Zetlin, Dokl. Acad. Nauk SSSR **71**,825 (1950)(in Russian); see also G.E.Baird and L.C.Biedenharn, J.Math.Phys.**4**,1449 (1963).

COMPUTERIZED SYMMETRIZATION OF QUANTUM STATES

Michael Ramek

Institut für Physikalische und Theoretische Chemie, Technische Universität Graz, A–8010 Graz, Austria

Bruno Gruber

Department of Physics and Molecular Science Program, Southern Illinois University, Carbondale, IL 62901, USA

Introduction

A package of computer programs was developed, which can be used for the calculation of the bases of irreducible representations of $SU(\ell+1)$ or direct products $SU(\ell+1)\times SU(\ell_1+1)\times SU(\ell_2+1)\times\cdots$, and of the bases of irreducible representations of the maximally embedded subalgebras $SU(\ell'+1)$, $SO(2\ell')$, $SO(2\ell'+1)$, $SP(2\ell')$, $\ell'\leq\ell$ or direct products of these. [1] The representations of $SU(\ell+1)$ (or the direct products $SU(\ell+1)\times SU(\ell_1+1)\times SU(\ell_2+1)\times\cdots$) can be of any symmetry, i. e. the programs are not limited to the completely symmetric case. Beyond obtaining the branching laws for the representations of $SU(\ell+1)$ with respect to the subalgebras listed above, the bases for the irreducible representations of the subalgebras are calculated explicitely in terms of the basis vectors of $SU(\ell+1)$.

The calculations start from the state vector corresponding to the highest weight of an irreducible representation of $SU(\ell+1)$ or any of its subalgebras. Repeated application of shift operators generates all states within this irreducible representation. The initial states of all other irreducible representations of the given subalgebra, which are contained in the given representation of $SU(\ell+1)$, are automatically generated using a precomputed list of all dominant subalgebra weights.

To avoid rounding errors, only integer arithmetic is used in all programs. Linear combination coefficients, and all quantities related with these, are treated in the explicit form $\pm\sqrt{p/q}$, p and q being integers which are stored and manipulated in portions of a few digits in several variables; the programs are therefore not restricted to any machine dependent integer arithmetic limitations.

The program package contains the three programs SPAN, STEP1, and STEP2. SPAN is designed to map an algebra (or algebra product) onto itself (e. g. $SU(4)\times SU(4)\mapsto SU(4)\times SU(4)$), STEP1 is designed to map an algebra (direct product) onto another algebra (direct product) (e. g. $SU(4)\times SU(4)\mapsto SU(4)$), and STEP2 is designed to add one more step in the algebra (direct product) embedding chain (e. g. $SU(4)\times SU(4)\mapsto SU(4)\mapsto SU(2)\times SU(2)$).

The programs are written in PASCAL [2]. Some of the unusual features are described in the following sections; a sample calculation for the algebra chain $SU(4)\times SU(4)\mapsto SU(4)$ is given in an appendix.

Input/Output

To keep the programs completely general, no specific algebra embeddings are coded in the programs. Instead, external files are used to store all relevant embedding data. This makes it necessary to enter the embedding symbol (from which all dimensions, the dominant weight conditions, and the number of operators are deduced), the projection matrix, and the operators in the first program execution referring to a specific embedding.

The programs support interactive use in two ways. First, the embedding definition is guided by an input dialogue; second, those parts of the programs, which do the actual calculation, write all results to external files and give only a few messages to the primary output channel. The programs require the following input/output devices:

SPAN: "INPUT", "OUTPUT", "SPANO", "TEX" , "ARCHIV";
STEP1: "INPUT", "OUTPUT", "SPANO", "OUTF", "TEX" , "ARCHIV";
STEP2: "INPUT", "OUTPUT", "SPANO", "INF", "OUTF", "TEX" , "MEMORY".

The use of these files is as follows:

INPUT and OUTPUT are the usual primary I/O channels; in an interactive execution they correspond to the terminal keyboard and screen. They are used to start the calculation, i. e. to specify the algebra chain, and the initial state.

SPANO is a textfile written by SPAN; it contains all results for later use by the other programs.

OUTF is a textfile written by STEP1 and STEP2; it contains all results for later use by STEP2.

INF is a textfile, which is read by STEP2; it must contain the OUTF output of either STEP1 or STEP2 .

TEX is a textfile which contains all results formatted as an input file for the text processing system TEX [3]. Generally speaking, the textfile TEX contains the same information as the files SPANO or OUTF (which are in readable form, too). However, the TEX output is the recommended final result file and should be printed on a high quality printer.

ARCHIV and MEMORY are the textfiles on which all data defining the algebra chains are stored.

Long Integer Handling

The most important computational feature of the program is the calculation of the linear combination coefficients, and all quantities related with these, in the explicit form $\pm\sqrt{p/q}$, p and q being integers. The advantage of this strategy is, that rounding errors occuring in real number calculations are completely avoided. As a consequence, however, intermediate results easily become larger than the largest integer allowed on any computer.

This problem is solved by using "long integers", which employ a polynomial expansion

$$\sum_{i=0}^{\text{USEDWORD}} c_i \cdot \text{BASIS}^i$$

instead of a single variable. This expansion is defined by the global constants "POWER" and "MAXWORD", and the global variables "BASIS" and "USEDWORD". In detail, BASIS is the basis of the expansion used; it is calculated from POWER as 10^{POWER}. To avoid malfunctions it is a necessary condition, that the square of BASIS is less than the largest integer the computer used can handle. Hence POWER is a machine dependent constant. MAXWORD is the maximum degree of the expansion; no principal upper limit applies for this constant (except the storage capacity of the computer used), but there is a logical lower limit: long integers must be capable of handling larger integer numbers than normal machine supplied integers; MAXWORD must therefore also be considered as machine dependent. USEDWORD is the highest degree of the expansion which ocurred in the calculation so far.

USEDWORD must never exceed MAXWORD; it was introduced to speed up the calculation as much as possible.

The concept of long integer storage and arithmetic is perhaps best demonstrated by an example:

$$12345678901234567890\,/\,2 = 6172839450617283945.$$

Assuming POWER = 4 and MAXWORD $\geq$ 5, the integer 12345678901234567890 is stored and manipulated in portions of four digits:

$$\begin{aligned}12345678901234567890 = 1234 \cdot 10000000000000000 \\ + 5678 \cdot 1000000000000 \\ + 9012 \cdot 100000000 \\ + 3456 \cdot 10000 \\ + 7890 \cdot 1\end{aligned}$$

Division by two is then performed by dividing each of the five portions by two:

$$\begin{aligned}617 \cdot 10000000000000000 \\ + 2839 \cdot 1000000000000 \\ + 4506 \cdot 100000000 \\ + 1728 \cdot 10000 \\ + 3945 \cdot 1 = 6172839450617283945\,.\end{aligned}$$

Output Organization

The result file TEX is organized in the following manner for all programs: each embedding starts on a new page, each representation starts on top of a new column. Page or column breaking is performed under the aspect of keeping most of the data for one state together.

Except for the initial state of a representation, all entries consist of the items LAYER NR, STATE, HISTOGRAM, WEIGHT, HISTORY, and MATRIX ELEMENT. The meaning of these items is:

STATE: the (normalized) linear combination of state vectors produced from a previous state by one of the lowering operators;

MATRIX ELEMENT: the quantity stripped off during the normalization of this state vector linear combination;

WEIGHT: the subalgebra weight corresponding to this state;

HISTORY: the sequence of operator actions necessary to create this state (starting with the initial state of the representation);

HISTOGRAM: the number of individual operator actions necessary to create this state (starting with the initial state of the representation);

LAYER NR: the total number of operator actions necessary to create this state (starting with the initial state of the representation).

References

[1] B. Gruber, M. Lorente, T. Nomura, and M. Ramek, *J. Phys. A*, in press

[2] K. Jensen and N. Wirth, *Pascal User Manual And Report*, Springer, New York, 1985

[3] D.E. Knuth, *The TEXbook*, Addison-Wesley, Amsterdam, 1984

Appendix

The results of a STEP1 calculation for one symmetry of the embedding SU(4) $\times$ SU(4) $\mapsto$ SU(4) for 4 particles are reproduced on the following pages. The corresponding results of SPAN, on which this calculation is based, are not given due to space limitations.

SU(4) × SU(4) ↦ SU(4)

[1, 1, 0, 0; 1, 1, 0, 0] ↦ [2, 2, 0, 0] + [2, 1, 1, 0] + [1, 1, 1, 1]

Representation 1:

Initial state:

$+\sqrt{1/4}\,\varphi^1_1(1)\varphi^1_2(2)\varphi^2_1(1)\varphi^2_2(2)$
$-\sqrt{1/4}\,\varphi^1_2(1)\varphi^1_1(2)\varphi^2_1(1)\varphi^2_2(2)$
$-\sqrt{1/4}\,\varphi^1_1(1)\varphi^1_2(2)\varphi^2_2(1)\varphi^2_1(2)$
$+\sqrt{1/4}\,\varphi^1_2(1)\varphi^1_1(2)\varphi^2_2(1)\varphi^2_1(2)$

Weight: [2, 2, 0, 0]

Layer 1

State:

$+\sqrt{1/8}\,\varphi^1_1(1)\varphi^1_3(2)\varphi^2_1(1)\varphi^2_2(2)$
$-\sqrt{1/8}\,\varphi^1_3(1)\varphi^1_1(2)\varphi^2_1(1)\varphi^2_2(2)$
$-\sqrt{1/8}\,\varphi^1_1(1)\varphi^1_3(2)\varphi^2_2(1)\varphi^2_1(2)$
$+\sqrt{1/8}\,\varphi^1_3(1)\varphi^1_1(2)\varphi^2_2(1)\varphi^2_1(2)$
$+\sqrt{1/8}\,\varphi^1_1(1)\varphi^1_2(2)\varphi^2_1(1)\varphi^2_3(2)$
$-\sqrt{1/8}\,\varphi^1_2(1)\varphi^1_1(2)\varphi^2_1(1)\varphi^2_3(2)$
$-\sqrt{1/8}\,\varphi^1_1(1)\varphi^1_2(2)\varphi^2_3(1)\varphi^2_1(2)$
$+\sqrt{1/8}\,\varphi^1_2(1)\varphi^1_1(2)\varphi^2_3(1)\varphi^2_1(2)$

Histogram: (0 1 0)

Weight: [2, 1, 1, 0]

History: 0 2)

Matrix element: $\sqrt{2}$

Layer 2

State:

$+\sqrt{1/8}\,\varphi^1_2(1)\varphi^1_3(2)\varphi^2_1(1)\varphi^2_2(2)$
$-\sqrt{1/8}\,\varphi^1_3(1)\varphi^1_2(2)\varphi^2_1(1)\varphi^2_2(2)$
$-\sqrt{1/8}\,\varphi^1_2(1)\varphi^1_3(2)\varphi^2_2(1)\varphi^2_1(2)$
$+\sqrt{1/8}\,\varphi^1_3(1)\varphi^1_2(2)\varphi^2_2(1)\varphi^2_1(2)$
$+\sqrt{1/8}\,\varphi^1_1(1)\varphi^1_2(2)\varphi^2_2(1)\varphi^2_3(2)$
$-\sqrt{1/8}\,\varphi^1_2(1)\varphi^1_1(2)\varphi^2_2(1)\varphi^2_3(2)$
$-\sqrt{1/8}\,\varphi^1_1(1)\varphi^1_2(2)\varphi^2_3(1)\varphi^2_2(2)$
$+\sqrt{1/8}\,\varphi^1_2(1)\varphi^1_1(2)\varphi^2_3(1)\varphi^2_2(2)$

Histogram: (1 1 0)

Weight: [1, 2, 1, 0]

History: 0 2 1)

Matrix element: $\sqrt{1}$

Layer 2

State:

$+\sqrt{1/4}\,\varphi^1_1(1)\varphi^1_3(2)\varphi^2_1(1)\varphi^2_3(2)$
$-\sqrt{1/4}\,\varphi^1_3(1)\varphi^1_1(2)\varphi^2_1(1)\varphi^2_3(2)$
$-\sqrt{1/4}\,\varphi^1_1(1)\varphi^1_3(2)\varphi^2_3(1)\varphi^2_1(2)$
$+\sqrt{1/4}\,\varphi^1_3(1)\varphi^1_1(2)\varphi^2_3(1)\varphi^2_1(2)$

Histogram: (0 2 0)

Weight: [2, 0, 2, 0]

History: 0 2 2)

Matrix element: $\sqrt{2}$

Layer 2

State:

$+\sqrt{1/8}\,\varphi^1_1(1)\varphi^1_4(2)\varphi^2_1(1)\varphi^2_2(2)$
$-\sqrt{1/8}\,\varphi^1_4(1)\varphi^1_1(2)\varphi^2_1(1)\varphi^2_2(2)$
$-\sqrt{1/8}\,\varphi^1_1(1)\varphi^1_4(2)\varphi^2_2(1)\varphi^2_1(2)$
$+\sqrt{1/8}\,\varphi^1_4(1)\varphi^1_1(2)\varphi^2_2(1)\varphi^2_1(2)$
$+\sqrt{1/8}\,\varphi^1_1(1)\varphi^1_2(2)\varphi^2_1(1)\varphi^2_4(2)$
$-\sqrt{1/8}\,\varphi^1_2(1)\varphi^1_1(2)\varphi^2_1(1)\varphi^2_4(2)$
$-\sqrt{1/8}\,\varphi^1_1(1)\varphi^1_2(2)\varphi^2_4(1)\varphi^2_1(2)$
$+\sqrt{1/8}\,\varphi^1_2(1)\varphi^1_1(2)\varphi^2_4(1)\varphi^2_1(2)$

Histogram: (0 1 1)

Weight: [2, 1, 0, 1]

History: 0 2 3)

Matrix element: $\sqrt{1}$

Layer 3

State:

$+\sqrt{1/8}\,\varphi^1_1(1)\varphi^1_3(2)\varphi^2_2(1)\varphi^2_3(2)$
$-\sqrt{1/8}\,\varphi^1_3(1)\varphi^1_1(2)\varphi^2_2(1)\varphi^2_3(2)$
$-\sqrt{1/8}\,\varphi^1_1(1)\varphi^1_3(2)\varphi^2_3(1)\varphi^2_2(2)$
$+\sqrt{1/8}\,\varphi^1_3(1)\varphi^1_1(2)\varphi^2_3(1)\varphi^2_2(2)$
$+\sqrt{1/8}\,\varphi^1_2(1)\varphi^1_3(2)\varphi^2_1(1)\varphi^2_3(2)$
$-\sqrt{1/8}\,\varphi^1_3(1)\varphi^1_2(2)\varphi^2_1(1)\varphi^2_3(2)$
$-\sqrt{1/8}\,\varphi^1_2(1)\varphi^1_3(2)\varphi^2_3(1)\varphi^2_1(2)$
$+\sqrt{1/8}\,\varphi^1_3(1)\varphi^1_2(2)\varphi^2_3(1)\varphi^2_1(2)$

Histogram: (1 2 0)

Weight: [1, 1, 2, 0]

History: 0 2 1 2)

Matrix element: $\sqrt{1}$

Layer 3

State:

$+\sqrt{1/8}\,\varphi^1_2(1)\varphi^1_4(2)\varphi^2_1(1)\varphi^2_2(2)$
$-\sqrt{1/8}\,\varphi^1_4(1)\varphi^1_2(2)\varphi^2_1(1)\varphi^2_2(2)$
$-\sqrt{1/8}\,\varphi^1_2(1)\varphi^1_4(2)\varphi^2_2(1)\varphi^2_1(2)$
$+\sqrt{1/8}\,\varphi^1_4(1)\varphi^1_2(2)\varphi^2_2(1)\varphi^2_1(2)$
$+\sqrt{1/8}\,\varphi^1_1(1)\varphi^1_2(2)\varphi^2_2(1)\varphi^2_4(2)$
$-\sqrt{1/8}\,\varphi^1_2(1)\varphi^1_1(2)\varphi^2_2(1)\varphi^2_4(2)$
$-\sqrt{1/8}\,\varphi^1_1(1)\varphi^1_2(2)\varphi^2_4(1)\varphi^2_2(2)$
$+\sqrt{1/8}\,\varphi^1_2(1)\varphi^1_1(2)\varphi^2_4(1)\varphi^2_2(2)$

Histogram: (1 1 1)

Weight: [1, 2, 0, 1]

History: 0 2 1 3)

Matrix element: $\sqrt{1}$

Layer 3

State:

$+\sqrt{1/8}\,\varphi^1_2(1)\varphi^1_3(2)\varphi^2_1(1)\varphi^2_3(2)$
$-\sqrt{1/8}\,\varphi^1_3(1)\varphi^1_2(2)\varphi^2_1(1)\varphi^2_3(2)$
$-\sqrt{1/8}\,\varphi^1_2(1)\varphi^1_3(2)\varphi^2_3(1)\varphi^2_1(2)$
$+\sqrt{1/8}\,\varphi^1_3(1)\varphi^1_2(2)\varphi^2_3(1)\varphi^2_1(2)$
$+\sqrt{1/8}\,\varphi^1_1(1)\varphi^1_3(2)\varphi^2_2(1)\varphi^2_3(2)$
$-\sqrt{1/8}\,\varphi^1_3(1)\varphi^1_1(2)\varphi^2_2(1)\varphi^2_3(2)$
$-\sqrt{1/8}\,\varphi^1_1(1)\varphi^1_3(2)\varphi^2_3(1)\varphi^2_2(2)$
$+\sqrt{1/8}\,\varphi^1_3(1)\varphi^1_1(2)\varphi^2_3(1)\varphi^2_2(2)$

Histogram: (1 2 0)

Weight: [1, 1, 2, 0]

History: 0 2 2 1)

Matrix element: $\sqrt{2}$

THIS STATE IS LINEARLY DEPENDENT!

Layer 3

State:

$+\sqrt{1/8}\,\varphi^1_1(1)\varphi^1_4(2)\varphi^2_1(1)\varphi^2_3(2)$
$-\sqrt{1/8}\,\varphi^1_4(1)\varphi^1_1(2)\varphi^2_1(1)\varphi^2_3(2)$
$-\sqrt{1/8}\,\varphi^1_1(1)\varphi^1_4(2)\varphi^2_3(1)\varphi^2_1(2)$
$+\sqrt{1/8}\,\varphi^1_4(1)\varphi^1_1(2)\varphi^2_3(1)\varphi^2_1(2)$
$+\sqrt{1/8}\,\varphi^1_1(1)\varphi^1_3(2)\varphi^2_1(1)\varphi^2_4(2)$
$-\sqrt{1/8}\,\varphi^1_3(1)\varphi^1_1(2)\varphi^2_1(1)\varphi^2_4(2)$
$-\sqrt{1/8}\,\varphi^1_1(1)\varphi^1_3(2)\varphi^2_4(1)\varphi^2_1(2)$
$+\sqrt{1/8}\,\varphi^1_3(1)\varphi^1_1(2)\varphi^2_4(1)\varphi^2_1(2)$

Histogram: (0 2 1)

Weight: [2, 0, 1, 1]

History: 0 2 2 3)

Matrix element: $\sqrt{2}$

Layer 3

State:

$+\sqrt{1/8}\,\varphi^1_2(1)\varphi^1_4(2)\varphi^2_1(1)\varphi^2_2(2)$
$-\sqrt{1/8}\,\varphi^1_4(1)\varphi^1_2(2)\varphi^2_1(1)\varphi^2_2(2)$
$-\sqrt{1/8}\,\varphi^1_2(1)\varphi^1_4(2)\varphi^2_2(1)\varphi^2_1(2)$
$+\sqrt{1/8}\,\varphi^1_4(1)\varphi^1_2(2)\varphi^2_2(1)\varphi^2_1(2)$
$+\sqrt{1/8}\,\varphi^1_1(1)\varphi^1_2(2)\varphi^2_2(1)\varphi^2_4(2)$
$-\sqrt{1/8}\,\varphi^1_2(1)\varphi^1_1(2)\varphi^2_2(1)\varphi^2_4(2)$
$-\sqrt{1/8}\,\varphi^1_1(1)\varphi^1_2(2)\varphi^2_4(1)\varphi^2_2(2)$
$+\sqrt{1/8}\,\varphi^1_2(1)\varphi^1_1(2)\varphi^2_4(1)\varphi^2_2(2)$

Histogram: (1 1 1)

Weight: [1, 2, 0, 1]

History: 0 2 3 1)

Matrix element: $\sqrt{1}$

THIS STATE IS LINEARLY DEPENDENT!

Layer 3

State:

$+\sqrt{1/8}\,\varphi^1_1(1)\varphi^1_3(2)\varphi^2_1(1)\varphi^2_4(2)$
$-\sqrt{1/8}\,\varphi^1_3(1)\varphi^1_1(2)\varphi^2_1(1)\varphi^2_4(2)$
$-\sqrt{1/8}\,\varphi^1_1(1)\varphi^1_3(2)\varphi^2_4(1)\varphi^2_1(2)$
$+\sqrt{1/8}\,\varphi^1_3(1)\varphi^1_1(2)\varphi^2_4(1)\varphi^2_1(2)$
$+\sqrt{1/8}\,\varphi^1_1(1)\varphi^1_4(2)\varphi^2_1(1)\varphi^2_3(2)$
$-\sqrt{1/8}\,\varphi^1_4(1)\varphi^1_1(2)\varphi^2_1(1)\varphi^2_3(2)$
$-\sqrt{1/8}\,\varphi^1_1(1)\varphi^1_4(2)\varphi^2_3(1)\varphi^2_1(2)$
$+\sqrt{1/8}\,\varphi^1_4(1)\varphi^1_1(2)\varphi^2_3(1)\varphi^2_1(2)$

Histogram: (0 2 1)

Weight: [2, 0, 1, 1]

History: 0 2 3 2)

Matrix element: $\sqrt{1}$

THIS STATE IS LINEARLY DEPENDENT!

Layer 4

State:

$+\sqrt{1/4}\,\varphi^1_2(1)\varphi^1_3(2)\varphi^2_2(1)\varphi^2_3(2)$
$-\sqrt{1/4}\,\varphi^1_3(1)\varphi^1_2(2)\varphi^2_2(1)\varphi^2_3(2)$
$-\sqrt{1/4}\,\varphi^1_2(1)\varphi^1_3(2)\varphi^2_3(1)\varphi^2_2(2)$
$+\sqrt{1/4}\,\varphi^1_3(1)\varphi^1_2(2)\varphi^2_3(1)\varphi^2_2(2)$

Histogram: (2 2 0)

Weight: [0, 2, 2, 0]

History: 0 2 1 2 1)

Matrix element: $\sqrt{2}$

Layer 4

State:

$+\sqrt{1/16}\,\varphi^1_1(1)\varphi^1_4(2)\varphi^2_2(1)\varphi^2_3(2)$
$-\sqrt{1/16}\,\varphi^1_4(1)\varphi^1_1(2)\varphi^2_2(1)\varphi^2_3(2)$
$-\sqrt{1/16}\,\varphi^1_1(1)\varphi^1_4(2)\varphi^2_3(1)\varphi^2_2(2)$
$+\sqrt{1/16}\,\varphi^1_4(1)\varphi^1_1(2)\varphi^2_3(1)\varphi^2_2(2)$
$+\sqrt{1/16}\,\varphi^1_2(1)\varphi^1_4(2)\varphi^2_1(1)\varphi^2_3(2)$
$-\sqrt{1/16}\,\varphi^1_4(1)\varphi^1_2(2)\varphi^2_1(1)\varphi^2_3(2)$
$-\sqrt{1/16}\,\varphi^1_2(1)\varphi^1_4(2)\varphi^2_3(1)\varphi^2_1(2)$
$+\sqrt{1/16}\,\varphi^1_4(1)\varphi^1_2(2)\varphi^2_3(1)\varphi^2_1(2)$
$+\sqrt{1/16}\,\varphi^1_1(1)\varphi^1_3(2)\varphi^2_2(1)\varphi^2_4(2)$
$-\sqrt{1/16}\,\varphi^1_3(1)\varphi^1_1(2)\varphi^2_2(1)\varphi^2_4(2)$
$-\sqrt{1/16}\,\varphi^1_1(1)\varphi^1_3(2)\varphi^2_4(1)\varphi^2_2(2)$
$+\sqrt{1/16}\,\varphi^1_3(1)\varphi^1_1(2)\varphi^2_4(1)\varphi^2_2(2)$
$+\sqrt{1/16}\,\varphi^1_2(1)\varphi^1_3(2)\varphi^2_1(1)\varphi^2_4(2)$
$-\sqrt{1/16}\,\varphi^1_3(1)\varphi^1_2(2)\varphi^2_1(1)\varphi^2_4(2)$
$-\sqrt{1/16}\,\varphi^1_2(1)\varphi^1_3(2)\varphi^2_4(1)\varphi^2_1(2)$
$+\sqrt{1/16}\,\varphi^1_3(1)\varphi^1_2(2)\varphi^2_4(1)\varphi^2_1(2)$

$SU(4) \times SU(4) \mapsto SU(4)$

$[1,1,0,0;1,1,0,0] \mapsto [2,2,0,0] + [2,1,1,0] + [1,1,1,1]$

Histogram: (1 2 1)

Weight: [1, 1, 1, 1]

History: 0 2 1 2 3)

Matrix element: $\sqrt{2}$

Layer 4

State:

$+ \sqrt{1/16}\,\varphi_3^1(1)\varphi_4^1(2)\varphi_1^2(1)\varphi_2^2(2)$
$- \sqrt{1/16}\,\varphi_4^1(1)\varphi_3^1(2)\varphi_1^2(1)\varphi_2^2(2)$
$- \sqrt{1/16}\,\varphi_3^1(1)\varphi_4^1(2)\varphi_2^2(1)\varphi_1^2(2)$
$+ \sqrt{1/16}\,\varphi_4^1(1)\varphi_3^1(2)\varphi_2^2(1)\varphi_1^2(2)$
$+ \sqrt{1/16}\,\varphi_1^1(1)\varphi_3^1(2)\varphi_2^2(1)\varphi_4^2(2)$
$- \sqrt{1/16}\,\varphi_3^1(1)\varphi_1^1(2)\varphi_2^2(1)\varphi_4^2(2)$
$- \sqrt{1/16}\,\varphi_1^1(1)\varphi_3^1(2)\varphi_4^2(1)\varphi_2^2(2)$
$+ \sqrt{1/16}\,\varphi_3^1(1)\varphi_1^1(2)\varphi_4^2(1)\varphi_2^2(2)$
$+ \sqrt{1/16}\,\varphi_2^1(1)\varphi_4^1(2)\varphi_1^2(1)\varphi_3^2(2)$
$- \sqrt{1/16}\,\varphi_4^1(1)\varphi_2^1(2)\varphi_1^2(1)\varphi_3^2(2)$
$- \sqrt{1/16}\,\varphi_2^1(1)\varphi_4^1(2)\varphi_3^2(1)\varphi_1^2(2)$
$+ \sqrt{1/16}\,\varphi_4^1(1)\varphi_2^1(2)\varphi_3^2(1)\varphi_1^2(2)$
$+ \sqrt{1/16}\,\varphi_1^1(1)\varphi_2^1(2)\varphi_3^2(1)\varphi_4^2(2)$
$- \sqrt{1/16}\,\varphi_2^1(1)\varphi_1^1(2)\varphi_3^2(1)\varphi_4^2(2)$
$- \sqrt{1/16}\,\varphi_1^1(1)\varphi_2^1(2)\varphi_4^2(1)\varphi_3^2(2)$
$+ \sqrt{1/16}\,\varphi_2^1(1)\varphi_1^1(2)\varphi_4^2(1)\varphi_3^2(2)$

Histogram: (1 2 1)
Weight: [1, 1, 1, 1]
History: 0 2 1 3 2)
Matrix element: $\sqrt{2}$

ORTHOGONALIZATION OF THIS STATE GIVES:

State:

$+ \sqrt{1/12}\,\varphi_3^1(1)\varphi_4^1(2)\varphi_1^2(1)\varphi_2^2(2)$
$- \sqrt{1/12}\,\varphi_4^1(1)\varphi_3^1(2)\varphi_1^2(1)\varphi_2^2(2)$
$- \sqrt{1/12}\,\varphi_3^1(1)\varphi_4^1(2)\varphi_2^2(1)\varphi_1^2(2)$
$+ \sqrt{1/12}\,\varphi_4^1(1)\varphi_3^1(2)\varphi_2^2(1)\varphi_1^2(2)$
$+ \sqrt{1/48}\,\varphi_1^1(1)\varphi_3^1(2)\varphi_2^2(1)\varphi_4^2(2)$
$- \sqrt{1/48}\,\varphi_3^1(1)\varphi_1^1(2)\varphi_2^2(1)\varphi_4^2(2)$
$- \sqrt{1/48}\,\varphi_1^1(1)\varphi_3^1(2)\varphi_4^2(1)\varphi_2^2(2)$
$+ \sqrt{1/48}\,\varphi_3^1(1)\varphi_1^1(2)\varphi_4^2(1)\varphi_2^2(2)$
$+ \sqrt{1/48}\,\varphi_2^1(1)\varphi_4^1(2)\varphi_1^2(1)\varphi_3^2(2)$
$- \sqrt{1/48}\,\varphi_4^1(1)\varphi_2^1(2)\varphi_1^2(1)\varphi_3^2(2)$
$- \sqrt{1/48}\,\varphi_2^1(1)\varphi_4^1(2)\varphi_3^2(1)\varphi_1^2(2)$
$+ \sqrt{1/48}\,\varphi_4^1(1)\varphi_2^1(2)\varphi_3^2(1)\varphi_1^2(2)$
$+ \sqrt{1/12}\,\varphi_1^1(1)\varphi_2^1(2)\varphi_3^2(1)\varphi_4^2(2)$
$- \sqrt{1/12}\,\varphi_2^1(1)\varphi_1^1(2)\varphi_3^2(1)\varphi_4^2(2)$
$- \sqrt{1/12}\,\varphi_1^1(1)\varphi_2^1(2)\varphi_4^2(1)\varphi_3^2(2)$
$+ \sqrt{1/12}\,\varphi_2^1(1)\varphi_1^1(2)\varphi_4^2(1)\varphi_3^2(2)$
$- \sqrt{1/48}\,\varphi_1^1(1)\varphi_4^1(2)\varphi_2^2(1)\varphi_3^2(2)$
$+ \sqrt{1/48}\,\varphi_4^1(1)\varphi_1^1(2)\varphi_2^2(1)\varphi_3^2(2)$
$+ \sqrt{1/48}\,\varphi_1^1(1)\varphi_4^1(2)\varphi_3^2(1)\varphi_2^2(2)$
$- \sqrt{1/48}\,\varphi_4^1(1)\varphi_1^1(2)\varphi_3^2(1)\varphi_2^2(2)$
$- \sqrt{1/48}\,\varphi_2^1(1)\varphi_3^1(2)\varphi_1^2(1)\varphi_4^2(2)$
$+ \sqrt{1/48}\,\varphi_3^1(1)\varphi_2^1(2)\varphi_1^2(1)\varphi_4^2(2)$
$+ \sqrt{1/48}\,\varphi_2^1(1)\varphi_3^1(2)\varphi_4^2(1)\varphi_1^2(2)$
$- \sqrt{1/48}\,\varphi_3^1(1)\varphi_2^1(2)\varphi_4^2(1)\varphi_1^2(2)$

Histogram: (1 2 1)

Weight: [1, 1, 1, 1]

History: 0 2 1 3 2)

Matrix element: $\sqrt{3/2}$

Layer 4

State:

$+ \sqrt{1/16}\,\varphi_2^1(1)\varphi_4^1(2)\varphi_1^2(1)\varphi_3^2(2)$
$- \sqrt{1/16}\,\varphi_4^1(1)\varphi_2^1(2)\varphi_1^2(1)\varphi_3^2(2)$
$- \sqrt{1/16}\,\varphi_2^1(1)\varphi_4^1(2)\varphi_3^2(1)\varphi_1^2(2)$
$+ \sqrt{1/16}\,\varphi_4^1(1)\varphi_2^1(2)\varphi_3^2(1)\varphi_1^2(2)$
$+ \sqrt{1/16}\,\varphi_2^1(1)\varphi_3^1(2)\varphi_1^2(1)\varphi_4^2(2)$
$- \sqrt{1/16}\,\varphi_3^1(1)\varphi_2^1(2)\varphi_1^2(1)\varphi_4^2(2)$
$- \sqrt{1/16}\,\varphi_2^1(1)\varphi_3^1(2)\varphi_4^2(1)\varphi_1^2(2)$
$+ \sqrt{1/16}\,\varphi_3^1(1)\varphi_2^1(2)\varphi_4^2(1)\varphi_1^2(2)$
$+ \sqrt{1/16}\,\varphi_1^1(1)\varphi_4^1(2)\varphi_2^2(1)\varphi_3^2(2)$
$- \sqrt{1/16}\,\varphi_4^1(1)\varphi_1^1(2)\varphi_2^2(1)\varphi_3^2(2)$
$- \sqrt{1/16}\,\varphi_1^1(1)\varphi_4^1(2)\varphi_3^2(1)\varphi_2^2(2)$
$+ \sqrt{1/16}\,\varphi_4^1(1)\varphi_1^1(2)\varphi_3^2(1)\varphi_2^2(2)$
$+ \sqrt{1/16}\,\varphi_1^1(1)\varphi_3^1(2)\varphi_2^2(1)\varphi_4^2(2)$
$- \sqrt{1/16}\,\varphi_3^1(1)\varphi_1^1(2)\varphi_2^2(1)\varphi_4^2(2)$
$- \sqrt{1/16}\,\varphi_1^1(1)\varphi_3^1(2)\varphi_4^2(1)\varphi_2^2(2)$
$+ \sqrt{1/16}\,\varphi_3^1(1)\varphi_1^1(2)\varphi_4^2(1)\varphi_2^2(2)$

Histogram: (1 2 1)

Weight: [1, 1, 1, 1]

History: 0 2 2 3 1)

Matrix element: $\sqrt{2}$

THIS STATE IS LINEARLY DEPENDENT!

Layer 4
State:

$+ \sqrt{1/4}\,\varphi_1^1(1)\varphi_4^1(2)\varphi_1^2(1)\varphi_4^2(2)$
$- \sqrt{1/4}\,\varphi_4^1(1)\varphi_1^1(2)\varphi_1^2(1)\varphi_4^2(2)$
$- \sqrt{1/4}\,\varphi_1^1(1)\varphi_4^1(2)\varphi_4^2(1)\varphi_1^2(2)$
$+ \sqrt{1/4}\,\varphi_4^1(1)\varphi_1^1(2)\varphi_4^2(1)\varphi_1^2(2)$

Histogram: (0 2 2)
Weight: [2, 0, 0, 2]
History: 0 2 2 3 3)
Matrix element: $\sqrt{2}$

Layer 5
State:

$+ \sqrt{1/8}\,\varphi_2^1(1)\varphi_4^1(2)\varphi_2^2(1)\varphi_3^2(2)$
$- \sqrt{1/8}\,\varphi_4^1(1)\varphi_2^1(2)\varphi_2^2(1)\varphi_3^2(2)$
$- \sqrt{1/8}\,\varphi_2^1(1)\varphi_4^1(2)\varphi_3^2(1)\varphi_2^2(2)$
$+ \sqrt{1/8}\,\varphi_4^1(1)\varphi_2^1(2)\varphi_3^2(1)\varphi_2^2(2)$
$+ \sqrt{1/8}\,\varphi_2^1(1)\varphi_3^1(2)\varphi_2^2(1)\varphi_4^2(2)$
$- \sqrt{1/8}\,\varphi_3^1(1)\varphi_2^1(2)\varphi_2^2(1)\varphi_4^2(2)$
$- \sqrt{1/8}\,\varphi_2^1(1)\varphi_3^1(2)\varphi_4^2(1)\varphi_2^2(2)$
$+ \sqrt{1/8}\,\varphi_3^1(1)\varphi_2^1(2)\varphi_4^2(1)\varphi_2^2(2)$

Histogram: (2 2 1)
Weight: [0, 2, 1, 1]
History: 0 2 1 2 1 3)
Matrix element: $\sqrt{2}$

Layer 5
State:

$+ \sqrt{1/8}\,\varphi_2^1(1)\varphi_4^1(2)\varphi_2^2(1)\varphi_3^2(2)$
$- \sqrt{1/8}\,\varphi_4^1(1)\varphi_2^1(2)\varphi_2^2(1)\varphi_3^2(2)$
$- \sqrt{1/8}\,\varphi_2^1(1)\varphi_4^1(2)\varphi_3^2(1)\varphi_2^2(2)$
$+ \sqrt{1/8}\,\varphi_4^1(1)\varphi_2^1(2)\varphi_3^2(1)\varphi_2^2(2)$
$+ \sqrt{1/8}\,\varphi_2^1(1)\varphi_3^1(2)\varphi_2^2(1)\varphi_4^2(2)$
$- \sqrt{1/8}\,\varphi_3^1(1)\varphi_2^1(2)\varphi_2^2(1)\varphi_4^2(2)$
$- \sqrt{1/8}\,\varphi_2^1(1)\varphi_3^1(2)\varphi_4^2(1)\varphi_2^2(2)$
$+ \sqrt{1/8}\,\varphi_3^1(1)\varphi_2^1(2)\varphi_4^2(1)\varphi_2^2(2)$

Histogram: (2 2 1)
Weight: [0, 2, 1, 1]
History: 0 2 1 2 3 1)
Matrix element: $\sqrt{2}$

THIS STATE IS LINEARLY DEPENDENT!

Layer 5

State:

$+ \sqrt{1/8}\,\varphi_3^1(1)\varphi_4^1(2)\varphi_1^2(1)\varphi_3^2(2)$
$- \sqrt{1/8}\,\varphi_4^1(1)\varphi_3^1(2)\varphi_1^2(1)\varphi_3^2(2)$
$- \sqrt{1/8}\,\varphi_3^1(1)\varphi_4^1(2)\varphi_3^2(1)\varphi_1^2(2)$
$+ \sqrt{1/8}\,\varphi_4^1(1)\varphi_3^1(2)\varphi_3^2(1)\varphi_1^2(2)$
$+ \sqrt{1/8}\,\varphi_1^1(1)\varphi_3^1(2)\varphi_3^2(1)\varphi_4^2(2)$
$- \sqrt{1/8}\,\varphi_3^1(1)\varphi_1^1(2)\varphi_3^2(1)\varphi_4^2(2)$
$- \sqrt{1/8}\,\varphi_1^1(1)\varphi_3^1(2)\varphi_4^2(1)\varphi_3^2(2)$
$+ \sqrt{1/8}\,\varphi_3^1(1)\varphi_1^1(2)\varphi_4^2(1)\varphi_3^2(2)$

Histogram: (1 3 1)

Weight: [1, 0, 2, 1]

History: 0 2 1 2 3 2)

Matrix element: $\sqrt{1/2}$

Layer 5

State:

$+ \sqrt{1/8}\,\varphi_1^1(1)\varphi_4^1(2)\varphi_2^2(1)\varphi_4^2(2)$
$- \sqrt{1/8}\,\varphi_4^1(1)\varphi_1^1(2)\varphi_2^2(1)\varphi_4^2(2)$
$- \sqrt{1/8}\,\varphi_1^1(1)\varphi_4^1(2)\varphi_4^2(1)\varphi_2^2(2)$
$+ \sqrt{1/8}\,\varphi_4^1(1)\varphi_1^1(2)\varphi_4^2(1)\varphi_2^2(2)$
$+ \sqrt{1/8}\,\varphi_2^1(1)\varphi_4^1(2)\varphi_1^2(1)\varphi_4^2(2)$
$- \sqrt{1/8}\,\varphi_4^1(1)\varphi_2^1(2)\varphi_1^2(1)\varphi_4^2(2)$
$- \sqrt{1/8}\,\varphi_2^1(1)\varphi_4^1(2)\varphi_4^2(1)\varphi_1^2(2)$
$+ \sqrt{1/8}\,\varphi_4^1(1)\varphi_2^1(2)\varphi_4^2(1)\varphi_1^2(2)$

Histogram: (1 2 2)

Weight: [1, 1, 0, 2]

History: 0 2 1 2 3 3)

Matrix element: $\sqrt{2}$

Layer 5

State:

$+ \sqrt{1/8}\,\varphi_3^1(1)\varphi_4^1(2)\varphi_1^2(1)\varphi_3^2(2)$
$- \sqrt{1/8}\,\varphi_4^1(1)\varphi_3^1(2)\varphi_1^2(1)\varphi_3^2(2)$
$- \sqrt{1/8}\,\varphi_3^1(1)\varphi_4^1(2)\varphi_3^2(1)\varphi_1^2(2)$
$+ \sqrt{1/8}\,\varphi_4^1(1)\varphi_3^1(2)\varphi_3^2(1)\varphi_1^2(2)$
$+ \sqrt{1/8}\,\varphi_1^1(1)\varphi_3^1(2)\varphi_3^2(1)\varphi_4^2(2)$
$- \sqrt{1/8}\,\varphi_3^1(1)\varphi_1^1(2)\varphi_3^2(1)\varphi_4^2(2)$
$- \sqrt{1/8}\,\varphi_1^1(1)\varphi_3^1(2)\varphi_4^2(1)\varphi_3^2(2)$
$+ \sqrt{1/8}\,\varphi_3^1(1)\varphi_1^1(2)\varphi_4^2(1)\varphi_3^2(2)$

Histogram: (1 3 1)

Weight: [1, 0, 2, 1]

History: 0 2 1 3 2 2)

Matrix element: $\sqrt{3/2}$

THIS STATE IS LINEARLY DEPENDENT!

Layer 5

State:

$+ \sqrt{1/8}\,\varphi_2^1(1)\varphi_4^1(2)\varphi_1^2(1)\varphi_4^2(2)$
$- \sqrt{1/8}\,\varphi_4^1(1)\varphi_2^1(2)\varphi_1^2(1)\varphi_4^2(2)$
$- \sqrt{1/8}\,\varphi_2^1(1)\varphi_4^1(2)\varphi_4^2(1)\varphi_1^2(2)$
$+ \sqrt{1/8}\,\varphi_4^1(1)\varphi_2^1(2)\varphi_4^2(1)\varphi_1^2(2)$
$+ \sqrt{1/8}\,\varphi_1^1(1)\varphi_4^1(2)\varphi_2^2(1)\varphi_4^2(2)$
$- \sqrt{1/8}\,\varphi_4^1(1)\varphi_1^1(2)\varphi_2^2(1)\varphi_4^2(2)$
$- \sqrt{1/8}\,\varphi_1^1(1)\varphi_4^1(2)\varphi_4^2(1)\varphi_2^2(2)$
$+ \sqrt{1/8}\,\varphi_4^1(1)\varphi_1^1(2)\varphi_4^2(1)\varphi_2^2(2)$

Histogram: (1 2 2)

Weight: [1, 1, 0, 2]

History: 0 2 2 3 3 1)

Matrix element: $\sqrt{2}$

THIS STATE IS LINEARLY DEPENDENT!

Layer 6
State:
$+ \sqrt{1/8}\,\varphi_3^1(1)\varphi_4^1(2)\varphi_2^2(1)\varphi_3^2(2)$
$- \sqrt{1/8}\,\varphi_4^1(1)\varphi_3^1(2)\varphi_2^2(1)\varphi_3^2(2)$
$- \sqrt{1/8}\,\varphi_3^1(1)\varphi_4^1(2)\varphi_3^2(1)\varphi_2^2(2)$
$+ \sqrt{1/8}\,\varphi_4^1(1)\varphi_3^1(2)\varphi_3^2(1)\varphi_2^2(2)$
$+ \sqrt{1/8}\,\varphi_2^1(1)\varphi_3^1(2)\varphi_3^2(1)\varphi_4^2(2)$
$- \sqrt{1/8}\,\varphi_3^1(1)\varphi_2^1(2)\varphi_3^2(1)\varphi_4^2(2)$
$- \sqrt{1/8}\,\varphi_2^1(1)\varphi_3^1(2)\varphi_4^2(1)\varphi_3^2(2)$
$+ \sqrt{1/8}\,\varphi_3^1(1)\varphi_2^1(2)\varphi_4^2(1)\varphi_3^2(2)$
Histogram: (2 3 1)
Weight: [0, 1, 2, 1]
History: 0 2 1 2 1 3 2)
Matrix element: $\sqrt{1}$

Layer 6
State:
$+ \sqrt{1/4}\,\varphi_2^1(1)\varphi_4^1(2)\varphi_2^2(1)\varphi_4^2(2)$
$- \sqrt{1/4}\,\varphi_4^1(1)\varphi_2^1(2)\varphi_2^2(1)\varphi_4^2(2)$
$- \sqrt{1/4}\,\varphi_2^1(1)\varphi_4^1(2)\varphi_4^2(1)\varphi_2^2(2)$
$+ \sqrt{1/4}\,\varphi_4^1(1)\varphi_2^1(2)\varphi_4^2(1)\varphi_2^2(2)$
Histogram: (2 2 2)
Weight: [0, 2, 0, 2]
History: 0 2 1 2 1 3 3)
Matrix element: $\sqrt{2}$

Layer 6
State:
$+ \sqrt{1/8}\,\varphi_2^1(1)\varphi_3^1(2)\varphi_3^2(1)\varphi_4^2(2)$
$- \sqrt{1/8}\,\varphi_3^1(1)\varphi_2^1(2)\varphi_3^2(1)\varphi_4^2(2)$
$- \sqrt{1/8}\,\varphi_2^1(1)\varphi_3^1(2)\varphi_4^2(1)\varphi_3^2(2)$
$+ \sqrt{1/8}\,\varphi_3^1(1)\varphi_2^1(2)\varphi_4^2(1)\varphi_3^2(2)$
$+ \sqrt{1/8}\,\varphi_3^1(1)\varphi_4^1(2)\varphi_2^2(1)\varphi_3^2(2)$
$- \sqrt{1/8}\,\varphi_4^1(1)\varphi_3^1(2)\varphi_2^2(1)\varphi_3^2(2)$
$- \sqrt{1/8}\,\varphi_3^1(1)\varphi_4^1(2)\varphi_3^2(1)\varphi_2^2(2)$
$+ \sqrt{1/8}\,\varphi_4^1(1)\varphi_3^1(2)\varphi_3^2(1)\varphi_2^2(2)$
Histogram: (2 3 1)
Weight: [0, 1, 2, 1]
History: 0 2 1 2 3 2 1)
Matrix element: $\sqrt{1}$

THIS STATE IS LINEARLY DEPENDENT!

Layer 6
State:
$+ \sqrt{1/8}\,\varphi_1^1(1)\varphi_4^1(2)\varphi_3^2(1)\varphi_4^2(2)$
$- \sqrt{1/8}\,\varphi_4^1(1)\varphi_1^1(2)\varphi_3^2(1)\varphi_4^2(2)$
$- \sqrt{1/8}\,\varphi_1^1(1)\varphi_4^1(2)\varphi_4^2(1)\varphi_3^2(2)$
$+ \sqrt{1/8}\,\varphi_4^1(1)\varphi_1^1(2)\varphi_4^2(1)\varphi_3^2(2)$
$+ \sqrt{1/8}\,\varphi_3^1(1)\varphi_4^1(2)\varphi_1^2(1)\varphi_4^2(2)$
$- \sqrt{1/8}\,\varphi_4^1(1)\varphi_3^1(2)\varphi_1^2(1)\varphi_4^2(2)$
$- \sqrt{1/8}\,\varphi_3^1(1)\varphi_4^1(2)\varphi_4^2(1)\varphi_1^2(2)$
$+ \sqrt{1/8}\,\varphi_4^1(1)\varphi_3^1(2)\varphi_4^2(1)\varphi_1^2(2)$
Histogram: (1 3 2)
Weight: [1, 0, 1, 2]
History: 0 2 1 2 3 2 3)
Matrix element: $\sqrt{1}$

Layer 6
State:
$+ \sqrt{1/4}\,\varphi_2^1(1)\varphi_4^1(2)\varphi_2^2(1)\varphi_4^2(2)$
$- \sqrt{1/4}\,\varphi_4^1(1)\varphi_2^1(2)\varphi_2^2(1)\varphi_4^2(2)$
$- \sqrt{1/4}\,\varphi_2^1(1)\varphi_4^1(2)\varphi_4^2(1)\varphi_2^2(2)$
$+ \sqrt{1/4}\,\varphi_4^1(1)\varphi_2^1(2)\varphi_4^2(1)\varphi_2^2(2)$
Histogram: (2 2 2)
Weight: [0, 2, 0, 2]
History: 0 2 1 2 3 3 1)
Matrix element: $\sqrt{2}$

THIS STATE IS LINEARLY DEPENDENT!

Layer 6
State:
$+ \sqrt{1/8}\,\varphi_3^1(1)\varphi_4^1(2)\varphi_1^2(1)\varphi_4^2(2)$
$- \sqrt{1/8}\,\varphi_4^1(1)\varphi_3^1(2)\varphi_1^2(1)\varphi_4^2(2)$
$- \sqrt{1/8}\,\varphi_3^1(1)\varphi_4^1(2)\varphi_4^2(1)\varphi_1^2(2)$
$+ \sqrt{1/8}\,\varphi_4^1(1)\varphi_3^1(2)\varphi_4^2(1)\varphi_1^2(2)$
$+ \sqrt{1/8}\,\varphi_1^1(1)\varphi_4^1(2)\varphi_3^2(1)\varphi_4^2(2)$
$- \sqrt{1/8}\,\varphi_4^1(1)\varphi_1^1(2)\varphi_3^2(1)\varphi_4^2(2)$
$- \sqrt{1/8}\,\varphi_1^1(1)\varphi_4^1(2)\varphi_4^2(1)\varphi_3^2(2)$
$+ \sqrt{1/8}\,\varphi_4^1(1)\varphi_1^1(2)\varphi_4^2(1)\varphi_3^2(2)$
Histogram: (1 3 2)
Weight: [1, 0, 1, 2]
History: 0 2 1 2 3 3 2)
Matrix element: $\sqrt{1}$

THIS STATE IS LINEARLY DEPENDENT!

Layer 7

State:

$+ \sqrt{1/8}\,\varphi_2^1(1)\varphi_4^1(2)\varphi_3^2(1)\varphi_4^2(2)$
$- \sqrt{1/8}\,\varphi_4^1(1)\varphi_2^1(2)\varphi_3^2(1)\varphi_4^2(2)$
$- \sqrt{1/8}\,\varphi_2^1(1)\varphi_4^1(2)\varphi_4^2(1)\varphi_3^2(2)$
$+ \sqrt{1/8}\,\varphi_4^1(1)\varphi_2^1(2)\varphi_4^2(1)\varphi_3^2(2)$
$+ \sqrt{1/8}\,\varphi_3^1(1)\varphi_4^1(2)\varphi_2^2(1)\varphi_4^2(2)$
$- \sqrt{1/8}\,\varphi_4^1(1)\varphi_3^1(2)\varphi_2^2(1)\varphi_4^2(2)$
$- \sqrt{1/8}\,\varphi_3^1(1)\varphi_4^1(2)\varphi_4^2(1)\varphi_2^2(2)$
$+ \sqrt{1/8}\,\varphi_4^1(1)\varphi_3^1(2)\varphi_4^2(1)\varphi_2^2(2)$

Histogram: (2 3 2)

Weight: [0, 1, 1, 2]

History: 0 2 1 2 1 3 2 3)

Matrix element: $\sqrt{1}$

Layer 7

State:

$+ \sqrt{1/8}\,\varphi_3^1(1)\varphi_4^1(2)\varphi_2^2(1)\varphi_4^2(2)$
$- \sqrt{1/8}\,\varphi_4^1(1)\varphi_3^1(2)\varphi_2^2(1)\varphi_4^2(2)$
$- \sqrt{1/8}\,\varphi_3^1(1)\varphi_4^1(2)\varphi_4^2(1)\varphi_2^2(2)$
$+ \sqrt{1/8}\,\varphi_4^1(1)\varphi_3^1(2)\varphi_4^2(1)\varphi_2^2(2)$
$+ \sqrt{1/8}\,\varphi_2^1(1)\varphi_4^1(2)\varphi_3^2(1)\varphi_4^2(2)$
$- \sqrt{1/8}\,\varphi_4^1(1)\varphi_2^1(2)\varphi_3^2(1)\varphi_4^2(2)$
$- \sqrt{1/8}\,\varphi_2^1(1)\varphi_4^1(2)\varphi_4^2(1)\varphi_3^2(2)$
$+ \sqrt{1/8}\,\varphi_4^1(1)\varphi_2^1(2)\varphi_4^2(1)\varphi_3^2(2)$

Histogram: (2 3 2)

Weight: [0, 1, 1, 2]

History: 0 2 1 2 1 3 3 2)

Matrix element: $\sqrt{2}$

THIS STATE IS LINEARLY DEPENDENT!

Layer 7
State:
$+ \sqrt{1/8}\,\varphi_2^1(1)\varphi_4^1(2)\varphi_3^2(1)\varphi_4^2(2)$
$- \sqrt{1/8}\,\varphi_4^1(1)\varphi_2^1(2)\varphi_3^2(1)\varphi_4^2(2)$
$- \sqrt{1/8}\,\varphi_2^1(1)\varphi_4^1(2)\varphi_4^2(1)\varphi_3^2(2)$
$+ \sqrt{1/8}\,\varphi_4^1(1)\varphi_2^1(2)\varphi_4^2(1)\varphi_3^2(2)$
$+ \sqrt{1/8}\,\varphi_3^1(1)\varphi_4^1(2)\varphi_2^2(1)\varphi_4^2(2)$
$- \sqrt{1/8}\,\varphi_4^1(1)\varphi_3^1(2)\varphi_2^2(1)\varphi_4^2(2)$
$- \sqrt{1/8}\,\varphi_3^1(1)\varphi_4^1(2)\varphi_4^2(1)\varphi_2^2(2)$
$+ \sqrt{1/8}\,\varphi_4^1(1)\varphi_3^1(2)\varphi_4^2(1)\varphi_2^2(2)$
Histogram: (2 3 2)
Weight: [0, 1, 1, 2]
History: 0 2 1 2 3 2 3 1)
Matrix element: $\sqrt{1}$

THIS STATE IS LINEARLY DEPENDENT!

Layer 8
State:
$+ \sqrt{1/4}\,\varphi_3^1(1)\varphi_4^1(2)\varphi_3^2(1)\varphi_4^2(2)$
$- \sqrt{1/4}\,\varphi_4^1(1)\varphi_3^1(2)\varphi_3^2(1)\varphi_4^2(2)$
$- \sqrt{1/4}\,\varphi_3^1(1)\varphi_4^1(2)\varphi_4^2(1)\varphi_3^2(2)$
$+ \sqrt{1/4}\,\varphi_4^1(1)\varphi_3^1(2)\varphi_4^2(1)\varphi_3^2(2)$
Histogram: (2 4 2)
Weight: [0, 0, 2, 2]
History: 0 2 1 2 1 3 2 3 2)
Matrix element: $\sqrt{2}$

---

20 linear independent states found in this representation.

Representation 2:

Initial state:
$+ \sqrt{1/8}\,\varphi_1^1(1)\varphi_3^1(2)\varphi_1^2(1)\varphi_2^2(2)$
$- \sqrt{1/8}\,\varphi_3^1(1)\varphi_1^1(2)\varphi_1^2(1)\varphi_2^2(2)$
$- \sqrt{1/8}\,\varphi_1^1(1)\varphi_3^1(2)\varphi_2^2(1)\varphi_1^2(2)$
$+ \sqrt{1/8}\,\varphi_3^1(1)\varphi_1^1(2)\varphi_2^2(1)\varphi_1^2(2)$
$- \sqrt{1/8}\,\varphi_1^1(1)\varphi_2^1(2)\varphi_1^2(1)\varphi_3^2(2)$
$+ \sqrt{1/8}\,\varphi_2^1(1)\varphi_1^1(2)\varphi_1^2(1)\varphi_3^2(2)$
$+ \sqrt{1/8}\,\varphi_1^1(1)\varphi_2^1(2)\varphi_3^2(1)\varphi_1^2(2)$
$- \sqrt{1/8}\,\varphi_2^1(1)\varphi_1^1(2)\varphi_3^2(1)\varphi_1^2(2)$
Weight: [2, 1, 1, 0]

Layer 1
State:
$+ \sqrt{1/8}\,\varphi_2^1(1)\varphi_3^1(2)\varphi_1^2(1)\varphi_2^2(2)$
$- \sqrt{1/8}\,\varphi_3^1(1)\varphi_2^1(2)\varphi_1^2(1)\varphi_2^2(2)$
$- \sqrt{1/8}\,\varphi_2^1(1)\varphi_3^1(2)\varphi_2^2(1)\varphi_1^2(2)$
$+ \sqrt{1/8}\,\varphi_3^1(1)\varphi_2^1(2)\varphi_2^2(1)\varphi_1^2(2)$
$- \sqrt{1/8}\,\varphi_1^1(1)\varphi_2^1(2)\varphi_2^2(1)\varphi_3^2(2)$
$+ \sqrt{1/8}\,\varphi_2^1(1)\varphi_1^1(2)\varphi_2^2(1)\varphi_3^2(2)$
$+ \sqrt{1/8}\,\varphi_1^1(1)\varphi_2^1(2)\varphi_3^2(1)\varphi_2^2(2)$
$- \sqrt{1/8}\,\varphi_2^1(1)\varphi_1^1(2)\varphi_3^2(1)\varphi_2^2(2)$
Histogram: (1 0 0)
Weight: [1, 2, 1, 0]
History: 0 1)
Matrix element: $\sqrt{1}$

Layer 1
State:
$+ \sqrt{1/8}\,\varphi_1^1(1)\varphi_4^1(2)\varphi_1^2(1)\varphi_2^2(2)$
$- \sqrt{1/8}\,\varphi_4^1(1)\varphi_1^1(2)\varphi_1^2(1)\varphi_2^2(2)$
$- \sqrt{1/8}\,\varphi_1^1(1)\varphi_4^1(2)\varphi_2^2(1)\varphi_1^2(2)$
$+ \sqrt{1/8}\,\varphi_4^1(1)\varphi_1^1(2)\varphi_2^2(1)\varphi_1^2(2)$
$- \sqrt{1/8}\,\varphi_1^1(1)\varphi_2^1(2)\varphi_1^2(1)\varphi_4^2(2)$
$+ \sqrt{1/8}\,\varphi_2^1(1)\varphi_1^1(2)\varphi_1^2(1)\varphi_4^2(2)$
$+ \sqrt{1/8}\,\varphi_1^1(1)\varphi_2^1(2)\varphi_4^2(1)\varphi_1^2(2)$
$- \sqrt{1/8}\,\varphi_2^1(1)\varphi_1^1(2)\varphi_4^2(1)\varphi_1^2(2)$
Histogram: (0 0 1)
Weight: [2, 1, 0, 1]
History: 0 3)
Matrix element: $\sqrt{1}$

Layer 2

State:

$- \sqrt{1/8}\,\varphi_1^1(1)\varphi_3^1(2)\varphi_2^2(1)\varphi_3^2(2)$
$+ \sqrt{1/8}\,\varphi_3^1(1)\varphi_1^1(2)\varphi_2^2(1)\varphi_3^2(2)$
$+ \sqrt{1/8}\,\varphi_1^1(1)\varphi_3^1(2)\varphi_3^2(1)\varphi_2^2(2)$
$- \sqrt{1/8}\,\varphi_3^1(1)\varphi_1^1(2)\varphi_3^2(1)\varphi_2^2(2)$
$+ \sqrt{1/8}\,\varphi_2^1(1)\varphi_3^1(2)\varphi_1^2(1)\varphi_3^2(2)$
$- \sqrt{1/8}\,\varphi_3^1(1)\varphi_2^1(2)\varphi_1^2(1)\varphi_3^2(2)$
$- \sqrt{1/8}\,\varphi_2^1(1)\varphi_3^1(2)\varphi_3^2(1)\varphi_1^2(2)$
$+ \sqrt{1/8}\,\varphi_3^1(1)\varphi_2^1(2)\varphi_3^2(1)\varphi_1^2(2)$

Histogram: (1 1 0)

Weight: [1, 1, 2, 0]

History: 0 1 2)

Matrix element: $\sqrt{1}$

$SU(4) \times SU(4) \mapsto SU(4)$

$[1,1,0,0;1,1,0,0] \mapsto [2,2,0,0]+[2,1,1,0]+[1,1,1,1]$

Layer 2

State:

$+\sqrt{1/8}\,\varphi_2^1(1)\varphi_4^1(2)\varphi_1^2(1)\varphi_2^2(2)$
$-\sqrt{1/8}\,\varphi_4^1(1)\varphi_2^1(2)\varphi_1^2(1)\varphi_2^2(2)$
$-\sqrt{1/8}\,\varphi_2^1(1)\varphi_4^1(2)\varphi_2^2(1)\varphi_1^2(2)$
$+\sqrt{1/8}\,\varphi_4^1(1)\varphi_2^1(2)\varphi_2^2(1)\varphi_1^2(2)$
$-\sqrt{1/8}\,\varphi_1^1(1)\varphi_2^1(2)\varphi_2^2(1)\varphi_4^2(2)$
$+\sqrt{1/8}\,\varphi_2^1(1)\varphi_1^1(2)\varphi_2^2(1)\varphi_4^2(2)$
$+\sqrt{1/8}\,\varphi_1^1(1)\varphi_2^1(2)\varphi_4^2(1)\varphi_2^2(2)$
$-\sqrt{1/8}\,\varphi_2^1(1)\varphi_1^1(2)\varphi_4^2(1)\varphi_2^2(2)$

Histogram: (1 0 1)

Weight: [1, 2, 0, 1]

History: 0 1 3)

Matrix element: $\sqrt{1}$

Layer 2

State:

$+\sqrt{1/8}\,\varphi_2^1(1)\varphi_4^1(2)\varphi_1^2(1)\varphi_2^2(2)$
$-\sqrt{1/8}\,\varphi_4^1(1)\varphi_2^1(2)\varphi_1^2(1)\varphi_2^2(2)$
$-\sqrt{1/8}\,\varphi_2^1(1)\varphi_4^1(2)\varphi_2^2(1)\varphi_1^2(2)$
$+\sqrt{1/8}\,\varphi_4^1(1)\varphi_2^1(2)\varphi_2^2(1)\varphi_1^2(2)$
$-\sqrt{1/8}\,\varphi_1^1(1)\varphi_2^1(2)\varphi_2^2(1)\varphi_4^2(2)$
$+\sqrt{1/8}\,\varphi_2^1(1)\varphi_1^1(2)\varphi_2^2(1)\varphi_4^2(2)$
$+\sqrt{1/8}\,\varphi_1^1(1)\varphi_2^1(2)\varphi_4^2(1)\varphi_2^2(2)$
$-\sqrt{1/8}\,\varphi_2^1(1)\varphi_1^1(2)\varphi_4^2(1)\varphi_2^2(2)$

Histogram: (1 0 1)

Weight: [1, 2, 0, 1]

History: 0 3 1)

Matrix element: $\sqrt{1}$

THIS STATE IS LINEARLY DEPENDENT!

Layer 2

State:

$-\sqrt{1/8}\,\varphi_1^1(1)\varphi_3^1(2)\varphi_1^2(1)\varphi_4^2(2)$
$+\sqrt{1/8}\,\varphi_3^1(1)\varphi_1^1(2)\varphi_1^2(1)\varphi_4^2(2)$
$+\sqrt{1/8}\,\varphi_1^1(1)\varphi_3^1(2)\varphi_4^2(1)\varphi_1^2(2)$
$-\sqrt{1/8}\,\varphi_3^1(1)\varphi_1^1(2)\varphi_4^2(1)\varphi_1^2(2)$
$+\sqrt{1/8}\,\varphi_1^1(1)\varphi_4^1(2)\varphi_1^2(1)\varphi_3^2(2)$
$-\sqrt{1/8}\,\varphi_4^1(1)\varphi_1^1(2)\varphi_1^2(1)\varphi_3^2(2)$
$-\sqrt{1/8}\,\varphi_1^1(1)\varphi_4^1(2)\varphi_3^2(1)\varphi_1^2(2)$
$+\sqrt{1/8}\,\varphi_4^1(1)\varphi_1^1(2)\varphi_3^2(1)\varphi_1^2(2)$

Histogram: (0 1 1)

Weight: [2, 0, 1, 1]

History: 0 3 2)

Matrix element: $\sqrt{1}$

Layer 3

State:

$-\sqrt{1/16}\,\varphi_1^1(1)\varphi_4^1(2)\varphi_2^2(1)\varphi_3^2(2)$
$+\sqrt{1/16}\,\varphi_4^1(1)\varphi_1^1(2)\varphi_2^2(1)\varphi_3^2(2)$
$+\sqrt{1/16}\,\varphi_1^1(1)\varphi_4^1(2)\varphi_3^2(1)\varphi_2^2(2)$
$-\sqrt{1/16}\,\varphi_4^1(1)\varphi_1^1(2)\varphi_3^2(1)\varphi_2^2(2)$
$+\sqrt{1/16}\,\varphi_2^1(1)\varphi_4^1(2)\varphi_1^2(1)\varphi_3^2(2)$
$-\sqrt{1/16}\,\varphi_4^1(1)\varphi_2^1(2)\varphi_1^2(1)\varphi_3^2(2)$
$-\sqrt{1/16}\,\varphi_2^1(1)\varphi_4^1(2)\varphi_3^2(1)\varphi_1^2(2)$
$+\sqrt{1/16}\,\varphi_4^1(1)\varphi_2^1(2)\varphi_3^2(1)\varphi_1^2(2)$
$-\sqrt{1/16}\,\varphi_1^1(1)\varphi_3^1(2)\varphi_2^2(1)\varphi_4^2(2)$
$+\sqrt{1/16}\,\varphi_3^1(1)\varphi_1^1(2)\varphi_2^2(1)\varphi_4^2(2)$
$+\sqrt{1/16}\,\varphi_1^1(1)\varphi_3^1(2)\varphi_4^2(1)\varphi_2^2(2)$
$-\sqrt{1/16}\,\varphi_3^1(1)\varphi_1^1(2)\varphi_4^2(1)\varphi_2^2(2)$
$+\sqrt{1/16}\,\varphi_2^1(1)\varphi_3^1(2)\varphi_1^2(1)\varphi_4^2(2)$
$-\sqrt{1/16}\,\varphi_3^1(1)\varphi_2^1(2)\varphi_1^2(1)\varphi_4^2(2)$
$-\sqrt{1/16}\,\varphi_2^1(1)\varphi_3^1(2)\varphi_4^2(1)\varphi_1^2(2)$
$+\sqrt{1/16}\,\varphi_3^1(1)\varphi_2^1(2)\varphi_4^2(1)\varphi_1^2(2)$

Histogram: (1 1 1)

Weight: [1, 1, 1, 1]

History: 0 1 2 3)

Matrix element: $\sqrt{2}$

Layer 3

State:

$+\sqrt{1/16}\,\varphi_3^1(1)\varphi_4^1(2)\varphi_1^2(1)\varphi_2^2(2)$
$-\sqrt{1/16}\,\varphi_4^1(1)\varphi_3^1(2)\varphi_1^2(1)\varphi_2^2(2)$
$-\sqrt{1/16}\,\varphi_3^1(1)\varphi_4^1(2)\varphi_2^2(1)\varphi_1^2(2)$
$+\sqrt{1/16}\,\varphi_4^1(1)\varphi_3^1(2)\varphi_2^2(1)\varphi_1^2(2)$
$-\sqrt{1/16}\,\varphi_1^1(1)\varphi_3^1(2)\varphi_2^2(1)\varphi_4^2(2)$
$+\sqrt{1/16}\,\varphi_3^1(1)\varphi_1^1(2)\varphi_2^2(1)\varphi_4^2(2)$
$+\sqrt{1/16}\,\varphi_1^1(1)\varphi_3^1(2)\varphi_4^2(1)\varphi_2^2(2)$
$-\sqrt{1/16}\,\varphi_3^1(1)\varphi_1^1(2)\varphi_4^2(1)\varphi_2^2(2)$
$+\sqrt{1/16}\,\varphi_2^1(1)\varphi_4^1(2)\varphi_1^2(1)\varphi_3^2(2)$
$-\sqrt{1/16}\,\varphi_4^1(1)\varphi_2^1(2)\varphi_1^2(1)\varphi_3^2(2)$
$-\sqrt{1/16}\,\varphi_2^1(1)\varphi_4^1(2)\varphi_3^2(1)\varphi_1^2(2)$
$+\sqrt{1/16}\,\varphi_4^1(1)\varphi_2^1(2)\varphi_3^2(1)\varphi_1^2(2)$
$-\sqrt{1/16}\,\varphi_1^1(1)\varphi_2^1(2)\varphi_3^2(1)\varphi_4^2(2)$
$+\sqrt{1/16}\,\varphi_2^1(1)\varphi_1^1(2)\varphi_3^2(1)\varphi_4^2(2)$
$+\sqrt{1/16}\,\varphi_1^1(1)\varphi_2^1(2)\varphi_4^2(1)\varphi_3^2(2)$
$-\sqrt{1/16}\,\varphi_2^1(1)\varphi_1^1(2)\varphi_4^2(1)\varphi_3^2(2)$

Histogram: (1 1 1)

Weight: [1, 1, 1, 1]

History: 0 1 3 2)

Matrix element: $\sqrt{2}$

ORTHOGONALIZATION OF THIS STATE GIVES:

State:

$+\sqrt{1/12}\,\varphi_3^1(1)\varphi_4^1(2)\varphi_1^2(1)\varphi_2^2(2)$
$-\sqrt{1/12}\,\varphi_4^1(1)\varphi_3^1(2)\varphi_1^2(1)\varphi_2^2(2)$
$-\sqrt{1/12}\,\varphi_3^1(1)\varphi_4^1(2)\varphi_2^2(1)\varphi_1^2(2)$
$+\sqrt{1/12}\,\varphi_4^1(1)\varphi_3^1(2)\varphi_2^2(1)\varphi_1^2(2)$
$-\sqrt{1/48}\,\varphi_1^1(1)\varphi_3^1(2)\varphi_2^2(1)\varphi_4^2(2)$
$+\sqrt{1/48}\,\varphi_3^1(1)\varphi_1^1(2)\varphi_2^2(1)\varphi_4^2(2)$
$+\sqrt{1/48}\,\varphi_1^1(1)\varphi_3^1(2)\varphi_4^2(1)\varphi_2^2(2)$
$-\sqrt{1/48}\,\varphi_3^1(1)\varphi_1^1(2)\varphi_4^2(1)\varphi_2^2(2)$
$+\sqrt{1/48}\,\varphi_2^1(1)\varphi_4^1(2)\varphi_1^2(1)\varphi_3^2(2)$
$-\sqrt{1/48}\,\varphi_4^1(1)\varphi_2^1(2)\varphi_1^2(1)\varphi_3^2(2)$
$-\sqrt{1/48}\,\varphi_2^1(1)\varphi_4^1(2)\varphi_3^2(1)\varphi_1^2(2)$
$+\sqrt{1/48}\,\varphi_4^1(1)\varphi_2^1(2)\varphi_3^2(1)\varphi_1^2(2)$
$-\sqrt{1/12}\,\varphi_1^1(1)\varphi_2^1(2)\varphi_3^2(1)\varphi_4^2(2)$
$+\sqrt{1/12}\,\varphi_2^1(1)\varphi_1^1(2)\varphi_3^2(1)\varphi_4^2(2)$
$+\sqrt{1/12}\,\varphi_1^1(1)\varphi_2^1(2)\varphi_4^2(1)\varphi_3^2(2)$
$-\sqrt{1/12}\,\varphi_2^1(1)\varphi_1^1(2)\varphi_4^2(1)\varphi_3^2(2)$
$+\sqrt{1/48}\,\varphi_1^1(1)\varphi_4^1(2)\varphi_2^2(1)\varphi_3^2(2)$
$-\sqrt{1/48}\,\varphi_4^1(1)\varphi_1^1(2)\varphi_2^2(1)\varphi_3^2(2)$
$-\sqrt{1/48}\,\varphi_1^1(1)\varphi_4^1(2)\varphi_3^2(1)\varphi_2^2(2)$
$+\sqrt{1/48}\,\varphi_4^1(1)\varphi_1^1(2)\varphi_3^2(1)\varphi_2^2(2)$
$-\sqrt{1/48}\,\varphi_2^1(1)\varphi_3^1(2)\varphi_1^2(1)\varphi_4^2(2)$
$+\sqrt{1/48}\,\varphi_3^1(1)\varphi_2^1(2)\varphi_1^2(1)\varphi_4^2(2)$
$+\sqrt{1/48}\,\varphi_2^1(1)\varphi_3^1(2)\varphi_4^2(1)\varphi_1^2(2)$
$-\sqrt{1/48}\,\varphi_3^1(1)\varphi_2^1(2)\varphi_4^2(1)\varphi_1^2(2)$

Histogram: (1 1 1)

Weight: [1, 1, 1, 1]

History: 0 1 3 2)

Matrix element: $\sqrt{3/2}$

Layer 3

State:

$-\sqrt{1/16}\,\varphi_2^1(1)\varphi_3^1(2)\varphi_1^2(1)\varphi_4^2(2)$
$+\sqrt{1/16}\,\varphi_3^1(1)\varphi_2^1(2)\varphi_1^2(1)\varphi_4^2(2)$
$+\sqrt{1/16}\,\varphi_2^1(1)\varphi_3^1(2)\varphi_4^2(1)\varphi_1^2(2)$
$-\sqrt{1/16}\,\varphi_3^1(1)\varphi_2^1(2)\varphi_4^2(1)\varphi_1^2(2)$
$+\sqrt{1/16}\,\varphi_2^1(1)\varphi_4^1(2)\varphi_1^2(1)\varphi_3^2(2)$
$-\sqrt{1/16}\,\varphi_4^1(1)\varphi_2^1(2)\varphi_1^2(1)\varphi_3^2(2)$
$-\sqrt{1/16}\,\varphi_2^1(1)\varphi_4^1(2)\varphi_3^2(1)\varphi_1^2(2)$
$+\sqrt{1/16}\,\varphi_4^1(1)\varphi_2^1(2)\varphi_3^2(1)\varphi_1^2(2)$
$-\sqrt{1/16}\,\varphi_1^1(1)\varphi_3^1(2)\varphi_2^2(1)\varphi_4^2(2)$
$+\sqrt{1/16}\,\varphi_3^1(1)\varphi_1^1(2)\varphi_2^2(1)\varphi_4^2(2)$
$+\sqrt{1/16}\,\varphi_1^1(1)\varphi_3^1(2)\varphi_4^2(1)\varphi_2^2(2)$
$-\sqrt{1/16}\,\varphi_3^1(1)\varphi_1^1(2)\varphi_4^2(1)\varphi_2^2(2)$
$+\sqrt{1/16}\,\varphi_1^1(1)\varphi_4^1(2)\varphi_2^2(1)\varphi_3^2(2)$
$-\sqrt{1/16}\,\varphi_4^1(1)\varphi_1^1(2)\varphi_2^2(1)\varphi_3^2(2)$
$-\sqrt{1/16}\,\varphi_1^1(1)\varphi_4^1(2)\varphi_3^2(1)\varphi_2^2(2)$
$+\sqrt{1/16}\,\varphi_4^1(1)\varphi_1^1(2)\varphi_3^2(1)\varphi_2^2(2)$

Histogram: (1 1 1)

Weight: [1, 1, 1, 1]

History: 0 3 2 1)

Matrix element: $\sqrt{2}$

ORTHOGONALIZATION OF THIS STATE GIVES:

State:

$-\sqrt{1/24}\,\varphi_2^1(1)\varphi_3^1(2)\varphi_1^2(1)\varphi_4^2(2)$
$+\sqrt{1/24}\,\varphi_3^1(1)\varphi_2^1(2)\varphi_1^2(1)\varphi_4^2(2)$
$+\sqrt{1/24}\,\varphi_2^1(1)\varphi_3^1(2)\varphi_4^2(1)\varphi_1^2(2)$
$-\sqrt{1/24}\,\varphi_3^1(1)\varphi_2^1(2)\varphi_4^2(1)\varphi_1^2(2)$
$+\sqrt{1/24}\,\varphi_2^1(1)\varphi_4^1(2)\varphi_1^2(1)\varphi_3^2(2)$
$-\sqrt{1/24}\,\varphi_4^1(1)\varphi_2^1(2)\varphi_1^2(1)\varphi_3^2(2)$
$-\sqrt{1/24}\,\varphi_2^1(1)\varphi_4^1(2)\varphi_3^2(1)\varphi_1^2(2)$
$+\sqrt{1/24}\,\varphi_4^1(1)\varphi_2^1(2)\varphi_3^2(1)\varphi_1^2(2)$
$-\sqrt{1/24}\,\varphi_1^1(1)\varphi_3^1(2)\varphi_2^2(1)\varphi_4^2(2)$
$+\sqrt{1/24}\,\varphi_3^1(1)\varphi_1^1(2)\varphi_2^2(1)\varphi_4^2(2)$
$+\sqrt{1/24}\,\varphi_1^1(1)\varphi_3^1(2)\varphi_4^2(1)\varphi_2^2(2)$
$-\sqrt{1/24}\,\varphi_3^1(1)\varphi_1^1(2)\varphi_4^2(1)\varphi_2^2(2)$
$+\sqrt{1/24}\,\varphi_1^1(1)\varphi_4^1(2)\varphi_2^2(1)\varphi_3^2(2)$
$-\sqrt{1/24}\,\varphi_4^1(1)\varphi_1^1(2)\varphi_2^2(1)\varphi_3^2(2)$
$-\sqrt{1/24}\,\varphi_1^1(1)\varphi_4^1(2)\varphi_3^2(1)\varphi_2^2(2)$
$+\sqrt{1/24}\,\varphi_4^1(1)\varphi_1^1(2)\varphi_3^2(1)\varphi_2^2(2)$
$-\sqrt{1/24}\,\varphi_3^1(1)\varphi_4^1(2)\varphi_1^2(1)\varphi_2^2(2)$
$+\sqrt{1/24}\,\varphi_4^1(1)\varphi_3^1(2)\varphi_1^2(1)\varphi_2^2(2)$
$+\sqrt{1/24}\,\varphi_3^1(1)\varphi_4^1(2)\varphi_2^2(1)\varphi_1^2(2)$
$-\sqrt{1/24}\,\varphi_4^1(1)\varphi_3^1(2)\varphi_2^2(1)\varphi_1^2(2)$
$+\sqrt{1/24}\,\varphi_1^1(1)\varphi_2^1(2)\varphi_3^2(1)\varphi_4^2(2)$
$-\sqrt{1/24}\,\varphi_2^1(1)\varphi_1^1(2)\varphi_3^2(1)\varphi_4^2(2)$
$-\sqrt{1/24}\,\varphi_1^1(1)\varphi_2^1(2)\varphi_4^2(1)\varphi_3^2(2)$
$+\sqrt{1/24}\,\varphi_2^1(1)\varphi_1^1(2)\varphi_4^2(1)\varphi_3^2(2)$

Histogram: (1 1 1)

Weight: [1, 1, 1, 1]

History: 0 3 2 1)

Matrix element: $\sqrt{4/3}$

Layer 4

State:

$+\sqrt{1/8}\,\varphi_3^1(1)\varphi_4^1(2)\varphi_1^2(1)\varphi_3^2(2)$
$-\sqrt{1/8}\,\varphi_4^1(1)\varphi_3^1(2)\varphi_1^2(1)\varphi_3^2(2)$
$-\sqrt{1/8}\,\varphi_3^1(1)\varphi_4^1(2)\varphi_3^2(1)\varphi_1^2(2)$
$+\sqrt{1/8}\,\varphi_4^1(1)\varphi_3^1(2)\varphi_3^2(1)\varphi_1^2(2)$
$-\sqrt{1/8}\,\varphi_1^1(1)\varphi_3^1(2)\varphi_3^2(1)\varphi_4^2(2)$
$+\sqrt{1/8}\,\varphi_3^1(1)\varphi_1^1(2)\varphi_3^2(1)\varphi_4^2(2)$
$+\sqrt{1/8}\,\varphi_1^1(1)\varphi_3^1(2)\varphi_4^2(1)\varphi_3^2(2)$
$-\sqrt{1/8}\,\varphi_3^1(1)\varphi_1^1(2)\varphi_4^2(1)\varphi_3^2(2)$

Histogram: (1 2 1)

Weight: [1, 0, 2, 1]

History: 0 1 2 3 2)

Matrix element: $\sqrt{1/2}$

SU(4) × SU(4) $\mapsto$ SU(4)

$[1,1,0,0;1,1,0,0] \mapsto [2,2,0,0]+[2,1,1,0]+[1,1,1,1]$

Layer 4

State:

$-\sqrt{1/8}\,\varphi^1_1(1)\varphi^1_4(2)\varphi^2_2(1)\varphi^2_4(2)$
$+\sqrt{1/8}\,\varphi^1_4(1)\varphi^1_1(2)\varphi^2_2(1)\varphi^2_4(2)$
$+\sqrt{1/8}\,\varphi^1_1(1)\varphi^1_4(2)\varphi^2_4(1)\varphi^2_2(2)$
$-\sqrt{1/8}\,\varphi^1_4(1)\varphi^1_1(2)\varphi^2_4(1)\varphi^2_2(2)$
$+\sqrt{1/8}\,\varphi^1_2(1)\varphi^1_4(2)\varphi^2_1(1)\varphi^2_4(2)$
$-\sqrt{1/8}\,\varphi^1_4(1)\varphi^1_2(2)\varphi^2_1(1)\varphi^2_4(2)$
$-\sqrt{1/8}\,\varphi^1_2(1)\varphi^1_4(2)\varphi^2_4(1)\varphi^2_1(2)$
$+\sqrt{1/8}\,\varphi^1_4(1)\varphi^1_2(2)\varphi^2_4(1)\varphi^2_1(2)$

Histogram: (1 1 2)

Weight: [1, 1, 0, 2]

History: 0 1 2 3 3)

Matrix element: $\sqrt{2}$

Layer 4

State:

$-\sqrt{1/8}\,\varphi^1_2(1)\varphi^1_3(2)\varphi^2_2(1)\varphi^2_4(2)$
$+\sqrt{1/8}\,\varphi^1_3(1)\varphi^1_2(2)\varphi^2_2(1)\varphi^2_4(2)$
$+\sqrt{1/8}\,\varphi^1_2(1)\varphi^1_3(2)\varphi^2_4(1)\varphi^2_2(2)$
$-\sqrt{1/8}\,\varphi^1_3(1)\varphi^1_2(2)\varphi^2_4(1)\varphi^2_2(2)$
$+\sqrt{1/8}\,\varphi^1_2(1)\varphi^1_4(2)\varphi^2_2(1)\varphi^2_3(2)$
$-\sqrt{1/8}\,\varphi^1_4(1)\varphi^1_2(2)\varphi^2_2(1)\varphi^2_3(2)$
$-\sqrt{1/8}\,\varphi^1_2(1)\varphi^1_4(2)\varphi^2_3(1)\varphi^2_2(2)$
$+\sqrt{1/8}\,\varphi^1_4(1)\varphi^1_2(2)\varphi^2_3(1)\varphi^2_2(2)$

Histogram: (2 1 1)

Weight: [0, 2, 1, 1]

History: 0 1 3 2 1)

Matrix element: $\sqrt{2/3}$

Layer 4

State:

$+\sqrt{1/8}\,\varphi^1_3(1)\varphi^1_4(2)\varphi^2_1(1)\varphi^2_3(2)$
$-\sqrt{1/8}\,\varphi^1_4(1)\varphi^1_3(2)\varphi^2_1(1)\varphi^2_3(2)$
$-\sqrt{1/8}\,\varphi^1_3(1)\varphi^1_4(2)\varphi^2_3(1)\varphi^2_1(2)$
$+\sqrt{1/8}\,\varphi^1_4(1)\varphi^1_3(2)\varphi^2_3(1)\varphi^2_1(2)$
$-\sqrt{1/8}\,\varphi^1_1(1)\varphi^1_3(2)\varphi^2_3(1)\varphi^2_4(2)$
$+\sqrt{1/8}\,\varphi^1_3(1)\varphi^1_1(2)\varphi^2_3(1)\varphi^2_4(2)$
$+\sqrt{1/8}\,\varphi^1_1(1)\varphi^1_3(2)\varphi^2_4(1)\varphi^2_3(2)$
$-\sqrt{1/8}\,\varphi^1_3(1)\varphi^1_1(2)\varphi^2_4(1)\varphi^2_3(2)$

Histogram: (1 2 1)

Weight: [1, 0, 2, 1]

History: 0 1 3 2 2)

Matrix element: $\sqrt{3/2}$

THIS STATE IS LINEARLY DEPENDENT!

Layer 4

State:

$-\sqrt{1/8}\,\varphi^1_2(1)\varphi^1_3(2)\varphi^2_2(1)\varphi^2_4(2)$
$+\sqrt{1/8}\,\varphi^1_3(1)\varphi^1_2(2)\varphi^2_2(1)\varphi^2_4(2)$
$+\sqrt{1/8}\,\varphi^1_2(1)\varphi^1_3(2)\varphi^2_4(1)\varphi^2_2(2)$
$-\sqrt{1/8}\,\varphi^1_3(1)\varphi^1_2(2)\varphi^2_4(1)\varphi^2_2(2)$
$+\sqrt{1/8}\,\varphi^1_2(1)\varphi^1_4(2)\varphi^2_2(1)\varphi^2_3(2)$
$-\sqrt{1/8}\,\varphi^1_4(1)\varphi^1_2(2)\varphi^2_2(1)\varphi^2_3(2)$
$-\sqrt{1/8}\,\varphi^1_2(1)\varphi^1_4(2)\varphi^2_3(1)\varphi^2_2(2)$
$+\sqrt{1/8}\,\varphi^1_4(1)\varphi^1_2(2)\varphi^2_3(1)\varphi^2_2(2)$

Histogram: (2 1 1)

Weight: [0, 2, 1, 1]

History: 0 3 2 1 1)

Matrix element: $\sqrt{4/3}$

THIS STATE IS LINEARLY DEPENDENT!

Layer 5

State:

$-\sqrt{1/8}\,\varphi^1_2(1)\varphi^1_3(2)\varphi^2_3(1)\varphi^2_4(2)$
$+\sqrt{1/8}\,\varphi^1_3(1)\varphi^1_2(2)\varphi^2_3(1)\varphi^2_4(2)$
$+\sqrt{1/8}\,\varphi^1_2(1)\varphi^1_3(2)\varphi^2_4(1)\varphi^2_3(2)$
$-\sqrt{1/8}\,\varphi^1_3(1)\varphi^1_2(2)\varphi^2_4(1)\varphi^2_3(2)$
$+\sqrt{1/8}\,\varphi^1_3(1)\varphi^1_4(2)\varphi^2_2(1)\varphi^2_3(2)$
$-\sqrt{1/8}\,\varphi^1_4(1)\varphi^1_3(2)\varphi^2_2(1)\varphi^2_3(2)$
$-\sqrt{1/8}\,\varphi^1_3(1)\varphi^1_4(2)\varphi^2_3(1)\varphi^2_2(2)$
$+\sqrt{1/8}\,\varphi^1_4(1)\varphi^1_3(2)\varphi^2_3(1)\varphi^2_2(2)$

Histogram: (2 2 1)

Weight: [0, 1, 2, 1]

History: 0 1 2 3 2 1)

Matrix element: $\sqrt{1}$

Layer 5

State:

$-\sqrt{1/8}\,\varphi^1_1(1)\varphi^1_4(2)\varphi^2_3(1)\varphi^2_4(2)$
$+\sqrt{1/8}\,\varphi^1_4(1)\varphi^1_1(2)\varphi^2_3(1)\varphi^2_4(2)$
$+\sqrt{1/8}\,\varphi^1_1(1)\varphi^1_4(2)\varphi^2_4(1)\varphi^2_3(2)$
$-\sqrt{1/8}\,\varphi^1_4(1)\varphi^1_1(2)\varphi^2_4(1)\varphi^2_3(2)$
$+\sqrt{1/8}\,\varphi^1_3(1)\varphi^1_4(2)\varphi^2_1(1)\varphi^2_4(2)$
$-\sqrt{1/8}\,\varphi^1_4(1)\varphi^1_3(2)\varphi^2_1(1)\varphi^2_4(2)$
$-\sqrt{1/8}\,\varphi^1_3(1)\varphi^1_4(2)\varphi^2_4(1)\varphi^2_1(2)$
$+\sqrt{1/8}\,\varphi^1_4(1)\varphi^1_3(2)\varphi^2_4(1)\varphi^2_1(2)$

Histogram: (1 2 2)

Weight: [1, 0, 1, 2]

History: 0 1 2 3 2 3)

Matrix element: $\sqrt{1}$

Layer 5

State:

$+\sqrt{1/8}\,\varphi^1_3(1)\varphi^1_4(2)\varphi^2_1(1)\varphi^2_4(2)$
$-\sqrt{1/8}\,\varphi^1_4(1)\varphi^1_3(2)\varphi^2_1(1)\varphi^2_4(2)$
$-\sqrt{1/8}\,\varphi^1_3(1)\varphi^1_4(2)\varphi^2_4(1)\varphi^2_1(2)$
$+\sqrt{1/8}\,\varphi^1_4(1)\varphi^1_3(2)\varphi^2_4(1)\varphi^2_1(2)$
$-\sqrt{1/8}\,\varphi^1_1(1)\varphi^1_4(2)\varphi^2_3(1)\varphi^2_4(2)$
$+\sqrt{1/8}\,\varphi^1_4(1)\varphi^1_1(2)\varphi^2_3(1)\varphi^2_4(2)$
$+\sqrt{1/8}\,\varphi^1_1(1)\varphi^1_4(2)\varphi^2_4(1)\varphi^2_3(2)$
$-\sqrt{1/8}\,\varphi^1_4(1)\varphi^1_1(2)\varphi^2_4(1)\varphi^2_3(2)$

Histogram: (1 2 2)

Weight: [1, 0, 1, 2]

History: 0 1 2 3 3 2)

Matrix element: $\sqrt{1}$

THIS STATE IS LINEARLY DEPENDENT!

Layer 5

State:

$+\sqrt{1/8}\,\varphi^1_3(1)\varphi^1_4(2)\varphi^2_2(1)\varphi^2_3(2)$
$-\sqrt{1/8}\,\varphi^1_4(1)\varphi^1_3(2)\varphi^2_2(1)\varphi^2_3(2)$
$-\sqrt{1/8}\,\varphi^1_3(1)\varphi^1_4(2)\varphi^2_3(1)\varphi^2_2(2)$
$+\sqrt{1/8}\,\varphi^1_4(1)\varphi^1_3(2)\varphi^2_3(1)\varphi^2_2(2)$
$-\sqrt{1/8}\,\varphi^1_2(1)\varphi^1_3(2)\varphi^2_3(1)\varphi^2_4(2)$
$+\sqrt{1/8}\,\varphi^1_3(1)\varphi^1_2(2)\varphi^2_3(1)\varphi^2_4(2)$
$+\sqrt{1/8}\,\varphi^1_2(1)\varphi^1_3(2)\varphi^2_4(1)\varphi^2_3(2)$
$-\sqrt{1/8}\,\varphi^1_3(1)\varphi^1_2(2)\varphi^2_4(1)\varphi^2_3(2)$

Histogram: (2 2 1)

Weight: [0, 1, 2, 1]

History: 0 1 3 2 1 2)

Matrix element: $\sqrt{1}$

THIS STATE IS LINEARLY DEPENDENT!

Layer 6

State:

$-\sqrt{1/8}\,\varphi^1_2(1)\varphi^1_4(2)\varphi^2_3(1)\varphi^2_4(2)$
$+\sqrt{1/8}\,\varphi^1_4(1)\varphi^1_2(2)\varphi^2_3(1)\varphi^2_4(2)$
$+\sqrt{1/8}\,\varphi^1_2(1)\varphi^1_4(2)\varphi^2_4(1)\varphi^2_3(2)$
$-\sqrt{1/8}\,\varphi^1_4(1)\varphi^1_2(2)\varphi^2_4(1)\varphi^2_3(2)$
$+\sqrt{1/8}\,\varphi^1_3(1)\varphi^1_4(2)\varphi^2_2(1)\varphi^2_4(2)$
$-\sqrt{1/8}\,\varphi^1_4(1)\varphi^1_3(2)\varphi^2_2(1)\varphi^2_4(2)$
$-\sqrt{1/8}\,\varphi^1_3(1)\varphi^1_4(2)\varphi^2_4(1)\varphi^2_2(2)$
$+\sqrt{1/8}\,\varphi^1_4(1)\varphi^1_3(2)\varphi^2_4(1)\varphi^2_2(2)$

Histogram: (2 2 2)

Weight: [0, 1, 1, 2]

History: 0 1 2 3 2 1 3)

Matrix element: $\sqrt{1}$

Layer 6

State:

$-\sqrt{1/8}\,\varphi^1_2(1)\varphi^1_4(2)\varphi^2_3(1)\varphi^2_4(2)$
$+\sqrt{1/8}\,\varphi^1_4(1)\varphi^1_2(2)\varphi^2_3(1)\varphi^2_4(2)$
$+\sqrt{1/8}\,\varphi^1_2(1)\varphi^1_4(2)\varphi^2_4(1)\varphi^2_3(2)$
$-\sqrt{1/8}\,\varphi^1_4(1)\varphi^1_2(2)\varphi^2_4(1)\varphi^2_3(2)$
$+\sqrt{1/8}\,\varphi^1_3(1)\varphi^1_4(2)\varphi^2_2(1)\varphi^2_4(2)$
$-\sqrt{1/8}\,\varphi^1_4(1)\varphi^1_3(2)\varphi^2_2(1)\varphi^2_4(2)$
$-\sqrt{1/8}\,\varphi^1_3(1)\varphi^1_4(2)\varphi^2_4(1)\varphi^2_2(2)$
$+\sqrt{1/8}\,\varphi^1_4(1)\varphi^1_3(2)\varphi^2_4(1)\varphi^2_2(2)$

Histogram: (2 2 2)

Weight: [0, 1, 1, 2]

History: 0 1 2 3 2 3 1)

Matrix element: $\sqrt{1}$

THIS STATE IS LINEARLY DEPENDENT!

---

15 linear independent states found in this representation.

Representation 3:

Initial state:

$+\sqrt{1/24}\,\varphi^1_3(1)\varphi^1_4(2)\varphi^2_1(1)\varphi^2_2(2)$
$-\sqrt{1/24}\,\varphi^1_4(1)\varphi^1_3(2)\varphi^2_1(1)\varphi^2_2(2)$
$-\sqrt{1/24}\,\varphi^1_3(1)\varphi^1_4(2)\varphi^2_2(1)\varphi^2_1(2)$
$+\sqrt{1/24}\,\varphi^1_4(1)\varphi^1_3(2)\varphi^2_2(1)\varphi^2_1(2)$
$-\sqrt{1/24}\,\varphi^1_2(1)\varphi^1_4(2)\varphi^2_1(1)\varphi^2_3(2)$
$+\sqrt{1/24}\,\varphi^1_4(1)\varphi^1_2(2)\varphi^2_1(1)\varphi^2_3(2)$
$+\sqrt{1/24}\,\varphi^1_2(1)\varphi^1_4(2)\varphi^2_3(1)\varphi^2_1(2)$
$-\sqrt{1/24}\,\varphi^1_4(1)\varphi^1_2(2)\varphi^2_3(1)\varphi^2_1(2)$
$+\sqrt{1/24}\,\varphi^1_2(1)\varphi^1_3(2)\varphi^2_1(1)\varphi^2_4(2)$
$-\sqrt{1/24}\,\varphi^1_3(1)\varphi^1_2(2)\varphi^2_1(1)\varphi^2_4(2)$
$-\sqrt{1/24}\,\varphi^1_2(1)\varphi^1_3(2)\varphi^2_4(1)\varphi^2_1(2)$
$+\sqrt{1/24}\,\varphi^1_3(1)\varphi^1_2(2)\varphi^2_4(1)\varphi^2_1(2)$
$+\sqrt{1/24}\,\varphi^1_1(1)\varphi^1_4(2)\varphi^2_2(1)\varphi^2_3(2)$
$-\sqrt{1/24}\,\varphi^1_4(1)\varphi^1_1(2)\varphi^2_2(1)\varphi^2_3(2)$
$-\sqrt{1/24}\,\varphi^1_1(1)\varphi^1_4(2)\varphi^2_3(1)\varphi^2_2(2)$
$+\sqrt{1/24}\,\varphi^1_4(1)\varphi^1_1(2)\varphi^2_3(1)\varphi^2_2(2)$
$-\sqrt{1/24}\,\varphi^1_1(1)\varphi^1_3(2)\varphi^2_2(1)\varphi^2_4(2)$
$+\sqrt{1/24}\,\varphi^1_3(1)\varphi^1_1(2)\varphi^2_2(1)\varphi^2_4(2)$
$+\sqrt{1/24}\,\varphi^1_1(1)\varphi^1_3(2)\varphi^2_4(1)\varphi^2_2(2)$
$-\sqrt{1/24}\,\varphi^1_3(1)\varphi^1_1(2)\varphi^2_4(1)\varphi^2_2(2)$
$+\sqrt{1/24}\,\varphi^1_1(1)\varphi^1_2(2)\varphi^2_3(1)\varphi^2_4(2)$
$-\sqrt{1/24}\,\varphi^1_2(1)\varphi^1_1(2)\varphi^2_3(1)\varphi^2_4(2)$
$-\sqrt{1/24}\,\varphi^1_1(1)\varphi^1_2(2)\varphi^2_4(1)\varphi^2_3(2)$
$+\sqrt{1/24}\,\varphi^1_2(1)\varphi^1_1(2)\varphi^2_4(1)\varphi^2_3(2)$

Weight: [1, 1, 1, 1]

---

1 linear independent state found in this representation.

36 linear independent states found in all representations.

SU(4) × SU(4) ⟼ SU(4)

[1, 1, 0, 0; 1, 1, 0, 0] ⟼ [2, 2, 0, 0] + [2, 1, 1, 0] + [1, 1, 1, 1]

Summary of dominant weights and their state vectors:

[2, 2, 0, 0] found in rep.1:

$+\sqrt{1/4}\varphi_1^1(1)\varphi_2^1(2)\varphi_1^2(1)\varphi_2^2(2)$
$-\sqrt{1/4}\varphi_2^1(1)\varphi_1^1(2)\varphi_1^2(1)\varphi_2^2(2)$
$-\sqrt{1/4}\varphi_1^1(1)\varphi_2^1(2)\varphi_2^2(1)\varphi_1^2(2)$
$+\sqrt{1/4}\varphi_2^1(1)\varphi_1^1(2)\varphi_2^2(1)\varphi_1^2(2)$

[2, 1, 1, 0] found in rep.1:

$+\sqrt{1/8}\varphi_1^1(1)\varphi_3^1(2)\varphi_1^2(1)\varphi_2^2(2)$
$-\sqrt{1/8}\varphi_3^1(1)\varphi_1^1(2)\varphi_1^2(1)\varphi_2^2(2)$
$-\sqrt{1/8}\varphi_1^1(1)\varphi_3^1(2)\varphi_2^2(1)\varphi_1^2(2)$
$+\sqrt{1/8}\varphi_3^1(1)\varphi_1^1(2)\varphi_2^2(1)\varphi_1^2(2)$
$+\sqrt{1/8}\varphi_1^1(1)\varphi_2^1(2)\varphi_1^2(1)\varphi_3^2(2)$
$-\sqrt{1/8}\varphi_2^1(1)\varphi_1^1(2)\varphi_1^2(1)\varphi_3^2(2)$
$-\sqrt{1/8}\varphi_1^1(1)\varphi_2^1(2)\varphi_3^2(1)\varphi_1^2(2)$
$+\sqrt{1/8}\varphi_2^1(1)\varphi_1^1(2)\varphi_3^2(1)\varphi_1^2(2)$

[2, 1, 1, 0] found in rep.2:

$+\sqrt{1/8}\varphi_1^1(1)\varphi_3^1(2)\varphi_1^2(1)\varphi_2^2(2)$
$-\sqrt{1/8}\varphi_3^1(1)\varphi_1^1(2)\varphi_1^2(1)\varphi_2^2(2)$
$-\sqrt{1/8}\varphi_1^1(1)\varphi_3^1(2)\varphi_2^2(1)\varphi_1^2(2)$
$+\sqrt{1/8}\varphi_3^1(1)\varphi_1^1(2)\varphi_2^2(1)\varphi_1^2(2)$
$-\sqrt{1/8}\varphi_1^1(1)\varphi_2^1(2)\varphi_1^2(1)\varphi_3^2(2)$
$+\sqrt{1/8}\varphi_2^1(1)\varphi_1^1(2)\varphi_1^2(1)\varphi_3^2(2)$
$+\sqrt{1/8}\varphi_1^1(1)\varphi_2^1(2)\varphi_3^2(1)\varphi_1^2(2)$
$-\sqrt{1/8}\varphi_2^1(1)\varphi_1^1(2)\varphi_3^2(1)\varphi_1^2(2)$

[1, 1, 1, 1] found in rep.1:

$+\sqrt{1/16}\varphi_1^1(1)\varphi_4^1(2)\varphi_2^2(1)\varphi_3^2(2)$
$-\sqrt{1/16}\varphi_4^1(1)\varphi_1^1(2)\varphi_2^2(1)\varphi_3^2(2)$
$-\sqrt{1/16}\varphi_1^1(1)\varphi_4^1(2)\varphi_3^2(1)\varphi_2^2(2)$
$+\sqrt{1/16}\varphi_4^1(1)\varphi_1^1(2)\varphi_3^2(1)\varphi_2^2(2)$
$+\sqrt{1/16}\varphi_2^1(1)\varphi_4^1(2)\varphi_1^2(1)\varphi_3^2(2)$
$-\sqrt{1/16}\varphi_4^1(1)\varphi_2^1(2)\varphi_1^2(1)\varphi_3^2(2)$
$-\sqrt{1/16}\varphi_2^1(1)\varphi_4^1(2)\varphi_3^2(1)\varphi_1^2(2)$
$+\sqrt{1/16}\varphi_4^1(1)\varphi_2^1(2)\varphi_3^2(1)\varphi_1^2(2)$
$+\sqrt{1/16}\varphi_1^1(1)\varphi_3^1(2)\varphi_2^2(1)\varphi_4^2(2)$
$-\sqrt{1/16}\varphi_3^1(1)\varphi_1^1(2)\varphi_2^2(1)\varphi_4^2(2)$
$-\sqrt{1/16}\varphi_1^1(1)\varphi_3^1(2)\varphi_4^2(1)\varphi_2^2(2)$
$+\sqrt{1/16}\varphi_3^1(1)\varphi_1^1(2)\varphi_4^2(1)\varphi_2^2(2)$
$+\sqrt{1/16}\varphi_2^1(1)\varphi_3^1(2)\varphi_1^2(1)\varphi_4^2(2)$
$-\sqrt{1/16}\varphi_3^1(1)\varphi_2^1(2)\varphi_1^2(1)\varphi_4^2(2)$
$-\sqrt{1/16}\varphi_2^1(1)\varphi_3^1(2)\varphi_4^2(1)\varphi_1^2(2)$
$+\sqrt{1/16}\varphi_3^1(1)\varphi_2^1(2)\varphi_4^2(1)\varphi_1^2(2)$

[1, 1, 1, 1] found in rep.1:

$+\sqrt{1/12}\varphi_3^1(1)\varphi_4^1(2)\varphi_1^2(1)\varphi_2^2(2)$
$-\sqrt{1/12}\varphi_4^1(1)\varphi_3^1(2)\varphi_1^2(1)\varphi_2^2(2)$
$-\sqrt{1/12}\varphi_3^1(1)\varphi_4^1(2)\varphi_2^2(1)\varphi_1^2(2)$
$+\sqrt{1/12}\varphi_4^1(1)\varphi_3^1(2)\varphi_2^2(1)\varphi_1^2(2)$
$+\sqrt{1/48}\varphi_1^1(1)\varphi_3^1(2)\varphi_2^2(1)\varphi_4^2(2)$
$-\sqrt{1/48}\varphi_3^1(1)\varphi_1^1(2)\varphi_2^2(1)\varphi_4^2(2)$
$-\sqrt{1/48}\varphi_1^1(1)\varphi_3^1(2)\varphi_4^2(1)\varphi_2^2(2)$
$+\sqrt{1/48}\varphi_3^1(1)\varphi_1^1(2)\varphi_4^2(1)\varphi_2^2(2)$
$+\sqrt{1/48}\varphi_2^1(1)\varphi_4^1(2)\varphi_1^2(1)\varphi_3^2(2)$
$-\sqrt{1/48}\varphi_4^1(1)\varphi_2^1(2)\varphi_1^2(1)\varphi_3^2(2)$
$-\sqrt{1/48}\varphi_2^1(1)\varphi_4^1(2)\varphi_3^2(1)\varphi_1^2(2)$
$+\sqrt{1/48}\varphi_4^1(1)\varphi_2^1(2)\varphi_3^2(1)\varphi_1^2(2)$
$+\sqrt{1/12}\varphi_1^1(1)\varphi_2^1(2)\varphi_3^2(1)\varphi_4^2(2)$
$-\sqrt{1/12}\varphi_2^1(1)\varphi_1^1(2)\varphi_3^2(1)\varphi_4^2(2)$
$-\sqrt{1/12}\varphi_1^1(1)\varphi_2^1(2)\varphi_4^2(1)\varphi_3^2(2)$
$+\sqrt{1/12}\varphi_2^1(1)\varphi_1^1(2)\varphi_4^2(1)\varphi_3^2(2)$
$-\sqrt{1/48}\varphi_1^1(1)\varphi_4^1(2)\varphi_2^2(1)\varphi_3^2(2)$
$+\sqrt{1/48}\varphi_4^1(1)\varphi_1^1(2)\varphi_2^2(1)\varphi_3^2(2)$
$+\sqrt{1/48}\varphi_1^1(1)\varphi_4^1(2)\varphi_3^2(1)\varphi_2^2(2)$
$-\sqrt{1/48}\varphi_4^1(1)\varphi_1^1(2)\varphi_3^2(1)\varphi_2^2(2)$
$-\sqrt{1/48}\varphi_2^1(1)\varphi_3^1(2)\varphi_1^2(1)\varphi_4^2(2)$
$+\sqrt{1/48}\varphi_3^1(1)\varphi_2^1(2)\varphi_1^2(1)\varphi_4^2(2)$
$+\sqrt{1/48}\varphi_2^1(1)\varphi_3^1(2)\varphi_4^2(1)\varphi_1^2(2)$
$-\sqrt{1/48}\varphi_3^1(1)\varphi_2^1(2)\varphi_4^2(1)\varphi_1^2(2)$

[1, 1, 1, 1] found in rep.2:

$-\sqrt{1/16}\varphi_1^1(1)\varphi_4^1(2)\varphi_2^2(1)\varphi_3^2(2)$
$+\sqrt{1/16}\varphi_4^1(1)\varphi_1^1(2)\varphi_2^2(1)\varphi_3^2(2)$
$+\sqrt{1/16}\varphi_1^1(1)\varphi_4^1(2)\varphi_3^2(1)\varphi_2^2(2)$
$-\sqrt{1/16}\varphi_4^1(1)\varphi_1^1(2)\varphi_3^2(1)\varphi_2^2(2)$
$+\sqrt{1/16}\varphi_2^1(1)\varphi_4^1(2)\varphi_1^2(1)\varphi_3^2(2)$
$-\sqrt{1/16}\varphi_4^1(1)\varphi_2^1(2)\varphi_1^2(1)\varphi_3^2(2)$
$-\sqrt{1/16}\varphi_2^1(1)\varphi_4^1(2)\varphi_3^2(1)\varphi_1^2(2)$
$+\sqrt{1/16}\varphi_4^1(1)\varphi_2^1(2)\varphi_3^2(1)\varphi_1^2(2)$
$-\sqrt{1/16}\varphi_1^1(1)\varphi_3^1(2)\varphi_2^2(1)\varphi_4^2(2)$
$+\sqrt{1/16}\varphi_3^1(1)\varphi_1^1(2)\varphi_2^2(1)\varphi_4^2(2)$
$+\sqrt{1/16}\varphi_1^1(1)\varphi_3^1(2)\varphi_4^2(1)\varphi_2^2(2)$
$-\sqrt{1/16}\varphi_3^1(1)\varphi_1^1(2)\varphi_4^2(1)\varphi_2^2(2)$
$+\sqrt{1/16}\varphi_2^1(1)\varphi_3^1(2)\varphi_1^2(1)\varphi_4^2(2)$
$-\sqrt{1/16}\varphi_3^1(1)\varphi_2^1(2)\varphi_1^2(1)\varphi_4^2(2)$
$-\sqrt{1/16}\varphi_2^1(1)\varphi_3^1(2)\varphi_4^2(1)\varphi_1^2(2)$
$+\sqrt{1/16}\varphi_3^1(1)\varphi_2^1(2)\varphi_4^2(1)\varphi_1^2(2)$

[1, 1, 1, 1] found in rep.2:

$+\sqrt{1/12}\varphi_3^1(1)\varphi_4^1(2)\varphi_1^2(1)\varphi_2^2(2)$
$-\sqrt{1/12}\varphi_4^1(1)\varphi_3^1(2)\varphi_1^2(1)\varphi_2^2(2)$
$-\sqrt{1/12}\varphi_3^1(1)\varphi_4^1(2)\varphi_2^2(1)\varphi_1^2(2)$
$+\sqrt{1/12}\varphi_4^1(1)\varphi_3^1(2)\varphi_2^2(1)\varphi_1^2(2)$
$-\sqrt{1/48}\varphi_1^1(1)\varphi_3^1(2)\varphi_2^2(1)\varphi_4^2(2)$
$+\sqrt{1/48}\varphi_3^1(1)\varphi_1^1(2)\varphi_2^2(1)\varphi_4^2(2)$
$+\sqrt{1/48}\varphi_1^1(1)\varphi_3^1(2)\varphi_4^2(1)\varphi_2^2(2)$
$-\sqrt{1/48}\varphi_3^1(1)\varphi_1^1(2)\varphi_4^2(1)\varphi_2^2(2)$
$+\sqrt{1/48}\varphi_2^1(1)\varphi_4^1(2)\varphi_1^2(1)\varphi_3^2(2)$
$-\sqrt{1/48}\varphi_4^1(1)\varphi_2^1(2)\varphi_1^2(1)\varphi_3^2(2)$
$-\sqrt{1/48}\varphi_2^1(1)\varphi_4^1(2)\varphi_3^2(1)\varphi_1^2(2)$
$+\sqrt{1/48}\varphi_4^1(1)\varphi_2^1(2)\varphi_3^2(1)\varphi_1^2(2)$
$-\sqrt{1/12}\varphi_1^1(1)\varphi_2^1(2)\varphi_3^2(1)\varphi_4^2(2)$
$+\sqrt{1/12}\varphi_2^1(1)\varphi_1^1(2)\varphi_3^2(1)\varphi_4^2(2)$
$+\sqrt{1/12}\varphi_1^1(1)\varphi_2^1(2)\varphi_4^2(1)\varphi_3^2(2)$
$-\sqrt{1/12}\varphi_2^1(1)\varphi_1^1(2)\varphi_4^2(1)\varphi_3^2(2)$
$+\sqrt{1/48}\varphi_1^1(1)\varphi_4^1(2)\varphi_2^2(1)\varphi_3^2(2)$
$-\sqrt{1/48}\varphi_4^1(1)\varphi_1^1(2)\varphi_2^2(1)\varphi_3^2(2)$
$-\sqrt{1/48}\varphi_1^1(1)\varphi_4^1(2)\varphi_3^2(1)\varphi_2^2(2)$
$+\sqrt{1/48}\varphi_4^1(1)\varphi_1^1(2)\varphi_3^2(1)\varphi_2^2(2)$
$-\sqrt{1/48}\varphi_2^1(1)\varphi_3^1(2)\varphi_1^2(1)\varphi_4^2(2)$
$+\sqrt{1/48}\varphi_3^1(1)\varphi_2^1(2)\varphi_1^2(1)\varphi_4^2(2)$
$+\sqrt{1/48}\varphi_2^1(1)\varphi_3^1(2)\varphi_4^2(1)\varphi_1^2(2)$
$-\sqrt{1/48}\varphi_3^1(1)\varphi_2^1(2)\varphi_4^2(1)\varphi_1^2(2)$

[1, 1, 1, 1] found in rep.2:

$-\sqrt{1/24}\varphi_2^1(1)\varphi_3^1(2)\varphi_1^2(1)\varphi_4^2(2)$
$+\sqrt{1/24}\varphi_3^1(1)\varphi_2^1(2)\varphi_1^2(1)\varphi_4^2(2)$
$+\sqrt{1/24}\varphi_2^1(1)\varphi_3^1(2)\varphi_4^2(1)\varphi_1^2(2)$
$-\sqrt{1/24}\varphi_3^1(1)\varphi_2^1(2)\varphi_4^2(1)\varphi_1^2(2)$
$+\sqrt{1/24}\varphi_2^1(1)\varphi_4^1(2)\varphi_1^2(1)\varphi_3^2(2)$
$-\sqrt{1/24}\varphi_4^1(1)\varphi_2^1(2)\varphi_1^2(1)\varphi_3^2(2)$
$-\sqrt{1/24}\varphi_2^1(1)\varphi_4^1(2)\varphi_3^2(1)\varphi_1^2(2)$
$+\sqrt{1/24}\varphi_4^1(1)\varphi_2^1(2)\varphi_3^2(1)\varphi_1^2(2)$
$-\sqrt{1/24}\varphi_1^1(1)\varphi_3^1(2)\varphi_2^2(1)\varphi_4^2(2)$
$+\sqrt{1/24}\varphi_3^1(1)\varphi_1^1(2)\varphi_2^2(1)\varphi_4^2(2)$
$+\sqrt{1/24}\varphi_1^1(1)\varphi_3^1(2)\varphi_4^2(1)\varphi_2^2(2)$
$-\sqrt{1/24}\varphi_3^1(1)\varphi_1^1(2)\varphi_4^2(1)\varphi_2^2(2)$
$+\sqrt{1/24}\varphi_1^1(1)\varphi_4^1(2)\varphi_2^2(1)\varphi_3^2(2)$
$-\sqrt{1/24}\varphi_4^1(1)\varphi_1^1(2)\varphi_2^2(1)\varphi_3^2(2)$
$-\sqrt{1/24}\varphi_1^1(1)\varphi_4^1(2)\varphi_3^2(1)\varphi_2^2(2)$
$+\sqrt{1/24}\varphi_4^1(1)\varphi_1^1(2)\varphi_3^2(1)\varphi_2^2(2)$
$-\sqrt{1/24}\varphi_3^1(1)\varphi_4^1(2)\varphi_1^2(1)\varphi_2^2(2)$
$+\sqrt{1/24}\varphi_4^1(1)\varphi_3^1(2)\varphi_1^2(1)\varphi_2^2(2)$
$+\sqrt{1/24}\varphi_3^1(1)\varphi_4^1(2)\varphi_2^2(1)\varphi_1^2(2)$
$-\sqrt{1/24}\varphi_4^1(1)\varphi_3^1(2)\varphi_2^2(1)\varphi_1^2(2)$
$+\sqrt{1/24}\varphi_1^1(1)\varphi_2^1(2)\varphi_3^2(1)\varphi_4^2(2)$
$-\sqrt{1/24}\varphi_2^1(1)\varphi_1^1(2)\varphi_3^2(1)\varphi_4^2(2)$
$-\sqrt{1/24}\varphi_1^1(1)\varphi_2^1(2)\varphi_4^2(1)\varphi_3^2(2)$
$+\sqrt{1/24}\varphi_2^1(1)\varphi_1^1(2)\varphi_4^2(1)\varphi_3^2(2)$

[1, 1, 1, 1] found in rep.3:

$+\sqrt{1/24}\varphi_3^1(1)\varphi_4^1(2)\varphi_1^2(1)\varphi_2^2(2)$
$-\sqrt{1/24}\varphi_4^1(1)\varphi_3^1(2)\varphi_1^2(1)\varphi_2^2(2)$
$-\sqrt{1/24}\varphi_3^1(1)\varphi_4^1(2)\varphi_2^2(1)\varphi_1^2(2)$
$+\sqrt{1/24}\varphi_4^1(1)\varphi_3^1(2)\varphi_2^2(1)\varphi_1^2(2)$
$-\sqrt{1/24}\varphi_2^1(1)\varphi_4^1(2)\varphi_1^2(1)\varphi_3^2(2)$
$+\sqrt{1/24}\varphi_4^1(1)\varphi_2^1(2)\varphi_1^2(1)\varphi_3^2(2)$
$+\sqrt{1/24}\varphi_2^1(1)\varphi_4^1(2)\varphi_3^2(1)\varphi_1^2(2)$
$-\sqrt{1/24}\varphi_4^1(1)\varphi_2^1(2)\varphi_3^2(1)\varphi_1^2(2)$
$+\sqrt{1/24}\varphi_2^1(1)\varphi_3^1(2)\varphi_1^2(1)\varphi_4^2(2)$
$-\sqrt{1/24}\varphi_3^1(1)\varphi_2^1(2)\varphi_1^2(1)\varphi_4^2(2)$
$-\sqrt{1/24}\varphi_2^1(1)\varphi_3^1(2)\varphi_4^2(1)\varphi_1^2(2)$
$+\sqrt{1/24}\varphi_3^1(1)\varphi_2^1(2)\varphi_4^2(1)\varphi_1^2(2)$
$+\sqrt{1/24}\varphi_1^1(1)\varphi_4^1(2)\varphi_2^2(1)\varphi_3^2(2)$
$-\sqrt{1/24}\varphi_4^1(1)\varphi_1^1(2)\varphi_2^2(1)\varphi_3^2(2)$
$-\sqrt{1/24}\varphi_1^1(1)\varphi_4^1(2)\varphi_3^2(1)\varphi_2^2(2)$
$+\sqrt{1/24}\varphi_4^1(1)\varphi_1^1(2)\varphi_3^2(1)\varphi_2^2(2)$
$-\sqrt{1/24}\varphi_1^1(1)\varphi_3^1(2)\varphi_2^2(1)\varphi_4^2(2)$

$+\sqrt{1/24}\varphi_3^1(1)\varphi_1^1(2)\varphi_2^2(1)\varphi_4^2(2)$
$+\sqrt{1/24}\varphi_1^1(1)\varphi_3^1(2)\varphi_4^2(1)\varphi_2^2(2)$
$-\sqrt{1/24}\varphi_3^1(1)\varphi_1^1(2)\varphi_4^2(1)\varphi_2^2(2)$
$+\sqrt{1/24}\varphi_1^1(1)\varphi_2^1(2)\varphi_3^2(1)\varphi_4^2(2)$
$-\sqrt{1/24}\varphi_2^1(1)\varphi_1^1(2)\varphi_3^2(1)\varphi_4^2(2)$
$-\sqrt{1/24}\varphi_1^1(1)\varphi_2^1(2)\varphi_4^2(1)\varphi_3^2(2)$
$+\sqrt{1/24}\varphi_2^1(1)\varphi_1^1(2)\varphi_4^2(1)\varphi_3^2(2)$

NOTE ON THE EXPERIMENTAL EVIDENCE FOR QUANTUM MECHANICAL COHERENCE IN RED BLOOD CELLS

S. Rowlands

Faculty of Medicine, University of Calgary
Calgary, Alberta, Canada, T2N 4N1

Present address: R.R.1, Site 160, Box 31
Bowser, British Columbia, Canada, V0R 1G0

Frölich's theory of coherent excitations in biological systems (Frölich, 1968, 1980) predicts interactions between living cells at distances greater than the range of chemical forces. To exhibit such an interaction the theory requires the cells to have, as is normal, a membrane potential, a supply of energy and, of course, an intact structure. Such an interaction between human red blood cells at a distance of several micrometres has been experimentally demonstrated by one method (Rowlands et al., 1982a, b) and confirmed by another (Fritz, 1984). The interaction disappears when the structure of the cells is disorganised and it disappears reversibly if either the membrane potential or the supply of energy is brought to zero.

In the original experiments (Rowlands et al., 1982a) the Brownian motion of human red blood cells in anticoagulated blood was recorded by time-lapse cinematography. When red cells touch they adhere and, as time goes on, unique aggregates form (named rouleaux). The observed rate of aggregation of normal cells is significantly greater than

(a) that predicted by Einstein's theory of the brownian motion,
(b) the rate of aggregation of plastic microspheres of comparable size,
(c) the rate of aggregation of red cells treated with a noxious agent,
(d) the rate of aggregation of red cells with either zero membrane potential or zero energy.

In this last case (d) restoration of the membrane potential or of the energy stores returns the rate of aggregation to the range for normal living cells. Smoluchowski's analysis of the kinetics of aggregation under Brownian motion (see Chandrasekhar, 1943) shows that this interaction of human red cells occurs at a distance of several micrometres.

Müller-Herold et al., (1987) assert that changes in the shape and size of red blood cells, when the membrane potential or the energy stores are lowered, invalidates the evidence outlined above. Reference to Rowlands et al. (1982a) will show that the effects of such shape or size changes were compensated for in the design of the study as follows: The motion of 150-200 cells in a precision optical chamber was recorded by time-lapse cinematography and their rate of aggregation was measured. In each experiment there were to be seen at least thirty cells which, by chance, remained well clear of all other cells and which therefore did not participate in the process of aggregation. Their motion was the unperturbed Brownian motion of isolated cells. In each experiment this basic Brownian motion of thirty isolated cells was measured separately. Then the cellular interaction was calculated from the measured rate of aggregation divided by the rate of aggregation which would have occurred as a result of the measured unperturbed Brownian motion of the same cells in the same experiment. This control therefore corrects for any hydrodynamical change produced by changes in cell size or shape.

The change in unperturbed Brownian motion on manipulation of the cells was small (Rowlands et al., 1982a, Table 1, p. 55) and not statistically significant whereas the changes in the interaction were highly significant ($p < 0.001$, Table 2, p. 56). Scrutiny of the cine films revealed no change in cell shape or size on treating the cells with a noxious agent (glutaraldehyde) nor when the membrane potential was lowered and raised. Cells were, however, crenated (echinocyte formation) in three of four experiments in which the energy stores were depleted. Though the difference was not statistically significant, energy-depleted cells had less hydrodynamical drag than normal, yet they aggregated at a significantly lower rate than energy-replete cells (Rowlands et al., 1982a, Table 1, p. 55).

The assertion by Müller-Herold et al., (1987) that hydrodynamical changes invalidate the evidence for a Frölich interaction in human red cells is therefore incorrect.

The specificity of the interaction and its mode of transmission have been investigated. A critical review of many experiments on this cellular interaction is to be published shortly (Rowlands, 1988).

REFERENCES

1. Chandrasekhar,S., 1943, Stochastic problems in physics and astronomy, Rev. Mod. Phys. 15:1.
2. Fritz, O.G., 1984, Anomalous diffusion of erythrocytes in the presence of polyvinylpyrrolidone, Biophys. J. 46:219.
3. Frölich, H., 1968, Long range coherence and energy storage in biological systems, Int. J. Quantum Chem. 2:641.
4. Frölich, H., 1980, The biological effects of microwaves and related questions, Adv. Electron. Electron Phys. 53:85.
5. Müller-Herold, U., Lutz, H. U., and Kedem, O., 1987, Quantum mechanical coherence in red blood cells: No experimental evidence, J. Theor. Biol. 126:251.
6. Rowlands, S., Sewchand, L. S., and Enns, E. G., 1982a, A quantum mechanical interaction of human erythrocytes, Can. J. Physiol. Pharmacol. 60:52.
7. Rowlands, S., Sewchand, L. S., and Enns, E. G., 1982b, Further evidence for a Frölich interaction of erythrocytes, Phys. Lett. 87A:256.
8. Rowlands, S., 1988, The interaction of living red blood cells, in "Biological Coherence and Response to External Stimuli," H. Frölich, ed., Springer, Berlin, Heidelberg, New York.

THE ANALYSIS OF UNITARY IRREDUCIBLE REPRESENTATIONS OF u(p, 1) ALGEBRA BY USING THE PROJECTION OPERATOR METHOD

Yu. F. Smirnov, L. Ya. Stotland, and V. A. Knyr
Nuclear Physics Institute
Moscow State University

ABSTRACT

It is shown that for each unitary irreducible representation (UIR) of the $\mathbf{u}(p,1)$ algebra there is a unique extremal vector determining the structure of corresponding UIR. The classification of UIRs belonging to discrete and continuous series is given.

This paper continues our previous work[1] concerning the unitary irreducible representations (UIR) of the $\mathbf{u}(2,1)$. Here we shall analyze the UIRS of the more general $\mathbf{u}(p,1)$ algebra by using the same approach as in ref. [1], namely the projection operator method. As for applications of noncompact Lie groups and algebras to physical problems there is a rather rich literature now (see for example the books [2,3]).

The generators A_{ik} $(i,k = 1,2,\ldots,p+1)$ of the $\mathbf{u}$ algebra, acting in the space R of the weight vectors, satisfy the standard commutation relations

$$[A_{ik}, A_{lm}] = \delta_{kl}A_{im} - \delta_{im}A_{lk}\,. \tag{1}$$

We shall use the canonical reduction

$$\mathbf{U}(p,1) \supset \mathbf{U}(p) \supset \mathbf{U}(p-1) \supset \cdots \supset \mathbf{U}(1)\,. \tag{2}$$

In this case every basic vector of UIR can be labelled by a Gelfand-Graev scheme[3,4]

$$\left|\begin{matrix} f_{1p+1}\, f_{2p+1} \cdots f_{p+1\,p+1} \\ f_{1p}\, f_{2p} \cdots f_{pp} \\ \cdots \\ f_{11} \end{matrix}\right\rangle \tag{3}$$

It means that this vector belongs to a definite UIR of each algebra entering into the chain.[2] The signatures of UIRs of these algebras are indicated in corresponding rows of the Gelfand-Graev scheme.[4] Since the structure of the UIR $[f_{1p}\, f_{2p} \cdots f_{pp}]$ of the compact Lie algebra $\mathbf{u}(p)$ is well known, it suffices to consider only the highest weight vectors with respect to this subalgebra. These vectors will be called rank-one vectors and denoted by the shortened scheme

$$\left| \begin{matrix} f_{1p+1}\, f_{2p+1} \cdots f_{p+1\,p+1} \\ f_{1p}\, f_{2p} \cdots f_{pp} \end{matrix} \right\rangle, \tag{4}$$

where $f_{ip} - f_{jp} \geq 0$, $(f_{ip} - f_{jp})$ is a non-negative integer (if $i < j$). The subspace of all rank-one vectors in R will be labelled as R_H.

For the rank-one vectors the following obvious relations are valid

$$A_{ik} \left| \begin{matrix} f_{1p+1}\, f_{2p+1} \cdots f_{p+1\,p+1} \\ f_{1p}\, f_{2p} \cdots f_{pp} \end{matrix} \right\rangle = 0, \tag{5}$$

if $i < k \leq$ p.

Let us introduce the raising (lowering) operators $\mathbf{z}_i$, $\mathbf{z}_{-i}$ $(i = 1, 2, \dots, p)$ belonging to the Mickelsson-Zhelobenko algebra[5] and acting in R_H. They are determined by formulas

$$\mathbf{z}_i = \mathbf{P} A_{ip+1} \mathbf{P}, \quad \mathbf{z}_{-i} = \mathbf{P} A_{p+1i} \mathbf{P}, \quad (\mathbf{z}_i)^+ = -\mathbf{z}_{-i}. \tag{6}$$

Here $\mathbf{P}$ is an extremal projection operator that projects any vector of R into the R_H subspace. According to ref. [6] the projector P is of a form

$$\begin{gathered} \mathbf{P} = (\mathbf{P}_{12}\, P_{13} \cdots \mathbf{P}_{1p})(\mathbf{P}_{23}\, \mathbf{P}_{24} \cdots \mathbf{P}_{2p}) \cdots \mathbf{P}_{p-1p}, \\ \mathbf{P}_{ik} = \sum_r \frac{(-1)^r}{r!} \frac{(A_{ii} - A_{kk} + 1)!}{(A_{ii} - A_{kk} + 1 + r)!} A_{ki}^r A_{ik}^r. \end{gathered} \tag{7}$$

The result of the action of the operators $\mathbf{z}_{\pm i}$ on the vectors (4) is

$$\mathbf{z}_{\pm i} \left| \begin{matrix} f_{1p+1} \cdots f_{p+1\,p+1} \\ f_{1p} \cdots f_{pp} \end{matrix} \right\rangle \sim \left| \begin{matrix} f_{1p+1} \cdots f_{p+1\,p+1} \\ f'_{1p} \cdots f'_{pp} \end{matrix} \right\rangle \tag{8}$$

i.e. the $\mathbf{z}_{\pm i}$ transform the rank-one vector (4) into another rank-one vector with $f'_{kp} = f_{kp} \pm \delta_{ik}$. Since the relations $f_{ip} \geq f_{jp}$ $(i < j)$ are valid for the rank-one vectors there exists among them always the vector $|f\rangle$ satisfying the conditions

$$\mathbf{z}_{-1} |f\rangle = \mathbf{z}_2 |f\rangle = \cdots = \mathbf{z}_p |f\rangle = 0. \tag{9}$$

We shall refer to this vector as the extremal vector (EV). Various types of the UIR of the $\mathbf{u}(p,1)$ algebra can be labelled by the weight of EV. (In the case of the UIR belonging to continuous series it is necessary to use some additional parameters to characterise the EV). Because of the cyclicity of the UIRs an arbitrary rank-one vector can be written in the form

$$\left| \begin{matrix} f_{1p+1}\, f_{2p+1} \cdots f_{p+1\,p+1} \\ f_{1p}\, f_{2p} \cdots f_{pp} \end{matrix} \right\rangle = N(r_1\, r_2 \cdots r_p)^{-1}\, \mathbf{z}_1^{r_1}\, \mathbf{z}_{-2}^{r_2} \cdots \mathbf{z}_{-p}^{r_p} |f\rangle \tag{10}$$

where $N(r_1 r_2 \cdots r_p)$ is a normalization factor, $r_1 = f_{1p} - f_{1p+1}$, $r_i = f_{ip+1} - f_{ip}$. The normalization factor is determined by the formula

$$|N(r_1 r_2 \cdots r_p)|^2 = (-1)^{\Sigma r_i} \langle f| \, \mathbf{z}_p^{r_p} \mathbf{z}_{p-1}^{r_{p-1}} \cdots \mathbf{z}_2^{r_2} \mathbf{z}_{-1}^{r_1} \mathbf{z}_1^{r_1} \mathbf{z}_{-2}^{r_2} \cdots \mathbf{z}_{-p}^{r_p} \, |f\rangle \,. \tag{11}$$

The Hermitian properties (6) of the $\mathbf{z}_{\pm i}$ operators are taken into account in Eq. (11). Each UIR contains a single EV. The weight $[f_{1p+1} f_{2p+1} \cdots f_{p+1\,p+1}]$ of the EV will be called the extremal weight (EW).

In order to classify the UIRs of the $\mathbf{u}(p,1)$ algebra it is necessary to calculate the normalization factor (11). It can be done by using the permutation relations for the operators (6) obtained in ref. [5]:

$$\mathbf{z}_i \mathbf{z}_j = \rho_{ij} \mathbf{z}_j \mathbf{z}_i \quad \text{if} \quad i + j \neq 0. \tag{12}$$

In particular $\rho_{ij} = 1$ if $sgn\, i \neq sgn\, j$;

$$\rho_{ij} = \rho_{-i-j} = \frac{\phi_{ij} + 1}{\phi_{ij}} \quad \text{at } 1 \leq i < j \leq p, \tag{13}$$

$$\phi_{ij} = A_{ii} - A_{jj} - i + j;$$
$$\mathbf{z}_i \mathbf{z}_{-i} = \sum_{j=1}^{p} \omega_{ij} \mathbf{z}_{-j} \mathbf{z}_j + \epsilon_i, \tag{14}$$

$$\omega_{ij} = (1 - \phi_{ij})^{-1} \delta_i^- \delta_j^+ ,$$
$$\epsilon_i = \delta_i^- (\phi_{ip+1} - 1) \,, \; \delta_i^{\pm} = \prod_{k=i+1}^{p} \frac{(\phi_{ik} \pm 1)}{\phi_{ik}} \,. \tag{15}$$

The relations (12)-(15) and the condition (9) allow to obtain the normalization factor (11) in all cases except for the case $f_{1p+1} = f_{2p+1}$. In fact we can obtain from the Eq. (15) the following relation

$$\mathbf{z}_{-1}\mathbf{z}_1 = \prod_{k=2}^{p} \frac{\phi_{1k}}{\phi_{1k}+1} \left\{ \prod_{k=2}^{p} \frac{\phi_{1k}}{\phi_{1k}-1} \mathbf{z}_1 \mathbf{z}_{-1} + \sum_{j=2}^{p} \frac{1}{\phi_{1j} - 1} \prod_{k=j+1}^{p} \frac{\phi_{1k} - j - 2}{\phi_{jk}} \mathbf{z}_{-1} \mathbf{z}_j - (\phi_{1p+1} - 1) \right\} . \tag{16}$$

Applying it to the extremal vector $|f\rangle$ and using its property (9) we can write

$$\mathbf{z}_{-1}\mathbf{z}_1 \,|f\rangle = (\mu_{p+1} - \mu_1 + 1) \prod_{k=2}^{p} \frac{(\mu_1 - \mu_k)}{(\mu_1 - \mu_k + 1)} \,|f\rangle \tag{17}$$

where $\mu_i = f_{ip+1} - i \; (i = 1, 2, \ldots, p+1)$. In the case $f_{1p+1} = f_{2p+1}$ the denominator $\phi_{12} - 1$ of Eq. (16) vanishes and it is necessary to give a special definition for the vector $\mathbf{z}_{-1}\mathbf{z}_1 \,|f\rangle$. Since the extremal vector $|f\rangle$ is unique the combination $\mathbf{z}_{-1}\mathbf{z}_1 \,|f\rangle$ is proportional to EV $|f\rangle$. We choose the proportionality coefficient in the form

$$\mathbf{z}_{-1}\mathbf{z}_1 \,|f\rangle = \prod_{k+2}^{p} \frac{(\mu_1 - \mu_k)}{(\mu_1 - \mu_k + 1)} (\mu_{p+1} - \mu_1 + 1 + \delta_{f_{1p+1,2p+1}} \rho) \,|f\rangle \,. \tag{18}$$

Here $\delta_{f_1 f_2}$ is Kronecker's symbol. The real parameter ρ is a labelling index for the UIR belonging to the continuous series. By using Eqs. (12)-(18) it is possible to derive the reciprocal relations for the normalizing factors

$$\begin{aligned} &|N(r_1+1, r_2, \ldots, r_p)|^2 \\ &\quad = |N(r_1, r_2, \ldots, r_p)|^2 \prod_{i=2}^{p} \frac{(\mu_1 - \mu_i + r_1)}{(\mu_1 - \mu_i + r_1 + r_i + 1)} \\ &\quad \times \left(r_1 + \lambda \delta_{f_{1p+1} f_{2p+1}}\right) \left(\mu_1 - \mu_{p+1} + r_1 + 1 - \lambda \delta_{f_{1p+1} f_{2p+1}}\right), \end{aligned} \tag{19}$$

$$\begin{aligned} &|N(0,0,\ldots, r_l + 1, r_{l+1}, \ldots, r_p)|^2 \\ &\quad = |N(0,0,\ldots, r_l, r_{l+1}, \ldots, r_p)|^2 \frac{r_{l+1}}{(\mu_1 - \mu_l + r_l + 1)} \\ &\quad \cdot \prod_{i=l+1}^{p} \frac{\mu_l - \mu_i - r_l - 1)}{(\mu_l - \mu_i - r_l + r_i)} \left(\mu_1 - \mu_l + r_l - \lambda \delta_{f_{1p+1} f_{2p+1}}\right) \\ &\quad \left(\mu_{p+1} - \mu_l + \lambda \delta_{f_{1p+1} f_{2p+1}} + r_l + 2\right). \end{aligned} \tag{20}$$

Here instead of ρ a new parameter λ is introduced. It is connected with ρ by equation

$$\lambda^2 + \lambda(\mu_{p+1} - \mu_1 + 2) = \rho .$$

Thus the normalizing factors for the rank-one vectors are determined unambiguously by $p+1$ components of the extremal weight $[f_{1p+1} f_{2p+1} \cdots f_{p+1\,p+1}]$ for $f_{1p+1} \neq f_{2p+1}$, or by p quantum numbers $f_{2p+1} \cdots f_{p+1\,p+1}$ and some additional parameter λ for $f_{1p+1} = f_{2p+1}$. These results permit the classification of possible types of UIRs for the $\mathbf{u}(p,1)$ algebra. The analysis is based upon the condition of reality and positivity of $|N|^2$. This condition implies some limitations on the parameter λ which can be real or complex. In the last case

$$\lambda = (\mu_{p+1} - \mu_1 + 2)/2 + i\nu, \quad \nu > 0. \tag{21}$$

To find the structure of a UIR of $\mathbf{u}(p,1)$ means to formulate the restrictions for the possible values of weights of the rank-one vectors which belong to this UIR. Let us introduce the quantities $l_i = f_{ip} - i$ and find out their possible values for the different types of UIR.

1. There are the UIRs $C(\mu_2, \ldots, \mu_{p+1}, \nu)$ with $\mu_1 = \mu_2 - 1$ and complex λ (see Eq. (21)). In this case $l_1 \geq \mu_2 + 1$, $\mu_{i+1} + 1 \leq l_i \leq \mu_i$ $(i = 2, \ldots, p+1)$, $l_p \leq \mu_p$. Such UIRs belong to the continuous series because they are characterized by the continuous index ν in addition to the $\mu_2 \cdots \mu_{p+1}$ which are integers.

2. For real λ the following notation is convenient

$$\mu_{min\,(max)} = min\,(max)\, \{\mu_1 - \delta_{f_1 f_2}\lambda - 1, \quad \mu_{p+1} + \delta_{f_1 f_2}\lambda + 1\}.$$

The symbol $R_{\alpha\beta}\,(j, k, \mu_{max}, \mu_{min}, \mu_2 \cdots \mu_p)$ denotes the UIR. The integer numbers j, k determine the interrelations between numbers $\mu_1, \mu_2, \ldots, \mu_p, \mu_{min}, \mu_{max}$. Namely the following variants are possible.

$$\begin{aligned} &\mu_2 > \mu_3 > \cdots > \mu_j \geq \mu_{max} + 1 > \mu_{j+1} > \cdots > \mu_k \\ &\quad > \mu_{min} + 1 \geq \mu_{k+1} > \mu_{k+2} > \cdots > \mu_p; \\ &\mu_l = \mu_{l+1} + 1, \quad l = j+1, \ldots, k \\ &(j \leq k = 1, 2, \cdots, p-1, \quad j = k = p). \end{aligned} \tag{22}$$

α, β will be auxiliary indices.

Let us describe the structure of these UIRs.

A) $j, k \neq 1, p;\quad \mu_i - \mu_{max},\quad \mu_i - \mu_{min}$ are integer.
$l_1 \geq \mu_2 + 1;\quad \mu_{i+1} < l_i \leq \mu_i\quad$ for $\quad i = 2, \ldots, j-1;\quad k+1, \ldots, p-1;$
$\mu_{max} + 1 \leq l_j \leq \mu_i;\quad l_i = \mu_i\quad$ for $\quad i = j+1, \ldots, k;$
$l_p \leq \mu_p$ (except for $k = p-1,\quad \mu_p = \mu_{min} + 1$ when $l_p = \mu_p$).

B) $j = 1,\ k \neq 1$.
$l_i = \mu_i,\quad i = 2, 3, \ldots, k;\quad \mu_i \geq l_i \geq \mu_{i+1} + 1,$
$i = k+1, k+2, \ldots, p-1$ (except for $\mu_{k+1} = \mu_{min} + 1$ when $l_{k+1} = \mu_{k+1}$),
$l_p \leq \mu_p$ (except for $k = p-1,\ \mu_{min} + 1 = \mu_{k+1}$ when $l_{k+1} = \mu_{k+1}$).
Two variants of limitations on l_1 giving two types of representation are possible.

1) $R_1\,(1, k, \mu_{min}, \mu_{max}, \mu_2 \cdots \mu_p) : \mu_1 = \mu_{max} + 1,\quad l_1 \geq \mu_{max} + 1.$

2) $R_2\,(1, k, \mu_{min}, \mu_{max}, \mu_2 \cdots \mu_p) : l_1 = \mu_{max} = \mu_2 + 1.$

C) $j = k = 1$.
$\mu_{i+1} < l_i \leq \mu_i,\quad i = 2, 3, \ldots, p-1;\quad l_p \leq \mu_p.$
The class of UIRs $R(1, 1, \cdots)$ includes the following types of representations:

1) $R_1(1, 1, \cdots) : \mu_1 = \mu_{max} + 1,\quad l_1 \geq \mu_{max} + 1;$

2) $R_2(1, 1, \cdots) : \mu_1 = \mu_{min} + 1 = \mu_{max} = l_1;$

3) $R_3(1, 1, \cdots) : \mu_1 = \mu_{min} + 1,\quad \mu_{max} - 1 < \mu_{min} < \mu_{max},$
$\mu_{min} - \mu_2$ is integer, $l_1 \geq \mu_{min} + 1;$

4) $\mu_1 = \mu_2 + 1$
$R_{41}(1, 1, \cdots) : \mu_{max} < \mu_1 < \mu_{max} + 2,\quad l_1 \geq \mu_1;$
$R_{42}(1, 1, \cdots) : l_1 = \mu_{max} = \mu_1;$
$R_{43}(1, 1, \cdots) : \mu_1 - \mu_{min}$ is integer, $\quad l_1 = \mu_1, \ldots, \mu_{min};$
$R_{44}(1, 1, \cdots)$: both of the numbers $\mu_{min} - \mu_2$ and $\mu_{max} - \mu_2$ are not integer, $l_1 \geq \mu_1;$
$R_{45}(1, 1, \cdots) : \mu_1 - \mu_{max}$ is integer, $\quad \mu_1 - \mu_{min}$ is noninteger,
$\mu_{max} - \mu_{min} < 1,\ l_1 = \mu_1, \ldots, \mu_{max};$
$R_{46}(1, 1, \cdots) : l_1 = \mu_1 = \mu_{min}.$

D) $j = k = p$
$l_1 \geq \mu_2 + 1,\quad \mu_i \geq l_i \geq \mu_{i+1} + 1,\quad i = 2, \ldots, p-1,$

$R_1(p, p, \cdots) :\ \mu_p \geq \mu_{max} + 1,\quad \mu_p - \mu_{max}$ is integer,
$l_p = \mu_p, \ldots, \mu_{max} + 1;$

$R_2(p, p, \cdots) :\ \mu_p > \mu_{max} + 1,\quad \mu_p - \mu_{max}$ is noninteger,
$\mu_{max} - \mu_{min} < 1,\quad \mu_p - \mu_{min}$ is integer,
$l_p = \mu_p, \ldots, \mu_{min} + 1;$

$R_3(p, p, \cdots) :\ \mu_p = \mu_{min} + 1, l_p = \mu_p.$

Since the exact form of the orthonormalized basic vectors is known it is now easy to calculate the matrix elements of the $\mathbf{u}(p, 1)$ algebra generators. We shall not discuss these calculations here but it should be noted that the resulting matrix elements for the discrete series will coincide with the Gelfand-Graev formulas[3,4]. The comparison of list of representations give above with the results of the paper[7] shows that all UIRs of the $\mathbf{u}(p, 1)$ algebra are obtained by our approach. Thus it is established that in each UIR of the $\mathbf{u}(p, 1)$ algebra there is a single EV which determines completely the structure of the corresponding UIR. In such a situation the projection operator method is rather effective for analysis of the structure of UIRs. Unfortunately for the general $\mathbf{u}(p, q)$ case the problem is more complicated and new efforts are necessary.

REFERENCES

1. Yu. F. Smirnov, V. N. Tolstoy, V. A. Knyr, and L. Ya. Stotland, in "Group Theoretical Methods in Physics," Proc. of Third Yurmala Seminar, VNU Science Press, Utrecht, The Netherlands, 2:223 (1986).
2. I. A. Malkin and V. I. Manko, "Dynamical symmetries and coherent states of the quantum systems," (Russian) Nauka, Moscow (1979).
3. A. O. Barut and R. Raczka, "The Theory of Group Representations and Applications," Warszawa (1977).
4. I. M. Gelfand and M. I. Graev, Izv. Acad. Nauk SSSR, Ser. Math., (Russian) 29:1329 (1965).
5. D. P. Zhelobenko, Dokl. Acad. Nauk SSSR, (Russian) 4:1317 (1984).
6. R. M. Asherova, Yu. F. Smirnov, and V. N. Tolstoy, Matem. Zam., (Russian) 26:15 (1979).
7. U. Ottoson, Comm. Math. Phys. 10:144 (1968).

CALCULATING PROPERTIES OF LIE GROUPS

B.G. Wybourne
Physics Department, University of Canterbury
Christchurch, New Zealand

ABSTRACT

The calculation of the properties of the Lie groups via personal computers is briefly reviewed with particular emphasis of the package SCHUR.

1. INTRODUCTION

In many areas of chemistry, physics and mathematics it is necessary to be able to compute the properties of representations of Lie groups. For simple cases calculations can be made by hand provided the person is familiar with the diversity of calculational procedures. Interest often involves large representations making hand calculation excessively laborious with a considerable propensity for errors.

The modern personal computer can lead to efficient interactive evaluation of many group properties without the user requiring a detailed knowledge of the relevant algorithms. In this way the computer screen becomes a notepad leaving the user free to develop real science.

Among the properties frequently required are:

(1) Properties of irreps such as dimensions, Casimir eigenvalues, the second order Dynkin index and transformations between Dynkin and partition based labelling schemes.

(2) Kronecker products of lists of irreps of all the compact Lie groups.

(3) Branching decompositions for nested Lie groups.

(4) Young tableaux or Schur function type calculations such as outer, inner, skew and plethysm operations.

(5) The isomorphisms and automorphisms for groups such as $Sp_4 \sim SO_5$, $SU_4 \sim SO_6$ and $SO_8 \sim SO_8$.

(6) Modification rules for standardising non-standard representations.

In addition it is useful to have the ability to:

(a) allow the user to define sequences of instructions as functions to implement his or her own formulas;

(b) save results as logfiles for subsequent use or editing;

(c) bring to the screen helpfiles describing any of the commands available.

SCHUR

We have developed a PASCAL code, known as SCHUR, for IBM compatible personal computers for carrying out all the above. This programme requires a Pc with preferably 640 K RAM and a hard disk. A SUN workstation version is being developed.

Central to SCHUR is the efficient implementation of the Littlewood-Richardson rule. A standard XT clone operating at 8 MHz evaluates as a sorted list the 930 terms in {4321}.{4321} in ~4.7 seconds or ~1 second on a 286 Pc. Needless to say small outer products are evaluated virtually instantly. The 260,278 representations that arise in the seventh power of the fundamental 248-dimensional irrep of E_8 are determined in ~4 minutes.

The computation of Kronecker products in SO_7, say [1]x[21], is evaluated by just two instructions (indicated by an arrow ⟶) as below:

```
->gr so7
  Group is SO(7)
  REP>
->p11,21
  [32] + [311] + [3] + [221] + [211] + 2[21] + [111] + [1]
```

It is also possible to handle products of groups as shown in the following example where three groups are set and then the product to two irreps computed. The group

Sp_4 is locally isomorphic to SO_5 and the command *autom* converts the Sp_4 irrep into the appropriate SO_5 irrep. Successive isomorphisms and the SO_8 automorphism are shown. Finally we see we can choose to branch the groups. In this case we have branched the third group in the product, SO_8, into irreps of $U_1 x SU_4$.

```
-> gr3sp4su4so8
Groups are  Sp(4) * SU(4) * SO(8)
DP>
p[1*1*1],[1*1*s0+]
    <2>{2}[s;1]+ + <2>{2}[s;0]- + <2>{11}[s;1]+ + <2>{11}[s;0]-
  + <11>{2}[s;1]+ + <11>{2}[s;0]- + <11>{11}[s;1]+ + <11>{11}[s;0]-
  + <0>{2}[s;1]+ + <0>{2}[s;0]- + <0>{11}[s;1]+ + <0>{11}[s;0]-
DP>
autom gr1,so5,last
Groups are  SO(5) * SU(4) * SO(8)
    [11]{2}[s;1]+ + [11]{2}[s;0]- + [11]{11}[s;1]+ + [11]{11}[s;0]-
  + [1]{2}[s;1]+ + [1]{2}[s;0]- + [1]{11}[s;1]+ + [1]{11}[s;0]-
  + [0]{2}[s;1]+ + [0]{2}[s;0]- + [0]{11}[s;1]+ + [0]{11}[s;0]-
DP>
autom gr2,so6,last
Groups are  SO(5) * SO(6) * SO(8)
    [11][111]+[s;1]+ + [11][111]+[s;0]- + [11][1][s;1]+ + [11][1][s;0]-
  + [1][111]+[s;1]+ + [1][111]+[s;0]- + [1][1][s;1]+ + [1][1][s;0]-
  + [0][111]+[s;1]+ + [0][111]+[s;0]- + [0][1][s;1]+ + [0][1][s;0]-
DP>
autom gr3,so8,last
Groups are  SO(5) * SO(6) * SO(8)
    [11][111]+[s;1]- + [11][111]+[s;0]+ + [11][1][s;1]- + [11][1][s;0]+
  + [1][111]+[s;1]- + [1][111]+[s;0]+ + [1][1][s;1]- + [1][1][s;0]+
  + [0][111]+[s;1]- + [0][111]+[s;0]+ + [0][1][s;1]- + [0][1][s;0]+
DP>
br5,8,gr3last
Groups are  SO(5) * SO(6) * U(1) * SU(4)
    [11][111]+{4}{211} + [11][111]+{4}{0} + [11][111]+{0}{222}
```

```
+ [11][111]+{0}{2} + 2[11][111]+{0}{11} + [11][111]+{-4}{211}
+ [11][111]+{-4}{0} + [11][1]{4}{211} + [11][1]{4}{0} + [11][1]{0}{222}
+ [11][1]{0}{2} + 2[11][1]{0}{11} + [11][1]{-4}{211} + [11][1]{-4}{0}
+ [1][111]+{4}{211} + [1][111]+{4}{0} + [1][111]+{0}{222}
+ [1][111]+{0}{2} + 2[1][111]+{0}{11} + [1][111]+{-4}{211}
+ [1][111]+{-4}{0} + [1][1]{4}{211} + [1][1]{4}{0} + [1][1]{0}{222}
+ [1][1]{0}{2} + 2[1][1]{0}{11} + [1][1]{-4}{211} + [1][1]{-4}{0}
+ [0][111]+{4}{211} + [0][111]+{4}{0} + [0][111]+{0}{222}
+ [0][111]+{0}{2} + 2[0][111]+{0}{11} + [0][111]+{-4}{211}
+ [0][111]+{-4}{0} + [0][1]{4}{211} + [0][1]{4}{0} + [0][1]{0}{222}
+ [0][1]{0}{2} + 2[0][1]{0}{11} + [0][1]{-4}{211} + [0][1]{-4}{0}
DP>
```

Approximately 100 commands are available and it is possible to create user defined functions involving these commands. As a non-trivial example consider the calculation ot plethysms of G_2 irreps $(\mu)\otimes\{\lambda\}$. These are not explicitly calculated in SCHUR but the user may evaluate them by creating a small function as shown in Table 1. These plethysms could be evaluated by first expressing (μ) in terms of characters of $SO_{\perp}$ using the inverse branching rule 32. The resulting $SO_{\perp}$ irreps can then be formed into character of SU_7 by effectively skewing each SO_7 irrep with compatible members of a series of C of S-functions [12]. The plethysms in SU_7 can be evaluated by using *scom* to convert the SU_7 characters into S functions evaluating the plethysms and then using *convert* to return them as SU_7 reps. These can now be rapidly reduced to rep of G_2 to yield the final results.

Table 1. Setting a function to compute plethysms for the exceptional group g2. N.B. The irreps of g2 have been labelled with respect to the labels of its maximal subgroup su3. These labels (a,b) are related to the Racah labels by (a,b) ⟶ (a-b,b). The function g2plethysm evaluates $(\mu)\otimes\{\lambda\}$.

rem g2plethysm	This function computes g2 plethysms
gr g2	This sets the group as g2
enter rv1	Asks ther user to enter a g2 irrep ( )
enter sv2	Asks the user to enter { }
dim[rv1]	Computes the dimension of ( )
br32gr1[rv1]	Branches g2 -> so7

```
rule last skew1 with c                 Skews last with sfns of the C series
gr su7                                 Resets the group as su7
[convert pleth scon last,sv2] Performs the plethysms and converts to su7
br1,7gr1last                           Branches su7 -> so7
br30gr1last                            Branches so7 -> g2
stop                                   End of function
DP>
fn1                                    This instruction runs the above function
Group is G(2)
enter rv1
->31                                   User input indicated by ->
enter sv2
->21
Dimension = 64
Group is SO(7)
Group is SU(7)
Group is O(7)
Group is G(2)
   (94) + (84) + 2(83) + 2(82) + (81) + 5(73) + 6(72) + 4(71) + (7) + 4(63)
+ 12(62) + 12(61) + 4(6) + 15(52) + 18(51) + 10(5) + 9(42) + 20(41)
+ 14(4) + 17(31) + 13(3) + 6(21) + 10(2) + 4(1)
DP>
->dim last                             User requests the total dimension
Dimension = 87360
DP>
->mc last                              User requests the sum of multiplicities
MultSum = 191
DP>
->tc last                              User requests the sum of distinct terms
TermSum = 24
DP>
->dim[31]                              User requests the g2 dimension of (31)
```

Dimension = 64	
DP>	
–>gr u64	User resets the group as u64
Group is U(64)	
DP>	
–>dim[21]	User requests the dimension of {21}
Dimension = 87360	Result agrees with the plethysm dimension
DP	

The results may be checked by computing its dimension and then comparing that figure with the dimension of the {21} irrep of U_{64}.

The above example indicates the diversity of calculations now possible and how the computer screen becomes a scratch pad for doing calculations allowing the user to get on with physical applications as opposed to mathematical manipulations requiring an extensive knowledge of the relevant algorithms.

ACKNOWLEDGEMENTS

I am appreciative of assistance from the conference organisers and the University of Canterbury for making my participation in a splendid conference in magnificent surroundings. A special thanks to Bruno Gruber who successfully put it altogether - a truly non-trivial task.

REFERENCES

1. R.C. King, J. Phys. A: Math. Gen. **A8**, 429 (1975)
2. G.R.E. Black, R.C. King & B.G. Wybourne, J. Phys. A: Math. Gen. **16**, 1555 (1983).

ON COMMUTATIVITY OF PRIME RINGS WITH (σ,τ)-DERIVATIONS

M. Serif Yenigül
Ege Üniversitesí, Fen Fakültesi, Matematik Bölümü
35100 Bornova-İzmir/Turkey

ABSTRACT

This work contains two sections. In the first section we tried to show that whether some commutativity properties, which are already satisfied in prime rings with ordinary derivations, are valid in prime rings with (σ,τ)-derivations. We see that some of those properties are satisfied in this case. So we give the generalizations of [6; Lemma 1, Lemma 2] and [5; Lemma 1] furthermore we proved some other properties and also we give a simple and short proof for a known property in prime rings (Lemma 1.6).

In the second section we consider the prime rings with (σ,σ)-derivations. We see that some stronger results can be obtained in addition to the results previously found. Here we give the generalizations of [3; Lemma 1, Lemma 2] and [1; Lemma 1].

Under the view of these results and some conditions it is shown that, if d is (σ,σ)-derivation of a prime ring R and semicentralizing on R, then R is commutative.

(σ,τ)-DERIVATIONS OF PRIME RINGS

Throughout in general, R will represent a prime ring with center C. Let σ, τ be ring automorphisms of R. Then the set $C_{(\sigma,\tau)} = \{c \in R \mid c\,\sigma(x) = \tau(x)c$ for all x in $R\}$ is called (σ,τ)-center of R. Moreover $C_{(1,1)} = C$ and $C_{(\sigma,\sigma)} = C_\sigma = C$. We write $[x,y]_{(\sigma,\tau)} = x\sigma(y) - \tau(y)x$, $(x,y)_{(\sigma,\tau)} = x\sigma(y) + \tau(y)x$.

Let $d : x \to x'$ be a (σ,τ)-derivation of R. That is an additive map of R satisfying $(xy)' = x'\sigma(y) + \tau(x)y'$ for all x in R.

Let U be a non-zero ideal of R. We set

$$[U] = \{u \in U \mid [u',u]_{(\sigma,\tau)} \text{ in } C_{(\sigma,\tau)}\}$$
$$(U) = \{u \in U \mid (u',u)_{(\sigma,\tau)} \text{ in } C_{(\sigma,\tau)}\}$$

If

a) $U = [U] \cup (U)$ we call that d is semicentralizing on U.

b) $U = [U]$ we call that d is centralizing on U.

c) $U = (U)$ we call that d is skew centralizing on U.

Q will represent Martindale quotient ring of R and $S = RC$ is central closure of R, which is a subring of Q, containing R.

Lemma 1.1. Let R be a prime ring and d is a non-zero (σ,τ)-derivation of R. Then left and right annihilators of R' are zero.

Proof. Let $aR' = 0$. Then for all x in R we have $ax' = 0$. Replace x by xy we get $0 = a(xy)' = ax'\sigma(y) + a\tau(x)y' = a\tau(x)y'$. Since R is prime, τ is an automorphism and d is a non-zero (σ,τ)-derivations of R $a\tau(x)y' = 0$ implies $a = 0$.

In the same way, we can prove that if $R'b = 0$, this implies $b = 0$.

Lemma 1.2. Let R be a ring. Then the following statements are equivalent.

a) R is prime.

b) The right annihilators of a non-zero subset I of R, such that $IR \subseteq I$, are zero.

c) The left annihilators of a non-zero subset L of R, such that $RL \subseteq L$, are zero.

d) Let I and L be two non-zero subsets of R, such that $IR \subseteq l$, $RI \subseteq I$ and $LR \subseteq L$, $RL \subseteq L$. Then $IL = 0$ implies $I = 0$ or $L = 0$.

Proof. Let I be a non-zero subset of R and denote the right and left annihilators of I by I_r and I_e respectively. First to show that a) $\Rightarrow$ b) since $IR \subseteq I$ and $IRI_r \subseteq II_r = 0$ implies $IRI_r = 0$. The primeness of R and $I \neq 0$ implies that $I_r = 0$. Similarly we have b) $\Rightarrow$ d), a) $\Rightarrow$ c), c) $\Rightarrow$ d). Now we show that d) $\Rightarrow$ a). Assume that for a,b in R, $aRb = 0$. Take $I = RaR$, $L = RbR$ then $IL = 0$ and so we have $I = 0$ or $L = 0$. Assume that $I = 0$ i.e. $RaR = 0$ since $R_eR = 0$ and by d) we have $R_e = 0$ so $Ra \subseteq R_e = 0$. This implies $Ra = 0$ similarly $R_r = 0$ implies $a = 0$. In the same way if we assume that $L = 0$, we get that $b = 0$. This shows that R is a prime ring.

Lemma 1.3. Let R be a prime ring, $(0) \neq U$ an ideal of R and d a (σ,τ)-derivation of R. If d is trivial on U, then d is trivial on R.

Proof. Take $0 \neq u \in U$. Then $0 = (ur)' = u'\sigma(r) + \tau(u)r' = \tau(u)r'$ for all r in R. Since $u \neq 0$ and by Lemma 1.2 $\tau(u)r' = 0$ implies $r' = 0$ so d is trivial on R.

Lemma 1.4. Let R be a prime ring and $(0) \neq U$ an ideal of R. Let $a, b \in U$ and for each $u \in U$ if $b[a,u]_{(\sigma,\tau)} = 0$ then $b = 0$ or $a \in C_{(\sigma,\tau)}$.

Proof. Let $b[a,u]_{(\sigma,\tau)} = 0$, $0 \neq u \in U$ and for all r in R we have $0 = b[a,ur]_{(\sigma,\tau)} = b[a,u]_{(\sigma,\tau)}\sigma(r) + b\,\tau(u)[a,r]_{(\sigma,\tau)} = b\,\tau(u)[a,r]_{(\sigma,\tau)}$ so by Lemma 1.2 we get $b = 0$ or $[a,r]_{(\sigma,\tau)} = 0$ i.e. $a \in C_{(\sigma,\tau)}$. In particular, $0 = b[a,u]_{(1,1)} = b[a,u] = 0$ implies $b = 0$ or $a \in C$.

Lemma 1.5. Let R be a prime ring. For a, b in R, if $ab = 0$ and $b \in C_{(\sigma,\tau)}$, then $a = 0$ or $b = 0$.

Proof. If $ab = 0$, then for all x in R we have $0 = ab\,\sigma(x) = a\,\tau(x)b$ and this implies $a = 0$ or $b = 0$.

Lemma 1.6. Let $(0) \neq U$ be a right ideal of a prime ring R. If U is commutative then R is commutative.

Proof. Take $0 \neq u \in U$ and for all r in R, z in U we have $0 = [ur, z] = u[r, z] + [u, z]r = u[r, z]$. By Lemma 1.2 we bet $[r, z] = 0$. This means $U \subseteq C$. Now for all x, y in R

$$0 = [ux, y] = u[x, y] + [u, y]x = u[x, y].$$

This implies $[x, y] = 0$. So R is commutative.

(σ,σ)-DERIVATIONS OF PRIME RINGS

From now on we shall consider the properties of prime rings with (σ,σ)-derivations. For this reason we write $C_{(\sigma,\sigma)} = C_\sigma = C$ and $[\ ,\ ]_{(\sigma,\sigma)} = [\ ,\ ]_\sigma$.

Lemma 2.1. Let $d : x \to x'$ be a (σ,σ)-derivation of R and U is an ideal of R such that d is semicentralizing on U.

If x, y in $[U]$ $((U))$, then

1) $x + y \in [U]\ ((U)) \leftrightarrow x - y \in [U]\ ((U))$
2) If $v \in (U)$ then $[v', v^2]_\sigma = 0$.

Proof. 1) If $x, y \in [U]$ we have $[x', x]_\sigma = x'\sigma(x) - \sigma(x)x' \in C$ and $[y', y]_\sigma = y'\sigma(y) - \sigma(y)y' \in C$ so 1) comes from the following equations.

$$[x' - y', x - y]_\sigma = -[x' + y', x + y] + 2\{[x', x]_\sigma + [y', y]_\sigma\}$$
$$(x' - y', x - y)_\sigma = -(x' + y', x + y) + 2\{(x', x)_\sigma + (y', y)_\sigma\}$$

2) For any $v \in (U)$ we have

$$\begin{aligned} 0 = [(v', v)_\sigma, v]_\sigma &= [v'\sigma(u) + \sigma(v)v', v]_\sigma \\ &= (v'\sigma(v) + \sigma(v)v')\sigma(v) - \sigma(v)(v'\sigma(v) + \sigma(v)v') \\ &= v'\sigma(v^2) - \sigma(v^2)v' = [v', v^2]_\sigma. \end{aligned}$$

Lemma 2.2. Let $d : x \to x'$ be a (σ,σ)-derivation of a prime ring R of characteristic $\neq 2$ which is semicentralizing on a non-zero ideal U of R. If $v \in U \backslash [U]$, then $(v^2)' = 0$ and

$$\sigma(v^2)v' = v'\sigma(v^2) = 0.$$

Proof. By Lemma 2.1 we have $r(v^2)v' = v'\sigma(v^2)$. Moreover

$$[(v^2)' + v', v^2 + v]_\sigma = [(v^2)', v^2]_\sigma + [(v^2)', v]_\sigma + [v', v^2]_\sigma$$
$$[v', v]_\sigma = [v', v]_\sigma \notin C.$$

so that $v^2 + v \notin [U]$. In the same way we can show that $v^2 - v \notin [U]$. By Lemma 2.1,1) $(v^2 + v) - (v^2 - v) = 2v \notin [U]$. This shows that $v^2 + v + v^2 - v = 2v^2 \in (U)$ so

$v^2 \in (U)$. This means $((v^2)', v^2)_\sigma = (v^2)'\sigma(v^2) + \sigma(v^2)(v^2)' = 2(v^2)'\sigma(v^2) \in C$. This implies $(v^2)'\sigma(v^2) \in C$. Again by Lemma 2.1,2) we get

$$\begin{aligned} 0 &= (v^2)' \left[(v^2+v)', (v^2+v)^2\right]_\sigma \\ &= (v^2)' \left[(v^2)' + v', v^4 + 2v^3 + v^2\right]_\sigma \\ &= (v^2)' \left[v', v^4 + 2v^3 + v^2\right]_\sigma = (v^2)' \left[v', 2v^3\right]_\sigma \\ &= 2(v^2)'\sigma(v^2)[v'v]_\sigma \end{aligned}$$

so

$$(v^2)'\sigma(v^2)[v', v]_\sigma = 0 .$$

Since $(v^2)'\sigma(v^2) \in C$, $(v^2)' \in C$ and $[v', v]_\sigma \neq 0$ we get that $(v^2)' = 0$. On the other hand, $v^2 + v \notin [U]$ implies $v^2 + v \in (U)$.

$$0 = \left((v^2+v)', (v^2+v)\right)_\sigma = \left((v^2)' + v', v^2 + v\right)_\sigma$$

$$(v', v^2)_\sigma + (v', v)_\sigma = (v', v^2)_\sigma$$

so $0 = v'\sigma(v^2) + \sigma(v^2)v' = 2\sigma(v^2)v'$ therefore we get $\sigma(v^2)v' = 0 = v'\sigma(v^2)$.

Lemma 2.3. **Let $d : x \to x'$ be a (σ,σ)-derivation of a prime ring R of characteristic $\neq 2$, which is semicentralizing on a non-zero ideal U of R. If $C \cap U = 0$ and $v \in U \backslash [U]$, then $\sigma(v^2) \notin 0$ and $v'^3 = 0$.**

Proof. Since $(v^2)' = (v', v)_\sigma = v'\sigma(v) + \sigma(v)v' = 0$, we have $v'\sigma(v) = -\sigma(v)v'$. So that for all u in U we have $u'\sigma(u) = +\sigma(u)u' + c$ where $c \in C$ we prove first that $\sigma(v^2) \neq 0$. Suppose $\sigma(v^2) = 0$. Since d is semicentralizing we have for any x in R,

$$\begin{aligned} 0 &= \sigma(v)\left\{\sigma(v+xv)(v+xv)' \mp (v+xv)'\sigma(v+xv) + c\right\}\sigma(v) \\ &= \sigma(v)\left\{(\sigma(v) + \sigma(x)\sigma(v))(v' + x'\sigma(v) + \sigma(x)v')\right\}\sigma(v) \\ &\quad \mp \sigma(v)\left\{(v' + x'\sigma(v) + \sigma(x)v')(\sigma(v) + \sigma(x)\sigma(v))\right\}\sigma(v) \\ &= \sigma(v)\sigma(x)\sigma(v)v'\sigma(v) + \sigma(v)\sigma(x)\sigma(v)\sigma(x)v'\sigma(v) . \end{aligned} \tag{1}$$

Since $\sigma(v)v' = -v'\sigma(v)$ and by (1) we get

$$\begin{aligned} 0 &= -\sigma(v)\sigma(x)v'\sigma(v^2) + \sigma(v)\sigma(x)\sigma(v)\sigma(x)v'\sigma(v) \\ &= \sigma(v)\sigma(x)\sigma(v)\sigma(x)v'\sigma(v). \end{aligned}$$

The primeness of R implies $v'\sigma(v) = 0 = \sigma(v)v'$. But this contradicts $v \notin [U]$. So our assumption $\sigma(v^2) = 0$ is not true. i.e. $\sigma(v^2) \neq 0$.

Next we claim that $\sigma(v)v'^2 = 0$. By Lemma 2.2 we know that $v'\sigma(v^2) = \sigma(v^2)v' = 0$. So we have

$$\begin{aligned} 0 &= \sigma(v)\left\{\sigma(v+vxv)(v+vxv)' + (v+vxv)'\sigma(v+vxv) + c\right\}\sigma(v)v' \\ &= \sigma(v)\left\{(\sigma(v) + \sigma(v)\sigma(x)\sigma(v))\left(v' + v'\sigma(xv) + \sigma(v)(xv)'\right)\right\}\sigma(v)v' \\ &= \sigma(v)\left\{(\sigma(v) + \sigma(v)\sigma(x)\sigma(v))\left(v' + v'\sigma(x)\sigma(v) + \sigma(v)x'\sigma(v)\right.\right. \\ &\qquad \left.\left. + \sigma(v)\sigma(x)v'\right)\sigma(v)v'\right\} \\ &= \left(\sigma(v^2) + \sigma(v^2)\sigma(x)\sigma(v)\right)\left(v'\sigma(v)v' + \sigma(v)\sigma(x)v'\sigma(v)v'\right) \\ &= \sigma(v^3)\sigma(x)v'\sigma(v)v' + \sigma(v^2)\sigma(x)\sigma(v^2)\sigma(x)v'\sigma(v)v'. \end{aligned} \tag{i}$$

Replace x by $-x$, we get

$$0 = -\sigma(v^3)\sigma(x)v'\sigma(v)v' + \sigma(v^2)\sigma(x)\sigma(v^2)\sigma(x)v'\sigma(v)v'. \qquad (ii)$$

From (i) and (ii) we get that

$$\sigma(v^2)\sigma(x)\sigma(v^2)\sigma(x)v'\sigma(v)v' = 0.$$

So $\sigma(v^2)\sigma(x)\sigma(v^2)\sigma(x)\sigma(v)(v')^2 = 0$. Since R is prime we have $\sigma(v)(v')^2 = 0$. So for any x in R we have

$$\begin{aligned}
0 &= \sigma(v)\left\{\sigma(v+xv)(v+xv)' \mp (v+xv)'\sigma(v+xv) + c\right\}(v')^2 \\
&= \sigma(v)\left\{(\sigma(v)+\sigma(x)\sigma(v))\left(v' + x'\sigma(v) + \sigma(x)v'\right)\right\}(v')^2 \\
&\quad \mp \sigma(v)\left\{\left(v' + x'\sigma(v) + \sigma(x)v'\right)(\sigma(v)+\sigma(x)\sigma(v))\right\}(v')^2 \\
&= \sigma(v)\left\{(\sigma(v)+\sigma(x)\sigma(v))\left(v'^3 + \sigma(x)v'^3\right)\right\} \\
&= \left(\sigma(v^2)+\sigma(v)\sigma(x)\sigma(v)\right)\left(v'^3 + \sigma(x)v'^3\right) \\
&= \sigma(v^2)\sigma(x)v'^3 + \sigma(v)\sigma(x)\sigma(v)\sigma(x)v'^3. \qquad \text{(iii)}
\end{aligned}$$

Again replace x by $-x$ we get

$$0 = -\sigma(v^2)\sigma(x)v'^3 + \sigma(v)\sigma(x)\sigma(v)\sigma(x)v'^3. \qquad (iv)$$

From (iii) and (iv) we get that $\sigma(v^2)\sigma(x)v'^3 = 0$. By primeness of R we get $v'^3 = 0$.

Lemma 2.4. Let $d : x \to x'$ be a (σ,σ)-derivation of R of characteristic $\neq 2$ which is semicentralizing on a non-zero ideal U of R. If $C \cap U \neq 0$, then d is centralizing on U.

Proof. Suppose that U contains an element v which is not contained in $[U]$, i.e. $[v',v]_\sigma \notin C$.

Choose an arbitrary $0 \notin c \in C \cap U$ and since $c' \in C$ we have

$$[v'+c', v+c]_\sigma = [v',v]_\sigma + [v',c]_\sigma + [c',v]_\sigma + [c',c]_\sigma = [v'v]_\sigma \notin C$$

so that $v + c \notin [U]$. By Lemma 2.2 we have $\left((v+c)^2\right)' = 0$.

$$\begin{aligned}
0 &= \left[\left((v+c)^2\right)', v\right]_\sigma = \left[\left(v^2 + 2vc + c^2\right)', v\right]_\sigma \\
&= \left[(v^2)' + (2vc)' + (c^2)', v\right]_\sigma = [(2vc)', v]_\sigma \\
&= 2\left[v'\sigma(c) + \sigma(v)c', v\right]_\sigma
\end{aligned}$$

so that

$$\begin{aligned}
0 &= [v'\sigma(c) + \sigma(v)c', v]_\sigma = [v'\sigma(c), v]_\sigma + [\sigma(v)c', v]_\sigma \\
&= [v'\sigma(c), v]_\sigma = v'\sigma(c)\sigma(v) - \sigma(v)v'\sigma(c) \\
&\qquad + v'\sigma(v)\sigma(c) - v'\sigma(v)\sigma(c) \\
&= v'\left(\sigma[c,v]\right) + [v',v]_\sigma\sigma(c).
\end{aligned}$$

Since $[c,v]_\sigma = 0$, we have that $[v',v]_\sigma \sigma(c) = 0$ and $\sigma(c) \neq 0$ implies that $[v',v]_\sigma = 0$. This contradiction proves that $[U] = U$. So that d is centralizing on U.

Lemma 2.5. Let R be a prime ring with an idempotent $e \neq 0,1$. If d is a (σ,σ)-derivation of R such that $(e + ex - exe)' = 0$ for all x in R. Then $d = 0$.

Proof. Since for all x in R, $(e + ex - exe)' = 0$ we have $0 = (e + ee - eee)' = e'$. On the other hand,

$$\begin{aligned} 0 &= (e + ex - exe)' = e' + (ex - exe)' \\ &= (ex - exe)' = (ex)' - (exe)' \\ &= e'\sigma(x) + \sigma(e)x' - e'\sigma(xe) - \sigma(e)(xe)' \\ &= \sigma(e)x' - \sigma(e)(x'\sigma(e) + \sigma(x)e') \\ &= \sigma(e)x' - \sigma(e)x'\sigma(e). \end{aligned}$$

So we get $0 = \sigma(e)R'\sigma(1-e)$. If we set $e_1 = e$, $1 - e = e_2$ and $R_{12} = e_1Re_2$, then $(R_{12})' = 0$.

Since $(R_{12})' = 0$ we have $(R_{12}R_{21}R_{12})' = 0$. So for all x in $R_{12}R_{21}R_{12}$, $x = e_1r_1e_2r_2e_1r_3e_2$

$$\begin{aligned} 0 &= (R_{12}R_{21}R_{12})' = (e_1r_1e_2r_2e_1r_3e_2)' \\ &= (e_1r_1e_2)'\sigma(r_2e_1r_3e_2) + \sigma(e_1r_1e_2)(r_2e_1r_3e_2)' \\ &= \sigma(e_1r_1e_2)\,(r_2'\sigma(e_1r_3e_2) + \sigma(r_2)(e_1r_3e_2)') \\ &= \sigma(e_1r_1e_2)r_2'\sigma(e_1r_3e_2). \end{aligned}$$

So we get $0 = \sigma(e_1)\sigma(r_1)\sigma(e_2)r_2'\,\sigma(e_1)\sigma(r_3)\sigma(e_2)$. Since $\sigma(e_1) \neq 0$ and R is prime

$$\sigma(e_2)r_2'\,\sigma(e_1)\sigma(r_3)\sigma(e_2) = 0,$$

and again since $\sigma(e_2) \neq 0$ and R is prime we get $\sigma(e_2)r_2'\sigma(e_1) = 0$. This means that $0 = (e_2r_2e_1)' = (R_{21})'$. Similarly from $0 = (R_{ii}R_{ij})' = (e_ir_ie_ise_j)' = \sigma(e_i)(re_ise_j)' = \sigma(e_i)r'\sigma(e_ise_j) = \sigma(e_i)r'\sigma(e_i)\sigma(s)\sigma(e_j)$ where $i,j = 1,2$. Since $\sigma(e_j) \neq 0$ and R is prime we have $\sigma(e_i)r'\sigma(e_i) = 0$. This means that $(R_{ii})' = 0$. For $r \in R$

$$\begin{aligned} &e_1re_1 + e_1re_2 + e_2re_1 + e_2re_2 \\ &\quad \in \sum_{i,j=1}^{2} R_{ij}\; e_1re_1 + e_1re_2 + e_2re_1 + e_2re_2 \\ &\quad = e_1re_1 + e_1r(1-e_1) + (1-e_1)re_1 + (1-e_1)r(1-e_1) \\ &\quad e_1re_1 + e_1r - e_1re_1 + re_1 - e_1re_1 + r - e_1r - re_1 + e_1re_1 = r. \end{aligned}$$

Since $R = \sum\limits_{i,j=1}^{2} R$ and d is an additive, we conclude that $d = 0$.

Lemma 2.6. Let R be a prime ring, U a non-zero ideal of R and Q be the Martindale quotient ring of R. Let p,q,r be elements of Q, if $puqur = 0$ for all $u \in U$. Then at least one of p,q,r is zero.

Proof. If x, y are elements of Q such that $x \cup y = 0$, then $x = 0$ or $y = 0$ making use of this fact we can prove it in the same way as in [7; Lemma 2].

Now we can prove the following theorem.

Theorem. Let R be a prime ring of characteristic $\neq 2$. If d is a non-zero (σ, σ)-derivation of R, which is semicentralizing on R. Then R is commutative.

Proof. If (σ, σ)-derivation d is centralizing on R, then R is commutative [3; Theorem 1] so if we can show that (σ, σ)-derivation d is centralizing then the theorem will be proved. In Lemma 2.4 we show that if $C \neq 0$ then (σ, σ)-derivation d is centralizing, so we can take the case $C = 0$. Since we can extend (σ, σ)-derivation d of R to a (σ, σ)-derivation of S. Let e be an arbitrary idempotent in S (in fact e is a mapping from μ to R). Then there exists a non-zero ideal A of R such that $eA \subseteq R$ and $Ae \subseteq R$.

Then for any a in A we have

$$\sigma(ea)(ea)' = (ea)'\sigma(ea) \text{ or } \sigma(ea)(ea)' = -(ea)'\sigma(ea)$$

for this two case we have

$$\begin{aligned}
&\sigma(e)(ea)'\sigma(ea) = (ea)'\sigma(ea)\\
&\sigma(e)(e'\sigma(a) + \sigma(e)a')\sigma(ea) = (e'\sigma(a) + \sigma(e)a')\sigma(ea)\\
&(\sigma(e)e'\sigma(a) + \sigma(e)a')\,\sigma(ea) = e'\sigma(a)\sigma(ea) + \sigma(e)a'\sigma(ea)\\
&\sigma(e)e'\sigma(a)\sigma(ea) + \sigma(e)a'\sigma(ea) = e'\sigma(a)\sigma(ea) + \sigma(e)a'\sigma(ea)\\
&\sigma(e)e'\sigma(a)\sigma(ea) - e'\sigma(a)\sigma(ea) = 0\\
&(\sigma(e)e' - e')\,\sigma(a)\sigma(ea) = 0.
\end{aligned}$$

By Lemma 2.6 $\sigma(e)e' - e' = 0$. This implies $\sigma(e)e' = e'$. Similarly we can prove that

$$e'\sigma(e) = e'.$$

So that $e' = (e^2)' = e'\sigma(e) + \sigma(e)e' = 2e' = 0$ this implies $e' = 0$.

If f is an idempotent element of S, then for all x in S, $(f + fx - f \times f)$ is also idemponent in S. Since (σ, σ)-derivation d is not equal to zero by Lemma 2.5 S has nontrivial ideomponent. Hence S has to be a division ring and R is integral domain [2; Theorem 1.3.4]. So by Lemma 2.3 (σ, σ)-derivation d is centralizing. Therefore R is commutative.

REFERENCES

1. B. Felzenswalb, Derivation in prime rings, Proc. Amer. Math. Soc., Vol. 84. No. 1 (1982).
2. I. N. Herstein, Non commutative rings, The Math. Assoc. Amer. (1968).
3. Y. Hirano and H. Tominaga, Some commutativity theorems for prime rings with derivations and differentially semi prime rings, Math J. of Okayama Univ., Vol. 26 (1984).

4. W. S. Martindale, Prime rings satisfying general polynomial identity, J. of Algebra, Vol. 12 (1969).
5. J. H. Mayne, Ideals and centralizing mappings of prime rings, Proc. Amer. Math. Soc., Vol. 86 No. 2 (1982).
6. J.H. Mayne, Centralizing mappings of prime rings, Canada Math. Bul., Vol. 27(1) (1984).
7. E. C. Posner, Derivations in prime rings, Proc. Amer. Math. Soc., Vol. 8 No. 5 (1957).

INDEX